Advances in Near Infrared Spectroscopy and Related Computational Methods

Advances in Near Infrared Spectroscopy and Related Computational Methods

Special Issue Editors

Christian Huck
Krzysztof B. Bec

MDPI • Basel • Beijing • Wuhan • Barcelona • Belgrade

MDPI

Special Issue Editors
Christian Huck
Leopold-Franzens University
Austria

Krzysztof B. Bec
University of Wrocław
Poland

Editorial Office
MDPI
St. Alban-Anlage 66
4052 Basel, Switzerland

This is a reprint of articles from the Special Issue published online in the open access journal *Molecules* (ISSN 1420-3049) from 2018 to 2019 (available at: https://www.mdpi.com/journal/molecules/special_issues/infrared_computational).

For citation purposes, cite each article independently as indicated on the article page online and as indicated below:

LastName, A.A.; LastName, B.B.; LastName, C.C. Article Title. *Journal Name* **Year**, *Article Number*, Page Range.

ISBN 978-3-03928-052-0 (Pbk)
ISBN 978-3-03928-053-7 (PDF)

Contents

About the Special Issue Editors

Christian Huck obtained his doctorate in chemistry in 1998 from the University in Innsbruck, Austria, where he continued to work as an assistant professor until the habilitation in 2006. In 2013, he received a call as a full professor to the University of Stuttgart, Germany and in 2015, another call back to the University of Innsbruck, where he is currently vice-head of the Institute of Analytical Chemistry and Radiochemistry and head of the spectroscopy unit. From 2014 until 2017 he was a visiting professor at Kwansei-Gakuin University in Sanda, Japan, in the laboratory of Professor Yukihiro Ozaki. Christian has published more than 300 peer-reviewed manuscripts resulting in an h-index of 48 based on 7900 citations. He is Editor-in-Chief of NIR News, Editor of Spectrochimica Acta Part A and Editorial Board Member of numerous other journals including Molecules journal. Beside several numerous awards he was also the receiver of 2018 Tomas Hirschfeld Award.

Krzysztof B. Bec obtained his PhD degree (2014) in Physical and Theoretical Chemistry from University of Wroclaw. His research focused on thin-film IR spectroscopy and computational methods. He then worked with Professor Yukihiro Ozaki as Postdoctoral Fellow and Research Assistant Professor at Kwansei Gakuin University, Japan; he focused on advancing NIR spectroscopy and contributed into the development of ATR-FUV-DUV spectroscopy and its applications. He presently continues his work as FWF Lise Meitner Senior Fellow in Prof. Christian W. Huck team at University of Innsbruck, Austria.

molecules

MDPI

Editorial

Advances in Near-Infrared Spectroscopy and Related Computational Methods

Krzysztof B. Beć * and Christian W. Huck *

Institute of Analytical Chemistry and Radiochemistry, CCB-Center for Chemistry and Biomedicine,
Leopold-Franzens University, Innrain 80/82, 6020 Innsbruck, Austria
* Correspondence: Krzysztof.Bec@uibk.ac.at (K.B.B.); Christian.W.Huck@uibk.ac.at (C.W.H.)

Received: 25 November 2019; Accepted: 26 November 2019; Published: 29 November 2019

Over the last few decades, near-infrared (NIR) spectroscopy has distinguished itself as one of the most rapidly advancing spectroscopic techniques [1]. Mainly known as an analytical tool useful for sample characterization and content quantification, NIR spectroscopy is essential in various other fields, e.g., NIR imaging techniques in biophotonics, medical applications, or in characterization of food products, to name the few [2]. Its contribution to basic science and physical chemistry should be noted as well, e.g., in exploration of the nature of molecular vibrations or intermolecular interactions [3]. One of the current development trends involves the miniaturization and simplification of instrumentation [4], creating prospects for the spread of NIR spectrometers at a consumer level, e.g., in the form of smartphone attachments—a breakthrough not yet accomplished by any other analytical technique. NIR spectroscopy has been developing in conjunction with advanced methods of data analysis; recent years have highlighted the role of anharmonic quantum mechanical computations in shedding light on the complex nature of NIR spectra as well [5].

The importance of NIR spectroscopy is well demonstrated by a remarkable interest it receives among scientific and professional communities. Such observation can be roughly quantified using the statistical data collected by Web of Science [6]. A query for "near infrared spectroscopy" returns over 2200 records for 2018 year alone, out of this number almost 1800 records being scientific articles (Figure 1). This clearly evidences the maturity level that NIR spectroscopy has achieved nowadays. At the same time, statistical data evidences a steady progress in popularity of the eponymous technique as the number of articles published annually almost doubled over the last decade comparing 2009 to 2018 (Figure 1). However, one may also notice some adverse effects of such popularity. As unveiled by Web of Science query, this technique is used throughout various fields of application in a true myriad of contexts (Figure 2) [6]. The mentioned trend also resulted in a growing diversity of the methods and applications related to NIR spectroscopy and has led to a dispersion of the contributions among disparate scientific communities.

For this reason, we recognized the need to propose the Special Issue "Advances in Near Infrared Spectroscopy and Related Computational Methods" in *Molecules* journal. Our aim was to bring together these diverse communities, which may perceive NIR spectroscopy from different perspectives. Besides, we welcomed research topics not directly focused on the NIR region, however, which remained relevant by employing the methodologies essential in NIR spectroscopy. A number of other spectroscopic methods of analysis share methods and tools common with NIR spectroscopy. We believe such scope of the Special Issue promoted the exchange of ideas and thus was helpful in pushing the frontier of this discipline of science. Moreover, we hoped to create a formidable opportunity for the readership to obtain a thorough overview of state-of-the-art NIR spectroscopy, current development trends, and future prospects with no artificial limits or strict categorization. This way, we also put faith in offering an appropriate opportunity to all the contributors to make their results and techniques more visible, and to present the most recent accomplishments in their respective fields that have become possible with the use of NIR spectroscopy.

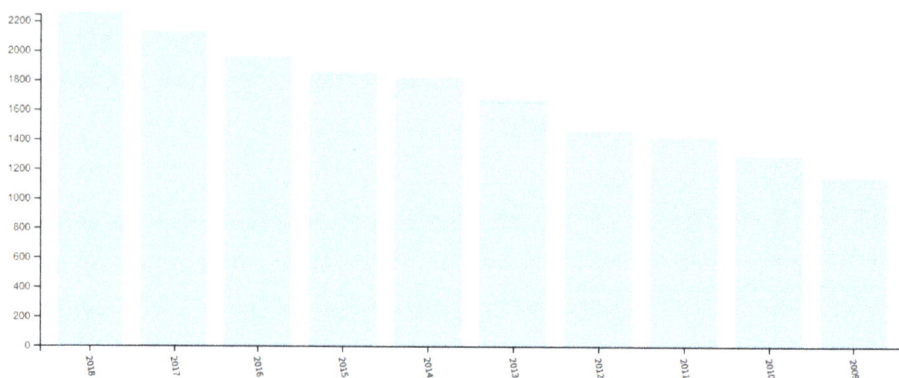

Figure 1. Results analysis for Web of Science query "near infrared spectroscopy" for publication years (2009–2018) [6].

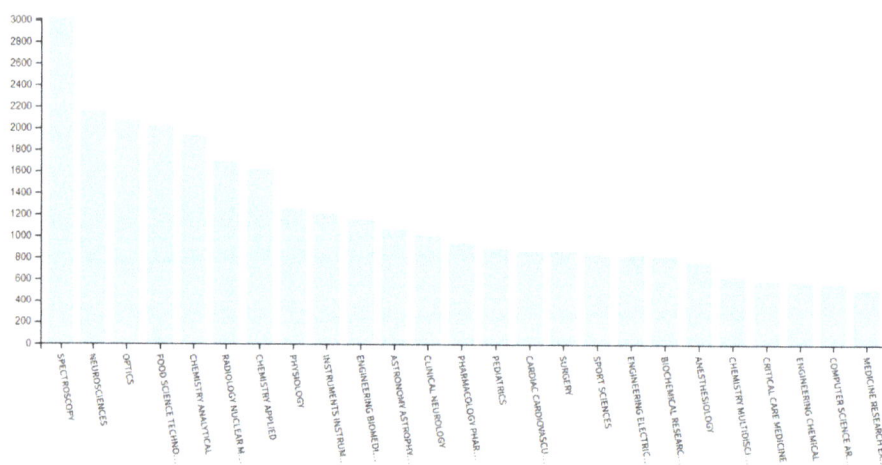

Figure 2. Results analysis for Web of Science query "near infrared spectroscopy" following the classification of Web of Science Categories. The figure presents only the 25 most significant categories [6].

The Special Issue has met a remarkably positive feedback with many contributions submitted by numerous scholars and professional spectroscopists performing their active research in academia and industry, resulting in a collection of 30 publications including two exhaustive review articles [7–36]. The diversity in the application field has been well represented by the submitted manuscripts. These articles discuss a variety of aspects relevant to NIR spectroscopy in a markedly broad context.

Many of these articles have a cross-field character and it would be difficult to ascribe them arbitrarily to certain disciplines of science. However, for sake of clarity a tentative and brief overview of these contributions may be helpful to present the Special Issue to the readership. The majority of the articles focuses on applied qualitative and quantitative analyses in a variety of fields [9,11,12,17–34]. Roughly, half of these may be associated with pharmaceutical and medical applications [17–24]. Most of the remaining applied studies were directed at agricultural applications [25–34], well reflecting the ever-growing significance of NIR spectroscopy in this area; a good perspective of this topic is included in a focused review article published in the Special Issue [30]. Modern strategies for food analysis also rely on this technique, and few contributions touched that field as well [9,12,33,34]. State-of-the-art analytical spectroscopy is based on sophisticated data-analytical methods. Development of new methods is, therefore, essential and benefits multiple applications [10,11,13–16]. Several articles focused on this direction, and the importance of research and development of calibration transfer methods is well reflected in this Special Issue [13,16]. Interestingly, Beganović *et al.* demonstrated that there exists room for improvement in fundamental aspects of analytical spectroscopy such as wavenumber region selection for subsequent calibration [3]. On the other hand, progress in technology and instrumentation is indispensable as well. The growing applicability and importance of miniaturized, portable NIR spectrometers is reflected by several focused articles [10–12]. The differences in design principles and emerging novel technologies that become applied in order to obtain affordable and ultra-miniaturized devices raise concerns about the resulting analytical performance of such spectrometers; therefore, comparative evaluation studies are critical [11,12]. Likewise, the potential of hyperspectral imaging can be recognized on the basis of the articles collected in this Special Issue as well [29–32]. The importance of NIR spectroscopy as a potent tool in exploring the complex nature of water, the elementary substance, is reflected in an exhaustive review article [36]. Finally, contributions focused on fundamental principles of NIR spectroscopy including theoretical NIR spectra simulation and physicochemical research should be mentioned, highlighting the significance of pushing the frontier of the underlying basic science [7,8]. One may note that these contributions reflect well the diversity and dynamics of contemporary development trends in NIR spectroscopy.

This special issue is accessible through the following link: https://www.mdpi.com/journal/molecules/special_issues/infrared_computational

As Guest Editors for this Special Issue, we would like to thank all the authors and co-authors for their contributions and all the reviewers for their effort in carefully evaluating the manuscripts. Last but not least, we would like to appreciate the editorial office of *Molecules* journal for their kind assistance in preparing this Special Issue.

Funding: This research received no external funding.

Conflicts of Interest: The authors declare no conflict of interest.

References

1. Siesler, H.W.; Ozaki, Y.; Kawata, S.; Heise, H.M. (Eds.) *Near Infrared Spectroscopy: Principles, Instruments, Applications*; Wiley-VCH: Weinheim, Germany, 2002.
2. Ozaki, Y.; Huck, C.W.; Beć, K.B. Near infrared spectroscopy and its applications. In *Molecular and Laser Spectroscopy*; Gupta, V.P., Ed.; Elsevier: Amsterdam, The Netherlands, 2017.
3. Czarnecki, M.A.; Morisawa, Y.; Futami, Y.; Ozaki, Y. Advances in molecular structure and interaction studies using near-infrared spectroscopy. *Chem. Rev.* **2015**, *115*, 9707–9744. [CrossRef] [PubMed]
4. Kirchler, C.G.; Pezzei, C.K.; Beć, K.B.; Mayr, S.; Ishigaki, M.; Ozaki, Y.; Huck, C.W. Critical evaluation of spectral information of benchtop vs. portable near-infrared spectrometers: Quantum chemistry and two-dimensional correlation spectroscopy for a better understanding of PLS regression models of the rosmarinic acid content in Rosmarini folium. *Analyst* **2017**, *142*, 455–464. [PubMed]
5. Beć, K.B.; Huck, C.W. Breakthrough potential in near-infrared spectroscopy: Spectra simulation. A review of recent developments. *Front. Chem.* **2019**, *7*, 48. [CrossRef] [PubMed]
6. Web of Science Database Query. Search for "Near Infrared Spectroscopy". Available online: https://wcs.webofknowledge.com/RA/analyze.do?product=WOS&SID=D4zaDudmNOG6Ax865O1&field=PY_PublicationYear_PublicationYear_en&yearSort=true (accessed on 5 November 2019).
7. Grabska, J.; Beć, K.B.; Kirchler, C.G.; Ozaki, Y.; Huck, C.W. Distinct difference in sensitivity of NIR vs. IR bands of melamine to inter-molecular interactions with impact on analytical spectroscopy explained by anharmonic quantum mechanical study. *Molecules* **2019**, *24*, 1402. [CrossRef]
8. Beć, K.B.; Grabska, J.; Huck, C.W.; Czarnecki, M.A. Spectra–structure correlations in isotopomers of ethanol (CX_3CX_2OX.; X = H, D): Combined near-infrared and anharmonic computational study. *Molecules* **2019**, *24*, 2189. [CrossRef]
9. Beganović, A.; Moll, V.; Huck, C.W. Comparison of multivariate regression models based on water—And carbohydrate-related spectral regions in the near-infrared for aqueous solutions of glucose. *Molecules* **2019**, *24*, 3696. [CrossRef]

10. Sun, L.; Hsiung, C.; Smith, V. Investigation of direct model transferability using miniature near-infrared spectrometers. *Molecules* **2019**, *24*, 1997. [CrossRef]

11. Wiedemair, V.; Langore, D.; Garsleitner, R.; Dillinger, K.; Huck, C.W. Investigations into the performance of a novel pocket-sized near-infrared spectrometer for cheese analysis. *Molecules* **2019**, *24*, 428. [CrossRef]

12. Neves, M.D.G.; Poppi, R.J.; Siesler, H.W. Rapid determination of nutritional parameters of pasta/sauce blends by handheld near-infrared spectroscopy. *Molecules* **2019**, *24*, 2029. [CrossRef]

13. Zhao, Y.; Zhao, Z.; Shan, P.; Peng, S.; Yu, J.; Gao, S. Calibration transfer based on affine invariance for NIR without transfer standards. *Molecules* **2019**, *24*, 1802. [CrossRef]

14. Chen, Y.; Wang, Z. Wavelength selection for NIR spectroscopy based on the binary dragonfly algorithm. *Molecules* **2019**, *24*, 421. [CrossRef] [PubMed]

15. Pei, Y.-F.; Zuo, Z.-T.; Zhang, Q.-Z.; Wang, Y.-Z. Data fusion of Fourier transform mid-infrared (MIR) and near-infrared (NIR) spectroscopies to identify geographical origin of wild *Paris polyphylla* var. *yunnanensis*. *Molecules* **2019**, *24*, 2559. [CrossRef] [PubMed]

16. Zhao, Y.; Yu, J.; Shan, P.; Zhao, Z.; Jiang, X.; Gao, S. PLS subspace-based calibration transfer for near-infrared spectroscopy quantitative analysis. *Molecules* **2019**, *24*, 1289. [CrossRef] [PubMed]

17. Delueg, S.; Kirchler, C.G.; Meischl, F.; Ozaki, Y.; Popp, M.A.; Bonn, G.K.; Huck, C.W. At-line monitoring of the extraction process of Rosmarini folium via wet chemical assays, UHPLC analysis, and newly developed near-infrared spectroscopic analysis methods. *Molecules* **2019**, *24*, 2480. [CrossRef] [PubMed]

18. Frosch, T.; Wyrwich, E.; Yan, D.; Domes, C.; Domes, R.; Popp, J.; Frosch, T. Counterfeit and substandard test of the antimalarial tablet Riamet®by means of Raman hyperspectral multicomponent analysis. *Molecules* **2019**, *24*, 3229. [CrossRef] [PubMed]

19. Marinelli, B.; Pluchinotta, F.; Cozzolino, V.; Barlafante, G.; Strozzi, M.C.; Marinelli, E.; Franchini, S.; Gazzolo, D. Osteopathic manipulation treatment improves cerebro–splanchnic oximetry in late preterm infants. *Molecules* **2019**, *24*, 3221. [CrossRef]

20. Chaber, R.; Arthur, C.J.; Łach, K.; Raciborska, A.; Michalak, E.; Bilska, K.; Drabko, K.; Depciuch, J.; Kaznowska, E.; Cebulski, J. Predicting ewing sarcoma treatment outcome using infrared spectroscopy and machine learning. *Molecules* **2019**, *24*, 1075. [CrossRef]

21. Zhang, X.; Yang, Y.; Wang, Y.; Fan, Q. Detection of the BRAF V600E mutation in colorectal cancer by NIR spectroscopy in conjunction with counter propagation artificial neural network. *Molecules* **2019**, *24*, 2238. [CrossRef]

22. Fang, M.; Xia, S.; Bi, J.; Wigstrom, T.P.; Valenzano, L.; Wang, J.; Tanasova, M.; Luck, R.L.; Liu, H. Detecting Zn(II) ions in live cells with near-infrared fluorescent probes. *Molecules* **2019**, *24*, 1592. [CrossRef]

23. Han, Y.; Jian, L.; Yao, Y.; Wang, X.; Han, L.; Liu, X. Insight into rapid DNA-specific identification of animal origin based on FTIR analysis: A case study. *Molecules* **2018**, *23*, 2842. [CrossRef]

24. Marotz, J.; Kulcke, A.; Siemers, F.; Cruz, D.; Aljowder, A.; Promny, D.; Daeschlein, G.; Wild, T. Extended parameter estimation from Hyperspectral imaging data for bedside diagnostic in medicine. *Molecules* **2019**, *24*, 4164. [CrossRef] [PubMed]

25. Fernández-Novales, J.; Tardáguila, J.; Gutiérrez, S.; Paz Diago, M. On-The-Go VIS + SW − NIR spectroscopy as a reliable monitoring tool for grape composition within the vineyard. *Molecules* **2019**, *24*, 2795. [CrossRef]

26. He, X.; Feng, X.; Sun, D.; Liu, F.; Bao, Y.; He, Y. Rapid and nondestructive measurement of rice seed vitality of different years using near-infrared hyperspectral imaging. *Molecules* **2019**, *24*, 2227. [CrossRef] [PubMed]

27. Xu, L.; Sun, W.; Ma, Y.; Chao, Z. Discrimination of *Trichosanthis fructus* from different geographical origins using near infrared spectroscopy coupled with chemometric techniques. *Molecules* **2019**, *24*, 1550. [CrossRef] [PubMed]

28. Toledo-Martín, E.M.; del Carmen García-García, M.; Font, R.; Moreno-Rojas, J.M.; Salinas-Navarro, M.; Gómez, P.; Del Río-Celestino, M. Quantification of total phenolic and carotenoid content in blackberries (*Rubus fructicosus* L.) using near infrared spectroscopy (NIRS) and multivariate analysis. *Molecules* **2018**, *23*, 3191. [CrossRef]

29. Zhang, J.; Dai, L.; Cheng, F. Classification of frozen corn seeds using hyperspectral VIS/NIR reflectance imaging. *Molecules* **2019**, *24*, 149. [CrossRef]

30. Feng, L.; Zhu, S.; Zhang, C.; Bao, Y.; Feng, X.; He, Y. Identification of maize kernel vigor under different accelerated aging times using hyperspectral imaging. *Molecules* **2018**, *23*, 3078. [CrossRef]

31. Feng, L.; Zhu, S.; Zhang, C.; Bao, Y.; Gao, P.; He, Y. Variety identification of raisins using near-infrared hyperspectral imaging. *Molecules* **2018**, *23*, 2907. [CrossRef]

32. Wu, N.; Zhang, C.; Bai, X.; Du, X.; He, Y. Discrimination of chrysanthemum varieties using hyperspectral imaging combined with a deep convolutional neural network. *Molecules* **2018**, *23*, 2831. [CrossRef]

33. Camps, C.; Camps, Z.-N. Optimized prediction of reducing sugars and dry matter of potato frying by FT-NIR spectroscopy on peeled tubers. *Molecules* **2019**, *24*, 967. [CrossRef]

34. Jiang, H.; Chen, Q. Determination of adulteration content in extra virgin olive oil using FT-NIR spectroscopy combined with the BOSS–PLS algorithm. *Molecules* **2019**, *24*, 2134. [CrossRef] [PubMed]

35. Muncan, J.; Tsenkova, R. Aquaphotomics—From innovative knowledge to integrative platform in science and technology. *Molecules* **2019**, *24*, 2742. [CrossRef] [PubMed]

36. Tan, J.Y.; Ker, P.J.; Lau, K.Y.; Hannan, M.A.; Hoon Tang, S.G. Applications of photonics in agriculture sector: A review. *Molecules* **2019**, *24*, 2025. [CrossRef] [PubMed]

molecules

MDPI

Article

Extended Perfusion Parameter Estimation from Hyperspectral Imaging Data for Bedside Diagnostic in Medicine

Jörg Marotz [1,2,*], Axel Kulcke [3], Frank Siemers [1], Diogo Cruz [4], Ahmed Aljowder [5], Dominik Promny [6], Georg Daeschlein [7] and Thomas Wild [2,4,5]

[1] Klinik für Plastische und Handchirurgie und Brandverletztenzentrum, BG-Klinikum Bergmannstrost, D-06002 Halle (Saale), Germany; frank.siemers@bergmannstrost.de

[2] Institute of Applied Bioscience and Process Management, University of Applied Science Anhalt, D-06366 Köthen (Anhalt), Germany; thomas.wild@woundconsulting.com

[3] Diaspective Vision GmbH, D-18233 Am Salzhaff, Germany; axel.kulcke@diaspective-vision.com

[4] Clinic of Plastic, Hand and Aesthetic Surgery, Medical Center Dessau, University of Applied Science Anhalt, D-06847 Dessau, Germany; diogo.cruz@klinikum-dessau.de

[5] Clinic of Dermatology, Immunology and Allergology, Medical Center Dessau, Medical University Brandenburg "Theodor Fontane" Medical Center Dessau, D-06847 Dessau, Germany; ahmedaljowder@hotmail.com

[6] Klinik für Plastische, Wiederherstellende und Handchirurgie, Zentrum für Schwerbrandverletzte, Klinikum Nürnberg, D-90471 Nürnberg, Germany; dpromny@gmail.com

[7] Klinik und Poliklinik für Hautkrankheiten, Universitätsmedizin Greifswald, D-17475 Greifswald, Germany; Georg.Daeschlein@med.uni-greifswald.de

* Correspondence: joerg.marotz@ipross.de; Tel.: +49-17696526456

Academic Editors: Christian Huck and Krzysztof B. Bec

Received: 12 September 2019; Accepted: 14 November 2019; Published: 17 November 2019

Abstract: Background: Hyperspectral Imaging (HSI) has a strong potential to be established as a new contact-free measuring method in medicine. Hyperspectral cameras and data processing have to fulfill requirements concerning practicability and validity to be integrated in clinical routine processes. **Methods:** Calculating physiological parameters which are of significant clinical value from recorded remission spectra is a complex challenge. We present a data processing method for HSI remission spectra based on a five-layer model of perfused tissue that generates perfusion parameters for every layer and presents them as depth profiles. The modeling of the radiation transport and the solution of the inverse problem are based on familiar approximations, but use partially heuristic methods for efficiency and to fulfill practical clinical requirements. **Results:** The parameter determination process is consistent, as the measured spectrum is practically completely reproducible by the modeling sequence; in other words, the whole spectral information is transformed into model parameters which are easily accessible for physiological interpretation. The method is flexible enough to be applicable on a wide spectrum of skin and wounds. Examples of advanced procedures utilizing extended perfusion representation in clinical application areas (flap control, burn diagnosis) are presented.

Keywords: hyperspectral image processing; perfusion measurements; clinical classifications

---

1. Introduction

Hyperspectral Imaging (HSI, imaging remission spectroscopy, or diffuse reflectance spectroscopy) as a non-contact, stressless imaging measuring method is currently an intensively developing area for diverse medical applications [1,2]. Despite the limited penetration depth in biological tissue in the visible (VIS) and near infrared (NIR) spectral range, the effect of the specific scattering and absorption by tissue components in this "diagnostic window" makes it possible to retrieve information of significant clinical value [3–11].

In perfused tissue like human skin, the remission spectra are mainly influenced by hemoglobin (oxygenated and reduced) absorption. Additional components such as the collagen matrix, melanin, fat, and water contribute by specific scattering and absorption processes. The main focus is the estimation of the perfusion-related parameters of skin or similar tissue systems. Those parameters make it possible to evaluate local (for instance wounds) and regional (for instance PAD, diabetic foot) perfusion quality, and often also systemic attributes of the blood supply and oxygen usage [12–14]. Normally in clinical practice, no other methods are available to gain such information in a quick and simplified manner.

For the estimation of perfusion parameters (volume fraction blood, oxygen saturation hemoglobin), sample one- or two-layer models of the tissue with infinite depth and homogeneous distribution of the components are frequently used. Additionally, from the NIR-part of the spectrum, the volume fraction of water can be estimated [15]. The drawback from this is the substantial simplification of the normally complex layered structure of skin and similar tissue systems. The penetration depth of the light, and therefore the measuring volume, depends on the spectral range (VIS: <1 mm, NIR: 4–6 mm in skin) caused by the specific spectral scattering and absorption in the tissue layers. In real layered tissue systems, different layers contribute to a remission signal, depending on the wavelength. Thus, the remission spectrum is a heterogeneous spectrum in relation to the measured volume. The perfusion parameters estimated by these models are values averaged over different layers with unknown weights. With these models and estimated model parameters, normally, the measured spectrum cannot be reproduced, indicating a loss of information.

Nevertheless, even those parameters have been proven to represent a considerable information profit, and to provide additional value for diverse clinical application areas [9–11,13,15–17].

Recently, compact and cost-efficient hyperspectral cameras for routine clinical practice have been made available. The practicability for clinical use is accomplished by means of a simple measuring process with laminar illumination, direct imaging by an integrated scanning process generating a "3D-data cube"(x-y-λ), fast acquisition of a large area (i.e., approximately 5 seconds for 20 × 30 cm), and no special measuring conditions (beside the avoidance of external light on the measuring area) [14].

In order to establish such easy-to-use cameras, and therefore, hyperspectral imaging technology in a clinical environment, the potentially high information content of the measurement has to be exploited and presented to the clinical user in an informative manner.

Only clinical applications concerning the skin are considered. Mainly quantitative information about the perfusion situation should be generated. In this context, the estimation of the delivered arterial blood quantity and oxygen saturation, as well as the oxygen consumption in the capillary system of the measured area, are of special interest. The imaging measurement additionally allows for the analysis of regional distributions of the perfusion quality and the identification of regional perfusion distortions.

Besides the assessment of the intact skin, perfusion analyses of wounds generally are of special interest, because their quality is an essential factor in wound healing processes. Therefore, the parameter estimation method should be applicable to a wide variety of perfused tissue systems.

Although analytical solutions of the light transport equations in the diffusion approximation in tissue systems are available [18,19], the measuring geometries often do not correspond to the use of a HSI-camera in a clinical environment [20,21]. Simple models like two- or three-layer systems cannot adequately represent the complexity and variability of real skin systems and generate parameters of limited comparability. Solution procedures of the inverse problem (calculation of the model parameters from the measured spectrum) for more realistic multilayer models are still computationally expensive [20]. Also, solution procedures based on artificial neural networks (for instance, seven-layer models require the reduction of the number of parameters) to be efficient, and therefore, do not always adequately describe physiological conditions [22].

We present an evaluation procedure for the remission spectra of skin and wounds, which transforms all the information about the spectrum consistently into model parameters, which are then easily accessible to physiological interpretation. However, we do not claim to generate an exact solution to the problem of the realistic modeling of actual tissue systems.

Consistent transformation means that the measured spectrum should be completely reproduced by the model and the determined model parameters. The consistently-determined model parameters should be used as a more interpretable basis for further clinical estimations, as, for instance, in classification procedures.

The objective is to exploit the information of HSI-measurements in consideration of clinical demands and to create data processing of high practicability.

2. Results

2.1. D-Physiological Perfusion Imaging

The model-based processing described in Section 4 provides "depth profiles" for perfusion parameters vHb and $xHbO_2$ (with six values in each case), one value for vH_2O, and one for vFat, calculated from Λ_5 and Λ_6. Furthermore the depth profiles of the intrinsic structure parameters (s0, s1), as well as the relative "layer thickness" d_i ($d_i = D_i - D_{i-1}$, relative to D_1^0), are available.

The profiles are presented independently from the layer thickness as a series of parameter values (bars) (see Figure 6). The values of vHb are scaled according to the layer thicknesses determined by the procedure. This form of presentation has been proved to be the most informative in practice.

From the perfusion profiles (for every image pixel), four survey images are generated, depicting vHb and $xHbO_2$ for the upper layers 1 and 2 (vHb_1, xHbO_2_1) and for the deeper layers 5 and 6 (vHb_2, xHbO_2_2). The values are color-coded in blue (low), via green (normal) to red (high).

To evaluate the physiological interpretation and validity of the model parameters, the spectra of normal perfused skin (healthy volunteers) and from patients in different clinical areas are recorded and the depth profiles analyzed and proved in terms of their physiological plausibility.

2.1.1. Example: Occlusion Test

As a first example, the data from an occlusion test with healthy volunteers are presented. The left arm has been occluded (venous and arterial) and the hands were measured with a HSI camera with the right hand as a reference (Figure 1). The occlusion test contains four phases: normal perfusion, venous occlusion, arterial occlusion, and reperfusion after arterial occlusion. The survey images are shown in Figure 2, and the depth profiles from the test areas in Figure 3.

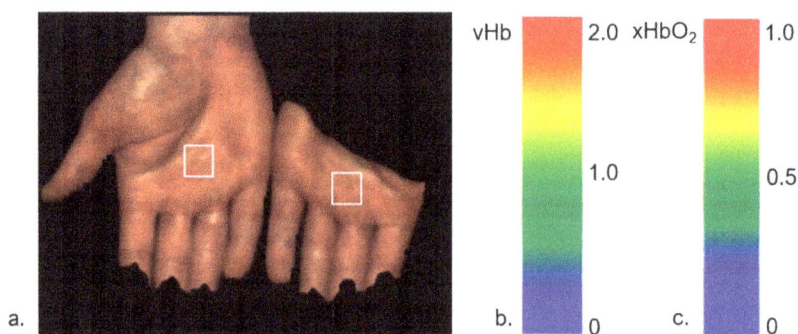

Figure 1. (**a**) Measurements of the hands in an occlusion test; right: hand of the occluded arm, left: reference hand; the white quadrates indicate the tested areas from which the profiles in Figure 3 were determined; (**b**) color scale vHb [0...2], c: color scale $xHbO_2$ [0 ... 1].

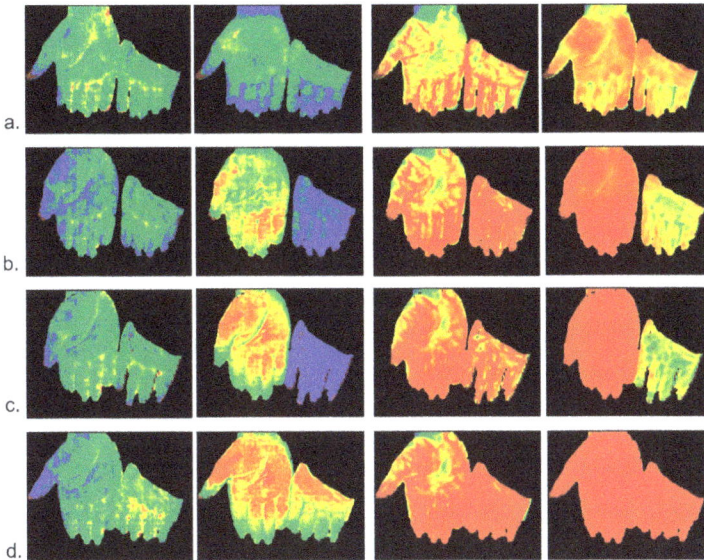

Figure 2. Survey images; color-coded parameters from left to right: vHb_1, xHbO$_2$_1 (superficial), vHb_2, xHbO$_2$_2 (deep); (**a**) normal perfusion, (**b**) venous occlusion, (**c**) arterial occlusion, (**d**) reperfusion after arterial occlusion. Color scales for vHb and xHbO$_2$, as depicted in Figure 1.

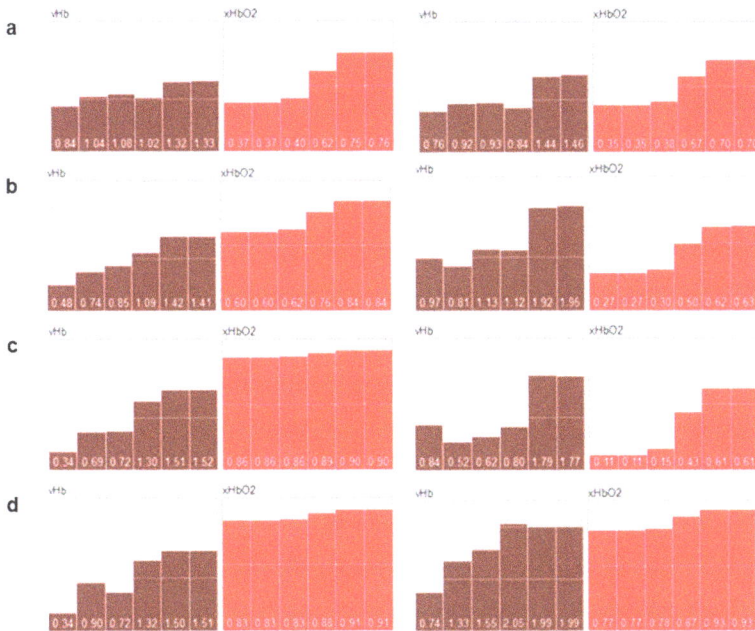

Figure 3. Perfusion profiles averaged over the test areas in Figure 1; the profiles depict the vHb- and HbO$_2$-values for 6 depth layers (layer 3 in Figure 8 has been split into 2 layers 3.1 and 3.2) from left to right; the thickness of the layers are not depicted; vHb are index values [0 . . . 2.5], xHbO$_2$: [0...1]; left column: reference hand; right column: test hand; (**a**) normal perfusion, (**b**) venous occlusion, (**c**) arterial occlusion, (**d**) reperfusion after arterial occlusion.

The survey images (Figure 2) clearly show the reaction of vHb and xHbO$_2$ in the different phases. The depth profiles of the parameters can be plausibly explained as follows:

"Normal" perfusion (a): the profiles from the reference and the test hand are similar; vHb shows the normal distribution over the layers 1–6; xHbO$_2$ in the superficial layers (1 and 2) is approx. 0.36 due to oxygen consumption in the capillary system; xHbO$_2$ in the deep layers (5 and 6) are a mixture of arterial (approx. 98%) and venous blood (0.36) from the capillary system; in the reticular system, the volume fraction of both arterial and venous blood are principally equal (in stationary states), so that xHbO$_2$ is the mean value of venous and arterial xHbO$_2$;

Venous occlusion (b): vHb increases in all layers, but mainly in 5 and 6, because blood cannot flow off; xHbO$_2$ decreases due to consumption and because the arterial supply is also hindered by venous occlusion, but there is still an arterial pressure in the capillary system;

Arterial occlusion (c): no blood flow; the available blood is gathered in the deeper vessels (layers 5 and 6); vHb in layers 1 and 2 is lower than for venous occlusion because of a lower arterial pressure; xHbO$_2$ strongly decreases due to consumption;

Reperfusion (d): expansion of all vessels, high blood flow; due to the high flow xHbO$_2$ increases in the capillary system because (stationary state) xHbO$_2$ in the superficial layers depends of the blood flow.

It is interesting to note the systemic reaction on the occlusion observable in the reference hand; the systemic blood flow increases in the deeper vessel system, while the superficial vHb (layers 1 and 2) decreases in the reference hand; due to the high flow, xHbO$_2$ increases.

After the reperfusion phase, the perfusion returns to normal values.

2.1.2. Example: Flap Transplant for Wound Coverage

In the following example, the perfusion evolution of a skin graft over twelve days is shown (measurement each second day).

The depth profiles are available for every point on the flap, and can be used to analyze the perfusion quality and distribution over the flap in detail over time. With close-meshed measurements, over time, developing perfusion problems can be detected and evaluated very early.

The survey images show the decreasing blood supply from the right side of the flap (Figure 4 vHb_2 and Figure 5a,b vHb_2), clearly indicating a distortion of the arterial conjunction already observable at day 5 (Figure 5b).

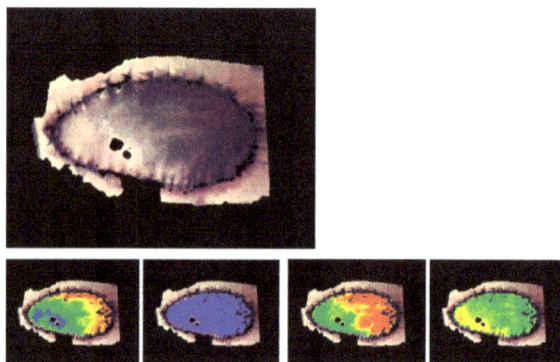

Figure 4. Transplant at day 1; survey images vHb_1, xHbO$_2$_1, vHb_2, xHbO$_2$_2; the arterial influx is on the right side of the flap; color scales for vHb and xHbO$_2$, as depicted in Figure 1.

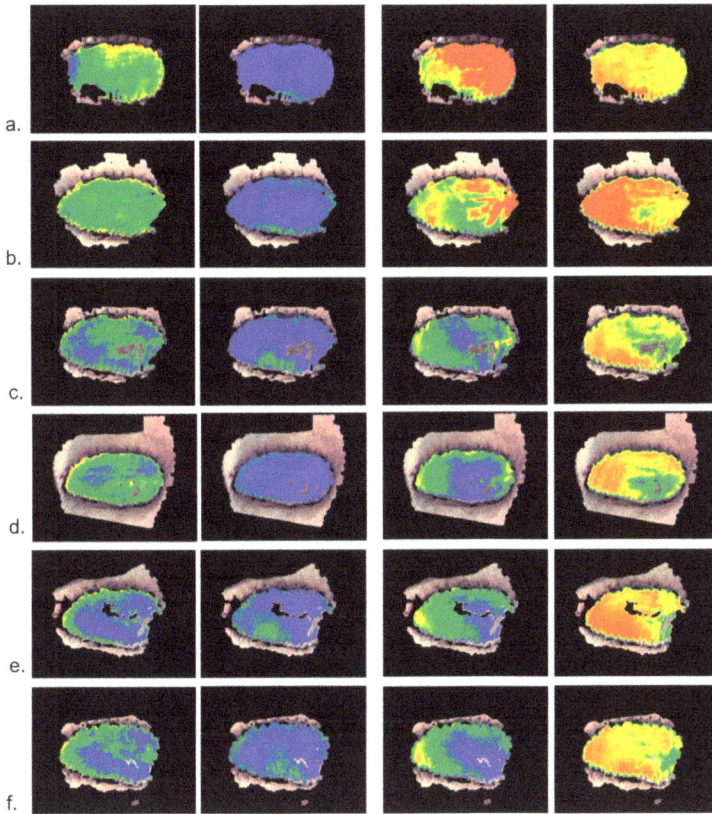

Figure 5. Same transplant measured at the following days (**a**) day 3, (**b**) day 5, (**c**) day 7, (**d**) day 9, (**e**) day 11, (**f**) day 13); vHb_1, xHbO$_2$_1, vHb_2, xHbO$_2$_2; color scales for vHb and xHbO$_2$, as depicted in Figure 1.

The survey images (Figure 5) clearly show the abated arterial blood supply over time.

The automated analysis is supported by image registration transforming the flap in every image to the same position and dimension. For automated analyses, the complete depth profiles are used.

With this methodology, an advanced procedure for describing and analyzing the perfusion dynamics in flaps is realizable.

2.1.3. Burn Wounds

The extended parameters have been used in a first attempt to generate a classification process for burn wounds.

Fundamental to the degree of skin damage by heat impact for the healing potential is the remaining perfusion quality in the wound area. With depth profiles, the perfusion situation can be depicted and evaluated on a new, higher level.

The example shows typical depth profiles of burn wounds with different degrees of damage (burn degrees: superficial, partial-thickness, full-thickness) (see Figure 6), as well as the classification of a burn wound on a hand, clinically assessed to be of partial-thickness (see Figure 7).

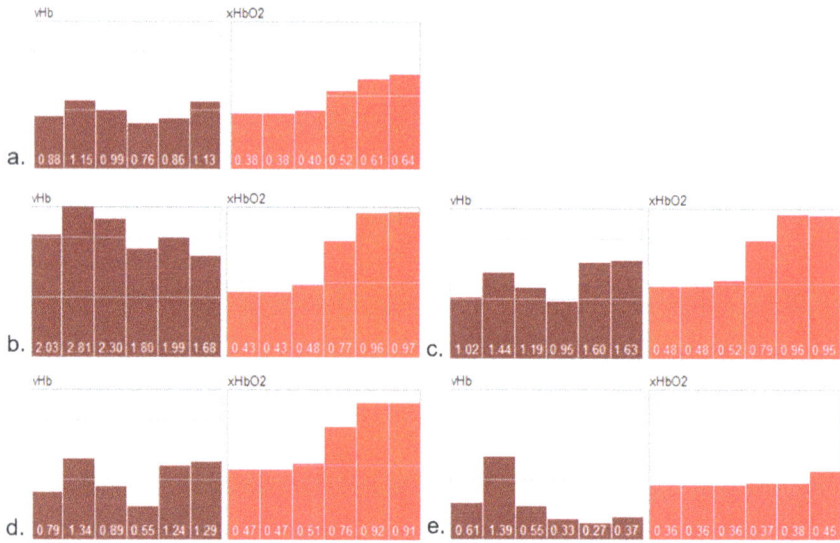

Figure 6. Depth profiles vHb, xHbO$_2$ for normal skin (**a**) and different burn degrees (**b**) superficial, (**c**) intermediate (superficial/partial), (**d**) partial-thickness, (**e**) full-thickness).

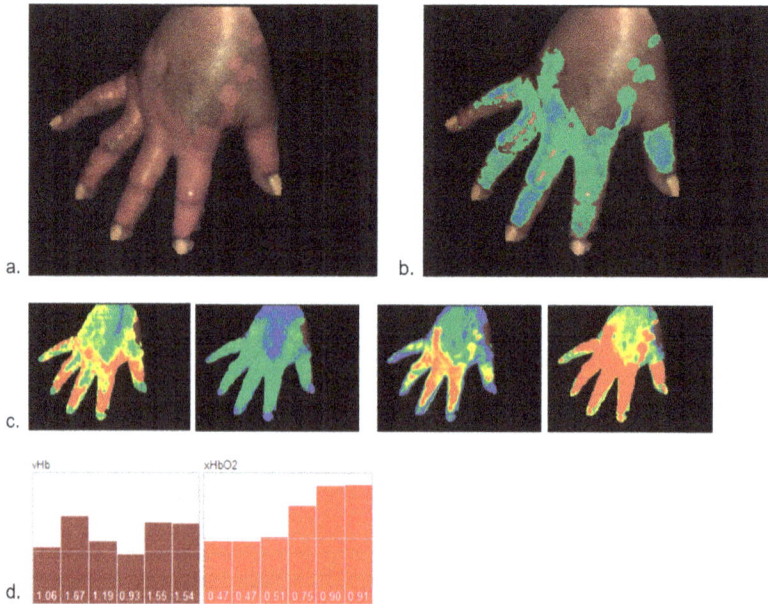

Figure 7. (**a**) Burn wound on a hand, (**c**) perfusion survey images (color scales for vHb and xHbO$_2$ as depicted in Figure 1) and (**b**) fuzzy classification (blue: superficial, green: partial-thickness, red: full-thickness); (**d**) perfusion profile from the burn area.

This first attempt of a classification process was constructed based on a small number of burn wounds (i.e., approx. 20). Additional to the perfusion parameters, the intrinsic structure parameters of the model were evaluated, and showed characteristic differences between the burn degrees. Also, spectral features, for instance, quantitatively describing the degree of tissue necrosis, are used in the classification process.

The perfusion profiles show significant differences between the burn degrees; for instance, a strong hyperemic reaction for superficial and intermediate partial degrees. With increasing degrees of damage, vHb decreases in all layers; the damage of the deeper vessels is obvious in Figure 6e (full thickness).

Theoretically, with these parameters, an efficient classification can be constructed, but the time after the burn has to be included as a fundamental factor. Especially for intermediate burn degrees, the development of the perfusion in the first 2–3 days is essential for the assessment of the healing potential.

With this methodology, a significant increase in terms of the quantitative and qualitative nature of descriptions and evaluations of burn wounds and wound processes seems to be achievable; therefore, a reliable diagnosis and treatment supporting procedure for burn medicine is foreseeable.

2.2. Comparison with Perfusion Parameters Based on a One-Layer Model

Actual standard data processing of hyperspectral imaging spectra involves the calculation of perfusion parameters based on a model consisting of a homogeneous, infinite, one-layer system with hemoglobin as the main component [14,15]. These parameters are comparable, and were validated with other parameters from standard tissue oximetry systems. The perfusion parameters are THI, StO2, and NIR-perfusion. THI (tissue hemoglobin index) denotes the relative volume content of hemoglobin/blood in the measuring volume, StO2 the oxygen saturation of the hemoglobin, and NIR-perfusion a measure of perfusion quality calculated from the NIR-spectral region. The algorithms for THI and StO2 as described in [14] and [15] use wavelength segments from 500 to 800 nm, restricting the depth sensitivity. In the NIR-region (NIR-perfusion), there is no separation of relative volume content and oxygen saturation.

Because the remission spectra can be completely reproduced by the five-layer model parameters, the THI-, StO2-, and NIR parameters can be principally calculated from the model parameters. THI and StO2 are related to a mixture of the vHb resp. $xHbO_2$ of layers 1–4, the NIR-perfusion parameter is a function of (vHb $\bullet$ $xHbO_2$), and vHb and $xHbO_2$ of layers 4–5.

Although these one-layer model parameters have been shown to be of high clinical value, the new five-layer model represents a description of the perfusion situation, especially differentiating between the superficial capillary blood volume and oxygen saturation and the parameters of the deeper vessel system. This gives rise to better clinical estimations of perfusion quality, or disturbances of the perfusion system. This additional clinical value will be described for different application areas in subsequent publications.

2.3. Wound Healing Disorders

Objective diagnostics in wound healing disorders is a long-term problem with no implications in daily life. The intra- and inter- observer difference is often discussed in the literature [23,24]. A validated, computer-assisted measurement tool based on conventional RGB-imaging has been available for some fifteen years [25,26]. Based on color segmentation, the software is able to quantify the surface, wound borderline, diameter, numeric and percentile part of necrotic tissue, and the fibrin and granulation tissue. Based on this quantification, we are able to analyze the progress of surface reduction, the progression of granulation tissue, and part of the fibrin and necrotic tissue. If the progress of granulation tissue is reduced, we have to check the local therapeutic concept or identify the underlying reasons, e.g., perfusion, edema, oxygenation, infection, etc.

Additionally, perfusion can be measured by ultrasound and oxygenation with $TcPO_2$; however, the examination is time consuming, and edema can only be measured with the circumferences of lower legs.

With hyperspectral imaging, all the following parameters are available within one measurement: perfusion, as described above; based on the characteristic features of the remission spectra, a detailed and quantified segmentation and classification of the wound area, providing portions of necrotic tissue, fibrin, granulation and epithelial tissue; and tissue which is endangered by insufficient perfusion.

The advanced and specific methodology for the clinical application areas addressed in the examples will be described in detail in separate articles.

3. Discussion

The requirements for the development of the data processing were:

- The tissue model should describe the physiological structure in a manner which is sufficiently detailed to enable information retrieval, especially concerning the perfusion situation with high clinical value (adequacy);
- The modeling should be able to reproduce real measured remission spectra from skin and wounds over the complete spectrum in detail; the variety of spectra is described in the confidence range, and should sufficiently cover a variety of clinical problems (consistency);
- The solution of the inverse problem should be practicable for imaging measurements with the described measuring geometry in clinical routine environment; the processing should be fast for imaging measurements (practicability).

The challenge is to find a reasonable compromise between the flexibility and adaptivity of the tissue model (many parameters), the physiological informative value, and the physical–mathematical correctness of the solution of the inverse problem.

The described tissue model seems to be sufficiently detailed to offer insights into the perfusion situation, and fulfills the adequacy requirement. The determined model parameters represent perfusion values of the capillary system and the deeper vessel system, and seem to be more informative concerning the perfusion situation.

The values have been proved to be physiologically and clinically plausible (up to now), and the multitude of parameters constitute a better basis for classification processes.

The processing also fulfills the consistency and the practicability requirement. A complete processing of a measurement image with a 50% tissue content needs approx. 10–15 seconds for the 3D physiological perfusion imaging result to be determined.

Many details of the spectrum forms are explainable by the modeling and the dynamics of measuring depth variation over the spectral range. In $\Lambda 3$, the measuring depth changes very dynamically with the wavelength. Structures such as those at 650nm and 715nm, the rise at 600nm, as well as in Λ_5 and Λ_6, are only explainable by the dynamic transition between different layers, and have to be distinguished from biochemical contributions to the spectrum.

The other side of the compromise is that the modeling of the radiation transfer through the system is not physically stringent:

The spectral segments are selected by simple plausibility arguments based on knowledge about the penetration depth in perfused human tissue. By this predetermined dependence upon measuring depth, the spectral segments define the layer thickness relative to the standard value D_1^0.

The heuristic visibility function, enabling the layer separation and differentiation in the successive procedure is based on a theoretical and simulative analysis. The specification of this function, as well as the dependence of the mean path length on the wavelength, is accessible to further refinement and optimization, for instance by expanding to higher orders (taking into account the nonlinearity of the "visibility"). The globally fixed function $D = f_D^A(A)$ could be empirically diversified for different spectrum forms and different layer structures.

The use of globalized heuristic functions does not sufficiently correspond to the variety of individual forms of skin systems; the interpretation of the results has to assessed with respect to these limitations.

The parameters have to be empirically validated concerning their physiological interpretation and clinical information content (in consideration of the different clinical context).

Clinically-validated classification processes based on the model parameters will account for gaining confidence in the usability and adequacy of the parameters.

The data processing of HSI remission spectra based on a five-layer model of perfused tissue generates perfusion parameters for every layer and presents them as depth profiles. The evaluation procedure transforms the whole information of the spectrum consistently into model parameters so that the measured spectrum can be completely reproduced.

For the first time, we present a complete system of powerful hyperspectral imaging data acquisition and data processing with high applicability in clinical practice. The main advance of the data processing method is its enhanced information content with highly plausible physiological interpretation and high clinical relevance, which is currently not available with other methods.

The data processing is integrated into a piece of software running on a computer which is associated with the hyperspectral camera. The data processing requires approx. 10–15 seconds, so that directly after data acquisition, the perfusion parameters are presented to the physician and the patient (bedside diagnostics).

4. Methods and Materials

4.1. Hyperspectral Measuring System

All measurements were performed with a HSI-camera TIVITA® Tissue (Diaspective Vision GmbH; Am Salzhaff, Germany) with written consent from volunteers. Data acquisition from patients was conducted in accordance with the Declaration of Helsinki, and the protocol was approved by the Ethics Committee of the Ärztekammer Sachsen-Anhalt, Germany (35/17). All patients gave informed consent.

The camera was a compact measuring system certified for clinical use [27]. Remission spectra were recorded in the spectral range of 500 to 1000 nm with a resolution of 5 nm; the measuring area was approx. 20×30 cm, standard image size was 640×480 Pixel, and the recording needed approx. 5 seconds.

4.2. Hyperspectral Imaging Data Analysis and Processing

To ensure good qualitative and undisturbed measuring data, the following tests were performed:

- Regular tests of the camera calibration and comparison of spectra from reference objects with corresponding reference spectra.

    By software:

- Quality tests of the spectra concerning wavelength-dependent noise to ensure that relevant spectral details for parameter estimation are presented in sufficient quality;
- Tests concerning disturbing influences on the spectra, such as reflection, external light, and strong inclination of parts of the measuring area.

To define adequate quality measures, experimental tests and numerical Monte Carlo simulations [28] were performed.

In the preprocessing procedure of the measuring data, the data quality was tested; data of insufficient quality were excluded from further processing.

4.2.1. Model-Based Analysis

The skin is modeled as a five-layer-system (Figure 8). Every layer is regarded as homogenous, and is provided with the relevant components:

- Layer 1 (stratum corneum, epidermis): melanin, vHb, and $xHbO_2$; vHb denotes the relative volume fraction of total hemoglobin, $xHbO_2$ the oxygen saturation of hemoglobin; layer 1 contains also blood and $xHbO_2$, because this layer cannot be sufficiently separated from the next;

- Layer 2 (upper dermis: papillary or capillary system): vHb, xHbO$_2$, and collagen structure;
- Layer 3 (reticular dermis): vHb, xHbO$_2$, and collagen structure;
- Layer 4 (deep dermis, subcutis): vHb, xHbO$_2$, vH$_2$O, vFat, collagen structure, and connective tissue; vH2O and vFat denote the volume fractions for water and fat;
- Layer 5 (subcutis): vHb, xHbO$_2$, vH$_2$O, vFat, and connective tissue.

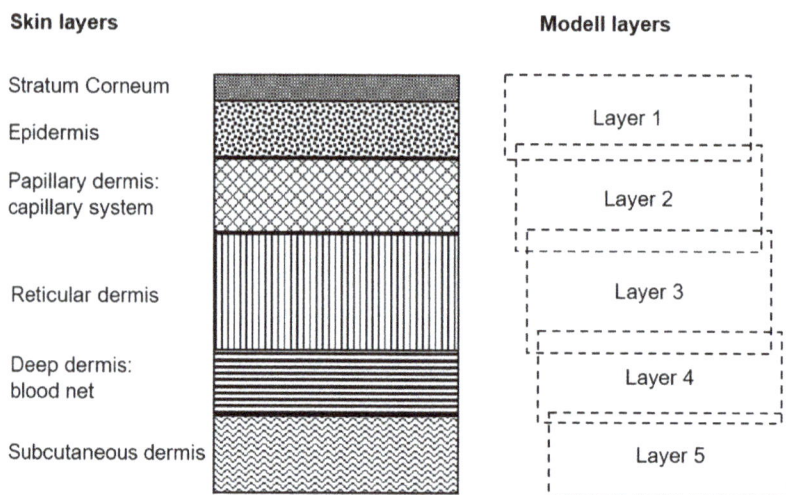

Figure 8. Five-layer skin model.

For every layer, the absorption of hemoglobin, water, and fat is explicitly described in a linear approximation; the background absorption and scattering by the collagen matrix, vessels, and connective tissue is jointly described by a linear function containing the so-called intrinsic structure parameters:

$$\text{"Absorbance"} \ A = \ln\left(\frac{R}{I_0}\right) = s_0 + s_1 l + L \sum_i \vartheta_i \, \varepsilon_i \tag{1}$$

where R: remission, I_0: incident intensity, $S(L) = s_0 + s_1(L)$: intrinsic contributions to absorption and scattering; ϑ_i: volume fraction, and ε_i extinction coefficient of component i; L denotes a mean path length, which could be calculated from the path length distribution [29].

Especially for hemoglobin, the derivates Hb ad HbO$_2$ are represented in the form

$$\vartheta_H \left(\varepsilon_{HbO2} + x \, \varepsilon_{Hb} \right) \tag{2}$$

with ϑ_H as the volume fraction of the total haemoglobin and x as the oxygen saturation of the haemoglobin.

The measuring geometry used with this HSI-camera with laminar illumination precludes the separation of different layers by technical control of the path length distribution. The remission at one measuring point is given by an integral over many path length distributions; the measuring volume defined by this distribution varies with the wavelength, depending on the scattering and absorption.

The form of the spectra is mainly determined by the absorption spectra of hemoglobin (mainly in the range 500–600nm and around 760 nm), as well as by the water and fat absorption spectra increasingly from approx. 700 nm (see Figure 9). The remission spectra contain contributions from the different layers in the measuring volume with a measuring depth depending on the wavelength.

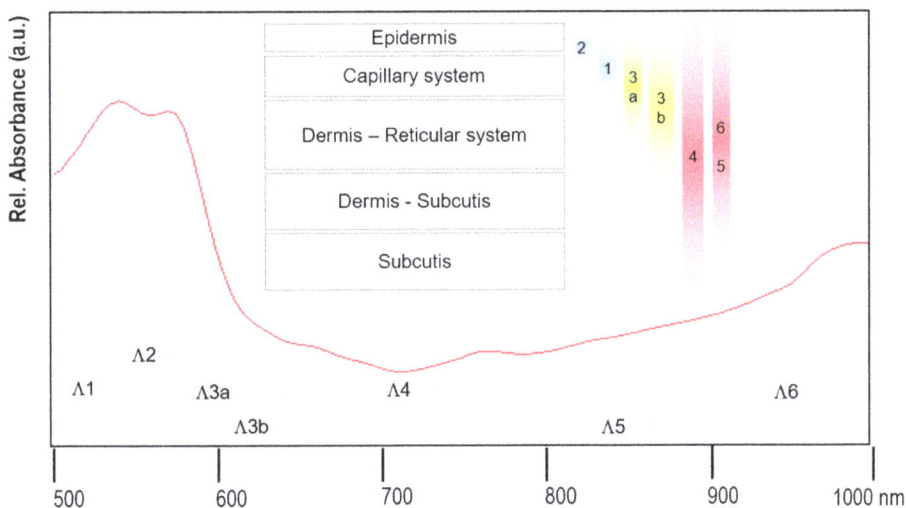

Figure 9. Remission spectrum represented in the "absorbance" mode. Spectral segments Λ_i and schematic: skin layers and approximate measuring depth in the spectral segments.

4.2.2. Transformation of the HSI-Remission Spectra

In this modeling, the parameters in L ϑ cannot be determined separately (the scattering is not described explicitly in the model), and depending on the wavelength, $L(\Lambda)$ may have different values.

Because layer thickness and path length distributions are not explicitly determinable parameters in this model, a depth scale cannot be defined. To obtain a depth profile, we have to make concrete statements about the measuring depth and the path length.

Basis is an analysis of the path length distribution which is dependent on the wavelength to estimate $L(\Lambda)$; therefore, the measuring depth D is defined as the maximal depth with minimal intensity I_{min}: path length distribution $h(l, L)$ Þ $L(L) = f(s(L), a(L))$; s: scattering; a: absorption; measuring depth D: $\frac{I}{I_0} = e^{-a \cdot L} = I_{min}$Þ $L_{min} = \frac{-\ln(I_{min})}{a}$; without further knowledge about the dependencies between D and L, the measuring depth D corresponding to the path length L_{min} is supposed to be $D = f_D^a \, L_{min}$.

Because the actual path length distribution is not known, as a first approach, a global function $D(L) = f_D^A(A(L)) \, (\approx \frac{1}{a_0(L)})$ is used, including a globally fixed $a_0(\lambda)$. Thereby, the measuring depth D for a spectral segment Λ becomes determinable using the total absorbance A of the system (corresponding to the assumption of a homogenous system and the dependence of the measuring depth on μ_a (absorption) and μ_s (scattering)).

Thereby, different measuring depths can be assigned to different spectral segments:

In the segment 535–585nm (Λ_2), the measuring depth is least and defines layer 1. The segment 500–535 (Λ_1) comprises layers 1 and 2, 585–595nm (Λ_3) layers 1, 2 and 3. The segment 595–690nm (Λ_4) additionally comprises layer 4. The segment 690–825nm (Λ_5) comprises all layers (1–5), and segment 825–1000nm (Λ_6) layers 1–4.

In Λ_3, the measuring depth changes very dynamically from layers 3 to 5; therefore, the segment is further subdivided in Λ_{3a} und Λ_{3b}.

Due to the higher absorption of H_2O and fat, the measuring depth is reduced in Λ_5 und Λ_6 in comparison to Λ_4. Water and fat fractions can only be determined in Λ_5, and especially Λ_6, with sufficient reliability.

It has to be emphasized once again that the layers are defined by the measuring depths of the spectral segments.

The spectral segments are not strictly fixed, but may be adapted to the actual form of the total absorbance spectrum, and are therefore overlapping.

The different layers contribute differently to the resulting remission spectrum depending on the spectral segments. The nonlinear relation between the layer contributions and the remission spectrum is not explicitly modeled in this framework. Instead, to achieve a separation of the layers, a heuristic function is introduced describing the "visibility" of a lower layer (layer 2) underlying an upper layer (layer 1) in a first order linear approximation:

$$\text{Visibility function } f_V = f_v^0 \left(\frac{D_2 - D_1}{D_1}\right)^\alpha R_1{}^\beta e^{-a_1 L_1} \tag{3}$$

where D_1 denotes the measuring depth (layer thickness) of layer 1 and D_2 of layers 1 and 2, L_1 is the mean path length corresponding to the measuring depth D_1, and R_1 the remission of layer 1.

In a first approximation, the pathways through layer 2 are only affected in layer 1 by absorption a_1, and therefore, by path length L_1 ($\beta = 0$). The exponent α, determining the volume portion of layer 2 relative to layer 1, is globally fixed.

The determination of the parameter of the visibility function is based on comparison with Monte Carlo simulations of two-layer systems.

Because D_1 resp. L_1 are not known for individual measurements, standard values are defined: $L_1^0 = \frac{D_1^0}{f_D^A}$.

In the first step, for every spectral segment, a numerical adaptation to a homogenous equivalent system (i.e., a homogeneous, one-layer model with the relevant components) is performed. The adaptation quality is a measure of the appropriateness of the segment selection, and therefore, for the layer structure. Inside the segment, the dependence on the measuring depth should be low.

For the approximate determination of the layer contributions, two-layer modeling is performed successively for the underlying layers:

1. The volume captured by Λ_2 is defined as layer 1. From the remission $R_1(\Lambda_2)$, the parameters $S_1(L_2)$ and $\vartheta_1 L_1(L_2)$ ($L_1(L_2) = L_1^0$) are determined.

2. In Λ_1, layer 2 is also captured; the combined remission $R_{12}(\Lambda_1)$ can be presented in the form $R_{12}(L_1) = R_1(L_1) + f_V(L_1) \cdot R_2(L_1)$, with f_V as the visibility function. $R_1(\Lambda_1)$ results from $S_1(L_1)$ and $\vartheta_1 L_1(L_1)$, with $L_1(L_1) = f_L(L) L_1^0$.

From the remission spectrum $R_2(L_1) = \frac{R_{12}(L_1) - R_1(L_1)}{f_V(L_1)}$, the parameters of layer 2 are determined. The parameters refer to the path length $\Delta L_2 = L_2(L_1) - L_1(L_1)$ (approximately), and are finally scaled with respect to the standard value L_1^0. $a_1 L_1$ in the visibility function is determined using the total absorbance $A(\Lambda_1)$.

3. In the further segments, i.e., Λ_3 etc., the further layers (3, etc.) are successively captured. The processing is analogue to 2. ($R_{123} = R_{12} + f_V(L_3) R_3$, etc.).

From this result, the component parameters of the layers are ($s_0, s_1, \{\vartheta_i L_1^0\}$) and the visibility function $f_V(L_i)$ (parameter D) for the actual spectral segments is Λ_i. D is the mean measuring depth of the actual layer. A depth range $[D_{min} \ldots D_{max}]$ for every layer is stored. With these values and the global function $f_D^A(A)$, the spectral segments can be reconstructed, with A as the absorbance of the total HES.

4.2.3. Reconstruction of the Spectrum

With the spectral segments, the layer parameters and the visibility function, the complete spectrum can be reconstructed successively:

1. With Λ_2 and the parameters of layer 1, $R_1(\Lambda_2)$ is calculated.

2. With Λ_1, the parameters of layer 2, $D_2(\Lambda_1)$, $R_2(\Lambda_1)$, and $A_1(\Lambda_1)$, $R12(\Lambda_1)$ is calculated.

   Based on R12, the parameters of HES12(Λ_1) are determined.

3. With Λ_3 and the parameters of layer 3, $R_3(\Lambda_3)$ is calculated from $D_2(\Lambda_3)$ and $D_3(\Lambda_3)$, as well as $R12(\Lambda_3)$ and $A12(\Lambda_3)$. With HES12, $R123(\Lambda_3)$ is calculated.

   Analogue processing for the further layers.

   Consistency: With the model parameters determined by this process, the measured spectrum can be reproduced nearly perfectly (see Figure 10). This means that the information contained in the spectrum is practically completely transformed into the model parameters.

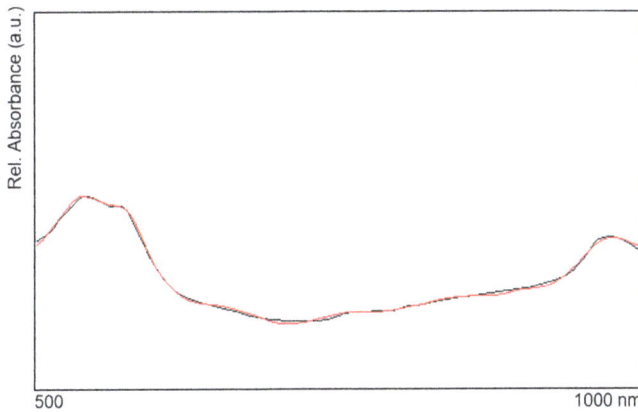

Figure 10. Measured spectrum (absorbance) (red) and reproduced spectrum by the model (black).

Uniqueness: Generally, there is no unique adaptation maximum in the (model) parameter space, especially for the spectral segments, except Λ_2 and partially Λ_5 and Λ_6. To reduce potential ambiguity, the pathway between the actual maximum and the successively following maximum for the next spectral segment is estimated by additional intermediate segments (not fulfilling the requirement of quasi-stationarity with respect to the measuring depth). Thereby, the actual valid maximum can be selected with a higher level of probability.

Principally, in each case, even the global maximum of adaptation cannot be regarded as the "true" solution due to the limited reality of the modeling system.

4.2.4. Parameters and Confidence Range of Modeling

The parameters vHb und xHbO$_2$ named in the layer model are related to the model parameters: $vHb = \vartheta_{H1}\, L_1^0$; $xHbO_2 \equiv x$ in formula (2).

Because L_1^0 is a globally fixed parameter, vHb represents an index value (range [0...2.5]); the x-values are in the range [0...1].

The physiologically-acceptable variation ranges of the model parameters define the variety of spectrum forms representable by the model. In the processing procedure, every real spectrum is proved to be within this confidence range before further processing.

The reproduction quality of the spectrum is a test of consistency.

5. Conclusions

Despite the aforementioned limitations, the presented processing method provides a more differentiated outcome in relation to the perfusion situation in the layered tissue structure, and comprehensively utilizes the information content of hyperspectral measuring data.

The examples show the potential for creating a new, valuable, clinical procedural and investigative category in different medical fields. The processing is a further step for establishing hyperspectral imaging in medicine, and considers the measuring conditions and essential requirements for clinical practicability.

To create a supporting powerful diagnosing system using hyperspectral imaging technology, model-based data processing has to be complemented by an efficient, knowledge-based method.

6. Further Validations and Developments

A fundamental problem is the lack of reliable and accurate reference methods for detailed perfusion values in the layers. A systematic comparison with a Monte-Carlo simulation is in progress, as well as a comparison with spectroscopic measurement methods, enabling control of the measuring depth, and other methods depicting the layer structure, e.g., OCT.

Methodical progressions concern the improvement of the modeling of radiation transport, the separation of the layers, and more generally, a reduction of the described limitations.

Author Contributions: J.M., writing—original draft preparation, methodology, formal analysis; A.K., methodology, resources; F.S. resources, validation; D.C. resources; A.A., validation, writing—review and editing; D.P., resources; G.D., resources, validation; T.W., writing—review and editing, validation.

Funding: This research received no external funding.

Conflicts of Interest: The author Axel Kulcke is affiliated with the company Diaspective Vision GmbH. The authors declare no conflict of interest.

References

1. Lu, G.; Fei, B. Medical hyperspectral imaging: A review. *J. Biomed. Opt.* **2014**, *19*, 1–23. [CrossRef] [PubMed]
2. Paul, D.W.; Ghassemi, P.; Ramella-Roman, J.C.; Prindeze, N.J.; Moffatt, L.T.; Alkhalil, A.; Shupp, J.W. Noninvasive imaging technologies for cutaneous wound assessment: A review. *Wound Repair Regen.* **2015**, *23*, 149–162. [CrossRef] [PubMed]
3. Best, S.L.; Thapa, A.; Jackson, N.; Olweny, E.; Holzer, M.; Park, S.; Wehner, E.; Zuzak, K.; Cadeddu, J.A. Renal oxygenation measurement during partial nephrectomy using hyperspectral imaging may predict acute postoperative renal function. *J. Endourol.* **2013**, *27*, 1037–1040. [CrossRef] [PubMed]
4. Kiyotoki, S.; Nishikawa, J.; Okamoto, T.; Hamabe, K.; Saito, M.; Goto, A.; Fujita, Y.; Hamamoto, Y.; Takeuchi, Y.; Satori, S.; et al. New method for detection of gastric cancer by hyperspectral imaging: A pilot study. *J. Biomed. Opt.* **2013**, *18*, 26010. [CrossRef] [PubMed]
5. Gerstner, A.O.H.; Martin, R.; Westermann, S.; Mahlein, A.K.; Schmidt, K.; Thies, B.; Laffers, W. Hyperspectral imaging in head and nek oncology. *Laryngo Rhino Otol.* **2014**, *92*, 453–457.
6. Chen, Y.; Shen, Z.; Shao, Z.; Yu, P.; Wu, J. Free Flap Monitoring Using Near-Infrared Spectroscopy: A Systemic Review. *Ann. Plast. Surg.* **2016**, *76*, 590–597. [CrossRef] [PubMed]
7. Calin, A.A.; Parasca, S.V.; Savastru, R.; Manea, D. Characterization of burns using hyperspectral imaging technique—A preliminary study. *Burns* **2015**, *41*, 118–124. [CrossRef]
8. Denstedt, M.; Pukstad, B.S.; Paluchowski, L.A.; Hernandez-Palacios, J.E.; Randeberg, L.L. Hyperspectral imaging as a diagnostic tool for chronic skin ulcers. In Proceedings of the Photonic Therapeutics and Diagnostics IX: SPIE BiOS, San Francisco, CA, USA, 8 March 2013.
9. Barberio, M.; Maktabi, M.; Gockel, I.; Rayes, N.; Jansen-Winkeln, B.; Köhler, H.; Rabe, S.M.; Seidemann, L.; Takoh, J.P.; Diana, M.; et al. Hyperspectral based discrimination of thyroid and parathyroid during surgery. *Curr. Dir. Biomed. Eng.* **2018**, *4*, 399–402. [CrossRef]
10. Jansen-Winkeln, B.; Holfert, N.; Köhler, H.; Moulla, Y.; Takoh, J.P.; Rabe, S.M.; Mehdorn, M.; Barberio, M.; Chalopin, C.; Neumuth, T.; et al. Determination of the transection margin during colorectal resection with hyperspectral imaging (HSI). *Int. J. Colorectal Dis.* **2019**, *34*, 731–739. [CrossRef]
11. Köhler, H.; Jansen-Winkeln, B.; Maktabi, M.; Barberio, M.; Takoh, J.; Holfert, N.; Moulla, Y.; Niebisch, S.; Diana, M.; Neumuth, T.; et al. Evaluation of hyperspectral imaging (HSI) for the measurement of ischemic conditioning effects of the gastric conduit during esophagectomy. *Surg. Endosc.* **2019**, *10*, 1–8. [CrossRef]

12. Sowa, M.G.; Kuo, W.C.; Ko, A.C.; Armstrong, D.G. Review of near infrared methods for wound assessment. *J. Biomed. Opt.* **2016**, *21*, 091304. [CrossRef] [PubMed]
13. Wild, T.; Becker, M.; Winter, J.; Schuschenk, N.; Daeschlein, G.; Siemers, F. Hyperspectral imaging of tissue perfusion and oxygenation in wounds: Assessing the impact of a micro capillary dressing. *J. Wound Care* **2018**, *27*, 38–51. [CrossRef] [PubMed]
14. Kulcke, A.; Holmer, A.; Wahl, P.; Siemers, F.; Wild, T.; Daeschlein, G. Compact hyperspectral camera for measurement of perfusion parameters in medicine. *Biomed. Eng. Biomed. Tech.* **2018**, *63*, 519–527. [CrossRef] [PubMed]
15. Holmer, A.; Marotz, J.; Wahl, P.; Dau, M.; Kämmerer, P.W. Hyperspectral imaging in perfusion and wound diagnostics—Methods and algorithms for the determination of tissue parameters. *Biomed. Tech.* **2018**, *63*, 547–556. [CrossRef]
16. Holmer, A.; Tetschke, F.; Marotz, J.; Malberg, H.; Markgraf, W.; Thiele, C.; Kulcke, A. Oxygenation and perfusion monitoring with a hyperspectral camera system for chemical based tissue analysis of skin and organs. *Physiol. Meas.* **2016**, *37*, 2064–2078. [CrossRef]
17. Daeschlein, G.; Langner, I.; Wild, T.; von Podewils, S.; Sicher, C.; Kiefer, T.; Juenger, M. Hyperspectral imaging as a novel diagnostic tool in microcirculation of wounds. *Clin. Hemorheol. Microcirc.* **2017**, *67*, 467–474. [CrossRef]
18. Kienle, A.; Patterson, M.S.; Dögnitz, N.; Bays, R.; Wagnieres, G.; van den Bergh, H. Noninvasive determination of the optical properties of two-layered turbid media. *Appl. Opt.* **1998**, *37*, 779–791. [CrossRef]
19. Liemert, A.; Kienle, A. Light diffusion in N-layered turbid media: Steady-state Domain. *J. Biomed. Opt.* **2010**, *15*, 025003. [CrossRef]
20. Naglic, P.; Vidovic, L.; Milanic, M.; Randeberg, L.L.; Majaron, B. Suitability of diffusion approximation for inverse analysis of diffuse reflectance spectra from human skin in vivo. *OSA Contin.* **2019**, *2*, 905–922. [CrossRef]
21. Bjorgan, A.M.; Milanic, M.; Randeberg, L. Estimation of skin optical parameters for real-time hyperspectral imaging applications. *J. Biomed. Opt.* **2014**, *19*, 066003. [CrossRef]
22. Zherebtsov, E.; Dremin, V.; Popova, A.; Doronin, A.; Kurakina, D.; Kirillin, M.; Meglinski, I.; Bykov, A. Hyperspectral imaging of human skin aided by artificial neural networks. *Biomed. Opt. Express* **2019**, *10*, 3545–3558. [CrossRef] [PubMed]
23. Mekkes, J.R.; Westerhof, W. Image processing in the study of wound healing. *Clin. Dermatol.* **1995**, *13*, 401–407. [CrossRef]
24. Stremitzer, S.; Wild, T.; Hoelzenbein, T. How precise is the evaluation of chronic wounds by health care professionals? *Int. Wound J.* **2007**, *4*, 156–161. [CrossRef] [PubMed]
25. Wild, T.; Prinz, M.; Fortner, N.; Krois, W.; Sahora, K.; Stremitzer, S.; Hoelzenbein, T. Digital measurement and analysis of wounds based on colour segmentation. *Eur. Surg.* **2008**, *40*, 325–329. [CrossRef]
26. Jelinek, H.F.; Prinz, M.; Wild, T. A digital assessment and documentation tool evaluated for daily podiatric wound practice. *Wounds* **2013**, *25*, 1–6.
27. Marotz, J.; Siafliakis, A.; Holmer, A.; Kulcke, A.; Siemers, F. First results of a new hyperspectral camera system for chemical based wound analysis. *Wound Med.* **2015**, *10*, 17–22. [CrossRef]
28. Herrmann, B.H.; Hornberger, C. Monte-Carlo Simulation of Light Tissue Interaction in Medical Hyperspectral Imaging Applications. *Curr. Dir. Biomed. Eng.* **2018**, *4*, 275–278. [CrossRef]
29. Stratonnikov, A.A.; Loschenov, V.B. Evaluation of blood oxygen saturation in vivo from diffuse reflectance spectra. *J. Biomed. Opt.* **2001**, *6*, 457–467. [CrossRef]

Sample Availability: Not available.

molecules

MDPI

Article

Distinct Difference in Sensitivity of NIR vs. IR Bands of Melamine to Inter-Molecular Interactions with Impact on Analytical Spectroscopy Explained by Anharmonic Quantum Mechanical Study

Justyna Grabska [1], Krzysztof B. Beć [1,2,*], Christian G. Kirchler [1], Yukihiro Ozaki [3] and Christian W. Huck [1]

[1]   Institute of Analytical Chemistry and Radiochemistry, Leopold-Franzens University, Innrain 80/82, CCB-Center for Chemistry and Biomedicine, 6020-Innsbruck, Austria; justyna.grabska7@gmail.com (J.G.); Christian.Kirchler@uibk.ac.at (C.G.K.); Christian.W.Huck@uibk.ac.at (C.W.H.)
[2]   Faculty of Chemistry, University of Wrocław, F. Joliot-Curie 14, 50-383 Wrocław, Poland
[3]   Department of Chemistry, School of Science and Technology, Kwansei Gakuin University, Sanda, Hyogo 669-1337, Japan; ozaki@kwansei.ac.jp
*   Correspondence: krzysztof.bec@chem.uni.wroc.pl

Received: 14 March 2019; Accepted: 7 April 2019; Published: 10 April 2019

Abstract: Melamine (IUPAC: 1,3,5-Triazine-2,4,6-triamine) attracts high attention in analytical vibrational spectroscopy due to its misuse as a food adulterant. Vibrational spectroscopy [infrared (IR) and Raman and near-infrared (NIR) spectroscopy] is a major quality control tool in the detection and quantification of melamine content. The physical background for the measured spectra is not interpreted in analytical spectroscopy using chemometrics. In contrast, quantum mechanical calculations are capable of providing deep and independent insights therein. So far, the NIR region of crystalline melamine has not been studied by quantum mechanical calculations, while the investigations of its IR spectra have remained limited. In the present work, we employed fully anharmonic calculation of the NIR spectrum of melamine based on finite models, and also performed IR spectral simulation by using an infinite crystal model—periodic in three dimensions. This yielded detailed and unambiguous NIR band assignments and revised the previously known IR band assignments. We found that the out-of-plane fundamental transitions, which are essential in the IR region, are markedly more sensitive to out-of-plane inter-molecular interactions of melamine than NIR transitions. Proper description of the chemical surrounding of the molecule of melamine is more important than the anharmonicity of its vibrations. In contrast, the NIR bands mostly arise from in-plane vibrations, and remain surprisingly insensitive to the chemical environment. These findings explain previous observations that were reported in IR and NIR analytical studies of melamine.

Keywords: melamine; FT-IR; NIR spectroscopy; quantum chemical calculation; anharmonic calculation; overtones; combination bands

---

1. Introduction

Melamine (IUPAC: 1,3,5-Triazine-2,4,6-triamine) has wide industrial importance, nowadays being used e.g., in the manufacture of polymers and resin [1], concrete [2], flame-resistant materials [3], and it may be utilized in the production of nanomaterials (e.g., *N*-doped carbon nanotubes) [4]. In the past, it was even more widely applied in industry and agriculture [5]. It was unfortunate that melamine has become infamous worldwide as a dairy adulterant after it caused a milk safety crisis in 2008 with severe casualties (290,000 people affected with 51,900 hospitalized in China only) [6]. That event had a global impact on the food industry, food production, supply chains, and corresponding legal regulations [7,8]. It strongly echoed in the field of food quality control, leading to a strong stimulus for development and the adaptation of adequate analytical routines [8–10]. A number of other food safety incidents in recent years have induced particular pressure on this area of analytical chemistry [11,12]. The methods that are based on vibrational spectroscopy [infrared (IR), Raman and near-infrared (NIR)] have become particularly important elements of this effort in controlling the food safety at every stage of its production and supply [13–15].

Vibrational spectroscopy stands out as a non-invasive, widely applicable, low-cost, and quick time-to-result analytical method. Therefore, it combines advantages that are highly valued in analytical chemistry. Despite being grouped together, the key differences among these three kinds of techniques should be noted. IR (4000–400 cm^{-1}) spectroscopy elucidates chemical information from the fundamental vibrational transitions. In contrast, the signal that was measured in NIR (10,000–4000 cm^{-1}) spectroscopy originates from the excitations of higher quanta transitions, mostly first overtones and binary combinations [16–18]. Raman spectroscopy also provides information regarding fundamental vibrations, but through a distinctly different working principle than IR spectroscopy. The differences in the wavelength regions and underlying physical background translate into distinct differences in the instrumentation and applicability of these methods. Each of these techniques offer unique advantages, but for the detection/quantification of melamine content in milk, NIR spectroscopy may be favored [9]. A number of factors contribute to this fact [19]. Higher sample volume resulting from low NIR absorptivity of matter in general, and water, in particular, allows for more straightforward measurement of transmittance or reflectance of the milk sample. The same may be achieved with ATR-IR (Attenuated Total Reflection IR) approach; however, the fiber probe compatibility of NIR instrumentation gives it superior flexibility in high-volume analysis. NIR spectroscopy also benefits from the largest tolerance for the sample inhomogeneity. Raman spectroscopy is suitable for the measurement of aqueous samples, but for similar analytical applications Raman instrumentation is often more expensive. Finally, in NIR spectroscopy, a strongly stimulated development for miniaturization [20,21] has culminated in highly affordable micro-spectrometers, which are available under 300 USD nowadays [22]. In a strict application to melamine detection/quantitation, all three techniques have been successfully used in the literature 9]. However, NIR spectroscopy demonstrates the best analytical performance in this case [9], on top of its practical advantages [23,24].

The nature of NIR spectra (overlaying overtones and combinations) [16–18] largely limits their interpretability [25,26]. Chemometric methods do not provide physical insights on the analyzed sample, and NIR spectroscopy is often used as a "black-box". In contrast, for IR and Raman spectroscopy, this limitation is less severe due to its more simple spectra with milder fundamental band overlapping [27,28]. Coincidently, quantum chemistry offers affordable methods (harmonic approximation) for the adequately accurate simulation of IR and Raman spectra [29,30]. In contrast, prediction of NIR bands require resource intensive anharmonic methods [31,32]. The difference in resource demand (harmonic vs. anharmonic approximation) is significant, and theoretical NIR studies of complex molecules have only recently appeared [33–38]. Lately, we have reported the quantum mechanical calculations of NIR spectra of various molecules in solution, liquid, and solid state, including short- [34], medium- [35], and long-chain [36] fatty acids. These studies could have been used, e.g., for the interpretation of the meaningful NIR bands that influence the chemometric models used in quantification of the content of phytopharmaceutical compounds in natural drugs [37–39]. On the other hand, simulations of IR spectra in crystalline phase that use a proper representation of infinite crystal lattice by a three-dimensional (3D) periodic model remain equally rare.

In the literature so far, quantum chemical calculation vibrational studies of melamine have been limited to finite models and harmonic approximation [40,41]. Mircescu et al. have harmonically calculated IR and Raman spectra of melamine [40]. They have used the single molecule model, and cluster of 10 melamine molecules. They have compared the spectra that were calculated with these two approaches and judged that the 10 molecule cluster model leads to a better quality of the calculated IR and Raman spectra of melamine. They concluded that the hydrogen-bonding of melamine in crystal needs to be taken into account in order to yield accurate calculated vibrational spectra. Accordingly, the calculations that are based on the 10 molecule cluster have led to much improved simulated IR and Raman spectra. Yuan et al. have drawn similar conclusions [41] in their quantum mechanical calculations of IR spectra of melamine. They have used a single molecule model, a four-molecule cluster with two hydrogen-bonds, and a large cluster consisting of 32 molecules of melamine featuring 30 intermolecular hydrogen-bonds. They have compared the IR spectra that were calculated on the basis of these models and concluded that proper representation of the hydrogen-bonded structure of melamine is essential in improving the quality of the calculated IR spectrum [40,41].

Therefore, the earlier studies [40,41] have recognized the importance for spectra calculation of the proper description of the chemical neighborhood in crystalline melamine, in particular, the hydrogen bonding network. However, the methodology in these studies has been limited to finite models, clusters of melamine molecules (10 to 32 molecules). Although improved vs. single molecule models, the finite boundary of these model clusters has led to the distortions of the molecular structure as compared with the structure of the crystal lattice of melamine. This has resulted in a number of "phantom bands" appearing in the calculated IR and Raman spectra, which could not be observed in the experimental spectra [40,41]. Additionally, these previous studies have been limited to harmonic approximation, which made any calculations of NIR bands unavailable. Accordingly, there are no NIR spectra simulations of melamine of any kind reported so far. On the other hand, melamine was intensively focused on in analytical near-infrared reflectance spectroscopy (NIRS) [9,42–46]; however, these studies have not been able to derive insightful physicochemical information on melamine.

The purpose and novelty of the present study is to explore NIR vs. IR spectra correspondences in crystalline melamine. To achieve this, for the first time, we employ anharmonic quantum mechanical calculations of NIR spectra of melamine, by using two different approaches, which we directly compare. Moreover, we improve the previous investigations of IR spectra in a well-defined crystalline lattice by employing an infinite three-dimensional (3D) periodic model of the crystalline melamine, for the first time as well. This yields more accurate calculated IR spectrum, but it also is essential in obtaining good comprehension of a number of relevant effects. In example, the impact of anharmonicity may become well separated from the influence of the chemical neighborhood. The distinct difference in the importance of inter-molecular interactions for the accurate reproduction of IR and NIR transitions of melamine is found and explained. This means that, in contrast to IR bands, the accurate reproduction of NIR bands requires significantly less attention in describing the long-range, and in particular, inter-plane, interactions in crystalline melamine.

2. Results and Discussion

2.1. Experimental and Simulated IR Spectra of Crystalline Melamine

The simulation of the IR spectrum of melamine in polycrystalline state requires a proper representation of the long-range ordered structure (Figure 1). There exists a decisive decrease in the accuracy for the IR spectrum calculated on the basis of finite model, even in anharmonic approximation (Figure 2). Such spectra are markedly poor and numerous bands are missing (Figure 2C,D). In contrast, the spectrum that was calculated for infinite (3D periodic) model (Figure 2B) correctly reproduces all of the major experimental bands (Figure 2A). The overestimation of the calculated peak positions, particularly noticeable above 3000 cm^{-1}, likely results foremost from neglecting the anharmonic effects that are typically strong in the X-H stretching region. However, this may be accurately corrected by employing wavenumber scaling. This observation may appear obvious; however, in light of NIR simulations (as discussed in Sections 2.2 and 2.3), it leads to farer-reaching conclusions. In Section 2.4, we will explore this topic in detail.

Therefore, the discussion of the IR bands will be based on harmonic periodic system calculations. The neglecting of anharmonicity in the case of periodic system calculations did not decrease noticeably the agreement with the experimental spectrum. We have carried out two separate calculations of IR spectra for the lattice model (B3LYP/Gatti and B3LYP/TZVP; Figures S2 and S3 in Supplementary Materials). The differences between these two simulated spectra are qualitatively negligible, with the exception of the low-lying bands in the region of 900–650 cm^{-1} (Figure S3 in Supplementary Materials). Accordingly, the overhead computing cost that is introduced by the larger TZVP basis set did not return any profit in the case of melamine. The accuracy of scaled B3LYP/Gatti allows for unambiguous band assignments in the entire 4000–650 cm^{-1} region of crystalline melamine (Figure 3A,B; Table 1).

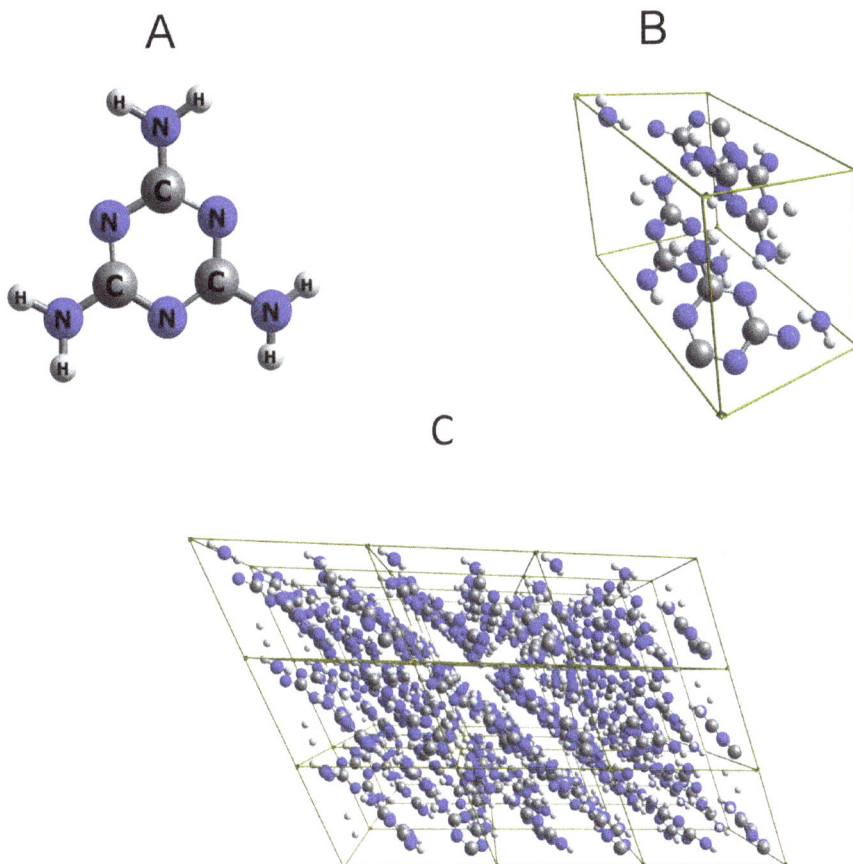

Figure 1. Molecular structure of melamine. (**A**) single molecule; (**B**) the content of a unit cell; (**C**) 3 × 3 × 3 supercell. The structure after optimization (B3LYP\Gatti) is presented in Figure S1 (Supplementary Materials).

A good comprehension of all IR bands in the crystalline melamine was accomplished; the resulting assignments are presented in Figure 3A,B and in Table 1. The upper IR region (X-H stretching region) is mostly populated by $\nu_{as}NH_2$ bands; three of them are separated, while the fourth one (at ca. 3188 cm^{-1}) overlaps with the neighboring strong $\nu_s NH_2$ peak. That single $\nu_s NH_2$ band at 3122 cm^{-1} has the highest intensity in this region. These features are very well reflected in the calculated spectrum. The broadening that was observable just below 3000 cm^{-1} in the experimental spectrum Figure 3A) originates from the strong anharmonic effects that occur because of the long-range ordering of hydrogen-bonded melamine molecules, which exists in the crystal lattice [47,48]. These effects have well-known impact on IR spectra [47–50].

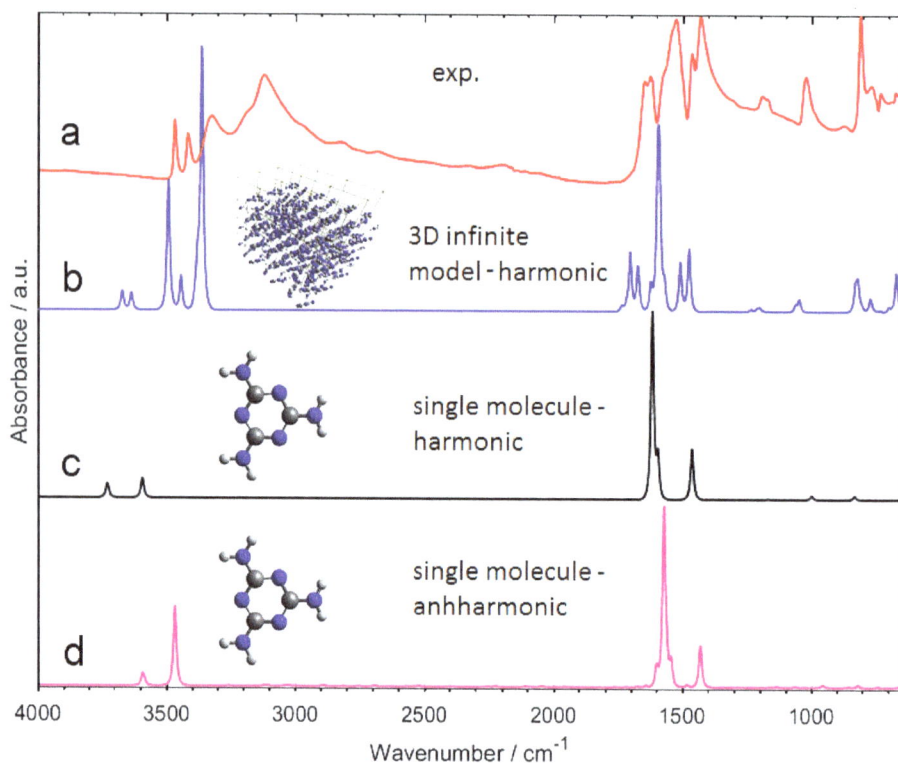

Figure 2. Comparison of the experimental Attenuated Total Reflection IR (ATR-IR) spectrum of polycrystalline melamine (**a**) with the calculated spectra (**b–d**). The spectra were calculated in harmonic ((**b**) B3LYP/Gatti for periodic (3D, infinite) model; (**c**) B3LYP-GD3BJ/SNST for single molecule model) and anharmonic ((**d**) GVPT2//DFT-B3LYP-GD3BJ/SNST) approximation. For the calculated spectra, no scaling was applied in this figure.

The lower IR region (fingerprint region; 1700–650 cm^{-1} in the present case; Figure 3B) of melamine features rather well separated bands. The most notable group of the intense bands in the region of 1650–1430 cm^{-1} primarily arises from in-plane NH$_2$ deformations (scissoring and rocking modes of NH$_2$; in-plane ring modes and C-N(H$_2$) stretching modes). The internal coordinates corresponding to these vibrations are rather highly mixed (Table 1). In contrast, the bands appearing at 1194 and 1174 cm^{-1} stems from relatively "clean" NH$_2$ rocking modes. However, the intensities of those two bands are weak. These observations will find good confirmation in the subsequent analysis of the NIR spectrum of melamine. The next band (1024 cm^{-1}) has a moderate intensity and it corresponds to two transitions; δ_{rock}NH$_2$ mixed with δ_{ip}ring (at calc. position of 1035 cm^{-1}) and δ_{rock}NH$_2$ mixed with νC-N(H$_2$) and δ_{ip}ring (at calc. position of 1021 cm^{-1}). The very strong band at 810 cm^{-1} originates from the out-of-plane deformations. The transitions corresponding to less mixed δ_{rock}NH$_2$ and δ_{rock}NH$_2$ give rise to weak bands at 768 and 675 cm^{-1} (calc. 755 and 661 cm^{-1}), respectively (Figure 3B and Table 1). Thus, the mixing of internal coordinates can consistently be noted for the corresponding bands with stronger intensities.

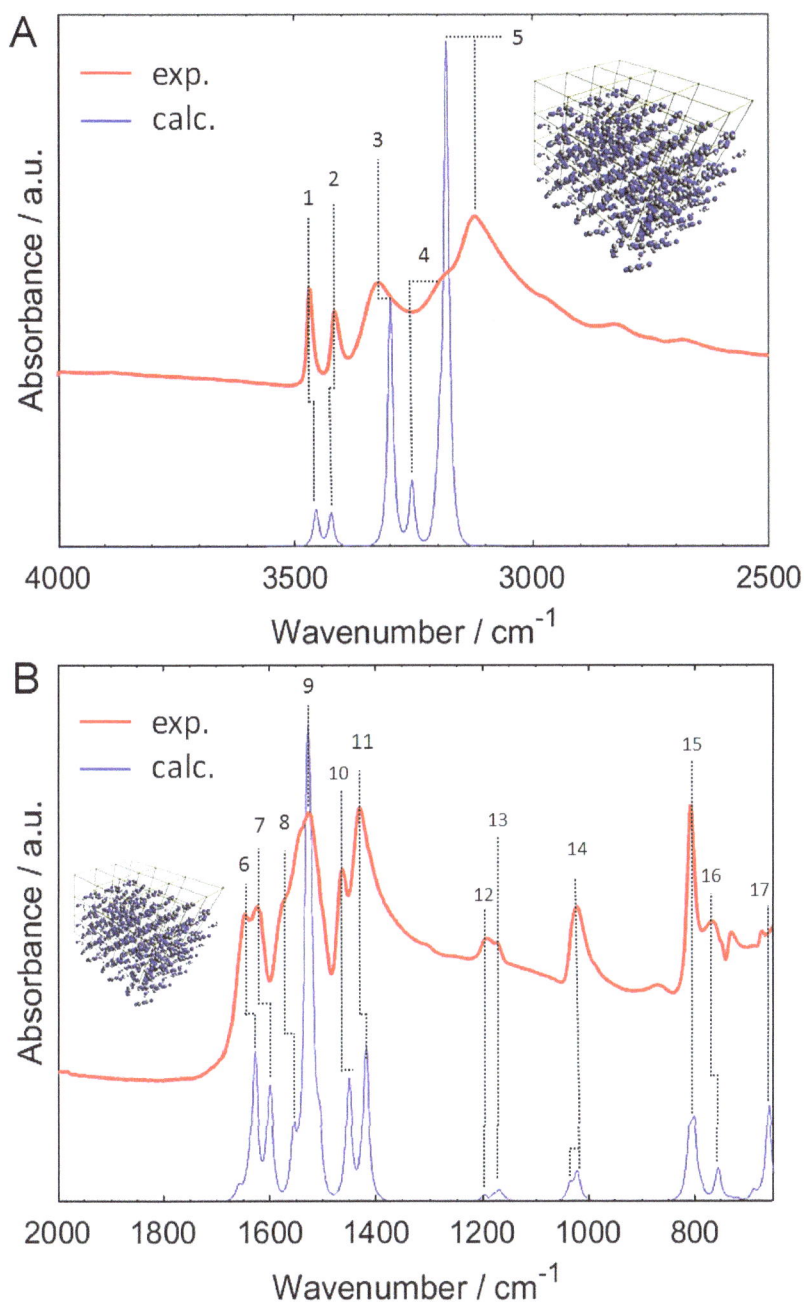

Figure 3. Band assignments in the ATR-IR spectrum of polycrystalline melamine based on scaled periodic//B3LYP/Gatti calculated spectrum. Band numbers correspond to those that are presented in Table 1. (**A**) 4000–2500 cm^{-1} region; (**B**) 2000–650 cm^{-1} region.

Table 1. Band assignments in the experimental ATR-IR spectrum of crystalline melamine. Band numbers correspond to those presented in Figure 3A,B.

Band Number	Wavenumber/cm^{-1}			Assignment
	Experimental	Scaled Calc.	Non-Scaled Calc.	
1	3468	3454	3669	$\nu_{as}NH_2$
2	3416	3423	3634	$\nu_{as}NH_2$
3	3324	3296	3491	$\nu_{as}NH_2$
4	~3188	3252	3441	$\nu_{as}NH_2$
5	3121.7	3179	3360	$\nu_{s}NH_2$
6	1647.7	1627	1705	$\delta_{sciss}NH_2$
7	1624.9	1599	1675	$\delta_{sciss}NH_2$
8	~1574	1553	1624	$\delta_{sciss}NH_2$; δ_{ip}ring
9	1527.7	1528	1597	$\delta_{sciss}NH_2$; $\delta_{rock}NH_2$; δ_{ip}ring
		1523	1591	$\delta_{rock}NH_2$; δ_{ip}ring
10	1465.6	1449	1511	$\delta_{sciss}NH_2$; νC-N(H$_2$)
11	1431.6	1417	1476	νC-N(H$_2$); δ_{ip}ring
12	1194.3	1197	1238	$\delta_{rock}NH_2$
13	1173.5	1177	1217	$\delta_{rock}NH_2$
		1167	1206	
14	1024.1	1035	1066	$\delta_{rock}NH_2$; δ_{ip}ring
		1021	1051	$\delta_{rock}NH_2$; νC-N(H$_2$); δ_{ip}ring
15	810.1	811	830	δ_{oop}ring; $\delta_{twist}NH_2$
		801	819	$\delta_{wagg}NH_2$
16	768.4	755	771	$\delta_{twist}NH_2$
17	674.5	661	673	$\delta_{wagg}NH_2$

2.2. Experimental and Simulated NIR Spectra of Crystalline Melamine

In decisive contrast to the IR region (Section 2.1), and as evidenced in Figure 4, the NIR bands of crystalline melamine are accurately reproduced on the basis of a finite model. In this case, even the calculations that are based on a single molecule of melamine provide good agreement between the calculated and experimental spectra (Figure 4). If further studies will allow for generalizing this observation, a lower requirement for the model complexity in modeling of NIR spectra could open other opportunities for refining the theoretical approach. For example, a hybrid approach combining higher-level harmonic computations augmenting DVPT2/GVPT2 anharmonic analysis could be used. Barone and co-workers have reported evidences of accurate and affordable hybrid B3LYP(harmonic)/B2PLYP(anharmonic) computations [51,52], as also seen in our previous studies [53].

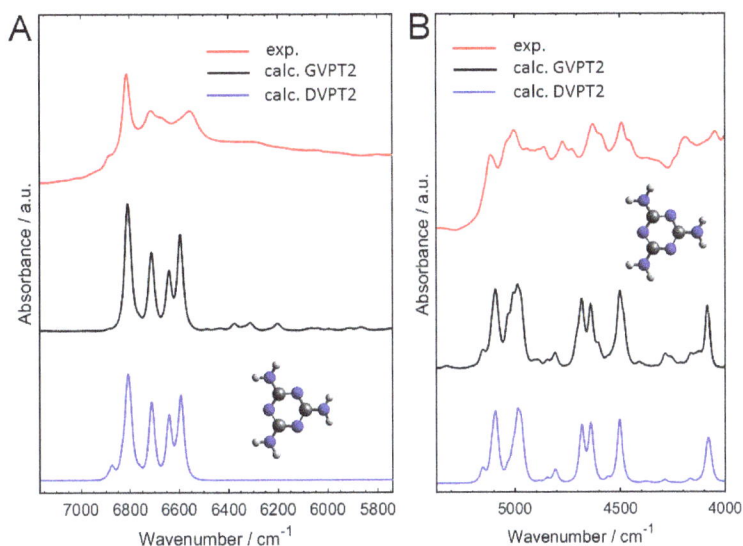

Figure 4. Comparison of the experimental diffuse reflectance near-infrared (NIR) spectrum of polycrystalline melamine with the calculated spectra. The spectra were calculated at B3LYP-GD3BJ/SNST level for an isolated molecule in two different anharmonic approximations (DVPT2 and GVPT2). In the GVPT2 calculation second overtones and ternary combinations were also included. (**A**) 7150–5750 cm^{-1} region; (**B**) 5400–4000 cm^{-1} region.

The principle difference between the DVPT2 and GVPT2 approaches is the treatment of vibrational resonances; the latter method features an advanced correction for Fermi and Darling-Dennison resonances by the variational method [54]. The improvement does not seem to significantly affect the major NIR bands of melamine (Figure 4). However, improved method allows for calculations of three quanta transitions (i.e., second overtones, 3ν; and ternary combinations, $2\nu_x + \nu_y$, and $\nu_x + \nu_y + \nu_z$). The addition of these minor bands slightly enhances the agreement, for example, below 6500 cm^{-1} and in the region of 4400–4150 cm^{-1} (Figures 4 and 5). This observation remains similar to our previous estimations that were based on the anharmonic study of methanol molecule and its deuterated isotopomers [55]. We have therein concluded that the bands due to three quanta transitions are only responsible for minority (ca. 20%) of the NIR spectra of methanol [55]. These higher order bands are weak, overlapping bands, and the corresponding spectral information is "diffused" along the wavenumber axis [55]. A similar case may be reported for melamine, and it is probably shared by other molecules. There are some exceptions, e.g., in the literature, 3νC=O has been reported to appear near 5150–5160 cm^{-1} in gas phase (and near 5122–5076 cm^{-1} in various solvents), as a well resolved band in some molecules with a C=O group [56,57]. No such exceptions were observed for melamine, and close inspection of the theoretical spectra confirms that the majority of the experimental bands were reproduced by both approximations (Figure 4). The more comprehensive, but also resource intensive, calculations of additional three quanta transitions of melamine allowed for explaining finer features that were observed in its NIR spectrum (Figure 4). Primarily, the influence of higher order bands may be seen throughout the 6500–5200 cm^{-1} region and in the vicinity of 4300–4100 cm^{-1} (Figure 4A,B). In the former case (6500–5200 cm^{-1} region), these bands remain very weak. For the latter (4300–4100 cm^{-1} region), there appears a marked band overlapping that gives rise to a broadened feature of moderate intensity at 4300–4200 cm^{-1}, and also a similar one appearing at the high-frequency wing of the ~4090 cm^{-1} band (Figure 4B). The present study of NIR spectra of crystalline melamine may be compared with our earlier calculations of NIR spectra of medium-chain fatty acids [35]. In contrast to the previously observed significance of the hydrogen-bonding interaction for the NIR region of crystalline sorbic acid [35], NIR bands of melamine reveal surprisingly low sensitivity to the hydrogen-bonding. This occurs despite the fact that melamine interacts strongly and it forms multiple hydrogen-bonds in crystalline states, as recently demonstrated by Yuan et al. [41]. We will discuss in detail the reasons for such distinctiveness of NIR region of melamine in Section 2.3.

The agreement with the experimental spectrum is slightly lower in the 5200–4000 cm^{-1} region, as the level of band overlapping is very high there. In contrast, the upper NIR region (ca. 6900–6450 cm^{-1}) was accurately reproduced in the calculation (Figure 4A). The first overtones and the binary combinations bands of NH_2 stretching modes populate this region (as the primary contributions). Similar to other kinds of X-H vibrations (e.g., OH), these modes are expected in the literature to be sensitive to the molecules' chemical neighborhood, e.g., prone to red-shifting in hydrogen-bonded complexes. This effect has been clearly observed for alcohols [58] or in our recent NIR investigation of thymol [37]. Surprisingly, in the case of melamine it may be concluded that the first overtones and binary combinations bands of NH_2 stretching modes are properly reproduced, even in the simplified case of an isolated molecule, for which the inter-molecular interactions are not reflected (Figure 4A).

2.3. An In-Depth Analysis of the Origin of NIR Bands of Crystalline Melamine

Numerous overlapping bands populate NIR spectra of even relatively simple molecules [16,31]. This makes their detailed analysis difficult, and the interpretation of the leading contributions is not as straightforward as it is for the corresponding IR spectra. To better elucidate these influences and to present them in a clear manner, we have developed a density map (colormap) of spectral contributions to highlight the modes of interest in a straightforward way (Figure 5). The color range corresponds to the square rooted intensity ratio of the selected simulated bands to the total intensity of modeled spectrum at any given wavenumber ν_i and additionally proportionalized to the calculated intensity at that point. The yielded value ranges from 0 (no contribution) to 1 (the NIR spectrum is influenced by the selected mode/modes entirely). The corresponding color varies from black to white, and it may be directly interpreted as the intensity of a given mode at a given wavenumber; the colortable is presented underneath the figure. The square root allows for elucidating less pronounced contributions. The density maps that are determined for various selections of modes-of-interest allow for unequivocal and thorough analysis of the influential determinants in the NIR spectrum of crystalline melamine (Figure 5), while keeping the figure compact and easy to read.

The upper NIR region (ca. 6900–6500 cm^{-1}) is almost entirely populated by the first overtones and the binary combination bands of NH_2 stretching modes (Figure 5). Other than this exception, the overtone bands of melamine may be described as non-essential for the other regions of NIR spectrum. Instead, the combination bands (binary combinations the most, ternary combinations to a bit lesser degree) play the primary role there. Again, the combinations that involve stretching NH_2 modes are the most important factor. The corresponding binary combinations provide very strong influence in the 6900–6500 cm^{-1} region and throughout a broad 5100–4000 cm^{-1} region. Ternary combinations that involve NH_2 stretching modes only give weak influence in the region of 6500–6000 cm^{-1}, where the spectral lineshape is rather flat. The highly populated region between 5100–4000 cm^{-1} is mostly influenced by the combinations of stretching modes with deformation modes of NH_2 groups. The other influential factor is νC-N mode (both in ring as well as between C atoms in ring and NH_2 groups). However, the wagging NH_2 modes are largely suppressed. The modes that do not influence NIR region are as follows; overtones of δ_{ip}ring, δ_{ip}C-N(H$_2$), δ_{oop}ring, δ_{oop}C-N(H$_2$), $\delta_{rock}NH_2$, $\delta_{twist}NH_2$, $\delta_{wagg}NH_2$.

The analysis of the NIR modes reveals a structural pattern. The modes that involve in-plane atoms displacements tend to impact NIR spectrum relatively stringer than those that involve out-of-plane motions. This observation will be discussed in detail in Section 2.4, together with the conclusions drawn from the analysis of IR spectrum of melamine (Section 2.1).

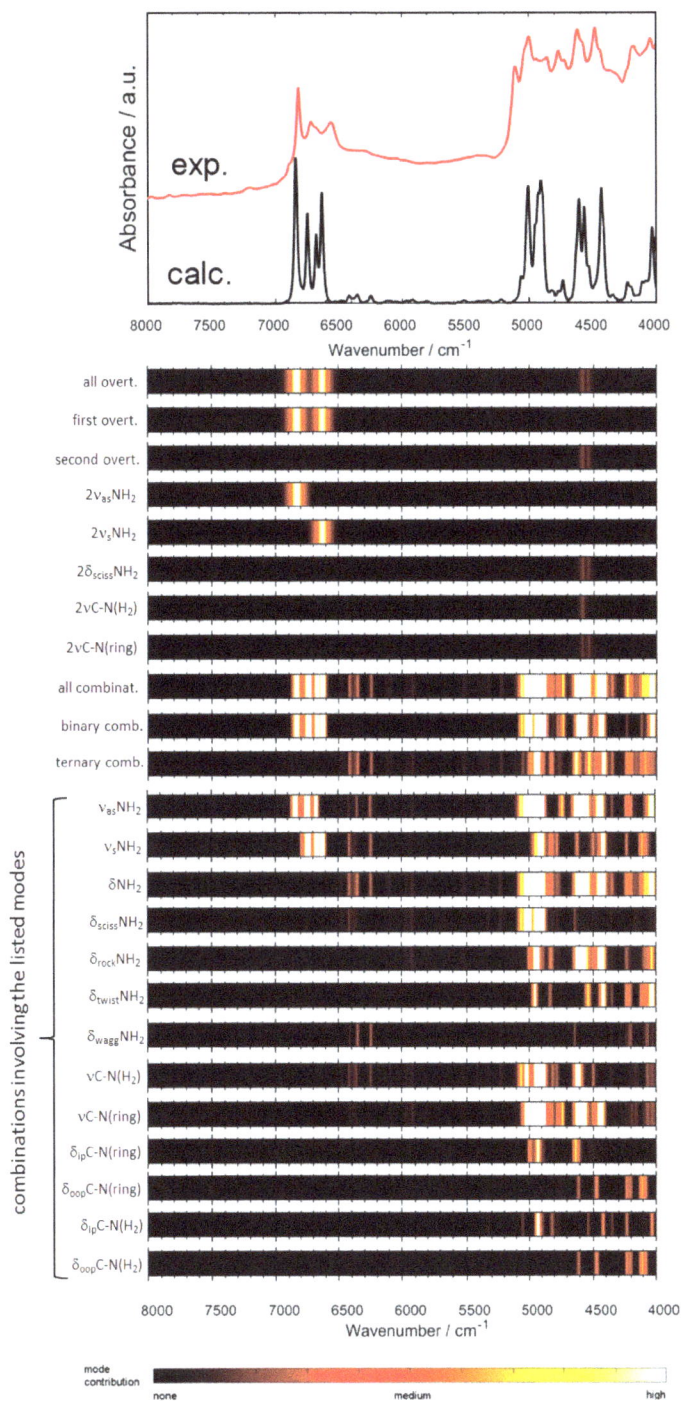

Figure 5. Analysis of the contributions to NIR spectrum of polycrystalline melamine based on the calculated spectrum (GVPT2//B3LYP-GD3BJ/SNST).

2.4. The Relationships Between IR and NIR Bands, and the Structural Features of Crystalline Melamine

Important conclusions may be drawn from the detailed comprehension of both IR and NIR spectra of crystalline melamine. The vibrational modes involving out-of-plane atomic motions are much more essential in the IR region (Figure 3 and Table 1), and thus a proper reflection of the inter-plane structure of crystalline melamine is required. Therefore, calculations that are based on a 3D infinite model of crystal lattice allows for the best description of these vibrations, even with harmonic approximation. However, in the NIR spectrum the most influential modes are those that involve in-plane atomic displacements. Thus, the neighboring planes of crystalline lattice do not impact these motions as much. Therefore, the NIR modes calculated for a single molecule are largely enough to accurately reflect the experimental spectrum.

This also makes perfect sense by comparing the IR vibrations calculated in periodic system to the ones calculated for a single molecule (also refer to Table S1 in Supporting Materials presenting the potential energy distribution calculated for melamine). In the calculations that are based on a single molecule (vs. the model of crystalline lattice), the out-of-plane deformation vibrations (e.g., $\delta_{twist}NH_2$, $\delta_{wagg}NH_2$, δ_{oop}ring) are positioned at higher calculated wavenumbers, while the in-plane deformation vibrations are positioned at the lower calculated wavenumbers. Thus, the oscillators with the out-of-plane atomic motions have lower calculated force constants because of neglecting the surrounding crystalline planes. The neglecting of intra-plane interactions of melamine lead to a surprisingly miniscule decrease of the accuracy of the calculated NIR modes. The corresponding vibrations with in-plane atomic motions (primarily νNH_2, but also δ_{ip}ring, $\nu C-N[H_2]$) were mostly overestimated in the calculated frequencies, which was easily corrected by applying scaling. These modes are seemingly less affected by the intermolecular interactions of melamine in crystal lattice, including the effects of hydrogen-bonding.

2.5. New Insights on the Quantitative Analytical Spectroscopy of Melamine

Melamine is known food adulterant, and for this reason, it has been frequently focused on analytical IR and NIR spectroscopy [9,42–46]. After reproducing the spectra of melamine, we can more deeply discuss some of the observations that were reported in these previous contributions.

NIR bands of melamine remain at relatively similar positions throughout various kinds of samples, unlike its IR bands, which are prone to shifting in response to the chemical environment [43]. This can be fully confirmed by our conclusions from quantum mechanical calculations. Lu et al. [42] have reported that the most relevant spectral region of milk powder useful for the detection of melamine is 5300–4900 cm^{-1}. This region contains the binary and ternary combinations of NH$_2$ stretching with NH$_2$ deformation modes of melamine; however, no overtone bands can be found there (Figure 5). Cantor et al. [9] have found that the most meaningful region for the analysis of melamine content in gelatin by IR spectroscopy is at around 800 cm^{-1}; the melamine bands therein stem from δ_{oop}ring, $\delta_{twist}NH_2$, and $\delta_{wagg}NH_2$ (Figure 3 and Table 1). On the other hand, the chemometric models that are based on NIR spectra of gelatin contaminated by melamine have been found to recognize the most the 7000–6500 cm^{-1} region. The NIR bands of melamine due to the first overtones and binary combinations of both symmetric and asymmetric NH$_2$ stretching modes can be exclusively found there (Figure 5). Our study allows for correcting Haughey et al. [45], who ascribed the NIR region of melamine in the vicinity of 6800 cm^{-1} to only overtone bands.

In more detail, the influential spectral regions in chemometric analysis of melamine content in milk powders based on NIR spectroscopy has been reported by Lim et al. [46]. They have found that the partial least squares regression (PLSR) models have recognized 6763 and 6808 cm^{-1} as the most meaningful in the analysis. They have concluded that these regions correspond to 6676 and 6529 cm^{-1} peaks of melamine observed in the spectrum of milk powder. We may now confirm that these two bands arise from the first overtones and binary combinations of $\nu_{as}NH_2$ and $\nu_s NH_2$ modes of melamine, respectively. These previous chemometric analyses consistently indicate that for the detection and quantification of melamine, its NH$_2$ stretching vibrations are the most meaningful;

however, we may now indicate that not only the first overtones, but also binary combinations equally contribute. On the other hand, the lower NIR region, where combination bands due to NH_2 stretching and deformation modes of melamine are present, is less essential.

Balabin et al. [43] have concluded that, for the samples with high content of melamine, NIR spectroscopy yields better analytical performance than IR spectroscopy. It may be now confirmed that this occurs due to relatively much lower sensitivity of NIR bands to intermolecular interactions with the matrix molecules when compared to IR bands. The hydrogen-bonding influences the IR region of melamine much stronger than the NIR region, in which only minor band shifts and negligible intensity variations occur due to this effect. In such sense, this feature of pure melamine in crystal evidenced here (lower sensitivity of NIR bands than IR bands to the chemical environment of molecules of melamine), seems to also be universally preserved in complex samples consisting of various other molecules and biomolecules, e.g., in milk powder [43]. Thus, analytical IR and NIR spectroscopy can evidence the structural and vibrational properties of melamine that were unveiled in our study to have a direct impact on the detection and quantification of melamine content.

3. Materials and Methods

3.1. Experimental

Melamine was purchased from Alfa Aesar (A11295; ≥99% purity) and used without further purification. The measurement of IR spectra was carried out on a Perkin Elmer Spectrum 100 FT-IR spectrometer that was equipped with an ATR accessory and a diamond prism (PerkinElmer, Inc. Waltham, MA, USA). The spectrometer was controlled by Perkin Elmer Spectrum software (version 10.4.00). The spectra were measured in the 4000–650 cm^{-1} region with a spectral resolution of 4 cm^{-1}. The number of accumulated scans was 16, and the measurement procedure was triplicated. The measurement was carried out at room temperature (~298 K). Sample preparation and spectra recording procedures were triplicated for each sample.

The measurement of NIR spectra was performed on a Büchi NIR Flex N-500 FT-NIR benchtop spectrometer that was controlled by the manufacturer's NIRWare 1.4.3010 software (BUCHI® AG, Flawil, Switzerland). The spectrometer was equipped with Büchi accessory for solid samples and operated in diffuse reflectance (DRIFT) mode and at room temperature (~298 K). The following recording parameters were selected; spectral resolution, spectral range, and scan number were 8 cm^{-1} (resulting in the interpolated 4 cm^{-1} of data spacing and 2 cm^{-1} of absolute accuracy), 10,000–4000 cm^{-1}, and 64, respectively. Sample preparation and spectra recording procedures were triplicated for each sample.

The experimental spectra in this work presented adequate quality for qualitative assessment, with no need for preprocessing of any kind.

3.2. Quantum Mechanical Calculations

3.2.1. IR spectrum Calculation in 3D Periodic Approximation

The simulation of IR spectra of crystalline melamine was based on harmonic analysis that was executed in three-dimensional periodic representation of crystal structure in Crystal 09 software (Aethia Srl, Italy) [59]. An infinite 3D model of crystal lattice of melamine was constructed by defining the primitive cell, in accordance with the experimental structural data obtained from the Cambridge Structural Database (CSD) [60,61]. An unconstrained and full geometry optimization was performed, in which both the atomic centers and cell parameters underwent the treatment prior to all subsequent calculations. The following procedural parameters were set throughout the geometry and vibrational computing steps. The Monkhorst–Pack reciprocal space was sampled over a shrinking factor that was equal to eight. The self-consistent field (SCF) direct procedure was iteratively converged with a tolerance of 10^{-13} atomic units per unit cell; the truncation of Coulomb and exchange sums in direct space was controlled by setting the Gaussian overlap tolerance criteria to 10^{-8}, 10^{-8}, 10^{-8}, 10^{-8}, and

10^{-16}. With the objective of accelerating the convergence of the SCF procedure, a linear mixing of Fock matrices by 25% between adjacent steps and an energy shifting of 0.8 hartree for the first SCF cycle were employed. The electron integrals were numerically calculated over a dense (XL) integration grid. The periodic Density Functional Theory (DFT) computations were performed with the use of B3LYP (Becke, three-parameter, Lee-Yang-Parr) [62] single-hybrid density functional, as implemented in Crystal 09 software.

Two separate calculations each using different basis sets were carried out. In the first one, we applied the following basis sets for the respective atomic centers: 3-1p1G for hydrogen, and 6-31d1G for carbon and nitrogen. Since Gatti et al. have jointly proposed these basis sets [63], for sake of clarity in the present work we will refer to them as "Gatti" basis sets. For the second calculation, we used triple-ζ valence basis set with polarization (TZVP), applied uniformly for all atomic centers (C, N, H). The employment of the TZVP basis set substantially increased the computational expense. Section 2.1 discusses the quality of IR simulation obtained at B3LYP/TZVP and B3LYP/Gatti levels.

Harmonic vibrational frequencies and intensities were obtained at the Gamma point in each case. The frequencies are numerically obtained in Crystal 09; to ensure the high stability of this procedure, numerical derivation (in the calculation of the second derivatives of the potential energy) was based on two-point finite difference scheme. The convergence criterion for the vibrational analysis was successfully achieved, as the sonic modes of the crystal lattice approached near zero values (not exceeding -0.5 cm^{-1}). The calculated band positions were overestimated; the overestimation was decreasing towards lower wavenumbers. Few factors contribute to this; the key ones are likely overestimation of bond strengths, imperfect molecular structure, and neglecting of anharmonicity. To account for this fact, the calculated wavenumbers were scaled while using linear scaling (Equation (1)).

$$\nu_{scal} = \nu_{calc} - s\nu_{calc}^2 \tag{1}$$

The best fit was achieved with the scaling parameter s equal to 2.7×10^{-5} (in the 4000–2500 cm^{-1} region) and 1.6×10^{-5} (in the 2500–650 cm^{-1} region). This scheme resembles the Wavenumber Linear Scaling (WLS), as developed by Yoshida et al. [64] Adjustments were necessary to fit the need of the study in a crystalline state. The calculated NIR spectra were scaled accordingly, while using scale factor s equal to 5.1×10^{-6} (Equation (1)). NIR scaling was significantly lesser; this is reasonable as the anharmonicity of the corresponding vibrations was already accounted for in the calculations. This is further explained in detail elsewhere (refer to Results and Discussion Section).

3.2.2. Anharmonic Calculation of NIR Spectra

The prediction of NIR bands requires multi-modal anharmonic vibrational analysis. The methods that were based on Vibrational Second-order Perturbation Theory (VPT2) offer superior cost/accuracy factor therein [65–67]. In this work, we employed and compared the deperturbed (DVPT2) and generalized (GVPT2) variants [66]. The implementation of the latter one allowed for simulating second overtones and ternary combination bands. For the determination of the basic electronic properties at the DFT level, B3LYP functional and triple-ζ SNST basis set [67] were selected; this method has repeatedly been evidenced to deliver good results [31,33,65]. Long-range interactions were refined by Grimme's D3 variant of empirical correction for dispersion with Becke–Johnson damping (GD3BJ) [68]. Prior to vibrational analysis, the molecular geometry optimization procedure was performed with very tight convergence criteria. The calculations were performed with the use of Gaussian 16 A.03 software [69].

The final spectra were simulated with the band broadening being obtained through a four-parameter Lorentz–Gauss product function [34,70] being applied to the quantum mechanically calculated IR and NIR band positions and intensities. The necessary data processing and assembly of figures were carried out in MATLAB R2016b [71].

Molecules **2019**, *24*, 1402

4. Conclusions

Quantum mechanical calculations of IR spectra of crystalline melamine were successfully carried out for an infinite (3D periodic) lattice model. All of the experimental IR bands were accurately reproduced, even when using reasonably affordable harmonic approximation. On the other hand, finite models gave very inaccurate calculated IR spectra. In contrast, the NIR spectrum of crystalline melamine was accurately modeled by anharmonic calculations that are based on the model of a single molecule. It is a striking difference in the dependence of the quality vs. model complexity, between these two spectral regions. This result may be explained based on direct comparison of the molecular motions corresponding to the vibrations being influential for the IR and NIR regions.

Vibrations involving out-of-plane atomic motions strongly affect the IR region of melamine. Therefore, it is essential to incorporate a proper description of the inter-plane interactions of melamine molecules as they appear in the crystalline lattice. The neglecting of the neighboring molecules in proximity planes leads to a completely incorrect calculated IR spectrum. In contrast, the in-plane vibrations of melamine are less affected by inter-molecular interactions. The vibrations involving in-plane atomic displacements are the most essential for the NIR spectrum of melamine. On the other hand, the out-of-plane motions are either suppressed in intensity or they are located outside of the NIR region. Hence, the calculated NIR spectrum of melamine is not significantly affected by a radical simplification of the molecular model. Even a single molecule model provides accurate reproduction of NIR spectrum of crystalline melamine. It may be concluded that the long-range ordering and, in particular, inter-plane interactions in the crystal lattice of melamine are significantly less important factors for NIR modes than for IR modes. From this observation, another important conclusion may be drawn. Due to the very significant computational cost of anharmonic calculations, the possibility to reduce the complexity of the molecular model (i.e., non-necessity to use 3D infinite model, which is also computationally costly) in the simulation of NIR spectra offers promising possibilities for similar studies of other crystalline materials.

Our findings shed light on the spectral features of melamine that have been reported in analytical spectroscopic studies of melamine as contaminant. In particular, the concluded in literature superiority of NIR spectroscopy in the analysis of the samples with relatively higher content of melamine was explained.

Supplementary Materials: The following are available online at http://www.mdpi.com/1420-3049/24/7/1402/s1, Figure S1: Optimized (B3LYP\Gatti) primitive cell of melamine. CRYSTAL output (A–D), and with atoms wrapped to the cell (E–H). Figure S2: Comparison of the experimental ATR-IR spectrum of polycrystalline melamine (4000–2500 cm^{-1} region) with the spectra calculated for periodic (3D, infinite) model of crystal lattice in harmonic approximation at B3LYP/Gatti and B3LYP/TZVP levels. Figure S3: Comparison of the experimental ATR-IR spectrum of polycrystalline melamine (2000–650 cm^{-1} region) with the spectra calculated for periodic (3D, infinite) model of crystal lattice in harmonic approximation at B3LYP/Gatti and B3LYP/TZVP levels. Table S1: Projection of normal coordinates of melamine onto natural internal coordinates. Based on Potential Energy Distribution (PED) analysis carried out for single molecule model vibrationally analyzed at B3LYP-GD3BJ/SNST level.

Author Contributions: Conceptualization, K.B.B.; methodology, J.G., K.B.B. and C.G.K.; formal analysis, J.G. and K.B.B.; investigation, J.G., K.B.B. and C.G.K.; writing—original draft preparation, J.G. and K.B.B.; writing—review and editing, all authors; supervision, Y.O. and C.W.H.

Funding: This work was supported by the Austrian Science Fund (FWF), P32004-N28. This work was supported by the National Science Center Poland (NCN), Grant 2017/27/B/ST4/00948.

Acknowledgments: Calculations have been carried out in Wrocław Centre for Networking and Supercomputing (http://www.wcss.pl), under grant no. 375.

Conflicts of Interest: The authors declare no conflict of interest. The funders had no role in the design of the study; in the collection, analyses, or interpretation of data; in the writing of the manuscript, or in the decision to publish the results.

References

1. Deim, H.; Matthias, G.; Wagner, R.A. Amino Resins. In *Ullmann's Encyclopedia of Industrial Chemistry*; Wiley-VCH: Weinheim, Germany, 2012.
2. Ogawa, A. Effect of a melamine resin admixture on the properties of concrete. *Concr. J. (Tokyo 1963)* **1973**, *11*, 12–21. [CrossRef]
3. Ashford, R.D. (Ed.) *Ashford's Dictionary of Industrial Chemicals*, 3rd ed.; Wavelength Publications Ltd.: London, UK, 2011; p. 5713.
4. Zhong, Y.; Jaidann, M.; Zhang, Y.; Zhang, G.; Liu, H.; Ionescu, M.; Li, R.; Sun, X.; Abou-Rachid, H.; Lussier, L.-S. Synthesis of high nitrogen doping of carbon nanotubes and modeling the stabilization of filled DAATO@CNTs (10,10) for nanoenergetic materials. *J. Phys. Chem. Solids* **2010**, *71*, 134–139. [CrossRef]
5. Hauck, R.D.; Stephenson, H.F. Nitrification of triazine nitrogen. *Fertilizer Nitrogen Sources* **1964**, *12*, 147–151. [CrossRef]
6. Chow, C.-Y. *Number of Melamine-Sickened Children Revised up Five-Fold*; South China Morning Post: Hong Kong, China, 2 December 2008; p. A9.
7. Ng, T.-W. *Lawyers Warned to Shun Milk Suits*; South China Morning Post: Hong Kong, China, 23 September 2008; p. A2, Archived from the original on 6 February 2009.
8. Pei, X.; Tandon, A.; Alldrick, A.; Giorgi, L.; Huang, W.; Yang, R. The China melamine milk scandal and its implications for food safety regulation. *Food Policy* **2011**, *6*, 412–420. [CrossRef]
9. Cantor, S.L.; Gupta, A.; Khan, M.A. Analytical methods for the evaluation of melamine contamination. *J. Pharm. Sci.* **2014**, *103*, 539–544. [CrossRef] [PubMed]
10. Henn, R.; Kirchler, C.G.; Grossgut, M.E.; Huck, C.W. Comparison of sensitivity to artificial spectral errors and multivariate LOD in NIR spectroscopy—Determining the performance of miniaturizations on melamine in milk powder. *Talanta* **2017**, *166*, 109–118. [CrossRef] [PubMed]
11. De Benedictis, L.; Huck, C.W. New approach to optimise near-infrared spectra with design of experiments and determination of milk compounds as influence factors for changing milk over time. *Food Chem.* **2016**, *212*, 552–560. [CrossRef] [PubMed]
12. Charlebois, S.; Schwab, A.; Henn, R.; Huck, C.W. An exploratory study for measuring consumer perception towards mislabelled food products and influence on self-authentication intentions. *Trends Food Sci. Technol.* **2016**, *50*, 211–218. [CrossRef]
13. *Near-Infrared Spectroscopy in Food Science and Technology*; Ozaki, Y.; McClure, W.F.; Christy, A.A. (Eds.) Wiley-Interscience: Hoboken, NJ, USA, 2007.
14. Su, W.-H.; Arvanitoyannis, I.-S.; Sun, D.-W. Chapter: 18 Trends in Food Authentication. In *Modern Techniques for Food Authentication*, 2nd ed.; Sun, D.-W., Ed.; Academic Press: Cambridge, MA, USA, 2018.
15. Marini, F. (Ed.) *Chemometrics in Food Chemistry*, 1st ed.; Elsevier: Amsterdam, The Netherlands, 2013; Volume 28.
16. Ozaki, Y.; Huck, C.W.; Beć, K.B. Near infrared spectroscopy and its applications. In *Molecular and Laser Spectroscopy*; Gupta, V.P., Ed.; Elsevier: Amsterdam, The Netherlands, 2017.
17. Huck, C.W. *Infrared Spectroscopy in Near-Infrared/Infrared Bioanalysis Including Imaging*; John Wiley & Sons, Encyclopedia of Analytical Chemistry: Hoboken, NJ, USA, 2016.
18. Ciurczak, E.W.; Drennen, J.K., III. *Pharmaceutical and Medical Applications of Near-Infrared Spectroscopy*; CRC Press: Boca Raton, FL, USA, 2002.
19. Chalmers, J.M.; Griffiths, P.R. (Eds.) *Handbook of Vibrational Spectroscopy*; John Wiley & Sons: Hoboken, NJ, USA, 2002; Volume 1.
20. Herberholz, L.; Kolomiets, O.; Siesler, H.W. Quantitative analysis by a portable near infrared spectrometer: Can it replace laboratory instrumentation for in situ analysis? *NIR News* **2010**, *21*, 1–8. [CrossRef]
21. Henn, R.; Schwab, A.; Huck, C.W. Evaluation of benchtop versus portable near-infrared spectroscopic method combined with multivariate approaches for the fast and simultaneous quantitative analysis of main sugars in syrup formulations. *Food Contr.* **2016**, *68*, 97–104. [CrossRef]
22. SCiO—Pocket molecular sensor. Available online: https://www.consumerphysics.com/scio-for-consumers/ (accessed on 10 April 2019).
23. Saranwong, S.; Kawano, S.; Ikehata, A.; Noguchi, G.; Park, S.; Sashida, K.; Okura, T.; Haff, R. Development of a low-cost NIR instrument for minced meat analysis: Part 1—Spectrophotometer and sample presentations. *Am. J. Agric. Sci. Technol.* **2013**, *2*, 61–68. [CrossRef]

24. Huck, C.W. Near-infrared (NIR) spectroscopy in natural product research. In *Handbook of Chemical and Biological Plant Analytical Methods*; Hostettman, K., Chen, S., Marston, A., Stuppner, H., Eds.; John Wiley & Sons: Hoboken, NJ, USA, 2014.

25. Siesler, H.W. Near-Infrared Spectra, Interpretation. In *Encyclopedia of Spectroscopy and Spectrometry*, 3rd ed.; Lindon, J.C., Tranter, G.E., Koppenaal, D.W., Eds.; Academic Press: Oxford, UK, 2017.

26. Weyer, L.G.; Lo, S.C. Spectra-structure correlations in the near-infrared. In *Handbook of Vibrational Spectroscopy*; Chalmers, J.M., Griffiths, P.R., Eds.; Wiley: Chichester, UK, 2002; Volume 3.

27. Beć, K.B.; Grabska, J.; Ozaki, Y.; Hawranek, J.P.; Huck, C.W. Influence of non-fundamental modes on mid-infrared spectra of aliphatic ethers. A fully anharmonic DFT study. *J. Phys. Chem. A* **2017**, *121*, 1412–1424. [CrossRef] [PubMed]

28. ydżba-Kopczyńska, B.I.; Beć, K.B.; Tomczak, J.; Hawranek, J.P. Optical constants of liquid pyrrole in the infrared. *J. Mol. Liq.* **2012**, *172*, 34–40. [CrossRef]

29. Beć, K.B.; Hawranek, J.P. Vibrational analysis of liquid *n*-butylmethylether. *Vib. Spectrosc.* **2013**, *64*, 164–171. [CrossRef]

30. Beć, K.B.; Kwiatek, A.; Hawranek, J.P. Vibrational analysis of neat liquid *tert*-butylmethylether. *J. Mol. Liq.* **2014**, *196*, 26–31. [CrossRef]

31. Beć, K.B.; Grabska, J.; Ozaki, Y. Advances in anharmonic methods and their applications to vibrational spectroscopies. In *Frontiers of Quantum Chemistry*; Wójcik, M.J., Nakatsuji, H., Kirtman, B., Ozaki, Y., Eds.; Springer: Singapore, 2017.

32. Beć, K.B.; Grabska, J.; Huck, C.W.; Ozaki, Y. Quantum mechanical simulation of NIR spectra. In *Applications in Physical and Analytical Chemistry*; Ozaki, Y., Wójcik, M.J., Popp, J., Eds.; Wiley: Hoboken, NJ, USA, 2019, in press.

33. Beć, K.B.; Futami, Y.; Wójcik, M.J.; Ozaki, Y. A spectroscopic and theoretical study in the near-infrared region of low concentration aliphatic alcohols. *Phys. Chem. Chem. Phys.* **2016**, *18*, 13666–13682. [CrossRef]

34. Grabska, J.; Ishigaki, M.; Beć, K.B.; Wójcik, M.J.; Ozaki, Y. Structure and near-infrared spectra of saturated and unsaturated carboxylic acids. An insight from anharmonic DFT calculations. *J. Phys. Chem. A* **2017**, *121*, 3437–3451. [CrossRef]

35. Grabska, J.; Beć, K.B.; Ishigaki, M.; Wójcik, M.J.; Ozaki, Y. Spectra-structure correlations of saturated and unsaturated medium-chain fatty acids. Near-infrared and anharmonic DFT study of hexanoic acid and sorbic acid. *Spectrochim. Acta A* **2017**, *185*, 35–44. [CrossRef]

36. Grabska, J.; Beć, K.B.; Ishigaki, M.; Huck, C.W.; Ozaki, Y. NIR spectra simulations by anharmonic DFT-saturated and unsaturated long-chain fatty acids. *J. Phys. Chem. B* **2018**, *122*, 6931–6944. [CrossRef]

37. Beć, K.B.; Grabska, J.; Kirchler, C.G.; Huck, C.W. NIR spectra simulation of thymol for better understanding of the spectra forming factors, phase and concentration effects and PLS regression features. *J. Mol. Liq.* **2018**, *268*, 895–902. [CrossRef]

38. Kirchler, C.G.; Pezzei, C.K.; Beć, K.B.; Mayr, S.; Ishigaki, M.; Ozaki, Y.; Huck, C.W. Critical evaluation of spectral information of benchtop vs. portable near-infrared spectrometers: Quantum chemistry and two-dimensional correlation spectroscopy for a better understanding of PLS regression models of the rosmarinic acid content in Rosmarini folium. *Analyst* **2017**, *142*, 455–464.

39. Kirchler, C.G.; Pezzei, C.K.; Beć, K.B.; Henn, R.; Ishigaki, M.; Ozaki, Y.; Huck, C.W. Critical evaluation of NIR and ATR-IR spectroscopic quantifications of rosmarinic acid in rosmarini folium supported by quantum chemical calculations. *Planta Med.* **2017**, *83*, 1076–1084. [CrossRef]

40. Mircescu, N.E.; Oltean, M.; Chis, V.; Leopold, N. FTIR, FT-Raman, SERS and DFT study on melamine. *Vib. Spectrosc.* **2012**, *62*, 165–171. [CrossRef]

41. Yuan, X.; Luo, K.; Zhang, K.; He, J.; Zhao, Y.; Yu, D. Combinatorial vibration-mode assignment for the FTIR spectrum of crystalline melamine: A strategic approach toward theoretical IR vibrational calculations of triazine-based compounds. *J. Phys. Chem. A* **2016**, *120*, 7427–7433. [CrossRef]

42. Lu, C.; Xiang, B.; Hao, G.; Xu, J.; Wang, Z.; Chen, C. Rapid detection of melamine in milk powder by near infrared spectroscopy. *J. Near Infrared Spectrosc.* **2009**, *17*, 59–67. [CrossRef]

43. Balabin, R.M.; Smirnov, S.V. Melamine detection by mid- and near-infrared (MIR/NIR) spectroscopy: A quick and sensitive method for dairy products analysis including liquid milk, infant formula, and milk powder. *Talanta* **2011**, *85*, 562–568. [CrossRef]

44. Abbas, O.; Lecler, B.; Dardenne, P.; Baeten, V. Detection of melamine and cyanuric acid in feed ingredients by near infrared spectroscopy and chemometrics. *J. Near Infrared Spectrosc.* **2013**, *21*, 183–194. [CrossRef]

45. Haughey, S.A.; Galvin-King, P.; Malechaux, A.; Elliott, C.T. The use of handheld near-infrared reflectance spectroscopy (NIRS) for the proximate analysis of poultry feed and to detect melamine adulteration of soya bean meal. *Anal. Methods* **2015**, *7*, 181–186. [CrossRef]

46. Lim, J.; Kim, G.; Mo, C.; Kim, M.S.; Chao, K.; Qin, J.; Fu, X.; Baek, I.; Cho, B.-K. Detection of melamine in milk powders using near-infrared hyperspectral imaging combined with regression coefficient of partial least square regression model. *Talanta* **2016**, *151*, 183–191. [CrossRef]

47. Boczar, M.; Boda, .; Wójcik, M.J. Theoretical modeling of infrared spectra of hydrogen-bonded crystals of salicylic acid. *Spectrochim. Acta A* **2006**, *64*, 757–760. [CrossRef]

48. Flakus, H.T.; Chelmecki, M. Infrared spectra of the hydrogen bond in benzoic acid crystals: Temperature and polarization effects. *Spectrochim. Acta A* **2002**, *58*, 179–196. [CrossRef]

49. Hanuza, J.; Godlewska, P.; Kucharska, E.; Ptak, M.; Kopacz, M.; Mączka, M.; Hermanowicz, K.; Macalik, L. Molecular structure and vibrational spectra of quercetin and quercetin-5′-sulfonic acid. *Vib. Spectrosc.* **2017**, *88*, 94–105. [CrossRef]

50. Kucharska, E.; Bryndal, I.; Lis, T.; Lorenc, J.; Hanuza, J. Influence of methyl and nitro group substitutions on the structure and vibrational characteristics of the hydrazo-bridge in 6,6′-dimethyl-3,3′,5,5′-tetranitro-2,2′-hydrazobipyridine. *Vib. Spectrosc.* **2016**, *83*, 70–77. [CrossRef]

51. Latouche, C.; Barone, V. Computational chemistry meets experiments for explaining the behavior of bibenzyl: A thermochemical and spectroscopic (Infrared, Raman, and NMR) investigation. *J. Chem. Theory Comput.* **2014**, *10*, 5586–5592. [CrossRef]

52. Vazart, F.; Latouche, C.; Cimino, P.; Barone, V. Accurate infrared (IR) spectra for molecules containing the C≡N moiety by anharmonic computations with the double hybrid B2PLYP density functional. *J. Chem. Theory Comput.* **2015**, *11*, 4364–4369. [CrossRef]

53. Beć, K.B.; Huck, C.W. Breakthrough potential in near-infrared spectroscopy: Spectra simulation. A review of recent developments. *Front. Chem.* **2019**, *7*, 48. [CrossRef]

54. Bloino, J.; Baiardi, A.; Biczysko, M. Aiming at an accurate prediction of vibrational and electronic spectra for medium-to-large molecules: An overview. *Int. J. Quantum Chem.* **2016**, *116*, 1543–1574. [CrossRef]

55. Grabska, J.; Czarnecki, M.A.; Beć, K.B.; Ozaki, Y. Spectroscopic and quantum mechanical calculation study of the effect of isotopic substitution on NIR spectra of methanol. *J. Phys. Chem. A* **2017**, *121*, 7925–7936. [CrossRef]

56. Chen, Y.; Morisawa, Y.; Futami, Y.; Czarnecki, M.A.; Wang, H.-S.; Ozaki, Y. Combined IR/NIR and density functional theory calculations analysis of the solvent effects on frequencies and intensities of the fundamental and overtones of the C=O stretching vibrations of acetone and 2-hexanone. *J. Phys. Chem. A* **2014**, *118*, 2576–2583. [CrossRef]

57. Workman, J., Jr.; Weyer, L. *Practical Guide and Spectral Atlas for Interpretive Near-Infrared Spectroscopy*, 2nd ed.; CRC Press: Boca Raton, FL, USA, 2012.

58. Grabska, J.; Beć, K.B.; Ozaki, Y.; Huck, C.W. Temperature drift of conformational equilibria of butyl alcohols studied by near-infrared spectroscopy and fully anharmonic DFT. *J. Phys. Chem. A* **2017**, *121*, 1950–1961. [CrossRef]

59. Dovesi, R.; Saunders, V.R.; Roetti, C.; Orlando, R.; Zicovich-Wilson, C.M.; Pascale, F.; Civalleri, B.; Doll, K.; Harrison, N.M.; Bush, I.J.; et al. *CRYSTAL09 User's Manual*; University of Torino: Torino, Italy, 2009.

60. Cambridge Structural Database. CSD Entry: MELAMI05. Available online: https://www.ccdc.cam.ac.uk/structures/search?id=doi:10.5517/cc7yptn&sid=DataCite (accessed on 10 April 2019).

61. Kooijman, H.; Beijer, F.H.; Sijbesma, R.P.; Meijer, E.W.; Spek, A.L. CCDC 237082: Experimental Crystal Structure Determination. *CSD Commun.* **2004**. [CrossRef]

62. Becke, A.D. Density-functional thermochemistry. III. The role of exact exchange. *J. Chem. Phys.* **1993**, *98*, 5648–5652. [CrossRef]

63. Gatti, C.; Saunders, V.R.; Roetti, C. Crystal-field effects on the topological properties of the electron-density in molecular-crystals. The case of urea. *J. Chem. Phys.* **1994**, *101*, 10686–10696. [CrossRef]

64. Yoshida, H.; Ehara, A.; Matsuura, H. Density functional vibrational analysis using wavenumber-linear scale factors. *Chem. Phys. Lett.* **2000**, *325*, 477–483. [CrossRef]

65. Beć, K.B.; Wójcik, M.J.; Nakajima, T. Quantum chemical calculations of basic molecules: Alcohols and carboxylic acids. *NIR News* **2016**, *27*, 15–21. [CrossRef]

66. Barone, V. Anharmonic vibrational properties by a fully automated second-order perturbative approach. *J. Chem. Phys.* **2005**, *122*, 014108. [CrossRef]

67. Barone, V.; Cimino, P.; Stendardo, E. Development and validation of the B3LYP/N07D computational model for structural parameter and magnetic tensors of large free radicals. *J. Chem. Theory Comput.* **2008**, *4*, 751–764. [CrossRef]

68. Grimme, S.; Antony, J.; Ehrlich, S.; Krieg, H. A consistent and accurate ab initio parameterization of density functional dispersion correction (DFT-D) for the 94 elements H-Pu. *J. Chem. Phys.* **2010**, *132*, 154104. [CrossRef]

69. Frisch, M.J.; Trucks, G.W.; Schlegel, H.B.; Scuseria, G.E.; Robb, M.A.; Cheeseman, J.R.; Scalmani, G.; Barone, V.; Mennucci, B.; Petersson, G.A.; et al. *Gaussian 16, Revision A.03*; Gaussian, Inc.: Wallingford, CT, USA, 2013.

70. Hawranek, J.P. On the numerical description of asymmetric absorption bands. *Acta Phys. Pol. B* **1971**, *40*, 811–814.

71. MATLAB. The MathWorks, Inc.: Natick, MA, USA. Available online: https://www.mathworks.com/products/matlab.html (accessed on 10 April 2019).

Sample Availability: Sample of the compound melamine is available from the authors.

molecules MDPI

Article

Spectra–Structure Correlations in Isotopomers of Ethanol (CX₃CX₂OX; X = H, D): Combined Near-Infrared and Anharmonic Computational Study

Krzysztof B. Beć [1,2,*], Justyna Grabska [1], Christian W. Huck [1] and Mirosław A. Czarnecki [2]

[1] Institute of Analytical Chemistry and Radiochemistry, Leopold-Franzens University, Innrain 80/82, CCB-Center for Chemistry and Biomedicine, 6020 Innsbruck, Austria; Justyna.Grabska@uibk.ac.at (J.G.); Christian.W.Huck@uibk.ac.at (C.W.H.)

[2] Faculty of Chemistry, University of Wrocław, F. Joliot-Curie 14, 50-383 Wrocław, Poland; miroslaw.czarnecki@chem.uni.wroc.pl

* Correspondence: Krzysztof.Bec@uibk.ac.at; Tel.: +43-512-507-5783

Academic Editor: Fabrizio Santoro
Received: 8 May 2019; Accepted: 8 June 2019; Published: 11 June 2019

Abstract: The effect of isotopic substitution on near-infrared (NIR) spectra has not been studied in detail. With an exception of few major bands, it is difficult to follow the spectral changes due to complexity of NIR spectra. Recent progress in anharmonic quantum mechanical calculations allows for accurate reconstruction of NIR spectra. Taking this opportunity, we carried out a systematic study of NIR spectra of six isotopomers of ethanol (CX₃CX₂OX; X = H, D). Besides, we calculated the theoretical spectra of two other isotopomers (CH_3CD_2OD and CD_3CH_2OD) for which the experimental spectra are not available. The anharmonic calculations were based on generalized vibrational second-order perturbation theory (GVPT2) at DFT and MP2 levels with several basis sets. We compared the accuracy and efficiency of various computational methods. It appears that the best results were obtained with B2PLYP-GD3BJ/def2-TZVP//CPCM approach. Our simulations included the first and second overtones, as well as binary and ternary combinations bands. This way, we reliably reproduced even minor bands in the spectra of diluted samples (0.1 M in CCl_4). On this basis, the effect of isotopic substitution on NIR spectra of ethanol was accurately reproduced and comprehensively explained.

Keywords: near-infrared spectroscopy; ethanol; anharmonic quantum mechanical calculations; isotopic substitution; overtones; combinations bands

1. Introduction

Near-infrared (NIR) spectra are appreciably more complex and difficult for interpretation than IR or Raman spectra [1–6]. This results from a large number of strongly overlapping overtones and combination bands, numerous resonances between different modes and anharmonicity of vibrations [1–6]. Interpretation of vibrational bands has been aided by studies of a series of similar compounds (including isotopomers), or by reconstruction of the spectra by using quantum mechanical calculations [5]. The former way may provide highly speculative assignments, while the latter method has limitations that prevent their common use in NIR spectroscopy. From the point of view of applied spectroscopy, there exists an essential difference in the applicability of quantum mechanical calculation of mid-infrared (MIR) [7–9] and NIR spectra. A simplistic and computationally inexpensive harmonic approximation fails to predict the overtones and combination modes [10]. Because of a considerable computational cost of anharmonic calculations, simulations of NIR spectra are rare. Nevertheless, in the literature one can find examples of application of different approaches used for calculation of overtones and combination bands, including variational approaches, which are expensive but useful in selected cases [11,12]. However, the studies using a vibrational self-consistent field (VSCF) approach [13,14] and its refined variants—e.g., PT2-VSCF—are more common [15–17]. Recently, a number of theoretical reconstructions of NIR spectra by means of efficient vibrational second-order perturbation (VPT2) method have been reported [18]; e.g., carboxylic acids [19], fatty acids [20–22], aminoacids [23], nucleobases [24], nitriles [25], azines [26], phenols [27,28], and alcohols [29,30]. Considerable efforts have been undertaken in order to develop anharmonic approaches applicable to even larger molecular systems [31–34]. On the other hand, meticulous probing of vibrational potential capable of yielding nearly-exact results is also available [35–38]. Recent advances in this field include the development of multi-dimensional approaches that provide complete information on mode couplings in linear triatomic molecules [39].

The isotopic effect appears to be helpful for the analysis of NIR spectra [10]. By shifting a part of overlapped contributions, one can reduce their complexity and reveal individual bands. Time-resolved NIR spectroscopy of deuterated alcohols has also been successfully used for elucidating the diffusion coefficients [40]. In our previous work, the effect of isotopic substitution on NIR spectra of methanol has been accurately reproduced by anharmonic calculations [41]. In particular, we were able to predict the vibrational contributions from non-uniformly substituted CX_3OX (X = H, D) species, which are not available from the experiment [41]. Further studies on molecules more complex than methanol are still necessary. A reasonable progress in this field is expected by examination of ethanol and all its isotopomers [42,43]. Ethanol has eight major isotopomers resulting from deuteration (CX_3CX_2OX; X = H, D), as compared to four in methanol. Moreover, the internal rotation around C-O(H) bond leads to rotational isomerism (*gauche*, *trans*) in molecules of ethanol, which adds additional origin of spectral variability in NIR spectra [42,43].

To enable detailed examination of the impact of various effects on NIR spectra of ethanol isotopomers, at first it is necessary to perform reliable theoretical reproduction of NIR spectra for eight isotopomers of ethanol (CX_3CX_2OX; X = H, D). We are interested in accurate reproduction of subtle effects observed in NIR spectra. Therefore, a combination of several electronic methods underlying VPT2 vibrational analysis will be useful for establishing the best approach capable of reproducing fine features in NIR spectra. The determination of electronic structure underlying the geometry optimization and harmonic analysis will be based on Møller–Plesset second-order perturbation (MP2) and density functional (DFT) theories. The efficiency and accuracy of reproduction of NIR bands by MP2 and DFT with single-hybrid B3LYP and double-hybrid B2PLYP density functionals will be overviewed. MP2 and DFT calculations included basis sets of increasing quality (6-31G(d,p), SNST, def2-TZVP, and aug-cc-pVTZ). Moreover, the impact of solvent cavity model will be evaluated. The anharmonic vibrational analysis will be carried out by means of generalized VPT2 (GVPT2). In our previous studies on methanol, it has been demonstrated that the relative contributions from the second overtones and ternary combinations are different for various isotopomers [41]. This work will provide detailed information on contributions from different vibrational modes and the trends observed with increasing of the alkyl chain length in going from methanol to ethanol. In addition, we will elucidate the accuracy of prediction of the three quanta transitions in NIR spectra.

2. Results and Discussion

2.1. Accuracy of Reproduction of NIR Spectra by Selected Approaches

Anharmonic calculations are significantly more challenging compared to harmonic approximation [1–4]. This holds even for efficient anharmonic approaches based on VPT2 method. At the same time, the higher quanta transitions are more prone to inaccuracies than the fundamental ones [44]. An insufficient accuracy of the ground state geometry and potential energy surface may easily propagate into inaccuracy of prediction in VPT2 calculation step. Thus, the theoretical prediction of NIR spectra is usually a compromise between the cost and accuracy. Effects like isotopic substitution [41] and conformational isomerism [29,30] may further complicate the vibrational analysis of NIR spectra.

One of the aims of this work was assessment of the efficiency of several combinations of the electronic theory methods and basis sets (Table 1). In addition, we examined the effect of the solvent model on the anharmonic vibrational energies. An efficient single-hybrid B3LYP functional is commonly used tool for spectroscopic studies [10]. Empirical correction for dispersion has been introduced to overcome one of the major weaknesses of DFT method [45]. In some cases, this approach markedly improves the robustness of calculation of primary parameters (i.e., energy). Therefore, recent literature suggests employing empirical correction for dispersion in DFT calculations [46]. Our previous studies have shown that, even for small and isolated molecules, this correction is advantageous in spectra modeling [29]. For molecules in solution an approximation of the solvent cavity often improves the quality of the simulated NIR spectra [28]. However, in the present work this advantage is less important (Table 1). We observed an improvement of RMSE from 45 to 35 cm^{-1} for CH_3CH_2OH, and from 27 to 24 cm^{-1} for CH_3CH_2OD. However, considering small additional cost of CPCM (ca. 10% of total CPU time in the case of GVPT2//B3LYP-GD3BJ/6-31G(d,p) calculations), it is advisable to include this correction step in the calculations.

Switching from B3LYP to B2PLYP density functional with a small basis set (6-31G(d,p)) leads to an increase in the RMSE value (Table 1). However, B2PLYP method overestimated the band positions in a systematic manner. In contrast, B3LYP approach provides irregular results. Some of the band positions (δCH) are blue-shifted, while the others (νOX and νCX; X = H, D) are red-shifted. Thus, more uniform band shift from B2PLYP method (Figures 1 and 2) resulted in better interpretability of NIR spectra as compared with B3LYP results. It has a peculiar effect in NIR spectra of aliphatic alcohols, as it reduces RMSE of NIR band positions, particularly for the νOX + δCH, and νCX + δCH combination bands. However, this apparent gain does not improve the true interpretability of the spectra. It is likely that simulated NIR spectra of larger molecules may suffer even more due to binary combinations involving the stretching and deformation of the C-H and O-H vibrations. Similar inconsistency of B3LYP as compared to B2PLYP has been noted before [29,41]. Due to electron correlation being computed effectively at the MP2 level, it is commonly accepted that B2PLYP functional requires larger basis sets. B2PLYP coupled with large def2-TZVP basis set noticeably improves the quality of simulated NIR spectra (Figures 1 and 2). This improvement is particularly evident in the reproduction of minor bands originating from the three quanta transitions (Figures 3 and 4). This effect is nicely illustrated by reduction of RMSE from 72–83 cm^{-1} for B2PLYP/6-31G(d,p) to 18–19 cm^{-1} for B2PLYP/def2-TZVP. SNST basis set, less complex than def2-TZVP but still of triple-ζ quality, leads to worse results. An exception was observed for the prominent doublet from the νCH combination bands (at ca. 4400 cm^{-1}), where B2PLYP/SNST calculations reproduced the peak shapes more resembling the experimental ones. However, the position of this doublet was also overestimated by this method. Hence, B2PLYP/def2-TZVP method appears to be the better tool for reliable reconstruction of NIR spectra. A similar conclusion was obtained for butyl alcohols [30].

In contrast, MP2 method does not appear to be particularly useful for anharmonic calculations of NIR spectra of ethanol isotopomers due to significant redshift of the νCX frequencies (Figures 1 and 2). It is an interesting observation, as the tendency of MP2 to describe incorrectly repulsive forces is known in the literature [47]. In this work, however, an insufficient basis set (6-31G(d,p)) may strongly deviate the results. Here, MP2 method seems to be more sensitive to the effect of a small basis set than DFT-B2PLYP method. As expected, an application of a larger basis set, e.g., aug-cc-pVTZ, improves the accuracy of calculations. However, these results are not as good as those obtained from B2PLYP/def2-TZVP method. Moreover, this improvement is accompanied by a substantial increase of computing time (by ca. 4 times). Nevertheless, selected spectral ranges (νOH + δCH $\approx$ 5050–4800 cm^{-1} and νCH + δCH doublet $\approx$ 4400 cm^{-1}) were better reproduced by MP2/aug-cc-pVTZ computations. Therefore, MP2 approach may be recommended as a reference method in selected cases. Our results demonstrate that different computational methods achieve different accuracy for particular regions of

NIR spectra, and these regions do not overlap. Thus, comparison of the spectra simulated by different methods (e.g., resulting from VPT2 calculations at DFT or MP2 levels) appears to be the best way for reliable interpretation of the experimental spectra.

Particular attention should be paid to 2νOH/OD band, which is the most characteristic peak for alcohols and the other important compounds like, e.g., phenols [27], terpenes [28], and polyphenols [48]. This peak is very sensitive to the chemical environment and inter- and intra-molecular interactions, and is frequently used for studies of the structure and physicochemical properties [49–53]. Hence, its proper theoretical reproduction is of essential importance. As shown in Figures 1–4, and Figures S1–S6 in SM, most of the methods did not reproduce correctly the shape of this band. The experimental band from 2νOH/OD vibration reveals a slight asymmetry. This asymmetry is a result of convolution of two components due to *trans* and *gauche* conformers. The lower-frequency *gauche* component has also the lower intensity. Among isotopomers of ethanol, this feature was reproduced correctly only by B2PLYP/def2-TZVP method. MP2/6-31G(d,p) and MP2/aug-cc-pVTZ methods predicted correct shape of the 2νOH/OD peak only for CH_3CD_2OH and CD_3CD_2OD. As can be seen (Table 1), the peak position was overestimated by all used methods, but B2PLYP calculations give the best agreement.

In comparison with other modes, large amplitude motions (LAMs)—e.g., torsion modes and hindered rotations—are more difficult for accurate description in harmonic approximation and also by anharmonic approaches that probe the potential curve relatively shallow (e.g., VPT2) [54]. We did not find any evidence that these low-frequency modes influence NIR bands directly (i.e., their overtone and combination modes do not appear in NIR region). However, a NIR spectrum provides some insights on LAMs as well. In our case, the shape of the 2νOH/OD band is an indirect probe of the accuracy of prediction of the low-frequency modes. The shape of this band results from two components due to *gauche* and *trans* rotational conformers. Unreliable theoretical abundances of these forms would result in biased relative intensities of the 2νOH/OD components (Table S1). Gibbs free energies may be affected by erroneous LAMs, which would propagate into incorrect relative abundances of *gauche* and *trans* conformers. Because it is an isolated band of strong intensity, the simulated 2νOH/OD may be used to assess the reliability of prediction of LAMs and the related Gibbs free energies. This kind of error would manifest itself as a distorted shape of simulated 2νOH/OD band. Above effect can be seen for some of the methods used in this study, e.g., for B3LYP (B3LYP-GD3BJ/6-31G(d,p); B3LYP-GD3BJ/6-31G(d,p)//CPCM; B3LYP-GD3BJ/SNST//CPCM;) and B2PLYP coupled with an insufficient basis set (B2PLYP-GD3BJ/6-31G(d,p)//CPCM;). However, the methods which yielded the most accurate spectra in the other regions (B2PLYP/def2-TZVP; MP2/6-31G(d,p); MP2/aug-cc-pVTZ) also reproduced 2νOH/OD peak accurately (Figures 1 and 2). Therefore, we conclude that the LAMs of ethanol and its derivatives were determined adequately by MP2 method. B2PLYP method also provides correct results, but it is more sensitive to the selection of a basis set. On the other hand, B3LYP tends to falsify the Gibbs free energies corrected by anharmonic ZPE. Further studies are needed to determine, whether this effect occurs because of an unreliable description of LAMs. On the other hand, inaccuracy of the 2νOH/OD frequencies prediction by VPT2 may also be considered as another contributing factor, as we have evidenced such occurrence in the case of the conformers of cyclohexanol [27]. Note that B3LYP functional coupled with a relatively simple basis set yields reasonable reproduction of NIR spectra and correctly predicts the effects of isotopic substitution at a relatively modest computational expense (Figures 1–4 and Figures S1–S6 in SM). However, a tendency to over- and underestimate the position and intensity of some bands may be unfavorable for the reliable interpretation of theoretical NIR spectra. For exploration of more subtle effects, B2PLYP functional seems to be more suitable. In the present study of isotopic substitution and the other effects (e.g., rotational isomerism) on NIR spectra of ethanol, we used B2PLYP/def2-TZVP method with additions of GD3BJ and CPCM.

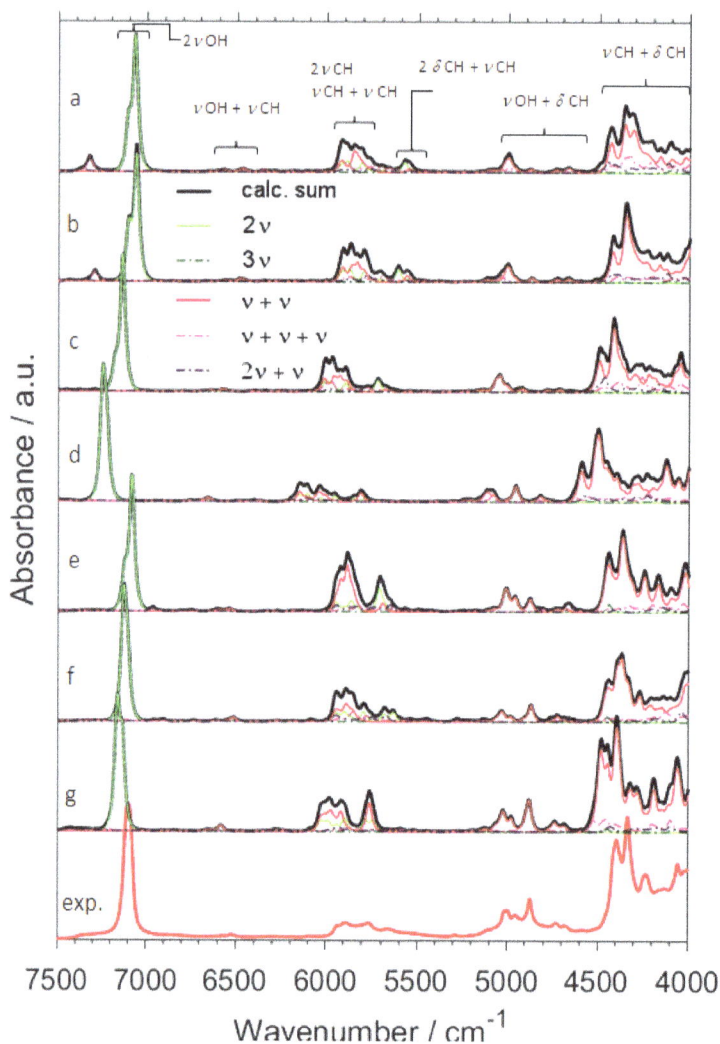

Figure 1. NIR spectra of CH_3CH_2OH calculated with GVPT2 method at different levels of electronic theory; (**a**) B3LYP-GD3BJ/6-31G(d,p); (**b**) B3LYP-GD3BJ/6-31G(d,p)//CPCM; (**c**) B2PLYP-GD3BJ/6-31G(d,p)//CPCM; (**d**) MP2/6-31G(d,p)//CPCM; (**e**) B3LYP-GD3BJ/SNST//CPCM; (**f**) B2PLYP-GD3BJ/def2-TZVP//CPCM; (**g**) MP2/aug-cc-pVTZ//CPCM; (**exp.**) Experimental spectrum of CH_3CH_2OH in CCl_4 (0.1 M).

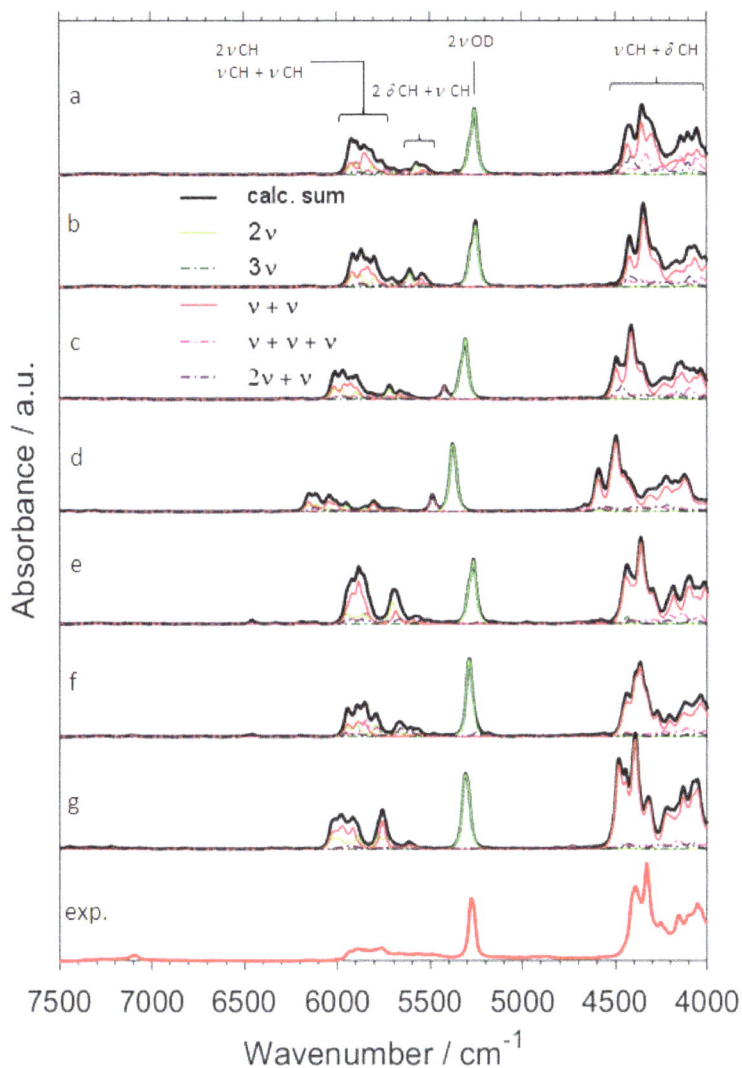

Figure 2. NIR spectra of CH_3CH_2OD calculated with GVPT2 method at different levels of electronic theory; (**a**) B3LYP-GD3BJ/6-31G(d,p); (**b**) B3LYP-GD3BJ/6-31G(d,p)//CPCM; (**c**) B2PLYP-GD3BJ/ 6-31G(d,p)//CPCM; (**d**) MP2/6-31G(d,p)//CPCM; (**e**) B3LYP-GD3BJ/SNST//CPCM; (**f**) B2PLYP-GD3BJ/def2-TZVP//CPCM; (**g**) MP2/aug-cc-pVTZ//CPCM; (**exp.**) Experimental spectrum of CH_3CH_2OD in CCl_4 (0.1 M).

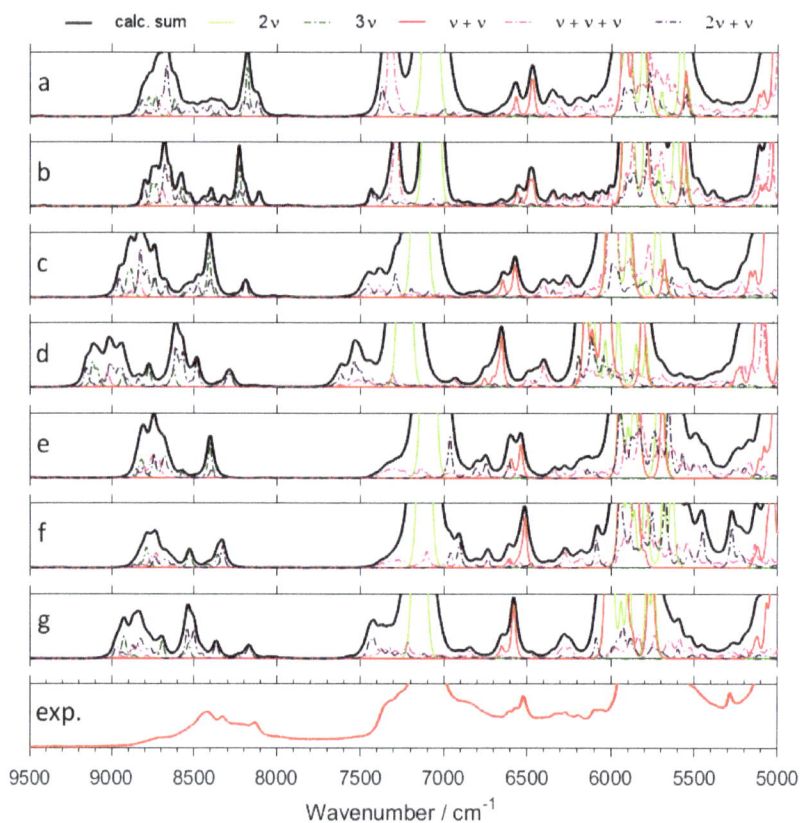

Figure 3. Contributions from minor bands in NIR spectra of CH_3CH_2OH calculated with GVPT2 method at different levels of electronic theory; (**a**) B3LYP-GD3BJ/6-31G(d,p); (**b**) B3LYP-GD3BJ/ 6-31G(d,p)//CPCM; (**c**) B2PLYP-GD3BJ/6-31G(d,p)//CPCM; (**d**) MP2/6-31G(d,p)//CPCM; (**e**) B3LYP-GD3BJ/SNST//CPCM; (**f**) B2PLYP-GD3BJ/def2-TZVP//CPCM; (**g**) MP2/aug-cc-pVTZ//CPCM; (**exp.**) Experimental spectrum of CH_3CH_2OH in CCl_4 (0.1 M).

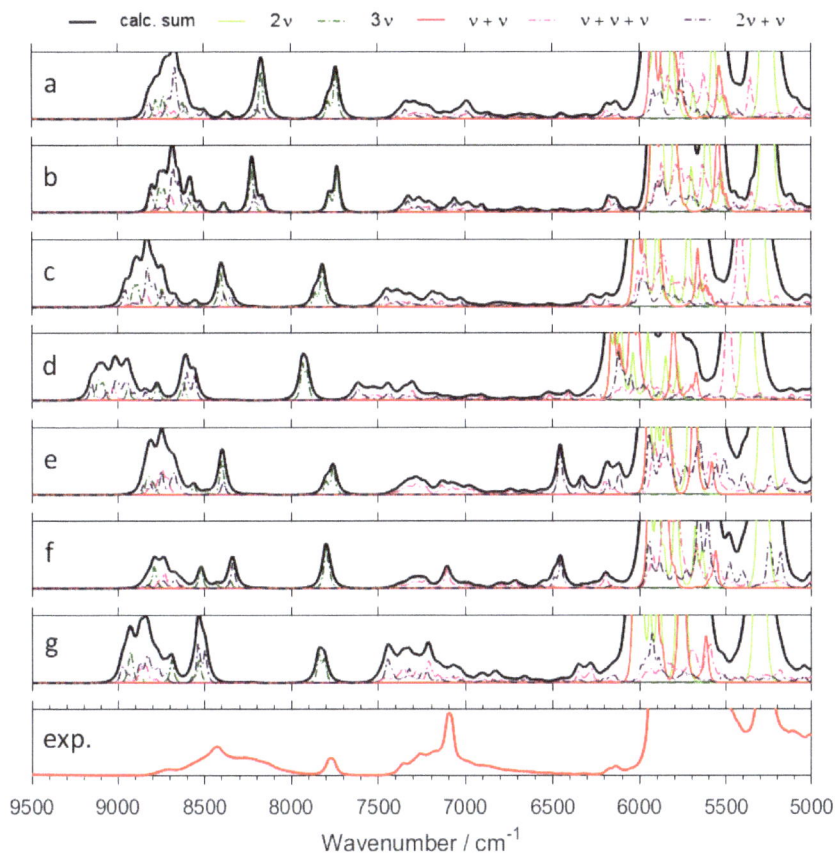

Figure 4. Contributions from minor bands in NIR spectra of CH_3CH_2OD calculated with GVPT2 method at different levels of electronic theory; (**a**) B3LYP-GD3BJ/6-31G(d,p); (**b**) B3LYP-GD3BJ/ 6-31G(d,p)//CPCM; (**c**) B2PLYP-GD3BJ/6-31G(d,p)//CPCM; (**d**) MP2/6-31G(d,p)//CPCM; (**e**) B3LYP-GD3BJ/SNST//CPCM; (**f**) B2PLYP-GD3BJ/def2-TZVP//CPCM; (**g**) MP2/aug-cc-pVTZ//CPCM; (**exp.**) Experimental spectrum of CH_3CH_2OD in CCl_4 (0.1 M).

Table 1. Positions of selected NIR bands (in cm^{-1}) from GVPT2 anharmonic vibrational analysis in CH_3CH_2OH and CH_3CH_2OD at different levels of electronic theory and corresponding RMSE values.

Assignment	Exp.	Calculated												
		MP2/aVTZ + CPCM	Diff.	B2PLYP-GD3BJ/def2-TZVP + CPCM	Diff.	B2PLYP-GD3BJ/SNST + CPCM	Diff.	MP2/6-31G(d,p) + CPCM	Diff.	B2PLYP-GD3BJ/6-31G(d,p) + CPCM	Diff.	B3LYP-GD3BJ/6-31G(d,p) + CPCM	Diff.	B3LYP-GD3BJ/6-31G(d,p)	Diff.
		CH_3CH_2OH												
$2\nu OH$	7099	7157	58	7125	26	7081	−18	7243	144	7134	35	7061	−38	7096	−3
$\nu_s CH_2 + \nu OH$	6520.4	6578	57.6	6513	−7.4	6536	15.6	6654	133.6	6576	55.6	6477	−43.4	6472	−48.4
$2b_{as}CH_3 + \nu OH$	5886.1	5980	93.9	5889	2.9	5922	35.9	6111	224.9	5970	83.9	5871	−15.1	5892	5.9
$2\nu_{as}CH_2$	5765.7	5897	131.3	5790	24.3	5864	98.3	5957	191.3	5898	132.3	5797	31.3	5812	46.3
$2\nu_s CH_2; 2\nu_s CH_3 + \delta_s CH_3$	5665.1	5759	93.9	5681	15.9	5708	42.9	5812	146.9	5722	56.9	5611	−54.1	5768	102.9
$[\delta_{wagg}CH_2, \delta_s CH_3] + \nu OH$	5013.8	5026	12.2	5029	15.2	5012	−1.8	5114	100.2	5049	35.2	5040	26.2	5019	5.2
$[\nu_{twist}CH_2, \delta_{ip}COH, \delta_{wagg}CH_2] + \nu OH$	4954.2	4978	23.8	4979	24.8	4959	4.8	5003	48.8	5009	54.8	5008	53.8	5003	48.8
$\delta_{ip}COH + \nu OH$	4873	4881	8	4868	−5	4877	4	4958	85	4926	53	4874	1	4883	10
$\delta_s CH_3 + \nu_{as}CH_3$	4394.8	4448	53.2	4366	−28.8	4443	48.2	4592	197.2	4473	78.2	4424	29.2	4436	41.2
$[\delta_{oop}COH, \tau CC] + \nu OH$	4333.5	4395	61.5	4331	−2.5	4364	30.5	4502	168.5	4416	82.5	4353	19.5	4357	23.5
		RMSE	70.0	**RMSE**	18.1	**RMSE**	40.8	**RMSE**	153.1	**RMSE**	72.2	**RMSE**	35.0	**RMSE**	44.6
		CH_3CH_2OD												
$2\nu_{as}CH_3$	5885.2	5978	92.8	5895	9.8	5921	35.8	6114	228.8	5967	81.8	5871	−14.2	5894	8.8
$2\nu_{as}CH_2$	5765.6	5917	151.4	5788	22.4	5862	96.4	6007	241.4	5896	130.4	5799	33.4	5815	49.4
$2\nu OD$	5277.1	5312	34.9	5289	11.9	5265	−12.1	5378	100.9	5306	28.9	5250	−27.1	5255	−22.1
$\delta_{scis}CH_2 + \nu_{as}CH_2$	4393.7	4445	51.3	4397	3.3	4439	45.3	4592	198.3	4493	99.3	4422	28.3	4424	30.3
$[\delta_s CH_3, \delta_{wagg}CH_2] + \nu_{as}CH_3$	4331.8	4392	60.2	4364	32.2	4364	32.2	4500	168.2	4415	83.2	4349	17.2	4355	23.2
$[\delta_{rock}CH_2, \delta_{rock}CH_3] + \nu_s CH_2$	4054.1	4057	2.9	4037	−17.1	4020	−34.1	4122	67.9	4036	−18.1	4068	13.9	4058	3.9
		RMSE	80.6	**RMSE**	18.6	**RMSE**	50.0	**RMSE**	179.4	**RMSE**	83.3	**RMSE**	23.6	**RMSE**	27.3

2.2. Origins of NIR Bands of CX_3CX_2OX (X = H, D)

The simulations of NIR spectra of ethanol isotopomers in CCl_4 solutions by GVPT2 anharmonic method at B2PLYP-GD3BJ/def2-TZVP//CPCM level accurately reproduced most of the experimental bands (Figures 5 and 6). On this basis, we performed detailed and reliable band assignments (Tables 2–9). The consistency of these assignments was positively verified by comparison with the experimental spectra of six isotopomers. High accuracy of simulations allows to analyze the theoretical spectra of CH_3CD_2OD and CD_3CH_2OD (Figure 6C,D and Tables 8 and 9) which are not available commercially. All assignments were supported by an analysis of the potential energy distributions (PEDs; Tables S2–S9).

NIR spectra of ethanol isotopomers mainly consist of the combinations of stretching and bending OX and CX (X = H, D) modes (Figures 1–6 and Tables 2–9). The region below 5500 cm^{-1} for CH_3CH_2OH is almost entirely contributed by the combination bands, while absorption from the overtones dominates above 5500 cm^{-1}. NIR spectrum of ethanol may be roughly divided into four regions, but only two of them contain meaningful contributions from overtones. These regions are contributed mainly by vibrations from: (1) $2\nu OX$; (2) $2\nu CX$ and $\nu CX + \nu CX$; (3) $\nu OX + \delta CX$; (4) $\nu CX + \delta CX$; (X = H, D). The other combination bands like $\nu OX + \nu CX$, and $2\delta CX + \nu CX$ have low intensity. Isotopic substitution introduces significant band shift, strongly affecting the appearance of NIR spectra (Figures 5 and 6 and Figures S1–S6). It should be noted, that the region of 5700–5400 cm^{-1} for ethanols containing CH_3 and CH_2 groups is strongly affected by the anharmonic effects. This effect is well seen for CH_3CH_2OH (Figure 1) and CH_3CH_2OD (Figure 2). The most meaningful contributions in this region originate from $\nu CH + \nu CH$ ($\nu_{as}CH_2 + \nu_s CH_2$), $2\delta CH + \nu CH$, and $2\nu_{as}CH_2$ vibrations as well.

One can notice the overestimated intensities of the $2\nu CH$ and $\nu CH + \nu CH$ bands appearing in the 6000–5500 cm^{-1} region (Figures 5 and 6 and Figures S1–S6). The magnitude of this effect varies between the different methods; however, it is present in all cases. A similar overestimation we have observed for butyl alcohols [30]. At present, we are unable to explain the reasons for these overestimations. Unexpectedly, B2PLYP functional (regardless of basis set; 6-31G(d,p), SNST, and def2-TZVP yielded similar results) significantly overestimates the frequencies of $2\delta CH + \delta CH$ transitions, shifting them to the 5500 cm^{-1} region. In contrast, the most of other transitions in NIR region is accurately reproduced by this approach. In the case of the $2\delta CH + \delta CH$ modes, large positive anharmonic constants appeared in GVPT2 vibrational analysis. Consequently, positions of the corresponding bands were predicted far from a simple combination of the harmonic frequencies. This shift has not been observed for the remaining approaches. Presently, the reason of this behavior is not clear. Because of very low intensity of $2\delta CH + \delta CH$ bands, these erroneous predictions do not provide meaningful contributions to NIR spectra. However, this occurrence demonstrates the need for using more than one method during examination of the fine spectral effects.

The deuteration of the OH group leads to a noticeable shift of the $\nu OD + \delta CH$ band. In contrast, the other bands do not shift meaningfully, as can be easily seen from comparison of CH_3CH_2OH and CH_3CH_2OD spectra (Figure 5A,B). In particular, the absorption from the $\nu CH + \delta CH$ in the 4600–4000 cm^{-1} region remains unaffected. This region can be used to monitor the isotopic substitutions of the CH_3 and CH_2 groups, as it leads to highly specific spectral changes. Obviously, simultaneous deuteration of both groups implies more significant changes. However, the most interesting effects result from the selective substitution of one of these groups. The presence of the CH_3 group gives rise to a prominent doublet near 4395 and 4330 cm^{-1}. This doublet has a complex structure resulting from overlapping of the contributions from the CH_2 (Figure S7 in Supplementary Material), leading to a broadening of the high-frequency wing of the doublet. As expected, this contribution is not present in the spectrum of CH_3CD_2OH (Figure S7B). On the other hand, the isotopic substitution of the CH_3 reveals a part of the overlapping contributions, as observed more clearly in the second derivative spectrum of CD_3CH_2OH (Figure S7C). This effect is well seen in the calculated spectra (Figure S7C).

The higher frequency NIR region (>7000 cm^{-1}) is also very sensitive to the isotopic effect. A weak absorption from the higher order overtones and combination bands creates difficulties in the analysis of this region. The deuteration of the OH group significantly reduces the number of the bands as a result of red-shift of the combination bands. The simulated spectra confirmed the high isotopic purity of the samples, except of CD_3CD_2OD which shows the $2\nu OH$ peak near 7100 cm^{-1} in the experimental spectrum (Figure 6B). This is in contrast to previously studied methanol, in which various non-uniform substitutions have been identified [41]. Contrary to -CX bonds, the H or D atoms in -OX bonds are labile, therefore, the OD group tends to exchange into the OH even by exposition of the deuterated alcohol to air. Since this band has a high absorptivity, therefore even small impurities due to the OH appear in NIR spectrum as a clear band at 7100 cm^{-1}. In contrast, no -CH bands are observed in the spectrum of CD_3CD_2OD (Figure 6B). NIR spectroscopy is particularly sensitive and selective for the isotopic effect, although the theoretical calculations are necessary for proper spectra interpretation. The spectral manifestations of the OH group in OD derivatives are obscured by ternary combinations from the CH vibrations that appear in the same region. For example, the $\delta_{as}'CH_3 + \nu_sCH_2 + \nu_{as}CH_2$ bands in CH_3CH_2OD (Figure 5B), and $\delta_{sciss}CH_2 + \nu_sCH_2 + \nu_{as}CH_2$ bands in CD_3CH_2OD (Figure 6D) are observed.

One can speculate that the isotopic substitution and conformational isomerism lead to convoluted spectral changes. This phenomenon will be a subject of our next paper (in preparation).

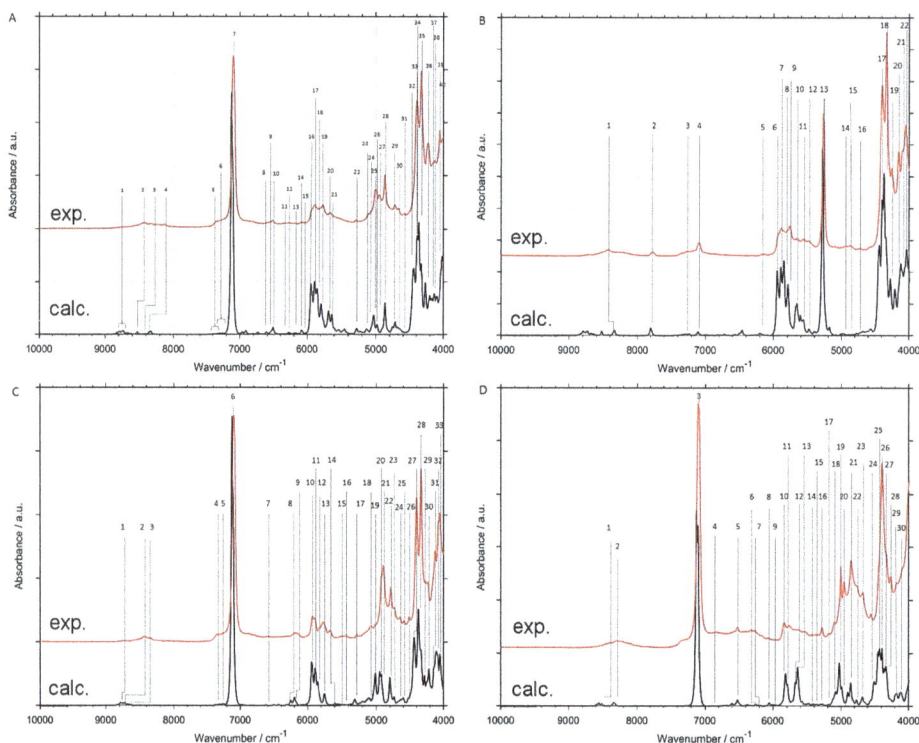

Figure 5. Band assignments in NIR spectra of deuterated ethanols based on GVPT2//B2PLYP-GD3BJ/ def2-TZVP//CPCM calculations. (**A**) CH_3CH_2OH; (**B**) CH_3CH_2OD; (**C**) CH_3CD_2OH; (**D**) CD_3CH_2OH. Band numbering corresponds to that presented in Tables 2–5.

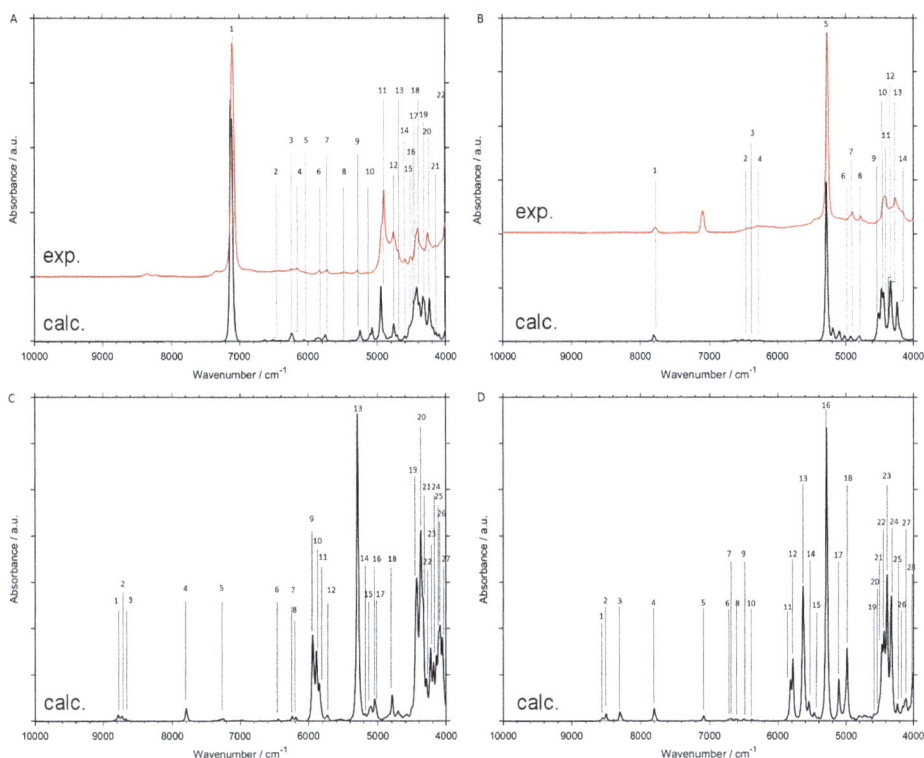

Figure 6. Band assignments in NIR spectra of deuterated ethanols based on GVPT2//B2PLYP-GD3BJ/ def2-TZVP//CPCM calculations. (**A**) CD_3CD_2OH; (**B**) CD_3CD_2OD; (**C**) CH_3CD_2OD; (**D**) CD_3CH_2OD. Band numbering corresponds to that presented in Tables 6–9.

Table 2. Band assignments in NIR spectra of CH_3CH_2OH based on GVPT2//B2PLYP-GD3BJ/ def2-TZVP//CPCM calculations. Band numbering corresponds to that presented in Figure 5A.

Peak Number	ν_{Exp}	ν_{Calc}	Assignment (Major Contribution)
1	8718.0	8739	$2\nu_{as}CH_2 + \nu_{as}CH_3$
2	8430.0	8526	$3\nu_{as}CH_2$
3	8329.0	8416	$3\nu_sCH_2$
4	8131.0	8329	$2\nu_sCH_2 + \nu_{as}CH_2$
5	7400–7300	7400–7300	$\delta_sCH_3 + \nu_{as}CH_3 + \nu_{as}'CH_3$ $[\delta_{as}'CH_3, \delta_{as}CH_3] + \nu_{as}CH_3 + \nu_{as}'CH_3$ $\delta_{sciss}CH_2 + \nu_{as}CH_2 + \nu_{as}'CH_3$ $[\delta_{rock}CH_2, \delta_{rock}CH_3] + \nu_{as}CH_2 + \nu OH$ $\delta_{twist}CH_2 + \nu_{as}CH_3 + \nu_{as}'CH_3$
6	7300–7200	7300–7200	$\delta_sCH_3 + \nu_sCH_3 + \nu_{as}'CH_3$ $[\delta_{as}CH_3, \delta_{as}'CH_3] + \nu_sCH_3 + \nu_{as}CH_3$ $[\delta_{as}CH_3, \delta_{as}'CH_3] + \nu_sCH_3 + \nu_{as}'CH_3$
7	7099.0	7125	$2\nu OH$
8	6610.0	6609	$\nu_sCH_3 + \nu OH$
9	6565.0	6540	$2\delta_{as}CH_3 + \nu OH$
10	6520.4	6513	$\nu_sCH_2 + \nu OH$
11	6331.0	6314	$2\delta_{twist}CH_2 + \nu OH$
12	6271.8	6275	$\delta_{ip}COH + \delta_{wagg}CH_2 + \nu OH$
13	6193.0	6178	$[\tau CC, \delta_{oop}COH] + \nu_{as}CH_3 + \nu_{as}'CH_3$

Table 2. *Cont.*

Peak Number	ν_{Exp}	ν_{Calc}	Assignment (Major Contribution)
14	6063.0	6085	$2\delta_{ip}COH + \nu OH$
15	6051.0	6021	$[\nu CC, \delta_{ip}COH] + \delta_{twist}CH_2 + \nu OH$
16	5936.0	5948	$2\nu_{as}'CH_3, \nu_{as}CH_3 + \nu_{as}'CH_3$
17	5886.1	5889	$2\delta_{as}'CH_3 + \nu OH$
18	5809.0	5846	$[\delta_{as}CH_3, \delta_{as}'CH_3] + \delta_{sciss}CH_2 + \nu_{as}CH_2$
19	5765.7	5790	$2\nu_{as}CH_2$
20	5665.1	5681	$2\nu_sCH_2; 2\nu_sCH_3 + \delta_sCH_3$
21	5634.0	5632	$\delta_{ip}COH + \delta_{wagg}CH_2 + \nu_sCH_3$
22	5287.6	5277	$\delta_{ip}OH + \delta CCO + \nu OH$
23	5111.0	5128	$\delta_{sciss}CH_2 + \nu OH$
24	5071.0	5118	$[\delta_{as}CH_3, \delta_{as}'CH_3] + \nu OH$
25	5013.8	5029	$[\delta_{wagg}CH_2, \delta_sCH_3] + \nu OH$
26	4996.2		
27	4954.2	4979	$[\delta_{twist}CH_2, \delta_{ip}COH, \delta_{wagg}CH_2] + \nu OH$
28	4873.0	4868	$\delta_{ip}COH + \nu OH$
29	4724.3	4763	$\delta_sCH_3 + 2[\delta_{as}CH_3, \delta_{as}'CH_3]$
30	4677.0	4726	$[\nu CO, \delta_{rock}'CH_3] + \nu OH$
31	4582.9	4648	$[\delta_{oop}COH, \tau CC] + \delta_{as}'CH_3 + \nu_sCH_3$
32	4454.0	4450	$3\delta_{sciss}CH_2$
33	4409.0	4396	$\delta_{sciss}CH_2 + \nu_{as}CH_2$
34	4394.8	4366	$\delta_sCH_3 + \nu_{as}'CH_3$
35	4333.5	4331	$[\delta_{oop}COH, \tau CC] + \nu OH$
36	4232.6	4269	$\delta_{twist}CH_2 + \nu_{as}'CH_3$
37	4162.0	4177	$\delta_{ip}COH + \nu_sCH_2$
38	4131.7	4137	$\delta_{twist}CH_2 + \nu_sCH_2$
39	4057.4	4020	$[\delta_{rock}CH_2, \delta_{rock}CH_3] + \nu_sCH_2$
40	4024.0	3997	$[\nu CO, \delta_{rock}'CH_3] + \nu_{as}CH_2$

Table 3. Band assignments in NIR spectra of CH_3CH_2OD based on GVPT2//B2PLYP-GD3BJ/ def2-TZVP//CPCM calculations. Band numbering corresponds to that presented in Figure 5B.

Peak Number	ν_{exp}	ν_{calc}	Assignment (Major Contribution)
1	8428.0	8334	$2\nu_sCH_2 + \nu_{as}CH_2$
2	7777.5	7796	$3\nu OD$
3	7260.0	7227	$\delta_{twist}CH_2 + \nu_{as}CH_3 + \nu_{as}'CH_3$
4	7099.1	7112	$\delta_{as}'CH_3 + \nu_sCH_2 + \nu_{as}CH_2$
5	6133.4	6200	$\tau CC + \nu_{as}CH_3 + \nu_{as}'CH_3$
6	5935.0	5946	$2\nu_{as}'CH_3$
7	5885.2	5895	$2\nu_{as}'CH_3$
8	5850.0	5847	$[\delta_{as}CH_3, \delta_{as}'CH_3] + \delta_{sciss}CH_2 + \nu_{as}CH_2$
9	5765.6	5788	$2\nu_{as}CH_2$
10	5665.7	5669	$2\delta_{wagg}CH_2 + \nu_sCH_2$
11	5564.3	5559	$\nu OD + \nu_sCH_2$
12	5494.3	5498	$\nu_{as}CH_2 + \delta_{twist}CH_2 + \delta_{wagg}CH_2$
13	5277.1	5289	$2\nu OD$
14	4947.0	4963	$[\delta_{rock}CH_2, \delta_{rock}CH_3] + \delta_{twist}CH_2 + \nu_sCH_2$
15	4873.0	4846	$[\delta_{rock}CH_2, \delta_{rock}CH_3] + [\delta_{rock}CH_2, \delta_{rock}CH_3] + \nu_sCH_2$
16	4720.8	4717	$[\nu CO, \delta_{rock}'CH_3, \delta_{ip}COD] + [\delta_{rock}'CH_3, \delta_{ip}COD, \delta_{sciss}CH_2CO] + \nu OD$
17	4393.7	4397	$\delta_{sciss}CH_2 + \nu_{as}CH_2$
18	4331.8	4364	$[\delta_sCH_3, \delta_{wagg}CH_2] + \nu_{as}CH_3$
19	4253.4	4275	$\delta_{twist}CH_2 + \nu_{as}'CH_3$
20	4155.1	4127	$[\delta_{rock}'CH_3, \delta_{ip}COD, \delta_{sciss}CH_2CO] + \nu_{as}CH_3$
21	4105.0	4063	$[\delta_{rock}'CH_3, \delta_{ip}COD, \delta_{sciss}CH_2CO] + \nu_sCH_3$
22	4054.1	4037	$[\delta_{rock}CH_2, \delta_{rock}CH_3] + \delta_sCH_2$

Table 4. Band assignments in NIR spectra of CH_3CD_2OH based on GVPT2//B2PLYP-GD3BJ/ def2-TZVP//CPCM calculations. Band numbering corresponds to that presented in Figure 5C.

Peak Number	ν_{exp}	ν_{calc}	Assignment (Major Contribution)
1	8717.0	8788	$3\nu_{as}CH_3$
2	8434.1	8728	$\nu_sCH_3 + \nu_{as}CH_3 + \nu_{as}'CH_3$
3	8337.0	8665	$2\nu_sCH_3 + \nu_{as}CH_3$
4	7345.0	7324	$[\delta_{rock}CD_2, \delta_{twist}CD_2] + \nu_{as}'CH_3 + \nu OH$
5	7251.0	7255	$[\delta_{as}CH_3, \delta_{as}'CH_3] + \nu_sCH_3 + \nu_{as}CH_3$
6	7098.2	7126	$2\nu OH$
7	6590.4	6618	$\nu_{as}'CH_3 + \nu OH$
8	6205.8	6257	$2\delta_{ip}COH + \nu OH$
9	6158.0	6198	$2\delta_{ip}COH + \nu OH$
10	5929.9	5943	$2\nu_{as}CH_3$
11	5895.4	5892	$2\nu_{as}'CH_3$
12	5828.0	5844	$\nu_sCH_3 + \nu_{as}'CH_3$
13	5763.8	5744	$\nu_sCD_2 + \nu OH$
14	5669.1	5595	$\nu_sCH_3 + 2\nu_{as}CH_3$
15	5449.0	5496	$\delta_{rock}CH_3 + \delta_sCH_3 + \nu_{as}'CH_3$
16	5439.4	5449	$\delta_{sciss}CD_2CO + 2\nu_{as}CH_3$
17	5282.0	5309	$\nu_{as}CD_2 + 2[\delta_{as}CH_3, \delta_{as}'CH_3]$
18	5070.1	5094	$\nu_sCD_2 + \nu_{as}CH_3$
19	5017.0	5018	$[\tau CC, \delta_{oop}COH] + 2[\delta_{as}CH_3, \delta_{as}'CH_3]$
20	4927.0	4954	$\delta_{ip}COH + \nu OH$
21	4898.3	4930	$\delta_{ip}COH + \nu OH$
22	4792.1	4806	$[\nu CO, \delta_{wagg}CD_2] + \nu OH$
23	4737.2	4744	$\delta_{sciss}CD_2 + \nu OH$
24	4650.2	4641	$[\nu CC, \delta_{rock}'CH_3] + \nu OH$
25	4591.7	4600	$[\nu CO, \delta_{wagg}CD_2] + \nu OH$
26	4513.3	4525	$\delta_{ip}COH + \nu OH$
27	4404.6	4437	$[\delta_{as}CH_3, \delta_{as}'CH_3] + \nu_{as}'CH_3$
28	4338.0	4373	$\delta_sCH_3 + \nu_{as}'CH_3$
29	4275.7	4285	$\delta_{ip}COH + \nu_{as}CH_3$
30	4238.1	4218	$\nu_sCD_2 + \nu_{as}CD_2$
31	4129.6	4147	$[\nu CO, \delta_{wagg}CD_2] + \nu_{as}'CH_3$
32	4079.7	4117	$\delta_{rock}CH_3 + \nu_{as}CH_3$
33	4056.2	4052	$\delta_{rock}CH_3 + \nu_sCH_3, \delta_{sciss}CD_2CO + \nu OH$

Table 5. Band assignments in NIR spectra of CD_3CH_2OH based on GVPT2//B2PLYP-GD3BJ/ def2-TZVP//CPCM calculations. Band numbering corresponds to that presented in Figure 5D.

Peak Number	ν_{exp}	ν_{calc}	Assignment (Major Contribution)
1	8405.0	8511	$3\nu_{as}CH_2$
2	8307.6	8323	$2\nu_sCH_2 + \nu_{as}CH_2$
3	7098.5	7124 (*t*) 7100 (*g*)	$2\nu OH$
4	6841.1	6897	$[\nu CO, \delta_{as}'CD_3] + \nu_sCH_2 + \nu_{as}CH_2$
5	6517.6	6510	$\nu_sCH_2 + \nu OH$
6	6324.0	6261	$[\delta_{ip}COH, \delta_{twist}CH_2] + \delta_{wagg}CH_2 + \nu OH$
7	6268.0	6111	$[\delta_{rock}'CD_3, \nu CC] + \nu_{as}'CD_3 + \nu OH$
8	6070.4	6059	$2[\delta_{ip}COH, \delta_{twist}CH_2] + \nu OH$
9	5966.3	5966	$[\delta_sCD_3, \nu CC] + \delta_{wagg}CH_2 + \nu OH$
10	5839.1	5825	$2\nu_{as}CH_2$
11	5772.1	5793	$2\nu_{as}CH_2$
12	5628.0	5686	$2\nu_sCH_2$
13	5533.0	5641	$2\nu_sCH_2$
14	5427.9	5419	$2\delta_{twist}CH_2 + \nu_{as}CH_2$
15	5358.0	5367	$[\nu CO, \delta_{as}'CD_3] + \nu_sCD_3 + \nu_{as}'CD_3$
16	5286.8	5287	$\delta_sCD_3 + \delta_{twist}CH_2 + \nu_{as}CH_2$

Table 5. *Cont.*

Peak Number	ν_{exp}	ν_{calc}	Assignment (Major Contribution)
17	5190.0	5188	$2[\delta_s CD_3, \nu CC] + \nu_s CH_2$
18	5102.7	5084	$\delta_{oop}COH + 2\delta_{wagg}CH_2$
19	5007.1	5028	$\delta_{wagg}CH_2 + \nu OH$
20	4955.8	4987	$[\delta_{twist}CH_2, \delta_{ip}COH, \delta_{wagg}CH_2] + \nu OH$
21	4853.8	4853	$[\delta_{ip}COH, \delta_{twist}CH_2] + \nu OH$
22	4764.7	4768	$[\delta_s CD_3, \nu CC] + \nu OH$
23	4676.7	4676	$\nu CO + \nu OH$
24	4558.4	4511	$\delta_{as}'CD_3 + [\delta_{twist}CH_2, \delta_{ip}COH] + \nu_{as}'CD_3$
25	4443.0	4429	$\delta_{sciss}CH_2 + \nu_{as}CH_2$
26	4390.5	4390	$\delta_{sciss}CH_2 + \nu_{as}CH_2$
27	4329.0	4356	$\delta_{sciss}CH_2 + \nu_s CH_2$
28	4263.8	4332	$\delta_{sciss}CH_2 + \nu_s CH_2$
29	4174.6	4180	$2[\delta_{twist}CH_2, \delta_{ip}COH, \delta_{wagg}CH_2] + \delta_{sciss}CH_2$
30	4100.0	4107	$\tau CC + \delta_{oop}COH + \nu OH$

Table 6. Band assignments in NIR spectra of CD_3CD_2OH based on GVPT2//B2PLYP-GD3BJ/ def2-TZVP//CPCM calculations. Band numbering corresponds to that presented in Figure 6A.

Peak Number	ν_{exp}	ν_{calc}	Assignment (Major Contribution)
1	7099.0	7126 (*t*) 7102 (*g*)	$2\nu OH$
2	6444.0	6495	$\nu_s CD_2 + [\nu_{as}CD_2, \nu_{as}CD_3] + [\nu_{as}'CD_3, \nu_{as}CD_2]$
3	6232.1	6244	$2\delta_{ip}COH + \nu OH$
4	6162.9	6224	$2[\nu CC, \delta_{wagg}CD_2] + \nu OH$
5	6063.0	6059	$[\delta_{sciss}CD_2, \nu CO] + [\nu CC, \delta_{wagg}CD_2] + \nu OH$
6	5838.3	5861	$[\nu_{as}CD_2, \nu_{as}CD_3] + \nu OH$
7	5732.3	5746	$[\delta_{rock}CD_2, \delta_{rock}CD_3] + [\delta_s CD_3, \delta_{wagg}CD_2] + \nu OH$
8	5478.7	5488	$[\nu CC, \delta_{wagg}CD_2] + \nu_s CD_3 + [\nu_{as}CD_2, \nu_{as}CD_3]$
9	5285.8	5247	$\nu_s CD_3 + \nu_{as}CD_2 + \nu CO$
10	5160.0	5103	$[\delta_{twist}CD_2, \delta_{rock}CD_3, \delta_{rock}CD_2] + 2[\nu CC, \delta_{wagg}CD_2]$
11	4903.7	4947	$\delta_{ip}COH + \nu OH$
12	4769.2	4766	$[\delta_{sciss}CD_2, \nu CO] + \nu OH$
13	4701.4	4714	$[\delta_{as}'CD_3, \delta_{as}CD_3] + \nu OH$
14	4598.4	4604	$[\nu CO, \delta_{wagg}CD_2] + \nu OH$
15	4525.0	4539	$[\delta_{twist}CD_2, \delta_{rock}CD_3, \delta_{rock}CD_2] + \delta_{ip}COH + [\nu_{as}'CD_3, \nu_{as}CD_2]$
16	4506.9	4499	$2\delta_{sciss}CD_2CO + \nu_{as}CD_3$
17	4430.0	4447	$[\nu_{as}CD_2, \nu_{as}CD_3] + \nu_{as}CD_3, 2[\nu_{as}'CD_3, \nu_{as}CD_2]$
18	4409.6	4420	$[\nu_{as}CD_2, \nu_{as}CD_3] + [\nu_{as}'CD_3, \nu_{as}CD_2]$
19	4332.0	4334	$2\nu_{as}CD_2$
20	4267.4	4235	$[\nu_s CD_2, \nu_s CD_3] + \nu_{as}CD_2$
21	4156.8	4140	$2\delta_{oop}COH + \nu OH$
22	4013.0	4012	$\delta_{sciss}CD_2CO + \nu OH$

Table 7. Band assignments in NIR spectra of CD_3CD_2OD based on GVPT2//B2PLYP-GD3BJ/ def2-TZVP//CPCM calculations. Band numbering corresponds to that presented in Figure 6B.

Peak Number	νExp	νCalc	Assignment (Major Contribution)
1	7771.0	7799	$3\nu OD$
2	6468.0	6525	$\nu_s CD_3 + [\nu_{as}'CD_3, \nu_{as}CD_2] + \nu_{as}CD_3$
3	6450.0	6447	$3\nu_{as}CD_2$
4	6290.1	6267	$2\nu_s CD_2 + \nu_{as}CD_2$
5	5276.0	5289	$2\nu OD$
6	4948.0	5020	$[\nu CO, \delta_{twist}CD_2] + 2[\delta_{sciss}CD_2, \nu CO]$
7	4902.2	4929	$[\delta_{twist}CD_2, \delta_{rock}CD_2] + 2[\delta_{sciss}CD_2, \nu CO]$
8	4779.2	4795	$\nu_s CD_2 + \nu OD$

Table 7. *Cont.*

Peak Number	νExp	νCalc	Assignment (Major Contribution)
9	4509.0	4510	$[\delta_{as}CD_3, \delta_{as}'CD_3] + [\nu CC, \delta_{wagg}CD_2] + \nu_{as}CD_3$
10	4437.4	4465	$2\nu_{as}'CD_3, \nu_{as}CD_3 + \nu_{as}'CD_3$
11	4409.3	4434	$2\nu_{as}CD_3, \nu_{as}CD_2 + \nu_{as}'CD_3$
12	4325.2	4381	$2[\delta_{sciss}CD_2, \nu CO] + \nu_s CD_3$
13	4269.5	4337	$2\nu_{as}CD_2$
14	4170.2	4241	$[\nu_s CD_2, \nu_s CD_3] + \nu_{as}CD_2$

Table 8. Band assignments in NIR spectra of CH_3CD_2OD based on GVPT2//B2PLYP-GD3BJ/ def2-TZVP//CPCM calculations. Band numbering corresponds to that presented Figure 6C.

Peak Number	νCalc	Assignment (Major Contribution)
1	8781	$3\nu_{as}'CH_3$
2	8722	$\nu_s CH_3 + \nu_{as}CH_3 + \nu_{as}'CH_3$
3	8666	$2\nu_s CH_3 + \nu_{as}CH_3, 2\nu_s CH_3 + \nu_{as}'CH_3$
4	7802	$3\nu OD$
5	7257	$[\delta_{as}CH_3, \delta_{as}'CH_3] + \nu_s CH_3 + \nu_{as}CH_3$
6	6445	$3\nu_{as}CD_2$
7	6188	$\tau CC + \nu_{as}CH_3 + \nu_{as}'CH_3$
8	5943	$2\nu_{as}CH_3, \nu_{as}CH_3 + \nu_{as}'CH_3$
9	5890	$2[\delta_{as}'CH_3, \delta_{as}CH_3] + \nu_{as}'CH_3, 2\nu_{as}'CH_3$
10	5845	$\nu_s CH_3 + \nu_{as}CH_3$
11	5728	$2\delta_s CH_3 + \nu_{as}CH_3$
12	5288	$2\nu OD$
13	5182	$\nu_{as}CD_2 + \nu_{as}'CH_3$
14	5120	$\nu_{as}CD_2 + \nu_s CH_3$
15	5097	$\nu_s CD_2 + \nu_{as}CH_3$
16	5046	$\tau CC + 2[\delta_{as}CH_3, \delta_{as}'CH_3]$
17	4796	$\delta_{oop}COD + 2[\delta_{as}'CH_3, \delta_{as}CH_3]$
18	4434	$\tau CC + [\delta_{wagg}CD_2, \nu CC] + \nu_{as}'CH_3$
19	4372	$\delta_s CH_3 + \nu_{as}CH_3, \delta_s CH_3 + \nu_{as}'CH_3$
20	4341	$2\nu_{as}CD_2$
21	4289	$\tau CC + \delta_{sciss}CD_2 + [\delta_{sciss}CD_2, \nu CO]$
22	4233	$\nu_s CD_2 + \nu_{as}CD_2$
23	4190	$2\nu_s CD_2, [\delta_{wagg}CD_2, \nu CC] + \nu_{as}CH_3$
24	4149	$[\delta_{sciss}CD_2, \nu CO] + \nu_{as}CH_3, [\delta_{sciss}CD_2, \nu CO] + \nu_{as}'CH_3$
25	4113	$\delta_{rock}CH_3 + \nu_{as}'CH_3$
26	4089	$[\delta_{sciss}CD_2, \nu CO] + \nu_s CH_3$
27	4056	$\delta_{rock}CH_3 + \nu_s CH_3$

Table 9. Band assignments in NIR spectra of CD_3CH_2OD based on GVPT2//B2PLYP-GD3BJ/ def2-TZVP//CPCM calculations. Band numbering corresponds to that presented Figure 6D.

Peak Number	ν_{calc}	Assignment (Major Contribution)
1	8560	$3\nu_{as}CH_2, 2\nu_s CH_2 + \nu_{as}CH_2$
2	8508	$3\nu_{as}CH_2$
3	8305	$2\nu_s CH_2 + \nu_{as}CH_2$
4	7802	$3\nu OD$
5	7088	$\delta_{sciss}CH_2 + \nu_s CH_2 + \nu_{as}CH_2$
6	6736	$\delta_{rock}CH_2 + \nu_s CH_2 + \nu_{as}CH_2$
7	6706	$[\nu CO, \delta_{as}'CD_3] + \nu_s CH_2 + \nu_{as}CH_2$
8	6614	$3\nu_{as}'CD_3$
9	6505	$\nu_s CD_3 + \nu_{as}CD_3 + \nu_{as}'CD_3$
10	6382	$2\nu_s CD_3 + \nu_{as}CD_3, 2\nu_s CD_3 + \nu_{as}'CD_3$
11	5821	$[\delta_{oop}COD, \tau CC] + \nu_{as}CH_2 + \nu_s CH_2$
12	5785	$2\nu_{as}CH_2, \delta_{wagg}CH_2 + \delta_{sciss}CH_2 + \nu_{as}CH_2$

Table 9. *Cont.*

Peak Number	ν_{calc}	Assignment (Major Contribution)
13	5638	$2\nu_s CH_2, \nu_s CH_2 + \nu_{as} CH_2$
14	5547	$\nu OD + \nu_s CH_2$
15	5473	$\delta_{rock} CH_2 + \delta_{sciss} CH_2 + \nu_{as} CH_2$
16	5288	$2\nu OD$
17	5106	$[\tau CC, \delta_{oop} COD] + 2\delta_{wagg} CH_2$
18	4986	$2[\nu CC, \delta_s CD_3] + \nu OD$
19	4585	$\delta_{rock}' CD_3 + 2\delta_{wagg} CH_2$
20	4503	$[\tau CC, \delta_{oop} COD] + \delta_{wagg} CH_2 + \nu_{as} CH_2$
21	4459	$2\nu_{as} CD_3$
22	4431	$[\delta_{as}' CD_3, \nu CO] + [\nu CC, \delta_s CD_3] + \nu_{as}' CD_3$
23	4385	$\delta_{sciss} CH_2 + \nu_{as} CH_2$
24	4324	$2\nu_{as}' CD_3, \delta_{sciss} CH_2 + \nu_s CH_2$
25	4237	$\delta_{wagg} CH_2 + \nu_s CH_2$
26	4167	$\delta_{twist} CH_2 + \nu_{as} CH_2$
27	4112	$\delta_{twist} CH_2 + \nu_s CH_2$
28	4015	$[\delta_{rock} CD_3, \delta_{rock} CH_2] + \delta_{rock} CH_2 + \nu_{as}' CD_3$

Another insight, which becomes possible only through theoretical simulation of NIR spectra, is estimation of the relative contributions from different kinds of vibrational transitions (Table 10). As compared with methanol [41], ethanol offers better opportunity to analyze these contributions, because of higher number of isotopomers and more complex NIR spectra. The effect of various kinds of isotopic substitution of the CH_3, CH_2, and OH groups on NIR spectra may be elucidated. In the $10,000$–$4000\ cm^{-1}$ region two quanta transitions, first overtones ($2\nu_x$) and binary combinations ($\nu_x + \nu_y$), are the most meaningful components of the spectra. In particular, binary combinations from the CH_3 group have significant contribution—e.g., for CH_3CH_2OH they are responsible for 47% of NIR intensity—while upon deuteration of the CH_3 group this contribution decreases to 32.6%. An even more pronounced effect is observed for OD derivatives, the analogous values for CH_3CH_2OD and CD_3CH_2OD are 51.2% and 35.5%, respectively. Simultaneously, the isotopic substitution of the methyl group increases the relative intensity of the first overtones, while the intensity of the second overtones remains insignificant. As expected, the importance of the second overtones increases in the upper NIR region ($10,000$–$7500\ cm^{-1}$). Interestingly, this trend is not observed for the ternary combinations ($\nu_x + \nu_y + \nu_z$ and $2\nu_x + \nu_y$), although for OD derivatives the $2\nu_x + \nu_y$ contribution increases and the $\nu_x + \nu_y + \nu_z$ contribution decreases upon deuteration of the CH_3 group. The isotopic substitution of the CH_2 group provides similar changes, but is noticeably less significant.

As can be seen (Table 10), the region above $7500\ cm^{-1}$ is contributed only by three and higher quanta transitions. Therefore, in this region the effect of isotopic substitution is even more visible. The deuteration of the CH_3 group increases the contributions from the second overtones at the expense of $\nu_x + \nu_y + \nu_z$ combinations, while the contributions from $2\nu_x + \nu_y$ remain similar. Interestingly, NIR spectrum of CD_3CD_2OD above $7500\ cm^{-1}$ includes the second overtones only.

These observations remain in agreement with our previous findings on methanol isotopomers [41]. However, the contributions from the three quanta transitions are more important for ethanol. For CH_3CH_2OH these transitions involve 25.9% of total intensity ($10,000$–$4000\ cm^{-1}$), while for CH_3OH this value was found to be 19.2%. The difference between CD_3OD and CD_3CD_2OD is even larger (23.5% vs. 36.7%).

Table 10. Contributions (in %) from the first and second overtones as well as binary and ternary combinations into NIR spectra of ethanol isotopomers based on GVPT2//B2PLYP-GD3BJ/ def2-TZVP//CPCM calculations. [a]

| | 10,000–4000 cm^{-1} | | | | |
	$2\nu_x$	$3\nu_x$	$\nu_x + \nu_y$	$\nu_x + \nu_y + \nu_z$	$2\nu_x + \nu_y$
CH_3CH_2OH	26.1	1.7	47.0	14.3	10.9
CH_3CH_2OD	18.0	2.2	51.2	17.4	11.1
CH_3CD_2OH	35.8	1.7	41.5	11.8	9.2
CD_3CH_2OH	40.9	1.2	32.6	15.1	10.1
CD_3CD_2OH	46.0	0.3	23.7	15.8	14.2
CD_3CD_2OD	43.1	2.0	19.2	17.8	17.9
CH_3CD_2OD	27.9	3.2	44.7	15.0	9.2
CD_3CH_2OD	36.0	2.5	35.5	10.8	15.2

| | 10,000–7500 cm^{-1} | | | | |
	$2\nu_x$	$3\nu_x$	$\nu_x + \nu_y$	$\nu_x + \nu_y + \nu_z$	$2\nu_x + \nu_y$
CH_3CH_2OH	0.0	39.7	0.0	22.3	38.0
CH_3CH_2OD	0.0	55.5	0.0	15.4	29.1
CH_3CD_2OH	0.0	43.9	0.0	30.5	25.6
CD_3CH_2OH	0.0	66.9	0.0	1.4	31.7
CD_3CD_2OH	0.0	0.0	0.0	43.0	57.0
CD_3CD_2OD	0.0	100.0	0.0	0.0	0.0
CH_3CD_2OD	0.0	69.9	0.0	16.7	13.4
CD_3CH_2OD	0.0	76.3	0.0	0.5	23.2

| | 7500–4000 cm^{-1} | | | | |
	$2\nu_x$	$3\nu_x$	$\nu_x + \nu_y$	$\nu_x + \nu_y + \nu_z$	$2\nu_x + \nu_y$
CH_3CH_2OH	26.5	1.2	47.6	14.2	10.5
CH_3CH_2OD	18.4	1.0	52.4	17.5	10.7
CH_3CD_2OH	36.0	1.4	41.8	11.7	9.1
CD_3CH_2OH	41.4	0.4	33.0	15.3	9.9
CD_3CD_2OH	46.0	0.3	23.7	15.8	14.2
CD_3CD_2OD	43.7	0.5	19.5	18.0	18.2
CH_3CD_2OD	28.4	2.0	45.5	15.0	9.1
CD_3CH_2OD	37.0	0.5	36.4	11.1	15.0

[a] The comparison is based on integrated intensity (cm^{-1}) summed over simulated bands, convoluted with the use of Cauchy−Gauss product function (details in the text) in relation to the total integrated intensity.

3. Experimental and Computational Methods

3.1. Materials and Spectroscopic Measurements

In Table 11 are collected the details on the samples used in this work. The experimental spectrum of CH_3CH_2OH was taken from our previous work [29]. All samples were used as received, while solvent (CCl_4) was distilled and additionally dried using freshly activated molecular sieves (Aldrich, 4A). All ethanols were measured in CCl_4 solution (0.1 mol dm^{-3}). NIR spectra were recorded on Thermo Scientific Nicolet iS50 spectrometer using InGaAs detector, with a resolution of 2 cm^{-1} (128 scans), in a quartz cells (Hellma QX, Hellma Optik GmbH, Jena, Germany) of 100 mm thicknesses at 298 K (25 °C).

Table 11. Samples used in this study

	Sample	Purity	D Atom Content	Other Remarks
1	CH_3CH_2OD	99%	≥99.5%	
2	CH_3CD_2OH	99%	98%	
3	CD_3CH_2OH	99%	99%	
4	CD_3CD_2OH	99%	99.5%	
5	CD_3CD_2OD	>99%	≥99.5%	anhydrous
6	CCl_4	>99%	-	

Samples were purchased from Sigma-Aldrich Chemie GmbH (Taufkirchen, Germany).

3.2. Computational Procedures

Our calculations were based on density functional theory (DFT) with double-hybrid B2PLYP density functional [55] (unfrozen core) coupled with Karlsruhe triple-ζ valence with polarization (def2-TZVP) [56] basis set. Grimme's third formulation of empirical correction for dispersion with Becke-Johnson damping (GD3BJ) was applied [57]. To better reflect solvation of molecules, CCl_4 cavity in solvent reaction field (SCRF) [58] was included at conductor-like polarizable continuum (CPCM) [59] level. Very tight criteria for geometry optimization and 10^{-10} convergence criterion in SCF procedure were set. Electron integrals and solving coupled perturbed Hartree-Fock (CPHF) equations were calculated over a superfine grid. The selected method provided good reproduction of NIR spectra of various molecules in CCl_4 solution [19,29,30].

We carried out the anharmonic vibrational analysis at generalized vibrational second-order perturbation theory (GVPT2) [60,61] level. In this approach, the anharmonic frequencies and intensities of the vibrational transitions up to three quanta were obtained. This allows to simulate fundamental, first and second overtones, as well as binary and ternary combination bands. Quantum mechanical calculations were carried out with Gaussian 16 (A.03) [62]. One of the major features implemented in GVPT2 approach is the automatic treatment of tight vibrational degenerations, i.e., resonances [63]. In this work the search for resonances included Fermi (i.e., 1-2) of type I ($\omega_i \approx 2\omega_j$) and type II ($\omega_i \approx \omega_j + \omega_k$), and Darling–Dennison (i.e., 2-2, 1-1, and 1-3) resonances. All possible resonant terms within search thresholds were included in the variational treatment. The resonance search thresholds (respectively, maximum frequency difference and minimum difference PT2 vs. variational treatment; in (cm^{-1})) were: 200 and 1 (for the search of 1–2 resonances), 100 and 10 (for 2-2, 1-1, and 1-3).

To display the simulated spectra we applied a four-parameter Cauchy–Gauss (Lorentz–Gauss) product function [20]. The theoretical bands were modelled with a_2 and a_4 parameters equal to 0.055 and 0.015, resulting with full-width at half-height (FWHH) of 25 cm^{-1}. Exception was made for better agreement with the weaker and broader experimental bands, which are presented in Figures 3 and 4. In this case the values were 0.075, 0.015, and 35 cm^{-1}, respectively. The final theoretical spectra were obtained by combining the spectra of *trans* and *gauche* conformers, mixed in accordance with the calculated abundances of each form [64]. The relative abundances of the *gauche* (n_g) and *trans* (n_t) conformers were determined as following equation [65].

$$\frac{n_g}{n_t} = \frac{A_t}{A_g} e^{\frac{-\Delta G^{298}}{RT}}$$

where Gibbs free energy (ΔG) corresponds to the value calculated at 298 K corrected by anharmonic (VPT2) zero-point energy (ZPE); A_t and A_g are the degeneracy prefactors of the Boltzmann term for the *gauche* (1) and *trans* (2) conformers.

The band assignments were aided by calculations of potential energy distributions (PEDs). PEDs were obtained with Gar2Ped software [66], using natural internal coordinate system defined in accordance with Pulay [67]. The numerical analysis of the theoretical results and the processing of the experimental spectra were performed with MATLAB R2016b (The Math Works Inc.) [68].

4. Conclusions

Isotopic substitution leads to much higher variability in NIR spectra as compared with IR spectra, due to significant contribution from the combination bands. The pattern of OH/OD, CH_3/CD_3, and CH_2/CD_2 groups in ethanol often leads to fine spectral changes, which may be monitored and explained in detail by anharmonic quantum mechanical simulations. Our studies were devoted to NIR spectra of eight isotopomers of ethanol (CX_3CX_2OX (X = H, D)) by using anharmonic GVPT2 vibrational analysis. The calculations were performed at several levels of electronic theory, including DFT and MP2 to find accurate and efficient theoretical approach for studies of isotopic effect in NIR spectra. Our results indicate that DFT approach using double-hybrid B2PLYP functional, coupled with def2-TZVP basis set, and supported by GD3BJ correction with CPCM solvent model yielded the best results. The theoretical spectra obtained by this approach enabled us to assign most of NIR bands, including two ($2\nu_x$ and $\nu_x + \nu_y$) and three quanta ($3\nu_x$, $\nu_x + \nu_y + \nu_z$, and $2\nu_x + \nu_y$) transitions. Accuracy of these calculations permitted us to analyze theoretical NIR spectra of CH_3CD_2OD and CD_3CH_2OD for which the experimental spectra are not available. The effect of the isotopic substitution of the OH, CH_3, and CH_2 groups was satisfactory reproduced and explained. Moreover, the relative contributions of selected groups and kinds of transitions were elucidated and discussed. The contributions from the CH_3 group appear to be more important than those from the CH_2 group. The isotopic substitution in the CH_3 group leads to the most prominent intensity changes in NIR spectra as compared to the changes due to the substitution of the other groups. The bands from the three quanta transitions are more important for isotopomers of ethanol than for derivatives of methanol.

Supplementary Materials: The following are available online, Figures S1–S35; Tables S1–S9.

Author Contributions: Conceptualization, K.B.B. and M.A.C.; Methodology K.B.B. and J.G.; Formal analysis, J.G. and K.B.B.; Investigation, all authors; Writing—original draft preparation, K.B.B.; Writing—review and editing, all authors; Supervision, C.W.H. and M.A.C.

Funding: This work was supported by the National Science Center Poland, Grant No. 2017/27/B/ST4/00948.

Acknowledgments: Calculations have been carried out in Wroclaw Centre for Networking and Supercomputing (http://www.wcss.pl), under grant no. 163.

Conflicts of Interest: The authors declare no conflict of interest. The founders had no role in the design of the study; in the collection, analyses, or interpretation of data; in the writing of the manuscript, or in the decision to publish the results.

References

1. Siesler, H.W.; Ozaki, Y.; Kawata, S.; Heise, H.M. (Eds.) *Near-Infrared Spectroscopy*; Wiley-VCH: Weinheim, Germany, 2002.
2. Beć, K.B.; Grabska, J.; Ozaki, Y. Advances in anharmonic methods and their applications to vibrational spectroscopies. In *Frontiers of Quantum Chemistry*; Wójcik, M.J., Nakatsuji, H., Kirtman, B., Ozaki, Y., Eds.; Springer: Singapore, 2017.
3. Beć, K.B.; Grabska, J.; Huck, C.W.; Ozaki, Y. Quantum mechanical simulation of NIR spectra. In *Applications in Physical and Analytical Chemistry*; Ozaki, Y., Wójcik, M.J., Popp, J., Eds.; Wiley: Hoboken, NJ, USA, 2019; in press.
4. Ozaki, Y.; Huck, C.W.; Beć, K.B. Near infrared spectroscopy and its applications. In *Molecular and Laser Spectroscopy*; Gupta, V.P., Ed.; Elsevier: Amsterdam, The Netherlands, 2017.
5. Siesler, H.W. Near-infrared spectra, interpretation. In *Encyclopedia of Spectroscopy and Spectrometry*, 3rd ed.; Lindon, J.C., Tranter, G.E., Koppenaal, D.W., Eds.; Academic Press: Oxford, UK, 2017.
6. Workman, J., Jr.; Weyer, L. *Practical Guide and Spectral Atlas for Interpretive Near-Infrared Spectroscopy*, 2nd ed.; CRC Press: Boca Raton, FL, USA, 2012.
7. Beć, K.B.; Hawranek, J.P. Vibrational analysis of liquid *n*-butylmethylether. *Vib. Spectrosc.* **2013**, *64*, 164–171. [CrossRef]
8. Beć, K.B.; Kwiatek, A.; Hawranek, J.P. Vibrational analysis of neat liquid *tert*-butylmethylether. *J. Mol. Liq.* **2014**, *196*, 26–31. [CrossRef]

9. Kirchler, C.G.; Pezzei, C.K.; Beć, K.B.; Henn, R.; Ishigaki, M.; Ozaki, Y.; Huck, C.W. Critical evaluation of NIR and ATR-IR spectroscopic quantifications of rosmarinic acid in rosmarini folium supported by quantum chemical calculations. *Planta Med.* **2017**, *83*, 1076–1084. [CrossRef] [PubMed]

10. Beć, K.B.; Huck, C.W. Breakthrough potential in near-infrared spectroscopy: Spectra simulation. A review of recent developments. *Front. Chem.* **2019**, *7*, 1–22. [CrossRef] [PubMed]

11. Bozzolo, G.; Plastino, A. Generalized anharmonic oscillator: A simple variational approach. *Phys. Rev. D* **1981**, *24*, 3113–3117.

12. Gribov, L.A.; Prokof'eva, N.I. Variational solution of the problem of anharmonic vibrations of molecules in the central force field. *J. Struct. Chem.* **2015**, *56*, 752–754. [CrossRef]

13. Bowman, J.M. Self-consistent field energies and wavefunctions for coupled oscillators. *J. Chem. Phys.* **1978**, *68*, 608–610. [CrossRef]

14. Gerber, R.B.; Chaban, G.M.; Brauer, B.; Miller, Y. *Theory and Applications of Computational Chemistry: The First 40 Years*; Elsevier: Amsterdam, The Netherlands, 2005; pp. 165–193.

15. Jung, J.O.; Gerber, R.B. Vibrational wave functions and spectroscopy of $(H_2O)_n$, n = 2, 3, 4, 5: Vibrational self-consistent field with correlation corrections. *J. Chem. Phys.* **1996**, *105*, 10332–10348. [CrossRef]

16. Norris, L.S.; Ratner, M.A.; Roitberg, A.E.; Gerber, R.B. Møller–Plesset perturbation theory applied to vibrational problems. *J. Chem. Phys.* **1996**, *105*, 11261–11267. [CrossRef]

17. Monteiro, J.G.S.; Barbosa, A.G.H. VSCF calculations for the intra- and intermolecular vibrational modes of the water dimer and its isotopologs. *Chem. Phys.* **2016**, *479*, 81–90. [CrossRef]

18. Barone, V.; Biczysko, M.; Bloino, J.; Borkowska-Panek, M.; Carnimeo, I.; Panek, P. Toward Anharmonic Computations of Vibrational Spectra for Large Molecular Systems. *Int. J. Quantum Chem.* **2012**, *112*, 2185–2200. [CrossRef]

19. Beć, K.B.; Futami, Y.; Wójcik, M.J.; Nakajima, T.; Ozaki, Y. Spectroscopic and computational study of acetic acid and its cyclic dimer in the near-infrared region. *J. Phys. Chem. A* **2016**, *120*, 6170–6183. [CrossRef]

20. Grabska, J.; Ishigaki, M.; Beć, K.B.; Wójcik, M.J.; Ozaki, Y. Structure and near-infrared spectra of saturated and unsaturated carboxylic acids. An insight from anharmonic DFT calculations. *J. Phys. Chem. A* **2017**, *121*, 3437–3451. [CrossRef]

21. Grabska, J.; Beć, K.B.; Ishigaki, M.; Wójcik, M.J.; Ozaki, Y. Spectra-structure correlations of saturated and unsaturated medium-chain fatty acids. Near-infrared and anharmonic DFT study of hexanoic acid and sorbic acid. *Spectrochim. Acta A* **2017**, *185*, 35–44. [CrossRef] [PubMed]

22. Grabska, J.; Beć, K.B.; Ishigaki, M.; Huck, C.W.; Ozaki, Y. NIR spectra simulations by anharmonic DFT-saturated and unsaturated long-chain fatty acids. *J. Phys. Chem. B* **2018**, *122*, 6931–6944. [CrossRef] [PubMed]

23. Biczysko, M.; Bloino, J.; Carnimeo, I.; Panek, P.; Barone, V. Fully ab initio IR spectra for complex molecular systems from perturbative vibrational approaches: Glycine as a test case. *J. Mol. Struct.* **2012**, *1009*, 74–82. [CrossRef]

24. Biczysko, M.; Bloino, J.; Brancato, G.; Cacelli, I.; Cappelli, C.; Ferretti, A.; Lami, A.; Monti, S.; Pedone, A.; Prampolini, G.; et al. Integrated computational approaches for spectroscopic studies of molecular systems in the gas phase and in solution: Pyrimidine as a test case. *Theor. Chem. Acc.* **2012**, *131*, 1201–1220. [CrossRef]

25. Beć, K.B.; Karczmit, D.; Kwaśniewicz, M.; Ozaki, Y.; Czarnecki, M.A. Overtones of νC≡N vibration as a probe of structure of liquid CH_3CN, CD_3CN and CCl_3CN. Combined IR, NIR, and Raman spectroscopic studies with anharmonic DFT calculations. *J. Phys. Chem. A* **2019**, *123*, 4431–4442. [CrossRef]

26. Grabska, J.; Beć, K.B.; Kirchler, C.G.; Ozaki, Y.; Huck, C.W. Distinct Difference in Sensitivity of NIR vs. IR Bands of Melamine to Inter-Molecular Interactions with Impact on Analytical Spectroscopy Explained by Anharmonic Quantum Mechanical Study. *Molecules* **2019**, *24*, 1402. [CrossRef]

27. Beć, K.B.; Grabska, J.; Czarnecki, M.A. Spectra-structure correlations in NIR region: Spectroscopic and anharmonic DFT study of *n*-hexanol, cyclohexanol and phenol. *Spectrochim. Acta A* **2018**, *197*, 176–184. [CrossRef]

28. Beć, K.B.; Grabska, J.; Kirchler, C.G.; Huck, C.W. NIR spectra simulation of thymol for better understanding of the spectra forming factors, phase and concentration effects and PLS regression features. *J. Mol. Liq.* **2018**, *268*, 895–902. [CrossRef]

29. Beć, K.B.; Futami, Y.; Wójcik, M.J.; Ozaki, Y. A spectroscopic and theoretical study in the near-infrared region of low concentration aliphatic alcohols. *Phys. Chem. Chem. Phys.* **2016**, *18*, 13666–13682. [CrossRef]

30. Grabska, J.; Beć, K.B.; Ozaki, Y.; Huck, C.W. Temperature drift of conformational equilibria of butyl alcohols studied by near-infrared spectroscopy and fully anharmonic DFT. *J. Phys. Chem. A* **2017**, *121*, 1950–1961. [CrossRef]

31. Pele, L.; Gerber, R.B. On the mean accuracy of the separable VSCF approximation for large molecules. *J. Phys. Chem. C* **2010**, *114*, 20603–20608.

32. Otaki, H.; Yagi, K.; Ishiuchi, S.; Fujii, M.; Sugita, Y. Anharmonic Vibrational Analyses of Pentapeptide Conformations Explored with Enhanced Sampling Simulations. *J. Phys. Chem. B* **2016**, *120*, 10199–10213. [CrossRef]

33. Yagi, K.; Otaki, H.; Li, P.-C.; Thomsen, B.; Sugita, Y. *Weight Averaged Anharmonic Vibrational Calculations: Applications to Polypeptide, Lipid Bilayers, and Polymer Materials*; Ozaki, Y., Wójcik, M.J., Popp, J., Eds.; Wiley: Hoboken, NJ, USA, 2019; in press.

34. Yagi, K.; Yamada, K.; Kobayashi, C.; Sugita, Y. Anharmonic Vibrational Analysis of Biomolecules and Solvated Molecules Using Hybrid QM/MM Computations. *J. Chem. Theory Comput.* **2019**, *15*, 1924–1938. [CrossRef]

35. Gonjo, T.; Futami, Y.; Morisawa, Y.; Wójcik, M.J.; Ozaki, Y. Hydrogen bonding effects on the wavenumbers and absorption intensities of the OH fundamental and the first, second and third overtones of phenol and 2,6-dihalogenated phenols studied by visible/near-infrared/infrared spectroscopy and density functional theory calculations. *J. Phys. Chem. A* **2011**, *115*, 9845–9853.

36. Futami, Y.; Ozaki, Y.; Hamada, Y.; Wójcik, M.J.; Ozaki, Y. Solvent dependence of absorption intensities and wavenumbers of the fundamental and first overtone of NH stretching vibration of pyrrole studied by near-infrared/infrared spectroscopy and DFT calculations. *J. Phys. Chem. A* **2011**, *115*, 1194–1198. [CrossRef]

37. Kuenzer, U.; Sorarù, J.-A.; Hofer, T.S. Pushing the limit for the grid-based treatment of Schrödinger's equation: A sparse Numerov approach for one, two and three dimensional quantum problems. *Phys. Chem. Chem. Phys.* **2016**, *18*, 31521–31533. [CrossRef]

38. Kuenzer, U.; Klotz, M.; Hofer, T.S. Probing vibrational coupling via a grid-based quantum approach-an efficient strategy for accurate calculations of localized normal modes in solid-state systems. *J. Comput. Chem.* **2018**, *39*, 2196–2209. [CrossRef]

39. Kuenzer, U.; Hofer, T.S. A four-dimensional Numerov approach and its application to the vibrational eigenstates of linear triatomic molecules—The interplay between anharmonicity and inter-mode coupling. *Chem. Phys.* **2019**, *520*, 88–89.

40. Wu, P.; Siesler, H.W. The diffusion of alcohols and water in polyamide 11: A study by FTNIR-spectroscopy. *Macromol. Symp.* **1999**, *143*, 323–336. [CrossRef]

41. Grabska, J.; Czarnecki, M.A.; Beć, K.B.; Ozaki, Y. Spectroscopic and quantum mechanical calculation study of the effect of isotopic substitution on NIR spectra of methanol. *J. Phys. Chem. A* **2017**, *121*, 7925–7936.

42. Wang, L.; Ishiyama, T.; Morita, A. Theoretical investigation of C−H vibrational spectroscopy. 1. Modeling of methyl and methylene groups of ethanol with different conformers. *J. Phys. Chem. A* **2017**, *121*, 6687–6700. [CrossRef]

43. Wang, L.; Ishiyama, T.; Morita, A. Theoretical investigation of C−H vibrational spectroscopy. 2. Unified assignment method of IR, Raman, and sum frequency generation spectra of ethanol. *J. Phys. Chem. A* **2017**, *121*, 6701–6712.

44. Bloino, J.; Baiardi, A.; Biczysko, M. Aiming at an accurate prediction of vibrational and electronic spectra for medium-to-large molecules: An overview. *Int. J. Quantum Chem.* **2016**, *116*, 1543–1574.

45. Grimme, S.; Antony, J.; Ehrlich, S.; Krieg, H. A consistent and accurate ab initio parameterization of density functional dispersion correction (DFT-D) for the 94 elements H-Pu. *J. Chem. Phys.* **2010**, *132*, 154104.

46. Fornaro, T.; Biczysko, M.; Monti, S.; Barone, V. Dispersion corrected DFT approaches for anharmonic vibrational frequency calculations: Nucleobases and their dimers. *Phys. Chem. Chem. Phys.* **2014**, *16*, 10112–10128.

47. Matczak, P.; Wojtulewski, S. Performance of Møller-Plesset second-order perturbation theory and density functional theory in predicting the interaction between stannylenes and aromatic molecules. *J. Mol. Model.* **2015**, *21*, 41. [CrossRef]

48. Kirchler, C.G.; Pezzei, C.K.; Beć, K.B.; Mayr, S.; Ishigaki, M.; Ozaki, Y.; Huck, C.W. Critical evaluation of spectral information of benchtop vs. portable near-infrared spectrometers: Quantum chemistry and two-dimensional correlation spectroscopy for a better understanding of PLS regression models of the rosmarinic acid content in Rosmarini folium. *Analyst* **2017**, *142*, 455–464.

49. Czarnecki, M.A. Effect of temperature and concentration on self-association of octan-1-ol studied by two-dimensional Fourier transform near-infrared correlation spectroscopy. *J. Phys. Chem. A* **2000**, *104*, 6356–6361. [CrossRef]

50. Czarnecki, M.A. Two-dimensional correlation analysis of the second overtone of the ν(OH) mode of octan-1-ol in the pure liquid phase. *Appl. Spectrosc.* **2000**, *54*, 1767–1770. [CrossRef]

51. Czarnecki, M.A.; Czarnik-Matusewicz, B.; Ozaki, Y.; Iwahashi, M. Resolution enhancement and band assignments for the first overtone of OH(D) stretching modes of butanols by two-dimensional near-infrared correlation spectroscopy. 3. Thermal dynamics of hydrogen bonding in butan-1-(ol-d) and 2-methylpropan-2-(ol-d) in the pure liquid states. *J. Phys. Chem. A* **2000**, *104*, 4906–4911.

52. Czarnecki, M.A.; Maeda, H.; Ozaki, Y.; Suzuki, M.; Iwahashi, M. Resolution enhancement and band assignments for the first overtone of OH stretching mode of butanols by two-dimensional near-infrared correlation spectroscopy. Part I: Sec-butanol. *Appl. Spectrosc.* **1998**, *52*, 994–1000. [CrossRef]

53. Czarnecki, M.A.; Morisawa, Y.; Futami, Y.; Ozaki, Y. Advances in molecular structure and interaction studies using near-infrared spectroscopy. *Chem. Rev.* **2015**, *115*, 9707–9744. [CrossRef]

54. Umer, M.; Kopp, W.A.; Leonhard, K. Efficient yet accurate approximations for ab initio calculations of alcohol cluster thermochemistry. *J. Chem. Phys.* **2015**, *143*, 214306. [CrossRef]

55. Grimme, S. Semiempirical hybrid density functional with perturbative second-order correlation. *J. Chem. Phys.* **2006**, *124*, 034108. [CrossRef]

56. Weigend, F.; Ahlrichs, R. Balanced basis sets of split valence, triple zeta valence and quadruple zeta valence quality for H to Rn: Design and assessment of accuracy. *Phys. Chem. Chem. Phys.* **2005**, *7*, 3297–3305. [CrossRef]

57. Grimme, S.; Ehrlich, S.; Goerigk, S.L. Effect of the damping function in dispersion corrected density functional theory. *J. Comput. Chem.* **2011**, *32*, 1456–1465. [CrossRef]

58. Miertuš, S.; Scrocco, E.; Tomasi, J. Electrostatic interaction of a solute with a continuum. A direct utilization of ab initio molecular potentials for the prevision of solvent effects. *Chem. Phys.* **1981**, *55*, 117–129.

59. Barone, V.; Cossi, M. Quantum calculation of molecular energies and energy gradients in solution by a conductor solvent model. *J. Phys. Chem. A* **1998**, *102*, 1995–2001. [CrossRef]

60. Barone, V. Anharmonic vibrational properties by a fully automated second-order perturbative approach. *J. Chem. Phys.* **2005**, *122*, 014108. [CrossRef]

61. Barone, V.; Biczysko, M.; Bloino, J. Fully anharmonic IR and Raman spectra of medium-size molecular systems: Accuracy and interpretation. *Phys. Chem. Chem. Phys.* **2014**, *16*, 1759–1787. [CrossRef]

62. Frisch, M.J.; Trucks, G.W.; Schlegel, H.B.; Scuseria, G.E.; Robb, M.A.; Cheeseman, J.R.; Scalmani, G.; Barone, V.; Mennucci, B.; Petersson, G.A.; et al. *Gaussian 16, Revision A.03*; Gaussian, Inc.: Wallingford, CT, USA, 2013.

63. Bloino, J.; Biczysko, M. IR and Raman Spectroscopies beyond the Harmonic Approximation: The Second-Order Vibrational Perturbation Theory Formulation. In *Reference Module in Chemistry, Molecular Sciences and Chemical Engineering*; Reedijk, J., Ed.; Elsevier: Waltham, MA, USA, 2015.

64. McQuarrie, D.A. *Statistical Mechanics*; Harper & Row: New York, NY, USA, 1976.

65. Robertson, M.B.; Klein, P.G.; Ward, I.M.; Packer, K.J. NMR study of the energy difference and population of the *gauche* and *trans* conformations in solid polyethylene. *Polymer* **2001**, *42*, 1261–1264. [CrossRef]

66. Martin, J.M.L.; Van Alsenoy, C. *GAR2PED*; University of Antwerp: Antwerp, Belgium, 1995.

67. Pulay, P.; Fogarasi, G.; Pang, F.; Boggs, J.E. Systematic ab initio gradient calculation of molecular geometries, force constants, and dipole moment derivatives. *J. Am. Chem. Soc.* **1979**, *101*, 2550–2560. [CrossRef]

68. *MATLAB*; The MathWorks, Inc.: Natick, MA, USA, 2016.

Sample Availability: Samples of the compounds (CH_3CH_2OD, CH_3CD_2OH, CD_3CH_2OH, CD_3CD_2OH, CD_3CD_2OD) are available from the authors.

molecules

MDPI

Article

Comparison of Multivariate Regression Models Based on Water- and Carbohydrate-Related Spectral Regions in the Near-Infrared for Aqueous Solutions of Glucose

Anel Beganović, Vanessa Moll and Christian W. Huck *

Institute of Analytical Chemistry and Radiochemistry, CCB-Center for Chemistry and Biomedicine, Innrain 80/82, 6020 Innsbruck, Austria; anel.beganovic@uibk.ac.at (A.B.); vanessa.moll@student.uibk.ac.at (V.M.)
* Correspondence: christian.w.huck@uibk.ac.at; Tel.: +43-512-507-57304

Received: 3 September 2019; Accepted: 10 October 2019; Published: 15 October 2019

Abstract: The predictive power of the two major water bands centered at 6900 cm^{-1} and 5200 cm^{-1} in the near-infrared (NIR) region was compared to carbohydrate-related spectral areas located in the first overtone (around 6000 cm^{-1}) and combination (around 4500 cm^{-1}) region using glucose in aqueous solutions as a model substance. For the purpose of optimal coverage of stronger as well as weaker absorbing NIR regions, cells with three different declared optical pathlengths were employed. The sample set consisted of multiple separately prepared batches in the range of 50–200 mmol/L. Moreover, the samples were divided into a calibration set for the construction of the partial least squares regression (PLS-R) models and a test set for the validation process with independent samples. The first overtone and combination region showed relative prediction errors between 0.4–1.6% with only one PLS-R factor required. On the other hand, the errors for the water bands were found between 1.6–8.3% and up to three PLS-R factors required. The best PLS-R models resulted from the cell with 1 mm optical pathlength. In general, the results suggested that the carbohydrate-related regions in the first overtone and combination region should be preferred over the regions of the two dominant water bands.

Keywords: FT-NIR spectroscopy; PLS-R; water; glucose; test set validation; RMSEP

1. Introduction

Glucose is of great importance in physiological systems, medicine, and health care, as well as in the food and beverage industry. Besides other analytical methods, glucose and other carbohydrates are often quantitatively determined enzymatically or via chromatographic methods such as high performance liquid chromatography (HPLC) or gas chromatography (GC) [1–3]. However, these methods are rather time-consuming and expensive as they usually require sample preparation, long measurement time (incubation of enzymes, separation on column, etc.), and qualified personnel. Considering these drawbacks of conventional analytical techniques, near-infrared spectroscopy (NIRS) is of increasing interest as an alternative method for the quantification of glucose and other carbohydrates in aqueous solutions. NIRS mostly does not require any sample preparation, offers fast and non-invasive analyses, and multiple sample characteristics are accessible with one single measurement. Moreover, NIR spectrometers are cheap to run and can be operated by relatively untrained personnel. Alongside the mentioned advantages, NIRS comes with a few downsides. One is that the information contained in NIR spectra often needs to be extracted using multivariate data analysis tools such as principal component analysis (PCA) or partial least squares regression (PLS-R). The other is, in order to establish a robust model for a reliable prediction of future samples, sufficient reference measurements with known target parameters need to be provided to the calibration. Both mentioned drawbacks make NIRS time-consuming and require highly skilled personnel in the calibration phase, and therefore might be cost-intensive initially [4–6].

Glucose in water and in various other water-based matrices like buffer and serum solutions, blood anticoagulants, fruit juices, and alcoholic beverages was extensively studied in the field of NIRS. Furthermore, non-invasive blood glucose monitoring and glucose quantifications in other body fluids (e.g., urine) using NIRS is a matter of great interest in literature. The aqueous NIR spectrum of glucose is dominated by intense water bands centered at around 6900 cm^{-1} and 5200 cm^{-1}, which are assigned to combinations of water OH stretching and bending modes [7–10]. Depending on the utilized optical cell pathlength (distance of interaction between sample and light), these bands can become unusable due to the complete absorption of NIR radiation by water [11–13]. Clearly visible glucose-related bands usually do not appear at lower concentrations in the NIR spectrum of aqueous glucose without spectral pre-treatments (e.g., derivative functions) [14], but become apparent at higher concentrations [15]. However, prominent absorption features of dissolved carbohydrates are located in the combination region around 4500 cm^{-1}, the first overtone region around 6000 cm^{-1}, as well as in the short wavelength NIR region around 9500 cm^{-1} [16–18]. In case of carbohydrates, vibrations in these regions are mostly due to C−H-based combination and overtone vibrations [8,19]. The available literature is primarily concerned with the combination and first overtone region.

Addition of glucose or other carbohydrates may have a similar effect on the liquid water structure (hydrogen bond network) as an addition of inorganic salts [20–22] or changes in temperature [10,23,24]. Thereby, changes in the appearance of the water bands in the NIR region are induced [13,25–27]. Furthermore, carbohydrates in aqueous solution can act on the water cluster as structure breakers or structure makers [26]. The extent of the changes in the water structure is connected to the concentration of the corresponding solute and thus the water bands can be utilized for quantitative analyses [13,25,27]. As already mentioned, glucose and other carbohydrates exhibit characteristic NIR spectral features in the combination, first overtone, and short wavelength NIR region, next to the two dominant water bands. The direct relation between the concentration of dissolved carbohydrates and the spectral response allowed successful applications of these regions in previous quantification studies [14,16–18,28,29].

Chen et al. [17] compared the predictive power of the combination and first overtone region for glucose and other biomolecules in aqueous solutions in the range of approximately 0–35 mM. The authors optimized the optical pathlength for each of the two regions and concluded that the combination region was superior relative to the first overtone NIR region. Beganović et al. [13] investigated the performance of the two major water bands in the NIR centered at around 6900 cm^{-1} and 5200 cm^{-1}. In order to overcome the issue of complete absorption of NIR light most commonly occuring at the water band located around 5200 cm^{-1}, they utilized a cell with 0.1 mm optical pathlength. By this, the authors demonstrated rich information content of the so-called combination band of water at 5200 cm^{-1}, which is often not taken into consideration in literature. Compared to the water band at 6900 cm^{-1}, the authors reported lower prediction errors for the more intense water band at 5200 cm^{-1}. To best of our knowledge, no previously conducted study compared the performance of all these NIR regions directly—the water-based, as well as the carbohydrate-based regions. The present study intends to point out the NIR regions with the maximum of relevant information content for the analysis of carbohydrate-based aqueous solutions. This is of particular interest for applications working close to the limits of detection (LOD) as well as limits of quantification (LOQ).

Therefore, this work aims at the comparison of the two dominating waterbands in the NIR centered at around 6900 cm^{-1} and 5200 cm^{-1} to the sugar-related spectral regions located in the first overtone and combination region around 6000 cm^{-1} and 4500 cm^{-1}, respectively. Glucose in aqueous solutions is used as a model substance. In addition, cells with different optical pathlengths are utilized in order to access the whole NIR region from 10,000–4000 cm^{-1} (thinner cell pathlength) and to account for the lower absorption in the first overtone region (thicker cell pathlength).

2. Material and Methods

2.1. Samples

D-(+)-glucose (≥99.5%) was purchased from Carl Roth (Karlsruhe, Germany) and Milli-Q water with a resistivity of 18.2 MΩ cm was used for the preparation of the glucose solutions. The calibration set was composed of pure Milli-Q water and glucose concentrations ranging from 50–200 mmol/L in steps of 30 mmol/L. For the test set, samples with glucose concentrations of 60.2, 130.5 and 186.0 mmol/L were prepared.

In order to avoid any effects of measurement time, multiple independent batches for both calibration and test samples were prepared. The calibration and test set consisted of three and two batches per sample, respectively. The preparation and measurement of the samples were randomized. Furthermore, all samples were measured on the day they were prepared.

2.2. FT-NIR Measurements

The Büchi NIRFlex N-500 FT-NIR spectrometer (Büchi, Flawil, Switzerland) equipped with the liquids measurement cell was used to acquire NIR spectra of the glucose solutions. The spectra were recorded in transmission mode in the range of 10,000–4000 cm^{-1} with a spectral resolution of 8 cm^{-1}, while each sample was scanned 64 times. In order to account for the lower absorption towards increasing wavenumbers, the measurements were performed using three cell types with different declared optical pathlengths. The cells were purchased from Hellma GmbH & Co. KG (Müllheim, Germany) and specified as follows: one 106-QS quartz SUPRASIL® cell with 0.1 mm optical pathlength and demountable cell windows, and multiple 100-QX quartz SUPRASIL® cells with 1 mm and 2 mm optical pathlength, respectively. Spectrometer reference measurements were performed in the beginning, as well as in the middle of each measurement day. Data acquisition was accomplished using the NIRWare 1.4.3010 software package (Büchi, Flawil, Switzerland).

Samples were always freshly prepared and measured randomly over a period of four weeks. In contrast to the 1 mm and 2 mm cells—which were simply filled with a certain amount of sample solution—the 0.1 mm cell is demountable and thus had to be filled differently. For the 0.1 mm cell, approximately 40 µL of glucose solution was applied onto the sample recess of one optical cell window, followed by the careful attachment of the second cell window. Excessive sample solution was displaced and collected with a tissue. Sticky glucose residues on the outside surface of the cell were removed.

The NIR measurements were performed at 35 °C due to the fact that the NIRFlex N-500 liquids measurement cell was subjected to significant fluctuations at lower temperatures. However, at 35 °C, the temperature fluctuation stabilized at ±0.1 °C [13]. In order to avoid the introduction of temperature-driven shifts in the NIR spectrum, each cell filled with sample solution was tempered to 35 °C before the measurements were started. The 0.1 mm cell was thermally equilibrated for 30 s, while the 1 mm and the 2 mm cells were thermally equilibrated for 1 min and for 2 min, respectively. To avoid water evaporation, the 1 mm and 2 mm cells were covered with a lid, whereas the 0.1 mm cell did not offer any cover possibility since the two cell windows were kept together by adhesion. The only possibility to prevent the evaporation of water out of the 0.1 mm cell was to minimize the time between the filling of the cell with sample solution and the actual sample measurement.

All samples were measured nine times, while the cells were refilled with fresh solution for each of the nine repeat measurements. The cells were cleaned thoroughly after every single measurement using Milli-Q water and ethanol. Lint-free tissues, as well as a conventional compressed air system, were used in order to dry the cells and remove potential dust particles.

2.3. Band Assignment and Division of Spectral Regions

In this study, the two water bands at around 6900 cm^{-1} and 5200 cm^{-1}, as well as two regions of glucose-related vibrations located at around 5900 cm^{-1} and 4400 cm^{-1}, were used for the comparison of the predictive power of each region separately at different cell pathlengths. Since water is known to be a strong absorber in the near-infrared [8,30], complete absorption of the NIR light can occur in certain spectral regions—depending on the cell pathlength, infrared source, and detector [30]. As a consequence, the water band centered at 5200 cm^{-1} could not be utilized for the purpose of any quantitative analysis using both the 1 mm and 2 mm cells. This region was only accessible using the 0.1 mm cell.

The NIR regions related to water were selected in such a way that they ranged from the beginning to the end of the corresponding NIR band while the regions related to glucose were chosen according to the spectral pattern after the application of the second derivative discussed later (Section 3). For the water band at around 6900 cm^{-1} a spectral range of 7692–6248 cm^{-1} was selected in order to match the region frequently used in aquaphotomics [27], and was labeled as W1 in this study. This band is commonly referred to as the first overtone of water [27,31], although it is actually a combination of symmetric and antisymmetric stretching vibration modes of water [8,9]. The spectral region for the second water band—labeled as W2—was set to 5400–4600 cm^{-1} and is assigned to the combination of bending and antisymmetric water stretching modes [7,8].

The two regions at around 5900 cm^{-1} and 4400 cm^{-1} related to glucose vibrations in water were labeled as G1 and G2, respectively. For G1, the spectral region was set to 6100–5800 cm^{-1} and is assigned to first overtone vibrations of C−H compounds [8,16,32]. The spectral region for G2 was set to 4520–4300 cm^{-1} and is assigned to combinations of C−H stretching and CH$_2$ deformation vibrations, as well as combinations of stretching vibrations of glucose-related O−H and C−O compounds [8,16]. Figure 1 shows an exemplary NIR spectrum of water containing glucose with the described division of spectral regions. Note that the small band around 4500 cm^{-1} (marked with an asterisk in Figure 1) is caused by O−H residues in the quartz windows of the 0.1 mm cell (probably due to water impurities [33]) and is assigned to a combination of an O−H stretching vibration and one of the SiO$_2$ fundamental vibrations [8,34].

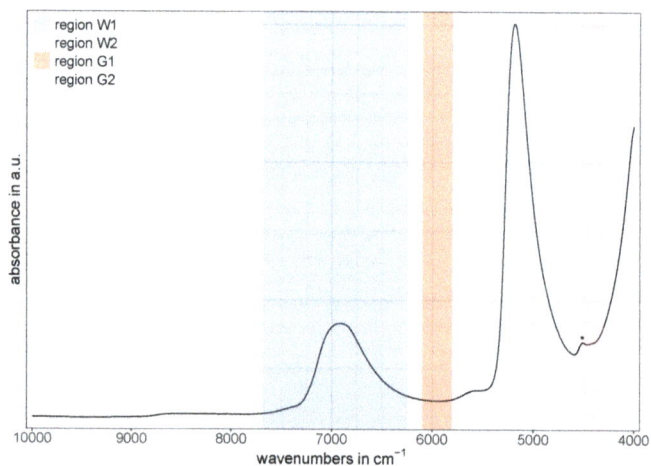

Figure 1. Illustration of the division of the spectral regions W1, W2, G1, and G2 using an exemplary spectrum collected with the 0.1 mm cell. The O−H residue in the cell's quartz windows at around 4500 cm^{-1} is marked with an asterisk. It is assigned to a combination of an O−H stretching vibration and one of the SiO$_2$ fundamental vibrations [8,34].

2.4. Multivariate Data Analysis

The Unscrambler X Ver. 10.5 (Camo Software AS, Oslo, Norway) was used for the pre-treatment of the NIR spectra as well as the construction and validation of the multivariate regression models. Due to the occurrence of interference fringes using the 0.1 mm cell, the frequency filtering technique fast Fourier transform filter (FFT-filter) [13] was applied to these NIR spectra using OriginPro Ver. 9.1G (OriginLab Corporation, Northampton, MA, USA). Thereby, the NIR spectra were first Fourier transformed, followed by the application of a filter function and finally retransformed by inverse Fourier transformation. In order to only eliminate the disturbing interferences and leave the regular spectra containing the targeted information untouched, the parabolic low-pass filter was chosen as an FFT filter function. This filter blocks all frequencies above a certain threshold value (cutoff frequency), while lower frequency elements are allowed to pass [13]. The cutoff frequency was set to 0.02625 Hz—all frequencies above were eliminated before the spectra were inverse Fourier transformed. The suitability of this approach has been validated before [13]. Afterward, the spectra of all three cell pathlengths were transformed from transmittance to absorbance.

The NIR spectra were reduced batchwise from nine spectra to one representative spectrum for each batch. All previously defined spectral regions were subjected to an individual optimization of pre-treatments (see Table 1). However, in case of the regions W1 and W2, the pre-treatments were chosen as proposed in aquaphotomics literature [27] with an additional application of a standard normal variate (SNV) transformation [35]. For the regions G1 and G2, it was found that a second order Savitzky–Golay derivative [36] with a second order polynomial and a varying number of smoothing points was optimal. Second order derivative spectra were also calculated for the two water-related regions W1 and W2. The results were inferior compared to the pre-treatments mentioned above and will therefore not be discussed any further.

Table 1. Details of spectral pre-treatments applied to each spectral region.

Cell	Region	Pre-Treatments
0.1 mm	W1	FFT-filter, SavGol smoothing (25 SP, second polynomial order), SNV
	W2	FFT-filter, SavGol smoothing (25 SP, second polynomial order), SNV
	G1	FFT-filter, SavGol smoothing (13 SP, second polynomial order), SNV, second SavGol derivative (15 SP, second polynomial order)
	G2	FFT-filter, second SavGol derivative (11 SP, second polynomial order)
1 mm	W1	SavGol smoothing (second polynomial order, 25 SP)
	G1	second SavGol derivative (19 SP, second polynomial order)
	G2	second SavGol derivative (25 SP, second polynomial order)
2 mm	W1	SavGol smoothing (25 SP, second polynomial order), SNV
	G1	second SavGol derivative (9 SP, second polynomial order)
	G2	second SavGol derivative (7 SP, second polynomial order)

SavGol—Savitzky-Golay; SP—smoothing points.

For each spectral region, regression models were calculated using partial least squares regression (PLS-R) along with the NIPALS algorithm. The calibration process incorporated 21 calibration samples from three batches, whereas the performance of the calibration models was evaluated with six completely independent samples from two batches (test set validation). Note that these samples were never employed in any calibration [37,38]. The performances of the PLS-R models were assessed using the root mean square error (RMSE), which was calculated according to Equation (1), where y_i and $\hat{y}_i$ represent the reference and predicted values, respectively. Furthermore, in order to enable a more straightforward interpretation of the RMSE's scale, a percentage error called normalized RMSE (NRMSE) was introduced, which refers to the calibration range of 0–200 mmol/L (see Equation (2)). The errors of the calibration (CAL) and test set validation (TSV) were referred to as root mean square error of calibration (RMSEC) and root mean square error of prediction (RMSEP), respectively:

$$RMSE = \sqrt{\frac{1}{n}\sum_{i=1}^{n}(y_i - \hat{y}_i)^2},$$ (1)

$$NRMSE = \frac{RMSE}{y_{max} - y_{min}} \times 100.$$ (2)

The RMSE's magnitude is closely associated with the number of PLS-R factors (or latent variables), which is a crucial parameter for a satisfactory performing PLS-R model [37,38]. Since the glucose-water system used in this study is rather simple, the number of PLS-R factors employed in the PLS-R models should be kept quite low in order to avoid modeling of noise and thus non-relevant spectral information (overfitting). However, using too few PLS-R factors can lead to poor model performance due to the lack of explained variance in the NIR spectra (underfitting). The optimal number of PLS-R factors was determined by the examination of the regression coefficients, the loadings and correlation loadings of each PLS-R factor as well as the explained variances.

3. Results and Discussion

The full-range raw NIR spectra of the calibration and test set for all three utilized cell pathlengths are depicted in Figure 2. The artifacts occuring in the region around 5200 cm^{-1} in the raw NIR spectra of the 1 mm and 2 mm cells in Figure 2 are caused by the complete absorption of NIR light in this spectral region [11].

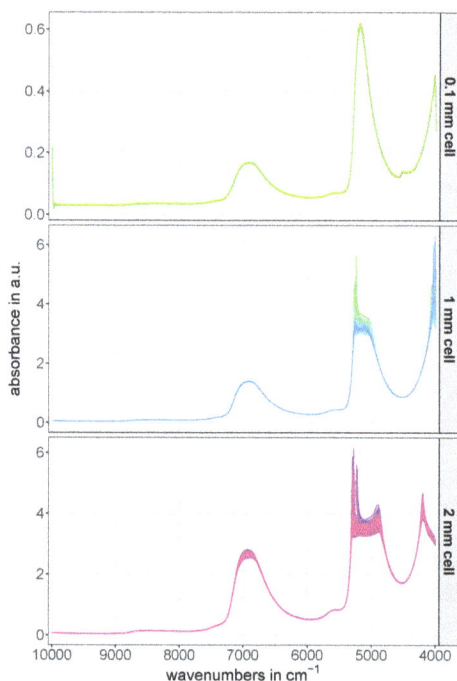

Figure 2. Raw NIR spectra (calibration and test set) of all three utilized cell pathlengths. The spectra were only transformed from transmittance to absorbance.

The pre-treated NIR calibration set spectra of the regions W1, W2, G1 and G2 for all three cell pathlengths are presented in Figure 3. Since each of the three batches per concentration was averaged from nine spectra to one representative spectrum, three spectra per concentration are shown in Figure 3. The glucose-related regions G1 and G2 showed an evident pattern towards increasing glucose concentrations (Figure 3e,f,j–l), while such an obvious pattern was missing in the NIR spectra of the water-related regions W1 and W2 at first glance (Figure 3a–c,g). However, a closer look revealed that there actually was a certain concentration dependent pattern, although it was not as pronounced as in the regions associated with glucose vibrations. An example of this is shown in Figure 4.

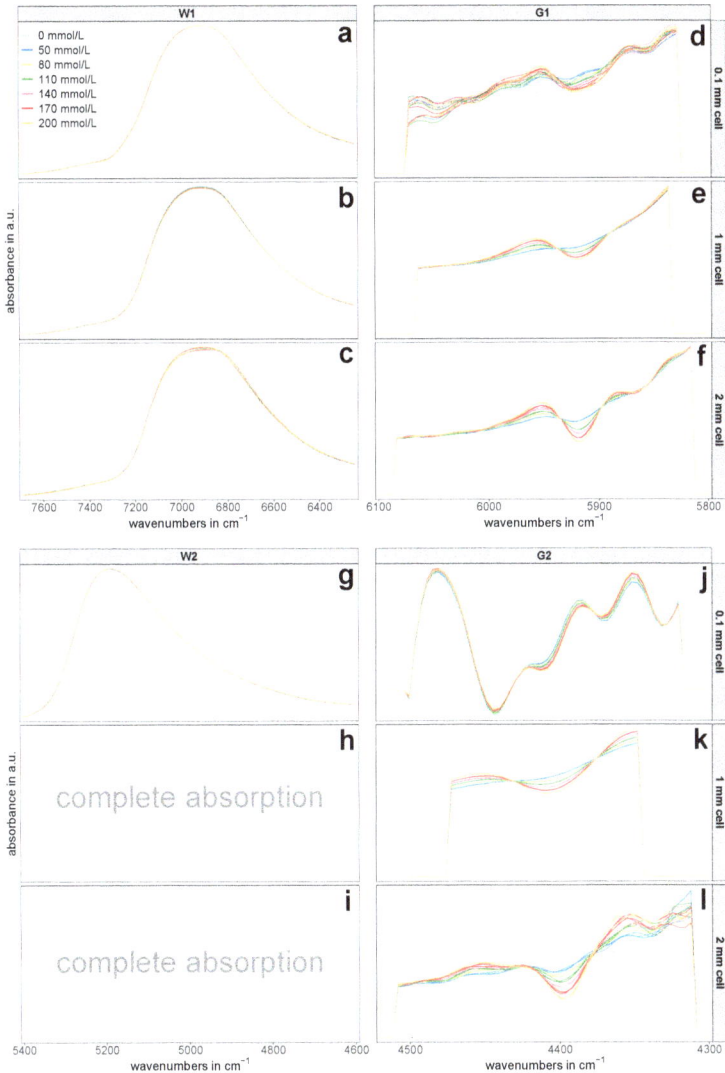

Figure 3. Pre-treated NIR spectra of the calibration set. The spectral regions W1 (**a–c**), W2 (**g**), G1 (**d–f**), and G2 (**j–l**) are shown for all three utilized cell pathlengths separately. Each concentration is represented by three representative NIR spectra (one per batch). No spectra were available in region W2 for the 1 mm and 2 mm cells (**h,i**) due to the complete absorption of the NIR light.

In the course of data analysis, the number of smoothing points for the second derivative in the first overtone and combination region was individually optimized prior to the PLS-R. As a consequence, the exact spectral range used for the PLS-R varied for each cell. Nevertheless, the spectral regions subjected to the calculation of the derivative spectra did not vary in between the three cell pathlengths.

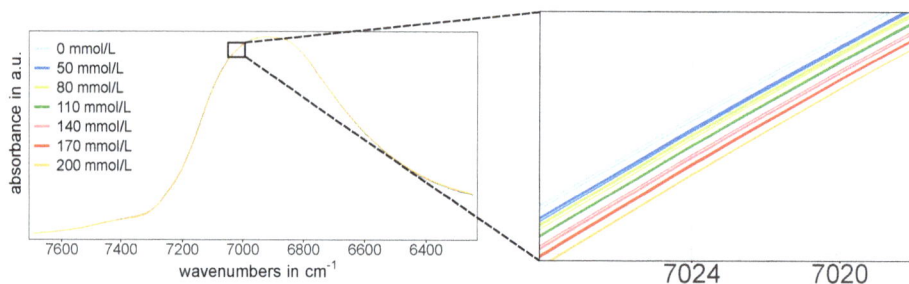

Figure 4. Illustration of the glucose concentration dependent pattern in the water band located at around 6900 cm^{-1} (region W1). NIR spectra of the calibration set recorded with the 0.1 mm cell are shown here exemplarily.

3.1. Measurements with 0.1 mm Cell Pathlength

The results of the PLS-R calibration and test set validation procedure is presented in Table 2. The 0.1 mm cell allowed the evaluation of the performance of all four investigated regions. Comparing the results for the 0.1 mm cell in Table 2, probably the most noticeable value is the relatively high prediction error of RMSEP = 22.6 mmol/L of the first overtone region G1. This error's magnitude of more than 11% employing three PLS-R factors was hardly surprising, considering the lack of a distinct concentration dependent pattern in Figure 3d. The reason for this was that a small amount of interference fringes was still present in this spectral region and that there was insufficient spectral information content due to the short pathlength [30]. These remaining fringes were hardly noticeable in the regular absorption spectrum but became evident after the application of the second derivative—despite previous smoothing of the NIR spectra. Considering the PLS-R scores of the calibration set of region G1 in Figure 5d–f, the first PLS-R factor mostly accounted for the changes in glucose concentration. Despite that, the PLS-R calibration model was not able to predict the test set adequately.

The models for the two water-related regions W1 and W2 both showed similar percentage errors of around NRMSEP = 4% for the prediction of unknown samples from the test set using two PLS-R factors, respectively. A consideration of more than two PLS-R factors for each model would have further reduced the prediction error; however, a closer look at the model statistics gave no justification for the use of a third PLS-R factor. In case of region W1, the validation model's explained Y-variance (variance in glucose concentration) comparably increased from PLS-R factor 1 to 2 and from PLS-R factor 2 to 3 (see Table 3), and therefore suggested a model based on three PLS-R factors. In contrast, the correlation loadings of PLS-R factor 3 showed very low values with a maximum of 0.2 (see Figure 6a), which led to the exclusion of PLS-R factor 3 from the PLS-R model due to the risk of modeling glucose-unrelated spectral information. For the combination band of water (region W2), two PLS-R factors were considered as optimal as the explained Y-variance in the PLS-R validation model increased by 2.4% from PLS-R factor 1 to 2 (see Table 3) and the correlation loadings indicated many X-variables with strong contributions to the second PLS-R factor (see Figure 6g).

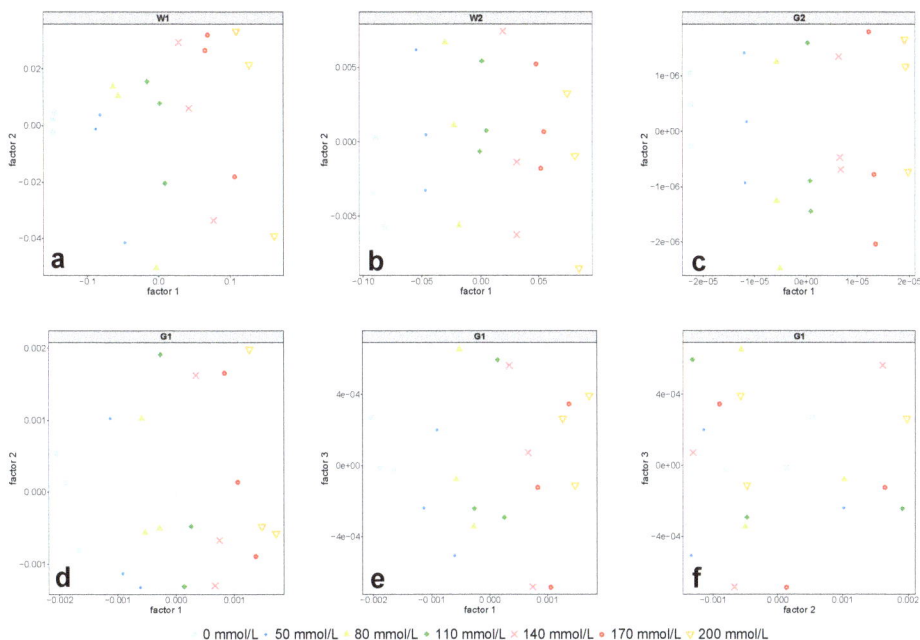

Figure 5. PLS-R score plots of the calibration set for the regions W1 (**a**), W2 (**b**), G2 (**c**), and G1 (**d–f**) obtained by measurements with the 0.1 mm cell. Each data point corresponds to one batch. The featured factors were chosen according to the required number of PLS-R factors in Table 2.

Among all prediction errors of the measurements conducted with the 0.1 mm cell, the glucose-related combination region G2 yielded by far the lowest prediction error. The model required only one PLS-R factor to yield an NRMSEP as low as 0.7% along with an R^2_{TSV} of 0.9993. The explained Y-variance of the test set already reached 99.9% in the first PLS-R factor (see Table 3), and thus made the use of more PLS-R factors invalid. This remarkable prediction performance of region G2 can be attributed to the very distinct concentration pattern in the second derivative NIR spectra (see Figure 3j), which was not observed in the regular (untreated) spectra. In addition to that, the concentrations in the PLS-R score plot of region G2 in Figure 5c were perfectly separated along PLS-R factor 1. This demonstrated that PLS-R factor 1 exclusively accounted for changes in glucose concentration and thus allowed the exclusion of further PLS-R factors.

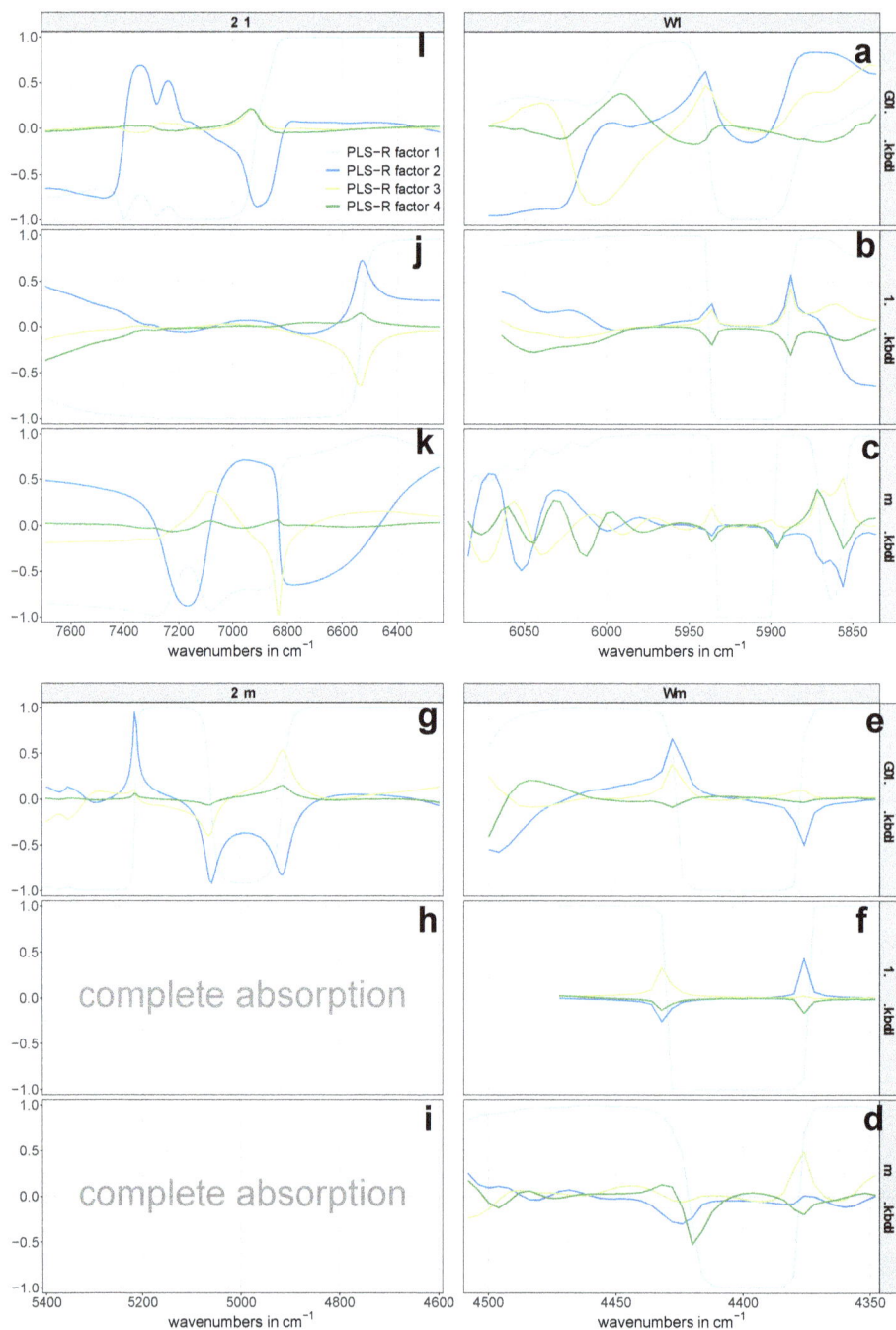

Figure 6. Correlation loadings of the first four PLS-R factors for the regions W1 (**a–c**), W2 (**g–i**), G1 (**d–f**), and G2 (**j–l**) with the corresponding cell pathlengths.

Table 2. Results of the PLS-R calibration and test set validation for all three cell pathlengths as well as for each spectral region.

Cell	Region	Type of Validation	PLS-R Factors	RMSEC in mmol/L	NRMSEC in %	RMSEP in mmol/L	NRMSEP in %	SEC in mmol/L	SEP in mmol/L	Bias in mmol/L	R^2
0.1 mm	W1	CAL	2	4.7	2.3	-	-	4.8	-	2.1×10^{-5}	0.995
		TSV		-	-	7.6	3.8	-	5.6	5.6	0.98
	W2	CAL	2	2.7	1.4	-	-	2.8	-	5.5×10^{-5}	0.998
		TSV		-	-	8.3	4.1	-	9.0	0.5	0.98
	G1	CAL	3	5.2	2.6	-	-	5.3	-	0	0.994
		TSV		-	-	22.6	11.3	-	23.1	−8.0	0.83
	G2	CAL	1	1.6	0.8	-	-	1.6	-	0	0.9994
		VAL		-	-	1.4	0.7	-	1.5	0.1	0.9993
1 mm	W1	CAL	2	2.1	1.0	-	-	2.1	-	2.3×10^{-5}	0.9989
		TSV		-	-	3.2	1.6	-	3.2	1.1	0.997
	G1	CAL	1	1.2	0.6	-	-	1.3	-	0	0.9996
		TSV		-	-	1.8	0.9	-	1.9	0.4	0.9989
	G2	CAL	1	0.6	0.3	-	-	0.6	-	0	0.99992
		TSV		-	-	0.7	0.4	-	0.8	0.2	0.9998
2 mm	W1	CAL	3	4.0	2.0	-	-	4.1	-	1.5×10^{-5}	0.996
		TSV		-	-	10.1	5.1	-	9.5	−5.3	0.97
	G1	CAL	1	1.1	0.6	-	-	1.1	-	0	0.9997
		TSV		-	-	1.5	0.8	-	1.6	−0.6	0.9992
	G2	CAL	1	3.1	1.6	-	-	3.2	-	0	0.998
		TSV		-	-	3.3	1.6	-	3.0	1.8	0.996

CAL—calibration; TSV—test set validation.

3.2. Measurements with 1 mm Cell Pathlength

The performance of the three exploitable regions of the 1 mm cell was remarkable. The PLS-R calibration model of the so-called first overtone of water (region W1) predicted the independent test set samples with an error of RMSEP = 3.2 mmol/L and a relative error of NRMSEP = 1.6% (see Table 2). These errors were achieved using the first two PLS-R factors and together accounted for 99.7% of the Y-variance (see Table 3), which, as a consequence, did not allow the consideration of further PLS-R factors in the model for region W1.

The pre-treated NIR spectra of the glucose-related regions G1 and G2 in Figure 3e,k, respectively, showed the same concentration dependent pattern from pure water towards increasing glucose content. This clearly evident pattern indicated that glucose in aqueous solution produces own NIR bands. This finding is in contrast to the frequently found view in the literature [26,39], according to which carbohydrates do not exhibit own NIR bands in aqueous solutions, but rather characteristically disturbs the water structure. Actually, at low concentrations, these bands are more like tiny changes in the untreated spectra's path line, which cannot be recognized by the eye, but are rather revealed and highlighted by calculating derivative spectra. The high predictive power of the two glucose-related regions is best represented by the low errors in the prediction of the independent test set: the PLS-R model for the first overtone region G1 yielded an NRMSEP value of 0.9%, whereas the NRMSEP for the combination region G2 was as low as 0.4% (see Table 2). The fact that for each PLS-R model only one PLS-R factor was necessary to achieve the mentioned prediction errors using the two regions associated with glucose vibrations showed the distinct glucose-related nature of these regions. The use of only one PLS-R factor for the two glucose-related regions was further confirmed by the fact that the concentrations in the PLS-R score plots in Figure 7a,b were clearly separated along the first PLS-R factor.

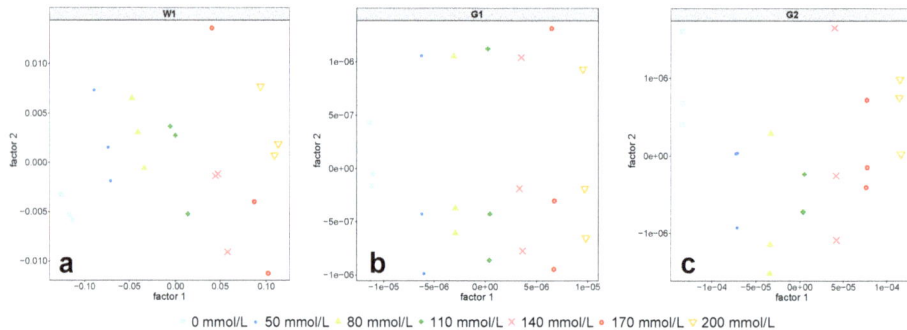

Figure 7. PLS-R score plots of the calibration set for the regions W1 (**a**), G1 (**b**), and G2 (**c**) obtained by measurements with the 1 mm cell. Each data point corresponds to one batch. The featured factors were chosen according to the required number of PLS-R factors in Table 2.

However, these findings allow for reconsidering the statement of Chen et al. [17], according to which a 1 mm cell pathlength is too thin for satisfactory glucose quantification from NIR spectra in the first overtone region in an aqueous matrix. The authors of the aforementioned study did not use derivative spectra. In contrast to Chen et al. [17], the results presented herein rather suggest that a cell pathlength of 1 mm is perfectly suitable. By applying a second derivative function to the first overtone region, a clear concentration dependent pattern becomes evident (see Figure 3e) and thus allows the construction of highly accurate PLS-R models for glucose quantifications. Our study did not investigate the exact cell pathlength at which it becomes too thin for high-quality NIR spectra. Nevertheless, considering the relatively high prediction error of the 0.1 mm cell in region G1, it can be concluded that this limit is below a cell pathlength of 1 mm.

3.3. Measurements with 2 mm Cell Pathlength

For the 2 mm cell, the test set validation for the so-called first overtone of water (region W1) yielded an NRMSEP of around 5% utilizing three PLS-R factors (see Table 2). An additional consideration of PLS-R factor 4 would have reduced the relative error by nearly half, but, from the interpretation of the model statistics, it was concluded that this would have led to the modeling of noise or glucose-unrelated spectral information. Although the explained Y-variance of the validation model increased by 2.4% from PLS-R factor 3 to PLS-R factor 4 (see Table 3), the correlation loadings showed negligibly small values for PLS-R factor 4 (see Figure 6c). This suggested that the X-variables modeled in PLS-R factor 4 were not of importance for the regression model and therefore might have contained non-relevant spectral information for the quantification of the target solute.

In direct comparison to the two glucose-related regions, the predictive power of region W1 was inferior. Using a pathlength of 2 mm, the PLS-R calibration models for the regions G1 and G2 predicted the independent test set samples with prediction errors of NRMSEP = 0.8% and NRMSEP = 1.6%, respectively, whereas both models required only one PLS-R factor (see Table 2). The second derivative spectra in Figure 3f,l showed a clear glucose concentration dependent pattern towards increasing glucose content, which was also reflected in the PLS-R score plots in Figure 8d,e. However, compared to the derivative spectra of the 1 mm cell in the combination region G2 (Figure 3k), the spectra in Figure 3l appeared noisy to some extent. This noisy pattern could be removed with a higher number of smoothing points in the second derivative, but the test set validation yielded poorer RMSEP values and required more PLS-R factors. It is conceivable that the high absorption of the adjacent water combination band and the associated spectral artifacts (see Figure 2) had an impact on region G2 and consequently led to the somewhat higher prediction error in this region.

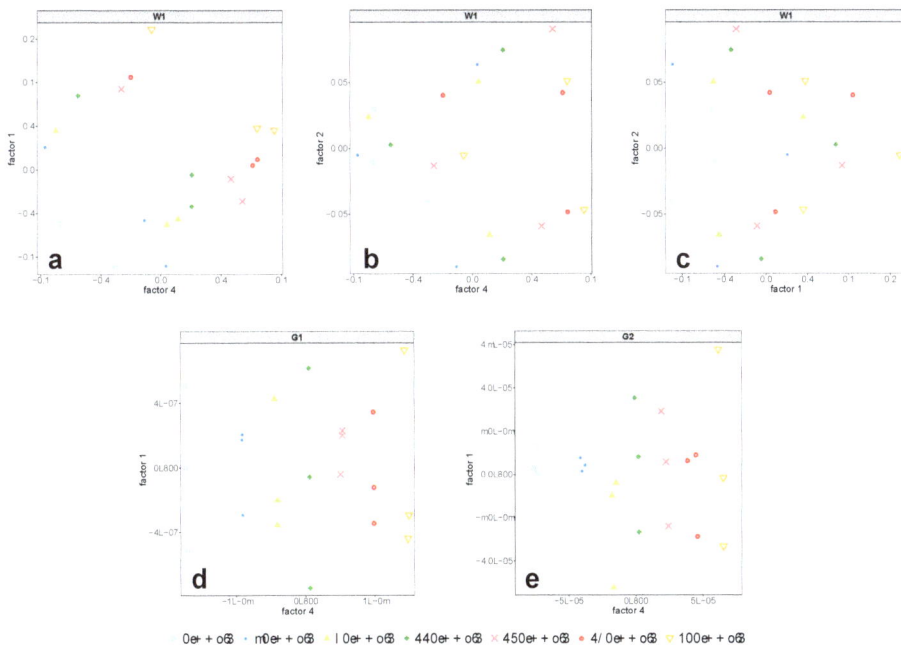

Figure 8. PLS-R score plots of the calibration set for the regions W1 (**a–c**), G1 (**d**), and G2 (**e**) obtained by measurements with the 2 mm cell. Each data point corresponds to one batch. The featured factors were chosen according to the required number of PLS-R factors in Table 2.

3.4. Comparison between Cell Pathlengths

An overall comparison between the predictive power of water- and glucose-based PLS-R models indicated that the glucose-related regions G1 and G2 considerably outperformed the two water bands W1 and W2. The glucose regions yielded far lower prediction errors with NRMSEP values as low as 0.4% along with the utilization of only one PLS-R factor. This emphasizes the dominant presence of glucose-related spectral information in these two NIR regions at around 5900 cm^{-1} and 4400 cm^{-1}. The only exception with poor predictive power was the 0.1 mm cell in region G1 due to the reasons described earlier. With regard to the pathlength, the 1 mm cell turned out to give the most accurate PLS-R models for both water and glucose related regions. This optical pathlength seemed to have a favorable ratio between transmitted and absorbed NIR light for quantitative analyses of aqueous glucose solutions and most probably for carbohydrate solutions in general.

Table 3. Explained Y-variances for each spectral region and utilized cell. The type of validation (calibration or test set validation) and the number of PLS-R factors are specified. The values for the explained variances are given in %.

Region	Cell	Type	PLS-R Factor			
			1	2	3	4
W1	0.1 mm	CAL	95.5	99.5	99.8	99.9
		TSV	96.4	98.1	99.7	99.7
	1 mm	CAL	96.8	99.9	99.9	100
		TSV	97.7	99.7	99.8	99.5
	2 mm	CAL	48.1	98.8	99.6	99.9
		TSV	22.7	93.8	96.6	99.0
W2	0.1 mm	CAL	99.4	99.8	99.9	100
		TSV	95.3	97.7	97.9	99.2
	1 mm	CAL	-	-	-	-
		TSV	-	-	-	-
	2 mm	CAL	-	-	-	-
		TSV	-	-	-	-
G1	0.1 mm	CAL	96.8	99.2	99.4	99.6
		TSV	79.2	71.7	83.0	80.3
	1 mm	CAL	100	100	100	100
		TSV	99.9	99.9	99.9	99.9
	2 mm	CAL	100	100	100	100
		TSV	99.9	99.9	99.9	99.9
G2	0.1 mm	CAL	99.9	100	100	100
		TSV	99.9	99.9	99.9	99.9
	1 mm	CAL	100	100	100	100
		TSV	100	99.9	100	99.9
	2 mm	CAL	99.8	99.9	99.9	99.9
		TSV	99.6	99.6	99.4	99.6

CAL—calibration; TSV—test set validation.

4. Conclusions

The good predictive performance of the PLS-R models with the water-related regions W1 and W2 confirmed the well documented fact, in which sugars (in this case glucose) in aqueous solutions affect the water bands in the NIR region by disturbing the structure of the hydrogen bond network of liquid water [26,39]. On the other hand, considering the significantly higher predictive power of the PLS-R models based on the regions G1 and G2, it must be concluded that the regions associated with carbohydrate vibrations (i.e., C–H, O–H, C–O) are even better suited for highly accurate quantifications. These vibrations cause rather small bands in the NIR spectrum, thus derivative functions need to be applied in order to reveal the concentration dependent patterns.

The validity of the results obtained herein was further confirmed by the fact that multiple batches were used for the construction of the PLS-R calibration models. Moreover, independent test set samples were utilized in the process of PLS-R model validation. Therefore, a certain robustness of the constructed PLS-R models can be assumed.

This study demonstrated the superiority of characteristic glucose bands over the dominant and intense water bands in the NIR spectrum in terms of quantitative predictive power. Relative prediction errors lower than 1% were obtained while only one PLS-R factor was required. Further investigations need to be carried out in order to determine reliable values for the limit of detection (LOD) and the limit of quantification (LOQ) of both the water- and glucose-related regions. However, despite the promising predictive power of the glucose bands, it has to be noted that the sample matrix employed in the present study was rather simple. The establishment of reliable PLS-R models based on NIR data obtained from more complex matrices like body fluids such as blood or urine, or food products like beverages, is undoubtedly much more challenging. Nevertheless, the findings reported herein can support the selection of the most informative NIR regions for investigations of aqueous carbohydrate systems.

Author Contributions: A.B. and C.W.H. conceived the study; A.B. and V.M. performed the experiments and data analysis; A.B. prepared the original draft (text, figures and tables); C.W.H. was in charge of supervision; and all authors reviewed the manuscript.

Funding: This research received no external funding.

Conflicts of Interest: The authors declare no conflict of interest.

Abbreviations

The following abbreviations are used in this manuscript:

CAL	Calibration
FFT	Fast Fourier-Transform
FT-NIR	Fourier-Transform Near-Infrared
LOD	Limit of Detection
LOQ	Limit of Quantification
NIPALS	Nonlinear Iterative Partial Least Squares
NIR	Near-Infrared
NIRS	Near-Infrared Spectroscopy
NRMSE	Normalized Root Mean Square Error
NRMSEC	Normalized Root Mean Square Error of Calibration
NRMSEP	Normalized Root Mean Square Error of Prediction
PCA	Principal Component Analysis
PLS-R	Partial Least Squares Regression
RMSE	Root Mean Square Error
RMSEC	Root Mean Square Error of Calibration
RMSEP	Root Mean Square Error of Prediction
SavGol	Savitzky-Golay
SEC	Standard Error of Calibration
SEP	Standard Error of Prediction
SNV	Standard Normal Variate
SP	Smoothing Points
TSV	Test Set Validation

References

1. Peterson, J.I.; Young, D.S. Evaluation of the hexokinase/glucose-6-phosphate dehydrogenase method of determination of glucose in urine. *Anal. Biochem.* **1968**, *23*, 301–316. [CrossRef]
2. Neeley, W.E. Simple automated determination of serum or plasma glucose by a hexokinase/glucose-6-phosphate dehydrogenase method. *Clin. Chem.* **1972**, *18*, 509–515.
3. Nielsen, S.S. (Ed.) *Food Analysis*, 4th ed.; Springer: New York, NY, USA, 2010.

4.   Siesler, H.W.; Ozaki, Y.; Kawata, S.; Heise, H.M. (Eds.) *Near-Infrared Spectroscopy: Principles, Instruments, Applications*; WILEY-VCH Verlag GmbH: Weinheim, Germany, 2002.

5.   Ozaki, Y.; McClure, W.F.; Christy, A.A. (Eds.) *Near-Infrared Spectroscopy in Food Science and Technology*; John Wiley & Sons, Inc.: Hoboken, NJ, USA, 2006.

6.   Burns, D.A.; Ciurczak, E.W. (Eds.) *Handbook of Near-Infrared Analysis*, 3rd ed.; CRC Press: Boca Raton, FL, USA, 2007.

7.   Buijs, K.; Choppin, G.R. Near-nfrared studies of the structure of water. I. Pure Water. *J. Chem. Phys.* **1963**, *39*, 2035–2041. [CrossRef]

8.   Workman, J., Jr.; Weyer, L. *Practical Guide and Spectral Atlas for Interpretive Near-Infrared Spectroscopy*; CRC Press: Boca Raton, FL, USA, 2012.

9.   Šašić, S.; Ozaki, Y. Band assignment of near-infrared spectra of milk by use of partial least-squares regression. *Appl. Spectrosc.* **2000**, *54*, 1327–1338. [CrossRef]

10. Segtnan, V.H.; Šašić, Š.; Isaksson, T.; Ozaki, Y. Studies on the structure of water using two-dimensional near-infrared correlation spectroscopy and principal component analysis. *Anal. Chem.* **2001**, *73*, 3153–3161. [CrossRef]

11. Rambla, F.J.; Garrigues, S.; De La Guardia, M. PLS-NIR determination of total sugar, glucose, fructose and sucrose in aqueous solutions of fruit juices. *Anal. Chim. Acta* **1997**, *344*, 41–53. [CrossRef]

12. Jung, Y.; Hwang, J. Near-infrared studies of glucose and sucrose in aqueous solutions: water displacement effect and red shift in water absorption from water-solute interaction. *Appl. Spectrosc.* **2013**, *67*, 171–180. [CrossRef] [PubMed]

13. Beganović, A.; Beć, K.B.; Henn, R.; Huck, C.W. Handling of uncertainty due to interference fringe in FT-NIR transmittance spectroscopy—Performance comparison of interference elimination techniques using glucose-water system. *Spectrochim. Acta Part A Mol. Biomol. Spectrosc.* **2018**, *197*, 208–215. [CrossRef] [PubMed]

14. Yano, T.; Funatsu, T.; Suehara, K.I.; Nakano, Y. Measurement of the concentrations of glucose and citric acid in the aqueous solution of a blood anticoagulant using near infrared spectroscopy. *J. Near Infrared Spectrosc.* **2001**, *9*, 43–48. [CrossRef]

15. Lanza, E.; Li, B.W. Application for near infrared spectroscopy for predicting the sugar content of fruit juices. *J. Food Sci.* **1984**, *49*, 995–998. [CrossRef]

16. Golic, M.; Walsh, K.; Lawson, P. Short-wavelength near-infrared spectra of sucrose, glucose, and fructose with respect to sugar concentration and temperature. *Appl. Spectrosc.* **2003**, *57*, 139–145. [CrossRef] [PubMed]

17. Chen, J.; Arnold, M.A.; Small, G.W. Comparison of combination and first overtone spectral regions for near-infrared calibration models for glucose and other biomolecules in aqueous solutions. *Anal. Chem.* **2004**, *76*, 5405–5413. [CrossRef] [PubMed]

18. Kasemsumran, S.; Du, Y.P.; Maruo, K.; Ozaki, Y. Improvement of partial least squares models for in vitro and in vivo glucose quantifications by using near-infrared spectroscopy and searching combination moving window partial least squares. *Chemom. Intell. Lab. Syst.* **2006**, *82*, 97–103. [CrossRef]

19. Beć, K.B.; Huck, C.W. Breakthrough potential in near-infrared spectroscopy: Spectra simulation. A review of recent developments. *Front. Chem.* **2019**, *7*, 48. [CrossRef] [PubMed]

20. Frost, V.J.; Molt, K. Analysis of aqueous solutions by near-infrared spectrometry (NIRS) III. Binary mixtures of inorganic salts in water. *J. Mol. Struct.* **1997**, *410*, 573–579. [CrossRef]

21. Inoue, A.; Kojima, K.; Taniguchi, Y.; Suzuki, K. Near-infrared spectra of water and aqueous electrolyte solutions at high pressures. *J. Solut. Chem.* **1984**, *13*, 811–823. [CrossRef]

22. Grant, A.; Davies, A.M.C.; Bilverstone, T. Simultaneous determination of sodium hydroxide, sodium carbonate and sodium chloride concentrations in aqueous solutions by near-infrared spectrometry. *Analyst* **1989**, *114*, 819–822. [CrossRef]

23. Maeda, H.; Ozaki, Y.; Tanaka, M.; Hayashi, N.; Kojima, T. Near infrared spectroscopy and chemometrics studies of temperature-dependent spectral variations of water: relationship between spectral changes and hydrogen bonds. *J. Near Infrared Spectrosc.* **1995**, *3*, 191–201. [CrossRef]

24. Fisher, H.F.; McCabee, W.C.; Subramanian, S. Near-infrared spectroscopic investigation of the effect of temperature on the structure of water. *J. Phys. Chem.* **1970**, *74*, 4360–4369. [CrossRef]

25. Berentsen, S.; Stolz, T.; Molt, K. Analysis of aqueous solutions by near-infrared spectrometry (NIRS) IV. One-and two-component systems of organic compounds in water. *J. Mol. Struct.* **1997**, *410*, 581–585. [CrossRef]

26. Giangiacomo, R. Study of water–sugar interactions at increasing sugar concentration by NIR spectroscopy. *Food Chem.* **2006**, *96*, 371–379. [CrossRef]

27. Bázár, G.; Kovacs, Z.; Tanaka, M.; Furukawa, A.; Nagai, A.; Osawa, M.; Itakura, Y.; Sugiyama, H.; Tsenkova, R. Water revealed as molecular mirror when measuring low concentrations of sugar with near infrared light. *Anal. Chim. Acta* **2015**, *896*, 52–62. [CrossRef] [PubMed]

28. Hazen, K.H.; Arnold, M.A.; Small, G.W. Measurement of glucose in water with first-overtone near-infrared spectra. *Appl. Spectrosc.* **1998**, *52*, 1597–1605. [CrossRef]

29. Jensen, P.S.; Bak, J. Measurements of urea and glucose in aqueous solutions with dual-beam near-infrared Fourier transform spectroscopy. *Appl. Spectrosc.* **2002**, *56*, 1593–1599. [CrossRef]

30. Jensen, P.S.; Bak, J. Near-infrared transmission spectroscopy of aqueous solutions: Influence of optical pathlength on signal-to-noise ratio. *Appl. Spectrosc.* **2002**, *56*, 1600–1606. [CrossRef]

31. Kasemsumran, S.; Du, Y.P.; Murayama, K.; Huehne, M.; Ozaki, Y. Simultaneous determination of human serum albumin, γ-globulin, and glucose in a phosphate buffer solution by near-infrared spectroscopy with moving window partial least-squares regression. *Analyst* **2003**, *128*, 1471–1477. [CrossRef]

32. Westad, F.; Schmidt, A.; Kermit, M. Incorporating chemical band-assignment in near infrared spectroscopy regression models. *J. Near Infrared Spectrosc.* **2008**, *16*, 265–273. [CrossRef]

33. Hellma Analytics Employee. (Hellma GmbH & Co. KG, Müllheim, Germany). Personal communication, 2016.

34. Malfait, W.J. The 4500 cm^{-1} infrared absorption band in hydrous aluminosilicate glasses is a combination band of the fundamental (Si, Al)-OH and OH vibrations. *Am. Mineral.* **2009**, *94*, 849–852. [CrossRef]

35. Barnes, R.J.; Dhanoa, M.S.; Lister, S.J. Standard normal variate transformation and de-trending of near-infrared diffuse reflectance spectra. *Appl. Spectrosc.* **1989**, *43*, 772–777. [CrossRef]

36. Savitzky, A.; Golay, M.J.E. Smoothing and differentiation of data by simplified least squares procedures. *Anal. Chem.* **1964**, *36*, 1627–1639. [CrossRef]

37. Næs, T.; Isaksson, T.; Fearn, T.; Davies, T. *A User Friendly Guide to Multivariate Calibration and Classification*; NIR Publications: Chichester, UK, 2004.

38. Esbensen, K.H.; Guyot, D.; Westad, F.; Houmoller, L.P. *Multivariate Data Analysis: In Practice: An Introduction to Multivariate Data Analysis and Experimental Design*, 5th ed.; CAMO Software AS: Oslo, Norway, 2009.

39. Workman, J., Jr.; Weyer, L. *Practical Guide to Interpretive Near-Infrared Spectroscopy*, 1st ed.; CRC Press: Boca Raton, FL, USA, 2007.

molecules

MDPI

Article

Investigation of Direct Model Transferability Using Miniature Near-Infrared Spectrometers

Lan Sun *, Chang Hsiung and Valton Smith

Viavi Solutions Inc., 1402 Mariner Way, Santa Rosa, CA 95407, USA; Chang.Hsiung@viavisolutions.com (C.H.); Valton.Smith@viavisolutions.com (V.S.)
* Correspondence: Lan.Sun@viavisolutions.com; Tel.: +1-707-525-7022

Academic Editors: Christian Huck and Krzysztof B. Bec
Received: 17 April 2019; Accepted: 23 May 2019; Published: 24 May 2019

Abstract: Recent developments in compact near infrared (NIR) instruments, including both handheld and process instruments, have enabled easy and affordable deployment of multiple instruments for various field and online or inline applications. However, historically, instrument-to-instrument variations could prohibit success when applying calibration models developed on one instrument to additional instruments. Despite the usefulness of calibration transfer techniques, they are difficult to apply when a large number of instruments and/or a large number of classes are involved. Direct model transferability was investigated in this study using miniature near-infrared (MicroNIR™) spectrometers for both classification and quantification problems. For polymer classification, high cross-unit prediction success rates were achieved with both conventional chemometric algorithms and machine learning algorithms. For active pharmaceutical ingredient quantification, low cross-unit prediction errors were achieved with the most commonly used partial least squares (PLS) regression method. This direct model transferability is enabled by the robust design of the MicroNIR™ hardware and will make deployment of multiple spectrometers for various applications more manageable.

Keywords: NIR; direct model transferability; MicroNIR™; SVM; hier-SVM; SIMCA; PLS-DA; TreeBagger; PLS; calibration transfer

---

1. Introduction

In recent years, compact near infrared (NIR) instruments, including both handheld and process instruments, have attracted considerable attention and received wider adoption due to their cost-effectiveness, portability, ease of use, and flexibility in installation. These instruments have been used for various applications in different industries, such as the pharmaceutical industry, agriculture, the food industry, the chemical industry, and so on. [1–5] They enable point-of-use analysis that brings advanced laboratory analysis to the field [6,7] and online and inline analysis that permits continuous process monitoring [8,9]. Moreover, scalability of NIR solutions has become possible. It is common that users of compact NIR instruments would desire more than one instrument to be used for their applications. Sometimes a large number of instruments are deployed.

Intrinsically, NIR solutions require multivariate calibration models for most applications due to the complexity of the spectra resulting from vibrational overtones and combination bands. Usually a calibration data set is collected using an NIR instrument to develop a calibration model. However, when multiple instruments are deployed for the same application, it is too time and labor consuming to collect calibration sets and develop calibration models for these instruments individually. It is also very inconvenient to manage different calibration models for different instruments. Therefore, it is highly desirable that calibration development is performed only once, and that the calibration model can be used on all these instruments successfully. In practice, when multiple instruments are involved for a particular application, the calibration model is often developed on one instrument and then applied to the rest of the instruments, especially when a project starts with one instrument for a feasibility test and then multiple instruments are procured. When a large number of instruments are involved, a global model approach can be taken in which calibration data from at least two to three instruments are pooled to develop the calibration model, in order to minimize noncalibrated variations from the instruments [10]. For any of the cases, model transferability from one or multiple instruments to the others is critical.

Historically, instrument-to-instrument variations could prohibit the success of the direct use of calibration models developed on one instrument with the other instruments. To avoid full recalibration, various calibration transfer methods have been developed to mathematically correct for instrument-to-instrument variations [10,11]. Common methods include direct standardization [12], piecewise direct standardization (PDS) [12–14], spectral space transformation [15], generalized least squares (GLS) [16], and so on. These methods have been extensively used to transfer quantitative calibration models [17–20], but very few studies were focused on the transfer of classification models [21,22]. Although these methods are very useful, they can only deal with calibration transfer from one instrument to another at a time and require transfer datasets to be collected from the same physical samples with both instruments. This is practical when there are only a few instruments involved. One instrument can be designated as the master instrument to develop the calibration model. Then data collected by the other instruments can be transformed into the master instrument's approximate space via the respective pair of transfer datasets. Thus, the master calibration model can be used by the other instruments. Alternatively, the master calibration data can be transferred to the other target instruments and calibration models can be developed on these target instruments. However, in the new era of handheld and process NIR instrumentation, a large number of instruments (e.g., > 20) could be deployed for one application. It would be difficult to perform calibration transfer in this way, especially when these instruments are placed in different locations. Other calibration transfer methods have been developed without using the transfer datasets from both instruments [23–25]. But unlike the commonly used methods, these methods have not been extensively studied and made easily available to general NIR users. Moreover, calibration transfer of classification models typically requires transfer data to be collected from every class. When a large number of classes are included in the model, the efforts required would be close to rebuilding a library on the secondary instrument. This may explain why very few studies have been conducted on transfer of classification models.

Considering all the advantages and potentials the handheld and process NIR instruments can offer and the challenges for calibration transfer when a large number of instruments and/or a large number of classes are involved, it is intriguing to understand if advances in instrumentation and modeling methods could make direct use of the master calibration model acceptable. However, to the best of our knowledge, little research has been done in this area.

The authors have demonstrated in the past that the use of miniature near-infrared (MicroNIR™) spectrometers with the aid of support vector machine (SVM) modeling can achieve very good direct transferability of models with a large number of classes for pharmaceutical raw material identification [26]. In the current study, using MicroNIR™ spectrometers, direct model transferability was investigated for polymer classification. Five classification methods were tested, including two conventional chemometric algorithms, partial least squares discriminant analysis (PLS-DA) [27] and soft independent modeling of class analogy (SIMCA) [28], and three machine learning algorithms that are burgeoning in chemometrics, bootstrap-aggregated (bagged) decision trees (TreeBagger) [29], support vector machine (SVM) [30,31] and hierarchical SVM (hier-SVM) [26]. High cross-unit prediction success rates were achieved. Direct transferability of partial least squares (PLS) regression models was also investigated to quantify active pharmaceutical ingredients (API). Low cross-unit prediction errors were obtained.

2. Results

2.1. Classification of Polymers

Polymers are encountered in everyday life and are of interest for many applications. In this study polymer classification was used as an example to investigate direct model transferability. Resin kits containing 46 materials representing the most important plastic resin used in industry today were used. Each material was treated as one class. Three resin kits were used to show prediction performance on different physical samples of the same material. The samples were measured by three randomly chosen MicroNIR™ OnSite spectrometers (labeled as Unit 1, Unit 2 and Unit 3).

2.1.1. Spectra of the Resin Samples

Spectra collected by the three spectrometers were compared in Figure 1. For clarity, example spectra of two samples were presented. The same observations were obtained for the other samples. The raw spectra in Figure 1a only show baseline shifts between measurements using different spectrometers for the same sample. These shifts were mainly due to different measurement locations, since these resin samples are injection molded and are not uniform in thickness and molecular orientation. In fact, baseline shifts were also observed when using the same spectrometer to measure different locations of the same sample. These shifts can be corrected by spectral preprocessing, and the preprocessed spectra from the same sample collected by different spectrometers were very similar as shown in Figure 1b.

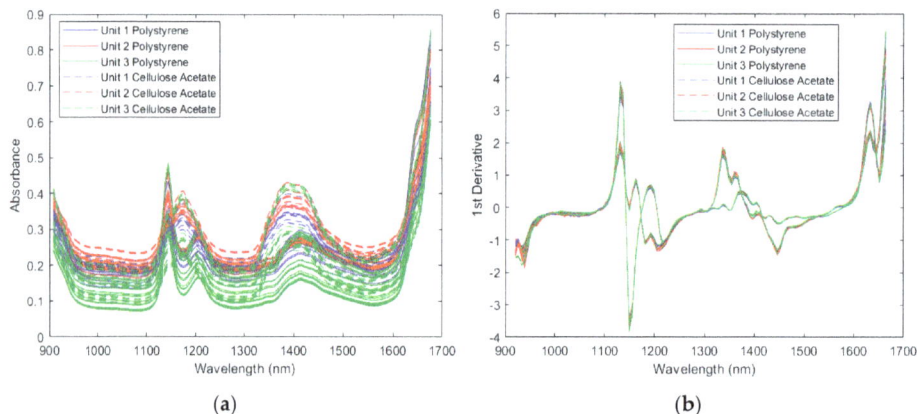

(a) (b)

Figure 1. Spectra of example polymer samples by three instruments: (**a**) raw spectra; (**b**) preprocessed spectra by Savitzky-Golay 1st derivative (5 smoothing points and 3rd polynomial order) and standard normal variate (SNV).

2.1.2. Direct Model Transferability of the Classification Models

The performance of the polymer classification models was evaluated at four levels, the same-unit-same-kit performance, the same-unit-cross-kit performance, the cross-unit-same-kit performance, and the cross-unit-cross-kit performance. To account for the most variation in sample shape and thickness, each resin sample was scanned in five specified locations. In addition, at each position the sample was scanned in two orientations with respect to the MicroNIR™ lamps to account for any directionality in the structure of the molding. For each position and orientation, three replicate scans were acquired, totaling thirty scans per sample, per spectrometer. Prediction was performed for every spectrum in the validation set. For the same-unit-same-kit performance, the models built with data collected from four locations on each sample in one resin kit by one spectrometer were used to predict data collected from the other location on each sample in the same resin kit by the same spectrometer. The total number of predictions was 276 for all 46 materials for each case. For the same-unit-cross-kit performance, the models built with all the data collected from one resin kit by one spectrometer were used to predict all the data collected from a different resin kit by the same spectrometer. The total number of predictions was 1380 for all 46 materials for each case. For the cross-unit-same-kit performance, the models built with all the data collected from one resin kit by one spectrometer were used to predict all the data collected from the same resin kit by a different spectrometer. The total number of predictions was 1380 for all 46 materials for each case. For the cross-unit-cross-kit performance, the models built with all the data collected from one resin kit by one spectrometer were used to predict all the data collected from a different resin kit by a different spectrometer. The total number of predictions was 1380 for all 46 materials for each case. Five different classification algorithms were used to build the models, which were PLS-DA, SIMCA, TreeBagger, SVM and hier-SVM.

The prediction performance was evaluated in terms of prediction success rates and the number of missed predictions. The representing results were summarized in Tables 1 and 2, respectively. The prediction success rates were calculated by dividing the number of correct predictions with the number of total predictions. The number of missed predictions is presented to make the difference clearer, since with a large number of total predictions a small difference in prediction success rate would mean a conceivable difference in the number of missed predictions. It should be noted that in a few cases the total number of predictions was not exactly 276 or 1380, because extra spectra were collected unintentionally during experiments and no spectra were excluded from analysis. To make the comparison consistent, in these tables all the models were developed using data from Kit 1 for different spectrometers. The prediction data were collected using different resin kits and different spectrometers for the four levels of performance.

The same-unit-same-kit cases were control cases and presented as the diagonal elements for each algorithm in the left three columns of the tables. As expected, 100% prediction success rates and 0 missed predictions were obtained for all algorithms except for one PLS-DA case (Unit 1 K1 for modeling and testing) where there was only 1 missed prediction. The same-unit-cross-kit cases showed the true prediction performance of the models for each spectrometer, since independent testing samples were used. The results are presented as the diagonal elements for each algorithm in the right three columns of the tables. All the models showed very good same-unit-cross-kit predictions. Although SIMCA showed the best performance, the differences in performance were very small between algorithms. It should be noted that samples made of the same type of material but with different properties are included in the resin kits, indicating that the MicroNIR™ spectrometers have the resolution to resolve minor differences between these polymer materials. For the cross-kit cases, Kit 2 was used for Unit 1 and Unit 2, while Kit 3 was used for Unit 3, because at the time of data collection using Unit 3, Kit 2 was no longer available. Nonetheless, conclusions about the cross-kit performance were not impacted by this.

Table 1. Prediction success rates (%) of polymer classification.

Algorithm	Unit# Kit# for Modeling	Unit# Kit# for Testing					
		Unit1 K1	Unit2 K1	Unit3 K1	Unit1 K2	Unit2 K2	Unit3 K3
	Unit 1 K1	99.64	89.68	83.99	95.87	88.91	82.39
PLS-DA	Unit 2 K1	91.96	100	81.52	90.87	99.57	84.49
	Unit 3 K1	76.74	75.32	100	75.07	73.12	99.20
	Unit 1 K1	100	99.42	96.45	99.35	97.32	96.81
SIMCA	Unit 2 K1	98.77	100	95.43	97.68	99.93	95.80
	Unit 3 K1	96.30	93.29	100	96.09	92.17	100
	Unit 1 K1	100	97.11	95.80	98.04	95.94	96.30
TreeBagger	Unit 2 K1	97.83	100	93.55	94.49	98.26	96.16
	Unit 3 K1	95.14	98.41	100	96.09	98.84	98.84
	Unit 1 K1	100	99.86	97.54	98.26	97.90	97.83
SVM	Unit 2 K1	98.70	100	97.03	94.93	98.26	98.26
	Unit 3 K1	97.83	96.18	100	96.30	95.00	99.57
	Unit 1 K1	100	100	97.97	97.83	97.83	97.25
Hier-SVM	Unit 2 K1	99.93	100	98.26	98.26	99.13	99.13
	Unit 3 K1	99.13	100	100	96.88	97.83	100

Table 2. Number of missed predictions of polymer classification in the format of number of missed predictions/total number of predictions.

Algorithm	Unit# Kit# for Modeling	Unit# Kit# for Testing					
		Unit1 K1	Unit2 K1	Unit3 K1	Unit1 K2	Unit2 K2	Unit3 K3
PLS-DA	Unit 1 K1	1/276	143/1386	221/1380	57/1380	153/1380	243/1380
	Unit 2 K1	111/1380	0/277	255/1380	126/1380	6/1380	214/1380
	Unit 3 K1	321/1380	342/1386	0/276	344/1380	371/1380	11/1380
SIMCA	Unit 1 K1	0/276	8/1386	49/1380	9/1380	37/1380	44/1380
	Unit 2 K1	17/1380	0/277	63/1380	32/1380	1/1380	58/1380
	Unit 3 K1	51/1380	93/1386	0/276	54/1380	108/1380	0/1380
TreeBagger	Unit 1 K1	0/276	40/1386	58/1380	27/1380	56/1380	51/1380
	Unit 2 K1	30/1380	0/277	89/1380	76/1380	24/1380	53/1380
	Unit 3 K1	67/1380	22/1386	0/276	54/1380	16/1380	16/1380
SVM	Unit 1 K1	0/276	2/1386	34/1380	24/1380	29/1380	30/1380
	Unit 2 K1	18/1380	0/277	41/1380	70/1380	24/1380	24/1380
	Unit 3 K1	30/1380	53/1386	0/276	51/1380	69/1380	6/1380
Hier-SVM	Unit 1 K1	0/276	0/1386	28/1380	30/1380	30/1380	38/1380
	Unit 2 K1	1/1380	0/277	24/1380	24/1380	12/1380	12/1380
	Unit 3 K1	12/1380	0/1386	0/276	43/1380	30/1380	0/1380

The direct model transferability was first demonstrated by the cross-unit-same-kit results, which are presented by the non-diagonal elements for each algorithm in the left three columns of the tables. Except the PLS-DA algorithm, all the other algorithms showed good performance. In general, the order of performance was Hier-SVM > SVM > SIMCA > TreeBagger >> PLS-DA. When the hier-SVM algorithm was used, the worst case only had 28 missed predictions out of 1380 predictions, and 1/3 of the cases showed perfect predictions.

The direct model transferability was further demonstrated by the most stringent cross-unit-cross-kit cases, which are often the real-world cases. The results are presented by the non-diagonal elements for each algorithm in the right three columns of the tables. Other than the PLS-DA algorithm, all the other algorithms showed good performance, but which was slightly worse than the cross-unit-same-kit results with some exceptions. In general, the order of performance was hier-SVM > SVM > TreeBagger ≈ SIMCA >> PLS-DA.

Besides the representing results shown in these tables, all possible combinations of datasets were analyzed, including 6 same-unit-same-kit cases, 6 same-unit-cross-kit cases, 8 cross-unit-same-kit cases, and 16 cross-unit-cross-kit cases in total for each algorithm. The conclusions were similar to those presented above. For the most stringent cross-unit-cross-kit cases, the mean prediction success rates of all the cases were 98.15%, 97.00%, 96.74%, 95.83%, and 80.19% for hier-SVM, SVM, TreeBagger, SIMCA, and PLS-DA, respectively. The high prediction success rates for hier-SVM, SVM, TreeBagger and SIMCA indicate good direct model transferability for polymer classification with MicroNIR™ spectrometers. To achieve the best result, hier-SVM should be used. But the conventional SIMCA algorithm that is available to most NIR users is also sufficient.

2.2. Quantification of Active Pharmaceutical Ingredients

Quantitative analysis of an active pharmaceutical ingredient is important in several different steps of a pharmaceutical production process and it was proved that NIR spectroscopy is a good alternative to other more time-consuming means of analysis [32]. As one of the process analytical technology (PAT) tools adopted by the pharmaceutical industry, compact NIR spectrometers can be installed for real-time process monitoring, enabling the quality by design (QbD) approach that is now accepted by most pharmaceutical manufacturers to improve manufacturing efficiency and quality [33,34]. In this context, multiple NIR spectrometers will be needed for the same application. It is important to understand the direct transferability of calibration models to determine APIs quantitatively.

To investigate this, a five-component pharmaceutical powder formulation including three APIs, acetylsalicylic acid (ASA), ascorbic acid (ASC), and caffeine (CAF), as well as two excipients, cellulose and starch, was used. A set of 48 samples was prepared by milling varying amounts of the three APIs in the concentration range of 13.77–26.43% *w/w* with equal amounts (40% *w/w*) of a 1:3 (*w/w*) mixture of cellulose and starch [4]. The set of samples was measured by three randomly chosen MicroNIR™ 1700ES spectrometers (labeled as Unit 1, Unit 2 and Unit 3).

2.2.1. Spectra of the Pharmaceutical Samples

The spectra were first compared across the three instruments. Raw spectra of two samples with the lowest ASA concentration and the highest ASA concentration collected by all three instruments are shown in Figure 2a. Only slight baseline shifts can be seen between spectra collected by different instruments. The preprocessed spectra collected by different instruments became almost identical, as shown in Figure 2b. However, spectral differences between the high concentration sample and the low concentration can be clearly seen. Similar observations were obtained for the other two APIs, ASC (Figure 2c,d) and CAF (Figure 2e,f). It should be noted the optimized preprocessing steps were chosen to generate the preprocessed spectra for each API, respectively.

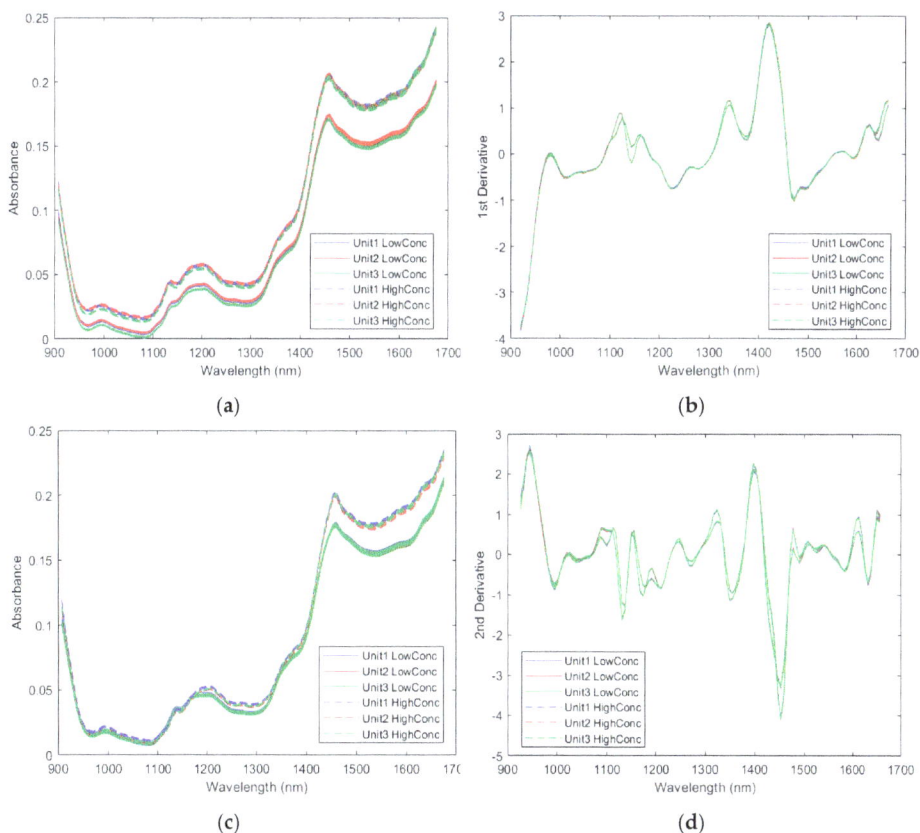

Figure 2. *Cont.*

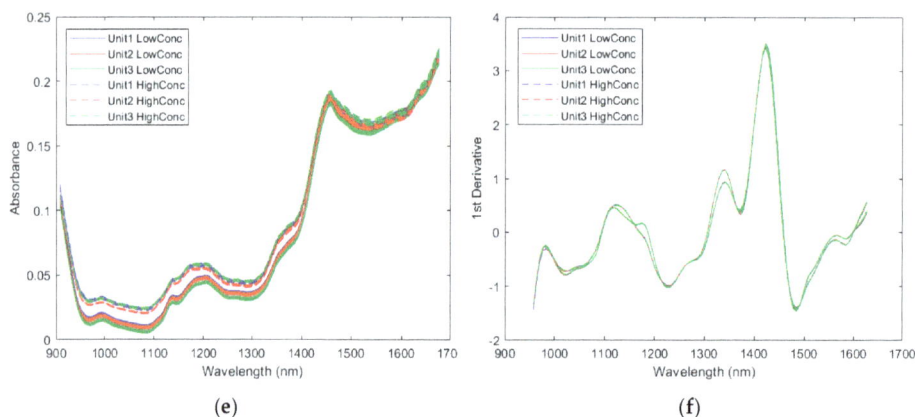

(e)   (f)

Figure 2. Spectra of samples with the highest and the lowest active pharmaceutical ingredient (API) concentrations measured by three instruments: (**a**) selected raw spectra based on the acetylsalicylic acid (ASA) concentration; (**b**) selected preprocessed spectra based on the ASA concentration by Savitzky-Golay 1st derivative (5 smoothing points and 2nd polynomial order) and SNV; (**c**) selected raw spectra based on the ascorbic acid (ASC) concentration; (**d**) selected preprocessed spectra based on the ASC concentration by Savitzky-Golay 2nd derivative (7 smoothing points and 3rd polynomial order) and SNV; (**e**) selected raw spectra based on the caffeine (CAF) concentration; (**f**) selected preprocessed spectra based on the CAF concentration by Savitzky-Golay 1st derivative (17 smoothing points and 3rd polynomial order) and SNV.

2.2.2. Direct Model Transferability of the Quantitative Models

To develop the quantitative calibration models, 38 out of the 48 samples were selected as the calibration samples via the Kennard-Stone algorithm [35], based on the respective API concentration, which was determined by the amount of API added to the powder sample. The remaining 10 samples were used as the validation samples. Twenty spectra were collected from each sample with every spectrometer. Thus, 760 spectra from the 38 calibration samples were used to build every model and 200 spectra from the 10 validation samples were used to validate each model. For each API, an individual model was developed on each instrument by partial least squares (PLS) regression. Different preprocessing procedures with different settings were tested and the optimal one was determined based on the cross-validation statistics using the calibration set. The same optimal preprocessing procedure was selected on all three instruments for the same API. The API models were developed using the corresponding preprocessed spectra.

The model performance was first evaluated in terms of normalized root mean square error of prediction (NRMSEP), which is root mean square error of prediction (RMSEP) normalized to the mean reference value of the validation set. NRMSEP was used to provide an estimate of how big the error was relative to the value measured. Since the mean reference value was the same for all the validation sets, it is equivalent to comparing RMSEP. Two types of prediction performance were examined, the same-unit performance and the cross-unit performance. Using a calibration model developed on one instrument, the same-unit performance was determined by predicting the validation set obtained with the same instrument, and the cross-unit performance was determined by predicting the validation set obtained with a different instrument. The cross-unit performance is the indicator of direct model transferability. The results were reported under the No Correction section in Tables 3–5 for ASA, ASC and CAF, respectively. The unit number in the row title represents which of the instruments was used to develop the calibration model, and the unit number in the column title represents which instrument was used to collect the validation data. Therefore, the NRMSEP values on the diagonal indicate the same-unit performance, while the other values indicate the cross-unit performance. The data show that cross-unit performance was close to the same-unit performance, all below 5%.

Table 3. The normalized root mean square error of prediction (NRMSEP, %) for ASA.

Test Sets	No Correction			Bias	PDS	GLS
	Unit 1	Unit 2	Unit 3	Unit 1	Unit 1	Unit 1
Unit 1	3.4	3.5	3.5	-	-	-
Unit 2	4.0	4.2	3.9	3.7	3.3	3.6
Unit 3	4.3	4.5	4.2	4.1	3.5	4.4

Table 4. The normalized root mean square error of prediction (NRMSEP, %) for ASC.

Test Sets	No Correction			Bias	PDS	GLS
	Unit 1	Unit 2	Unit 3	Unit 1	Unit 1	Unit 1
Unit 1	3.0	2.6	2.7	-	-	-
Unit 2	2.7	2.7	2.6	2.3	3.5	2.6
Unit 3	2.5	2.5	2.7	2.2	3.1	2.4

Table 5. The normalized root mean square error of prediction (NRMSEP, %) for CAF.

Test Sets	No Correction			Bias	PDS	GLS
	Unit 1	Unit 2	Unit 3	Unit 1	Unit 1	Unit 1
Unit 1	4.0	4.6	3.7	-	-	-
Unit 2	4.1	4.7	4.2	4.2	4.3	3.2
Unit 3	4.2	4.9	4.0	4.1	6.2	3.9

In another independent study, the same samples were measured by a benchtop Bruker Vector 22/N FT-NIR spectrometer. The reported mean absolute bias based on 3 validation samples was 0.28, 0.62 and 0.11 for ASA, ASC and CAF, respectively [36]. In the current study, the mean absolute bias of the three same-unit cases based on 10 validation samples was 0.21, 0.35 and 0.22 for ASA, ASC and CAF, respectively. The mean absolute bias of the six cross-unit cases based on 10 validation samples was 0.14, 0.30 and 0.25, respectively. These results indicate that both the same-unit and the cross-unit MicroNIR™ performance is comparable with the benchtop instrument performance. However, it should be noted in the current study 38 samples were used for calibration and 10 samples were used for validation, while in the other study 45 samples were used for calibration and 3 samples were used for validation.

The model performance was further examined by the predicted values of the validation set versus the reference values. Using calibration models developed on Unit 1, the same-unit predicted results and the cross-unit predicted results for ASA, ASC and CAF are shown in Figure 3. It can be seen that most of the predicted values stay close to the 45-degree lines, explaining the good model performance. Moreover, the cross-unit results (red circles) are very close to the same-unit results (blue circles), explaining the similar cross-unit performance to the same-unit performance.

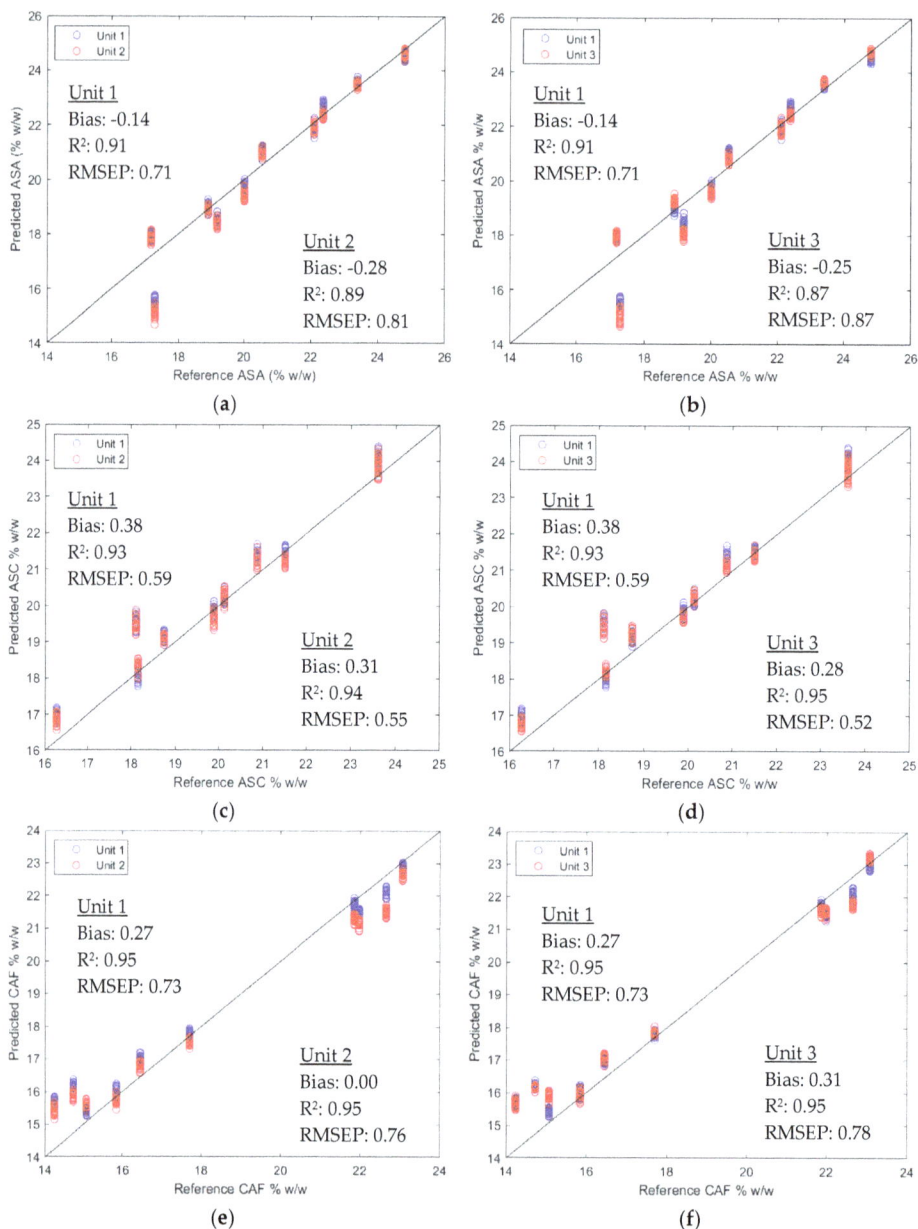

Figure 3. Predicted values versus reference values using models developed on Unit 1: (**a**) validation sets by Unit 1 and Unit 2 for ASA prediction; (**b**) validation sets by Unit 1 and Unit 3 for ASA prediction; (**c**) validation sets by Unit 1 and Unit 2 for ASC prediction; (**d**) validation sets by Unit 1 and Unit 3 for ASC prediction; (**e**) validation sets by Unit 1 and Unit 2 for CAF prediction; (**f**) validation sets by Unit 1 and Unit 3 for CAF prediction. The corresponding bias, R^2 for prediction, and root mean square error for prediction (RMSEP) are presented in each plot.

The corresponding Bland-Altman plots were used to illustrate the agreement between the cross-unit prediction results and the same-unit prediction results in Figure 4. The Bland-Altman analysis is a well-accepted technique for method comparison in highly regulated clinical sciences [37] and shows good visual comparison between two instruments [11]. The x-axis shows the mean predicted value and the y-axis shows the difference between the cross-unit predicted value and the same-unit predicted value. The limits of agreement (LOA) were calculated by Equation (1):

$$LOA = \bar{d} \pm 1.96 \times SD \tag{1}$$

where $\bar{d}$ is the bias or the mean difference, and SD is the standard deviation of the differences. It can be seen from Figure 4 that with only a few exceptions, all data points stayed within the LOA, indicating that at a 95% confidence level, the cross-unit prediction results agreed well with the same-unit prediction results. LOA relative to the mean of the mean predicted values (x-axis) was below 3% for all three APIs.

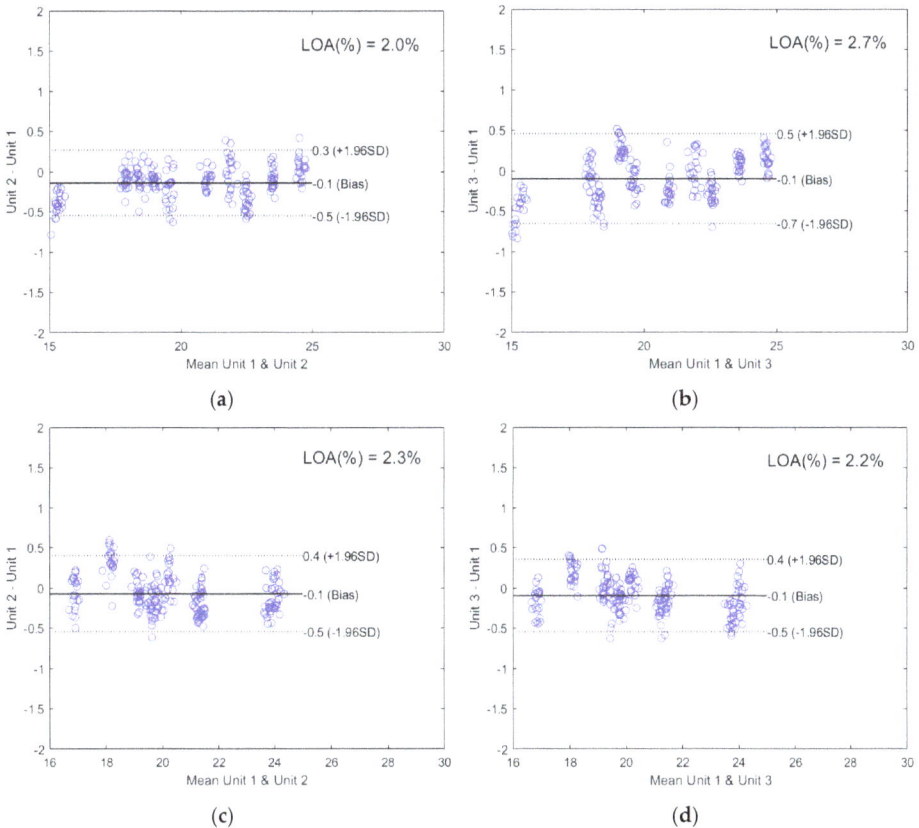

Figure 4. *Cont.*

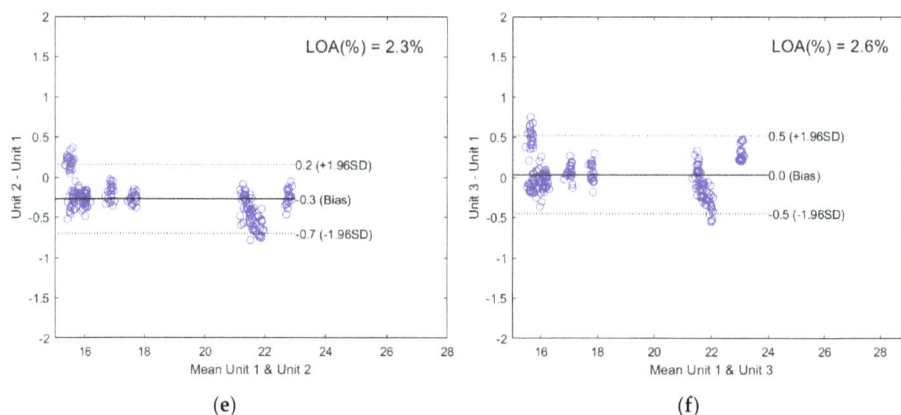

Figure 4. The Bland-Altman plots comparing the cross-unit prediction results and the same-unit prediction results using models developed on Unit 1: (**a**) validation sets by Unit 1 and Unit 2 for ASA prediction; (**b**) validation sets by Unit 1 and Unit 3 for ASA prediction; (**c**) validation sets by Unit 1 and Unit 2 for ASC prediction; (**d**) validation sets by Unit 1 and Unit 3 for ASC prediction; (**e**) validation sets by Unit 1 and Unit 2 for CAF prediction; (**f**) validation sets by Unit 1 and Unit 3 for CAF prediction.

The corresponding reduced Hotelling's T^2 and reduced Q residuals are shown in Figure 5. The reduced statistics were calculated by normalizing Hotelling's T^2 and Q residuals to their respective 95% confidence limit. The black circles represent the calibration data, the blue circles represent the same-unit validation data, and the red circles represent the cross-unit validation data. It can be clearly seen that the cross-unit validation data stayed close to the same-unit validation data, further explaining the similar cross-unit performance to the same-unit performance. It was noticed that 20 calibration data points (from the same physical sample) and 20 cross-unit validation data points (from another physical sample) are in the high reduced Hotelling's T^2 and high reduced Q residuals quadrant for ASA (Figure 5a,b). These explained why the prediction results of one sample significantly deviated from the 45-degree lines in Figure 3a,b. However, to keep the analysis consistent with the other two APIs and data available in literature [4,36] for comparison, no sample was excluded from calibration or validation.

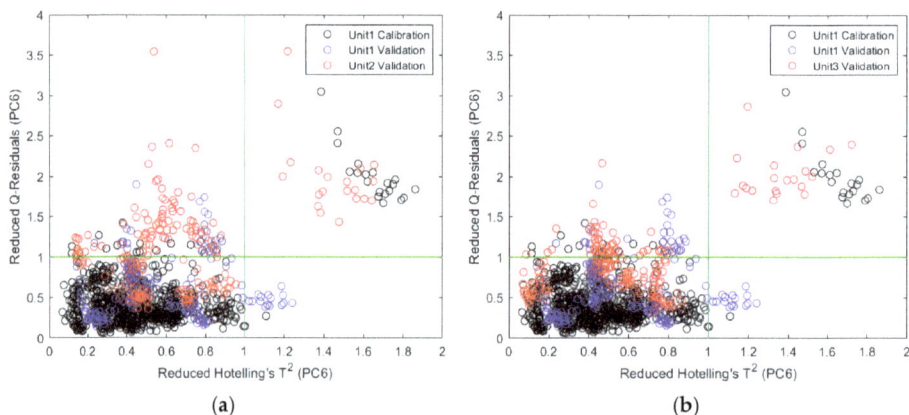

Figure 5. *Cont.*

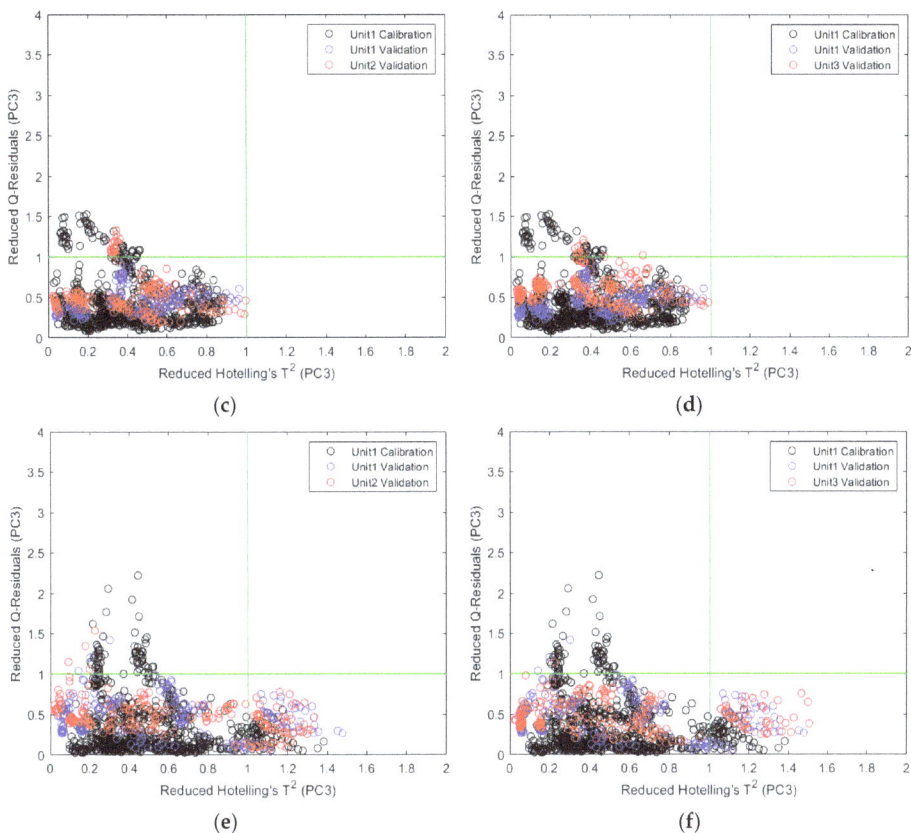

Figure 5. Reduced Q residuals versus reduced Hotelling's T^2 for models developed on Unit 1: (**a**) validation sets by Unit 1 and Unit 2 for ASA prediction; (**b**) validation sets by Unit 1 and Unit 3 for ASA prediction; (**c**) validation sets by Unit 1 and Unit 2 for ASC prediction; (**d**) validation sets by Unit 1 and Unit 3 for ASC prediction; (**e**) validation sets by Unit 1 and Unit 2 for CAF prediction; (**f**) validation sets by Unit 1 and Unit 3 for CAF prediction.

2.2.3. Calibration Transfer

To check how direct model transfer compared with calibration transfer, three types of calibration transfer methods were tested. The first method was bias correction by standardizing the predicted values, which is probably the simplest method. The second method was PDS by mapping spectral responses of the slave instrument to the master instrument, which is probably the most commonly used method. The third method was GLS by removing the differences between instruments from both instruments. To perform the calibration transfer, 8 transfer samples were selected from the calibration samples with the Kennard-Stone algorithm. The calibration transfer results using Unit 1 as the master instrument were summarized in Tables 3–5 for ASA, ASC and CAF, respectively. It should be noted that different settings for PDS and GLS were tested. The results presented were obtained under the best settings based on RMSEP. By comparing these results with the corresponding same-unit and cross-unit results (Column 1 under No Correction), there was not a single method that could improve cross-unit results for all three APIs. Choosing the best method for individual API, only slight improvement (decrease of 0.3–0.9% in RMSEP%) of cross-unit performance was observed. Calibration transfer could sometimes damage the performance when a certain method was applied to a certain API. In addition, for ASC and CAF, the cross-unit performance was already close to or slightly better than the same-unit performance. For ASA, although the same-unit performance was better than the cross-unit performance using the calibration model on Unit 1 (Column 1 under No Correction in Table 3), it was similar to the cross-unit performance using calibration models on Unit 2 and Unit 3 (Row 1 under No Correction in Table 3). All these observations indicate that the instrument-to-instrument difference was small. Therefore, calibration transfer may not be necessary for this application.

3. Discussion

The good direct model transferability demonstrated in this study was enabled by the minimal instrument-to-instrument differences owing to the robust design of the MicroNIR™ hardware. The MicroNIR™ spectrometer utilizes a wedged linear variable filter (LVF) as the dispersive element on top of an InGaAs array detector, which results in an extremely compact and rugged spectral engine with no moving parts [4]. The operation of the on-board illumination allows for a steady output of optical power and an extended lamp-life. Thus, a very stable performance can be achieved without the need for realignment of hardware over time. In addition to the hardware design, the performance of every MicroNIR™ spectrometer is evaluated and calibrated at the production level. The accuracy of the MicroNIR™ wavelength calibration enables precise spectral alignments from instrument to instrument. The repeatability of the photometric response ensures the consistency of signal amplitude from instrument to instrument. The unit-specific temperature calibration stabilizes the MicroNIR™ response over the entire operating temperature range. In the Supplementary Material, the wavelength reference plots and the photometric response plots are shown for the MicroNIR™ OnSite units used for the polymer classification example (Figure S1) and the MicroNIR™ ES units used for the API quantification example (Figure S2), respectively. Very small instrument-to-instrument differences were observed. It should be noted that findings from the handheld MicroNIR™ OnSite and ES units could be extended to the MicroNIR™ PAT units for process monitoring, since the spectral engine and the calibration protocol at the production level are the same.

In this study, both a classification example and a quantification example were investigated. For the quantification example, the good direct model transferability was demonstrated with the most commonly used regression method, PLS. For the classification example, the good direct model transferability was demonstrated with both the commonly used chemometric algorithm, SIMCA, and the machine learning algorithms, SVM, hier-SVM and TreeBagger. It should be noted the PLS-DA performance could be improved to about 90% prediction success rate by manually optimizing the number of PLS factors. The results presented in Tables 1 and 2 were based on automatically selected PLS factors. This automatic selection procedure sometimes causes overfitting. However, since all the other algorithms were also using automatic model building, which may not always generate the best results, for a fair comparison no manual intervention was introduced to PLS-DA. In fact, even with the improved performance, PLS-DA still didn't perform as well as the other algorithms for this specific application. Although the direct model transferability was good with conventional SIMCA, it can be further improved with the use of SVM algorithms. SVM has found increasing interest in chemometrics in recent years, since it is such a sound methodology, where geometric intuition, elegant mathematics, theoretical guarantees, and practical algorithms meet [38]. Among SVM's many appealing features, generalization ability, that is the ability to accurately predict outcome values for previously unseen data, can help minimize cross-unit prediction errors. The basic principle of SVM is to construct the maximum margin hyperplanes to separate data points into different classes. Maximizing the margin reduces complexity of the classification function, thus minimizing the possibility of overfitting. Therefore, better generalization can be achieved intrinsically for SVM [38]. When many classes are involved, like the polymer classification example in this study, the hier-SVM algorithm was shown to be beneficial, because this multilevel classification scheme facilitates refined classification for chemically similar materials to achieve more accurate prediction [26]. In addition, the TreeBagger algorithm is based on random forest, which is one of the most powerful classifiers in machine learning [39]. However, for the current study, the cross-unit performance of TreeBagger was not as good as the SVM algorithms.

The combination of the hardware design and implementation of advanced calibration techniques results in a repeatable and reproducible performance between different MicroNIR™ spectrometers, allowing effective direct model transferability. However, it is not intended to say that this will be the ultimate solution that eliminates all problems that necessitate calibration transfer. The scope of the current study was limited to model transferability only involving instrument-to-instrument differences, not very heterogeneous samples, and data collected with sound sampling and measurement protocols. For example, when different instruments are placed in different environments, environmental changes may have to be corrected for the model via calibration transfer. Very heterogeneous samples, such as biological samples, will be more difficult to handle in general. Even very small instrument-to-instrument differences could cause unsatisfactory cross-unit prediction results. A global model approach using data from samples with all expected sources of variance and/or measured with multiple instruments for calibration could significantly minimize prediction errors. Model updating techniques will also be very helpful [40]. Direct model transferability will be evaluated for very heterogeneous materials in our future studies. In addition, poor cross-unit model performance often results from nonqualified calibration data that are not collected with a careful sampling plan and a proper measurement protocol. The success of a multi-instrument NIR project must start with reliable NIR data that are collected with best practices in sampling [41,42] and measurement [43,44].

The current study demonstrated the possibility of direct model transfer from instrument to instrument for both classification and quantification problems, which has laid a good foundation for the use of a large number of compact NIR instruments. More studies should be encouraged in wider applications and using all kinds of instruments from various manufacturers. Scalability of handheld and process NIR solutions can become more manageable when the number of times that calibration transfer has to be performed between instruments can be minimized.

4. Materials and Methods

4.1. Materials

For the polymer classification study, 46 injection molded resins were obtained from The ResinKit™ (The Plastics Group of America, Woonsocket, RI, USA). The set of resins contains a variety of polymer materials, as well as various properties within the same type of material (for example different densities or strengths). Each resin was treated as an individual class in this study. All the resins used in this study are listed in Table 6 and detailed properties of these materials are available upon request. To evaluate the cross-kit prediction performance, three resin kits were used.

For the API quantification study, 48 pharmaceutical powders consisting of different concentrations of three crystalline active ingredients, as well as two amorphous excipients were provided by Prof. Heinz W. Siesler at University of Duisburg-Essen, Germany [4]. The active ingredients used were acetylsalicylic acid (ASA, Sigma-Aldrich Chemie GmbH, Steinheim, Germany), ascorbic acid (ASC, Acros Organics, NJ, USA), and caffeine (CAF, Sigma-Aldrich Chemie GmbH, Steinheim, Germany), and the two excipients used were cellulose (CE, Fluka Chemie GmbH, Buchs, Switzerland) and starch (ST, Carl Roth GmbH, Karlsruhe, Germany). The concentration of the active ingredients ranged from 13.77–26.43% (*w/w*), and all samples consisted of 40% (*w/w*) of a 3:1 (*w/w*) mixture of cellulose and starch.

Table 6. Polymer materials used for the classification study.

No.	Polymer Type	No.	Polymer Type
1	PolyStyrene-General Purpose	24	Polyethylene-High Density
2	PolyStyrene-High Impact	25	Polypropylene-Copolymer
3	Styrene-Acrylonitrile (SAN)	26	Polypropylene-Homopolymer
4	ABS-Transparent	27	Polyaryl-Ether
5	ABS-Medium Impact	28	Polyvinyl Chloride-Flexible
6	ABS-High Impact	29	Polyvinyl Chloride-Rigid
7	Styrene Butadiene	30	Acetal Resin-Homopolymer
8	Acrylic	31	Acetal Resin-Copolymer
9	Modified Acrylic	32	Polyphenylene Sulfide
10	Cellulose Acetate	33	Ethylene Vinyl Acetate
11	Cellulose Acetate Butyrate	34	Urethane Elastomer (Polyether)
12	Cellulose Acetate Propionate	35	Polypropylene-Flame Retardant
13	Nylon-Transparent	36	Polyester Elastomer
14	Nylon-Type 66	37	ABS-Flame Retardant
15	Nylon-Type 6 (Homopolymer)	38	Polyallomer
16	Thermoplastic Polyester (PBT)	39	Styrenic Terpolymer
17	Thermoplastic Polyester (PETG)	40	Polymethyl Pentene
18	Phenylene Oxide	41	Talc-Reinforced Polypropylene
19	Polycarbonate	42	Calcium Carbonate-Reinforced Polypropylene
20	Polysulfone	43	Nylon (Type 66–33% Glass)
21	Polybutylene	44	Thermoplastic Rubber
22	Ionomer	45	Polyethylene (Medium Density)
23	Polyethylene-Low Density	46	ABS-Nylon Alloy

4.2. Spectra Collection

4.2.1. Resin Samples

Three MicroNIR™ OnSite spectrometers (Viavi Solutions Inc., Santa Rosa, CA, USA) in the range of 908–1676 nm were randomly picked to collect the spectra of the resin samples. The spectral bandwidth is ~1.1% of a given wavelength. Three kits of samples were measured in the diffuse reflection mode. A MicroNIR™ windowless collar was used to interface with the samples, which optimized the sample placement relative to the spectrometer. Each sample was placed between the windowless collar of the MicroNIR™ spectrometer and a 99% diffuse reflection standard (Spectralon®, LabSphere, North Sutton, NH, USA). The reason for using the Spectralon® behind each sample was to return signal back to the spectrometer, particularly for very transparent samples, in order to improve the signal-to-noise ratio.

Each sample was scanned in five specified locations to account for the most variation in sample shape and thickness. In addition, at each position the sample was scanned in two orientations with respect to the MicroNIR™ lamps to account for any directionality in the structure of the molding. For each position and orientation, three replicate scans were acquired, totaling thirty scans per sample, per spectrometer. The MicroNIR™ spectrometer was re-baselined after every ten samples, using a 99% diffuse reflectance reference scan (Spectralon®), as well as a lamps-on dark scan, in which nothing was placed in front of the spectrometer. Each sample was measured by all three spectrometers following the same protocol.

4.2.2. Pharmaceutical Samples

Each of the 48 samples were placed in individual glass vials, and their spectra were collected by three randomly picked MicroNIR™ 1700ES spectrometers in the range of 908–1676 nm using the MicroNIR™ vial-holder accessory. The spectral bandwidth is ~1.1% of a given wavelength. In this measurement setup, the samples were scanned from the bottom of the vial in the diffuse reflection mode.

Each sample was scanned twenty times using each MicroNIR™ spectrometer. The sample was rotated in the vial-holder between every scan to account for sample placement variation, as well as the non-uniform thickness of the vial. Before every new sample, the MicroNIR™ spectrometer was re-baselined by scanning a 99% diffuse reflectance reference (Spectralon®), as well as a lamps-on dark scan, which consisted of an empty vial in place of a sample. Each sample was measured by all three spectrometers following the same protocol.

4.3. Data Processing and Multivariate Analysis

4.3.1. Polymer Classification

All steps of spectral processing and chemometric analysis were performed using MATLAB (The MathWorks, Inc., Natick, MA). All spectra collected were pretreated using Savitzky-Golay first derivative followed by standard normal variate (SNV).

PLS-DA, SIMCA, TreeBagger, SVM and hier-SVM were applied to preprocessed datasets. Autoscaling was performed when running these algorithms. To implement PLS-DA, the number of PLS factors was chosen by training set cross validation and the same number was used for all classes. To implement SIMCA, the number of principal components (PC) was optimized for each class by training set cross validation. No optimization was performed for TreeBagger, SVM and hier-SVM, and the default settings were used. For TreeBagger, the number of decision trees in the ensemble was set to be 50. Since random selection of sample subsets and variables is involved when running TreeBagger, there are small differences in the results from run to run. To avoid impacts from these differences, all the TreeBagger results were based on the mean of 10 runs. For SVM algorithms, the linear kernel with parameter C of 1 was used.

For the same-unit-same-kit performance, the models built with data collected from four locations on each sample in one resin kit by one spectrometer were used to predict data collected from the other location on each sample in the same resin kit by the same spectrometer. For the same-unit-cross-kit performance, the models built with all the data collected from one resin kit by one spectrometer were used to predict all the data collected from a different resin kit by the same spectrometer. For the cross-unit-same-kit performance, the models built with all the data collected from one resin kit by one spectrometer were used to predict all the data collected from the same resin kit by a different spectrometer. For the cross-unit-cross-kit performance, the models built with all the data collected from one resin kit by one spectrometer were used to predict all the data collected from a different resin kit by a different spectrometer.

4.3.2. API Quantification

All steps of spectral processing and chemometric analysis were performed using MATLAB. Some functions in PLS_Toolbox (Eigenvector Research, Manson, WA, USA) were called in the MATLAB code. To develop the calibration models, 38 out of the 48 samples were selected as the calibration samples via the Kennard-Stone algorithm based on the respective API concentration. The remaining 10 samples were used as the validation samples. The preprocessing procedure was optimized for each API separately based on the calibration set cross validation. The same preprocessing procedure was used on all three instruments for the same API. PLS models were developed using the corresponding preprocessed datasets for each API.

To evaluate the same-unit performance, the model built on one instrument was used to predict the validation set collected by the same instrument. To evaluate the cross-unit performance without calibration transfer, the model built on one instrument was used to predict the validation set collected by the other instruments.

For calibration transfer demonstration, Unit 1 was used as the master instrument, and Unit 2 and Unit 3 were used as the slave instruments. Eight transfer samples were selected from the calibration samples with the Kennard-Stone algorithm. To perform bias correction, bias was determined using the transfer data collected by the slave instrument, and the bias was applied to the predicted values using the validation data collected by the slave instrument. To perform PDS, the window size was optimized based on RMSEP, and the corresponding lowest RMSEP was reported in this study. To perform GLS,

Molecules **2019**, *24*, 1997

parameter *a* was optimized based on RMSEP, and the corresponding lowest RMSEP was reported in this study.

5. Conclusions

In this study, direct model transferability was investigated when multiple MicroNIR™ spectrometers were used. As demonstrated by the polymer classification example, high prediction success rates can be achieved for the most stringent cross-unit-cross-kit cases with multiple algorithms including the widely used SIMCA method. Better performance was achieved with SVM algorithms, especially when a hierarchical approach was used (hier-SVM). As demonstrated by the API quantification example, low prediction errors were achieved for the cross-unit cases with PLS models. These results indicate that the direct use of a model developed on one MicroNIR™ spectrometer on the other MicroNIR™ spectrometers is possible. The successful direct model transfer is enabled by the robust design of the MicroNIR™ hardware and will make deployment of multiple spectrometers for various applications more manageable and economical.

Supplementary Materials: The supplementary materials on reproducibility of MicroNIR™ products are available online http://www.mdpi.com/1420-3049/24/10/1997/s1. Figure S1: MicroNIR™ OnSite manufacturing data demonstrating instrument-to-instrument reproducibility, Figure S2: MicroNIR™ 1700ES manufacturing data demonstrating instrument-to-instrument reproducibility.

Author Contributions: Conceptualization, L.S.; data curation, V.S.; formal analysis, L.S. and C.H.; investigation, L.S., C.H. and V.S.; methodology, L.S., C.H. and V.S; software, L.S. and C.H.; validation, L.S. and C.H.; visualization, L.S.; writing—original draft, L.S. and V.S.; writing—review & editing, L.S., C.H. and V.S.

Funding: This research received no external funding.

Acknowledgments: The authors would like to thank Heinz W. Siesler at University of Duisburg-Essen, Germany, for providing the pharmaceutical samples used in this study.

Conflicts of Interest: The authors declare no conflicts of interest.

References

1. Yan, H.; Siesler, H.W. Hand-held near-infrared spectrometers: State-of-the-art instrumentation and practical applications. *NIR News* **2018**, *29*, 8–12. [CrossRef]
2. Dos Santos, C.A.T.; Lopo, M.; Páscoa, R.N.M.J.; Lopes, J.A. A Review on the Applications of Portable Near-Infrared Spectrometers in the Agro-Food Industry. *Appl. Spectrosc.* **2013**, *67*, 1215–1233. [CrossRef]
3. Santos, P.M.; Pereira-Filho, E.R.; Rodriguez-Saona, L.E. Application of Hand-Held and Portable Infrared Spectrometers in Bovine Milk Analysis. *J. Agric. Food Chem.* **2013**, *61*, 1205–1211. [CrossRef]
4. Alcalà, M.; Blanco, M.; Moyano, D.; Broad, N.; O'Brien, N.; Friedrich, D.; Pfeifer, F.; Siesler, H. Qualitative and quantitative pharmaceutical analysis with a novel handheld miniature near-infrared spectrometer. *J. Near Infrared Spectrosc.* **2013**, *21*, 445. [CrossRef]
5. Paiva, E.M.; Rohwedder, J.J.R.; Pasquini, C.; Pimentel, M.F.; Pereira, C.F. Quantification of biodiesel and adulteration with vegetable oils in diesel/biodiesel blends using portable near-infrared spectrometer. *Fuel* **2015**, *160*, 57–63. [CrossRef]
6. Risoluti, R.; Gregori, A.; Schiavone, S.; Materazzi, S. "Click and Screen" Technology for the Detection of Explosives on Human Hands by a Portable MicroNIR–Chemometrics Platform. *Anal. Chem.* **2018**, *90*, 4288–4292. [CrossRef]
7. Pederson, C.G.; Friedrich, D.M.; Hsiung, C.; von Gunten, M.; O'Brien, N.A.; Ramaker, H.-J.; van Sprang, E.; Dreischor, M. Pocket-size near-infrared spectrometer for narcotic materials identification. In *Proceedings Volume 9101, Proceedings of the Next-Generation Spectroscopic Technologies VII, SPIE Sensing Technology + Applications, Baltimore, MD, USA, 10 June 2014*; Druy, M.A., Crocombe, R.A., Eds.; International Society for Optics and Photonics: Bellingham, WA, USA, 2014; pp. 91010O-1–91010O-11. [CrossRef]
8. Wu, S.; Panikar, S.S.; Singh, R.; Zhang, J.; Glasser, B.; Ramachandran, R. A systematic framework to monitor mulling processes using Near Infrared spectroscopy. *Adv. Powder Technol.* **2016**, *27*, 1115–1127. [CrossRef]

9. Galaverna, R.; Ribessi, R.L.; Rohwedder, J.J.R.; Pastre, J.C. Coupling Continuous Flow Microreactors to MicroNIR Spectroscopy: Ultracompact Device for Facile In-Line Reaction Monitoring. *Org. Process Res. Dev.* **2018**, *22*, 780–788. [CrossRef]
10. Feudale, R.N.; Woody, N.A.; Tan, H.; Myles, A.J.; Brown, S.D.; Ferré, J. Transfer of multivariate calibration models: A review. *Chemom. Intell. Lab. Syst.* **2002**, *64*, 181–192. [CrossRef]
11. Workman, J.J. A Review of Calibration Transfer Practices and Instrument Differences in Spectroscopy. *Appl. Spectrosc.* **2018**, *72*, 340–365. [CrossRef]
12. Wang, Y.; Veltkamp, D.J.; Kowalski, B.R. Multivariate instrument standardization. *Anal. Chem.* **1991**, *63*, 2750–2756. [CrossRef]
13. Wang, Y.; Lysaght, M.J.; Kowalski, B.R. Improvement of multivariate calibration through instrument standardization. *Anal. Chem.* **1992**, *64*, 562–564. [CrossRef]
14. Wang, Z.; Dean, T.; Kowalski, B.R. Additive Background Correction in Multivariate Instrument Standardization. *Anal. Chem.* **1995**, *67*, 2379–2385. [CrossRef]
15. Du, W.; Chen, Z.-P.; Zhong, L.-J.; Wang, S.-X.; Yu, R.-Q.; Nordon, A.; Littlejohn, D.; Holden, M. Maintaining the predictive abilities of multivariate calibration models by spectral space transformation. *Anal. Chim. Acta* **2011**, *690*, 64–70. [CrossRef]
16. Martens, H.; Høy, M.; Wise, B.M.; Bro, R.; Brockhoff, P.B. Pre-whitening of data by covariance-weighted pre-processing. *J. Chemom.* **2003**, *17*, 153–165. [CrossRef]
17. Cogdill, R.P.; Anderson, C.A.; Drennen, J.K. Process analytical technology case study, part III: Calibration monitoring and transfer. *AAPS Pharm. Sci. Tech.* **2005**, *6*, E284–E297. [CrossRef]
18. Shi, G.; Han, L.; Yang, Z.; Chen, L.; Liu, X. Near Infrared Spectroscopy Calibration Transfer for Quantitative Analysis of Fish Meal Mixed with Soybean Meal. *J. Near Infrared Spectrosc.* **2010**, *18*, 217–223. [CrossRef]
19. Salguero-Chaparro, L.; Palagos, B.; Peña-Rodríguez, F.; Roger, J.M. Calibration transfer of intact olive NIR spectra between a pre-dispersive instrument and a portable spectrometer. *Comput. Electron. Agric.* **2013**, *96*, 202–208. [CrossRef]
20. Krapf, L.C.; Nast, D.; Gronauer, A.; Schmidhalter, U.; Heuwinkel, H. Transfer of a near infrared spectroscopy laboratory application to an online process analyser for in situ monitoring of anaerobic digestion. *Bioresour. Technol.* **2013**, *129*, 39–50. [CrossRef]
21. Myles, A.J.; Zimmerman, T.A.; Brown, S.D. Transfer of Multivariate Classification Models between Laboratory and Process Near-Infrared Spectrometers for the Discrimination of Green Arabica and Robusta Coffee Beans. *Appl. Spectrosc.* **2006**, *60*, 1198–1203. [CrossRef]
22. Milanez, K.D.T.M.; Silva, A.C.; Paz, J.E.M.; Medeiros, E.P.; Pontes, M.J.C. Standardization of NIR data to identify adulteration in ethanol fuel. *Microchem. J.* **2016**, *124*, 121–126. [CrossRef]
23. Ni, W.; Brown, S.D.; Man, R. Stacked PLS for calibration transfer without standards. *J. Chemom.* **2011**, *25*, 130–137. [CrossRef]
24. Lin, Z.; Xu, B.; Li, Y.; Shi, X.; Qiao, Y. Application of orthogonal space regression to calibration transfer without standards. *J. Chemom.* **2013**, *27*, 406–413. [CrossRef]
25. Kramer, K.E.; Morris, R.E.; Rose-Pehrsson, S.L. Comparison of two multiplicative signal correction strategies for calibration transfer without standards. *Chemom. Intell. Lab. Syst.* **2008**, *92*, 33–43. [CrossRef]
26. Sun, L.; Hsiung, C.; Pederson, C.G.; Zou, P.; Smith, V.; von Gunten, M.; O'Brien, N.A. Pharmaceutical Raw Material Identification Using Miniature Near-Infrared (MicroNIR) Spectroscopy and Supervised Pattern Recognition Using Support Vector Machine. *Appl. Spectrosc.* **2016**, *70*, 816–825. [CrossRef] [PubMed]
27. Ståhle, L.; Wold, S. Partial least squares analysis with cross-validation for the two-class problem: A Monte Carlo study. *J. Chemom.* **1987**, *1*, 185–196. [CrossRef]
28. Wold, S. Pattern recognition by means of disjoint principal components models. *Pattern Recognit.* **1976**, *8*, 127–139. [CrossRef]
29. Breiman, L. Random Forrest. *Mach. Learn.* **2001**, *45*, 5–32. [CrossRef]
30. Cortes, C.; Vapnik, V. Support-Vector Networks. *Mach. Learn.* **1995**, *20*, 273–297. [CrossRef]
31. Boser, B.E.; Guyon, I.M.; Vapnik, V.N. A training algorithm for optimal margin classifiers. In Proceedings of the Fifth Annual Workshop on Computational Learning Theory-COLT'92, Pittsburgh, PA, USA, 27 July 1992; pp. 144–152.
32. Blanco, M.; Bautista, M.; Alcalà, M. API Determination by NIR Spectroscopy Across Pharmaceutical Production Process. *AAPS Pharm. Sci. Tech.* **2008**, *9*, 1130–1135. [CrossRef]

33. Swarbrick, B. The current state of near infrared spectroscopy application in the pharmaceutical industry. *J. Near Infrared Spectrosc.* **2014**, *22*, 153–156. [CrossRef]
34. Gouveia, F.F.; Rahbek, J.P.; Mortensen, A.R.; Pedersen, M.T.; Felizardo, P.M.; Bro, R.; Mealy, M.J. Using PAT to accelerate the transition to continuous API manufacturing. *Anal. Bioanal. Chem.* **2017**, *409*, 821–832. [CrossRef]
35. Kennard, R.W.; Stone, L.A. Computer Aided Design of Experiments. *Technometrics* **1969**, *11*, 137–148. [CrossRef]
36. Sorak, D.; Herberholz, L.; Iwascek, S.; Altinpinar, S.; Pfeifer, F.; Siesler, H.W. New Developments and Applications of Handheld Raman, Mid-Infrared, and Near-Infrared Spectrometers. *Appl. Spectrosc. Rev.* **2011**, *47*, 83–115. [CrossRef]
37. Bland, J.M.; Altman, D.G. Measuring agreement in method comparison studies. *Stat. Methods Med. Res.* **1999**, *8*, 135–160. [CrossRef]
38. Bennett, K.P.; Campbell, C. Support Vector Machines: Hype or Hallelujah? *Sigkdd Explor. Newslett.* **2000**, *2*, 1–13. [CrossRef]
39. Briand, B.; Ducharme, G.R.; Parache, V.; Mercat-Rommens, C. A similarity measure to assess the stability of classification trees. *Comput. Stat. Data Anal.* **2009**, *53*, 1208–1217. [CrossRef]
40. Wise, B.M.; Roginski, R.T. A Calibration Model Maintenance Roadmap. *IFAC-PapersOnLine* **2015**, *48*, 260–265. [CrossRef]
41. Petersen, L.; Esbensen, K.H. Representative process sampling for reliable data analysis—A tutorial. *J. Chemom.* **2005**, *19*, 625–647. [CrossRef]
42. Romañach, R.; Esbensen, K. Sampling in pharmaceutical manufacturing—Many opportunities to improve today's practice through the Theory of Sampling (TOS). *TOS Forum* **2015**, *4*, 5–9. [CrossRef]
43. The Effects of Sample Presentation in Near-Infrared (NIR) Spectroscopy. Available online: https://www.viavisolutions.com/en-us/literature/effects-sample-presentation-near-infrared-nir-spectroscopy-application-notes-en.pdf (accessed on 12 March 2019).
44. MicroNIRTM Sampling Distance. Available online: https://www.viavisolutions.com/en-us/literature/micronir-sampling-distance-application-notes-en.pdf (accessed on 12 March 2019).

Sample Availability: Not available.

molecules

MDPI

Article

Investigations into the Performance of a Novel Pocket-Sized Near-Infrared Spectrometer for Cheese Analysis

Verena Wiedemair [1], Dominik Langore [1], Roman Garsleitner [2], Klaus Dillinger [2] and Christian Huck [1],*

[1] CCB—Center for Chemistry and Biomedicine, Institute of Analytical Chemistry and Radiochemistry, University of Innsbruck, Innrain 80/82; 6020 Innsbruck, Austria; verena.wiedemair@uibk.ac.at (V.W.); dominik.langore@student.uibk.ac.at (D.L.)

[2] Chemical devision, HBLFA für Landwirtschaft und Ernährung, Lebensmittel und Biotechnologie Tirol, Rotholz 50a, 6200 Strass im Zillertal, Austria; roman.garsleitner@hblfa-tirol.at (R.G.); klaus.dillinger@hblfa-tirol.at (K.D.)

* Correspondence: christian.w.huck@uibk.ac.at; Tel.: +43-512-507-57304

Received: 6 December 2018; Accepted: 22 January 2019; Published: 24 January 2019

Abstract: The performance of a newly developed pocket-sized near-infrared (NIR) spectrometer was investigated by analysing 46 cheese samples for their water and fat content, and comparing results with a benchtop NIR device. Additionally, the automated data analysis of the pocket-sized spectrometer and its cloud-based data analysis software, designed for laypeople, was put to the test by comparing performances to a highly sophisticated multivariate data analysis software. All developed partial least squares regression (PLS-R) models yield a coefficient of determination (R^2) of over 0.9, indicating high correlation between spectra and reference data for both spectrometers and all data analysis routes taken. In general, the analysis of grated cheese yields better results than whole pieces of cheese. Additionally, the ratios of performance to deviation (RPDs) and standard errors of prediction (SEPs) suggest that the performance of the pocket-sized spectrometer is comparable to the benchtop device. Small improvements are observable, when using sophisticated data analysis software, instead of automated tools.

Keywords: NIR; SCiO; pocket-sized spectrometer; cheese; fat; moisture; multivariate data analysis

1. Introduction

A wide variety of cheese and cheese products can be purchased in stores around the world. Cheese is an important source of nutrients, such as fat and protein [1], and is consumed worldwide with an annual production of approximately 23 million tonnes in 2014 [2]. Quality control is vital, to ensure food safety and to protect consumer interest. Hence, a wide variety of physico-chemical analyses were developed to determine pH, fat, nitrogen fractions, volatile fatty acids and others [3,4]. Traditional methods to determine these nutrients have some disadvantages. They are often time consuming, expensive and have a limited sample throughput [3,5]. Furthermore, trained personnel is needed to operate the machines and execute the analyses [6].

That is why, fast and non-destructive spectral methods were developed to measure main components, such as fat, protein and moisture, of cheese [3,7–9]. More recent approaches also focus on minor components, like vitamins, minerals and carotenoids [10,11]. But lately, near-infrared spectroscopy (NIRS) was also used to identify sensory properties, as well as the origin of cheese [12–16]. Cheese is not spatially homogeneous, which poses a challenge for spectral analysis [10].

NIRS is widely applied in food analyses [17–20], as well as in other fields including pharmaceutical sciences and petrochemistry [21–24]. NIR excites molecular vibrations and thus overtones and combination bands can be observed in a NIR spectrum [15]. The main advantages of NIRS are its comparatively low cost, fast measurements and easy handling [25,26]. Furthermore, it is possible to miniaturize NIR technology and thus reduce costs and weight, as well as improve consumer-friendliness [27,28]. With the miniaturization process beginning already a decade ago, its main challenge was to preserve spectral performance in terms of wavenumber range and resolution [28]. Thanks to technical advances, like the micro electro-mechanical system (MEMS) and the linear variable filter (LVF) technology, the performance of miniaturized devices increased greatly [29]. These technologies are implemented in various miniaturized devices designed for pharmaceutical and chemical industries. But currently, the miniaturization process is even more sophisticated, which makes a pocket-sized NIR spectrometer possible for the first time. Multiple companies (e.g., Tellspec, Consumer Physics and others) have launched NIR spectrometers so small, they fit into the palm of a hand. Additionally, the implementation of those spectrometers into mobile phones is not a fantasy of the future, but already possible now [30]. Small spectrometers like that are usually not targeted to industries, but to consumers, who want to make educated, science-based food choices. But this device could also be used at food competitions, where the judges usually have to rely on the claims of producer and who were, until now, unable to verify those claims. Hence, the operation of this new generation of miniaturized spectrometers is easy and intuitive. The devices can be operated without any knowledge in the field of chemistry or physics.

But the question of the performance of those pocket-sized spectrometers remains. Since they are not targeted at scientists and researchers, there is little knowledge about the performance of these spectrometers. This is why the present study aims at shedding light upon some aspects of the performance of a pocket-sized molecular sensor—The SCiO (Consumer Physics, Tel Aviv, Israel). Only few studies have been published using this device and thus more information is needed [31,32].

SCiO is operated with a smartphone via Bluetooth. SCiO can be controlled using either the SCiO or the SCiO Lab app. The SCiO app contains a set of pre-established calibration models for, for example, calories and water content of fruits and vegetables; and sugar, fat and calories for chocolates. It also contains a pre-established app for water, protein and fat content of dairy products like cheese, yoghurts and puddings. Hence, once a sample is measured all values are directly given to the user, without presenting a spectrum or a model. The main advantage of this app is its easy handling and intuitiveness. A disadvantage of the SCiO app is that only products for which a calibration model already exists can be measured. If the user wants to measure another product, the SCiO Lab app has to be used. There, spectra can be recorded and later be analysed using the cloud-based web application, SCiO Lab.

The current study investigates different cheese samples in terms of their fat and moisture content using a benchtop NIR spectrometer, as well as SCiO. The performance of both devices is compared using statistical parameters, such as the standard error of prediction and the coefficient of determination (R^2).

2. Results

2.1. Near-Infrared Spectroscopy

The NIR region is often divided into three sub-regions: Region I ranges from 800–1200 nm (12500–8500 cm^{-1}) and is also called the "Herschel" region. It is the only region where electronic transitions can be observed. Furthermore, the Herschel region contains overtones and combination bands. Region II is located from 1200–1800 nm (8500–5500 cm^{-1}) and mainly comprises first overtones. Region III ranges from 1800–2500 nm (5500–4000 cm^{-1}), where the combination band can be found [23].

As shown in Figure 1, the spectral range of the benchtop NIRFlex N-500 is 2500–1000 nm (10000–4000 cm^{-1}) and thus it mainly includes vibrations in region II and III. The benchtop device also reaches into the Herschel region; however, it does not cover it completely. SCiO on the other hand exclusively records spectra in region I, as its wavelength range reaches from 740–1070 nm (13514–9346 cm^{-1}). This also means that SCiO reaches into the visible part of the electromagnetic spectrum. This makes a comparison between the two devices interesting, as the recorded vibrations influence the performance of each spectrometer. The NIR region was chosen as the reference range for various reasons. First of all, molecular vibrations corresponding to water and fat are quite pronounced in this region. Secondly, the different colours of the cheese are irrelevant in this region. Thirdly, using SCiO and the NIRFlex N-500, the same measurement setup and mode could be used. Table 1 lists important peaks in the respective spectra and the corresponding vibrations.

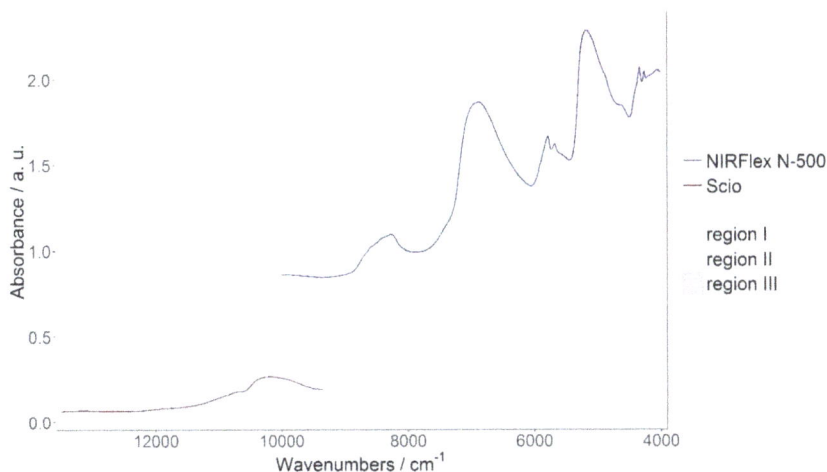

Figure 1. Averaged spectra of whole pieces of cheese of spectra recorded with SCiO (red) and NIRFlex N-500 (blue).

Table 1. Important peaks and their respective vibrations in the spectra recorded with the NIRFlex N-500 and SCiO [5,33].

Device	Vibration	Wavenumber/cm^{-1}	Wavelength/nm
NIRFlex N-500	C-H str. 2nd overtones	8888–8068	1125–1240
	O-H str. 1st overtones N-H str. 1st overtones	7264–6068	1377–1648
	C-H str. 1st overtones	5856–5604	1708–1784
	Combination of O-H str. and O-H def., C=O str. 2nd overtones	5404–4784	1850–2090
SCiO	C-H str. 3rd overtones N-H str. 2nd overtones	10,834–10,660	923–938
	and O-H str. 2nd overtones	10,616–9506	942–1052

2.2. Multivariate Data Analysis

First, all models developed using The Unscrambler X version 10.5 are presented and evaluated. Then the data generated with the SCiO web-application will be examined in detail and, lastly, the performance of the pre-established models from the SCiO app will be discussed.

2.2.1. Regression Models Established with The Unscrambler X Version 10.5

For spectral data pre-treatment, the descriptive statistics tool, which is implemented in The Unscrambler X, was consulted and then a fitting pre-treatment was applied (see Section 4.3.1). Table 2 lists the important statistical parameters for the cross- and test set-validated models, calculated using the respective calibration sets for fat content. Figure 2 shows the PLS regressions for whole pieces of cheese for SCiO and the NIRFlex N-500.

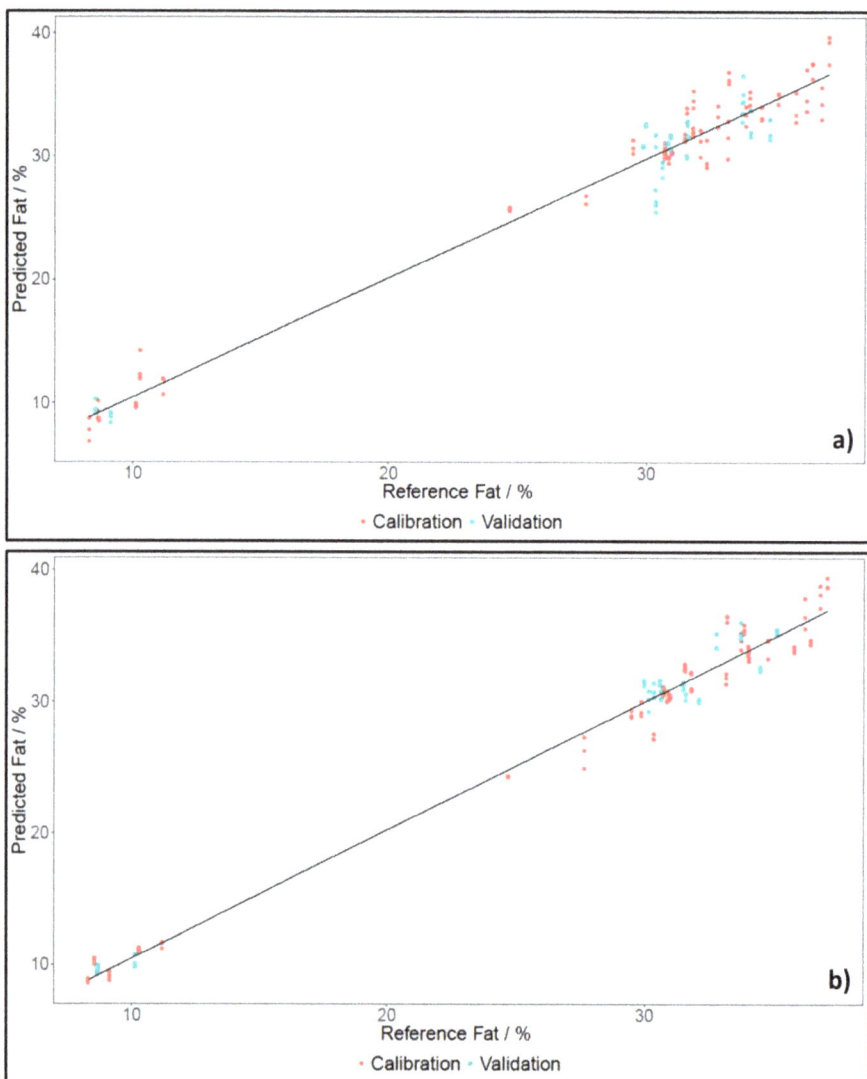

Figure 2. PLS regression of fat content of whole pieces of cheese, established using data of NIRFlex N-500 (**a**) and SCiO (**b**).

Table 2. Parameters of the established PLS-R models for fat content. CV denotes cross-validated models, whereas TV refers to test set-validated regressions.

Spectrometer	State of the Cheese	R^2 (CV)	RMSECV/%	PC (CV)	R^2 (TV)	RMSEP/%	Bias (TV)	PC (TV)	RPD
NIRFlex N-500	Whole pieces	0.9726	1.5711	2	0.9431	1.8964	−0.3369	2	5.109
	Grated cheese	0.9930	0.7845	2	0.9913	0.7676	0.3719	2	14.022
SCiO	Whole pieces	0.9801	1.2466	2	0.9838	1.1874	0.1634	2	7.754
	Grated cheese	0.9838	1.0527	2	0.9940	0.8194	0.1776	2	10.398

R^2—Coefficient of determination; RMSECV—Root mean square error of cross validation; PC—Principle component; RMSEP—Root mean square error of prediction; RPD—Ratio of performance to deviation.

For whole pieces of cheese and grated cheese measured with the NIRFlex N-500 and SCiO, two principle components (PCs) were needed to develop calibration models for fat content. Cheese has a high fat content, which is why a low number of principle components for calibration models can be expected. The number of PCs used was determined using the variance plot, which showed how much of the original information is included in the respective PC. Since the same number of PCs were used for calibrations established with SCiO and the benchtop device, as well as for grated and whole pieces of cheese, the results had a high comparability.

The coefficients of determination for all models were between 0.9431 and 0.9940. This indicated a high correlation between the spectra and the reference data. The errors of the cross-validated calibration models, called root mean square errors of cross validation (RMSECV), were between 0.78% and 1.57%, which was acceptable, considering that the average of the reference data was 29.17%. When investigating the test set-validated models, it immediately meets the eye that the Biases were not zero, which is to be expected when using an independent test set. Furthermore, the root mean square error of prediction (RMSEP) values were between 0.77% and 1.90%. All but one RMSEPs were lower than their respective RMSECV. This was most likely due to the Kennard–Stone sample selection, whereby a set of samples was chosen to best represent the multivariate space of the data. Because of that, high value samples, as well as low value samples, were selected in order to best describe the multivariate space. Hence, the error of the calibration set was higher, since it included more extreme values. The independent test set, on the other hand, could then easily be fitted into the well-described multivariate space. Furthermore, all RMSECV values were close to their respective RMSEP values, indicating a robust model.

In general, the best model for the estimation of fat content was the model developed for NIRFlex N-500 data when measuring grated pieces of cheese. The cross-validation and the independent test set-validation both had a coefficient of determination of over 0.99, and the RMSECV and RMSEP were both under 1%. Furthermore, the RMSECV and RMSEP values were close, indicating a robust model. Additionally, the statistical parameters suggested that the analysis of fat content for grated pieces of cheese was slightly better. This was most likely due to the fact that whole pieces of cheese can be spatially inhomogeneous.

Table 3 lists the results for moisture content for grated cheese and whole pieces measured with the NIRFlex N-500 and SCiO. Figure 3 shows the PLS regressions for whole pieces of cheese for SCiO and the NIRFlex N-500.

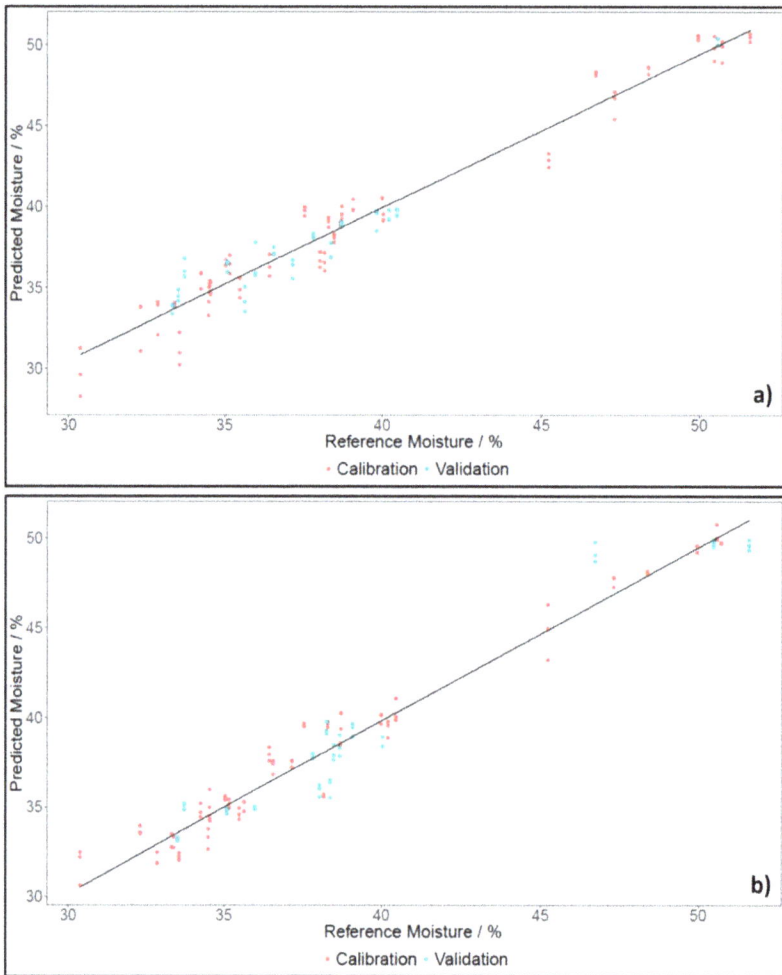

Figure 3. PLS regression of moisture content of whole pieces of cheese, established using data of NIRFlex N-500 (**a**) and SCiO (**b**).

Table 3. Statistical parameterd of the established PLS-R models for moisture content. CV denotes cross-validated models, whereas TV refers to test set validated regressions.

Spectrometer	State of the Cheese	R^2 (CV)	RMSECV/%	PC (CV)	R^2 (TV)	RMSEP/%	Bias (TV)	PC (TV)	RPD
NIRFlex N-500	Whole pieces	0.9598	1.2239	3	0.9376	1.0960	0.0408	3	5.597
	Grated cheese	0.9873	0.6868	3	0.9561	0.9337	−0.1843	3	6.697
SCiO	Whole pieces	0.9659	1.0407	2	0.9394	1.1357	−0.3763	2	4.341
	Grated cheese	0.9637	1.0400	2	0.9327	1.7147	0.1297	2	3.208

R^2—Coefficient of determination; RMSECV—Root mean square error of cross validation; PC—Principle component; RMSEP—Root mean square error of prediction; RPD—Ratio of performance to deviation.

For measurements with the NIRFlex N-500, three principle components were needed to develop models for moisture content for whole pieces and grated cheese. For data recorded with SCiO, two principle components were needed to develop models for moisture content. The benchtop device recorded spectra in a broader range than the SCiO, which is why multiple O-H vibrations could be found. However, these vibrations did not only stem from water, but also from other components. The SCiO on the other hand, recorded spectra in a very narrow range, hence only one O-H vibration was visible. Since the spectra of the NIRFlex N-500 were more complex, more principle components had to be used. The number of PCs did not differ greatly though, hence the results could still be compared.

The coefficients of determination for all models were between 0.9327 and 0.9873. This indicated a high correlation between the spectra and the moisture content. The RMSECV values of the cross-validated models were around 1%, with the highest being 1.22%, and the lowest being 0.69%. The RMSEP values for the test set-validated models were also around 1%, with the exception of RMSEP for data recorded with SCiO for grated cheese, which gave a value of 1.71%. Only one RMSEP value was lower than the respective RMSECV value. In general, an error of about 1% for moisture content was quite low, considering that the mean of the reference data was 38.42%.

The difference in prediction accuracy for moisture content of grated cheese and whole pieces of cheese was smaller than for fat content. The prediction of moisture content seemed to work better on grated cheese when working with the NIRFlex N-500; however, when using SCiO, whole pieces showed better prediction results.

2.2.2. Regression Models Established with the SCiO Lab Web Application

SCiO only offers a limited amount of statistical properties. It does not list coefficients of determination for test set-validated models or the Bias of any model. The algorithm for the validation process is proprietary and is; therefore, not accessible for users. The calibration model was validated using leave-one-out cross-validation (LOOCV) and then a test set was also used to validate the model. Table 4 lists the result for the analysis of SCiO data, recorded with the SCiO Lab app, using the SCiO Lab web application.

Table 4. Statistical parameters of the established PLS-R models for moisture and fat content. CV denotes cross-validated models.

Content	State of the Cheese	PC (CV)	R^2 (CV)	RMSE/%	SEP/%	RPD
Moisture	Whole pieces	4	0.972	0.949	1.050	5.453
	Grated cheese	4	0.977	0.834	1.102	5.034
Fat	Whole pieces	4	0.988	0.950	0.785	11.448
	Grated cheese	4	0.982	1.118	0.779	10.779

PC—Principle components; R^2—Coefficient of determination; RMSE—Root mean square error; SEP—Standard error of prediction; RPD—Ratio of performance to deviation.

The coefficients of determination for all cross-validated models were between 0.972 and 0.988, indicating a high correlation between spectra and reference values. Additionally, all root mean square errors (RMSEs) were around 1%, with the lowest being 0.834% and the highest being 0.950%. All but one standard errors of prediction (SEPs) were lower than their respective RMSEs. The lowest SEP was achieved for the prediction of fat content in grated cheese, at 0.779%. The highest SEP was 1.102%. These results were in accordance with the analysis of the same data set using the software The Unscrambler X version 10.5. However, interestingly, the prediction of the fat content with the benchtop device yielded a much higher SEP. Furthermore, different spectral pre-treatments were needed. This is most likely due to the limited options in SCiO Lab. Additionally, SCiO Lab suggested four principle components for all models. Conducting a manual analysis with the Unscrambler X; however, showed that only two PCs explain over 90% of the variance of the original data set. This suggests that the automated algorithm in SCiO Lab takes in too many PCs. This might also be the reason why some of the SEPs were lower for the SCiO Lab web application.

Additionally, it was attempted to imitate the results from the SCiO Lab web application with the Unscrambler X. However, since it was unknown how many smoothing points the SCiO Lab web application applies, results could not be duplicated. The SEP for fat content for whole pieces of cheese was 0.844%, and for water content, 0.825%. This means that for moisture the imitated results, using Unscrambler X, yielded better results, but for fat content the SCiO Lab web application yielded better results. Looking at grated cheese, the SEP for fat content was 0.771% and for moisture content it was 1.400%. This means that the SEP for moisture content was higher than predicted with the SCiO Lab web application, but the fat content was almost the same.

2.2.3. Results from the Pre-Established Model in the SCiO App

For the analysis of the performance of the pre-established model in the SCiO app, all values were registered in Microsoft Excel and then SEPs and Biases were calculated (Table 5).

Table 5. SEPs and Biases for the pre-established "dairy products" model in the SCiO app.

Content	State of the Cheese	SEP/%	Bias/%	RPD
Moisture	Whole pieces	1.349	2.266	4.021
	Grated cheese	1.159	2.443	4.681
Fat	Whole pieces	1.064	−0.826	7.832
	Grated cheese	1.218	−0.789	6.844

SEP—Standard error of prediction; RPD—Ratio of performance to deviation.

The Biases for moisture content were quite high, indicating a systematic shift between the data in the pre-established model and the recorded data. The SEPs were all around 1%, but slightly higher than for the self-developed models. This is most likely due to the fact that the model of the SCiO app was developed for "dairy products", including yoghurt, cheese and puddings.

3. Discussion

Reviewing all SEPs and RPDs in Tables 2–5, the qualities of the calibration model can be estimated. In the Supplementary Material, a table that compresses the most important information can be found. With the help of RPD, the quality of the model can be estimated. RPDs below 2 are not sufficient, whereas an RPD between 2 and 3 is adequate for screening. RPDs between 3 and 5 are satisfactory. If a value over 5 is achieved, the model is estimated to be good. A RPD over 10 indicates an excellent model [34].

Four models reached an RPD value of over 10: The model developed using the SCiO data for data analysis with The Unscrambler X for fat content of grated cheese (10.398); and the models developed using SCiO data for data analysis with the SCiO Lab web application for fat content of whole pieces and grated cheese (11.448 and 10.799, respectively). The highest RPD was yielded for data of fat content collected with the NIRFlex N-500 of grated cheese (14.022). Furthermore, four models had a value between 3 and 5: The models developed using data recorded with SCiO for moisture content of whole pieces of cheese (4.341) and grated cheese (3.208). Additionally, the RPDs for the SCiO data, recorded and evaluated using the SCiO App, was around 4 for moisture content of whole pieces of cheese (4.021) and grated cheese (4.681). All remaining models had an RPD of 5 or higher, indicating that the performances were satisfactory or even good.

Regarding the SEP values, the results for fat content were, overall, better than for moisture. Figure 2 in Section 2.2.1 shows a gap between cheeses with a low and a high fat content. This may have contributed to an enhancement of results because the range of fat content was broadened, which was beneficial, considering how the SEP was calculated. Unfortunately, cheese with a fat content between 12% and 25% were unavailable.

Overall, the measurements of grated cheese seemed to work better, most likely due to the more homogenous nature of grated cheese. However, results did not improve too much, which leads to the question as whether or not the sample preparation is really necessary.

4. Materials and Methods

4.1. Sample Management and Reference Data

Forty-six cheese samples were analysed for their fat and moisture content using NIRS, as well as traditional wet chemical methods. Twenty samples were classified as hard cheese according to their water content of the fat-free mass, the other 26 samples were classified as semi-hard cheese [35]. Upon receipt, a part of each sample was grated and prepared for reference analysis. The remaining whole piece, as well as some grated cheese, was passed on for spectral analysis. The reference data for fat was collected using the Van Gulik method [36], and the moisture content was calculated from dry mass, which was determined using gravimetry [37]. Great care was taken to always uphold the cold chain and thus keep the samples fresh and unaltered. NIR measurements were conducted for grated cheese, as well as whole pieces using two different NIR spectrometers.

4.2. Near-Infrared Spectroscopy

All samples were measured, as grated cheese and whole pieces, with the NIRFlex N-500 (Büchi, Flawil, Switzerland) and SCiO (Consumer Physics, Tel Aviv, Isreal). The latter is a pocket-sized molecular sensor and was launched in 2016. The former is a modular benchtop device.

For measurements of the whole pieces of cheese, the NIRFlex N-500 was operated with the Fibre Optics Solids module. The fibre had an outer diameter of 4 mm and a spectral resolution of 8 cm^{-1}. The digital resolution was 4 cm^{-1}. All samples were measured at four spots three times, in the range 800–2500 nm (10000–4000 cm^{-1}), with 64 scans in diffuse reflection mode. Hence, a total of twelve spectra were received for each sample.

With SCiO, the pieces of cheese were measured at six spots one time in diffuse reflection mode, in the wavelength range 740–1070 nm (13514–9346 cm^{-1}), with the SCiO Lab app. The medium resolution of SCiO was 13 cm^{-1}, with the lowest resolution (18 cm^{-1}) being found at high wavenumbers and the highest resolution (9 cm^{-1}) at low wavenumbers.

For measurements of grated cheese, a cylindrical quartz cuvette (h = 25 mm, inner diameter = 31.6 mm) was filled with cheese and used for measurements. Thus, a constant measuring angle and method was assured. Grated cheese was analysed with the NIRFlex N-500 in diffuse reflection mode, with 64 scans in the wavelength range 800–2500 nm (10000–4000 cm^{-1}). The spectral resolution was again 8 cm^{-1} and the digital was 4 cm^{-1}. The cuvette was constantly rotated during measurement and each sample was analysed six times.

For analysis of grated cheese using SCiO, the quartz cuvette was put onto SCiO and each sample was measured six times using the SCiO Lab app. The cuvette was manually rotated between measurements. Spectra were recorded in the range 740–1070 nm (13514–9346 cm^{-1}) and the medium resolution was again 13 cm^{-1}. All measurements were taken in diffuse reflection mode.

Additionally, grated cheese and whole pieces were also measured with SCiO, using the pre-establish model from the SCiO app. Samples were again measured six times in the same manner as before.

4.3. Multivariate Data Analysis

Spectra recorded with the benchtop NIRFlex N-500 were analysed using the external multivariate data analysis software, The Unscrambler X version 10.5 (Camo Software, Oslo, Norway). Spectra recorded with SCiO, using the SCiO Lab app, were also evaluated using The Unscrambler X version 10.5; however, in addition, they were also analysed using the cloud-based web-application SCiO Lab (Consumer Physics, Tel Aviv, Israel). It offers a limited amount of multivariate data analysis options and was developed for laypeople. SCiO Lab, itself, suggests a certain set of pre-treatments; however, the user can also, him- or her-self, decide which pre-treatments to use when turning on the expert mode. In SCiO Lab it is possible to conduct a standard normal variate (SNV) and to calculate first and second derivatives. Furthermore, the average spectrum can be subtracted and the logarithm can be calculated. It is also possible to select a wavelength range. Additionally, the results given by the SCiO app were transferred to Microsoft Excel (Microsoft Corporation, Redmond, WA, USA) for statistical analysis.

4.3.1. Spectral Pre-Treatments

Spectra recorded with the benchtop device were evaluated using The Unscrambler X version 10.5. For whole pieces of cheese, the twelve spectra of one sample were first averaged by a factor of four, in order to obtain three representative spectra per sample. Next, descriptive statistics, which is an implemented tool in the software, was applied to identify necessary spectral pre-treatments. Standard normal variate (SNV) was applied to reduce multiplicative scatter effects. No other pre-treatments were necessary as they did not improve the model much. Regression models for fat and moisture content were then established. The regression model for fat content considered only the regions in the spectrum where C-H vibrations occur. The two regions considered reached from 6040 to 5440 cm^{-1} (1656–1838 nm) and from 8548 to 8076 cm^{-1} (1170–1238 nm). For moisture content, no wavenumber range was selected, as this did not improve results.

Evaluating spectra of grated cheese recorded with the benchtop NIRFlex N-500, the same route of identifying necessary spectral pre-treatments was taken. First, the six spectra of one sample were averaged by a factor of two, in order to again receive three spectra per sample. Next, descriptive statistics was applied and SNV was again used to remove scatter effects. The same wavenumber region as before was used to establish a regression model for fat content. For moisture content, the whole spectral range was used.

Spectra of whole pieces of cheese and grated cheese recorded with SCiO were first analysed with The Unscrambler X version 10.5. All spectra were first averaged by a factor of two, in order to obtain three spectra per sample. Next, descriptive statistics was applied. Like before, SNV was identified as necessary spectral pre-treatment to remove scatter noise. For fat and moisture content, the whole spectral range (740–1070 nm, 13514–9346 cm^{-1}) was used.

When using SCiO Lab for multivariate data analysis, different combinations of the provided spectral pre-treatment options were tried. For moisture and fat content for whole pieces and grated cheese, the same spectral pre-treatments were used, namely 1st derivative followed by SNV.

4.3.2. Regression Models

Using the software, The Unscrambler X version 10.5, for model development, the data sets for whole pieces of cheese and grated cheese, recorded with the NIRFlex N-500 and SCiO, were split into a calibration and a test set using Kennard–Stone sample selection [38], respectively. All calibration sets comprised two thirds of the data (31 samples, 93 spectra), and the test sets consisted of the remaining third (15 samples, 45 spectra). As SCiO Lab does not offer sophisticated data selection algorithms, the same samples as given in The Unscrambler X version 10.5 were selected in the cloud-based web application.

In The Unscrambler X version 10.5, partial least squares regression (PLS-R) with cross-validation was applied to the respective calibration sets to establish calibration models. The models were then validated using test set-validation with the pre-established test sets.

In the SCiO Lab web application, the calibration set was also used to build a PLS-R. Therefore, the data set was manually split into a calibration and validation set. Samples selected corresponded to the sample set created using the Kennard–Stone algorithm in The Unscrambler X version 10.5. Afterwards the developed model was validated using the test set.

For the evaluation of the established PLS-R models, different statistical quality parameters were consulted. R^2 is a measure of the linearity, RMSECV and RMSEP are indicators of the accuracy of the established model and the Bias can point to methodical errors. The RMSECV is similar to a standard deviation, showing how great the differences between expected and actual values are. The RMSEP denotes the difference between the actual reference value and the predicted value by the

Molecules **2019**, *24*, 428

established calibration model. Additionally, the RPDs [34] can be used to evaluate the applicability of established models. All mentioned parameters are listed in the result section for all developed PLS-R models.

5. Conclusions

This study examines the performance of a new hand-held NIR spectrometer, called SCiO, in comparison to a benchtop device. Forty-six different cheese samples—Grated cheese and whole pieces—were investigated in terms of their moisture and fat content. Additionally, different data analysis routes were taken in order to investigate if a deeper knowledge of chemometrics is necessary for the operation of SCiO, or if the limited tools implemented in the SCiO app and the SCiO Lab web application are sufficient to yield acceptable results.

In general, the analysis of cheese worked better when investigating grated cheese, instead of whole pieces. Furthermore, all calibration models showed high correlation between spectra and reference data. All RPDs indicated that the developed models for whole pieces of cheese were satisfactory to excellent.

This study shows that although overall results do improve by applying more sophisticated multivariate data analysis, the difference is only marginal. This implies that, in the near future, companies could easily use small and cheap NIR devices, with pre-established apps, for quality analysis. This is especially important for small businesses, as often appear in the cheese industry, because they are often unable to afford large instruments or an expert in data analysis.

Supplementary Materials: The supplementary materials are available online.

Author Contributions: Conceptualization, K.D. and C.H.; data curation, V.W., D.L. and R.G.; formal analysis, V.W., D.L. and R.G.; investigation, V.W., D.L. and R.G.; methodology, V.W., D.L. and R.G.; project administration, V.W.; resources, K.D. and C.H.; supervision, K.D. and C.H.; validation, V.W. and D.L.; visualization, V.W.; writing—Original draft, V.W.; writing—Review and editing, V.W., R.G. and C.H.

Funding: This research received no external funding.

Acknowledgments: The authors want to thank HBLFA für Landwirtschaft und Ernährung, Lebensmittel und Biotechnologie Tirol for sample management and preparation, as well as for reference measurements.

Conflicts of Interest: The authors declare no conflicts of interest.

References

1. United States Department of Agriculture. USDA Food Composition Database. 2016. Available online: https://ndb.nal.usda.gov/ndb/ (accessed on 8 January 2018).
2. Food and Agriculture Organization of the United Nations. Livestock Processed 2014. 2017. Available online: http://www.fao.org/faostat/en/#data/QP (accessed on 8 January 2018).
3. Karoui, R.; Mouazen, A.M.; Dufour, É.; Pillonel, L.; Schaller, E.; De Baerdemaeker, J.; Bosset, J.-O. Chemical characterisation of European Emmental cheeses by near infrared spectroscopy using chemometric tools. *Int. Dairy J.* **2006**, *16*, 1211–1217. [CrossRef]
4. Woodcock, T.; Fagan, C.C.; O'Donnell, C.P.; Downey, G. Application of Near and Mid-Infrared Spectroscopy to Determine Cheese Quality and Authenticity. *Food Bioprocess. Technol.* **2008**, *1*, 117–129. [CrossRef]
5. Lénárt, J.; Szigedi, T.; Dernovics, M.; Fodor, M. Application of FT-NIR spectroscopy on the determination of the fat and protein contents of lyophilized cheeses. *Acta Aliment.* **2012**, *41*, 351–362. [CrossRef]
6. Tao, F.; Ngadi, M. Applications of spectroscopic techniques for fat and fatty acids analysis of dairy foods. *Curr. Opin. Food Sci.* **2017**, *17*, 100–112. [CrossRef]
7. McQueen, D.H.; Wilson, R.; Kinnunen, A.; Jensen, E.P. Comparison of two infrared spectroscopic methods for cheese analysis. *Talanta* **1995**, *42*, 2007–2015. [CrossRef]
8. Rodriguez-Otero, J.L.; Hermida, M.; Cepeda, A. Determination of fat, protein, and total solids in cheese by near-infrared reflectance spectroscopy. *J. AOAC Int.* **1995**, *78*, 802–806.

9. Blazquez, C.; Downey, G.; O'Donnell, C.; O'Callaghan, D.; Howard, V. Prediction of Moisture, Fat and Inorganic Salts in Processed Cheese by near Infrared Reflectance Spectroscopy and Multivariate Data Analysis. *J. Near Infrared Spectrosc.* **2017**, *12*, 149–157. [CrossRef]

10. Holroyd, S.E. The Use of near Infrared Spectroscopy on Milk and Milk Products. *J. Near Infrared Spectrosc.* **2013**, *21*, 311–322. [CrossRef]

11. Lucas, A.; Andueza, D.; Rock, E.; Martin, B. Prediction of dry matter, fat, pH, vitamins, minerals, carotenoids, total antioxidant capacity, and color in fresh and freeze-dried cheeses by visible-near-infrared reflectance spectroscopy. *J. Agric. Food Chem.* **2008**, *56*, 6801–6808. [CrossRef]

12. González-Martín, I.; González-Pérez, C.; Hernández-Hierro, J.M.; González-Cabrera, J.M. Use of NIRS technology with a remote reflectance fibre-optic probe for predicting major components in cheese. *Talanta* **2008**, *75*, 351–355. [CrossRef]

13. Karoui, R.; Pillonel, L.; Schaller, E.; Bosset, J.-O.; De Baerdemaeker, J. Prediction of sensory attributes of European Emmental cheese using near-infrared spectroscopy: A feasibility study. *Food Chem.* **2007**, *101*, 1121–1129. [CrossRef]

14. Downey, G.; Sheehan, E.; Delahunty, C.; O'Callaghan, D.; Guinee, T.; Howard, V. Prediction of maturity and sensory attributes of Cheddar cheese using near-infrared spectroscopy. *Int. Dairy J.* **2005**, *15*, 701–709. [CrossRef]

15. Huck-Pezzei, V.A.; Seitz, I.; Karer, R.; Schmutzler, M.; De Benedictis, L.; Wild, B.; Huck, C.W. Alps food authentication, typicality and intrinsic quality by near infrared spectroscopy. *Food Res. Int.* **2014**, *62*, 984–990. [CrossRef]

16. Ottavian, M.; Facco, P.; Barolo, M.; Berzaghi, P.; Segato, S.; Novelli, E.; Balzan, S. Near-infrared spectroscopy to assist authentication and labeling of Asiago d'allevo cheese. *J. Food Eng.* **2012**, *113*, 289–298. [CrossRef]

17. Schmutzler, M.; Huck, C.W. Simultaneous detection of total antioxidant capacity and total soluble solids content by Fourier transform near-infrared (FT-NIR) spectroscopy: A quick and sensitive method for on-site analyses of apples. *Food Control* **2016**, *66*, 27–37. [CrossRef]

18. Guelpa, A.; Marini, F.; Du Plessis, A.; Slabbert, R.; Manley, M. Verification of authenticity and fraud detection in South African honey using NIR spectroscopy. *Food Control* **2017**, *73*, 1388–1396. [CrossRef]

19. Prieto, N.; Pawluczyk, O.; Dugan, M.E.R.; Aalhus, J.L. A Review of the Principles and Applications of Near-Infrared Spectroscopy to Characterize Meat, Fat, and Meat Products. *Appl. Spectrosc.* **2017**, *71*, 1403–1426. [CrossRef] [PubMed]

20. Čurda, L.; Kukačková, O. NIR spectroscopy: A useful tool for rapid monitoring of processed cheeses manufacture. *J. Food Eng.* **2004**, *61*, 557–560. [CrossRef]

21. Koide, T.; Yamamoto, Y.; Fukami, T.; Katori, N.; Okuda, H.; Hiyama, Y. Analysis of Distribution of Ingredients in Commercially Available Clarithromycin Tablets Using Near-Infrared Chemical Imaging with Principal Component Analysis and Partial Least Squares. *Chem. Pharm. Bull.* **2015**, *63*, 663–668. [CrossRef] [PubMed]

22. Ariyasu, A.; Hattori, Y.; Otsuka, M. Non-destructive prediction of enteric coating layer thickness and drug dissolution rate by near-infrared spectroscopy and X-ray computed tomography. *Int. J. Pharm.* **2017**, *525*, 282–290. [CrossRef] [PubMed]

23. Ozaki, Y. Near-infrared spectroscopy—Its versatility in analytical chemistry. *Anal. Sci. Int. J. Jpn. Soc. Anal. Chem.* **2012**, *28*, 545–563. [CrossRef]

24. Yan, H.; Siesler, H.W. Quantitative analysis of a pharmaceutical formulation: Performance comparison of different handheld near-infrared spectrometers. *J. Pharm. Biomed. Anal.* **2018**, *160*, 179–186. [CrossRef] [PubMed]

25. Henn, R.; Kirchler, C.G.; Grossgut, M.-E.; Huck, C.W. Comparison of sensitivity to artificial spectral errors and multivariate LOD in NIR spectroscopy—Determining the performance of miniaturizations on melamine in milk powder. *Talanta* **2017**, *166*, 109–118. [CrossRef] [PubMed]

26. Rodriguez-Saona, L.E.; Koca, N.; Harper, W.J.; Alvarez, V.B. Rapid Determination of Swiss Cheese Composition by Fourier Transform Infrared/Attenuated Total Reflectance Spectroscopy. *J. Dairy Sci.* **2006**, *89*, 1407–1412. [CrossRef]

27. Pederson, C.G.; Friedrich, D.M.; Hsiung, C.; von Gunten, M.; O'Brien, N.A.; Ramaker, H.-J.; van Sprang, E.; Dreischor, M. Pocket-size near-infrared spectrometer for narcotic materials identification. In *Next-Generation Spectroscopic Technologies VII*; International Society for Optics and Photonics: Bellingham, WA, USA, 2014.

28. Zontov, Y.V.; Balyklova, K.S.; Titova, A.V.; Rodionova, O.Y.; Pomerantsev, A.L. Chemometric aided NIR portable instrument for rapid assessment of medicine quality. *J. Pharm. Biomed. Anal.* **2016**, *131*, 87–93. [CrossRef] [PubMed]

29. Friedrich, D.M.; Hulse, C.A.; von Gunten, M.; Williamson, E.P.; Pederson, C.G.; O'Brien, N.A. Miniature near-infrared spectrometer for point-of-use chemical analysis. *SPIE Proc.* **2014**, *8992*, 899203.

30. McGonigle, A.J.S.; Wilkes, T.C.; Pering, T.D.; Willmott, J.R.; Cook, J.M.; Mims, F.M.; Parisi, A.V. Smartphone Spectrometers. *Sensors* **2018**, *18*, 223. [CrossRef]

31. Kaur, H.; Künnemeyer, R.; McGlone, A. Comparison of hand-held near infrared spectrophotometers for fruit dry matter assessment. *J. Near Infrared Spectrosc.* **2017**, *25*, 267–277. [CrossRef]

32. Wilson, B.K.; Kaur, H.; Allan, E.L.; Lozama, A.; Bell, D. A New Handheld Device for the Detection of Falsified Medicines: Demonstration on Falsified Artemisinin-Based Therapies from the Field. *Am. J. Trop. Med. Hyg.* **2017**, *96*, 1117–1123. [CrossRef]

33. Workman, J. *Handbook of Organic Compounds: Methods and Interpretations*; NIR, IR, Raman, and UV-Vis Spectra Featuring Polymers and Surfactants; Academic Press: San Diego, CA, USA, 2001.

34. Williams, P. Variables Affecting Near-Infared Reflectance Spectroscopic Analysis. In *Near Infrared Technology in the Agriculture and Food Industries*; Williams, P., Norris, K., Eds.; American Association of Cereal Chemists: St. Paul, MN, USA, 1987; Chapter 8; pp. 143–167.

35. Bundesministerium für Arbeit, Soziales, Gesundheit und Konsumentenschutz. In *Österreichisches Lebensmittelbuch:Codexkapitel/B32/Milch und Milchprodukte*, 6th ed.; Universty of Applied science: Vienna, Austria, 2017.

36. *ISO 3433:2008, 2008-01: Cheese - Determination of fat content - Van Gulik method*; International Organization for Standardization: Geneva, Switzerland, 2008.

37. *ISO 5534:2004, 2004-05: Cheese and processed cheese - Determination of the total solids content (Reference method)*; International Organization for Standardization: Geneva, Switzerland, 2004.

38. Kennard, R.W.; Stone, L.A. Computer Aided Design of Experiments. *Technometrics* **1969**, *11*, 137–148. [CrossRef]

Sample Availability: Not available.

molecules

MDPI

Article

Rapid Determination of Nutritional Parameters of Pasta/Sauce Blends by Handheld Near-Infrared Spectroscopy

Marina D. G. Neves [1,2], Ronei J. Poppi [1,*] and Heinz W. Siesler [2]

1   Institute of Chemistry, University of Campinas, Campinas CP 6154, Brazil; marina.de.gea.n@gmail.com
2   Department of Physical Chemistry, University of Duisburg-Essen, D 45117 Essen, Germany; hw.siesler@uni-due.de
*   Correspondence: rjpoppi@unicamp.br; Tel.: +55-19-3521-3126

Academic Editor: Christian Huck
Received: 6 May 2019; Accepted: 25 May 2019; Published: 28 May 2019

Abstract: Nowadays, near infrared (NIR) spectroscopy has experienced a rapid progress in miniaturization (instruments < 100 g are presently available), and the price for handheld systems has reached the < $500 level for high lot sizes. Thus, the stage is set for NIR spectroscopy to become the technique of choice for food and beverage testing, not only in industry but also as a consumer application. However, contrary to the (in our opinion) exaggerated claims of some direct-to-consumer companies regarding the performance of their "food scanners" with "cloud evaluation of big data", the present publication will demonstrate realistic analytical data derived from the development of partial least squares (PLS) calibration models for six different nutritional parameters (energy, protein, fat, carbohydrates, sugar, and fiber) based on the NIR spectra of a broad range of different pasta/sauce blends recorded with a handheld instrument. The prediction performance of the PLS calibration models for the individual parameters was double-checked by cross-validation (CV) and test-set validation. The results obtained suggest that in the near future consumers will be able to predict the nutritional parameters of their meals by using handheld NIR spectroscopy under every-day life conditions.

Keywords: handheld near-infrared spectroscopy; pasta/sauce blends; partial least squares calibration; nutritional parameters

---

1. Introduction

The miniaturization of vibrational spectrometers has started more than two decades ago, but only within the last decade have real hand-held Raman, MIR (mid-infrared) and near infrared (NIR) scanning spectrometers become commercially available and been utilized for a broad range of analytical applications [1–6]. While the weight of the majority of Raman and MIR spectrometers is still in the 1 kg range, the miniaturization of NIR spectrometers has advanced down to the < 100 g level, and developments are under way to integrate them into mobile phones [7,8]. Furthermore, most of the Raman and MIR handheld spectrometers are still in the price range of several ten thousand US$, whereas miniaturized NIR systems have reached the < 500 US$ level. In view of the high price level of Raman and MIR instruments in the near future, only the acquisition of NIR systems can be taken into consideration for private use, whereas handheld Raman and MIR spectrometers will be restricted to industrial, military and homeland security applications, as well as public use, by first responders, customs or environmental institutions.

Because vibrational spectroscopy is a non-invasive technique that allows a rapid and non-destructive analysis [9,10], its use is increasing in analytical applications of food science [11]. In recent years, primarily handheld near-infrared spectroscopy has demonstrated an immense potential in this respect for different purposes such as authentication [12–14], classification [15–17], quality control [18–21], the detection of adulteration [22–24], and the determination of food parameters [25] such as the preliminary investigations of pasta/sauce mixtures [26,27].

Over the last years public health awareness has grown strongly, and the control of nutritional parameters of everyday life food is just one aspect of this issue. Beyond body weight control, nutritional parameters are directly related to quality of life and disease control, such as as obesity, high cholesterol, gastritis, diabetes and high blood pressure. Thus, in the present study the quantitative analysis of nutritional parameters by handheld NIR spectroscopy is exemplarily demonstrated in detail for different pasta/sauce blends in combination with a chemometric data evaluation. The objective of these investigations is to prove how feasible it will be for consumers in the near future to be able to predict the nutritional parameters of their meals by using handheld NIR spectroscopy [8].

2. Experimental Section

2.1. Experimental Set-Up

For each pasta/sauce-type blend five different combinations (ranging from a 0% to 100% (*w/w*) sauce addition) were investigated. Each pasta/sauce mixture was prepared "ready-to-eat" on a plate, and the NIR spectra were recorded at room temperature (22 ± 1 °C) at a distance of 1–2 mm above the sample surface at five different positions of the plate in order to compensate inevitable compositional and surface heterogeneities (Figure 1). Previous investigations have shown that the effective pathlength of NIR radiation for diffuse reflection measurements varies (wavelength and material dependent) from several hundred micrometers to millimeters [28–30].

Figure 1. Different morphologies of the investigated pastas and a typical experimental set-up for the measurement of a pasta (here without sauce) with the handheld NIR spectrometer.

2.2. Instrumentation

Near-infrared spectra were measured in diffuse reflection with a Viavi MicroNIR 1700 (formerly JDSU, Santa Rosa, CA, USA) handheld spectrometer, based on a linear variable filter (LVF) monochromator.

The five replicate spectra were recorded with an integration time of 8.8 ms by averaging 1000 scans in the wavelength range of 908–1676 nm with an uncooled 128 pixel InGaAs array detector at a spectral resolution of 12.5 nm at 1000 nm. The S/N ratio derived from the 100% line, recorded with the parameters given above, was 5067:1. As reference, a 99% Spectralon reflectance standard (Labsphere Inc., North Sutton, NH, USA) was used.

2.3. Materials

Five different commercial pastas (Farfalle—Edeka, Italy; Tortiglioni—Birkel, Germany; Penne—GutBio, Germany; Fusilli de lentilles corail—Barilla, Italy; Casarecce de pois chiches—Barilla, Italy) and five different commercial tomato sauces (Ricotta—Barilla, Italy; Gorgonzola—Barilla, Italy; Zucchini & Aubergine—Barilla, Italy; Siciliana—Bertolli, Italy; Kräuter—Knorr, Germany) were used for the preparation of the samples. Both the pastas and the sauces were carefully selected to represent a large variation of nutritional parameters and morphologies, in order to develop representative chemometric PLS [31,32] models for the individual parameters of energy, fat, protein, carbohydrates, sugar and fiber. The nutritional parameter values of the calibration mixtures were calculated from the package labels of the pastas and sauces according to the mixture compositions and are summarized in Table 1. The mutual assignment of the five sauces to the five pastas established 25 basic combinations, and for each combination five different proportions of pasta and sauce were prepared by mixing 75 g of dry pasta with five different weights of sauce (0.00 g, 18.75 g, 37.50 g, 56.25 g and 75.00 g). These proportions correspond to pasta/sauce blend ratios (%(w/w)) of 100/0, 100/25, 100/50, 100/75, and 100/100. Before mixing, the dry pastas were cooked by boiling in water for 10 min, and after draining for a defined time period of 5 min they were put on the plate, and the sauces were added and mixed with the pastas. Thus, 125 plates in total were prepared, and five replicate spectra were measured for each plate, yielding 625 NIR spectra for further processing and analysis.

Table 1. Nutritional parameter values calculated for 100 g of dry pasta and 100 g of sauce.

Sample	Energy (kcal)	Carbohydrate (g)	Fat (g)	Fiber (g)	Protein (g)	Sugar (g)
			Pasta			
1	374.0	75.0	1.8	3.0	13.5	3.0
2	347.0	69.0	2.1	4.0	12.0	6.0
3	360.0	61.0	2.8	6.5	21.0	3.4
4	335.0	47.4	2.9	12.0	25.0	1.8
5	348.0	45.1	7.3	14.0	21.0	2.9
			Sauce			
1	97.0	6.8	7.7	1.8	3.4	5.0
2	136.0	8.6	11.3	2.0	3.0	6.5
3	74.0	6.6	4.7	2.1	1.5	4.8
4	91.0	6.0	7.4	1.4	1.5	5.4
5	33.0	4.8	0.9	1.0	1.0	3.9

2.4. Spectral Preprocessing Treatment

In Figure 2, the sample preparation and spectra acquisition scheme is exemplarily demonstrated with specific reference to the pasta-1/sauce-1 blends. Thus, in a first step, the average spectra of the replicate measurements were calculated, and the resulting 125 spectral datasets were then concatenated in a matrix. In the matrix containing the average spectra, Savitzky-Golay (SG) smoothing [33] was applied by using a window-size 7 and 2nd degree polynomial, followed by an extended multiplicative scatter correction (EMSC) [34–36]. Finally, the spectral range was truncated to 950–1350 nm. The effects of the subsequent pretreatment steps on the original 625 raw spectra are demonstrated in detail in Figure 3.

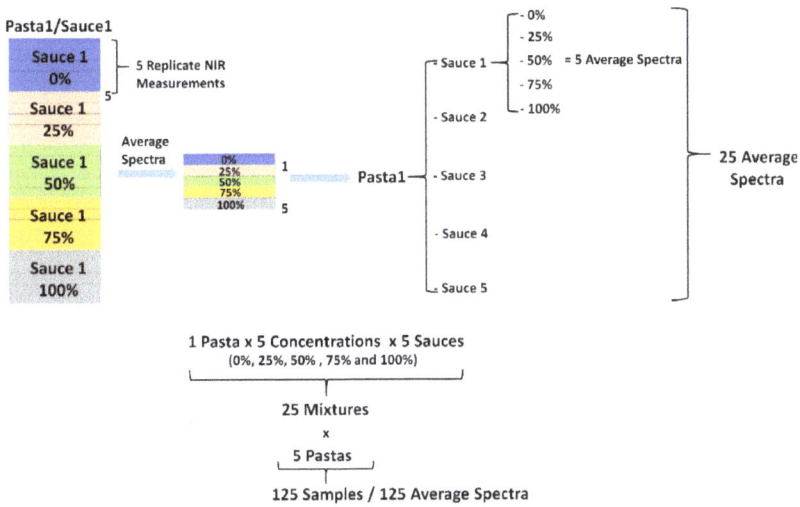

Figure 2. Sample preparation and spectra acquisition scheme demonstrated exemplarily for Pasta 1 and Sauce 1.

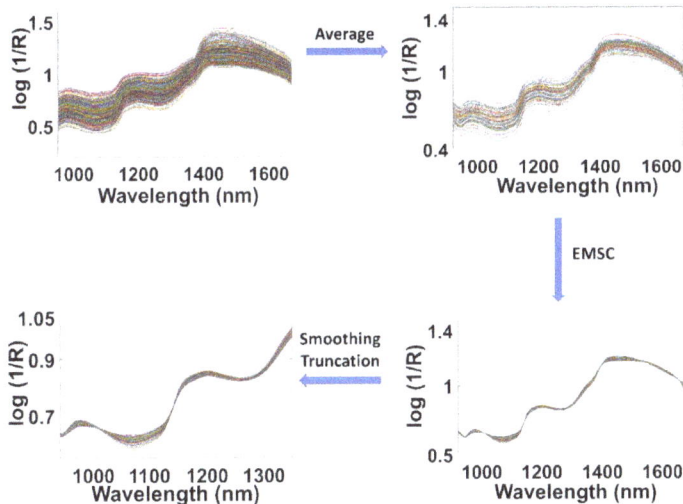

Figure 3. Pretreatments applied to the NIR spectra recorded for the pasta/sauce mixtures.

2.5. Chemometric Data Analysis

Individual PLS calibrations with mean centering and leave-one-out cross validation (CV) were developed for the different nutritional parameters with MatLab software (version R2016a, The MathWorks, Inc., Natick, MA, USA) and the PLS toolbox (version 8.6., Eigenvector Inc., Manson, WA, USA).

For the separation of the available pasta/sauce mixtures into calibration and test samples for the different nutritional parameters, the 125 samples were arranged by increasing order of the respective parameter, and one sample was removed randomly from each consecutive group of five samples. The 100 remaining samples were used as the calibration set, whereas the 25 removed samples were used as the test set. The test set samples were finally used for an additional validation step and the demonstration of the predictive capability for "unknown" samples.

3. Results and Discussion

The choice of the number of latent variables (factors) is a critical point in the PLS model development and should be based on the relation to other statistical parameters such as RMSEC and RMSECV [37]. Figure 4 shows plots of the RMSEC/RMSECV values versus the latent variable number for the individual calibrations of the nutritional parameters. Basically, the selection is a compromise between the magnitude of error, robustness of calibration and overfitting. In the present case, eight factors were chosen for energy, carbohydrate, sugar, fiber and protein, respectively, and only seven factors for fat, because the graphs of the RMSEs versus the number of latent variables flatten out beyond these numbers of latent variables (red arrows in Figure 4).

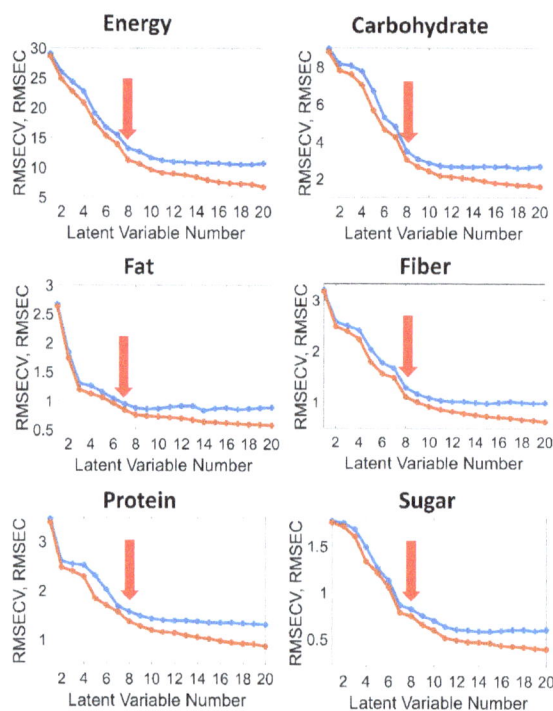

Figure 4. RMSEC (red) and RMSECV (blue) versus the latent variable number for the individual calibrations of the nutritional parameters.

The comparatively high number of factors can be readily explained by the complexity of the samples under investigation. Apart from the fact that six parameters are determined, the samples were prepared with five different types of pastas with varying morphologies and sauces, with considerable variations of ingredients (vegetables, cheese, etc.). Furthermore, residual amounts of water lead to hydrogen bonding interactions with carbohydrates, sugars, fibers, and proteins. In Table 2, the content ranges and selected calibration parameters such as root mean square error of calibration (RMSEC), root mean square error of cross validation (RMSECV), root mean square error of prediction (RMSEP), bias, slope, offset and correlation, have been summarized. The residual predictive deviation (RPD) was also included to estimate how well the calibration model can predict the compositional data [37,38]. Generally, the RMSEs and RPDs shown in Table 2 furnish evidence that, at best, medium quality calibrations have been achieved that can be used for the screening purposes of the nutritional parameters under investigation. In Figure 5, the predicted versus actual concentration graphs are shown for the calibration and test set samples for all nutritional parameters, with a linear regression fit.

As an additional feature, this figure also reflects two classes of calibration samples for the parameters of carbohydrate, protein and fiber.

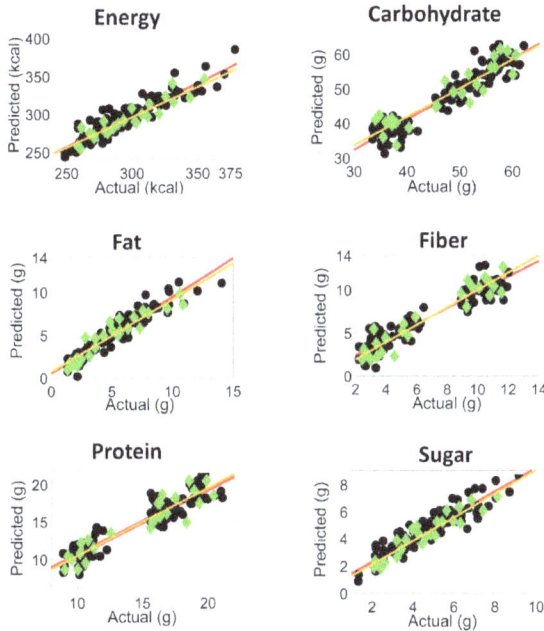

Figure 5. Graphs of the predicted versus actual content of the respective nutritional parameter per serving (calibration fit (◆), prediction fit (—), calibration samples (—) and predicted test set samples (●)).

Table 2. Content Range and statistical parameters obtained for the individual PLS models of the nutritional parameters.

Parameter	Energy	Carbohydrate	Fat	Fiber	Protein	Sugar
# LVs	8	8	7	8	8	8
RMSEC	11.15 [a]	2.97 [b]	0.83 [b]	1.10 [b]	1.36 [b]	0.65 [b]
RMSECV	13.10 [a]	3.43 [b]	0.94 [b]	1.27 [b]	1.56 [b]	0.74 [b]
RMESEP	10.64 [a]	3.59 [b]	0.95 [b]	1.11 [b]	1.39 [b]	0.61 [b]
Content Range	248.67–378.54 [a]	33.55–62.13 [b]	1.34–14.06 [b]	2.23–12.03 [b]	8.89–21.67 [b]	1.34–9.15 [b]
R^2 Cal	0.85	0.89	0.91	0.89	0.87	0.86
R^2 CV	0.80	0.85	0.88	0.85	0.83	0.82
R^2 Pred	0.86	0.85	0.89	0.90	0.86	0.88
RPD	2.02	2.54	2.77	2.45	2.26	2.19
Slope CV	0.85	0.89	0.91	0.89	0.87	0.86
Offset CV	43.12 [a]	5.21 [b]	0.46 [b]	0.73 [b]	1.92 [b]	0.62 [b]
Slope Pred	0.80	0.83	0.84	0.99	0.91	0.87
Offset Pred	55.0 [a]	8.81 [b]	0.75 [b]	0.37 [b]	1.40 [b]	0.38 [b]

[a] = kcal; [b] = g.

An overview of the prediction results for the test set samples is provided in Tables 3 and 4. The predictions for energy and carbohydrate of the test set samples were obtained with R^2Cal 0.85 and 0.89, respectively, and average relative prediction errors of 2.7 and 6.4 %(w/w), respectively. Protein had an R^2Cal of 0.87 and an average relative prediction error of 8.3 %(w/w). The calibrations for sugar and fat led to R^2Cal values of 0.86 and 0.91, respectively, and average relative prediction errors of 11.4 and 16.1 %(w/w), respectively. With an R^2Cal of 0.89, the largest average relative prediction error of 18.2 %(w/w) was obtained for the fiber calibration model. The comparatively large relative prediction errors for fat, sugar and fiber are not really unexpected and are partly due to the much lower content of these components and, for sugar and fiber, they are a consequence of the structural similarity with the main component carbohydrate. A comparison of the regression vectors of carbohydrate and sugar (not shown here), for example, highlighted an almost identical pattern of important wavelength variables for their calibration models. However, although the NIR spectra contain overlapping features, the PLS method takes into account both the spectral information and the reference nutritional values when building the quantification models. Thus, despite the addressed structural similarity, it is still possible to reasonably quantify the sugar and fiber parameters, as shown in Tables 2–4.

Table 3. The actual and predicted nutritional parameter content and relative error obtained for the test set samples per serving via the individual PLS models developed for energy, carbohydrates and fat.

Energy (kcal)			Carbohydrate (g)			Fat (g)		
Actual	Predicted	Relative Error (%)	Actual	Predicted	Relative Error (%)	Actual	Predicted	Relative Error (%)
299.8	307.9	2.7	60.1	54.1	9.9	1.4	1.0	25.3
355.2	346.3	2.5	59.1	61.1	3.4	6.8	7.1	4.4
281.4	275.7	2.0	60.7	59.8	1.3	4.8	6.5	35.1
331.9	315.8	4.9	57.8	61.3	6.2	1.6	1.8	16.7
285.7	282.0	1.3	56.0	56.2	0.3	5.7	5.5	2.8
260.8	254.4	2.5	51.7	54.0	4.5	3.2	2.6	17.9
279.3	277.7	0.6	56.6	59.3	4.7	7.3	5.8	20.5
330.9	340.5	2.9	54.5	50.0	8.3	1.7	1.6	9.2
289.4	287.4	0.7	56.7	56.4	0.5	3.6	4.7	27.9
258.6	258.4	0.1	53.3	51.6	3.2	5.1	5.6	10.0
271.2	272.2	0.4	45.5	44.2	3.0	4.3	5.0	15.2
307.5	299.7	2.5	47.2	52.0	10.3	2.1	2.3	8.4
344.2	321.6	6.6	48.8	49.0	0.4	2.2	1.5	32.8
325.6	323.5	0.7	50.4	48.3	4.2	3.6	3.6	1.0
303.1	288.7	4.8	51.9	45.7	12.0	10.6	9.8	8.1
320.6	300.2	6.4	45.7	44.7	2.3	5.6	7.0	23.7
267.7	274.8	2.7	50.1	45.6	8.9	2.2	2.2	1.8
293.6	291.8	0.6	35.5	36.2	2.0	7.6	7.8	3.8
302.3	290.5	3.9	37.2	39.4	5.8	2.5	2.3	9.4
278.5	290.8	4.4	40.3	39.1	3.1	2.8	4.7	67.4
271.2	268.7	0.9	37.9	40.2	6.1	6.3	5.8	8.4
311.6	296.4	4.9	36.1	39.9	10.6	8.0	7.6	5.2

Table 3. *Cont.*

Energy (kcal)			Carbohydrate (g)			Fat (g)		
Actual	Predicted	Relative Error (%)	Actual	Predicted	Relative Error (%)	Actual	Predicted	Relative Error (%)
313.0	313.3	0.1	38.4	45.4	18.1	5.5	5.0	8.9
261.4	282.3	8.0	37.2	40.7	9.4	9.6	8.1	15.5
286.0	282.5	1.2	34.1	41.3	21.1	10.9	8.5	21.6
Average Relative Error (%)		2.7	Average Relative Error (%)		6.4	Average Relative Error (%)		16.1

Table 4. The actual and predicted nutritional parameter content and relative error obtained for the test set samples per serving via the individual PLS models developed for fiber, protein and sugar.

Fiber (g)			Protein (g)			Sugar (g)		
Actual	Predicted	Relative Error (%)	Actual	Predicted	Relative Error (%)	Actual	Predicted	Relative Error (%)
3.2	5.5	70.2	12.6	13.5	7.1	3.2	2.6	17.9
3.4	3.8	12.3	10.0	8.7	13.4	3.4	3.2	5.6
2.6	3.1	22.2	10.6	9.5	10.3	6.3	5.1	18.6
2.4	1.9	20.7	11.3	11.5	2.3	2.2	2.1	5.6
2.8	2.4	14.8	10.8	12.6	16.4	8.1	7.1	12.4
4.0	4.4	12.0	10.3	12.2	18.6	4.5	3.6	18.6
3.7	2.7	26.2	9.3	10.5	12.5	5.6	6.1	8.1
3.0	1.9	36.7	9.6	10.0	4.4	6.9	6.9	0.1
3.4	2.2	35.0	9.1	8.6	4.7	7.4	6.1	18.2
4.6	2.3	49.7	10.2	10.8	6.0	5.4	5.3	3.0
5.9	6.9	16.4	15.8	14.9	5.4	6.3	6.3	1.5
5.1	6.1	17.9	18.4	14.9	19.1	3.6	3.9	7.9
5.0	5.2	3.9	16.9	16.1	4.4	4.6	4.1	10.9
5.4	4.7	12.9	16.0	15.3	4.2	2.6	2.2	14.3
10.1	10.3	1.4	15.6	14.2	9.3	3.2	3.9	21.0
10.5	11.4	8.8	19.3	17.5	9.4	4.9	4.3	11.2
9.4	11.1	18.8	19.9	20.5	3.2	2.4	1.8	25.3
8.9	9.6	7.5	20.4	18.5	9.5	2.1	1.9	7.9
9.3	8.0	13.6	18.6	20.5	10.3	4.3	4.4	2.9
9.7	10.4	7.2	18.9	19.2	1.2	6.0	5.3	10.6
10.8	11.0	2.2	16.3	15.7	4.2	2.2	2.3	8.7
11.1	10.4	6.0	17.5	18.2	4.0	4.0	4.7	18.8
11.7	12.6	8.1	16.2	17.5	8.1	4.2	4.9	18.0
10.8	9.4	13.4	16.3	17.5	7.4	5.2	5.3	1.8
11.6	9.7	16.2	16.5	18.6	12.7	5.1	4.2	16.8
Average Relative Error (%)		18.2	Average Relative Error (%)		8.3	Average Relative Error (%)		11.4

4. Conclusions

In combination with chemometric evaluation routines, NIR spectroscopy has proved a powerful analytical tool for authentication, adulteration and quality control in food science. The presented method, using a miniaturized spectrometer and PLS calibration models to quantify nutritional parameters of pasta/sauce mixtures, is simple, fast and non-destructive. The achieved calibration results provide an overview of the realistically expectable prediction accuracy for quantifying energy, carbohydrate, fat, fiber, protein and sugar via the application of handheld instruments. However, the results also demonstrate that the "cloud-derived" concentration data reported by several direct-to-consumer companies in commercial videos and advertising papers are beyond any realistic accuracy that is achievable with their relatively simple food-scanners.

Author Contributions: M.D.G.N. carried out the experimental works and wrote the original draft. H.W.S. and R.J.P. supervised the methodology used. All authors conceived and designed the experiments, as well as read and approved the final manuscript.

Funding: This research was funded by Conselho Nacional de Desenvolvimento Científico e Tecnológico (CNPq) (proc. 142387/2016-9 and 303994/2017-7) and by Coordenação de Aperfeiçoamento de Pessoal de Nível Superior–Brasil (CAPES) (Finance Code 001).

Acknowledgments: M.D.G.N. gratefully acknowledges financial support from PDSE/CAPES—Edital 47/2017, University of Campinas and helpful discussions with Frank Pfeifer (University of Duisburg—Essen).

Conflicts of Interest: The authors declare no conflict of interest.

References

1. Sorak, D.; Herberholz, L.; Iwascek, S.; Altinpinar, S.; Pfeifer, F.; Siesler, H.W. New developments and applications of handheld raman, mid-infrared, and near-infrared spectrometers. *Appl. Spectrosc. Rev.* **2012**, *47*, 83–115.
2. Crocombe, R.A. Portable Spectroscopy. *Appl. Spectrosc.* **2018**, *72*, 1701–1751. [CrossRef]
3. Guillemain, A.; Dégardin, K.; Roggo, Y. Performance of NIR handheld spectrometers for the detection of counterfeit tablets. *Talanta* **2017**, *165*, 632–640. [PubMed]
4. Soriano-Disla, J.M.; Janik, L.J.; McLaughlin, M.J. Assessment of cyanide contamination in soils with a handheld mid-infrared spectrometer. *Talanta* **2018**, *178*, 400–409. [CrossRef]
5. Yakes, B.J.; Brückner, L.; Karunathilaka, S.R.; Mossoba, M.M.; He, K. First use of handheld Raman spectroscopic devices and on-board chemometric analysis for the detection of milk powder adulteration. *Food Control* **2018**, *92*, 137–146.
6. Jentzsch, P.V.; Gualpa, F.; Ramos, L.A.; Ciobotă, V. Adulteration of clove essential oil: Detection using a handheld Raman spectrometer. *Flavour Fragr. J.* **2018**, *33*, 184–190. [CrossRef]
7. Pügner, T.; Knobbe, J.; Grüger, H. Near-Infrared Grating Spectrometer for Mobile Phone Applications. *Appl. Spectrosc.* **2016**, *70*, 734–745. [CrossRef]
8. BASF Hertzstueck™ Smartphone NIR. *Spectrosc. Eur.* **2018**, *30*, 11.
9. Lohumi, S.; Lee, S.; Lee, H.; Cho, B.K. A review of vibrational spectroscopic techniques for the detection of food authenticity and adulteration. *Trends Food Sci. Technol.* **2015**, *46*, 85–98. [CrossRef]
10. Lohumi, S.; Mo, C.; Cho, B.-K.; Hong, S.-J.; Kang, J.-S. Nondestructive Evaluation for the Viability of Watermelon (Citrullus lanatus) Seeds Using Fourier Transform Near Infrared Spectroscopy. *J. Biosyst. Eng.* **2014**, *38*, 312–317. [CrossRef]
11. Dos Santos, C.A.T.; Lopo, M.; Páscoa, R.N.M.J.; Lopes, J.A. A review on the applications of portable near-infrared spectrometers in the agro-food industry. *Appl. Spectrosc.* **2013**, *67*, 1215–1233. [CrossRef]
12. Grassi, S.; Casiraghi, E.; Alamprese, C. Handheld NIR device: A non-targeted approach to assess authenticity of fish fillets and patties. *Food Chem.* **2018**, *243*, 382–388. [CrossRef]
13. Karunathilaka, S.R.; Yakes, B.J.; He, K.; Chung, J.K.; Mossoba, M. Non-targeted NIR spectroscopy and SIMCA classification for commercial milk powder authentication: A study using eleven potential adulterants. *Heliyon* **2018**, *4*, e00806. [CrossRef] [PubMed]

14. Fardin-Kia, A.R.; Mossoba, M.M.; Chung, J.K.; Srigley, C.; Karunathilaka, S.R. Rapid screening of commercial extra virgin olive oil products for authenticity: Performance of a handheld NIR device. *NIR News* **2017**, *28*, 9–14.

15. Santos, P.M.; Pereira-Filho, E.R.; Rodrigues-Saona, L.E. Application of Hand-Held and Portable Infrared Spectrometers in Bovine Milk Analysis. *J. Agric. Food Chem.* **2015**, *61*, 1205–1211. [CrossRef] [PubMed]

16. Liu, N.; Parra, H.A.; Pustjens, A.; Hettinga, K.; Mongondry, P.; van Ruth, S.M. Evaluation of portable near-infrared spectroscopy for organic milk authentication. *Talanta* **2018**, *184*, 128–135. [CrossRef]

17. De Lima, G.F.; Andrade, S.A.C.; da Silva, V.H.; Honorato, F.A. Multivariate Classification of UHT Milk as to the Presence of Lactose Using Benchtop and Portable NIR Spectrometers. *Food Anal. Methods* **2018**, *11*, 2699–2706. [CrossRef]

18. Modroño, S.; Soldado, A.; Martínez-Fernández, A.; de la Roza-Delgado, B. Handheld NIRS sensors for routine compound feed quality control: Real time analysis and field monitoring. *Talanta* **2017**, *162*, 597–603. [CrossRef] [PubMed]

19. González Arrojo, A.; Cuevas Valdés, M.; Garrido-Varo, A.; Maroto, F.; Pérez-Marín, D.; Soldado, A.; de la Roza-Delgado, B. Matching portable NIRS instruments for in situ monitoring indicators of milk composition. *Food Control* **2017**, *76*, 74–81.

20. Wiedemair, V.; Huck, C.W. Evaluation of the performance of three hand-held near-infrared spectrometer through investigation of total antioxidant capacity in gluten-free grains. *Talanta* **2018**, *189*, 233–240. [CrossRef]

21. Pérez-Marín, D.; Paz, P.; Guerrero, J.E.; Garrido-Varo, A.; Sánchez, M.T. Miniature handheld NIR sensor for the on-site non-destructive assessment of post-harvest quality and refrigerated storage behavior in plums. *J. Food Eng.* **2010**, *99*, 294–302. [CrossRef]

22. Basri, K.N.; Hussain, M.N.; Bakar, J.; Sharif, Z.; Khir, M.F.A.; Zoolfakar, A.S. Classification and quantification of palm oil adulteration via portable NIR spectroscopy. *Spectrochim. Acta Part A Mol. Biomol. Spectrosc.* **2017**, *173*, 335–342. [CrossRef]

23. Basri, K.N.; Laili, A.R.; Tuhaime, N.A.; Hussain, M.N.; Bakar, J.; Sharif, Z.; Abdul Khir, M.F.; Zoolfakar, A.S. FT-NIR, MicroNIR and LED-MicroNIR for detection of adulteration in palm oil via PLS and LDA. *Anal. Methods* **2018**, *10*, 4143–4151. [CrossRef]

24. Rodrigues, R.R.T.; Tosato, F.; Correia, R.M.; Domingos, E.; Aquino, L.F.M.; Romão, W.; Lacerda, V.; Filgueiras, P.R. Portable near infrared spectroscopy applied to quality control of Brazilian coffee. *Talanta* **2017**, *176*, 59–68.

25. Sánchez, M.T.; Entrenas, J.A.; Torres, I.; Vega, M.; Pérez-Marín, D. Monitoring texture and other quality parameters in spinach plants using NIR spectroscopy. *Comput. Electron. Agric.* **2018**, *155*, 446–452. [CrossRef]

26. Yan, H.; Siesler, H.W. Hand-held near-infrared spectrometers: State-of-the-art instrumentation and practical applications. *NIR News* **2018**, *29*, 8–12. [CrossRef]

27. Yan, H.; Siesler, H.W. Handheld Raman, Mid-Infrared and Near Infrared Spectrometers: State-of-the-Art Instrumentation and Useful Applications. *Spectroscopy* **2018**, *33*, 6–16.

28. Kolomiets, O.; Hoffmann, U.; Geladi, P.; Siesler, H.W. Quantitative Determination of Pharmaceutical Drug Formulations by Near-Infrared Spectroscopic Imaging. *Appl. Spectrosc.* **2008**, *62*, 1200–1208. [CrossRef] [PubMed]

29. Hudak, S.J.; Haber, K.; Sando, G.; Kidder, L.H.; Lewis, E.N. Practical limits of spatial resolution in diffuse reflectance NIR chemical imaging. *NIR News* **2007**, *18*, 6–8. [CrossRef]

30. Bashkatov, A.N.; Genina, E.A.; Kochubey, V.I.; Tuchin, V.V. Optical properties of human skin, subcutaneous and mucous tissues in the wavelength range from 400 to 2000 nm. *J. Phys. D Appl. Phys.* **2005**, *38*, 2543–2555. [CrossRef]

31. Sjöström, M.; Eriksson, L.; Wold, S. PLS-regression: A basic tool of chemometrics. *Chemom. Intell. Lab. Syst.* **2001**, *58*, 109–130.

32. Geladi, P.; Kowalski, B.R. Partial Least-Squares regression: A tutorial. *Analytica Chimica Acta* **1986**, *185*, 1–17. [CrossRef]

33. Savitzky, A.; Golay, M.J.E. Smoothing and Differentiation of Data by Simplified Least Squares Procedures. *Anal. Chem.* **1964**, *36*, 1627–1639. [CrossRef]

34. Afseth, N.K.; Kohler, A. Extended multiplicative signal correction in vibrational spectroscopy, a tutorial. *Chemom. Intell. Lab. Syst.* **2012**, *117*, 92–99. [CrossRef]

35. Rinnan, Å.; van den Berg, F.; Engelsen, S.B. Review of the most common pre-processing techniques for near-infrared spectra. *TrAC Trends Anal. Chem.* **2009**, *28*, 1201–1222. [CrossRef]

36. Martens, H.; Stark, E. Extended multiplicative signal correction and spectral interference subtraction: New preprocessing methods for near infrared spectroscopy. *J. Pharm. Biomed. Anal.* **1991**, *9*, 625–635. [CrossRef]

37. Fearn, T. Assessing calibrations: SEP, RPD, RER and R2. *NIR News* **2002**, *6*, 12–13. [CrossRef]

38. Williams, P.C.; Sobering, D.C. Comparison of Commercial near Infrared Transmittance and Reflectance Instruments for Analysis of Whole Grains and Seeds. *J. Near Infrared Spectrosc.* **1993**, *1*, 25–32. [CrossRef]

Sample Availability: Samples of the compounds are not available from the authors.

molecules

MDPI

Article

Calibration Transfer Based on Affine Invariance for NIR without Transfer Standards

Yuhui Zhao [1], Ziheng Zhao [1], Peng Shan [2,*], Silong Peng [3], Jinlong Yu [1] and Shuli Gao [1]

[1] School of Computer Science and Engineering, Northeastern University, Shenyang 110819, China;
     yuhuizhao@neuq.edu.cn (Y.Z.); 13081850350@163.com (Z.Z.); jianren_d@163.com (J.Y.);
     15238247216@163.com (S.G.)
[2] College of Information Science and Engineering, Northeastern University, Shenyang 110819, China
[3] Institute of Automation, Chinese Academy of Sciences, Beijing 100190, China; silong.peng@ia.ac.cn
*   Correspondence: peng.shan@neuq.edu.cn; Tel.: +86-156-0337-1089

Academic Editors: Christian Huck and Krzysztof B. Bec
Received: 26 March 2019; Accepted: 6 May 2019; Published: 9 May 2019

Abstract: Calibration transfer is an important field for near-infrared (NIR) spectroscopy in practical applications. However, most transfer methods are constructed with standard samples, which are expensive and difficult to obtain. Taking this problem into account, this paper proposes a calibration transfer method based on affine invariance without transfer standards (CTAI). Our method can be utilized to adjust the difference between two instruments by affine transformation. CTAI firstly establishes a partial least squares (PLS) model of the master instrument to obtain score matrices and predicted values of the two instruments, and then the regression coefficients between each of the score vectors and predicted values are computed for the master instrument and the slave instrument, respectively. Next, angles and biases are calculated between the regression coefficients of the master instrument and the corresponding regression coefficients of the slave instrument, respectively. Finally, by introducing affine transformation, new samples are predicted based on the obtained angles and biases. A comparative study between CTAI and the other five methods was conducted, and the performances of these algorithms were tested with two NIR spectral datasets. The obtained experimental results show clearly that, in general CTAI is more robust and can also achieve the best Root Mean Square Error of test sets (RMSEPs). In addition, the results of statistical difference with the Wilcoxon signed rank test show that CTAI is generally better than the others, and at least statistically the same.

Keywords: near-infrared (NIR) spectroscopy; calibration transfer; affine invariance; multivariate calibration; partial least squares (PLS)

1. Introduction

With the characteristics of high efficiency, low cost and non-destructivity, near-infrared (NIR) spectroscopy has been widely used in control of food and pharmaceutical quality [1–4]. Multivariate calibration methods are commonly used to obtain quantitative or qualitative information from near-infrared spectra, such as principal component regression (PCR) [5,6] and partial least squares (PLS) [7–10]. Since changes of the instruments and measurement conditions may result in poor applicability of the model. Recalibration can be utilized to solve this problem, but recalibration is time consuming and takes an immense amount of work. In order to reduce consumption of the recalibration, calibration transfer has been widely studied and applied [11]. There are two main situations about calibration transfer: (1) The uniform calibration model is used to predict spectra being measured on multiple instruments; (2) the new spectra are measured on the same instrument after a period of time.

A number of related methods for calibration model transfer have been proposed, which are divided into two categories. Ones require transfer standards and ones not require transfer standards. The first category of methods has the characteristic that a set of samples are separately measured on the master and slave instrument. A great variety of transfer methods with standard samples have been proposed. For examples, SBC [12,13] assumes a linear relationship between predicted values of different instruments. First, the regression coefficient between the spectra and the response values on the master instrument is calculated. Then the predicted values of the master and slave setting are computed based on the regression coefficient. Finally, a linear equation is fitted between the predicted values. PDS proposed by Wang et al. is employed to correct the spectral differences [14]. In PDS [15–18], each wavelength of the master instrument is related to the wavelength window of the slave instrument, and a band transfer matrix is finally formed based on the regression coefficients of each window. The observation is consistent with this assumption that in various transfer methods the spectral correlation between master and slave is limited to smaller regions. The keys to PDS are the selection of window size and the number of standard samples. Due to the construction of multiple regression models, a huge amount of calculations are desired. The calibration model transfer for near-infrared spectra based on canonical correlation analysis [19] is proposed by Liang et al. The PLS model is built using the master instrument calibration set, and a part of the calibration set of master and slave instrument is taken as standard samples. Then, the features extracted respectively by canonical correlation analysis (CCA) [20,21]. The relationship between master and slave data is established with ordinary least squares (OLS) [22,23], and the test set is finally corrected. For CCA, SBC and PDS, a good result can be achieved with standard samples, but standard samples are difficult to obtain in some cases. For the transfer methods such as calibration transfer via extreme learning machine auto-encoder (TEAM) [24] method, calibration transfer by generalized least squares (GLSW) [25] method and spectral space transform (SST) [26,27] and so on, standard samples are also required, although the principles of these methods are different.

The second category is the methods without transfer standards. For examples, multiplicative scatter correction (MSC) [28–30] proposed by Bouveresse et al. first calculates the mean spectra of the calibration set as the reference spectra, then the linear relationship is found between every spectra and the reference spectra, and the slope and bias are obtained; finally, the slope and bias are utilized to correct slave spectra. While the standard samples are not required in MSC, it is difficult to handle complex situations. MSC is a transfer method using pre-processing techniques, and more pre-processing approaches include finite impulse response (FIR) [31] filtering and multivariate filtering via orthogonal signal correction (OSC) [32,33], etc. TCR [34] is also a standard-free method which combines transfer component analysis (TCA) [35] and ordinary least squares (OLS). The basic idea of TCA is to project the data of two instruments in a Reproducing Kernel Hilbert Space, where the data are distributed as close as possible at the same time preserving the key attributes of the original data. TCR is a robust model with good generalization abilities, but does not achieve more accurate predictions. Other techniques belonging to this category include kernel principal component analysis (KPCA) [36,37], domain generalization via invariant feature representation (DICA) [38] and so on.

Different from the above methods, this paper studies the relationship of regression coefficients between the feature vector and predicted values on two spectrometers. Samples of the calibration transfer method based on affine invariance without transfer standards (CTAI) are shown in Figure 1A. The response values of the slave spectrometer are not required, and the map is not necessary between master and slave samples. The samples are further processed under the PLS model. The spectral features and prediction values are respectively obtained, and the processed samples are shown in Figure 1B. We obtain the linear models between the feature vector and the predicted values respectively. According to the linear models of two instruments, the relationship between the predicted values is further obtained. Firstly, the PLS model is built on the master instrument; secondly, the score matrices and predicted values are extracted according to the PLS model, respectively; further, the angles and biases are calculated between two regression coefficients; finally, the prediction values are corrected by affine transformation. If the concentration information of the master spectra and the slave spectra are in the same range, CTAI can achieve more accurate predicted results and more robust model even without standard samples compared with other methods. The predictive performance of CTAI is verified by two near-infrared (NIR) datasets.

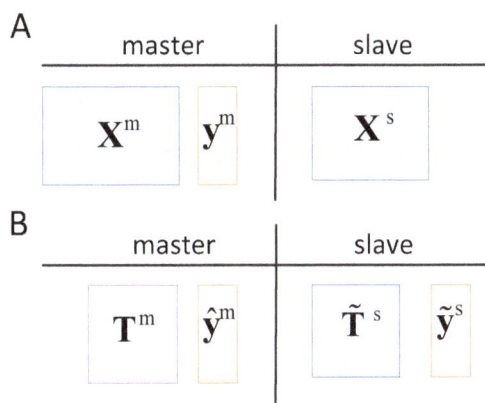

Figure 1. Data setting of the calibration transfer method based on affine invariance without transfer standards (CTAI). We assume the data to be available in (**A**), and the data after being processed based on PLS model of the master instrument is shown in (**B**).

2. Results and Discussion

2.1. Analysis of the Corn Dataset

The training errors, prediction errors, cross-validation errors, biases and the correlation coefficients for the predicted vs. actual results about the PLS model of the corn dataset are shown in Table 1. Large correlation coefficients and small biases can be seen in all results. The results reflect a good linear relationship between the spectra and measured values of the corn dataset. There are no significant differences between Root Mean Square Error of calibration set (RMSEC), Minimum Root Mean Square Error of Cross-Validation (RMSECV) and Root Mean Square Error of test set (RMSEP), indicating that there is no over-fitting and under-fitting phenomenon, which can explain the reasonable selection of the number of latent variables. Moreover, we can see that $RMSEP^m$ of the PLS on the instrument m5spec are smaller than the $RMSEP^m$ of the instrument mp6spec. For most calibration transfer methods, it is important that the master instrument has more accurate prediction results. Thus, m5spec as the master instrument and mp6spec as the slave instrument is a more reasonable choice.

In order to more fully assess the predicted performance of CTAI, the methods MSC, TCR, CCA, SBC and PDS are tested. In this work, when PDS was performed, PLS was utilized to compute the transformation function. For the PLS model, the optimal number of latent variables is shown in Table 1. The optimal dimensionality of the subspace in TCR is 4, 6, 10 and 10. In addition, optimal window sizes of PDS are all 3. We set the standard samples in range [5,30]. When the model is stable, the number of standard samples is selected for modeling based on the smallest RMSEC criteria.

As shown in Table 2, we can see the correlation coefficients r_{pre} and corresponding p_{pre} values, which indicate the prediction values between the master instrument and the slave instrument are linearly correlated. We can also see that the t_{pre} is greater than the t critical value. We then know the bias adjustment in predicted results should be implemented. Furthermore, the RMSE of prediction without any correction for the slave instrument shows more error of prediction than the master instrument. The corrected results of CTAI result in a significant reduction in RMSE of prediction. The same situation can be found between y^m and $\bar{y}^n$ in Table 2. The absolute value of t in each component is 15.437, 19.657, 19.408 and 8.762, respectively. The critical value of t is 2.131, and all results are greater than it.

It is further proved that the adjustment of bias is very important. For the corn dataset, the effect of correction in CTAI is vividly described in Figure 2. It can be seen that the corrected predicted values of CTAI more close to the straight line, and RMSEP is greatly reduced.

Table 1. Summary of the partial least squares (PLS) models and properties.

Instrument	Reference Values	RMSEC^m	RMSEP^m	RMSECV$_{min}$ (LV)	Bias^m	r^m	p^m
m5spec	moisture	0.00599	0.00764	0.01066(14)	0.0008	0.99973	2.6×10^{-24}
m5spec	oil	0.02686	0.05664	0.05049(15)	−0.01327	0.9332	1.3×10^{-7}
m5spec	protein	0.0507	0.10066	0.11012(15)	0.02814	0.97632	1×10^{-10}
m5spec	starch	0.09539	0.18993	0.19227(15)	0.01789	0.97464	1.6×10^{-10}
mp6spec	moisture	0.09991	0.15637	0.14775(10)	−0.02678	0.92083	4.2×10^{-7}
mp6spec	oil	0.06052	0.09098	0.09872(12)	0.01868	0.87697	8.2×10^{-6}
mp6spec	protein	0.10101	0.13338	0.15043(12)	0.02128	0.96659	1.1×10^{-9}
mp6spec	starch	0.27636	0.26723	0.35978(9)	0.02124	0.93136	1.6×10^{-7}
B1	protein	0.3288	0.33254	0.50337(15)	0.00906	0.98508	2.3×10^{-38}
B2	protein	0.21636	0.83755	0.32441(15)	−0.13124	0.8485	7.2×10^{-15}
B3	protein	0.30288	0.51567	0.43896(15)	−0.034	0.96009	3.2×10^{-28}

RMSEC^m: Root Mean Square Error of calibration set; RMSEP^m: Root Mean Square Error of test set; RMSECV$_{min}$: Minimum Root Mean Square Error of Cross-Validation; LV: The optimal number of latent variables is selected only with the lowest RMSECV; r^m: Pearson correlation coefficient for predicted vs. actual values; p^m: p values corresponding to the Pearson correlation coefficient is obtained by test.

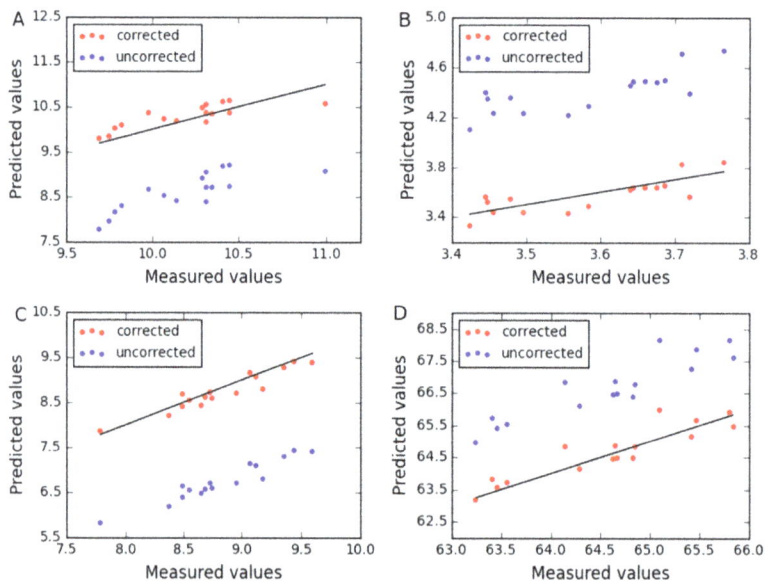

Figure 2. The relationship between the uncorrected and the corrected predict values for corn dataset by (**A**) moisture, (**B**) oil, (**C**) protein and (**D**) starch. The blue and red dots represent the uncorrected and the corrected predicted results for each sample, respectively.

Moreover, the results listed in Tables 3 and 4 show the difference between the 16 predictive corn samples by different methods. In general, the results of CTAI exhibit the best performance for prediction compared to other five methods. When moisture is used as the property, CTAI achieves the lowest RMSEP (0.21095). More specifically, the RMSEP improvements provided by CTAI with respect to MSC, TCR, CCA, SBC and PDS are as high as 87.35%, 46%, 9.48%, 50.45% and 12.96%, respectively. Though there are no statistically significant differences, CTAI is greatly improved in predictive accuracy compared with CCA and TCR. There is a significant difference at the 95%

confidence level between CTAI and MSC, SBC and PDS. When oil is used as the property, it can be seen that there is no significant difference between RMSEC and RMSEP in different transfer methods, so the over-fitting phenomenon does not appear. CTAI also produces the lowest RMSECV (0.08141) and RMSEC (0.08233). The results by Wilcoxon signed rank test reveal that CTAI is significantly different from MSC and TCR and has similar performance compared with CCA, SBC and PDS. It is noticeable that the RMSEP improvement rates of CTAI compared with CCA, SBC and PDS are 27.98%, 1.52% and 13.28%, respectively. Other properties are similar with the property of oil; CTAI achieves better predictive performance.

Table 2. Summary of the relevant results between uncorrected and CTAI corrected.

Instrument Reference Values		m5spec*-mp6spec				B1*-B2	B1*-B3	B3*-B2
		Moisture	Oil	Protein	Starch		Protein	
$\hat{y}^m$ vs $\tilde{y}^s$	$RMSEP^u_{pre}$	1.60705	0.7989	2.06797	2.11743	0.69894	2.92541	1.23368
	$RMSEP_{pre}$	0.21255	0.06922	0.13195	0.33358	0.31537	0.62632	0.65398
	k_{pre}	0.6498	0.77129	0.94553	0.82527	0.88809	0.76290	0.86909
	r_{pre}	0.81644	0.89598	0.96286	0.92197	0.97594	0.87695	0.93715
	p_{pre}	1.1×10^{-4}	2.6×10^{-6}	2.3×10^{-9}	3.8×10^{-7}	2×10^{-33}	6.8×10^{-17}	1.3×10^{-23}
	t_{pre}	−15.429	19.335	−19.147	8.838	2.292	10.684	−3.826
y^m vs $\tilde{y}^s$	$RMSEP^u$	1.60762	0.81532	2.09665	2.10291	0.71977	2.90011	1.08008
	RMSEP	0.21095	0.08233	0.16614	0.34714	0.41419	0.68215	0.38446
	k	0.65191	0.53297	0.98736	0.79329	0.96898	0.85693	0.93896
	r	0.81922	0.78858	0.95844	0.91487	0.96770	0.89517	0.97796
	p	1.0×10^{-4}	2.8×10^{-4}	5.1×10^{-9}	6.9×10^{-7}	2.2×10^{-30}	1.8×10^{-18}	2.5×10^{-34}
	t	−15.437	19.657	−19.408	8.762	2.256	10.649	−3.701
$t_{critical\_value}$		2.131	2.131	2.131	2.131	2.01	2.01	2.01

*: The master instrument; $RMSEP^u_{pre}$: RMSEP of uncorrected slave instrument relative to primary instrument prediction; $RMSEP_{pre}$: RMSEP of CTAI corrected slave instrument relative to primary instrument prediction; k_{pre}: The slope between predicted values of uncorrected slave instrument and primary prediction; r_{pre}: Correlation coefficient of uncorrected slave prediction relative to master prediction; p_{pre}: p values corresponding to the Pearson correlation coefficient are obtained by test; t_{pre}: The result of One-Sample *t*-Test between uncorrected slave prediction and master prediction; $RMSEP^u$: RMSEP of uncorrected slave instrument relative to primary actual values; RMSEP: RMSEP of CTAI corrected slave instrument relative to primary actual values; k: The slope between predicted values of uncorrected slave instrument and primary actual values; r: Pearson correlation coefficient of uncorrected slave prediction relative to primary actual values; p: p values corresponding to the Pearson correlation coefficient are obtained by test; t: The result of One-Sample *t*-Test between uncorrected slave prediction and master actual values; $t_{critical\_value}$: The t critical value for n–1 degrees of freedom at the significance level alpha = 0.05.

Table 3. Summary of Root Mean Square Error of test set (RMSEP) and Root Mean Square Error of calibration set (RMSEC) of different methods. The m5spec was used as the master spectra, and the mp6spec was used as the secondary spectra for corn dataset. The protein content was chosen as the property for wheat dataset.

	Method	CTAI	MSC	TCR	CCA	SBC	PDS
moisture	RMSEC	0.22646	1.92839	0.61873	0.15996(14[a])	0.18506(5[a])	0.14742(17[a])
	RMSEP	0.21095	1.6689	0.39066	0.23304(14[a])	0.42574(5[a])	0.24238(17[a])
oil	RMSEC	0.08141	1.21647	0.14543	0.15764(6[a])	0.08423(23[a])	0.10794(28[a])
	RMSEP	0.08233	1.23209	0.14225	0.11432(6[a])	0.08361(23[a])	0.09495(28[a])
protein	RMSEC	0.17247	1.77294	0.28297	0.27860(14[a])	0.17422(6[a])	0.24662(23[a])
	RMSEP	0.16614	1.80087	0.35223	0.39535(14[a])	0.19101(6[a])	0.28193(23[a])
starch	RMSEC	0.39517	1.89165	1.21093	0.33937(10[a])	0.38426(23[a])	0.62099(23[a])
	RMSEP	0.34714	1.93129	0.79852	0.85704(10[a])	0.36969(23[a])	0.78977(23[a])
B1*-B2	RMSEC	0.55682	1.31153	0.99246	1.11889(5[a])	0.48509(6[a])	1.3676(7[a])
	RMSEP	0.41419	0.92194	0.86881	2.68469(5[a])	0.4677(6[a])	4.09019(7[a])
B1*-B3	RMSEC	0.81895	2.91695	0.84682	1.00007(15[a])	1.00007(8[a])	0.57858(5[a])
	RMSEP	0.68215	2.40587	0.72996	1.10564(15[a])	0.79294(8[a])	1.33547(5[a])
B3*-B2	RMSEC	0.54753	1.25096	0.76972	1.57073(14[a])	0.56236(5[a])	2.1039(8[a])
	RMSEP	0.38446	1.38468	0.63689	2.29856(14[a])	0.53534(5[a])	1.83564(8[a])

[a]: Number of standard samples; the number of samples for slave instrument with labels is 20 in TCR.

Table 4. RMSEP comparison of CTAI and other methods, RMSEP improvements and p values by the Wilcoxon signed rank test ($\alpha = 0.05$). The m5spec was used as the master spectra, and the mp6spec was used as the secondary spectra for corn dataset. The protein content was chosen as the property for wheat dataset.

	MSC		TCR		CCA		SBC		PDS	
	$h(\%)$	p	$h(\%)$	p	$h(\%)$	p	$h(\%)$	p	$h(\%)$	p
moisture	87.35	4.3×10^{-4}	46	0.53	9.48	0.43	50.45	**0.01**	12.96	**0.04**
oil	93.31	4.3×10^{-4}	42.12	**0.01**	27.98	0.32	1.52	0.23	13.28	0.46
protein	90.77	4.3×10^{-4}	52.83	0.09	57.97	**0.03**	13.02	0.23	41.06	**0.01**
starch	82.02	4.3×10^{-4}	56.52	0.23	59.49	0.83	6.09	**0.02**	56.04	0.75
B1*-B2	55.07	0.11	52.32	0.79	84.57	$\mathbf{5.3 \times 10^{-9}}$	11.44	$\mathbf{2.6 \times 10^{-9}}$	89.87	$\mathbf{9.2 \times 10^{-3}}$
B1*-B3	71.64	$\mathbf{7.5 \times 10^{-10}}$	6.55	0.11	38.3	$\mathbf{1.8 \times 10^{-5}}$	13.97	$\mathbf{1 \times 10^{-5}}$	48.92	$\mathbf{9.8 \times 10^{-5}}$
B3*-B2	72.23	$\mathbf{3.1 \times 10^{-9}}$	39.63	$\mathbf{4.6 \times 10^{-3}}$	83.27	**0.02**	28.18	$\mathbf{7.5 \times 10^{-10}}$	79.05	0.06

In order to compare the predictive stability of various methods, Figures 3–6 show the plots of measured vs. predicted values for the calibration set and the test set. If the model predicts better, the point will be closer to the straight line. When moisture is used as the property, it is observed from Figure 3 that CTAI is in general closer to the straight line than the other models. It confirms that the CTAI achieves the best overall performance. When oil is used as the property, it is clear that CTAI provides satisfactory results not only in the calibration set but also in the test set. It reconfirmed that CTAI achieves more accurate prediction results. In addition, the standard error has also achieves good results in CTAI compared with others. From the discussion above, one can easily conclude that CTAI can achieve the best performance in all models and has better generalization ability.

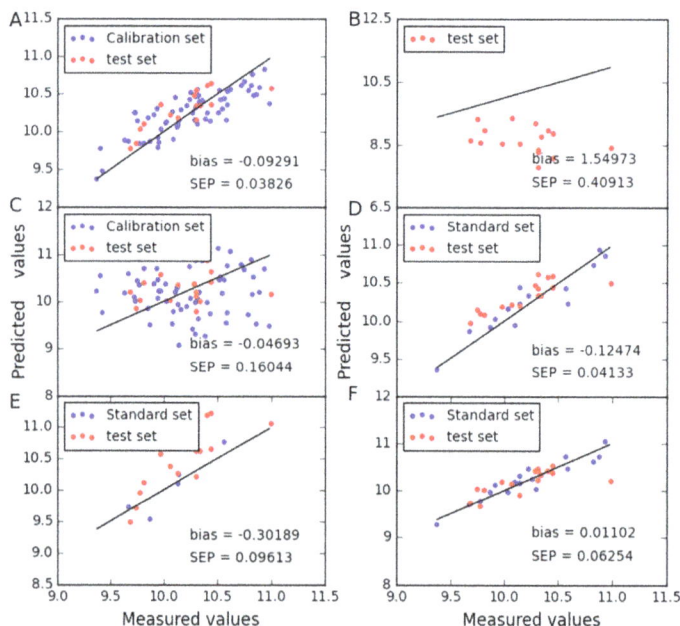

Figure 3. Moisture content predicted for corn dataset as determined by (**A**) CTAI, (**B**) MSC, (**C**) TCR, (**D**) CCA, (**E**) SBC and (**F**) PDS. The blue and red dots represent the results for each sample in the train set and test set, respectively.

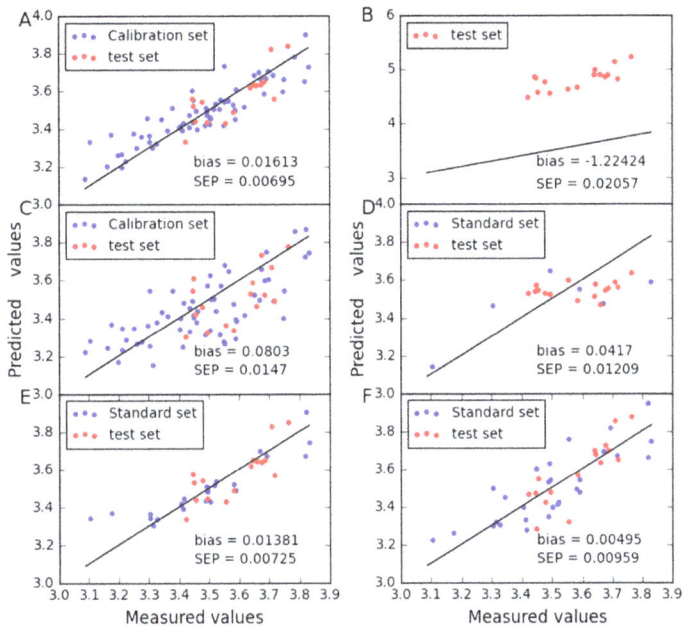

Figure 4. Oil content predicted for corn dataset as determined by (**A**) CTAI, (**B**) MSC, (**C**) TCR, (**D**) CCA, (**E**) SBC and (**F**) PDS. The blue and red dots represent the results for each sample in the train set and test set, respectively.

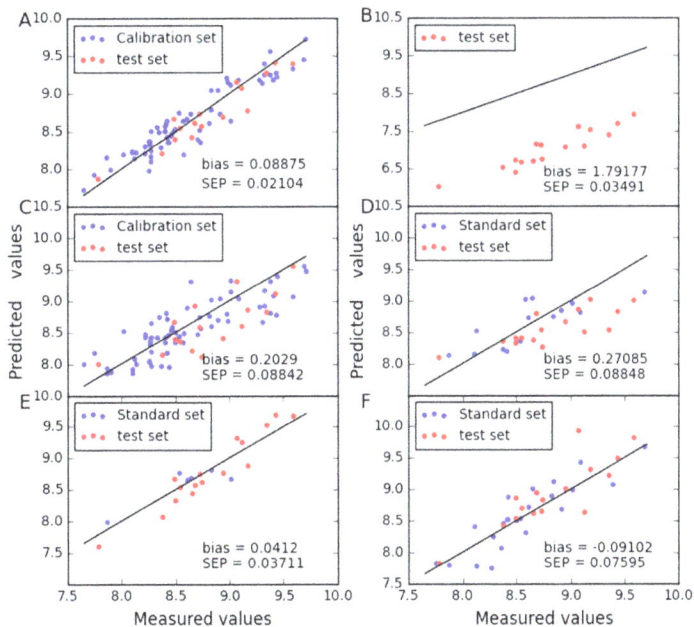

Figure 5. Protein content predicted for corn dataset as determined by (**A**) CTAI, (**B**) MSC, (**C**) TCR, (**D**) CCA, (**E**) SBC and (**F**) PDS. The blue and red dots represent the results for each sample in the train set and test set, respectively.

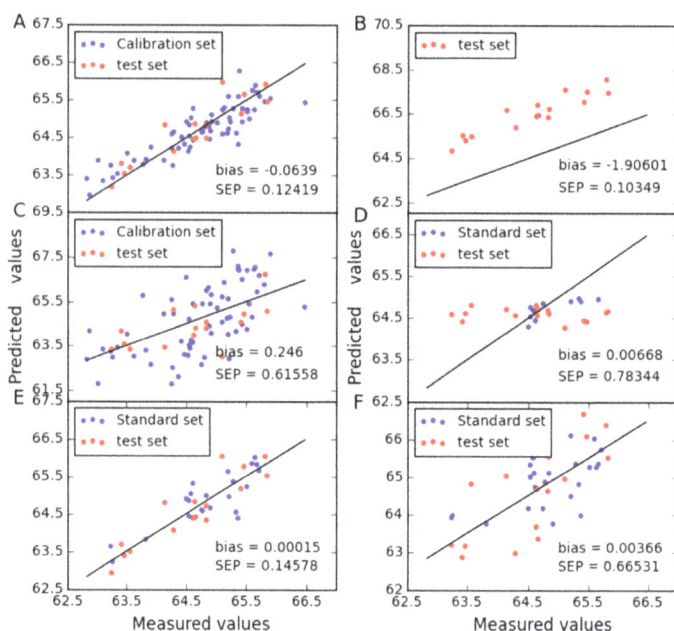

Figure 6. Starch content predicted for corn dataset as determined by (**A**) CTAI, (**B**) MSC, (**C**) TCR, (**D**) CCA, (**E**) SBC and (**F**) PDS. The blue and red dots represent the results for each sample in the train set and test set, respectively.

2.2. Analysis of the Wheat Dataset

The RMSEP of the PLS model is listed in Table 1. We can see that the predicted performance of the instrument B1 is better than B3 and the instrument B3 is better than B2. Thus, three combinations (B1-B2; B1-B3; B3-B2) of the instruments B1, B2 and B3 are used to analyze the wheat dataset. The first instrument of every combination stands for master instrument and the second instrument stands for slave instrument. For PLS model, the optimal number of latent variables is 14, 15 and 15, respectively, and the corresponding optimal dimensionality of the subspace in TCR is 17, 12 and 17, respectively. Moreover, the optimal number of window sizes for B1-B2, B1-B3 and B3-B2 is 3, 9 and 13, respectively.

For the three combinations of instruments (B1-B2; B1-B3; B3-B2), we can see between y^m and $\widetilde{y}^n$ the correlation coefficients r_{pre} are large and p_{pre} are close to zero in Table 2. Hence, there is a linear relationship between the predicted values of the two instruments for wheat dataset. For all combinations, the absolute value of t is greater than $t_{critical\_value}$. So there is a significant bias between uncorrected predicted values of the slave instrument and predicted values of the master instrument. So we can correct the predicted values of the slave instrument by affine transformation. The experimental results show that the prediction performance of CTAI is significantly enhanced. We found the same phenomenon for the uncorrected prediction values of the slave instrument relative to the master instrument actual values. Furthermore, for the predicted performance of CTAI, Figure 7 shows the difference between uncorrected and corrected predicted values for B1-B2, B1-B3 and B3-B2. It can be seen that CTAI plays an important role in the correction of predicted values.

In addition, Table 3 lists the results of different methods for calibration set and test set. For the B1-B2, CTAI produces the lowest RMSEP (0.41419) and the second lowest RMSEC (0.55682). For PDS and CCA, it is worth noting that RMSEP is significantly larger than RMSEC. Therefore, the predictive performance of PDS and CCA are poor under this setting. Further, a statistical testing is utilized to evaluate the RMSEP difference between the CTAI and other methods for the wheat dataset. The Wilcoxon signed rank sum test was performed and at the significance level alpha = 0.05. It can be seen from Table 4 that there is a statistically significant difference compared with CCA, SBC and PDS. In addition, the improvement rates of prediction provided by CTAI for MSC and TCR are up to 55.07% and 52.32%, respectively. For the combination (B1-B3), CTAI displays the lowest RMSEP (0.68215), followed by TCR (0.72996) and SBC (0.79294). For PDS, we can see that under-fitting still existed under this setting, and for CCA, this phenomenon also exists, but it is not particularly serious. The results by Wilcoxon signed rank test show that CTAI is significantly different from MSC, CCA, SBC and PDS (shown in Table 4). Compared with TCR, RMSEP improvement rates of CTAI can reach 6.55%. For the last combination, both RMSEP and RMSEC achieve the best predicted results. Further, except for PDS, the differences between CTAI and other models are statistically significant at the 95% confidence level. Compared with PDS, the RMSEP improvements of CTAI are as high as 79.05%. It is also worth noting that there is no under-fitting phenomenon in PDS under the current setting, but the predicted results are still poor. Therefore, the predictive performance of PDS is worse for wheat datasets under the current model.

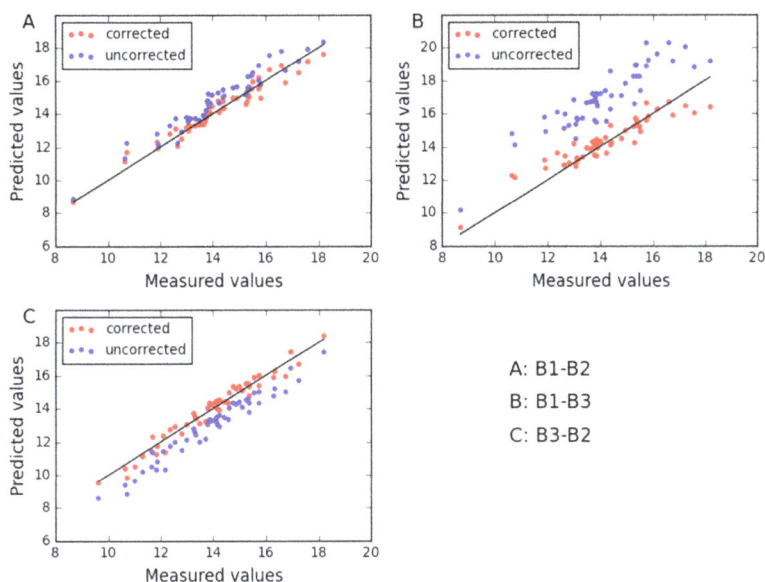

Figure 7. The relationship between the uncorrected and the corrected predict values for wheat dataset by (**A**) B1-B2, (**B**) B1-B3 and (**C**) B3-B2. The blue and red dots represent the uncorrected and the corrected predicted results for each sample, respectively.

To further display the predictive abilities of different models, the correlation between measured and predicted values obtained in Figures 8–10. Zero differences between measured and predicted values result in points over the straight line of the plot. It can be seen that good correlations are found between expected and predicted concentrations, which confirm the good performance of CTAI. CTAI achieved the lowest standard error for three combinations. Moreover, the predictive abilities of PDS and CCA are poor for wheat dataset. For SBC, PDS and CCA, they require standard samples and TCR requires reference values of the slave instrument samples, both of which are expensive and difficult to obtain. Obviously, this means that CTAI shows much more outstanding performance.

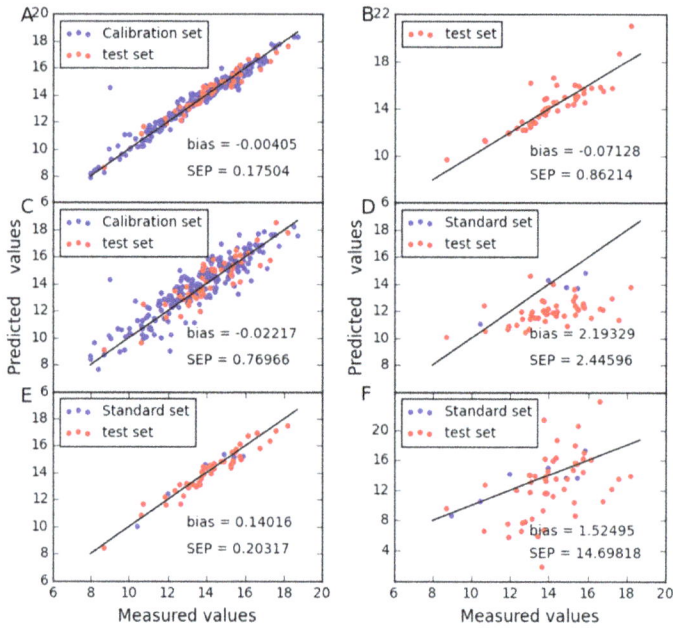

Figure 8. Protein content predicted between instruments B1 and B2 for wheat dataset as determined by (**A**) CTAI, (**B**) MSC, (**C**) TCR, (**D**) CCA, (**E**) SBC and (**F**) PDS. The blue and red dots represent the results for each sample in the train set and test set, respectively.

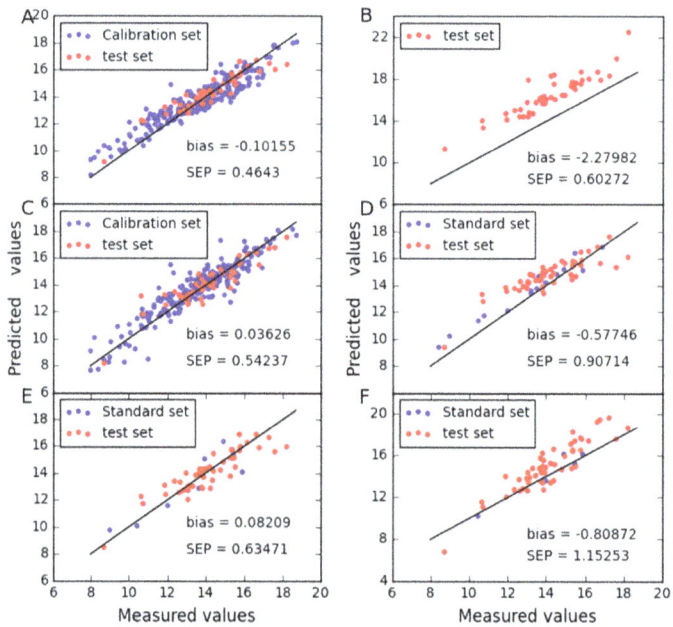

Figure 9. Protein content predicted between instruments B1 and B3 for wheat dataset as determined by (**A**) CTAI, (**B**) MSC, (**C**) TCR, (**D**) CCA, (**E**) SBC and (**F**) PDS. The blue and red dots represent the results for each sample in the train set and test set, respectively.

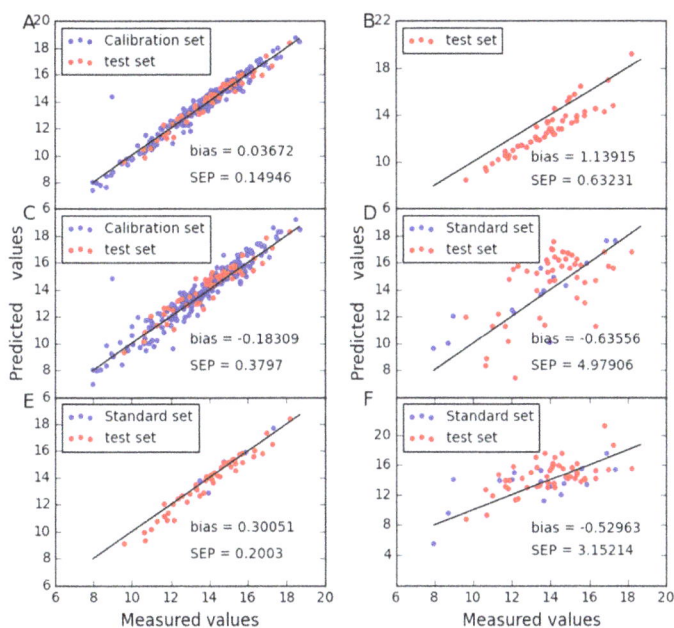

Figure 10. Protein content predicted between instruments B3 and B2 for wheat dataset as determined by (**A**) CTAI, (**B**) MSC, (**C**) TCR, (**D**) CCA, (**E**) SBC and (**F**) PDS. The blue and red dots represent the results for each sample in the train set and test set, respectively.

3. Materials and Methods

3.1. Dataset Description

3.1.1. Corn Dataset

The corn dataset, which contains 80 samples, was measured on three NIR spectrometers (m5, mp5 and mp6). Each sample consists of four components: Moisture, oil, protein, and starch. The wavelength range is 1100–2400 nm with interval 2 nm (700 channels). The spectra measured in m5spec were used as the master spectra, and the spectra measured by mp6spec were used as the secondary spectra. The data can be obtained from http://www.eigenvector.com/data/Corn/. The dataset was divided into a calibration set of 64 samples and a test set of 16 samples based on Kennard-Stone (KS) algorithm. The NIR spectra are shown in Figure 11A, which represents the difference between m5 and mp6.

3.1.2. Wheat Dataset

The wheat dataset was used as the shootout data for the International Diffuse Conference 2016, and the protein content was chosen as the property. Related information about the wheat dataset at http://www.idrc-chambersburg.org/content.aspx?page_id=22&club_id=409746&module_id=191116 can be easily accessed. 248 samples of the wheat dataset from three different NIR instrument manufacturers (B1, B2 and B3) were analyzed. According to KS algorithm, 198 samples were chosen as the calibration set and the remainder of samples formed the test set. The wavelength range is 570–1100 nm with an interval of 0.5 nm. The spectral difference between B1 and B2 is shown in Figure 11B. The spectral difference between B1 and B3 is shown in Figure 11C. The spectral difference between B2 and B3 is shown in Figure 11D.

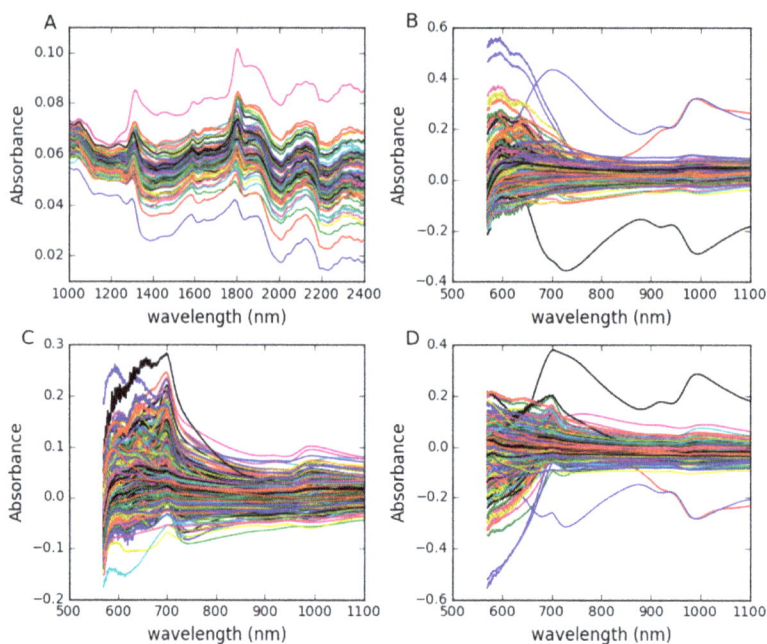

Figure 11. (**A**) Spectral differences between m5 and mp6 of corn samples; (**B**) spectral differences between B1 and B2 of wheat samples; (**C**) spectral differences between B1 and B3 of wheat samples; (**D**) spectral differences between B2 and B3 of wheat samples.

3.2. Determination of the Optimal Parameters

Latent variables of PLS in CTAI are allowed to take values in the set [1,15], and it is determined by the 10-fold cross-validation. The optimal number of latent variables is selected only when the lowest RMSECV.

Five methods were used for comparison, where the latent variable range and parameter optimization all of SBC, CCA, PDS and MSC in PLS are consistent with CTAI. In particular, the window size in PDS is searched for from 3 to 16 in increments of 2, and is selected by 5-fold cross-validation. In addition, the dimensionality of the TCA space in TCR is estimated in the range [1,24] and the optimization criteria are consistent as described in [24].

3.3. Model Performance Evaluation

In this experiment, root mean squared error RMSE is employed as indicators for parameter selection and model evaluation. Furthermore, RMSEC is the training error, RMSECV denotes the cross-validation error and RMSEP indicates the prediction error of the test set. The RMSE calculation method is written as:

$$\text{RMSE} = \sqrt{(\mathbf{y} - \hat{\mathbf{y}})^{\text{T}}(\mathbf{y} - \hat{\mathbf{y}})/n} \tag{1}$$

where $\hat{y}$ is the predict value, $\mathbf{y}$ is the measured value and n represents the number of samples.

Bias and standard error (SE) are also utilized as reference indicators for model evaluation. The bias and SE are as follows:

$$\begin{cases} \text{bias} = \sum_i^n (y_i - \hat{y}_i)/n \\ \text{SE} = \sqrt{((\mathbf{y} - \hat{\mathbf{y}})^{\text{T}}(\mathbf{y} - \hat{\mathbf{y}}) - \text{bias})/n} \end{cases} \tag{2}$$

Moreover, the Pearson correlation coefficient and corresponding test is used to determine if there is a linear relationship between the master instrument and the slave instrument. One-Sample *t*-Test is also utilized to determine whether a bias adjustment in predicted results should be implemented [11].

In order to compare CTAI and other methods further, another important parameter (*h*) is cited in order to compare the rate of improvement, defined as follows:

$$h = \left(1 - \frac{\text{RMSEP}}{\text{RMSEP}_{\text{other}}}\right) \times 100\% \tag{3}$$

where RMSEP represents the prediction error of CTAI and $\text{RMSEP}_{\text{other}}$ represents the others.

In addition, the Wilcoxon signed rank sum test at the 95% confidence level is used to determine whether there is a significant difference between CTAI and the others.

3.4. Computational Environment

All experimental procedures were implemented on a personal computer by python language, software version python 2.7, and run on an acer notebook with a 2.60 GHz Intel (R) Core (TM) i5-3230M CPU, 8 GB RAM and a Microsoft Windows 7 operating system (Acer Incorporated, Taiwan, China). Normalization and cross-validation are performed using the sklearn package. The Wilcoxon signed rank test is implemented using the scipy package and other programs are implemented by the individual.

3.5. Calibration Transfer

3.5.1. Notation

In the following text, matrices are represented by bold capital letters (e.g., **X**), column vectors by bold lower case letters (e.g., **y**) and scalars by italic letters (e.g., *empha*). The transposition operation is indicated by superscript ^{T}.

3.5.2. Overview of PLS

PLS is used to establish the linear relationship between the input space and the response space. The purpose of the PLS model is to ensure the optimal number of latent variables. The latent variables are linear combinations of the primitive variables. The latent variables are calculated in this way so that they contain a maximum of relevant information concerning the relation between **X** and **y**. Mathematically, this is shown by the following objective function.

$$H = \underset{\mathbf{w}}{\operatorname{argmax}} \quad \text{cov}\langle \mathbf{Xw}, \mathbf{y}\rangle$$
$$\text{subject to } \|\mathbf{w}\|_2 = 1 \tag{4}$$

where **w** represents the weight vector. This objective is a maximization problem under one constraint, which can be settled in virtue of the Lagrange multiplier method.

Assuming a PLS model is built between spectral matrix $\mathbf{X} \in \mathcal{R}^{n \times p}$ and concentration vector $\mathbf{y} \in \mathcal{R}^{n \times 1}$, the model is named PLS1 (*n* denotes the number of samples and *p* represents the optimal numbers of latent variables). In the algorithm, the first weighting vector must be the primary eigenvector of the matrix $\mathbf{X}^{\text{T}}\mathbf{yy}^{\text{T}}\mathbf{X}$. From the second latent variable on, it requires the following latent variables to be orthogonal (uncorrelated) to the former ones. Hence, the following weighting vectors will be the dominant eigenvectors of the matrix $\mathbf{X}^{\text{T}}\mathbf{yy}^{\text{T}}\mathbf{X}$; also, repeat a sequence of the steps until convergence. The PLS1 is built using the following model:

$$\begin{cases} \mathbf{X}^{n \times p} = \mathbf{T}^{n \times A}\left(\mathbf{P}^{p \times A}\right)^{\text{T}} + \mathbf{E}^{n \times p} \\ \mathbf{y}^{n \times 1} = \mathbf{T}^{n \times A}\left(\mathbf{Q}^{1 \times A}\right)^{\text{T}} + \mathbf{F}^{n \times 1} \end{cases} \tag{5}$$

where **T** is the score matrix and **P** and **Q** represent the **X**-loading matrix and **y**-loading vector, respectively; **E** and **F** denote the matrix of residuals; A is the optimal number of principal components over the master instrument PLS model.

Finally, the regression coefficient β of the model can be written as follows:

$$\beta = \mathbf{W}\left(\mathbf{P}^{\mathsf{T}}\mathbf{W}\right)^{-1}\mathbf{Q}^{\mathsf{T}} \tag{6}$$

where $\mathbf{W} = [\mathbf{w}_1, \mathbf{w}_2, \dots, \mathbf{w}_A]$ represents the weight matrix.

3.5.3. Affine Transformation

This paper focuses on the rotation and translation properties of two-dimensional affine transformation [39]. After transformation, the original line is still a straight line and the original parallel line is still parallel. Affine transformation is a transformation of coordinates. Based on Figure 12, the derivation is written as follows:

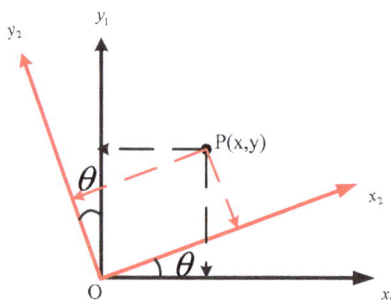

Figure 12. Derivation of affine transformation. In the coordinate system, the counterclockwise rotation of P is equivalent to the clockwise rotation of the coordinate system.

Point P in the original coordinate system (black) is (x, y). A counterclockwise rotation of the point P is equivalent to clockwise rotation of the coordinate system. Thus, the point P in the black coordinate system is equivalent with the point P in the red coordinate system after the rotation. Based on this conclusion, we can determine the coordinates of the point P by simple stereo geometry, and then add the offset of the X-axis and the Y-axis based on this position; the formula is as follows:

$$\begin{cases} x' = x\cos\theta - y\sin\theta + \Delta x \\ y' = y\cos\theta + x\sin\theta + \Delta y \end{cases} \tag{7}$$

where θ is the angle of rotation, Δx is the offset on the X axis and Δy is the offset on the Y axis; x' and y' are coordinate in the new coordinate system.

3.5.4. Calibration Transfer Method based on Affine Transformation

Based on the inputs and outputs $\{\mathbf{X}^m, \mathbf{y}^m\}$ from the master instrument, and the inputs $\{\mathbf{X}^s\}$ from the slave instrument, our task is to predict the unknown outputs $\{\hat{y}^s\}$ in the slave instrument. We assume that $\mathbf{X}^m$ and $\mathbf{X}^s$ are the spectra of two similar substances, and $\mathbf{y}^m$ and $\hat{y}^s$ are in the same range. Due to the difference between two instruments, the observed spectral data are different. The observations from the perspective of the master instrument model are as follows:

$$\begin{cases} \hat{y}^m = F(\mathbf{X}^m, \beta^m) = \sum_{i=1}^{A} t_i^m q_i^m \\ \overline{y}^s = F(\mathbf{X}^s, \beta^m) = \sum_{i=1}^{A} \overline{t}_i^s q_i^m \end{cases} \tag{8}$$

where F is the linear prediction function, which is obtained by partial least squares in this paper; β^m is the coefficient of the master model and $\hat{y}^m$, t_i^m and q_i^m are the predicted values, the i-th column score vector and the loading vector, respectively. Accordingly, $\widetilde{y}^s$ and $\widetilde{t}_i^s$ are the biased predicted values and the i-th biased column score vector for the slave instrument, respectively.

Therefore, the score vectors and predicted values both of the two instruments are different. As a result, there is a certain bias that needs to be corrected in the coefficient between the score vector and predicted values.

When correcting the bias, direct calculation will produce large errors. In order to solve this problem, we need to transform the score vectors and predicted values of the master and slave instrument into the range [0, 1] and thus keep the same scale between different values. The corresponding equations are given as follows:

$$
\begin{cases}
t^{m-norm} = (t_i^m - \min(t_i^m))/(\max(t_i^m) - \min(t_i^m)) \\
\hat{y}^{m-norm} = (\hat{y}^m - \min(\hat{y}^m))/(\max(\hat{y}^m) - \min(\hat{y}^m)) \\
\widetilde{t}^{s-norm} = (\widetilde{t}^s - \min(\widetilde{t}_i^s))/(\max(\widetilde{t}_i^s) - \min(\widetilde{t}_i^s)) \\
\widetilde{y}^{s-norm} = (\widetilde{y}^s - \min(\widetilde{y}^s))/(\max(\widetilde{y}^s) - \min(\widetilde{y}^s))
\end{cases}
\tag{9}
$$

where t_i^{m-norm} and $\hat{y}^{m-norm}$ are the normalized score vector and the predicted values of the master instrument, respectively; $\widetilde{t}_i^{s-norm}$ and $\widetilde{y}^{s-norm}$ are the normalized and biased score vector and predicted values, respectively.

Two linear regression equations between score vector and predicted values are as follows:

$$
\begin{cases}
\hat{y}^{m-norm} = t_i^{m-norm} \tan \theta_i^m + b_i^m \\
\widetilde{y}^{s-norm} = \widetilde{t}_i^{s-norm} \tan \widetilde{\theta}_i^s + \widetilde{b}_i^s
\end{cases}
\tag{10}
$$

where $\tan \theta_i^m$ and $\tan \widetilde{\theta}_i^s$ are the regression coefficients (slopes) computed on the two instrument; b_i^m and $\widetilde{b}_i^s$ are the intercepts.

In order to more intuitively reflect the difference between two instruments, it can be better understood from Figure 13. The blue line is the regression coefficient between the score vector and predicted values. The black and red coordinate systems are the observations of the master and slave instrument, and there is a difference from different observations.

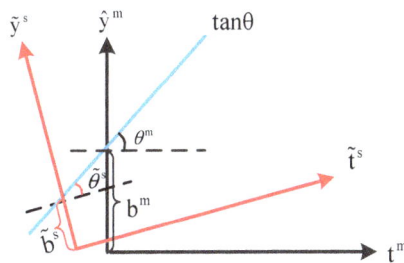

Figure 13. Theory of CTAI. $\tan \theta$ is the coefficient between the feature vector and the predicted values. The angles and deviations observed under different instruments are different. We correct the predicted value of the slave instrument with the rotation and translation of affine transformation.

The unknown angles and biases between two instruments are solved as follows:
Firstly, the regression coefficient β^m, the weight $\mathbf{W}^m$ and loading $\mathbf{P}^m$ matrix of PLS are obtained.
Secondly, a linear regression both of master and slave instrument is performed and slopes and intercepts are determined, respectively.

On the grounds of the PLS model, the score matrices and predicted values are calculated as shown below:

$$\begin{cases} \mathbf{T}^m = \mathbf{X}^m \mathbf{W}^m (\mathbf{P}^m \mathbf{W}^m)^{-1}, & \hat{\mathbf{y}}^m = \mathbf{X}^m \boldsymbol{\beta}^m \\ \widetilde{\mathbf{T}}^s = \mathbf{X}^s \mathbf{W}^m (\mathbf{P}^m \mathbf{W}^m)^{-1}, & \widetilde{\mathbf{y}}^s = \mathbf{X}^s \boldsymbol{\beta}^m \end{cases} \tag{11}$$

where $\mathbf{T}^m$ and $\widetilde{\mathbf{T}}^s$ represent the score matrices of two instruments.

The score matrix $\mathbf{T}^m$, predicted values $\hat{\mathbf{y}}^m$, the score matrix $\widetilde{\mathbf{T}}^s$ and predicted values $\widetilde{\mathbf{y}}^s$ are pre-processed using Equation (9).

According to score vector of each column and predicted values, the least square is used to compute the corresponding slopes and intercepts, respectively. The equations are as follows:

$$\begin{cases} \min\limits_{\theta_i^m, b_i^m} \|\hat{\mathbf{y}}^{m-norm} - \mathbf{T}_{aug}^m * \begin{bmatrix} \tan \theta_i^m \\ b_i^m \end{bmatrix} \|^2 \\ \min\limits_{\widetilde{\theta}_i^s, \widetilde{b}_i^s} \|\widetilde{\mathbf{y}}^{s-norm} - \mathbf{T}_{aug}^s * \begin{bmatrix} \tan \widetilde{\theta}_i^s \\ \widetilde{b}_i^s \end{bmatrix} \|^2 \end{cases} \tag{12}$$

where $\mathbf{T}_{aug}^m$ is an augmented matrix $\begin{bmatrix} \mathbf{t}_i^{m-norm}, & 1 \end{bmatrix}$; $\mathbf{T}_{aug}^s$ is an augmented matrix $\begin{bmatrix} \widetilde{\mathbf{t}}_i^{s-norm}, & 1 \end{bmatrix}$; $\mathbf{1}$ is the column vector with all ones.

Finally, the angle and biases between the two instruments are obtained. The equations for calculating the angles and biases are as follows:

$$\begin{cases} \Delta \theta_i = \theta_i^m - \widetilde{\theta}_i^s \\ \Delta b_i = b_i^m - \widetilde{b}_i^s \end{cases} \tag{13}$$

where $\Delta \theta_i$ is the angle of the two coefficients; Δb_i is the corresponding bias.

The score matrix and predicted values of the test set are extracted by Equation (11).

The angles and biases obtained by Equation (13) are brought into the affine transformation to correct the predicted values. Since the rotation angle is relative to the origin of the coordinate, each sample needs to be adjusted before rotation. The equation is shown as follows:

$$\hat{\mathbf{U}}_i = \widetilde{\mathbf{U}}_i \mathbf{M}_i \tag{14}$$

where the matrix $\mathbf{M}_i = \begin{bmatrix} \lambda_t \cos \Delta \theta_i & \lambda_t \sin \Delta \theta_i & 0 \\ -\lambda_y \sin \Delta \theta_i & \lambda_y \cos \Delta \theta_i & 0 \\ 0 & b_i^m & 1 \end{bmatrix}, \widetilde{\mathbf{U}}_i = \begin{bmatrix} \widetilde{\mathbf{t}}_i^{s-test}, \widetilde{\mathbf{y}}^{s-test}, 1 \end{bmatrix}$

and $\hat{\mathbf{U}}_i = \begin{bmatrix} \hat{\mathbf{t}}_i^{s-test}, \hat{\mathbf{y}}^{s-test}, 1 \end{bmatrix}$. In addition, $\lambda_t = \begin{bmatrix} (\widetilde{\mathbf{t}}_i^{s-test} - \min(\widetilde{\mathbf{t}}_i^s))/(\max(\widetilde{\mathbf{t}}_i^s) - \min(\widetilde{\mathbf{t}}_i^s)) + \min(\widetilde{\mathbf{t}}_i^s) \end{bmatrix} \times$ $(\max(\widetilde{\mathbf{t}}_i^s) - \min(\widetilde{\mathbf{t}}_i^s))$ and $\lambda_y = \begin{bmatrix} (\widetilde{\mathbf{y}}_i^{s-test} - \min(\widetilde{\mathbf{y}}^s))/(\max(\widetilde{\mathbf{y}}^s) - \min(\widetilde{\mathbf{y}}^s)) + \min(\widetilde{\mathbf{y}}^s) \end{bmatrix} \times (\max(\widetilde{\mathbf{y}}^s) - \min(\widetilde{\mathbf{y}}^s))$ represent the corresponding scaling factors for feature vector and predicted values, respectively; $\widetilde{\mathbf{t}}_i^{s-test}$ and $\widetilde{\mathbf{y}}^{s-test}$ are biased score vector and predicted values of the test set, respectively; $\hat{\mathbf{y}}^{s-test}$ is corrected predicted values; $\hat{\mathbf{t}}_i^{s-test}$ is corrected score vector.

Each column score vector and predicted values are solved separately, and a prediction matrix is obtained. The mean of the prediction matrix is the final predicted values.

Therefore, according to the expansion of the predicted values, $\boldsymbol{\beta}^s$ is as follows:

$$\boldsymbol{\beta}^s = ((\mathbf{X}^s)^T \mathbf{X}^s)^{-1} (\mathbf{X}^s)^T (\sum_i^A (\widetilde{\mathbf{t}}_i^{s-test} * \lambda_t * \sin \Delta \theta_i + (\widetilde{\mathbf{y}}^{s-test} - \widetilde{b}_i^s * 1) * \lambda_y * \cos \Delta \theta_i + b_i^m)/A) \tag{15}$$

3.5.5. Summary of CTAI

Given calibration set of the master $(\mathbf{X}_{cal}^m, \mathbf{y}_{cal}^m)$, calibration set of the slave $\mathbf{X}_{cal}^s$ and test set $(\mathbf{X}_{test}^s, \mathbf{y}_{test}^s)$.

1. The PLS model is built on the calibration set $(\mathbf{X}_{cal}^m, \mathbf{y}_{cal}^m)$ and the coefficient $\boldsymbol{\beta}^m$; the weight matrix $\mathbf{W}^m$ and the loading matrix $\mathbf{P}^m$ can be obtained.

2. Modeling of affine transformation; it consists of the two datasets $(\mathbf{X}_{cal}^m, \mathbf{y}_{cal}^m)$ and $\mathbf{X}_{cal}^s$.

    (a) Computing $(\mathbf{T}_{cal}^m, \hat{\mathbf{y}}_{cal}^m)$ and $(\widetilde{\mathbf{T}}_{cal}^s, \widetilde{\mathbf{y}}_{cal}^s)$ of master and slave instrument by Equation (11).

    (b) $(\mathbf{T}_{cal}^m, \hat{\mathbf{y}}_{cal}^m)$ and $(\widetilde{\mathbf{T}}_{cal}^s, \widetilde{\mathbf{y}}_{cal}^s)$ are normalized separately by Equation (9).

    (c) $(\tan\theta_i^m, b_i^m)$ and $(\tan\widetilde{\theta}_i^s, \widetilde{b}_i^s)$ are calculated by Equation (12).

    (d) Computing $\Delta\theta_i$ angle and Δb_i bias between master and slave instrument by Equation (13).

3. Prediction.

    (a) $(\widetilde{\mathbf{T}}_{test}^s, \widetilde{\mathbf{y}}_{test}^s)$ is obtained by Equation (11).

    (a) The matrix $\mathbf{M}_i$ is introduced to correct predicted values by Equation (14).

    (c) The corrected prediction values are accumulated. The mean values are the last result.

4. Conclusions

In this study, the relationship of regression coefficients between feature vector and predicted values on different instruments was investigated and CTAI was proposed for calibration transfer based on affine invariance without transfer standards (CTAI). Based on the PLS model of the master instrument, the score matrix and the predicted values of the master spectra, the pseudo score matrix and the pseudo predicted values of the slave spectra are obtained. Then, angles and biases between the coefficients of the master instrument and the corresponding coefficients of the slave instrument are computed. Finally, new samples are corrected by affine transformation. Different transfer methods are tested with two NIR datasets, CTAI achieves the lowest RMSEP and standard error, and the results of statistical difference indicate that CTAI is generally better than other methods, which proves that CTAI is successfully used to correct the difference on different instruments. Hence, the proposed method may provide an efficient way for calibration transfer when standard samples are unavailable in practical applications.

Author Contributions: Conceptualization, Y.Z. and P.S.; methodology, Y.Z. and P.S.; software, Z.Z.; validation, Z.Z.; formal analysis, P.S., Y.Z., Z.Z. and J.Y.; data curation S.G.; writing—original draft preparation, Z.Z.; writing—review and editing, J.Y., P.S. and S.P.; visualization, S.P.; supervision, P.S. and Y.Z.; project administration, Z.Z.; funding acquisition, Y.Z.

Funding: This research was funded by National Natural Science Foundation of China (Grant no. 61601104), Natural Science Foundation of Hebei Province (Grant no. F2017501052) and the Basic Science Research Fund of Northeast University at Qin Huang Dao (Grant no. XNB201611).

Conflicts of Interest: No conflict of interest exits in the submission of this manuscript, and the manuscript is approved by all authors for publication. I would like to declare on behalf of my co-authors that the work described was original research that has not been published previously, and not under consideration for publication elsewhere, in whole or in part. All the authors listed have approved the manuscript that is enclosed.

References

1. Huang, H.; Yu, H.; Xu, H.; Ying, Y. Near infrared spectroscopy for on/in-line monitoring of quality in foods and beverages: A review. *J. Food Eng.* **2008**, *87*, 303–313. [CrossRef]
2. Roggo, Y.; Chalus, P.; Maurer, L.; Lema-Martinez, C.; Edmond, A.; Jent, N. A review of near infrared spectroscopy and chemometrics in pharmaceutical technologies. *J. Pharm. Biomed. Anal.* **2007**, *44*, 683–700. [CrossRef]

3. Martinez, J.C.; Guzmán-Sepúlveda, J.R.; Bolañoz Evia, G.R.; Córdova, T.; Guzmán-Cabrera, R. Enhanced Quality Control in Pharmaceutical Applications by Combining Raman Spectroscopy and Machine Learning Techniques. *Int. J. Thermophys.* **2018**, *39*, 79. [CrossRef]

4. Porep, J.U.; Kammerer, D.R.; Carle, R. On-line application of near infrared (NIR) spectroscopy in food production. *Trends Food Sci. Tech.* **2015**, *46*, 211–230. [CrossRef]

5. Geladi, P.; Esbensen, K. Regression on multivariate images: Principal component regression for modeling, prediction and visual diagnostic tools. *J. Chemom.* **1991**, *5*, 97–111. [CrossRef]

6. Næs, T.; Martens, H. Principal component regression in NIR analysis: View-points, background details and selection of components. *J. Chemom.* **1988**, *2*, 155–167. [CrossRef]

7. Wold, S.; Sjöström, M.; Eriksson, L. PLS-regression: A basic tool of chemometrics. *Chemom. Intell. Lab. Syst.* **2001**, *58*, 109–130. [CrossRef]

8. Sijmen, D.J. SIMPLS: An alternative approach to partial least squares regression. *Chemom. Intell. Lab. Syst.* **1993**, *18*, 251–263.

9. Geladi, P.; Kowalski, B.R. Partial least-squares regression: A tutorial. *Anal. Chim. Acta* **1986**, *185*, 1–17. [CrossRef]

10. Matthew, B.; Rayens, W. Partial least squares for discrimination. *J. Chemometrics.* **2012**, *30*, 446–452.

11. Workman, J.J. A Review of Calibration Transfer Practices and Instrument Differences in Spectroscopy. *Appl. Spectrosc.* **2018**, *72*, 340–365. [CrossRef]

12. Bouveresse, E.; Hartmann, C.; Massart, D.L.; Last, I.R.; Prebble, K.A. Standardization of near-infrared spectrometric instruments. *Anal. Chem.* **1996**, *68*, 982–990. [CrossRef]

13. Feudale, R.N.; Woody, N.A.; Tan, H.; Myles, A.J.; Brown, S.D.; Ferré, J. Transfer of multivariate calibration models: A review. *Chemom. Intell. Lab. Syst.* **2002**, *64*, 181–192. [CrossRef]

14. Wang, Y.; Veltkamp, D.J.; Kowalski, B.R. Multivariate instrument standardization. *Anal. Chem.* **1991**, *63*, 2750–2756. [CrossRef]

15. Wang, Y.; Michael, J.L.; Kowalski, B.R. Improvement of multivariate calibration through instrument standardization. *Anal. Chem.* **1992**, *64*, 562–564. [CrossRef]

16. Bouveresse, E.; Massart, D. Improvement of the piecewise direct standardisation procedure for the transfer of NIR spectra for multivariate calibration. *Chemom. Intell. Lab. Syst.* **1996**, *32*, 201–213. [CrossRef]

17. Wang, Z.; Thomas, D.; Kowalski, B.R. Additive background correction in multivariate instrument standardization. *Anal. Chem.* **1995**, *67*, 2379–2385. [CrossRef]

18. Tan, H.-W.; Brown, S.D. Wavelet hybrid direct standardization of near-infrared multivariate calibrations. *J. Chemometrics.* **2001**, *15*, 647–663. [CrossRef]

19. Fan, W.; Liang, Y.; Yuan, D.; Wang, J. Calibration model transfer for near-infrared spectra based on canonical correlation analysis. *Anal. Chim. Acta* **2008**, *623*, 22–29. [CrossRef] [PubMed]

20. Zheng, K.; Zhang, X.; Iqbal, J.; Fan, W.; Wu, Ti.; Du, Y.; Liang, Y. Calibration transfer of near-infrared spectra for extraction of informative components from spectra with canonical correlation analysis. *J. Chemometrics.* **2014**, *28*, 773–784. [CrossRef]

21. Melzer, T.; Reiter, M.; Bischof, H. Appearance models based on kernel canonical correlation analysis. *Pattern Recognit.* **2003**, *36*, 1961–1971. [CrossRef]

22. Leng, L.; Zhang, T.; Kleinman, L.; Zhu, W. Ordinary least square regression, orthogonal regression, geometric mean regression and their applications in aerosol science. *J. Phys. Conf. Ser.* **2007**, *78*, 012084. [CrossRef]

23. Donald, C.; Orcutt, G.H. Application of least squares regression to relationships containing auto-correlated error terms. *J. Amer. Stat. Assoc.* **1949**, *44*, 32–61.

24. Chen, W.-R.; Bin, J.; Lu, H.-M.; Zhang, Z.-M.; Liang, Y.-Z. Calibration transfer via an extreme learning machine auto-encoder. *Analyst* **2016**, *141*, 1973–1980. [CrossRef] [PubMed]

25. Wise, B.M.; Martens, H.; Høy, M.; Bro, R.; Brockhoff, P.B. Calibration Transfer by Generalized Least Squares. In Proceedings of the Seventh Scandinavian Symposium on Chemometrics (SSC7), Copenhagen, Denmark, 19–23 August 2001.

26. Du, W.; Chen, Z.-P.; Zhong, L.-J.; Wang, S.-X.; Yu, R.-Q.; Nordon, A.; Littlejohn, D.; Holden, M. Maintaining the predictive abilities of multivariate calibration models by spectral space transformation. *Anal. Chim. Acta.* **2011**, *690*, 64–70. [CrossRef] [PubMed]

27. Chen, Z.P.; Li, L.M.; Yu, R.Q.; Littlejohn, D.; Nordon, A.; Morris, J.; Dann, A.S.; Jeffkins, P.A.; Richardson, M.D.; Stimpson, S.L. Systematic prediction error correction: A novel strategy for maintaining the predictive abilities of multivariate calibration models. *Analyst.* **2010**, *136*, 98–106. [CrossRef] [PubMed]

28. Kramer, K.E.; Morris, R.E.; Rose-Pehrsson, S.L. Comparison of two multiplicative signal correction strategies for calibration transfer without standards. *Chemom. Intell. Lab. Syst.* **2008**, *92*, 33–43. [CrossRef]

29. Preisner, O.; Lopes, J.A.; Guiomar, R.; Machado, J.; José, C.M. Fourier transform infrared (FT-IR) spectroscopy in bacteriology: towards a reference method for bacteria discrimination. *Anal. Bioanal. Chem.* **2007**, *387*, 1739–1748. [CrossRef] [PubMed]

30. Isaksson, T.; Næs, T. The effect of multiplicative scatter correction (MSC) and linearity improvement in NIR spectroscopy. *Appl. Spectrosc.* **1988**, *42*, 1273–1284. [CrossRef]

31. Blank, T.B.; Sum, S.T.; Brown, S.D.; Monfre, S.L. Transfer of near-infrared multivariate calibrations without standards. *Anal. Chem.* **1996**, *68*, 2987–2995. [CrossRef]

32. Wold, S.; Antti, H.; Lindgren, F.; Öhman, J. Orthogonal signal correction of near infrared spectra. *Chemom. Intell. Lab. Syst.* **1998**, *44*, 175–185. [CrossRef]

33. Sjöblom, J.; Svensson, O.; Josefson, M.; Kullberg, H.; Wold, S. An evaluation of orthogonal signal correction applied to calibration transfer of near infrared spectra. *Chemom. Intell. Lab. Syst.* **1998**, *44*, 229–244. [CrossRef]

34. Malli, B.; Birlutiu, A.; Natschläger, T. Standard-free calibration transfer-An evaluation of different techniques. *Chemom. Intell. Lab. Syst.* **2017**, *161*, 49–60. [CrossRef]

35. Pan, S.J.; Tsang, I.; Kwok, J.; Yang, Q. Domain adaptation via transfer component analysis. *IEEE Trans. Neural Netw.* **2011**, *22*, 199–210. [CrossRef] [PubMed]

36. Schölkopf, B.; Smola, A.; Müller, K.R. Kernel principal component analysis. In *Artificial Neural Networks — ICANN'97, Proceeding of 7th International Conference Lausanne, Lausanne, Switzerland, 8–10 October 1997*; Springer: Berlin/Heidelberg, Germany, 1997; pp. 583–588.

37. Schölkopf, B.; Smola, A.; Müller, K.-R. Nonlinear component analysis as a kernel eigenvalue problem. *Neural Comput.* **1998**, *10*, 1299–1319. [CrossRef]

38. Muandet, K.; Balduzzi, D.; Schölkopf, B. Domain generalization via invariant feature representation. In Proceedings of the 30th International Conference on Machine Learning (ICML-13), Atlanta, GA, USA, 16–21 June 2013; pp. 10–18.

39. Bloomenthal, J.; Jon, R. Homogeneous coordinates. *Visual Computer.* **1994**, *11*, 15–26. [CrossRef]

Sample Availability: Samples are not available from the authors.

molecules

Article

Wavelength Selection for NIR Spectroscopy Based on the Binary Dragonfly Algorithm

Yuanyuan Chen [1,2,*] and Zhibin Wang [2,3]

[1] School of Information and Communication Engineering, North University of China, Taiyuan 030051, China
[2] Engineering Technology Research Center of Shanxi Province for Opto-Electronic Information and Instrument, North University of China, Taiyuan 030051, China; wangzhibin@nuc.edu.cn
[3] School of Science, North University of China, Taiyuan 030051, China
* Correspondence: chenyy@nuc.edu.cn; Tel.: +86-139-3424-7637

Academic Editor: Christian Huck
Received: 29 December 2018; Accepted: 22 January 2019; Published: 24 January 2019

Abstract: Wavelength selection is an important preprocessing issue in near-infrared (NIR) spectroscopy analysis and modeling. Swarm optimization algorithms (such as genetic algorithm, bat algorithm, etc.) have been successfully applied to select the most effective wavelengths in previous studies. However, these algorithms suffer from the problem of unrobustness, which means that the selected wavelengths of each optimization are different. To solve this problem, this paper proposes a novel wavelength selection method based on the binary dragonfly algorithm (BDA), which includes three typical frameworks: single-BDA, multi-BDA, ensemble learning-based BDA settings. The experimental results for the public gasoline NIR spectroscopy dataset showed that: (1) By using the multi-BDA and ensemble learning-based BDA methods, the stability of wavelength selection can improve; (2) With respect to the generalized performance of the quantitative analysis model, the model established with the wavelengths selected by using the multi-BDA and the ensemble learning-based BDA methods outperformed the single-BDA method. The results also indicated that the proposed method is not limited to the dragonfly algorithm but can also be combined with other swarm optimization algorithms. In addition, the ensemble learning idea can be applied to other feature selection areas to obtain more robust results.

Keywords: wavelength selection; NIR spectroscopy; binary dragonfly algorithm; ensemble learning; quantitative analysis modeling

1. Introduction

Over the past decades, near-infrared (NIR) spectroscopy has been successfully applied in many areas, such as agriculture, medicine, environment [1]. Compared with traditional laboratory methods, NIR has the advantages of rapid speed and noninvasiveness. Usually, the NIR spectra collected from samples have the following several characteristics. Firstly, the number of necessary wavelengths is far higher than the number of samples, which, from the point of view of multivariate equations, means that the number of X factors (wavelengths) is far higher than the number of equations (samples). Hence, it is impossible to obtain a unique solution. Secondly, the contribution of each wavelength to quantitative analysis is different, since some wavelengths may be strongly correlated with the target content, while others may show little or no correlation. Obviously, it is not possible to consider the whole range of NIR spectra to perform subsequent qualitive or quantitative analysis. Hence, wavelength selection (also called "feature selection" and "variable selection") is an essential preprocessing step to find the most representative wavelengths and eliminate uninformative wavelengths.

In recent years, many researchers have focused on the wavelength selection issue and proposed a series of algorithms which have been proven effective in many areas. For example, Norgaard et al. [2]

proposed the interval PLS (iPLS) method, which first divides the whole range of spectrum into several intervals and then adopts a forward/backward stepwise selection algorithm to choose the most effective interval combinations. However, in how many intervals should the whole range of the spectrum be divided? Should it be divided equally or non-equally? These factors have a great influence on the wavelength selection results. Centner et al. [3] proposed a novel feature selection method called uninformative variable elimination (UVE), which brings in some random variables as the criterion of evaluating the correlation between wavelength and output target component. If the correlation coefficient of a certain wavelength is smaller than the random variables, the wavelength variable can be eliminated as an uninformative variable. The idea of the UVE algorithm is ingenious and intuitionistic. However, the disadvantage of this method is that the number of selected wavelengths is commonly large because this approach can only eliminate the uninformative wavelengths without selecting the most representative wavelengths [4]. Our group previously proposed an L1 regularization-based wavelength selection method [5], which considered the wavelength selection problem as a sparsity optimization issue. By adjusting the value of the sparsity parameter λ, we can freely control the number of selected wavelengths, which is reduced while the value of λ increases. Detailed information can be found in reference [5].

Besides the above-mentioned methods, evolution and swarm optimization methods (such as genetic algorithm [6], bat algorithm [7], particle swarm optimization [1], etc.) have also been applied to solve the wavelength selection problem. The main idea at the basis of these methods is similar. Firstly, they imply the generation of an initial population which is comprised of some individuals, each of which is a binary sequence. The length of each binary sequence is equal to the number of wavelengths, and the value of each point in the binary sequence is "1" or "0", indicating if the corresponding wavelength is selected or not. Secondly, the iterative searching is implemented by a series of heuristic strategies (for example, selection, crossover, mutation operators in genetic algorithm). However, due to the fact that there are some random mechanisms in swarm optimization methods (such as roulette wheel selection, Lévy flight, etc.), the wavelength selection results of each optimization are variable; hence, it is often confusing which wavelength variables should be taken into account. In 2016, Mirjalili et al. [8,9] proposed a novel swarm optimization method called "dragonfly algorithm (DA)" based on the observation, summary, and abstraction of the behaviors of the dragonfly in nature. Hence, the main contribution of this paper is to apply the binary dragonfly algorithm to solve the wavelength selection problem and improve its stability on the basis of the ensemble learning method.

The paper is organized as follows. Section 2 introduces the principles of the dragonfly algorithm and the basic idea of the transition dragonfly algorithm from continuous domain to binary domain. The proposed algorithm is introduced in detail in Section 3. The experimental results and discussion are presented in Section 4. Finally, Section 5 summarizes the contribution of this work and suggests some directions for future studies.

2. Dragonfly Algorithm

In the following section, we will introduce the principles of the dragonfly algorithm in continuous and discrete domains.

2.1. Continuous Dragonfly Algorithm

Biologists found that dragonflies have two interesting swarming behaviors: static and dynamic. This observation inspired the design of the dragonfly algorithm because there are two similar phases (called exploration and exploitation) in traditional swarm optimization methods. In the static swarm mode, dragonflies fly over different directions in a small area, which corresponds to the exploration phase; in the dynamic swarm mode, dragonflies fly in a bigger area along one direction, which is the main objective of the exploitation phase.

The main task of dragonflies is trying their best to survive; hence, they should be attracted towards food sources while avoiding enemies. Biologists observed that dragonflies usually change their position through five main strategies: separation, alignment, cohesion, attraction to food, and distraction from enemy, as shown in Figure 1.

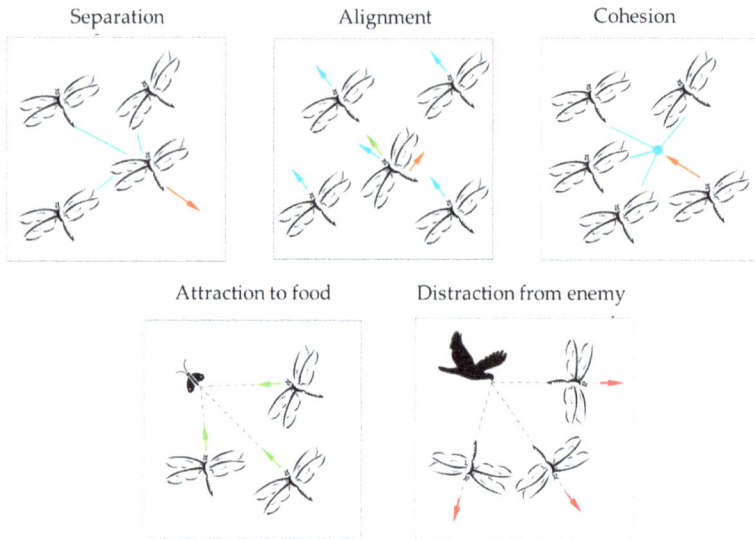

Separation   Alignment   Cohesion

Attraction to food   Distraction from enemy

Figure 1. Five main strategies by which dragonflies change their position [8].

The strategies to change position are mathematically modeled, as shown in Table 1.

Table 1. Mathematical modeling of the five main position-changing strategies of dragonfly [8].

Position Updating Strategies	Equations	Description
Separation	$S_i = -\sum_{j=1}^{N} X - X_j$	X: position of the current individual X_j: position of the j-th neighboring individual
Alignment	$A_i = \frac{\sum_{j=1}^{N} V_j}{N}$	N: number of the neighboring individuals
Cohesion	$C_i = \frac{\sum_{j=1}^{N} X_j}{N} - X$	V_j: velocity of the j-th neighboring individual
Attraction to food	$F_i = X^{+} - X$	X^{+}: position of the food source
Distraction from enemy	$E_i = X^{-} + X$	X^{-}: position of the enemy
Position updating	$\Delta X_{t+1} = (sS_i + aA_i + cC_i + fF_i + eE_i) + w\Delta X_t$ $X_{t+1} = X_t + \Delta X_{t+1}$	(s, a, c, f, e): separation, alignment, cohesion, food, and enemy factors w: inertia weight t: current iteration
Position updating with Lévy flight	$X_{t+1} = X_t + Levy(d) \times X_t$ $Levy(x) = 0.10 \times \frac{r_1 \times \sigma}{\|r_2\|^{\frac{1}{\beta}}}$ $\sigma = \left(\frac{\Gamma(1+\beta) \times sin\left(\frac{\pi\beta}{2}\right)}{\Gamma\left(\frac{1+\beta}{2}\right) \times \beta \times 2^{\left(\frac{\beta-1}{2}\right)}} \right)^{\frac{1}{\beta}}$ $\Gamma(x) = (x-1)!$	d: dimension of the position vectors r_1, r_2: random numbers in [0, 1] β: constant

As seen from the above table, dragonflies tend to align their flying while maintaining proper separation and cohesion in the dynamic swarm manner. However, while in the static swarm manner, alignments are very low, and cohesion is high to attack preys. Hence, we can assign high alignment and low cohesion weights to dragonflies exploring the search space, and low alignment and high cohesion to dragonflies exploiting the search space. In the transition between exploration and exploitation, the radius of the neighborhood is increased proportionally to the number of iterations. Another way to balance exploration and exploitation is to adaptively tune the swarming factors (s, a, c, f, *e*, and *w*) during the optimization. Additionally, to improve the randomness, stochastic behavior, and exploration of artificial dragonflies, dragonflies are required to fly around the search space using a random walk (*Lévy* flight) when there are no neighboring solutions. More detailed information can be found in reference [8].

2.2. Binary Dragonfly Algorithm (BDA)

In the continuous domain, dragonflies are able to update their positions through adding the step vector ΔX to the position vector X. However, in the discrete domain, the position of dragonflies cannot be updated in this way, because the position vectors X can only be 0 or 1. There are many transition methods described in the literature mapping the transition from continuous domain to discrete domain. Among them, the easiest and most effective method is to employ a transfer function which receives velocity values as inputs and returns a number in [0, 1], representing the probability of the changing positions.

There are two types of transfer functions: s-shaped and v-shaped. According to Saremi et al. [10], the v-shaped transfer functions outperform the s-shaped transfer functions because they do not force particles to take values of 0 or 1. In this paper, the following transfer function is utilized [11]:

$$T(\Delta x) = \left| \frac{\Delta x}{\sqrt{\Delta x^2 + 1}} \right| \tag{1}$$

After calculating the probability of changing position for all dragonflies, the following position-updating formula is employed to update the position of dragonflies in binary search spaces:

$$X_{t+1} = \begin{cases} -X_t, & r < T(\Delta x_{t+1}) \\ X_t, & r \geq T(\Delta x_{t+1}) \end{cases} \tag{2}$$

where r is a number in [0, 1].

3. Wavelength Selection Framework Based on BDA and Ensemble Learning

The flowchart of BDA-based wavelength selection method is illustrated in Figure 2. The main procedures are as follows.

Step 1: Mapping the wavelength selection problem to the BDA optimization problem. More precisely, initialize the binary dragonfly population. Without loss of generality, suppose there are K wavelengths in the whole range, hence the individuals in the initialized population are in binary series whose length is equal to K, in which the values of "1" and "0" indicate if the corresponding wavelength is selected or not.

Step 2: Evaluating the "goodness" or "badness" of each individual in the initialized population. For each individual, find the selected wavelength combinations and then establish the quantitative analysis model to predict the content of octane by using the partial least-squares (PLS) method. In this paper, the cost function (similar to the fitness function in the genetic algorithm) is defined as the root-mean-square error of cross validation (RMSECV) of the quantitative analysis model, which means that the individual with the smallest RMSECV has a good wavelength combination.

Step 3: Judging whether the stop criteria are satisfied or not. If yes, output the best individual in the population as the selected wavelength. If no, go to **step 4**. Generally, there are two approaches to design the stop criteria. The first takes the maximum iterations as the stop criteria, while the second takes the absolute error between two successive iterations as the stop criteria.

Step 4: According to Table 1, each individual in the population updates its position through five strategies, including separation, alignment, cohesion, attraction to food, and distraction from enemy. Finally, a new binary dragonfly population will be generated.

Step 5: Go back to **Step 2**, calculate the cost function of each individual in the new population, then execute the loop until the stop criteria are satisfied.

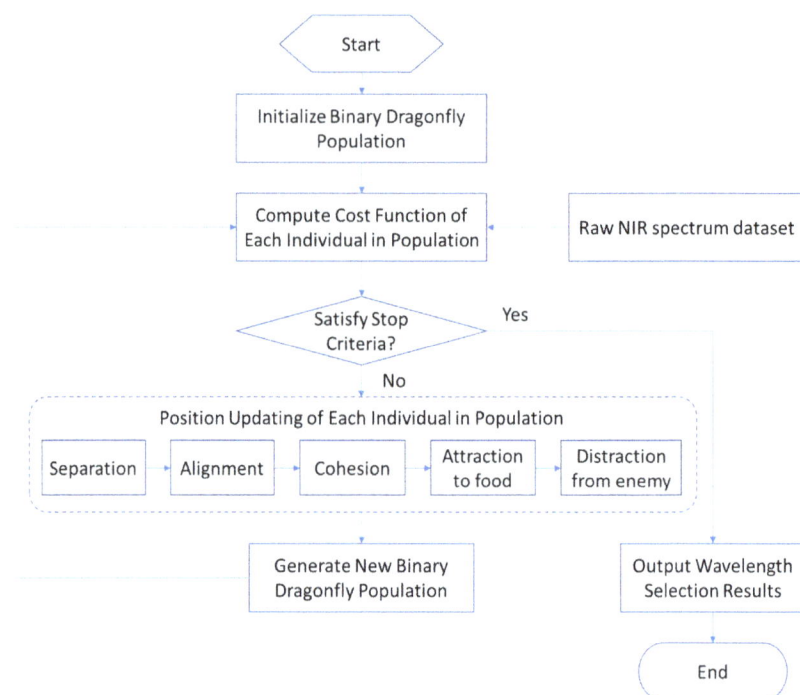

Figure 2. The flowchart of the binary dragonfly algorithm (BDA)-based wavelength selection method.

In this paper, we propose three typical wavelength selection methods based on BDA, as illustrated in Figure 3.

Similar to traditional swarm optimization methods, the wavelength section algorithm was designed on the basis of the single BDA. Figure 3a shows the procedure, which consists in using the BDA only once to search the most representative wavelengths. As mentioned above, because there are some random mechanisms in BDA, the wavelength selection results of each search are different.

To solve this problem, Figure 3b illustrates a possible solution which was designed on the basis of the multi-BDA. Differently from the single-BDA method, in this case the BDA is used many times, and then all the resulting selected wavelengths are aggregated by a voting strategy; those wavelengths with high votes are selected as the final wavelengths. However, an issue needs to be considered, that is, what is the quantitative criterion of "high votes"? Should it be 60%, 70%, 80%, or higher? The higher the vote percentage, the lower the number of selected wavelengths. If the number of selected wavelengths is too small, the performance of the subsequent quantitative analysis model may be reduced. Therefore, reaching a tradeoff between them is a challenge.

Furthermore, we propose a novel wavelength selection framework based on BDA and ensemble learning, as shown in Figure 3c. Traditionally, ensemble learning is often used to combine several models to improve the generalized performance of a model. However, in this paper, ensemble learning was not used to improve the model's performance. Instead, the idea of traditional ensemble learning was introduced to improve the stability of the wavelength selection results. Hence, only one PLS model was used during the wavelength selection period. Compared with the multi-BDA method, before the BDA searching procedure, Bootstrap sampling is added to randomly generate a series of different subsets. Suppose there are N samples in the raw dataset: N samples are drawn from the raw dataset with replacement, and then replicated samples are eliminated; hence, usually, the size of the subset becomes smaller because of the elimination.

(a)

(b)

(c)

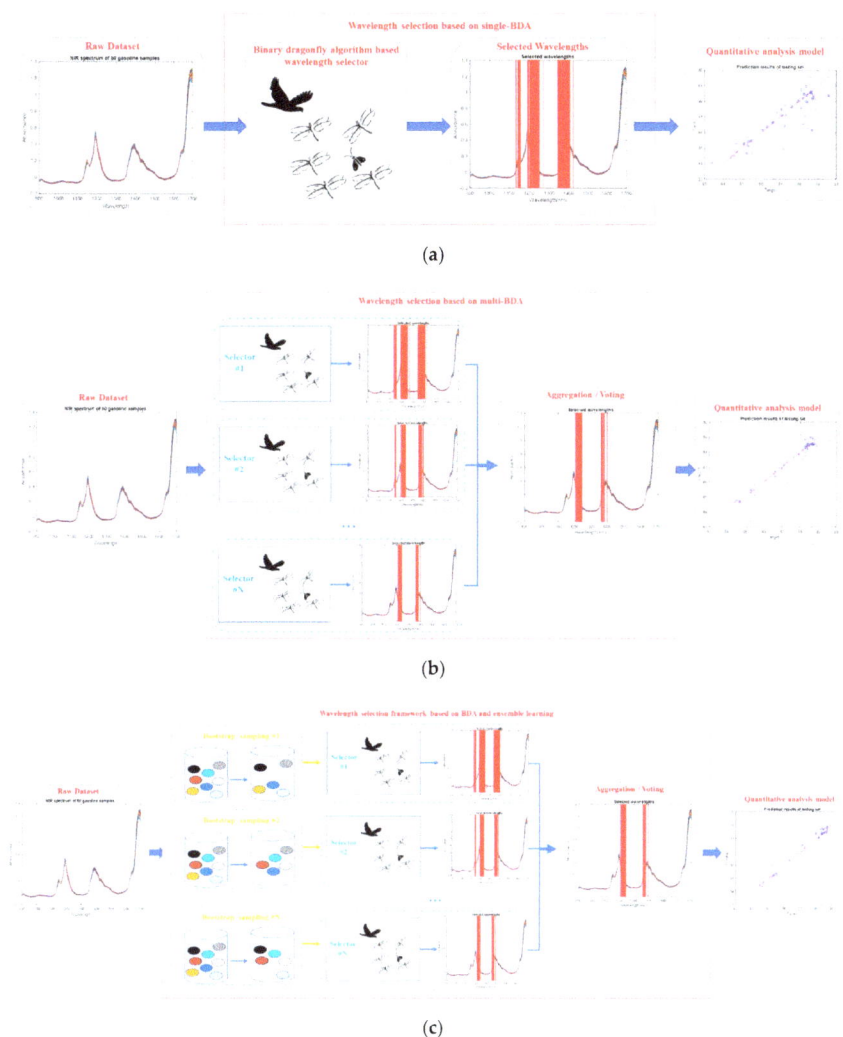

Figure 3. Three typical wavelength selection methods based on BDA. (**a**) Wavelength selection based on the single-BDA method; (**b**) wavelength selection based on the multi-BDA method; (**c**) wavelength selection framework based on the BDA and ensemble learning method.

Similar to the traditional swarm optimization algorithms, the single-BDA method also suffers from instability. In contrast, the multi-BDA and ensemble learning-based BDA methods can improve the stability. The key technical skill is to reduce the randomness inherent in the BDA. Even if both multi-BDA and ensemble learning-based BDA can achieve this, their principles are different. Although the multi-BDA method has many BDA selectors, they are built on the whole raw dataset. In contrast, the BDA selectors in the ensemble leaning method are built on different subsets. The variety of subsets leads to the different wavelength selection results from each BDA. This is the essence of ensemble learning. Besides, the computational complexity of the ensemble learning-based BDA method is smaller than that of the multi-BDA method, which mainly reflects in the computation of the cost (fitness) function RMSECV.

First of all, let us look at the scenario of the multi-BDA method. For example, consider a raw dataset with 100 samples and suppose the raw dataset is uniformly divided into five folds, each of which contains 20 samples. Establish five PLS models, each of which is trained on the basis of four folds (80) samples and tested on the remaining (20) samples.

Next, let us look at the scenario of the ensemble learning-based BDA method. According to the theoretical analysis, the size of subsets generated by Bootstrap sampling is approximately 63.2% that of the raw dataset (the whole theoretical analysis process can be found in reference [12]). Hence, there are about 63 samples in the subsets, and, similar to the multi-BDA method, the subset is uniformly divided into five folds, each of which contains about 12 samples. Establish five PLS models, each of which is trained on the basis of four folds (about 48) samples and tested on the remaining (12) samples.

The experimental results proved that, though the size of the samples in the subset became smaller, the performance was close to that of the multi-BDA method.

4. Experimental Results and Discussion

To validate the performance of the proposed wavelength selection methods, we applied the above-mentioned three typical algorithms to the public gasoline NIR spectroscopy dataset. The main aim was to find the most representative wavelengths to predict the content of octane by NIR spectrometry on the basis of the binary dragonfly algorithm.

4.1. Dataset Description

This dataset consists of 60 samples of gasoline measured by NIR spectrometry. The wavelength range was 900–1700 nm at 2 nm intervals (401 channels). The octane value for each of the samples was also included [13]. The raw whole range of the NIR spectrum of 60 samples is illustrated in Figure 4. Because this dataset had been embedded into MATLAB software (MathWorks Inc., Natick, MA, USA), the results were implemented in MATLAB R2017b. The basic source code of the BDA could be downloaded from the inventor's personal website (http://www.alimirjalili.com/DA.html) or MathWorks official website (https://www.mathworks.com/matlabcentral/fileexchange/51032-bda-binary-dragonfly-algorithm). On the basis of this source code, we implemented the single-BDA, multi-BDA, and ensemble learning-based wavelength selection codes.

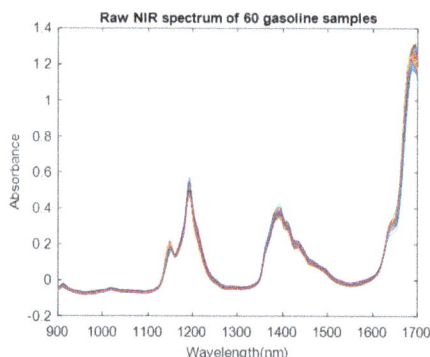

Figure 4. Raw spectrum of 60 gasoline samples.

4.2. Experimental Results

Firstly, the single-BDA method was applied to select the most representative wavelengths. In this paper, the PLS method was adopted to establish a quantitative analysis model. With respect to each binary individual, firstly, those wavelengths with value "1" were selected, and then a quantitative analysis model was established on the basis of these wavelengths. The values of parameters in Table 1 were adaptively tuned according to references [8–10]; the values of the remaining parameters of the BDA are listed in Table 2; we determined them by trial and error. The influence of different parameter values on the performance of the quantitative analysis model will be introduced in detail in the discussion section. Additionally, the BDA was repeated 10 times in the multi-BDA method; also, 10 parallel BDA wavelength selectors were present in the ensemble learning method. The influence of the BDA repetition on the wavelength selection results will also be introduced in the discussion section.

Table 2. Values of the BDA parameters.

Parameters	Values
Maximum number of iterations	50
Number of dragonflies	10
Number of wavelengths	401
Separation, alignment, cohesion, food, and enemy factors	adaptive tuning
Number of principal components	2
Number of folds of cross validation	5

As mentioned above, because of the random mechanism in the BDA, the wavelength selection results of each search were different. Figure 5 illustrates the wavelength selection results by using the single-BDA method. We implemented it for 10 times, and the wavelength selection results of each search are shown in Figure 5a, from which it is evident that the results were different. The corresponding generalized performances of the quantitative analysis model established with these selected wavelengths were also different, and the determined coefficient R^2 varied from 0.912 to 0.932, as shown in Figure 5b. These results indicated that, similar to traditional swarm optimization methods, the BDA can also be applied to solve the wavelength selection problem. However, it is often confusing which wavelength variables should be selected.

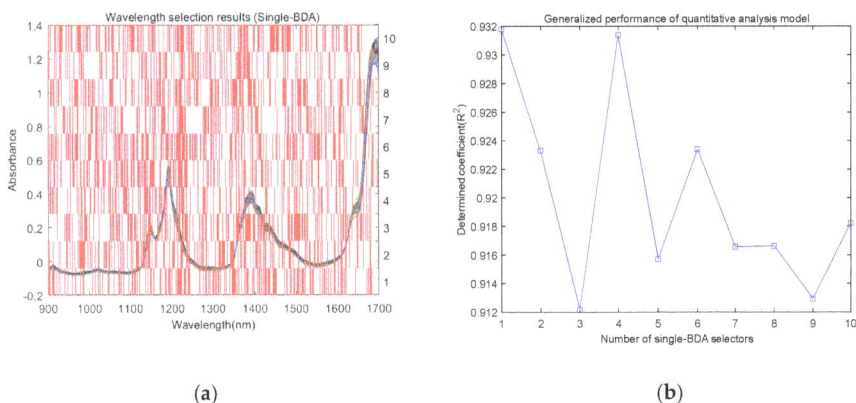

(a)   (b)

Figure 5. Wavelength selection results using the single-BDA method. (**a**) Wavelength selection results by applying a 10 times search; (**b**) corresponding generalized performance of the quantitative analysis model.

As mentioned above, the difference between the multi-BDA and the single-BDA methods is that there as a voting strategy to aggregate the wavelength selection results of the multiple-time search. The experimental results of the multi-BDA method are shown in Figure 6, from which it is obvious that by adjusting the value of the votes percentage (VP), we can control the number of selected wavelengths. Actually, if VP is equal to or smaller than 40%, the number of selected wavelengths is so large that the dimension is still high; on the contrary, if the VP is equal to or greater than 80%, there might be no wavelength satisfying this criterion. Hence, here we only show the wavelength selection results with VP between 50% and 80%. By comparing Figure 6 to Figure 5b, we can see that the determined coefficient R^2 of the quantitative analysis model established on the basis of the wavelengths selected with the multi-BDA method outperformed that obtained with the single-BDA method. However, the case of VP $\geq$ 80% was an exception, as only 7 wavelengths were selected. This means that if the number of selected wavelengths is too small, the generalized performance of the quantitative analysis model may decrease. Hence, a tradeoff has to be reached.

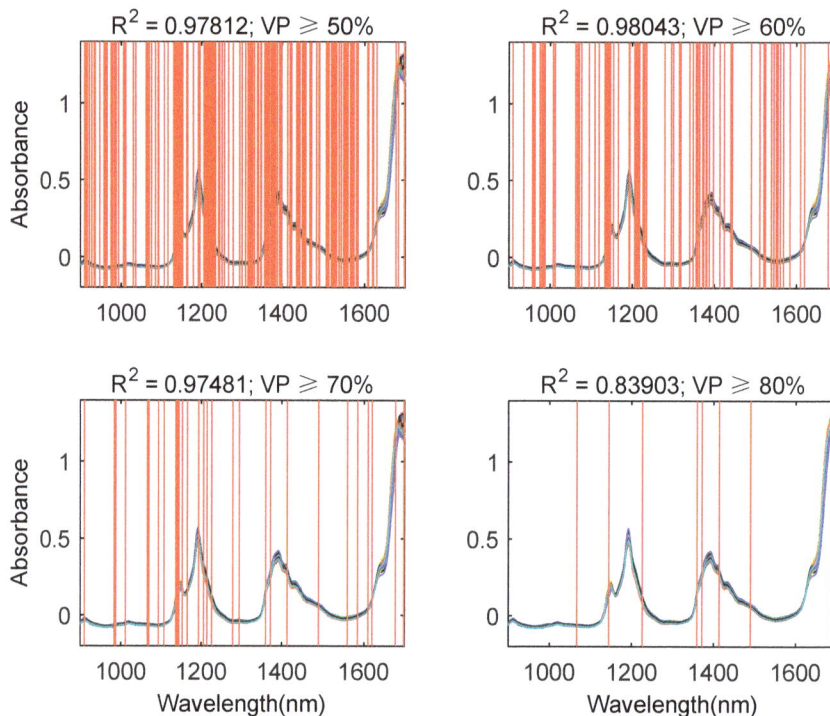

Figure 6. Wavelength selection results by using the multi-BDA method. VP: votes percentage. Upper left, upper right, lower left, and lower right represent selected wavelengths while the votes percentage is equal to or greater than 50%, 60%, 70%, and 80%, respectively.

Figure 7 describes the wavelength selection results by using the BDA and the ensemble learning method. The difference between this method and the multi-BDA method is that a series of Bootstrap sampling generators were added before the BDA. Because Bootstrap sampling is a method of random sampling with replacement, the size of the subset is smaller than the whole dataset. As mentioned above, the subset only contained about 63.2% samples of the whole set. By comparing Figure 7 to Figure 6, it is easy to see that, although the sample size was reduced through Bootstrap sampling, the generalized performance of the quantitative analysis models with selected features was close to that of the multi-BDA method. Hence, by using this method, we could reduce the computational complexity. However, similar to the multi-BDA method, if the votes percentage is equal to or greater than 80%, the generalized performance of the quantitative analysis model decreases. In summary, we suggest that the votes percentage should be set between 60% and 70%.

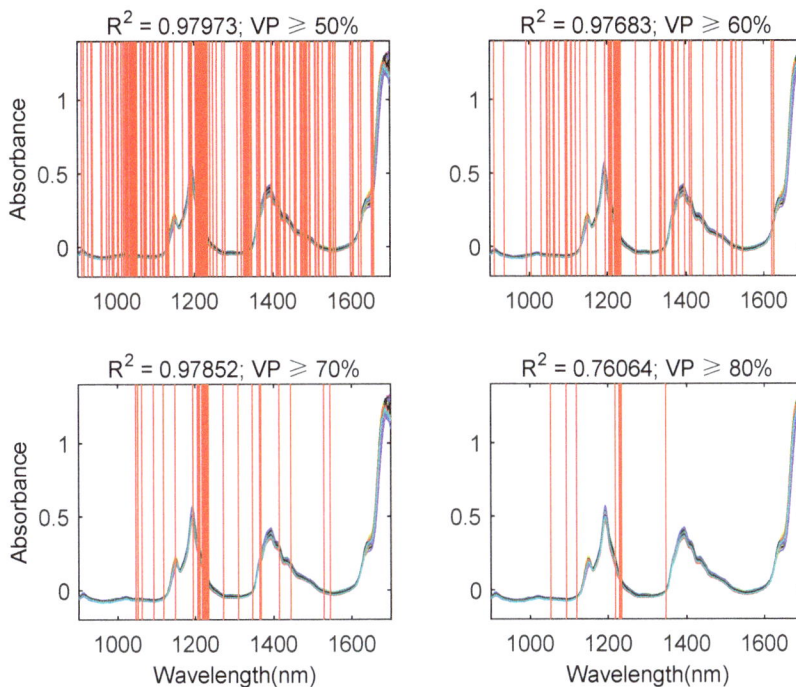

Figure 7. Wavelength selection results by using the BDA and ensemble learning method. VP: votes percentage. Upper left, upper right, lower left, and lower right represent selected wavelengths with votes percentage equal to or greater than 50%, 60%, 70%, and 80%, respectively.

4.3. Discussion

4.3.1. Influence of the Values of the BDA Parameters on the Generalized Performance of the Model

Similar to other evolution and swarm optimization algorithms, there are some parameters (as listed in Tables 1 and 2) inherent in BDA and while the values of these parameters are different, the corresponding wavelength selection results will different too. Hence, we want to quantitatively analyze the influence of the values of these parameters on the wavelength selection results.

As mentioned above, in this paper the values of parameters in Table 1 were adaptively tuned as follows:

$$w = 0.9 - iter \times \frac{0.9 - 0.4}{max\_iter} \tag{3}$$

where *iter* and *max_iter* is the current and maximum iteration, respectively. It is obviously to see that w is linearly decreased from 0.9 to 0.4.

With respect to parameter *s*, *a* and *c*, the adaptive tuning formula is:

$$x = 2 \times \gamma \times \theta, \quad x \in \{s, a, c\} \tag{4}$$

where γ is a random between 0 and 1.

$$\theta = \begin{cases} 0.1 - iter \times \frac{0.1 - 0}{max\_iter/2}, & iter \leq max\_iter/2 \\ 0, & iter > max\_iter/2 \end{cases} \tag{5}$$

From Equations (4) and (5), we can find that the range of parameter *s*, *a* and *c* is between 0 and 0.2. Parameter f is a random between 0 and 2, and parameter e is equal to θ.

The adaptive tuning procedure of parameters (w, s, a, c, f, e) is illustrated in Figure 8. We can easily find that while the current iteration period is larger than half of maximum iteration, values of parameters s, a, c and e become zero. In other words, the separation, alignment, cohesion and detraction from enemy strategies were not included in the position updating procedure, and only inertia weight w and attraction to food f were considered.

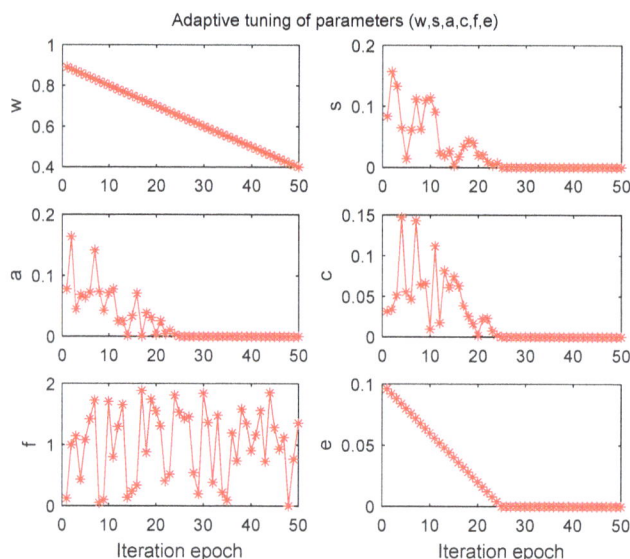

Figure 8. Adaptive tuning of parameters (w, s, a, c, f, e).

Table 3 shows the influence of different parameter values on the generalized performance (RMSECV) of quantitative analysis model. The second column means that all of the parameters were adaptively tuned, and the third and fourth columns mean that the corresponding parameter's value were fixed to its maximum or minimum respectively while other parameters were adaptively tuned. We can find that there is a trend that while these parameters were not adaptively tuned, the generalized performance (here we use mean RMSECV of 10 times repeated) of quantitative analysis model will decrease, which is consisting with reference [8].

Table 3. The values of related parameters of BDA.

Parameters	Mean RMSECV of 10 Times Repeated PLS Models with Selected Wavelengths		
	Adaptive Tuning	Fixed (Maximum)	Fixed (Minimum)
w (0.4–0.9)		0.5196	0.2583
s (0–0.2)		0.5500	0.3700
a (0–0.2)	0.3801	0.4833	0.4186
c (0–0.2)		0.4955	0.4071
f (0–2)		0.4871	0.5224
e (0–0.1)		0.3796	0.4570

Further, we consider the parameters in Table 2. Previously principal component analysis (PCA) showed that the cumulative contribution of the first two principal components was higher than 90%, hence in this paper we take the first two principal components to establish the PLS regression model. Additionally, 5-fold cross validation is often used so that we can ignore its influence on the model's performance. Hence, here we put focus on two parameters: maximum number of iterations and number of dragonflies. We implemented a two-dimensional grid to evaluate the influence of these two parameters, in which the maximum number of iterations and number of dragonflies were range from 10 to 100 and 10 to 50, respectively. Because of the limitation of space, here we only take single-BDA as an example. At each parameter value pairs, mean RMSECV of 10 times repeated PLS models with selected wavelengths was computed as the evaluation index. The experimental results were illustrated in Figure 9, from which we can obviously find that with the increase of maximum number of iteration and number of dragonflies (population size), mean RMSECV of 10 times repeated PLS models showed a decrease trend. while the computational complexity increases a lot. Hence, there is a trade-off between them.

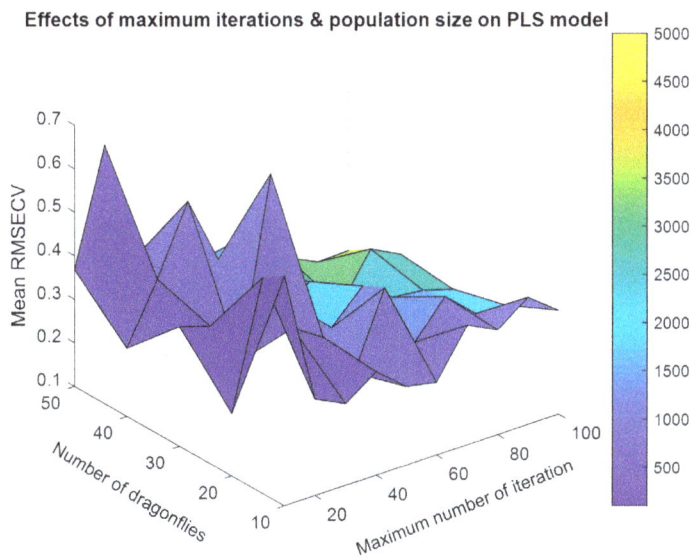

Figure 9. Influence of maximum number of iterations and number of dragonflies on the generalized performance of quantitative analysis model.

4.3.2. Comparison Between Proposed Methods and Traditional Methods

In order to validate the efficiency of proposed methods, here we compared the proposed single-BDA, multi-BDA and ensemble learning based BDA methods with traditional methods. Considering the fact that it is difficult to find a common standard to implement the comparison between proposed methods with interval PLS, hence here we limit the comparison between proposed methods with evolution and swarm optimization methods. Because our previous studies have validated the efficiency of binary bat algorithm and genetic algorithm, here we implemented the comparison between proposed BDA and binary bat algorithm and genetic algorithm. We set the population size and maximum of iteration of these methods were all same. As mentioned above, for each method, mean RMSECV of 10 times repeated PLS models with selected wavelengths was computed as the evaluation index. The comparison results were listed in Table 4, from which we can obviously find that compared with genetic algorithm and binary bat algorithm, the performance of single-BDA method is a little lower. However, there is no significant statistical difference between them. Additionally, the performance of multi-BDA and ensemble learning based BDA methods were similar, both outperforms traditional methods and single-BDA.

Table 4. Comparison between proposed BDA based methods and traditional methods.

Methods	Population Size	Maximum of Iteration	RMSECV of 10 Times Repeated PLS Models with Selected Wavelengths	
			Mean	Std
Genetic algorithm			0.4016	0.0624
Binary bat algorithm			0.3672	0.0482
Single-BDA	20	50	0.3801	0.0549
Multi-BDA			0.3265	0.0215
Ensemble learning based BDA			0.3294	0.0168

5. Conclusions

In terms of the wavelength selection problem in NIR spectroscopy, this paper proposes a novel method based on binary dragonfly algorithms, which includes three typical frameworks: single-BDA, multi-BDA, and ensemble learning-based BDA framework. The experimental results for a public gasoline NIR dataset showed that by using the proposed method, we could improve the stability of the wavelength selection results through the multi-BDA and ensemble learning-based BDA methods. With respect to the subsequent quantitative analysis modeling, the wavelengths selected with the multi-BDA and ensemble learning BDA methods outperformed those selected with the single-BDA method. When comparing the ensemble learning-based BDA and the multi-BDA methods, it can be seen that they can provide similar wavelength selection results with lower computational complexity. The results also indicated that the proposed method is not limited to the dragonfly algorithm but it can be combined with other swarm optimization algorithms. In addition, the ensemble learning idea can be applied to other feature selection areas to obtain more robust results.

Besides, during the experimental results analysis, we found that, for the same selected wavelengths, there were great changes of the generalized performance (RMSECV) between different subsets. This means that not only the wavelength variables, but also the samples influence the generalized performance of the quantitative analysis model. Currently, the majority of studies are focused on the wavelength selection problem; we suggest that sample selection using swarm optimization methods is an open problem that needs to be studied further.

Additionally, in this study, we found that the votes percentage should be set between 60% and 70%. However, we do not know whether this range is also suitable for other datasets. This should be validated further. Actually, in general, we should not be concerned by which wavelengths are selected. Instead, we may want to know how good the quantitative analysis model will be by using the selected wavelengths. Hence, improving the generalized performance of the quantitative analysis model is a hot topic in NIR spectroscopy. With the development of artificial intelligence, ensemble learning methods (such as random forest, adaboost, etc.) and deep learning (such as convolutional neural networks [14–19]) will be mainstream in the future.

Author Contributions: conceptualization, Y.C. and Z.W.; methodology, Y.C.; software, Y.C.; validation, Z.W.; formal analysis, Y.C.; investigation, Y.C.; resources, Z.W.; data curation, Y.C.; writing—original draft preparation, Y.C.; writing—review and editing, Z.W.; visualization, Y.C.; supervision, Y.C.; project administration, Y.C.; funding acquisition, Y.C.

Funding: This research was funded by the NATIONAL NATURAL SCIENCE FOUNDATION OF CHINA, grant number 61605176 and INTERNATIONAL COOPERATION PROJECTS OF SHANXI PROVINCE, CHINA, grant number 201803D421033.

Acknowledgments: The authors would like to give thanks to the many researchers who contributed previous remarkable works in this field.

Conflicts of Interest: The authors declare no conflict of interest.

Molecules **2019**, *24*, 421

References

1. Zou, X.; Zhao, J.; Povey, M.J.; Holmes, M.; Mao, H. Variables selection methods in near-infrared spectroscopy. *Anal. Chim. Acta* **2010**, *667*, 14–32. [CrossRef] [PubMed]

2. Norgaard, L.; Saudland, A.; Wagner, J.; Nielsen, J.P.; Munck, L.; Engelsen, S.B. Interval partial least-squares regression (iPLS): A comparative chemometric study with an example from near-infrared spectroscopy. *Appl. Spectrosc.* **2000**, *54*, 413–419. [CrossRef]

3. Centner, V.; Massart, D.L.; de Noord, O.E.; de Jong, S.; Vandeginste, B.M.; Sterna, C. Elimination of uninformative variables for multivariate calibration. *Anal. Chem.* **1996**, *68*, 3851–3858. [CrossRef] [PubMed]

4. Moros, J.; Kuligowski, J.; Quintás, G.; Garrigues, S.; de la Guardia, M. New cut-off criterion for uninformative variable elimination in multivariate calibration of near-infrared spectra for the determination of heroin in illicit street drugs. *Anal. Chim. Acta* **2008**, *6*, 150–160. [CrossRef] [PubMed]

5. Zhang, R.; Chen, Y.; Wang, Z.; Li, K. A novel ensemble L1 regularization based variable selection framework with an application in near infrared spectroscopy. *Chemom. Intell. Lab. Syst.* **2017**, *163*, 7–15.

6. Givianrad, M.H.; Saber-Tehrani, M.; Zarin, S. Genetic algorithm-based wavelength selection in multicomponent spectrophotometric determinations by partial least square regression: Application to a sulfamethoxazole and trimethoprim mixture in bovine milk. *J. Serb. Chem. Soc.* **2013**, *78*, 555–564. [CrossRef]

7. Chen, Y.; Wang, Z. Feature Selection of Infrared Spectrum Based on Improved Bat Algorithm. Available online: http://journal02.magtech.org.cn/Jwk_irla/CN/Y2014/V43/I8/2715 (accessed on 15 December 2018).

8. Mirjalili, S. Dragonfly algorithm: A new meta-heuristic optimization technique for solving single-objective, discrete, and multi-objective problems. *Neural Comput. Appl.* **2016**, *27*, 1053–1073. [CrossRef]

9. Mafarja, M.M.; Eleyan, D.; Jaber, I. Binary dragonfly algorithm for feature selection. In Proceedings of the 2017 International Conference on New Trends in Computing Sciences (ICTCS), Amman, Jordan, 11–13 October 2017.

10. Saremi, S.; Mirjalili, S.; Lewis, A. How important is a transfer function in discrete heuristic algorithms. *Neural Comput. Appl.* **2015**, *26*, 625–640. [CrossRef]

11. Mirjalili, S.; Lewis, A. S-shaped versus V-shaped transfer functions for binary particle swarm optimization. *Swarm Evol. Comput.* **2013**, *9*, 1–14. [CrossRef]

12. Yuanyuan, C.; Zhibin, W. Quantitative analysis modeling of infrared spectroscopy based on ensemble convolutional neural networks. *Chemom. Intell. Lab. Syst.* **2018**, *181*, 1–10. [CrossRef]

13. Kalivas, J.H. Two data sets of near infrared spectra. *Chemom. Intell. Lab. Syst.* **1997**, *37*, 255–259. [CrossRef]

14. Le, B.T.; Xiao, D.; Mao, Y.; He, D. Coal analysis based on visible-infrared spectroscopy and a deep neural network. *Infrared Phys. Technol.* **2018**, *93*, 34–40. [CrossRef]

15. Wang, C.; Wu, X.-H.; Li, L.Q.; Wang, Y.-S.; Li, Z.-W. University, Convolutional Neural Network Application in Prediction of Soil Moisture Content. Available online: http://www.gpxygpfx.com/article/2018/1000-0593-38-1-36.html (accessed on 15 December 2018).

16. Acquarelli, J.; Laarhoven, T.V.; Gerretzen, J.; Tran, T.N.; Buydens, L.M.C.; Marchiori, E. Convolutional neural networks for vibrational spectroscopic data analysis. *Anal. Chim. Acta* **2017**, *954*, 22–31. [CrossRef] [PubMed]

17. Afara, I.O.; Sarin, J.K.; Ojanen, S.; Finnilä, M.; Herzog, W.; Saarakkala, S.; Töyräs, J. Deep Learning Classification of Cartilage Integrity Using Near Infrared Spectroscopy. In Proceedings of the Biomedical Optics Congress 2018, Hollywood, FL, USA, 3–6 April 2018.

18. Malek, S.; Melgani, F.; Bazi, Y. One-dimensional convolutional neural networks for spectroscopic signal regression. *J. Chemom.* **2018**, *32*, e2977. [CrossRef]

19. Zhang, J.; Liu, W.; Hou, Y.; Qiu, C.; Yang, S.; Li, C.; Nie, L. Sparse Representation Classification of Tobacco Leaves Using Near-Infrared Spectroscopy and a Deep Learning Algorithm. *Anal. Lett.* **2018**, *51*, 1029–1038. [CrossRef]

molecules

MDPI

Article

Data Fusion of Fourier Transform Mid-Infrared (MIR) and Near-Infrared (NIR) Spectroscopies to Identify Geographical Origin of Wild *Paris polyphylla* var. *yunnanensis*

Yi-Fei Pei [1,2], Zhi-Tian Zuo [1], Qing-Zhi Zhang [2,*] and Yuan-Zhong Wang [1,*]

1   Institute of Medicinal Plants, Yunnan Academy of Agricultural Sciences, Kunming 650200, China
2   College of Traditional Chinese Medicine, Yunnan University of Chinese Medicine, Kunming 650500, China
*   Correspondence: ynkzqz@126.com (Q.-Z.Z.); boletus@126.com (Y.-Z.W.)

Academic Editors: Christian Huck and Krzysztof B. Bec
Received: 16 May 2019; Accepted: 11 July 2019; Published: 13 July 2019

Abstract: Origin traceability is important for controlling the effect of Chinese medicinal materials and Chinese patent medicines. *Paris polyphylla* var. *yunnanensis* is widely distributed and well-known all over the world. In our study, two spectroscopic techniques (Fourier transform mid-infrared (FT-MIR) and near-infrared (NIR)) were applied for the geographical origin traceability of 196 wild *P. yunnanensis* samples combined with low-, mid-, and high-level data fusion strategies. Partial least squares discriminant analysis (PLS-DA) and random forest (RF) were used to establish classification models. Feature variables extraction (principal component analysis—PCA) and important variables selection models (recursive feature elimination and Boruta) were applied for geographical origin traceability, while the classification ability of models with the former model is better than with the latter. FT-MIR spectra are considered to contribute more than NIR spectra. Besides, the result of high-level data fusion based on principal components (PCs) feature variables extraction is satisfactory with an accuracy of 100%. Hence, data fusion of FT-MIR and NIR signals can effectively identify the geographical origin of wild *P. yunnanensis*.

Keywords: origin traceability; data fusion; *Paris polyphylla* var. *yunnanensis*; Fourier transform mid-infrared spectroscopy; near-infrared spectroscopy

1. Introduction

The rhizome of *Paris polyphylla* var. *yunnanensis* (Franch.) Hand. -Mazz (*P. yunnanensis*) and *P. polyphylla* Smith var. *chinensis* (Franch.) Hara (*P. chinensis*), named as "chonglou" in Chinese, is a renowned and traditional herb with a history of thousands of years in China and plants belong to *Paris* genus, Liliaceae family. As an ancient history ethnobotanical medicinal plant, it is used to treat snake bite and insect sting, innominate toxin swelling, and a variety of inflammatory and traumatic in the folk in China. Various phytochemical researches have demonstrated that steroidal saponin, phytosterol, molting hormone, flavone, and pentacyclic triterpene are the major chemical components in *Paris* [1,2]. Additionally, *Paris* is considered to have anti-bacterial, anti-myocardial ischemia, anti-tumor, analgesia, immune-regulation according to numerous pharmacological studies [3–7]. As an important and precious medicinal plant, the raw material plants of *P. yunnanensis* are widely spread over southern China, especially in Yunnan Province [8]. Our previous studies have shown that there are significant differences in the content of wild *P. yunnanensis* samples from different geographical sources, and the saponin content in southern Yunnan is relatively higher than other regions [9,10]. Hence, it is crucial to the traceable geographical origin of wild *P. yunnanensis* samples to ensure effective medicinal values, which helps to ensure the effectiveness of the medication.

Some techniques have been applied to identify the authenticity and quality of various herbal medicines, including Fourier transform mid-infrared (FT-MIR), near-infrared (NIR), ultraviolet-visible (UV-Vis), Raman, liquid chromatography-mass spectrometry (LC-MS), high performance liquid chromatography (HPLC), etc. [11–16]. In recent years, several researches in classification of *P. yunnanensis* have widely used chemometrics models combined with various analytical techniques, including partial least squares discriminant analysis (PLS-DA), principal component analysis (PCA), hierarchical cluster analysis (HCA), random forest (RF), support vector machine (SVM), etc. [17–21]. Among them, spectroscopic techniques are fast, lossless, and efficient for the analysis of herbal medicines. Besides, quality of *P. yunnanensis* is difficult to identify by one or several chemical components due to the synergistic effect of TCMs, while the integral chemical information of medicinal plants can be provided by chromatographic or spectral fingerprints. For example, Yang et al. applied PCA and cluster analysis combined with FT-MIR to classify the small quantity wild and cultivated *P. yunnanensis* samples [21]. PLS-DA and RF models combined with FT-MIR have successfully traced the cultivated *P. yunnanensis* samples from Yunnan Province with different cultivation years [17]. Besides, our previous study has shown that the PLS-DA model combined with different parts (rhizomes and leaves) FT-MIR information can effectively distinguish the samples of cultivated *P. yunnanensis* collected in different cities of Yunnan Province [19].

Only a kind of chemical profile was obtained by the single technique, the relative complete chemical information would be provided by multiple platforms. The data fusion strategy contains low-, middle- and high-level, which can effectively fuse the chemical information obtained by different platforms of samples into one dataset to identification and classification researches [22]. As a case study, Li et al. found that the discriminant model established by FT-MIR and NIR spectral data combined with high-level data fusion strategy can effectively identify *Panax notoginseng* from different cultivation regions [23]. Wu et al. demonstrated that FT-MIR combined with UV-Vis by data fusion strategy could obtain a reliable and good result to trace the geographical origins of wild *P. yunnanensis* samples [24]. Hence, it is vital to effectively combine multiple techniques datasets of *P. yunnanensis* to obtain excellently chemical information to identification analysis and the effective results.

In this study, to obtain further realization of the similarities and differences among wild *P. yunnanensis* samples from central, western, northwest, southeast, and southwest Yunnan, we studied the collected samples using two spectroscopic techniques (FT-MIR and NIR), and fused with data fusion strategy (low-, mid- and high-level) combined with chemometrics including PCA, PLS-DA, and RF. The important variables regions among each classification models and the fast-quality assessment effects for geographical origins of *P. yunnanensis* were compared. Additionally, the chemical fingerprint of wild *P. yunnanensis* samples from different areas at Yunnan Province for FT-MIR and NIR spectra were analyzed. The results of our study may provide some basis for comprehensive utilization of *P. yunnanensis* resources.

2. Results and Discussion

2.1. Macroscopic Chemistry Components in IR Spectra

Averaged raw FT-MIR and NIR spectra of wild *P. yunnanensis* samples from central, northwest, western, southeast, and southwest Yunnan Province are shown in Figure 1. Twenty-five major common peaks were contained in FT-MIR spectra from these five regions, as shown in Figure 1a. The 4000 to 3700 cm^{-1} and 2620 to 1800 cm^{-1} absorptions were the FT-MIR spectral baseline area and diamond crystal spectral region, respectively, which areas provide invalid spectral information for this study [25]. Besides, regions of 3700 to 2620 cm^{-1} and 1800 to 1300 cm^{-1} were defined as characteristic areas in our study, which mainly contained C=O, C=C, and C–H stretching vibration as well as C–H bending vibration mode [21,26]. In addition, the region of 1300 to 650 cm^{-1} was the fingerprint region, which greatly contained C–O stretching vibration, C–C stretching vibration, C–OH bending vibration mode, as well as sugar skeleton vibration [27,28]. For all above these useful regions, FT-MIR spectra can be divided into five distinct ranges, including 3700 to 2000, 1800 to 1500, 1500 to 1200, 1200 to 900, and 900 to 650 cm^{-1} [29]. In the region of 3700 to 2000 cm^{-1}, the broad absorption band in the range 3700–3000 cm^{-1} corresponds to the stretching vibrations of free hydroxyl groups ν(OH) and the groups involved in intra- and intermolecular hydrogen bonds [29]. Absorption at the peaks of 2928 and 2852 cm^{-1} were assigned to normal vibration mode such as the CH$_3$ asymmetric normal vibration mode at 2960 to 2920 cm^{-1}, CH$_2$ asymmetric normal vibration mode at 2930 to 2900 cm^{-1}, CH$_3$ symmetric normal vibration mode at 2900–2880 cm^{-1}, and CH$_2$ symmetric normal vibration mode at 2860 to 2850 cm^{-1} [29]. Additionally, the region two to the region four were useful to deconvolute the bands into Lorentzian components, and the detailed information can be observed in Table 1. Moreover, amide I band is observed in the region of 1800 to 1500 cm^{-1} too, which corresponds to the C=C stretching mode of fatty acids and flavonoids [29].

Table 1. Peak assignments on the FT-MIR and NIR spectra of wild *P. yunnanensis*.

Spectral Type	Wavenumber (cm^{-1})	Base Group and Vibration Mode	Contribution
NIR	8347	C–H, N–H and O–H stretching vibration mode	CH$_2$, saccharides, and glycosides
	7256	C–H stretching and deformation vibration mode	CH$_2$
	6950	C–H, N–H and O–H stretching vibration mode	CH$_2$, saccharides, and glycosides
	6324	C–H, N–H and O–H stretching vibration mode	CH$_2$, saccharides, and glycosides
	5686	C–H, N–H and O–H stretching vibration mode	CH$_2$, saccharides, and glycosides
	5169	C–H, N–H and O–H and hydrogen bond stretching vibration mode	CH$_2$, saccharides, glycosides, and water molecule
FT-MIR	3382	O–H asymmetric and hydrogen bond stretching vibration mode	Saccharides, glycosides, and water molecule
	3334	O–H asymmetric and hydrogen bond stretching vibration mode	Saccharides, glycosides, and water molecule
	2930	C–H asymmetric stretching vibration mode	CH$_2$ and CH$_3$
	1743	C=O stretching vibration mode	Free carboxyl groups of pectins or/and fatty acids
	1653	asymmetric stretching vibrations of carboxyl groups participating in the hydrogen bonds and hydrogen bond scissoring vibration mode	Flavonoids, saccharides, steroid saponin, and water molecules
	1610	COO symmetric normal vibrations mode	The carboxyl group present in pectin
	1456	CH$_3$ asymmetric deformation and CH$_2$ scissoring vibration	CH$_2$ and CH$_3$
	1414	C–H symmetric bending vibration mode and OH–O in-plane bending mode	CH$_2$
	1370	C–H symmetric deformation vibration mode	CH$_3$
	1242	C–O stretching vibration mode	Saccharides and oils
	1150	C–C and C–O stretching and C–OH bending vibration mode	Saccharides and glycosides
	1078	C–C and C–O stretching and C–OH bending vibration mode	Saccharides and glycosides
	1020	C–C and C–O stretching and C–OH bending vibration mode	Saccharides and glycosides
	922	Sugar skeleton vibration mode	Saccharides

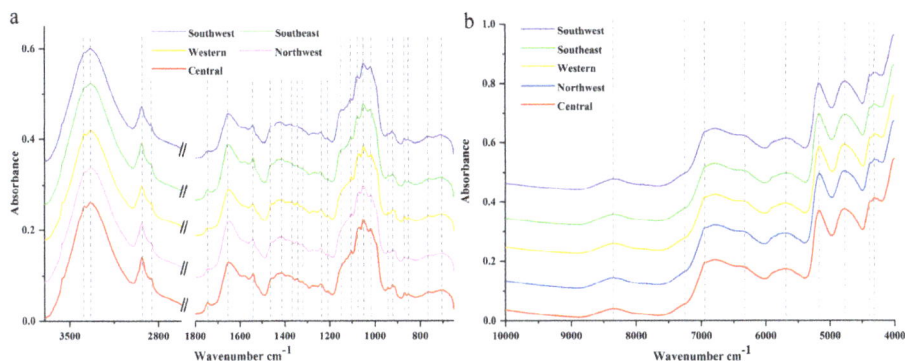

Figure 1. Averaged spectra of *P. yunnanensis* samples collected from five regions: (**a**) Fourier transform mid-infrared (FT-MIR) spectra; (**b**) near-infrared (NIR) spectra.

Nine major common peaks are obtained in average NIR spectra from five geographical origins, as shown in Figure 1b. The bands in the region of 9000 to 4500 cm^{-1} be associated with the first or second overtones [30]. The peaks in the region 4500 to 4000 cm^{-1} are so narrow that it was difficult to provide detailed information [31]. Besides, the peaks at 5169, 3382, 3334, and 1653 cm^{-1} were considered, which also may correspond to hydrogen bond stretching and scissoring vibration mode attributed to water molecules [29]. The fingerprint of FT-MIR and NIR spectra characteristics of *P. yunnanensis* from different geographical origins were similar, as shown in Figure 1, which indicated similar chemical composition among these samples. Additionally, detailed peak positions and assignments were applied in Table 1.

2.2. Single Block Models

Raw FT-MIR and NIR spectra were pretreated by standard normal variate (SNV), first derivative (FD), second derivative (SD), SNV-FD, and SNV-SD, and parameters of these pretreatment algorithms are applied in Table S1, including parameters of cumulative interpretation ability (R^2), cumulative prediction ability (Q^2), the root mean square error of estimation (RMSEE), the root mean square error of cross-validation (RMSECV), and accuracy. For the FT-MIR spectra dataset, the worst classification ability was by FD algorithms, which also had accuracy worse than the raw dataset. But for the NIR spectra dataset, the SNV pretreatment algorithm was the worst preprocessing algorithm. However, SD was the best preprocessing algorithm, both for FT-MIR and NIR, whereby the accuracy even reached 100%. Among all preprocessing algorithms, the best pretreatment algorithm (SD) for each kind of spectroscopy should be selected and used to establish geographical classification models.

PLS-DA and RF classification models were established on FT-MIR and NIR SD datasets, respectively. The efficiency for each class and accuracy of the calibration set and validation set of these geographical origin models are shown in Table 2. The parameter of the root means square error of prediction (RMSEP) was one important parameter for evaluating model classification ability. The values of RMSEP were 0.203 and 0.236 of PLS-DA based on FT-MIR and NIR spectra datasets, respectively. With the higher accuracy and lower RMSEP, the PLS-DA model effect of using FT-MIR data to classify geographical origins was better than that of NIR.

Table 2. The classification efficiency values and total accuracy of independent decision making with Partial least squares discriminant analysis (PLS-DA) and random forest (RF) models. RFE: Recursive feature elimination.

Model		Calibration Set						Validation Set					
		Class1	Class2	Class3	Class4	Class5	Accuracy	Class1	Class2	Class3	Class4	Class5	Accuracy
FT-MIR	PLS-DA	0.961	1.000	0.995	0.981	0.990	97.66%	1.000	0.991	0.913	1.000	0.991	97.06%
	RF	0.772	0.888	0.801	0.829	0.790	71.88%	0.886	0.964	0.973	1.000	0.946	92.65%
NIR	PLS-DA	1.000	1.000	1.000	1.000	1.000	100%	0.870	0.964	0.940	0.936	0.964	89.71%
	RF	0.803	0.854	0.775	0.837	0.834	72.66%	0.813	0.917	0.491	0.923	0.955	76.47%
FT-MIR (RFE)	PLS-DA	0.911	0.990	0.881	0.961	0.975	91.41%	0.794	1.000	0.694	0.955	0.923	82.35%
	RF	0.947	0.951	0.853	0.876	0.942	86.72%	0.845	0.917	0.964	0.964	0.936	88.24%
FT-MIR (Bo)	PLS-DA	0.951	0.961	0.995	0.961	0.985	95.31%	0.886	0.991	0.905	0.964	0.962	91.18%
	RF	0.890	0.942	0.829	0.881	0.922	83.59%	0.886	0.964	0.973	1.000	0.946	92.65%
FT-MIR (PCs)	PLS-DA	0.906	0.911	0.868	0.927	0.863	83.59%	0.926	0.991	0.843	0.926	0.972	89.71%
	RF	0.780	0.922	0.730	0.764	0.772	68.75%	0.964	1.000	0.991	0.917	0.991	95.59%
NIR (RFE)	PLS-DA	0.807	0.922	0.926	0.902	0.966	85.16%	0.779	0.845	0.675	0.891	0.953	75%
	RF	0.733	0.888	0.791	0.888	0.942	77.34%	0.813	0.964	0.567	0.889	0.955	77.94%
NIR (Bo)	PLS-DA	0.807	0.922	0.858	0.906	0.947	82.81%	0.779	0.837	0.551	0.962	0.962	75%
	RF	0.729	0.878	0.764	0.893	0.922	75.78%	0.772	0.878	0.486	0.870	0.955	72.06%
NIR (PCs)	PLS-DA	0.860	0.937	0.974	0.915	0.990	89.84%	0.927	0.926	0.991	0.878	0.955	90%
	RF	0.745	0.922	0.881	0.881	0.951	81.25%	0.955	0.964	1.000	1.000	0.991	97.06%

Bo: Boruta, PCs: Principal components.

The permutation test can be used to determine whether the established PLS-DA model is at risk of overfitting [32,33]. The intercepts generated by 200 random permutation tests for the FT-MIR PLS-DA model permutation test with the selected-class 1 variables were $R^2 = 0.395$ and $Q^2 = -0.861$, the values of R^2 and Q^2 of permutation tests with the remaining four categories variables are shown in Figure S1. All these results showed that this model was robust without overfitting. Additionally, the PLS-DA models involved in this paper were subjected to permutation tests and there was no overfitting.

For the RF model based on the FT-MIR spectra data matrix, the initial number of trees (n_{tree}) and number of variables (m_{try}) were set as 2000 and the square root of the number of variables. The optimal value of n_{tree} was selected based on the lowest total out-of-bag (OOB) classification error value, meanwhile assured of the lower OOB classification error values of the most classes. Besides, the optimal n_{tree} should be selected from the smooth region of the curve when the minimum OOB value was obtained in multiple regions. The optimal m_{try} was selected according to the lowest OOB classification error value. As shown in Figure 2a,b, the most suitable values 1780 and 33 were selected as the best n_{tree} and m_{try}, respectively, for the RF model based on the FT-MIR spectra data matrix. Based on the same principle, the optimal values 434 and 39 were the best n_{tree} and m_{try}, respectively, for the RF model based on the NIR spectra dataset, as shown in Figure 2c,d.

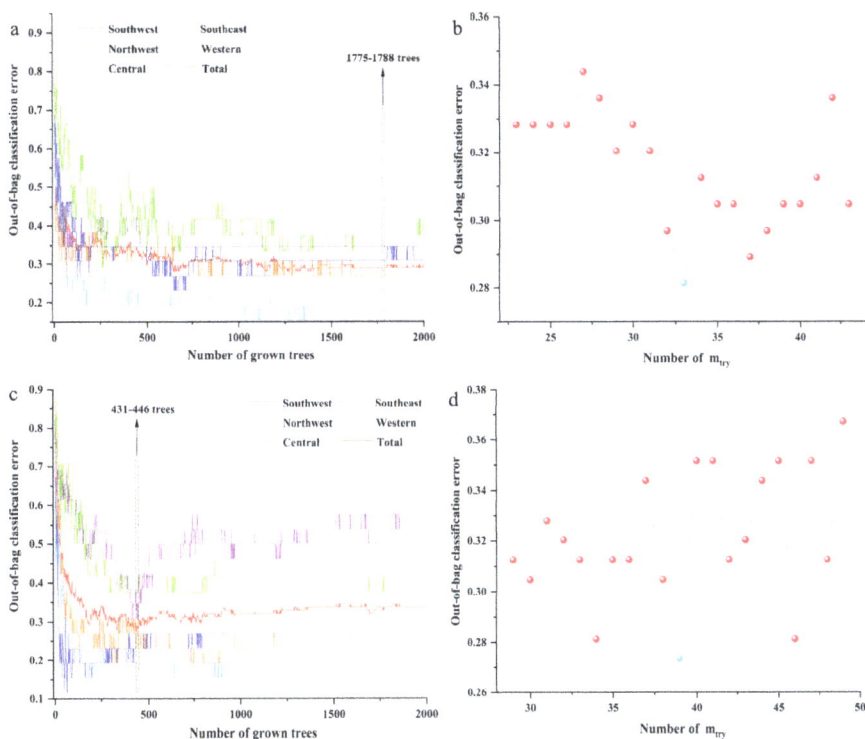

Figure 2. The parameter optimization of random forest models of independent decision making: (a) number of trees (n_{tree}) of the FT-MIR dataset; (b) number of variables (m_{try}) of the FT-MIR dataset; (c) n_{tree} of the NIR dataset; (d) m_{try} of the NIR dataset.

2.3. Important Variable Datasets Selected for Mid-Level Data Fusion

Mid-level on principal components (Mid-level-PCs), mid-level on recursive feature elimination (Mid-level-RFE) and mid-level on Boruta (Mid-level-Bo) dataset matrixes needed to be established to complete mid-level data fusion. The mid-level-RFE dataset was established as follows:

RF models were established on FT-MIR and NIR spectra datasets of wild *P. yunnanensis* samples. For the two RF models, the initial n_{tree} for both was defined as 2000, and the initial m_{try} was set as 33 for the FT-MIR dataset and 39 for the NIR dataset. A total of 1033 trees and 529 trees were the optimal values for n_{tree} of FT-MIR and NIR datasets, respectively, which are shown in Figure S2. Based on the optimal n_{tree}, the number of m_{try} was calculated to be 36 and 32 for FT-MIR and NIR spectra data matrixes, respectively. Next, the optimal n_{tree} and m_{try} were used to further obtain the importance of each variable of individual spectra matrix. All variables of FT-MIR and NIR datasets were arranged from small importance to large importance, respectively. The 10-fold cross-validation error rates of the RF model, based on FT-MIR and NIR data matrixes, are shown in Figure 3a,b. It was reduced sequentially by five variables for each step for the sorted important variables of FT-MIR and NIR spectra data matrixes, respectively. For the FT-MIR dataset (Figure 3a), all variables were divided into region 1, region 2, and region 3, which represent irrelevant variables, interference variables, and important variables, respectively [23]. Among them, region 3 remained for further research. In other words, 45 of the most important variables of the FT-MIR dataset were selected to prepare for mid-level data fusion. However, all variables of the NIR data matrix were divided into region 1 and region 2 (Figure 3b), which represent irrelevant variables and important variables. Variables of region 1 were excluded and the other 145 NIR variables of region 2 were used to fuse with important variables of the FT-MIR dataset to establish Mid-level-RFE models. In other words, these two block datasets straightforwardly concatenated and reconstituted the independent data matrix named Mid-level-RFE.

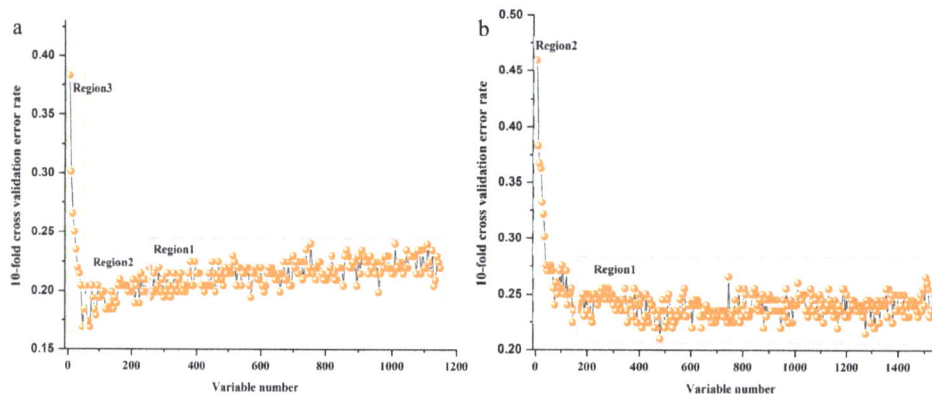

Figure 3. The 10-fold cross-validation error rates of the Random Forest (RF) model (sequentially reduced every five variables) based on total *P. yunnanensis* samples: (**a**) FT-MIR dataset; (**b**) NIR dataset.

Basing on the optimal n_{tree} and m_{try}, important (confirmed and tentative) variables were calculated to be 304 and 343 for NIR and FT-MIR spectra datasets, respectively. These two block datasets (important variables) reconstituted the independent data matrix named Mid-level-Bo.

Similarly, two block datasets of 25 PCs NIR variables and 17 PCs FT-MIR variables straightforwardly concatenated and reconstituted the independent data matrix named Mid-level-PCs.

Besides, the selected variables for establishing mid-level data fusion classification models using RFE and Bo algorithms are shown in Figure 4. The important (confirmed) and tentative variables are represented by blue lines and yellow lines, respectively.

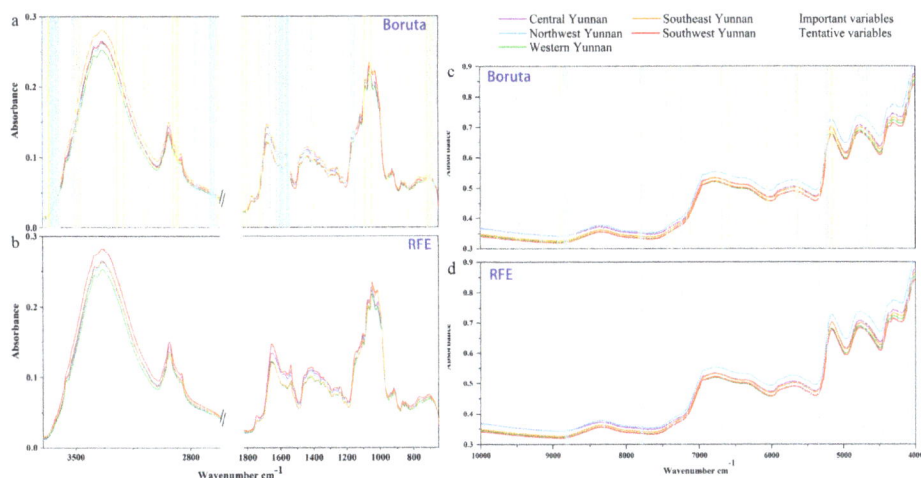

Figure 4. The important variables of Boruta algorithm and RFE algorithm of random forest models based on total *P. yunnanensis* samples: (**a**,**b**) the FT-MIR dataset; (**c**,**d**) the NIR dataset. RFE: Recursive feature elimination.

2.4. Important Variables Datasets Selected for High-Level Data Fusion

High-level data fusion uses the same PCs as mid-level data fusion as the PCs were selected from the unsupervised PCA model. Hence, 17 PCs of the FT-MIR spectra dataset (FT-MIR-PCs) and 25 PCs of the NIR spectra data matrix (NIR-PCs) were used to establish PLS-DA and RF models, respectively. For FT-MIR-PCs and NIR-PCs RF models, the two parameters of n_{tree} and m_{try} needed to be optimized first. As shown in Figure S3, the optimal values 535 and 1030 trees were selected and; furthermore, calculated the suitable m_{try} to be 4 and 3 for FT-MIR-PCs and NIR-PCs RF models, respectively. Then, final RF models were established on the optimal parameters and the most important variables. Besides, vote results of validation sets and calibration sets of PLS-DA and RF models based on the FT-MIR-PCs and NIR-PCs datasets were obtained as shown in Tables S2–S5.

For the RF model based on the FT-MIR and NIR spectra datasets, the initial n_{tree} was set as 2000, and the initial m_{try} were set as 33 (FT-MIR) and 39 (NIR), respectively. The 1780 trees and 434 trees are the optimal values for n_{tree} of FT-MIR and NIR datasets (Figure 2). The numbers of m_{try} were defined to be 33 and 39 for FT-MIR and NIR spectra data matrixes, respectively, based on the optimal n_{tree}. Furthermore, all variables of FT-MIR and NIR datasets were sorted, respectively. The 10-fold cross validation error rates of the RF model, based on the FT-MIR and NIR data matrixes are shown in Figure S4a,b. For the FT-MIR dataset (Figure S4a), all variables were divided into three regions. Among them, 80 of the most important variables (region 3) of the FT-MIR dataset were selected to establish FT-MIR-RFE PLS-DA and RF models. All variables of the NIR data matrix were divided into two regions (Figure S4b) and 200 NIR variables of region 2 were used to establish NIR-RFE PLS-DA and RF models. PLS-DA and RF models were based on two important variables datasets (FT-MIR-RFE and NIR-RFE), respectively. For FT-MIR-RFE and NIR-RFE RF models, 293 and 1459, respectively, were selected as the suitable trees, as shown in Figure S5. Furthermore, the optimal m_{try} values were calculated to be 8 and 14, respectively. Based on the optimal parameters, FT-MIR-RFE and NIR-RFE datasets were used to establish the RF model. Similarly, vote results of validation sets and calibration sets of PLS-DA and RF models based on FT-MIR-RFE and NIR-RFE data matrixes could be obtained, respectively, as shown in Tables S6–S9.

The FT-MIR-Bo and NIR-Bo datasets were obtained by selecting important variables using the Bo algorithm with FT-MIR and NIR spectra datasets. These two data matrixes were used to establish PLS-DA and RF models, respectively. As shown in Figure S6, 1143 and 875 trees were selected to be the optimal n_{tree} for FT-MIR-Bo and NIR-Bo datasets, respectively. Besides, the suitable m_{try} values of the two datasets were calculated to be 12 and 20, respectively. Similarly, vote results of validation sets and calibration sets of two models were obtained as shown in Tables S10–S13.

Like mid-level data fusion, the comparison of selected variables for establishing high-level data fusion classification models using two algorithms is shown in Figure S7. In addition, it was found that both the fingerprint region and characteristic region variables contribute to the classification of *P. yunnanensis* from different origins.

The results of PLS-DA and RF models based on RFE, Bo, and PCs selection algorithms are shown in Table 2. For FT-MIR datasets containing three-variable selection algorithms, the validation set classification results of the PLS-DA and RF models were similar and slightly lower than that obtained by models based on the raw FT-MIR data matrix. Among these models, the RF model using the PCs selection algorithm showed the best ability. Besides, the classification ability of the PLS-DA and RF models based on NIR (RFE) and NIR (Bo) data matrixes were significantly lower than that based on the NIR (PCs) dataset. However, the accuracy for the validation sets of the PLS-DA and RF models based on the NIR (PCs) data matrix were both higher than that based on the FT-MIR (PCs) dataset, which is contrary to the results based on original FT-MIR and NIR datasets. Additionally, it is not hard to see that the classification ability of the PLS-DA and RF models based on the RFE algorithm were like the Bo algorithm, no matter whether based on the FT-MIR or the NIR spectral data. Distribution of their important variables was almost the same (Figure S7), which may be the reason for the similar classification results. Bo variables selection algorithm would be the preferred one than RFE algorithm because it used lesser operation time.

2.5. Low-Level Data Fusion Models

In this case, the low-level data matrix (196 × 2695) was used to establish PLS-DA and RF models. For the low-level data fusion RF model, 2000 and 51were set as original n_{tree} and m_{try} values, respectively. As shown in Figure S8, 802 (n_{tree}) and 51 (m_{try}) were selected as the optimal parameters to establish the final low-level data fusion RF model. The results of calibration and validation sets for the two kinds of models are reported in Table 3. Although the accuracy of the calibration sets of the two models was quite different, the accuracy of the validation sets was only 3% different. Compared with the accuracy of individual data source models, the accuracy of low-level data fusion by both models was similar to that of the FT-MIR dataset, which was higher than that of the NIR data matrix. Besides, the samples of the second class (Northwest Yunnan) were correctly predicted by both low-level data fusion models. The low-level data fusion strategy enhanced (to 100%) the accuracy of samples from Northwest Yunnan.

Variables whose VIP (Variable importance in the projection) score was greater than 1 were selected as important variables to establish PLS-DA and RF models. The selected variables distribution is shown in Figure S9 and results of classification models are displayed in Table 3. The important variables were lesser dispersed in regions 3580 to 3530 cm^{-1}, 3460 to 3150 cm^{-1}, 3000 to 2950 cm^{-1}, 2750 to 2720 cm^{-1}, 1400 to 1200 cm^{-1}, and 1000 to 800 cm^{-1} of FT-MIR spectra and regions 6540 to 6200 cm^{-1} and 5070 to 4000 cm^{-1} of NIR spectra. Many interference and unrelated variables were eliminated by selecting VIP values. The efficiency and accuracy of each class in the low-level (VIP) data fusion model was unchanged and even enhanced.

Table 3. The classification efficiency values and total accuracy of low-, mid-, and high-level data fusion strategies decision making with PLS-DA and RF models. RFE: Recursive feature elimination, Bo: Boruta, PCs: Principal components, VIP: Variable importance in the projection.

	Model	Calibration Set						Validation Set					
		Class1	Class2	Class3	Class4	Class5	Accuracy	Class1	Class2	Class3	Class4	Class5	Accuracy
Low-level	PLS-DA	1.000	1.000	1.000	1.000	1.000	100%	0.926	1.000	0.949	1.000	0.981	95.59%
	RF	0.872	0.885	0.858	0.897	0.932	82.81%	0.886	1.000	0.905	0.972	0.991	92.65%
Low-level (VIP)	PLS-DA	1.000	1.000	1.000	1.000	1.000	100%	0.926	1.000	0.991	1.000	0.991	97.06%
	RF	0.927	0.927	0.849	0.922	0.947	86.72%	0.926	1.000	0.991	1.000	0.991	97.06%
Mid-level (RFE)	PLS-DA	0.946	0.712	0.764	0.864	0.878	75%	0.705	0.794	0.000	0.727	0.798	55.88%
	RF	0.966	0.888	0.868	0.894	0.881	84.38%	0.764	0.861	0.491	0.900	0.854	69.12%
Mid-level (Bo)	PLS-DA	0.961	1.000	1.000	0.995	0.995	98.44%	0.926	1.000	0.991	0.991	1.000	97.06%
	RF	0.947	0.942	0.885	0.922	0.951	89.06%	0.926	1.000	0.949	0.991	0.991	95.59%
Mid-level (PCs)	PLS-DA	0.951	0.961	0.974	0.995	0.995	96.09%	1.000	1.000	0.957	0.991	1.000	98.53%
	RF	0.927	0.951	0.922	0.922	0.981	90.63%	0.886	1.000	0.991	0.981	0.955	94.12%
High-level (RFE)	PLS-DA	0.951	0.981	0.953	1.000	0.990	96.09%	0.870	1.000	0.802	0.991	0.981	89.71%
	RF	0.976	0.976	0.904	0.922	0.995	91.21%	0.926	1.000	0.991	1.000	0.991	97.06%
High-level (Bo)	PLS-DA	0.976	0.981	0.979	1.000	0.990	97.66%	0.926	0.991	0.850	1.000	0.981	92.65%
	RF	0.966	0.976	0.904	0.902	0.951	90.63%	0.917	0.964	0.850	0.972	1.000	91.18%
High-level (PCs)	PLS-DA	0.981	1.000	0.990	0.981	1.000	98.44%	0.964	1.000	1.000	0.991	1.000	98.53%
	RF	0.881	0.971	0.872	0.911	0.961	87.5%	0.966	1.000	1.000	0.991	1.000	100%

2.6. Mid-Level Data Fusion Models

In our study, Mid-level-PCs, Mid-level-RFE, and Mid-level-Bo three kinds of data matrixes were used to establish PLS-DA and RF classification models to prepare for mid-level data fusion. As shown in Figure S10a,b, 427 (n_{tree}) and 6 (m_{try}) were selected as optimal parameters to establish the final mid-level-PCs RF model and obtained a validation set accuracy of 94.12%. The total accuracy of the validation set of the PLS-DA model was 98.53% (Table 3).

Similarly, 2000 and 13 were set as raw n_{tree} and m_{try} values for the mid-level-RFE RF model. According to the principle of parameter optimization of the RF model, the lowest values 262 (n_{tree}) and 13 (m_{try}) were defined as the suitable parameters to establish the RF model, as shown in Figure S10c,d. However, the validation set accuracy of the mid-level-RFE was only 69.12%, which was lesser than that of each individual spectral RF model. Besides, the PLS-DA model based on the mid-level-RFE had a worse classification ability with an accuracy of only 55.88% (Table 3).

The numbers of 2000 and 25 were set as raw n_{tree} and m_{try} values, respectively, for the mid-level-Bo RF model. As shown in Figure S10e,f, the suitable parameters were calculated to be 800 and 15, to establish the mid-level-Bo RF model, and the obtained validation set accuracy was 95.59%. The difference of accuracy for both the PLS-DA model and the RF model based on mid-level-RFE and mid-level-Bo was about 30%. By comparing the regions of important variables between mid-level-RFE and mid-level-Bo data matrixes (Figure 4), we could find that the number of important FT-MIR variables for the RFE variable selection algorithm was less than that of the Bo variable selection algorithm, and there was little difference in the important variables of the NIR spectroscopy selected by the two algorithms. Hence, we can infer that the RFE selection algorithm had excluded some of the important variables that may be enhancing the accuracy of classification models. In addition, we can also extrapolate that the FT-MIR dataset was more important for geographical origin classification of wild *P. yunnanensis* samples than the NIR data matrix, which provides more effective information.

2.7. High-Level Data Fusion Models

For high-level data fusion, four fuzzy aggregation operators were chosen as the voting rule for the voting decision, including minimum, maximum, product, and average [23]. The category that has the maximum value in each operator is considered to be the selected class. It is worth mentioning that when the difference between the maximum value and the second largest value is less than 0.01, both values are considered maximum. Three kinds of vote results would be obtained by high-level data fusion including correct, false, and multiple discriminated. As shown in Table S14, the true Class of sample NO. 4 belongs to Class 1 and the four fuzzy aggregation operators were fully accorded with the true Class. For example, for No. 31 the true category is Class 1, while three voting results are distinguished into Class 3 and one voting result is defined as Class 5. The high-level data fusion voting results of this sample is Class 3. Besides, NO. 114 was voted into Class 2 and Class 3 by FT-MIR-PCs and NIR-PCs RF models. This sample truly belongs to Class 3, while the final data fusion voting result is distinguished into Class 2. Besides, sample 6 truly belongs to Class 1 while pertained to Class 1 and Class 4 by voting, which was multiple discriminated. Although multivariate discrimination does not affect the accuracy of the model, it influences efficiency values of each Class. This explains that the accuracy of the validation set of RF model is 1, while the efficiency of Class 1 and Class 4 does not reach 1 (Table S14).

As shown in Table 3, the accuracy of the validation set of the High-level-PCs RF model was reached at 100% and that of the High-level-PCs PLS-DA model was 98.53%. The high-level data fusion classification ability based on the PCs selection variables algorithm was better than that based on RFE and Bo algorithms. Besides, there was little difference between the PLS-DA model and the RF model in identifying the origin of *P. yunnanensis* samples based on the same data set.

3. Materials and Methods

3.1. Samples Preparation

The 196 rhizomes of wild *P. yunnanensis* samples were obtained from five different origins at central, western, northwest, southeast, and southwest areas of Yunnan Province, as shown in Figure 5. The detail collection information is shown in Table S15. All wild samples were identified as *P. polyphylla* var. *yunnanensis* (Franch.) Hand. -Mazz. by Professor Hang Jin (Institute of Medicinal Plants, Yunnan Academy of Agricultural Sciences, Kunming, China). All rhizome samples were washed with tap water and were dried in a drying oven at 50 °C, then sifted through 100 mesh sieves. Additionally, all samples were preserved in polyethylene zip-lock bags and kept in a dark and dry environment for further analysis.

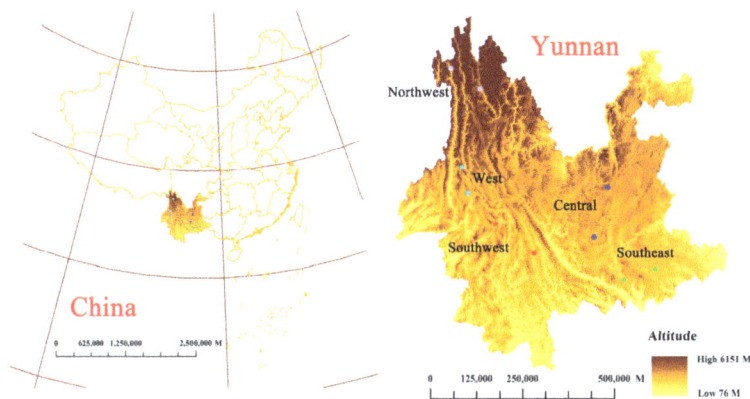

Figure 5. Location distribution of wild *P. yunnanensis* samples in central, western, northwest, southeast, and southwest areas, Yunnan Province.

3.2. Fourier Transform Mid-Infrared Spectroscopy (FT-MIR)

FT-MIR spectra were collected with an FTIR spectrometer equipped with a deuterated triglycine sulfate (DTGS) detector and a ZnSe ATR (attenuated total reflection) accessory (Perkin Elmer, Norwalk, CT, USA). All spectra recorded ranges of 4000 to 650 cm^{-1} with 4 cm^{-1} resolution, and 16 scans were averaged. Three analytical replicates of FT-MIR spectral data of all wild *P. yunnanensis* samples were obtained.

3.3. Near-Infrared Spectroscopy (NIR)

NIR analysis was conducted with an Antaris II spectrometer (Thermo Fisher Scientific, Madison, WI, USA) equipped, combined with a diffuse reflection module. All spectra recorded ranges of 10,000 to 4000 cm^{-1} with a spectral resolution of 4 cm^{-1}, and 16 scans were averaged for wild samples. Three scans were repeated for all wild samples.

3.4. Spectral Data Analysis and Software

FT-MIR spectra were converted from transmittance to absorbance and the advanced ATR correction was completed by OMNIC 9.7.7 software (Thermo Fisher Scientific, Madison, WI, USA). The spectral linear relation was greatly disturbed by high-frequency random noise, the interference of light scattering, baseline drift, etc. [34]. Hence, FT-MIR and NIR spectra were processed using SNV, FD, SD, and their combination (SNV-FD and SNV-SD), to decrease a part of the irrelevant interferences [10,35,36]. All these pretreatment procedures were performed by SIMAC-P^{+} (Version 13.0, Umetrics, Umeå, Sweden). Additionally, the spectral regions of 4000 to 3700 cm^{-1} and 2620 to 1800 cm^{-1} were excluded for all FT-MIR spectra before establishing classification models due to a mass of interference information.

Hence, the regions of 3700 to 2620 cm^{-1} and 1800 to 650 cm^{-1} formed a data matrix for constructing classification models.

All samples for each class were separated into calibration sets and validation sets as a rate of 2 to 1 with Kennard-Stone algorithm using MATLAB (Version R2017a, Mathworks, Natick, MA, USA) [37,38]. In other words, the number of calibration sets (128 samples) in one to five categories was 26, 26, 24, 26, and 26, respectively, and the number of validation sets (68 samples) in one to five classes was 14, 14, 12, 14, and 14, respectively. The preprocessing algorithms were estimated by parameters of R^2, Q^2, RMSEE, RMSECV, and accuracy of the calibration set [19,39]. The better pretreatment algorithm required higher values of R^2, Q^2, and accuracy as well as lower values of RMSEE and RMSECV. Hence, the best preprocessing algorithm would be selected for identification analysis to establish PLS-DA and RF classification models.

The groundwork of PLS-DA is the PLS algorithm and it belongs to the binary classification algorithm from 0 to 1, which has been widely applied to resolve the classification problems for geographical origins, growth years, and others [40]. For each sample, the probability of being assigned to each class could be obtained, and the category with the highest probability was seen as the category of this sample. For the validation samples, RF is based on the assembly classification or regression trees algorithm and has a better ability to handle the nonlinear and high-order interaction effects data matrixes [41]. Both of these kinds of class-modeling methods belong to supervised pattern recognition and require a calibration set for each class in order to establish an individual model to explore their similarities between samples from the one class and the differences among all classes. Besides, the validation sets were used to validate the identification ability of supervised models. In our study, all PLS-DA were completed by SIMAC-P$^+$ (Version 13.0, Umetrics, Umeå, Sweden) and RF were established by RStudio (version 3.5.2, Boston, MA, USA). The operation of RF was roughly divided into the following three steps: Firstly, 2000 and the square root of the number of variables were set as the initial values of n_{tree} and m_{try}, respectively. Secondly, these two parameters were optimized according to the lowest OOB classification error values. Thirdly, the RF model was established with the selected optimal values of n_{tree} and m_{try}.

The RMSEP and accuracy of the validation set were the two parameters used to estimate the classification ability of PLS-DA. The lower values of RMSEP shows the better prediction ability of PLS-DA models. Besides, for the PLS-DA and RF models established by the best preprocessing algorithm, indices of true positives (TP), true negatives (TN), false positives (FP), and false negatives (FN) were calculated for each class. The sensitivity (SEN) and specificity (SPE) were obtained by the above indices for each class. The efficiency of values was calculated by the geometric mean of SEN and SPE to evaluate the effectiveness of each class of PLS-DA and RF models. All formulas for the above parameters were as follow:

$$\text{SEN} = \frac{\text{TP}}{\text{TP} + \text{FN}} \tag{1}$$

$$\text{SPE} = \frac{\text{TN}}{\text{TN} + \text{FP}} \tag{2}$$

$$\text{efficiency} = \sqrt{\text{SEN} \times \text{SPE}} \tag{3}$$

The map of sample collection information in our study was obtained by Arc Map (version 10), and all figures were drawn by Origin (version 2018, OriginLab Corporation, Northampton, MA, USA) and Adobe Photoshop CC (version 2019, Adobe Systems Incorporated, San Jose, CA, USA).

3.5. Data Fusion Strategy

Data fusion strategy was applied in this study, as the comprehensive information of *P. yunnanensis* samples were unable to be provided by individual data sources. To compare and select the best data fusion strategy to trace geographical origins, the low-, mid-, and high- level data fusion strategies and three algorithms for variable selection were considered. The best preprocessing FT-MIR and NIR datasets were used to finish data fusion approaches. The schemes for these three strategies combined with two kinds of spectral signals are shown in Figure 6.

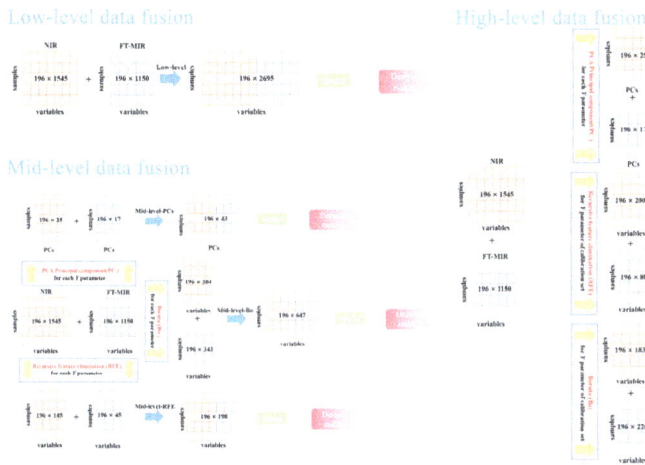

Figure 6. Scheme of the low-, mid-, and high-level data fusion approaches used to combine the FT-MIR signals and NIR signals.

In low-level data fusion strategy, the FT-MIR and NIR spectral signals are straightforwardly concatenated and reconstitute an independent data matrix. This new dataset (low-level data matrix) was equal to 196 rows and 2695 columns, namely 196 samples and 2695 spectral variables (= 1545 NIR variables + 1150 FT-MIR variables). Finally, the low-level dataset was used to establish the PLS-DA and RF models.

Mid-level data fusion strategy, namely feature-level data fusion, was made up of feature important variables from single data sources including FT-MIR and NIR spectra. PLS-DA and RF classification models were established by new data matrixes, which were formed by concatenating the feature important variables from FT-MIR and NIR by different variable selection algorithms. In this case, important variables were selected based on each Y parameter (spectral datasets of total samples). More in detail, three different variable selection algorithms were as follows:

- Mid-level-PCs consisted of principal components, which were selected by PCA of FT-MIR and NIR spectral datasets, respectively. PCs were selected based on values of eigenvalue greater than 1. Hence, the mid-level-PCs data matrix was obtained with 196 rows and 42 columns, namely 196 samples and 42 PCs variables (= 25 NIR PCs variables + 17 PCs FT-MIR variables).
- Mid-level-RFE, which consisted of merging together the important variables of FT-MIR and NIR spectral datasets, was selected by the recursive feature elimination algorithm based on the RF model. The mid-level-REF dataset size was equal to 196 rows and 190 columns (= 145 NIR REF variables + 45 FT-MIR REF variables).
- Mid-level-Bo, which consisted of merging together the important (confirmed and tentative) variables of FT-MIR and NIR spectral datasets, was selected by the Boruta algorithm based on the RF model. The mid-level-Bo data matrix consisted of 196 rows and 647 columns (= 153 NIR Bo confirmed variables + 151 NIR Bo tentative variables + 207 FT-MIR Bo confirmed variables + 136 FT-MIR Bo tentative variables).

High-level data fusion strategy, namely decision-data fusion, fused the vote results from the models of the FT-MIR and NIR datasets. Additionally, individual spectral matrices were formed by feature or important variables using variable selection algorithms (PCs, RFE, and Bo) before establishing models. In our study, important variables were selected by RFE and Bo for the high-level data fusion strategy based on the Y parameter of the calibration set, which was different from the mid-level data fusion variables selection method.

- NIR-PCs data matrix obtained 196 rows and 25 columns (25 NIR PCs variables) and the FT-MIR-PCs data matrix obtained 196 rows and 17 columns (17 FT-MIR PCs variables).
- NIR-RFE data matrix obtained 196 rows and 200 columns (200 NIR RFE variables) and the FT-MIR-RFE data matrix obtained 196 rows and 80 columns (80 FT-MIR RFE variables).
- NIR-Bo data matrix obtained 196 rows and 183 columns (83 NIR Bo confirmed variables + 108 NIR Bo tentative variables) and the FT-MIR-Bo data matrix obtained 196 rows and 226 columns (117 FT-MIR Bo confirmed variables + 109 FT-MIR Bo tentative variables).

Besides, PCs were obtained using MATLAB (Version R2017a, Mathworks, Natick, MA, USA) and RFE and Bo were completed by RStudio (version 3.5.2).

4. Conclusions

In our study, the use of low-, mid-, and high-level data fusion strategies, combined with feature extraction and important variable selection algorithms, were researched to fuse the chemical information from FT-MIR and NIR spectroscopies for the identification and classification of geographical origins of wild *P. yunnanensis* samples.

In fact, PCs was the feature extraction algorithm of three kinds of important variable selection algorithms, which obtained a better ability for establishing classification models no matter whether in mid- or high-level data fusion. Between the two important variable selection algorithms of RFE and Bo, the latter can obtain important variables that are similar, or more accurate, to the former and can complete the calculation in a shorter time.

Besides, the two kinds of IR spectroscopies bring complementary chemical information profiles about multiple geographical sources of *P. yunnanensis*. While FT-MIR provides chemical information among 4000 to 650 cm^{-1}, NIR describes the chemical information from 10,000 to 4000 cm^{-1}. The data fusion strategy improved the geographical traceability ability of models for *P. yunnanensis*, while FT-MIR spectra data provided more contributions than NIR spectra. Besides, thanks to the application of the high-level data fusion strategy, the identification effect based on the random forest model reached the best performance level.

Supplementary Materials: The Supplementary Materials are available online.

Author Contributions: Y.-F.P., Q.-Z.Z. and Y.-Z.W. developed the concept of the manuscript, Y.-F.P. and Z.-T.Z. performed the experiments, analyzed the data, and discussed the results, and Y.-Z.W. corrected this manuscript finally.

Funding: This research was funded by National Natural Science Foundation of China, grant number 81460584.

Acknowledgments: This work was sponsored by the National Natural Science Foundation of China (Grant number: 81460584).

Conflicts of Interest: The authors declare no conflict of interest.

References

1. Jing, S.; Wang, Y.; Li, X.; Man, S.; Gao, W. Chemical constituents and antitumor activity from *Paris polyphylla* Smith var. *yunnanensis*. *Nat. Prod. Res.* **2017**, *31*, 660–666. [CrossRef] [PubMed]
2. Kang, L.P.; Huang, Y.Y.; Zhan, Z.L.; Liu, D.H.; Peng, H.S.; Nan, T.G.; Zhang, Y.; Hao, Q.X.; Tang, J.F.; Zhu, S.D.; et al. Structural characterization and discrimination of the *Paris polyphylla* var. *yunnanensis* and *Paris vietnamensis* based on metabolite profiling analysis. *J. Pharm. Biomed.* **2017**, *142*, 252–261.
3. Wu, X.; Wang, L.; Wang, G.C.; Wang, H.; Dai, Y.; Yang, X.X.; Ye, W.C.; Li, Y.L. Triterpenoid saponins from rhizomes of *Paris polyphylla* var. *yunnanensis*. *Carbohydr. Res.* **2013**, *368*, 1–7. [CrossRef] [PubMed]

4. Deng, D.; Lauren, D.R.; Cooney, J.M.; Jensen, D.J.; Wurms, K.V.; Upritchard, J.E.; Cannon, R.D.; Wang, M.Z.; Li, M.Z. Antifungal saponins from *Paris polyphylla* Smith. *Planta Med.* **2008**, *74*, 1397–1402. [CrossRef]

5. Li, P.; Fu, J.H.; Wang, J.K.; Ren, J.G.; Liu, J.X. Extract of *Paris polyphylla* Smith protects cardiomyocytes from anoxia-reoxia injury through inhibition of calcium overload. *Chin. J. Integr. Med.* **2011**, *17*, 283–289. [CrossRef]

6. Hwang, S.J.; Lin, H.C.; Chang, C.F.; Lee, F.Y.; Lu, C.W.; Hsia, H.C.; Wang, S.S.; Lee, S.D.; Tsai, Y.T.; Lo, K.J. A randomized controlled trial comparing octreotide and vasopressin in the control of acute esophageal variceal bleeding. *J. Hepatol.* **1992**, *16*, 320. [CrossRef]

7. Wang, Y.; Gao, W.; Li, X.; Wei, J.; Jing, S.; Xiao, P. Chemotaxonomic study of the genus *Paris* based on steroidal saponins. *Biochem. Syst. Ecol.* **2013**, *48*, 163–173. [CrossRef]

8. Cunningham, A.B.; Brinckmann, J.A.; Bi, Y.F.; Pei, S.J.; Schippmann, U.; Luo, P. *Paris* in the spring: A review of the trade, conservation and opportunities in the shift from wild harvest to cultivation of *Paris polyphylla* (Trilliaceae). *J. Ethnopharmacol.* **2018**, *222*, 208–216. [CrossRef]

9. Zhao, Y.; Zhang, J.; Yuan, T.; Shen, T.; Li, W.; Yang, S.; Hou, Y.; Wang, Y.; Jin, H. Discrimination of wild *Paris* based on near-infrared spectroscopy and high-performance liquid chromatography combined with multivariate analysis. *PLoS ONE* **2014**, *9*, e89100. [CrossRef]

10. Yang, Y.G.; Jin, H.; Zhang, J.; Zhang, J.Y.; Wang, Y.Z. Quantitative evaluation and discrimination of wild *Paris polyphylla* var. *yunnanensis* (Franch.) Hand.-Mazz from three regions of Yunnan Province using UHPLC-UV-MS and UV spectroscopy couple with partial least squares discriminant analysis. *J. Nat. Med.* **2017**, *71*, 148–157. [CrossRef]

11. Wang, Y.Z.; Liu, E.H.; Li, P. Chemotaxonomic studies of nine *Paris* species from China based on ultra-high performance liquid chromatography tandem mass spectrometry and Fourier transform infrared spectroscopy. *J. Pharm. Biomed.* **2017**, *140*, 20–30. [CrossRef] [PubMed]

12. Li, J.; Zhang, J.; Zhao, Y.L.; Huang, H.Y.; Wang, Y.Z. Comprehensive Quality Assessment Based Specific Chemical Profiles for Geographic and Tissue Variation in *Gentiana rigescens* Using HPLC and FTIR Method Combined with Principal Component Analysis. *Front. Chem.* **2017**, *5*, 125. [CrossRef] [PubMed]

13. Ballabio, D.; Robotti, E.; Grisoni, F.; Quasso, F.; Bobba, M.; Vercelli, S.; Gosetti, F.; Calabrese, G.; Sangiorgi, E.; Orlandi, M.; et al. Chemical profiling and multivariate data fusion methods for the identification of the botanical origin of honey. *Food Chem.* **2018**, *266*, 79–89. [CrossRef] [PubMed]

14. Gad, H.A.; Bouzabata, A. Application of chemometrics in quality control of Turmeric (*Curcuma longa*) based on Ultra-violet, Fourier transform-infrared and ^{1}H NMR spectroscopy. *Food Chem.* **2017**, *237*, 857–864. [CrossRef] [PubMed]

15. Qi, L.; Liu, H.; Li, J.; Li, T.; Wang, Y. Feature Fusion of ICP-AES, UV-Vis and FT-MIR for Origin Traceability of *Boletus edulis* Mushrooms in Combination with Chemometrics. *Sensors* **2018**, *18*, 241. [CrossRef] [PubMed]

16. Ma, L.H.; Zhang, Z.M.; Zhao, X.B.; Zhang, S.F.; Lu, H.M. The rapid determination of total polyphenols content and antioxidant activity in *Dendrobium officinale* using near-infrared spectroscopy. *Anal. Methods* **2016**, *8*, 4584–4589. [CrossRef]

17. Pei, Y.F.; Wu, L.H.; Zhang, Q.Z.; Wang, Y.Z. Geographical traceability of cultivated *Paris polyphylla* var. *yunnanensis* using ATR-FTMIR spectroscopy with three mathematical algorithms. *Anal. Methods* **2019**, *11*, 113–122. [CrossRef]

18. Wu, X.M.; Zhang, Q.Z.; Wang, Y.Z. Traceability the provenience of cultivated *Paris polyphylla* Smith var. *yunnanensis* using ATR-FTIR spectroscopy combined with chemometrics. *Spectrochim. Acta A* **2019**, *212*, 132–145. [CrossRef]

19. Pei, Y.F.; Zhang, Q.Z.; Zuo, Z.T.; Wang, Y.Z. Comparison and Identification for Rhizomes and Leaves of *Paris yunnanensis* Based on Fourier Transform Mid-Infrared Spectroscopy Combined with Chemometrics. *Molecules* **2018**, *23*, 3343. [CrossRef]

20. Yang, Y.G.; Zhang, J.; Jin, H.; Zhang, J.Y.; Wang, Y.Z. Quantitative Analysis in Combination with Fingerprint Technology and Chemometric Analysis Applied for Evaluating Six Species of Wild *Paris* Using UHPLC-UV-MS. *J. Anal. Methods Chem.* **2016**, 1–9. [CrossRef]

21. Yang, L.F.; Ma, F.; Zhou, Q.; Sun, S.Q. Analysis and identification of wild and cultivated Paridis Rhizoma by infrared spectroscopy. *J. Mol. Struct.* **2018**, *1165*, 37–41. [CrossRef]

22. Biancolillo, A.; Bucci, R.; Magrì, A.L.; Magrì, A.D.; Marini, F. Data-fusion for multiplatform characterization of an italian craft beer aimed at its authentication. *Anal. Chim. Acta* **2014**, *820*, 23–31. [CrossRef] [PubMed]

23. Li, Y.; Zhang, J.Y.; Wang, Y.Z. FT-MIR and NIR spectral data fusion: A synergetic strategy for the geographical traceability of *Panax notoginseng*. *Anal. Bioanal. Chem.* **2018**, *410*, 91–103. [CrossRef]

24. Wu, X.M.; Zhang, Q.Z.; Wang, Y.Z. Traceability of wild *Paris polyphylla* Smith var. *yunnanensis* based on data fusion strategy of FT-MIR and UV-Vis combined with SVM and random forest. *Spectrochim. Acta A* **2018**, *205*, 479–488. [CrossRef] [PubMed]

25. Horn, B.; Esslinger, S.; Pfister, M.; Fauhl-Hassek, C.; Riedl, J. Non-targeted detection of paprika adulteration using mid-infrared spectroscopy and one-class classification-Is it data preprocessing that makes the performance? *Food Chem.* **2018**, *257*, 112–119. [CrossRef] [PubMed]

26. Chen, J.B.; Guo, B.L.; Yan, R.; Sun, S.Q.; Zhou, Q. Rapid and automatic chemical identification of the medicinal flower buds of *Lonicera* plants by the benchtop and hand-held Fourier transform infrared spectroscopy. *Spectrochim. Acta A* **2017**, *182*, 81–86. [CrossRef]

27. Xu, C.H.; Chen, J.B.; Zhou, Q.; Sun, S.Q. Classification and identification of TCM by macro-interpretation based on FT-IR combined with 2DCOS-IR. *Biomed. Spectrosc. Imaging* **2015**, *4*, 139–158.

28. Türker-Kaya, S.; Huck, C. A Review of Mid-Infrared and Near-Infrared Imaging: Principles, Concepts and Applications in Plant Tissue Analysis. *Molecules* **2017**, *22*, 168. [CrossRef]

29. Socrates, G. *Infrared and Raman Characteristic Group Frequencies*, 3rd ed.; John Wiley & Sons, Ltd.: Chichester, UK; New York, NY, USA, 2001.

30. Fu, H.Y.; Huang, D.C.; Yang, T.M.; She, Y.B.; Zhang, H. Rapid recognition of Chinese herbal pieces of Areca catechu by different concocted processes using Fourier transform mid-infrared and near-infrared spectroscopy combined with partial least-squares discriminant analysis. *Chin. Chem. Lett.* **2013**, *24*, 639–642. [CrossRef]

31. Wang, Y.; Zuo, Z.T.; Shen, T.; Huang, H.Y.; Wang, Y.Z. Authentication of *Dendrobium* Species Using Near-Infrared and Ultraviolet-Visible Spectroscopy with Chemometrics and Data Fusion. *Anal. Lett.* **2018**, *51*, 2792–2821. [CrossRef]

32. Ma, N.; Liu, X.W.; Kong, X.J.; Li, S.H.; Jiao, Z.H.; Qin, Z.; Yang, Y.J.; Li, J.Y. Aspirin eugenol ester regulates cecal contents metabolomic profile and microbiota in an animal model of hyperlipidemia. *BMC Vet. Res.* **2018**, *14*, 405. [CrossRef]

33. Rodrigues, D.; Pinto, J.; Araújo, A.M.; Monteiro-Reis, S.; Jerónimo, C.; Henrique, R.; de Lourdes Bastos, M.; de Pinho, P.G.; Carvalho, M. Volatile metabolomic signature of bladder cancer cell lines based on gas chromatography-mass spectrometry. *Metabolomics* **2018**, *14*, 62. [CrossRef]

34. Chen, D.; Shao, X.G.; Hu, B.; Su, Q.D. A Background and noise elimination method for quantitative calibration of near-infrared spectra. *Anal. Chim. Acta* **2004**, *511*, 37–45. [CrossRef]

35. Li, X.L.; Xu, K.L.; Li, H.; Yao, S.; Li, Y.F.; Liang, B. Qualitative analysis of chiral alanine by UV-visible-shortwave near-infrared diffuse reflectance spectroscopy combined with chemometrics. *RSC Adv.* **2016**, *6*, 8395–8450. [CrossRef]

36. Wang, L.; Sun, D.W.; Pu, H.; Cheng, J.H. Quality analysis, classification, and authentication of liquid foods by near-infrared spectroscopy: A review of recent research developments. *Crit. Rev. Food Sci. Nutr.* **2017**, *57*, 1524–1538. [CrossRef]

37. Saptoro, A.; Tadé, M.O.; Vuthaluru, H. A Modified Kennard-Stone Algorithm for Optimal Division of Data for Developing Artificial Neural Network Models. *Chem. Prod. Proc. Mode.* **2012**, *7*. [CrossRef]

38. Rajer-Kanduč, K.; Zupan, J.; Majcen, N. Separation of data on the training and test set for modelling: A case study for modelling of five colour properties of a white pigment. *Chemom. Intel. Lab.* **2003**, *65*, 221–229. [CrossRef]

39. Xie, L.J.; Ye, X.Q.; Liu, D.H.; Ying, Y.B. Quantification of glucose, fructose and sucrose in bayberry juice by NIR and PLS. *Food Chem.* **2009**, *114*, 1135–1140. [CrossRef]

40. Ståle, L.; Wold, S. Partial least squares analysis with cross-validation for the two-class problem: A Monte Carlo study. *J. Chemometr.* **1987**, *1*, 185–196. [CrossRef]

41. Breiman, L. Random forests. *Mach Learn.* **2001**, *45*, 5–32. [CrossRef]

Sample Availability: Not available.

molecules

MDPI

Article

PLS Subspace-Based Calibration Transfer for Near-Infrared Spectroscopy Quantitative Analysis

Yuhui Zhao [1,*], Jinlong Yu [1], Peng Shan [2], Ziheng Zhao [1], Xueying Jiang [1] and Shuli Gao [1]

[1] School of Computer Science and Engineering, Northeastern University, Shenyang 110819, China;
    jianren_d@163.com (J.Y.); 13081850350@163.com (Z.Z.); xueying.jiang@163.com (X.J.);
    15238247216@163.com (S.G.)
[2] College of Information Science and Engineering, Northeastern University, Shenyang 110819, China;
    peng.shan@neuq.edu.cn
*   Correspondence: yuhuizhao@neuq.edu.cn; Tel.: +86-186-3039-0553

Academic Editors: Christian Huck and Krzysztof B. Bec
Received: 15 March 2019; Accepted: 28 March 2019; Published: 2 April 2019

Abstract: In order to enable the calibration model to be effectively transferred among multiple instruments and correct the differences between the spectra measured by different instruments, a new feature transfer model based on partial least squares regression (PLS) subspace (PLSCT) is proposed in this paper. Firstly, the PLS model of the master instrument is built, meanwhile a PLS subspace is constructed by the feature vectors. Then the master spectra and the slave spectra are projected into the PLS subspace, and the features of the spectra are also extracted at the same time. In the subspace, the pseudo predicted feature of the slave spectra is transferred by the ordinary least squares method so that it matches the predicted feature of the master spectra. Finally, a feature transfer relationship model is constructed through the feature transfer of the PLS subspace. This PLS-based subspace transfer provides an efficient method for performing calibration transfer with only a small number of standard samples. The performance of the PLSCT was compared and assessed with slope and bias correction (SBC), piecewise direct standardization (PDS), calibration transfer method based on canonical correlation analysis (CCACT), generalized least squares (GLSW), multiplicative signal correction (MSC) methods in three real datasets, statistically tested by the Wilcoxon signed rank test. The obtained experimental results indicate that PLSCT method based on the PLS subspace is more stable and can acquire more accurate prediction results.

Keywords: calibration transfer; NIR spectroscopy; PLS; quantitative analysis model

---

1. Introduction

In the past few decades, near-infrared spectroscopy (NIR) has been widely used in various fields, because of its fast speed and the fact that it does not cause damage to sample characteristics. These areas include pharmaceutical [1–3], biomedical [4], petrochemical [5], agricultural [6,7], food [8–10]. In the NIR analysis, the most frequently used multivariate calibration techniques are partial least squares regression (PLS) [11,12] and principal component regression (PCR) [13,14]. However, the established calibration model is often outdated or unsuitable for new samples due to factors of the diversity of measuring instruments and measuring environments, as well as the variability of the materials being measured. New samples refer to any samples not included in the calibration model, such as those samples collected at different times or with different instruments. Frequent calibration is not desirable because a large amount of time and resources are devoted to establishing calibration models. One advisable option would be to carry out the calibration transfer.

Numerous relevant calibration transfer methods have been proposed in articles. In general, these methods can be divided into two types: transfer standard and non-standard. The transfer standard requires the same standard samples to be measured on the master instrument and the slave instrument. In this type of method, according to the stages in which the adjustment occurs are further divided into four types.

The first type is the method of correcting the slave spectra. In the standard samples, the slave spectra are made as close as possible to the corresponding master spectra by a transfer matrix. The most widely used are direct standardization (DS) and piecewise direct standardization (PDS) methods [15,16]. In the PDS method, the transfer relationship between the master spectra and the slave spectra from the sliding window is established at each wavelength of the master spectra, and finally a band-shaped transfer matrix is formed for correcting the slave spectra.

The second type is the method of simultaneously correcting the master spectra and the slave spectra. Commonly used is calibration transfer by the generalized least squares (GLSW) method [17,18]. GLSW uses the difference between the standard set of the master instrument and the slave instrument to build the weight matrix, and then uses the weight matrix to reduce the weight of spectral feature to be suppressed. A detailed description of the weight matrix is provided in [17] and [18].

The third type is the method of correcting the predicted values. Mainly the slope and bias correction (SBC) method [19], this method considers that there is a linear relationship between the predicted values of the slave spectra obtained by the master spectral model and the response variable, usually using ordinary least squares method to calculate this relationship. The predicted values are then corrected using this relationship.

The fourth type is the projection method. For example, calibration transfer method based on canonical correlation analysis (CCACT) [20], which uses CCA to find the set of canonical variables that are maximally correlated between the standard set of the master instrument and the slave instruments. Further explore the transfer relationship between the two canonical variables.

In practical applications, it is difficult or even impossible to measure the same samples on two instruments due to the position of the measuring instrument and the stability of the samples, etc. At this time, it is necessary to use a method that does not require measurement of the same standard samples, that is, a non-standard method. These methods are mainly divided into two types.

One is the signal preprocessing method, which removes the baseline offset and the linearly sloped baselines by simple mathematical operations of the first derivative and the second derivative. Common methods include multiplicative signal correction (MSC) [21], finite impulse response (FIR) filtering [22], generalized moving window MSC (W-MSC) [21], OSC [23,24], etc., wherein FIR and MW-MSC are variants of MSC. However, it must be noted that these simple preprocessing methods do not handle complex changes between the master spectra and the slave spectra.

The other is the projection method. It includes transfer component analysis (TCA) [25] and kernel principal component analysis (KPCA) [26]. TCA projects the master spectra and the slave spectra into a common feature space in which the distribution of the master spectra and the slave spectra are as similar as possible while retaining the key properties of the spectra. TCA and KPCA use different kernels, so they can cope with nonlinear and more complex changes in the spectra.

In this paper, a novel projection method is proposed, which is a feature transfer model based on PLS subspace (PLSCT). PLSCT establishes the PLS model of the calibration set of the master instrument firstly, constructing a low-dimensional PLS subspace, which is a feature space constructed by the spectral feature vectors. The PLS model is then used to extract the predicted features of the master spectra and the pseudo predicted features of the slave spectra, that is, to project all spectra of the master instrument and slave instrument into this PLS subspace. Then, the ordinary least squares method is used to explore the relationship between the two features in the identical PLS subspace, the relationship will then be resorted to construct a feature transfer relationship model.

Notice that the pseudo predicted feature of the slave spectra is acquired by the PLS model established by the master instrument rather than the PLS model of the slave instrument. And PLSCT does not need the response variable corresponding to the standard set. In addition, compared with PDS, PLSCT corrects the feature of the spectra rather than the spectra. In contrast to CCACT, PLSCT uses

PLS to find the covariance between the spectra and the response variable, instead of using CCA to find the correlation between the master spectra and the slave spectra.

In order to validate the performance of the PLSCT model, we not only compare its prediction results against those of the SBC, PDS, CCACT, GLSW, and MSC methods, but also apply the Wilcoxon signed rank test [27] to determine whether PLSCT is statistically significantly superior to other models. The experiment was conducted in three real near-infrared datasets. By analyzing all the experimental results, we conclude that the PLSCT can significantly reduce the prediction error.

2. Results and Discussion

2.1. The Analysis of the Corn Dataset

First of all, Table 1 lists the latent variables (LVs) and the root mean square error of prediction (RMSEP) of Calibration, Direct transfer and Recalibration. The RMSEP was 0.010156 when using the calibration model of the master instrument to predict the spectra of the test set measured on the master instrument. However, when directly using the calibration model of the master instrument to predict the spectra of the test set measured on the slave instrument, the RMSEP was 1.41931, which indicates that if the model of the master instrument is directly applied to the slave instrument, a large prediction error will be generated.

Table 1. Root mean square error of prediction (RMSEP) obtained by Calibration, Direct transfer, and Recalibration on three spectra datasets.

Instrument	Methods	LVs	RMSEP
Corn	Calibration [1]	13	0.010156
	Direct transfer [2]		1.41931
	Recalibration [3]	5	0.208522
Wheat	Calibration [1]	12	0.258014
	Direct transfer [2]		0.85131
	Recalibration [3]	8	0.530799
Pharmaceutical tablet	Calibration [1]	7	3.123115
	Direct transfer [2]		4.514284
	Recalibration [3]	2	3.31598

[1] Calibration: the calibration model of the calibration set of the master instrument; [2] Direct transfer: the calibration model of master instrument is used on the slave instrument without modification; [3] Recalibration: the calibration model of the calibration set of the slave instrument.

The number of the factors for constructing the pseudo predicted feature matrix from the standard set of the slave spectra ($\widetilde{\mathbf{T}}_{std}^{s\_m}$) and the predicted feature matrix from the standard set of the master spectra ($\hat{\mathbf{T}}_{std}^{m}$), which is a key parameter in the PLSCT model, was determined by leave-one-out cross-validation. Figure 1A,B illustrates the effects of selecting the number of factors used to build $\widetilde{\mathbf{T}}_{std}^{s\_m}$ and $\hat{\mathbf{T}}_{std}^{m}$ on the cross-validation error when the number of the samples in the standard set is set to 25 and 30. From the results in Figure 1A,B, inferring that when the number of the samples in the standard set is set to 25 and 30, the number of factors should be set to 3. At this time, the root mean square error of cross-validation (RMSECV) reached the minimum and PLSCT achieves the best performance.

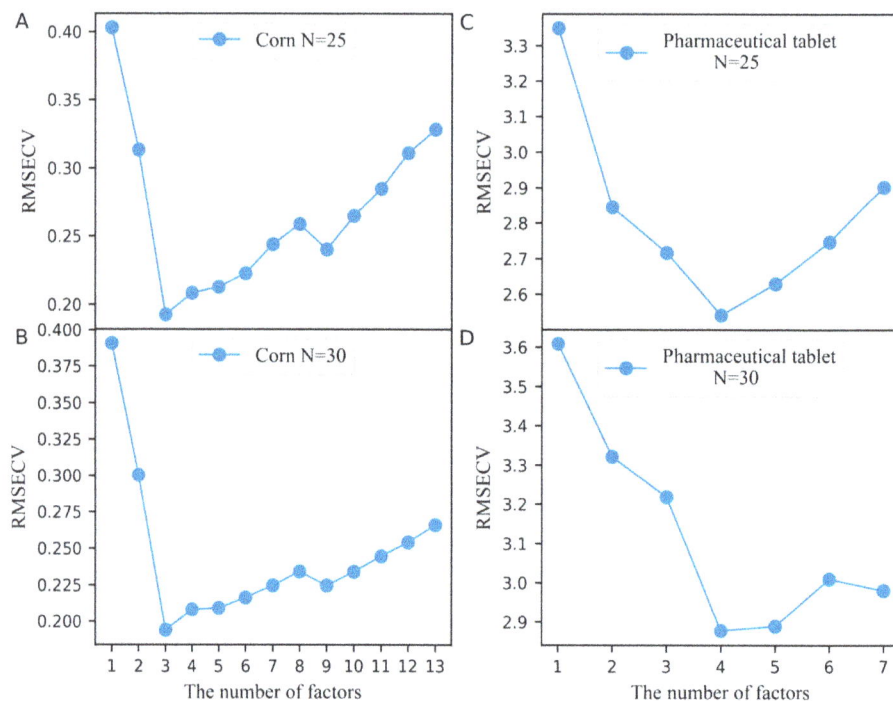

Figure 1. The effect of selecting the number of factors when building $\widetilde{\mathbf{T}}_{std}^{s\_m}$ and $\widehat{\mathbf{T}}_{std}^{m}$ on the cross-validation error. (**A**) Corn dataset and the number of the samples in the standard set is 25, (**B**) Corn dataset and the number of the samples in the standard set is 30, (**C**) Pharmaceutical tablet dataset and the number of the samples in the standard set is 25, (**D**) Pharmaceutical tablet dataset and the number of the samples in the standard set is 30.

In addition, the measured values of the moisture content of the corn dataset obtained from different models are compared with the predicted values when the number of the samples in the standard set is set to 30 are shown in Figure 2. In this case, the slope of the line was equal to 1. A point on the line indicates that the predicted value was equal to the measured value. As shown in Figure 2, PLSCT exhibited the smallest differences between the measured values and predicted values. This is attributed to the implementation of the feature transfer in the PLS subspace. The detailed description is shown in Figure 3.

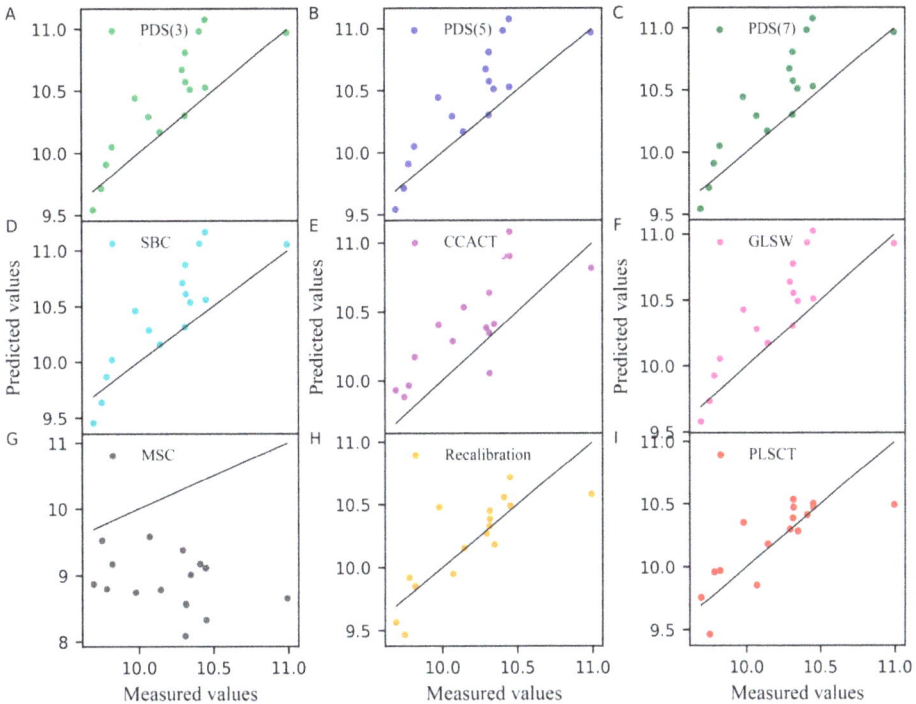

Figure 2. Measured values versus predicted values of water content for corn dataset as determined by (**A**) piecewise direct standardization with a window size of 3 (PDS(3)), (**B**) piecewise direct standardization with a window size of 5 (PDS(5)), (**C**) piecewise direct standardization with a window size of 7 (PDS(7)), (**D**) slope and bias correction (SBC), (**E**) calibration transfer method based on canonical correlation analysis (CCACT), (**F**) generalized least squares (GLSW), (**G**) multiplicative signal correction (MSC), (**H**) Recalibration and (**I**) partial least squares regression subspace based calibration transfer (PLSCT).

For comparison, the differences between the feature before and after transfer in the PLS subspace, the relationship between the first pseudo predicted feature of the slave instrument and the first predicted feature of the master instrument is displayed in Figure 3. In these two plots, the blue dots represent the feature before transfer, and the red dots represent the feature after transfer. The closer the dots are to a straight line, the smaller the differences between the pseudo predicted feature of the slave instrument and the predicted feature of the master instrument. Figure 3A,B depicts the differences between features in the standard set and the test set, respectively. Obviously, after transfer, the differences between the first pseudo predicted feature of the slave instrument and the first predicted feature of the master instrument was significantly reduced, not only in the standard set, but also in the test set.

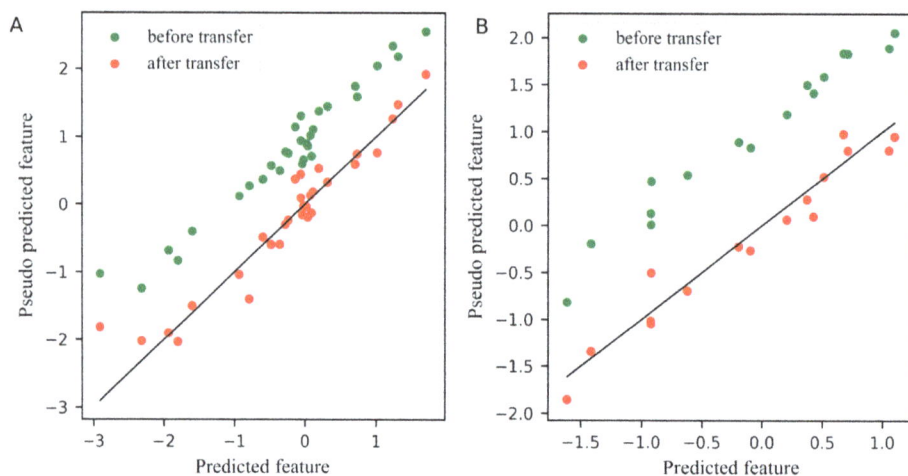

Figure 3. Plot for the differences between the feature before and after transfer in the partial least squares regression (PLS) subspace. (**A**) The differences of the first pseudo predicted feature of slave instrument standard set before and after transfer in PLS subspace, (**B**) The differences of the first pseudo predicted feature of slave instrument test set before and after transfer in PLS subspace.

In order to evaluate the effect of the number of the samples in the standard set on different calibration methods, 5, 10, 15, 20, 25, and 30, standard samples were considered in the experiment. As can be seen from Table A1 in the Appendix A, the RMSEP of MSC was relatively large, and the predictability of CCACT and GLSW were better than that of PDS, SBC and MSC. From 5 samples to 30 samples, the RMSEP of PLSCT was smaller than the RMSEP of PDS, SBC, CCACT, GLSW and MSC. Moreover, the RMSEP of PLSCT had been gradually stabilized when the number of the samples in the standard set from 20 to 30. So, we conclude that PLSCT had significantly better predictive performance than other models.

To further compare PLSCT with other models, the RMSEP improvement and p-value by Wilcoxon signed rank test are listed in Table A2 in the Appendix A. The RMSEP improvement of PLSCT to PDS(3), PDS(5), PDS(7), SBC, CCACT, GLSW, MSC, Recalibration2 and Recalibration were as high as 35.00575%, 34.99841%, 34.98937%, 41.95097%, 37.18537%, 30.21822%, 85.7502%, 8.610493% and 2.26298%, respectively. The Wilcoxon signed rank test shows statistically significant differences between PLSCT and other models (include Recalibration) at the 95% confidence level.

2.2. The Analysis of the Wheat Dataset

In Table 1, we can note that when no calibration transfer method was used, the difference between the RMSEP of directly using Calibration and the RMSEP of Recalibration was much smaller than the difference in corn dataset, in part because the difference between the two instruments in wheat dataset was relatively small.

Figure 4 displays the comparison of the measured values and the predicted values from different models. From these plots, it is worth noting that the differences between measured values and predicted values in PLSCT were only slightly larger than Recalibration and smaller than any other methods.

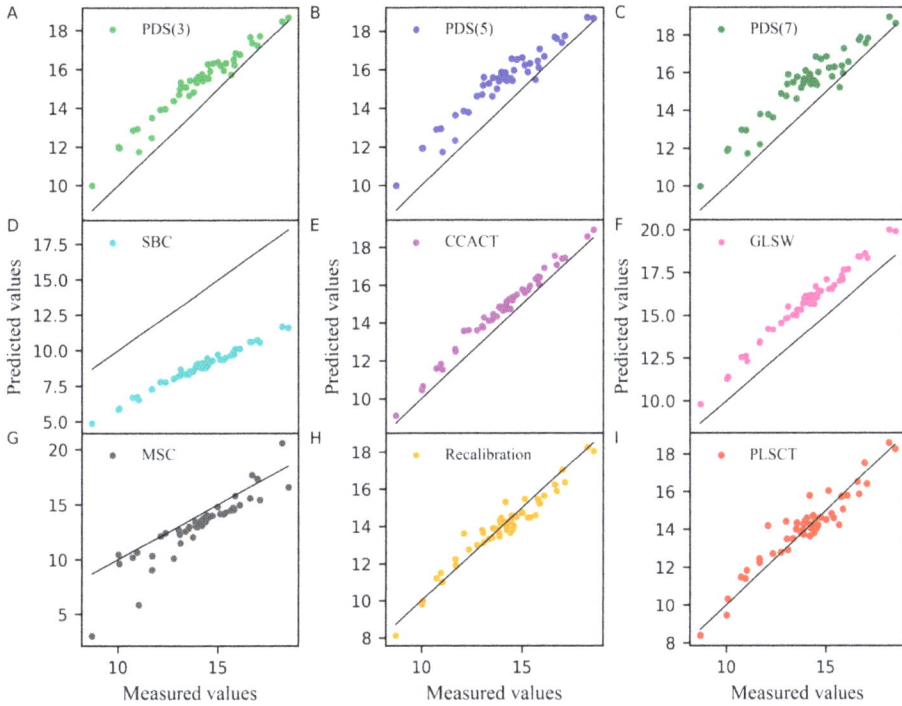

Figure 4. Measured values versus predicted values of protein content for wheat dataset as determined by (**A**) PDS(3), (**B**) PDS(5), (**C**) PDS(7), (**D**) SBC, (**E**) CCACT, (**F**) GLSW, (**G**) MSC, (**H**) Recalibration and (**I**) PLSCT.

Since the spectra difference between the master instrument and the slave instrument was small in the wheat dataset, the effect of feature transfer was not obvious in the PLS subspace from Figure 5. However, the difference between the first pseudo predicted feature after transfer and the first predicted feature is still slightly smaller. The number of samples of the standard set in Figure 5A was 30.

The performances of the different methods on wheat samples are also shown in Appendix A Table A1. The Table A2 shows clearly that PLSCT has much lower prediction error than PDS, SBC, GLSW and MSC when the number of the samples in the standard set is 10, 25 and 30. When the number of the samples in the standard set was 30, the minimum RMSEP obtained by PLSCT was 0.6604. The RMSEP of Recalibration2 fluctuated greatly, probably because there were outliers in the standard set of the slave instrument. These outliers also affect the performance of the SBC as shown in Figure 4D.

The results by Wilcoxon signed rank test reveal that PLSCT is significantly different from PDS(3), PDS(5), PDS(7), SBC, CCACT, GLSW, MSC and Recalibration2 at 95% confidence level. The RMSEP improvement resulting from PLSCT compared with these models were 51.77389%, 54.35396%, 57.02112%, 87.45319%, 42.18862%, 61.34526%, 56.43832% and 69.98222%, respectively (shown in Appendix A Table A2).

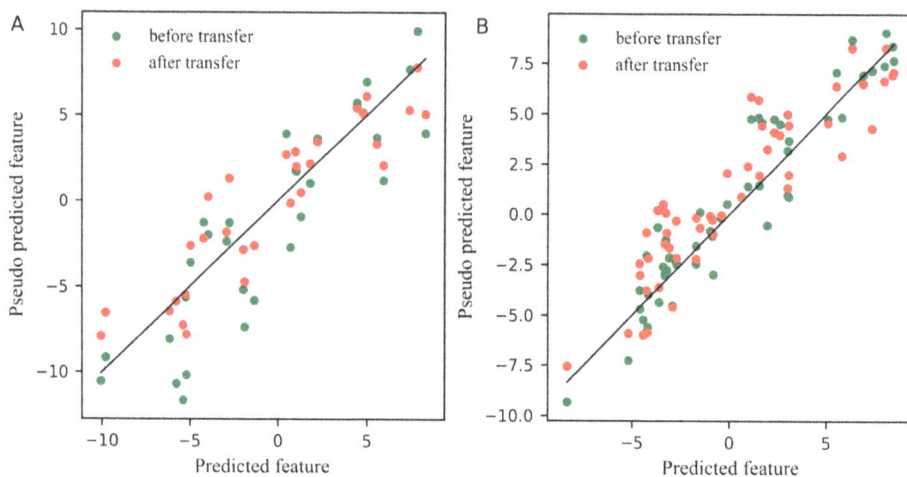

Figure 5. Plot for the differences between the feature before and after transfer in the PLS subspace. (**A**) The differences of the first pseudo predicted feature of slave instrument standard set before and after transfer in PLS subspace. (**B**) The differences of the first pseudo predicted feature of slave instrument test set before and after transfer in PLS subspace.

2.3. The Analysis of the Pharmaceutical Tablet Dataset

As in the previous cases, the LVs and RMSEP of Calibration, Direct transfer and Recalibration are shown in Table 1. The RMSEP of Calibration is 3.123115, the RMSEP of direct transfer is 4.514284, the RMSEP of Recalibration was 3.31598.

In the PLSCT model, the number of factors for constructing $\widetilde{\mathbf{T}}^{s\_m}_{std}$ and $\widehat{\mathbf{T}}^{m}_{std}$ was 4 when the number of the samples in the standard set was set to 25 and 30, as shown in Figure 1C,D. When the number of the samples in the standard set was set to 30, the comparison between the predicted values and measured values is shown in Figure 6. The results show that PLSCT has achieved the best performance.

Figure 7 displays the comparison of the first pseudo predicted feature of the slave instrument standard set and test set before and after transfer in the PLS subspace, where the number of samples of the standard set in Figure 7A was 30. From the two plots in Figure 7, the first pseudo predicted feature after transfer was significantly closer to the predicted feature of the master instrument, whether in the standard set or in the test set of the slave instrument.

From Appendix A Table A1, as the number of the samples in the standard set increases, the performance of PLSCT gradually got better. The RMSEP of PLSCT gradually became stable when the number of samples in the standard set was 25 and 30, which were outperformed than PDS, SBC, CCACT, GLSW and MSC significantly. From the results in Table A2, when the number of the samples in the standard set was greater than 20, the RMSEP of PLSCT was already less than that of Recalibration.

Compared with other models, the RMSEP improvement of PLSCT over them can reach up 16.3743%, 15.12146%, 14.35178%, 40.04516%, 16.81376%, 41.83697%, 24.21448%, 23.82937% and 2.908651%, respectively. Furthermore, the differences between PLSCT and other models are all statistically significant at the 95% confidence level (shown in Appendix A Table A2).

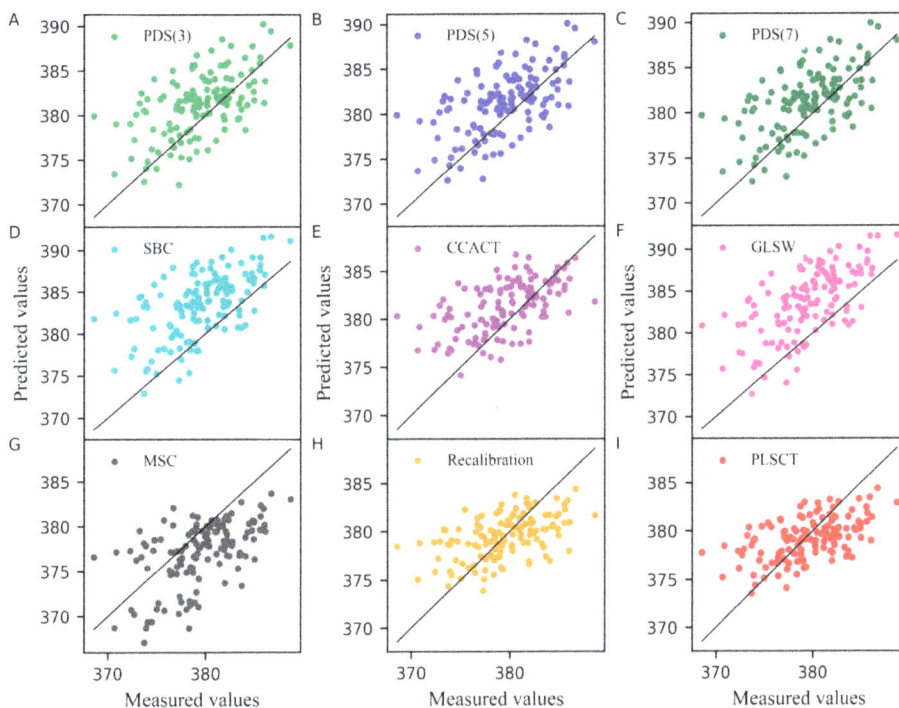

Figure 6. Measured values versus predicted values of pharmaceutical tablet dataset as determined by (**A**) PDS(3), (**B**) PDS(5), (**C**) PDS(7), (**D**) SBC, (**E**) CCACT, (**F**) GLSW, (**G**) MSC, (**H**) Recalibration and (**I**) PLSCT.

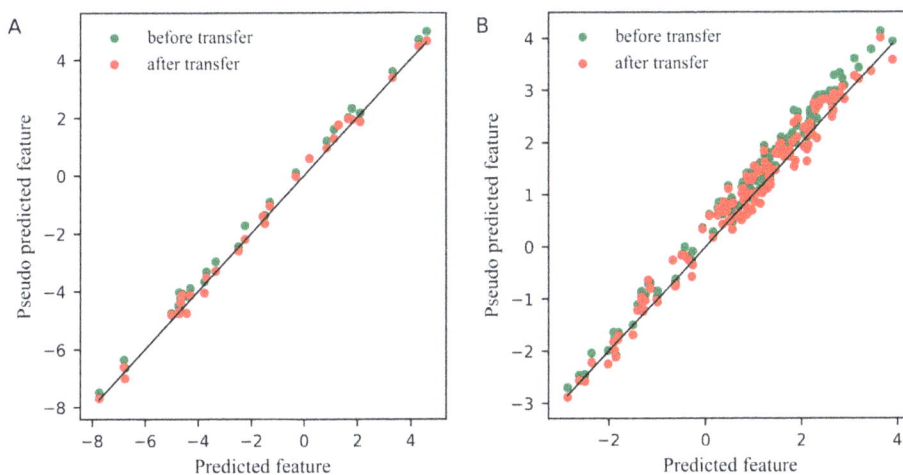

Figure 7. Plot for the differences between the feature before and after transfer in the PLS subspace. (**A**) The differences of the first pseudo predicted feature of slave instrument standard set before and after transfer in PLS subspace, (**B**) The differences of the first pseudo predicted feature of slave instrument test set before and after transfer in PLS subspace.

3. Materials and Methods

3.1. Dataset Description

3.1.1. Corn Dataset

The first dataset was corn dataset. We can conveniently access to obtain it at http://www.eigenvector.com/data/Corn/. The dataset is composed of 80 corn samples. Three near-infrared spectrometers were used to measure these samples, with wavelength range from 1100 nm to 2498 nm at 2 nm intervals (700 channels). The property of moisture, oil, protein and starch of corn is contained in the dataset. In this paper, the moisture content was chosen as the property of interest. We choose M5 as 'master instrument', MP5 as 'slave instrument'. The difference between the spectra measured on M5 instrument and MP6 instrument can be observed in Figure 8A.

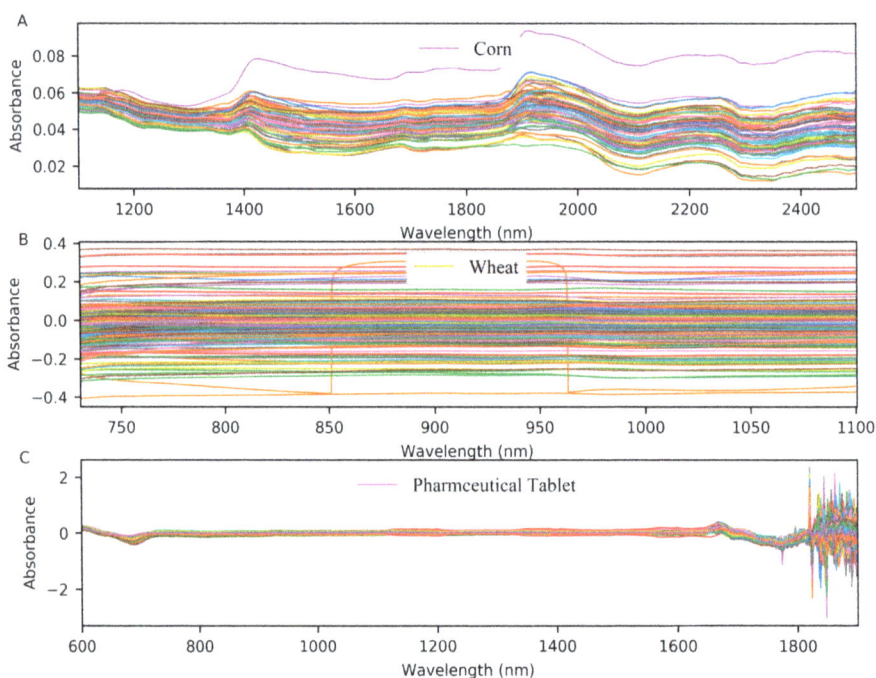

Figure 8. (**A**) The difference between the spectra of corn samples measured on M5 and MP5; (**B**) the difference between the spectra of wheat samples measured on A1 and A2; (**C**) the difference between the spectra of pharmaceutical tablet dataset.

3.1.2. Wheat Dataset

The second dataset was the wheat dataset, which consisted of 248 samples measured by three instruments of manufacturer A. This dataset was the shootout data of the International Diffuse Reflectance Conference (IDRC) in 2016. We can obtain it from http://www.idrc-chambersburg.org/content.aspx?page_id=22&club_id=409746&module_id=191116. The wavelength range of the manufacturer A was 730 nm–1100 nm and the interval was 0.5 nm. The dataset only provides the reference protein values. In this paper, we take the first instrument of manufacturer A as 'master instrument' and the second instrument as 'slave instrument'. Figure 8B shows the difference between the spectra measured on the A1 and A2 instruments.

3.1.3. Pharmaceutical Tablet Dataset

The third dataset came from the IDRC shootout 2002, which contains 655 pharmaceutical tablets measured on two spectrometers, with the range from 600 to 1898 nm, and the interval was 2 nm. We can obtain it from http://www.eigenvector.com/data/tablets/index.html. There are three reference values associated with this dataset, but we were only interested in weight content for each sample. The difference between the spectra in the pharmaceutical tablet dataset is shown in Figure 8C.

3.2. Dataset Division

We adopt the Kennard and Stone algorithm [28] to split the dataset. Firstly, the entire samples were split into the calibration set and the test set. The test set accounted for 20% of the total samples, and the remaining 80% was used as the calibration set. The corn dataset was divided into 64 samples for calibration set and 16 samples for the test set. The wheat dataset was divided into 198 samples for calibration set and 50 samples for the test set. For the pharmaceutical tablets dataset, we first integrated the three parts that have been divided, and then divided it into 524 samples for calibration sets and 131 samples for test sets. The standard samples were selected from the calibration set via the Kennard and Stone algorithm.

It must be noted that the Kennard and Stone algorithm was applied to the master spectra when splitting the calibration set and test set, while the Kennard and Stone algorithm was applied to the slave spectra when extracting the standard samples.

3.3. Determination of the Optimal Parameters

The number of latent variables used in the PLS model was selected by a 10-fold cross-validation. In order to avoid over-fitting caused by the inclusion of redundant latent variables, the optimal number of latent variables was achieved based on the statistical F-test [29] ($\alpha = 0.05$).

The predicted feature from the standard set of slave instrument is a pseudo predicted feature $\widetilde{\mathbf{T}}_{std}^{s\_m}$ constructed by the PLS model of the master instrument. Compared with the predicted feature $\widetilde{\mathbf{T}}_{std}^{s}$ constructed by the PLS model of the slave instrument, the $\widetilde{\mathbf{T}}_{std}^{s\_m}$ may contain some noise, which has a great influence on the solution of the transfer matrix ξ, further affecting the performance of the PLSCT model. In order to optimize the model, we used leave-one-out cross-validation to select the best number of factors in the standard set based on the minimum root mean square error of cross-validation (RMSECV) criterion. The response variable of the standard set used in cross-validation was the predicted value of the master instrument standard set obtained by the PLS model of the master instrument.

For the PDS method, its window sizes were set to 3, 5, and 7, respectively.

3.4. Model Performance Evaluation

In order to verify the prediction performance of different calibration models, we calculated the root mean square error of prediction (RMSEP). The calculation of RMSEP is as follows:

$$RMSEP = \sqrt{\frac{\sum\limits_{i=1}^{n}(\mathbf{y}_i - \hat{\mathbf{y}}_i)^2}{n}} \tag{1}$$

where $\mathbf{y}_i$ represents the measured value associated to the *i*-th test sample, $\hat{\mathbf{y}}_i$ is its final predicted value, while *n* is the number of samples in the test set.

In order to compare the prediction performance difference between the proposed model and other models more directly, Equation (2) was used to calculate the RMSEP improvement of the PLSCT method compared with other methods:

$$h = \left(1 - \frac{\text{RMSEP}_{\text{PLSCT}}}{\text{RMSEP}_{\text{other}}}\right) \times 100\% \tag{2}$$

where $\text{RMSEP}_{\text{PLSCT}}$ represents the prediction error of the PLSCT method, $\text{RMSEP}_{\text{other}}$ represents the prediction error of other comparison methods.

In addition, by comparing prediction error of the different models, the Wilcoxon signed rank test at the 95% confidence level was utilized to point out whether there was a significant difference between PLSCT and other methods. In python, we used the wilcoxon function in the scipy package to directly calculate the p-value between the two prediction errors. If $p > 0.05$, there is no significant difference between the two methods. Otherwise, there is significant difference.

3.5. Calibration Transfer Method

3.5.1. Notation

In this paper, we define the spectral matrix as $\mathbf{X}$, $n \times p$ represents the size of the matrix, n represents the number of samples, p represents the number of variables, and x_i represents the spectral variables corresponding to the i-th sample of the matrix. The response variables are defined as $\mathbf{y}$ and the predicted values are defined as $\hat{y}$. In order to distinguish the spectra collected on the two instruments, we added a superscript to the back of the matrix, such as defining the spectra from the master instrument as $\mathbf{X}^m$, defining the spectra from the slave instrument as $\mathbf{X}^s$, the predicted feature matrix of the master spectra obtained by the master instrument calibration model is $\hat{\mathbf{T}}^m$, the pseudo predicted feature matrix of the slave spectra obtained by the master instrument calibration model is $\widetilde{\mathbf{T}}^{s\_m}$. At the same time, a subscript was added to the back of the matrix to distinguish different data sets. For instance, $\mathbf{X}^m_{\text{cal}}$, $\mathbf{X}^m_{\text{std}}$, and $\mathbf{X}^m_{\text{test}}$ represent the calibration set, standard set and test set of the master instrument, respectively. $\mathbf{X}^s_{\text{cal}}$, $\mathbf{X}^s_{\text{std}}$, and $\mathbf{X}^s_{\text{test}}$ represent the calibration set, standard set and test set of the slave instrument, respectively.

3.5.2. Overview of PLS

PLS is a widely used multivariate calibration technique. PLS applies score vectors model the relationship between $\mathbf{X}$ and $\mathbf{y}$. It projects $\mathbf{X}$ and $\mathbf{y}$ into a PLS subspace, a low-dimensional space defined by a small number of the score vectors. The mean-centered $\mathbf{X}$ and $\mathbf{y}$ are decomposed as follows:

$$\begin{cases} \mathbf{X} = \mathbf{T}\mathbf{P}^T + \mathbf{E} \\ \mathbf{y} = \mathbf{T}\mathbf{q}^T + \mathbf{F} \end{cases} \tag{3}$$

where $\mathbf{T}$ is the score matrix, $\mathbf{P}$ and $\mathbf{q}$ represent loadings matrix for $\mathbf{X}$ and $\mathbf{y}$, respectively. $\mathbf{E}$ and $\mathbf{F}$ are the matrices of residuals corresponding to $\mathbf{X}$ and $\mathbf{y}$.

The matrix of regression coefficients is:

$$\beta = \mathbf{W}\left(\mathbf{P}^T\mathbf{W}\right)^{-1}\mathbf{q}^T \tag{4}$$

where $\mathbf{W}$ is the weight matrix.

With the regression coefficient matrix β, we can have the predicted values:

$$\hat{y} = \mathbf{X}\beta \tag{5}$$

3.5.3. Proposed PLSCT method

In the PLSCT, the PLS model was built on the calibration set of the master instrument to construct the PLS subspace, which is also the feature space constructed by the feature vectors of the spectra of the master instrument calibration set. The number of latent variables (LVs) in the PLS model is determined by cross-validation.

$$\beta^{m} = \mathbf{W}^{m}\left((\mathbf{P}^{m})^{T}\mathbf{W}^{m}\right)^{-1}(\mathbf{q}^{m})^{T} \tag{6}$$

On the basis of this PLS model, the predicted feature matrix of standard set in the master instrument $\mathbf{X}_{std}^{m}$ can be calculated via it, that is, the spectra of the master instrument can be projected into the PLS subspace:

$$\hat{\mathbf{T}}_{std}^{m} = \mathbf{X}_{std}^{m}\mathbf{W}^{m}\left((\mathbf{P}^{m})^{T}\mathbf{W}^{m}\right)^{-1} \tag{7}$$

Similarly, the pseudo predicted feature matrix of standard set in the slave instrument $\mathbf{X}_{std}^{s}$ can be calculated via this PLS model as well as $\mathbf{X}_{std}^{m}$, in other words, the spectra of the slave instrument can be projected into this PLS subspace:

$$\widetilde{\mathbf{T}}_{std}^{s\_m} = \mathbf{X}_{std}^{s}\mathbf{W}^{m}\left((\mathbf{P}^{m})^{T}\mathbf{W}^{m}\right)^{-1} \tag{8}$$

The two predicted feature matrices obtained are derived from the same PLS model of the master instrument, that is to say, all spectra are projected into the identical PLS subspace constructed by the master instrument. In the identical PLS subspace, there should be a linear relationship between the two feature matrices. So $\widetilde{\mathbf{T}}_{std}^{s\_m}$ and $\hat{\mathbf{T}}_{std}^{m}$ can be built as:

$$\widetilde{\mathbf{T}}_{std}^{s\_m}\boldsymbol{\xi} = \hat{\mathbf{T}}_{std}^{m} \tag{9}$$

The linear relationship between the two feature matrices can be solved through the ordinary least squares method, by the following equation:

$$\boldsymbol{\xi} = \left(\left(\widetilde{\mathbf{T}}_{std}^{s\_m}\right)^{T}\widetilde{\mathbf{T}}_{std}^{s\_m}\right)^{-1}\left(\widetilde{\mathbf{T}}_{std}^{s\_m}\right)^{T}\hat{\mathbf{T}}_{std}^{m} \tag{10}$$

Once $\boldsymbol{\xi}$ is computed, for the test set from the slave instrument $\mathbf{X}_{test}^{s}$, applying Equation (11) to calculate the predicted values corresponding to the spectra:

$$\hat{y}_{test} = \mathbf{X}_{test}^{s}\mathbf{W}^{m}\left((\mathbf{P}^{m})^{T}\mathbf{W}^{m}\right)^{-1}\boldsymbol{\xi}(\mathbf{q}^{m})^{T} \tag{11}$$

4. Conclusions

In this paper, an ingenious calibration transfer method based on PLS subspace is proposed. PLSCT uses the same PLS model to project the spectra into the identical PLS subspace. In the identical subspace, a feature transfer model is constructed by narrowing the differences between the predicted feature of master instrument and the pseudo predicted feature of the slave instrument via an ordinary least squares method. Additional, PLSCT does not need the response variable corresponding to the standard set. As expected, experimental results on three real datasets show that compared with PDS, SBC, CCACT, GLSW, and MSC, the PLSCT model is more stable and can obtain more accurate prediction results. The reason why the PLSCT model can achieve such remarkable results is that while the spectra of the slave instrument are projected into this subspace, some noise effects such as scattering that are unrelated to the response variable will be removed from the spectra, and then the feature transfer in the identical PLS subspace can more accurately narrow the differences between the predicted feature of master instrument and the pseudo predicted feature of slave instrument.

Molecules **2019**, *24*, 1289

Author Contributions: Conceptualization, P.S. and Y.Z.; methodology, P.S. and Y.Z.; software, J.Y.; validation, J.Y.; formal analysis, P.S., Y.Z., J.Y. and Z.Z.; data curation S.G.; writing—original draft preparation, J.Y.; writing—review and editing, J.Y., P.S. and X.J.; visualization, J.Y.; supervision, P.S. and Y.Z.; project administration, J.Y.; funding acquisition, Y.Z.

Funding: This research was funded by National Natural Science Foundation of China, grant number 61601104; Natural Science Foundation of Hebei Province, grant number F2017501052 and the Basic Science Research Fund of Northeast University at Qin Huang Dao, grant number XNB201611.

Conflicts of Interest: The authors declare no conflict of interest.

Appendix A

Table A1. RMSEP for three datasets using different transfer methods.

	PDS			SBC	CCACT	GLSW	MSC	PLSCT	Recalibration2 [2]	Recalibration [3]
	W[1] = 3	W[1] = 5	W[1] = 7							
Corn dataset										
N = 5	0.4142	0.4336	0.4354	0.5370	0.2411	0.4056	1.4302	0.1991	0.3538	0.2085
N = 10	0.3753	0.3617	0.3729	0.4440	0.2545	0.3696	1.4302	0.1980	0.2237	
N = 15	0.3507	0.3495	0.3357	0.4307	0.3663	0.3535	1.4302	0.2127	0.2425	
N = 20	0.3440	0.3440	0.3440	0.3900	0.2841	0.3208	1.4302	0.2087	0.2379	
N = 25	0.3373	0.3372	0.3366	0.3720	0.3528	0.3106	1.4302	0.2082	0.2314	
N = 30	0.3136	0.3135	0.3135	0.3511	0.3245	0.2921	1.4302	0.2038	0.2230	
Wheat dataset										
N = 5	8.2434	9.1587	8.4226	14.3731	1.6248	4.0835	1.5160	1.8478	2.7176	0.5308
N = 10	8.5844	9.3534	10.8927	10.5310	1.2496	3.5824	1.5160	0.8588	2.2233	
N = 15	2.1373	2.8513	3.2012	8.7159	1.5315	2.9205	1.5160	1.8280	1.3985	
N = 20	1.9586	2.0927	2.2380	7.0482	0.9688	2.4743	1.5160	1.8263	0.4520	
N = 25	1.5656	1.6480	1.7445	6.1945	1.0437	1.9804	1.5160	0.6850	2.3661	
N = 30	1.3694	1.4468	1.5366	5.2635	0.7735	1.7085	1.5160	0.6604	2.2000	
Pharmaceutical tablet dataset										
N = 5	4.7971	4.2899	4.4594	5.9983	4.1302	6.5988	4.2482	3.3202	5.8027	3.3160
N = 10	4.1431	4.0098	4.0444	5.4720	4.1112	5.6721	4.2482	3.5821	5.5904	
N = 15	3.9698	3.8314	3.8347	5.7069	3.9357	6.2284	4.2482	3.3834	5.8043	
N = 20	3.9787	3.8789	3.9190	5.2838	3.8979	5.6511	4.2482	3.2794	5.0811	
N = 25	3.9263	3.8416	3.7789	5.2514	4.0549	5.4809	4.2482	3.2765	4.9428	
N = 30	3.8499	3.7931	3.7590	5.3699	3.8703	5.5354	4.2482	3.2195	4.2267	

[1] w stands for window size of PDS method; [2] Recalibration2: the calibration model of the standard set of the slave instrument; [3] Recalibration: the calibration model of the calibration set of the slave instrument.

Table A2. RMSEP comparison of PLSCT and other methods, RMSEP improvement and *p*-values by the Wilcoxon signed rank test (α = 0.05) (the number of samples in the standard set is 30).

			PLSCT	
		Corn	Wheat	Pharmaceutical Tablet
PDS(3) [1]	*h* (%) [2]	35.00575	51.77389	16.3743
	p [3]	**0.00717**	**3.17 × 10⁻⁹**	**3.2 × 10⁻¹⁹**
PDS(5) [1]	*h* (%)	34.99841	54.35396	15.12146
	p	**0.00717**	**2.23 × 10⁻⁹**	**1.78 × 10⁻¹⁹**
PDS(7) [1]	*h* (%)	34.98937	57.02112	14.35178
	p	**0.00717**	**2.23 × 10⁻⁹**	**1.2 × 10⁻¹⁸**
SBC	*h* (%)	41.95097	87.45319	40.04516
	p	**0.011286**	**7.56 × 10⁻¹⁰**	**4.84 × 10⁻²³**
CCACT	*h* (%)	37.18537	42.18862	16.81376
	p	**0.004455**	**0.001161**	**4.37 × 10⁻²¹**
GLSW	*h* (%)	30.21822	61.34526	41.83697
	p	**0.00717**	**8.53 × 10⁻¹⁰**	**6.82 × 10⁻²³**
MSC	*h* (%)	85.7502	56.43832	24.21448
	p	**0.000531**	**1.57 × 10⁻⁶**	**1.51 × 10⁻¹⁸**
Recalibration2	*h* (%)	8.610493	69.98222	23.82937
	p	**0.017378**	**0.000231**	**3.05 × 10⁻²³**
Recalibration	*h* (%)	2.26298	−24.4164	2.908651
	p	0.876722	**9.06 × 10⁻⁵**	**0.000198**

[1] The number in brackets stands for window size of PDS method; [2] *h*: the RMSEP improvement; [3] *p*: *p*-value by Wilcoxon signed rank test.

References

1. Roggo, Y.; Chalus, P.; Maurer, L.; Lema-Martinez, C.; Edmond, A.; Jent, N. A review of near infrared spectroscopy and chemometrics in pharmaceutical technologies. *J. Pharm. Biomed. Anal.* **2007**, *44*, 683–700. [CrossRef]

2. Kumar, M.; Bhatia, R.; Rawal, R.K. Applications of Various Analytical Techniques in Quality Control of Pharmaceutical Excipients. *J. Pharm. Biomed. Anal.* **2018**, *157*, 122–136. [CrossRef] [PubMed]

3. Martinez, J.C.; Guzmán-Sepúlveda, J.R.; Bolaňoz Evia, G.R.; Córdova, T.; Guzmán-Cabrera, R. Enhanced Quality Control in Pharmaceutical Applications by Combining Raman Spectroscopy and Machine Learning Techniques. *Int. J. Thermophys.* **2018**, *39*, 79. [CrossRef]

4. Heesang, A.; Hyerin, S.; Dong-Myeong, S.; Kyujung, K.; Jong-ryul, C. Emerging optical spectroscopy techniques for biomedical applications—A brief review of recent progress. *Appl. Spectrosc. Rev.* **2017**, *53*, 264–278. [CrossRef]

5. Morris, R.E.; Hammond, M.H.; Cramer, J.A.; Johnson, K.J.; Giordano, B.C.; Kramer, K.E.; Rose-Pehrsson, S.L. Rapid fuel quality surveillance through chemometric modeling of near-infrared spectra. *Energy Fuels* **2009**, *23*, 1610–1618. [CrossRef]

6. López, A.; Arazuri, S.; García, I.; Mangado, J.S.; Jarén, C. A review of the application of near-infrared spectroscopy for the analysis of potatoes. *J. Agric. Food Chem.* **2013**, *61*, 5413–5424. [CrossRef] [PubMed]

7. Hernández-Hierro, J.M.; Valverde, J.; Villacreces, S.; Reilly, K.; Gaffney, M.; González-Miret, M.L.; Heredia, F.J.; Downey, G. Feasibility study on the use of visible–near-infrared spectroscopy for the screening of individual and total glucosinolate contents in Broccoli. *J. Agric. Food Chem.* **2012**, *60*, 7352–7358. [CrossRef]

8. Cen, H.; He, Y. Theory and application of near infrared reflectance spectroscopy in determination of food quality. *Trends Food Sci. Technol.* **2007**, *18*, 72–83. [CrossRef]

9. Huang, H.; Yu, H.; Xu, H.; Ying, Y. Near infrared spectroscopy for on/in-line monitoring of quality in foods and beverages:a review. *J. Food Eng.* **2008**, *87*, 303–313. [CrossRef]

10. Lukacs, M.; Bazar, G.; Pollner, B.; Henn, R.; Kirchler, C.G.; Huck, C.W.; Kovacs, Z. Near infrared spectroscopy as an alternative quick method for simultaneous detection of multiple adulterants in whey protein-based sports supplement. *Food Control* **2018**, *94*, 331–340. [CrossRef]

11. Geladi, P.; Kowalski, B.R. Partial least-squares regression: A tutorial. *Anal. Chim. Acta* **1986**, *185*, 1–17. [CrossRef]

12. Wold, S.; Sjöström, M.; Eriksson, L. PLS-regression: A basic tool of chemometrics. *Chemom. Intell. Lab. Syst.* **2001**, *58*, 109–130. [CrossRef]

13. Næs, T.; Martens, H. Principal component regression in NIR analysis: View-points, background details and selection of components. *J. Chemom.* **1988**, *2*, 155–167. [CrossRef]

14. Geladi, P.; Esbensen, K. Regression on multivariate images: Principal component regression for modeling, prediction and visual diagnostic tools. *J. Chemom.* **1991**, *5*, 97–111. [CrossRef]

15. Wang, Y.; Veltkamp, D.J.; Kowalski, B.R. Multivariate instrument standardization. *Anal. Chem.* **1991**, *63*, 2750–2756. [CrossRef]

16. Bouveresse, E.; Massart, D. Improvement of the piecewise direct standardisation procedure for the transfer of NIR spectra for multivariate calibration. *Chemom. Intell. Lab. Syst.* **1996**, *32*, 201–213. [CrossRef]

17. Wise, B.M.; Martens, H.; Høy, M.; Bro, R.; Brockhoff, P.B. Calibration transfer by generalized least squares. 2001. Available online: http://www.eigenvector.com/Docs/GLS_Standardization.pdf (accessed on 31 March 2019).

18. Wise, B.M.; Martens, H.; Høy, M. Generalized least squares for calibration transfer. Available online: http://www.eigenvector.com/Docs/GLS_Calibration_Trans.pdf (accessed on 22 October).

19. Bouveresse, E.; Hartmann, C.; Massart, D.L.; Last, I.R.; Prebble, K.A. Standardization of Near-Infrared Spectrometric Instruments. *Anal. Chem.* **1996**, *68*, 982–990. [CrossRef]

20. Fan, W.; Liang, Y.; Yuan, D.; Wang, J. Calibration model transfer for near-infrared spectra based on canonical correlation analysis. *Anal. Chim. Acta* **2008**, *623*, 22–29. [CrossRef]

21. Kramer, K.E.; Morris, R.E.; Rose-Pehrsson, S.L. Comparison of two multiplicative signal correction strategies for calibration transfer without standards. *Chemom. Intell. Lab. Syst.* **2008**, *92*, 33–43. [CrossRef]

22. Blank, T.B.; Sum, S.T.; Brown, S.D.; Monfre, S.L. Transfer of near-infrared multivariate calibrations without standards. *Anal. Chem.* **1996**, *68*, 2987–2995. [CrossRef]

23. Wold, S.; Antti, H.; Lindgren, F.; Öhman, J. Orthogonal signal correction of near infrared spectra. *Chemom. Intell. Lab. Syst.* **1998**, *44*, 175–185. [CrossRef]

24. Sjöblom, J.; Svensson, O.; Josefson, M.; Kullberg, H.; Wold, S. An evaluation of orthogonal signal correction applied to calibration transfer of near infrared spectra. *Chemom. Intell. Lab. Syst.* **1998**, *44*, 229–244. [CrossRef]
25. Pan, S.J.; Tsang, I.; Kwok, J.; Yang, Q. Domain adaptation via transfer component analysis. *IEEE Trans. Neural Netw.* **2011**, *22*, 199–210. [CrossRef] [PubMed]
26. Schölkopf, B.; Smola, A.; Müller, K.R. Kernel principal component analysis. *Artificial Neural Networks—ICANN'97* **1997**, 583–588. [CrossRef]
27. Wilcoxon, F. Individual comparisons by ranking methods. *Biometrics Bull.* **1945**, *1*, 80–83. [CrossRef]
28. Kennard, R.W.; Stone, L.A. Computer aided design of experiments. *Technometrics* **1969**, *11*, 137–148. [CrossRef]
29. Haaland, D.M.; Thomas, E.V. Partial Least-Squares Methods for Spectral Analyses. 1. Relation to Other Quantitative Calibration Methods and the Extraction of Qualitative Information. *Anal. Chem.* **1988**, *60*, 1193–1202. [CrossRef]

Sample Availability: Samples are not available from the authors.

molecules

MDPI

Article

At-Line Monitoring of the Extraction Process of Rosmarini Folium via Wet Chemical Assays, UHPLC Analysis, and Newly Developed Near-Infrared Spectroscopic Analysis Methods

Stefanie Delueg [1,†], Christian G. Kirchler [1,†], Florian Meischl [1], Yukihiro Ozaki [1,2], Michael A. Popp [3], Günther K. Bonn [1,4] and Christian W. Huck [1,*]

[1] Institute of Analytical Chemistry and Radiochemistry, University of Innsbruck, Innrain 80/82, 6020 Innsbruck, Austria
[2] School of Science and Technology, Kwansei Gakuin University, Gakuen, Sanda, Hyogo 669-1337, Japan
[3] Michael Popp Research Institute for New Phyto Entities, University of Innsbruck, Mitterweg 24, 6020 Innsbruck, Austria
[4] ADSI–Austrian Drug Screening Institute GmbH, Innrain 66a, 6020 Innsbruck, Austria
[*] Correspondence: Christian.W.Huck@uibk.ac.at; Tel.: +43-512-507-57304
[†] Authors contributed equally to this work.

Received: 13 June 2019; Accepted: 3 July 2019; Published: 6 July 2019

Abstract: The present study demonstrates the applicability of at-line monitoring of the extraction process of *Rosmarinus officinalis* L. leaves (Rosmarini folium) and the development of near-infrared (NIR) spectroscopic analysis methods. Therefore, whole dried Rosmarini folium samples were extracted by maceration with 70% (*v/v*) ethanol. For the experimental design three different specimen-taking plans were chosen. At first, monitoring was carried out using three common analytical methods: (a) total hydroxycinnamic derivatives according to the European Pharmacopoeia, (b) total phenolic content according to Folin–Ciocalteu, and (c) rosmarinic acid content measured by UHPLC-UV analysis. Precision validation of the wet chemical assays revealed a repeatability of (a) 0.12% relative standard deviation (RSD), (b) 1.1% RSD, and (c) 0.28% RSD, as well as an intermediate precision of (a) 4.1% RSD, (b) 1.3% RSD, and (c) 0.55% RSD. The collected extracts were analyzed with a NIR spectrometer using a temperature-controlled liquid attachment. Samples were measured in transmission mode with an optical path length of 1 mm. The combination of the recorded spectra and the previously obtained analytical reference values in conjunction with multivariate data analysis enabled the successful establishment of partial least squares regression (PLSR) models. Coefficients of determination (R^2) were: (a) 0.94, (b) 0.96, and (c) 0.93 (obtained by test-set validation). Since Pearson correlation analysis revealed that the reference analyses correlated with each other just one of the PSLR models is required. Therefore, it is suggested that PLSR model (b) be used for monitoring the extraction process of Rosmarini folium. The application of NIR spectroscopy provides a fast and non-invasive alternative analysis method, which can subsequently be implemented for on- or in-line process control.

Keywords: ultra-high performance liquid chromatography; Folin–Ciocalteu; total hydroxycinnamic derivatives; phytoextraction; near-infrared spectroscopy

---

1. Introduction

Plants have been the main source of traditional medicine systems over millennia and are still of great importance in healthcare today [1,2]. The demand for pharmaceuticals based on natural sources has even increased in recent times [3,4]. In Europe, herbal substances, preparations, and combinations are assessed and regulated by the Committee on Herbal Medicinal Products (HMPC), which is part of the

European Medicines Agency (EMA), and the European Pharmacopoeia (Ph. Eur.) [5,6]. Nevertheless, chemically complex plant-based preparations are in constant competition with chemically defined products. Therefore, quality assurance and analytics of these so-called "phytopharmaceuticals" is a big challenge for the manufacturers. Besides the incoming goods, inspection and extraction control of medicinal plants play an important role in the yield and purity of the product [7]. Furthermore, resource and cost efficiency can be increased by extraction optimization. Near-infrared (NIR) spectroscopy and Raman spectroscopy represent attractive analysis techniques for the research demand regarding the at-line, on-line, or in-line analysis of phytoextraction processes [3,8–14]. In contrast to common off-line reference analyses, NIR spectroscopic process monitoring as process analytical technology (PAT) has convincing advantages since its operation is non-destructive, contact-free, pollution-free, does not require any additional solvents, saves energy, and is highly cost-effective. The recorded NIR spectra include multiple physical and chemical parameters which can be determined simultaneously. The use of optical light fibers facilitates a distance of up to several hundred meters between the measurement probe and the analyzer. Furthermore, NIR spectroscopy fulfills the requirements of fast real-time process control. Nevertheless, the development of a NIR spectroscopic analysis method is time- and resource-consuming and has to be undertaken by experienced professionals [15]. As for reference analytics, the quantification of the total phenolic compound is specified by the European Pharmacopoeia. The antioxidant properties of certain phytogenic substances are attributed to the presence of phenol terpens in rosemary [16]. The analysis described in the European Pharmacopoeia is principally for the analysis of cinnamic acid derivatives. The assay is complicated and another wet chemical assay (Folin–Ciocalteu) has to be executed to verify the results. The Folin–Ciocalteu analysis is not that specific but is more reproducible. However, HPLC analysis is currently the method of choice. It is state-of-the-art, since the analyses can be measured without any major work-up and the measurement can be automated, in contrast to the wet-chemical investigations [17]. In order to meet the requirements of the EMA and still be up to date, all three analyses were carried out, calibrated into the system, and checked for reproducibility, traceability and comparability. Thus, a holistic view of the system and the determination of the saturation of the extraction could be determined. The present feasibility study reports the monitoring of the phytoextraction process of *Rosmarinus officinalis* L. leaves using common analytical methods as well as newly developed NIR spectroscopic methods applying partial least squares regression (PLSR) models as multivariate data analysis (MVA) tools. This analysis was used as the basis for an online fixation of NIR measurements in phytochemical extractions.

2. Results and Discussion

2.1. Wet Chemical Assays (European Pharmacopoeia and Folin–Ciocalteu)

The wet chemical assays for the determination of total hydroxycinnamic derivatives (THCD) according to Ph. Eur. and gallic acid equivalents (GAE%) referred to as Folin–Ciocalteu (FC) have similar reaction mechanisms. The chemical background is very complex and not yet fully understood. Both wet chemical assays are based on the reduction of a mixture composed of tungsten and molybdenum oxides [18]. In the fully oxidized valence state the isopolyphosphotungstates are colorless and the molybdenum compounds are yellow. The reagent mixture consists of heteropolyphosphotungstates-molybdates. In an acidic solution a hydrated octahedral complex of metal oxides, which is coordinated around a central phosphate, appears. Due to the reversible reduction of one or two electrons the color of the solution changes. In the case of the Ph. Eur. assay the solution turns red and in the case of the FC assay it turns blue [19]. The more intense the color the higher the concentration of the phenolic compounds is in the samples.

The Ph. Eur. assay, which can be assigned to the THCD, is more substance-specific than the FC assay. This is based on the different chemicals which are added for the assays. FC targets hydroxy groups, whereas the Ph. Eur. assay targets carboxyl groups which are not as common as hydroxy groups in the chemistry of natural products [20]. In the present study both assays were applied for monitoring the extraction process of Rosmarini folium. Correlation analyses of the two assays revealed a Pearson correlation of 0.966 (Table 1). This means that substances which were assessed by measurement via the Ph. Eur. assay were highly correlated with those measured by the FC assay, and vice versa.

Table 1. Pearson correlations of the reference analyses.

	Ph. Eur.	FC	UHPLC
Ph. Eur.	1	-	-
FC	0.966	1	-
UHPLC	0.955	0.953	1

Looking at the results of the precision studies in Table 2, the repeatability confirmed the high performance of the Ph. Eur. assay. Nevertheless, determination of the intermediate precision revealed the superiority of GAE% quantification via the FC assay, with a 1.3% relative standard deviation (RSD), compared to THCD quantification via the Ph. Eur. assay with a 4.1% RSD. The easier handling of the FC assay compared to the Ph. Eur. assay could be the reason for the better repeatability of the results on different days.

Table 2. Parameters of the precision studies of the reference analysis.

	Ph. Eur.	FC	UHPLC
Repeatability in % RSD	0.12	1.1	0.28
Intermediate precision in % RSD	4.1	1.3	0.55
Repeatability (absolute)	16 *	0.028 **	0.0017 ***
Intermediate precision (absolute)	593 *	0.033 **	0.0028 ***

* THCD mg/kg, ** GAE%, *** RA%

Both assays were used as reference analyses for the establishment of NIR spectroscopic methods.

2.2. Ultra-High Performance Liquid Chromatography

Nowadays, automatable methods like UHPLC-UV measurements are more common than wet chemical assays. This is because the sample preparation for UHPLC-UV measurement is often easier than for a wet chemical assay. Furthermore, fewer mistakes and variations in the analyses occur in UHPLC-UV. Also, in the present case, precision studies of the UHPLC-UV measurements of Rosmarini folium extracts obtained good repeatability (0.28% RSD) and excellent intermediate precision (0.55% RSD) for the determination of rosmarinic acid (RA) compared to the wet chemical assays (see Table 2). An example of a Rosmarini folium extract chromatogram compared to a RA reference solution, which was used for external calibration, is illustrated in Figure 1. Although the RA quantification showed such good results it is important to note that biological extracts are multi- substance mixtures of secondary metabolites. This is the reason that the Pearson correlations (Table 1) between the UHPLC-UV measurements and the wet chemical assays (0.955 and 0.953) were lower than the Pearson correlation between the wet chemical assays (0.966). Nevertheless, a high correlation between all three reference analysis methods was observed. Based on this fact, the reference analytical method of choice for the establishment of a NIR spectroscopic method should either be the UHPLC-UV analysis for the quantification of the single substance, RA, or the FC assay which was the better performing wet chemical assay (see Section 2.1) representing the plant extract as multi-substance mixture.

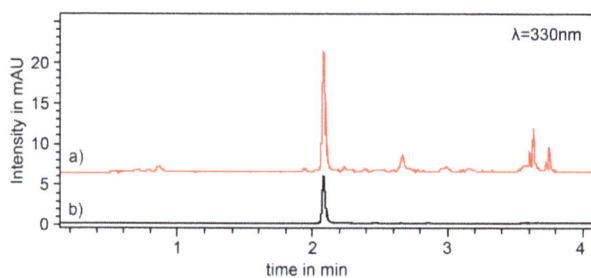

Figure 1. Chromatograms of (**a**) rosemary extract (red line) in 70% *v/v* ethanol (50 g/L) after 3 h continued stirred extraction, and (**b**) rosmarinic acid reference solution (black line), measured at 330 nm.

2.3. Near-Infrared Spectroscopy

Raw NIR spectra of all 90 samples are shown in Figure 2a. For the establishment of the PLSR models, uninformative and interfering spectral regions were excluded. Therefore, the best PLSR models were obtained by using the wavenumber region from 6028 to 5424 cm^{-1}, which is illustrated in Figure 2b.

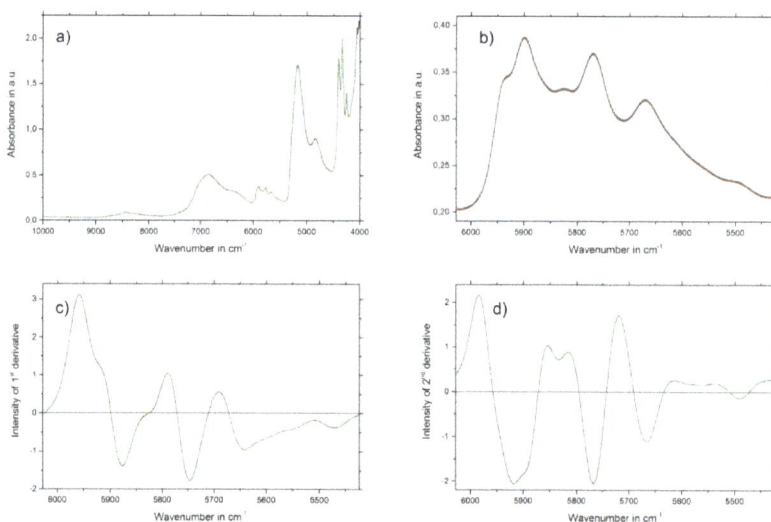

Figure 2. (**a**) Raw near-infrared (NIR) spectra of all 90 samples; (**b**) section from the raw NIR spectra showing the wavenumber region used for PLSR model calculation; (**c**) first derivate (13 smoothing points) and standard normal variate (SNV)-transformed NIR spectra region used for total hydroxycinnamic derivatives (THCD) in mg/kg and gallic acid equivalents (GAE)% PLSR model calculation; and (**d**) second derivate (23 smoothing points) and SNV-transformed NIR spectra region used for rosmarinic acid (RA)% PLSR model calculation.

The results of the best PLSR models for THCD in mg/kg, GAE% and RA% are given in Table 3. The best spectral pretreatment for THCD in mg/kg and GAE% was the first derivative, using 13 smoothing points followed by applying standard normal variate (SNV) transformation to the selected wavenumber region (see Figure 2c). The best spectral pretreatment for the determination of RA% was the second derivative, with 23 smoothing points followed by applying SNV transformation to the selected wavenumber region (see Figure 2d). Predicted versus reference plots, and the regression coefficient plots for the three PLSR models are shown in Figure 3a,b for THCD in mg/kg, Figure 3c,d for GAE%, and Figure 3e,f for RA%.

Table 3. Parameters of the established partial least squares regression PLSR models.

Reference Analysis	Ph. Eur.		FC		UHPLC	
Samples	90		90		90	
Outliers	0		0		0	
	CV	TSV	CV	TSV	CV	TSV
$R^2_{calibration}$	0.95	0.95	0.97	0.97	0.94	0.95
$R^2_{validation}$	0.94	0.94	0.96	0.96	0.94	0.93
RMSEC [a]	1425 *	1308 *	0.14 **	0.13 **	0.11 ***	0.09 ***
RMSECV [b] or RMSEP [c]	1527 *	1632 *	0.16 **	0.18 **	0.12 ***	0.13 ***
Factor	3	4	3	4	4	4
Calibration range	1975–25378 *		0.494–3.660 **		−1.810 ***	

* THCD mg/kg, ** GAE%, *** RA%, [a] root mean square error of calibration, [b] root mean square error of cross validation, [c] root mean square error of prediction.

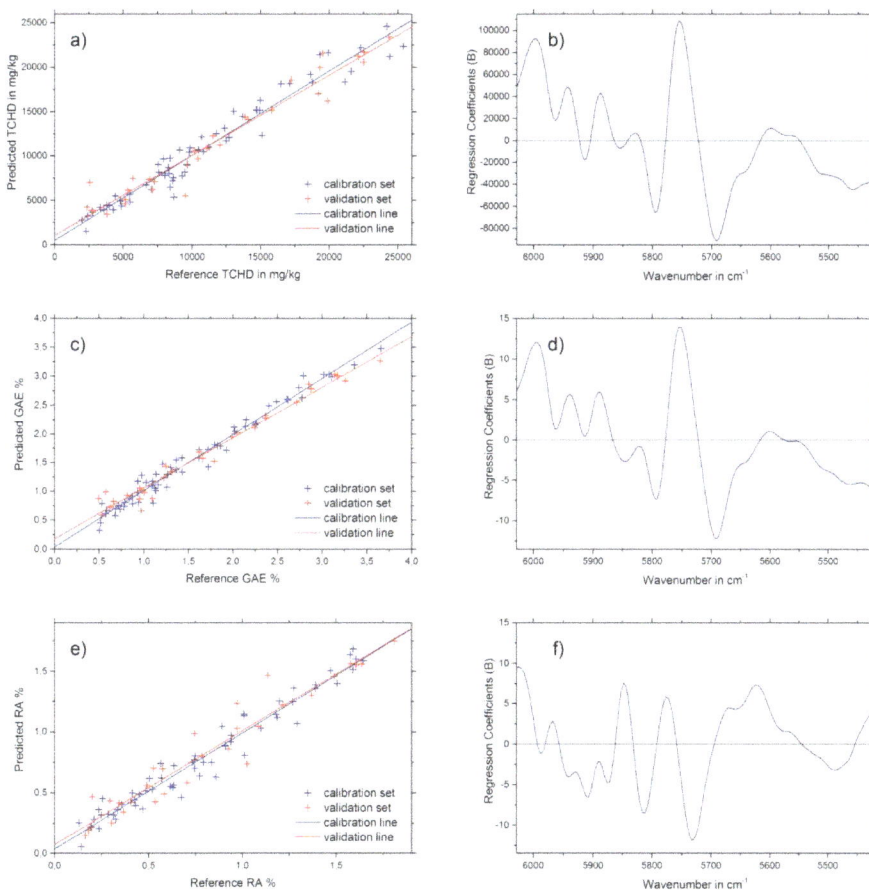

Figure 3. Predicted versus reference plots (left column) and regression coefficient plots (right column) for the best test-set validated PLSR models for: (**a**) and (**b**) THCD in mg/kg, (**c**) and (**d**) GAE%, and (**e**) and (**f**) RA%.

NIR bands of the wavenumber region used (see Figure 2b) which have an influence on the PLSR model calculations can be considered mainly as aromatic and unsaturated 2νCH. This is due to the diverse, but nevertheless chemically similar, structures of the THCD, the total phenolic content, and the RA content. Therefore, other bands which emerge from overtones or combinations of OH, CC, and CO vibrations can be excluded for the establishment of the PLSR model [21] The criteria for the successful end of the extraction process of Rosmarini folium is to access the extraction plateau. This can be easily achieved by the reference analyses methods (see Figure 4a). However, these need analysis time and manpower, as well as chemicals, and are therefore not suited for real-time at-line monitoring. All three NIR spectroscopic PLSR models also showed satisfactory results for the monitoring of the extraction process of Rosmarini folium. Since the reference analyses were all correlated (see Table 1) the application of just one of the PSLR models was required to obtain the desired result. Therefore, it is suggested that the best PLSR model should be applied. The model for GAE% showed the best performance, as indicated by comparing the values for root mean square error of cross validation (RMSECV) or root mean square error of prediction (RMSEP) to the given calibration ranges in Table 3. These values were almost in the range of the lower edge of the calibration line for THCD in mg/kg and RA%. For GAE%, the RMSECV or RMSEP were much smaller than the minimum value of the calibration range. Therefore, it is suggested that the PLSR model for GAE% be applied for monitoring the extraction process of Rosmarini folium. Figure 4b illustrates the extraction monitoring using NIR spectroscopy for GAE% prediction via the PLSR model. In direct comparison, Figure 4a shows monitoring via the reference analysis method.

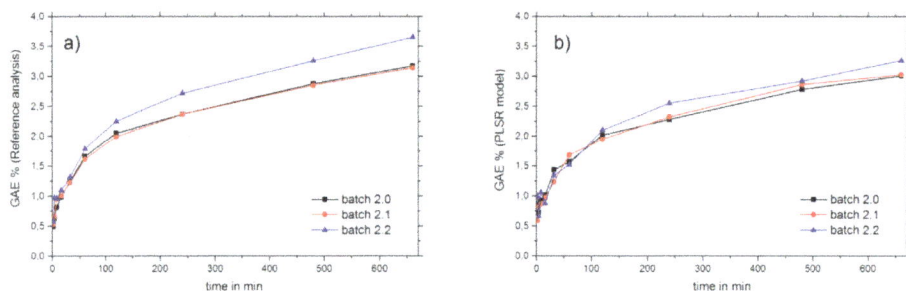

Figure 4. Monitoring of the extraction process of Rosmarini folium via (**a**) Folin–Ciocalteu reference analysis and (**b**) NIR spectroscopy.

Although the reference method has better intermediate precision with 0.033 GAE%, the PLSR model with a RMSEP of 0.18 GAE% is absolutely satisfactory for monitoring the extraction process of Rosmarini folium. Furthermore, in contrast to the common off-line reference analyses, NIR spectroscopic process monitoring has convincing advantages since its operation is non-destructive, contact-free, pollution-free, does not require any additional solvents, saves energy and is highly cost-effective [15].

3. Materials and Methods

3.1. Chemicals

Ethanol (99.9%, LiChrosolv for liquid chromatography), acetonitrile (99.9%, LiChrosolv Reag. Ph. Eur., gradient grade for liquid chromatography) were purchased from Merck Millipore (Darmstadt, Germany). Hydrochloric acid (0.5 N), sodium nitrite (>98%), sodium molybdate dihydrate (>99.5%) and sodium hydroxide tablets (>98%) were bought from Carl Roth GmbH + Co. KG (Karlsruhe, Germany). Rosmarinic acid (>99%), Folin–Ciocalteu's phenol reagent (2N), gallic acid (>97.5%), formic acid (98–100%, Suprapur for trace analysis) and sodium carbonate anhydrous (>99.8%) were obtained from Sigma Aldrich Handels GmbH (Vienna, Austria). H_2O was purified using a Mili-Q® reference water purification system from Merck Millipore. *Rosmarinus officinalis* L. leaves (Rosmarini folium) were collected in the wild at Lake Garda (Italy).

3.2. Extraction and Sampling

Dried *Rosmarinus officinalis* L. leaves were weighed (25 g ± 1 g) and extracted with 500 mL 70% (*v/v*) ethanol. The extraction was done in a 500 mL glass vessel with constant stirring using a color squid (IKA, Staufen im Breisgau, Germany). The extraction time lasted a maximum of 12 h. Three different sampling schedules were planned, and each was conducted three times (total: nine batches). For each sampling 1.5 mL were taken. The specimen-taking schedules are presented in Table 4. The numbering of the batches was done in following way: #(sampling). #(batch). Therefore, the three batches for each sampling schedule were denoted as 1.0, 1.1 and 1.2 or 2.0, 2.1 and 2.2 or 3.0, 3.1 and 3.2 for sampling 1, sampling 2, or sampling 3, respectively.

Table 4. Sampling schedule for extraction experiments.

Sampling 1	Sampling 2	Sampling 3
1.5 min	2 min	2.5 min
3 min	4 min	5 min
6 min	8 min	10 min
12 min	16 min	20 min
24 min	32 min	40 min
45 min	60 min	50 min
90 min	120 min	80 min
180 min	240 min	150 min
360 min	480 min	300 min
720 min	660 min	600 min

3.3. Wet Chemical Assays

3.3.1. European Pharmacopoeia

The THCD of plant extracts were determined according to the procedure reported by the Ph. Eur. [6] with some modifications: 1.0 mL of sample solution was taken to which 2.0 mL of 0.5 M hydrochloric acid, 2 mL of nitrite–molybdate solution (10 g of sodium nitrite and 10 g of sodium molybdate in 100 mL water) and 2 mL of 1 M sodium hydroxide solution were added. The mixture was made up to 10 mL with water. Absorbance was measured with a Jenway Genova Plus Life Science Spectrophotometer (Cole-Parmer, Stone, United Kingdom) at 505 nm and quantification was performed with RA as an external standard calibration. Every extraction sample and calibration sample was prepared in the same way as described above. The repeatability and intermediate precision were determined according to ICH (international council for harmonization of technical requirements for pharmaceuticals for human use) guidelines [22,23]. Therefore, three samples with low, medium, and high THCD content were analyzed for five days, three times per day.

3.3.2. Folin–Ciocalteu

The total phenolic content of plant extracts in GAE% was determined using FC reagent according to the procedure reported by Singleton, Orthofer, and Lamuela-Raventos [24] with some modifications. First, 1.5 mL of H_2O was placed in a macrocuvette (PMMA, Brand, Germany). Next, 100 µL of sample solution, 100 µL of FC's phenol reagent, and 1.3 mL of Na_2CO_3 were added. The mixture was heated to 60 °C for 30 min. After the heating procedure the samples were cooled for 20 min to room temperature. Absorbance was measured with the spectrophotometer at 750 nm. Quantification of the total phenolic content was performed by an external calibration with gallic acid. All extraction samples and calibration samples were prepared in the same way as described above. The repeatability and intermediate precision were determined according to ICH guidelines [22,23]. Therefore, three samples with low, medium, and high GAE% were analyzed for five days, three times per day.

3.4. Ultra-High Performance Liquid Chromatography

UHPLC analysis of RA was performed with an Agilent 1290 Infinity II LC Systems (Agilent Technologies, Santa Clara, CA, USA) equipped with a binary pump (G7120A), an autosampler (G7167B), a column oven (G7116B), and a DAD (diode array detector) (G7117A). Separation of RA was achieved by using an Agilent ZORBAX Eclipse Plus C18, Rapid Resolution HD 2.1 × 50 mm, 1.8 μm (Agilent Technologies, CA, USA) as the analytical column. The mobile phase was a composition of 0.5% formic acid in water (*v/v*, eluent A) and 0.5% formic acid in acetonitrile (*v/v*, eluent B). A gradient program was performed using the following steps (min/% eluent B): 0/15, 6/70, 6.1/100, 8/100, 8.1/15, and 10/15. The temperature of the column oven was set to 40 °C and detection of RA was performed at 330 nm. The flowrate was 1 mL/min and the injection volume was 1 μL. Quantification was performed using the external standard method. The repeatability and intermediate precision were determined according to ICH guidelines [22,23]. Therefore, three samples with low, medium and high RA% were analyzed for five days, three times per day.

3.5. Near-Infrared Spectroscopy

NIR spectra were measured using the NIRFlex N-500 FT-NIR spectrometer (Buchi, Flawil, Switzerland) with the NIRFLex Liquids cell and the cuvette cell add-on. The operating software was NIR Ware 1.4.3010 (Buchi, Flawil, Switzerland). Spectra of the extracts were recorded using precision cells (Hellma GmbH & Co. Kg., Müllheim, Germany) made of Quartz SUPRASIL 300 with a light path of 1 mm at a cell temperature of 35 °C. The spectral resolution was set to 8 cm^{-1} and the measurements were carried out in the wavenumber region from 10,000 to 4000 cm^{-1}. Three replicates for each sample were recorded, with 32 scans each. Spectra were averaged to one representative spectrum per sample.

3.6. Multivariate Data Analysis

MVA was performed using The Unscrambler X Version: 10.5 software (CAMO Software, Oslo, Norway). First, transmittance spectra were transformed to absorbance spectra in order to establish PLSR models. The following spectral pretreatments were applied, alone or in combination, to identify the best PLSR model: baseline correction, SNV transformation [25], multiplicative scatter correction (MSC) [26], and first or second derivative. Savitzky-Golay derivatives [27] (quadratic polynomial) were optimized by variation of the smoothing points. Furthermore, spectral regions which contained no relevant information or even worsened the PLSR models were excluded. The NIPALS algorithm [26] was applied for calculating the PLSR models. For each of the three reference methods (Ph. Eur., FC, and UHPLC) an optimized model was established. The models were validated by full cross validation (CV), also known as leave one out cross validation (LOOCV) [28], and test-set validation (TSV). For TSV, batches 2.0, 2.1, and 2.2 (30 samples) were set as the independent test-set, and batches 1.0, 1.1, 1.2, 3.1, 3.2 and 3.3 (60 samples which included the extreme values) were used as the calibration set. The number of factors was chosen at the suggestion of the software, The Unscrambler X, and examined by expert reviewing. The calculated PLSR models were evaluated with the following parameters: root mean square error of calibration (RMSEC), RMSECV for CV, RMSEP for TSV, coefficient of determination (R^2), regression coefficients, and the number of factors.

4. Conclusions

A fast analysis of the extraction process for the production of phytopharmaceuticals is indispensable in terms of economic viability and quality assurance. Common analytical methods which can be used for Rosmarini folium extraction monitoring are time- and resource-intensive and do not fulfill the requirement for real-time process control. However, the NIR spectroscopic analysis method provides a fast and non-invasive alternative analysis method, which can subsequently be implemented for on- or in-line process monitoring of the phytoextraction of Rosmarini folium.

Author Contributions: Conceptualization, S.D., C.G.K., G.K.B., and C.W.H.; Data curation, S.D., C.G.K., and F.M.; Formal analysis, S.D., C.G.K., and F.M.; Funding acquisition, M.A.P., G.K.B., and C.W.H.; Investigation, S.D., C.G.K., F.M., and Y.O.; Methodology, S.D., C.G.K., and F.M.; Project administration, S.D., C.G.K., G.K.B., and C.W.H.; Resources, M.A.P., G.K.B., and C.W.H.; Supervision, S.D., C.G.K., G.K.B., and C.W.H.; Validation, S.D., C.G.K., and F.M.; Visualization, S.D., C.G.K., and F.M.; Writing—original draft, S.D. and C.G.K; Writing—review & editing, S.D., C.G.K., and F.M.

Funding: This research was funded by the Austrian Federal Ministry of Education, Science and Research (Vienna, Austria)—Project: Novel analytical tools for the quality assessment of Chinese herbs with metabolic, immune related neuromodulatory effects (BMBWF—402.000/0017-WF/V/6/2016).

Acknowledgments: Florian Meischl is the recipient of a DOC Fellowship of the Austrian Academy of Sciences at the Institute of Analytical Chemistry and Radiochemistry, University of Innsbruck.

Conflicts of Interest: The authors declare no conflicts of interest.

References

1. Cragg, G.M.; Newman, D.J. Natural products: A continuing source of novel drug leads. *Biochim. Et Biophys. Acta (BBA) - Gen. Subj.* **2013**, *1830*, 3670–3695. [CrossRef]

2. Veeresham, C. Natural products derived from plants as a source of drugs. *J. Adv. Pharm. Technol. Res.* **2012**, *3*, 200–201. [CrossRef] [PubMed]

3. Uhlenbrock, L.; Sixt, M.; Tegtmeier, M.; Schulz, H.; Hagels, H.; Ditz, R.; Strube, J. Natural Products Extraction of the Future—Sustainable Manufacturing Solutions for Societal Needs. *Processes* **2018**, *6*, 177. [CrossRef]

4. Kirchler, C.G.; Pezzei, C.K.; Beć, K.B.; Henn, R.; Ishigaki, M.; Ozaki, Y.; Huck, C.W. Critical Evaluation of NIR and ATR-IR Spectroscopic Quantifications of Rosmarinic Acid in Rosmarini folium Supported by Quantum Chemical Calculations. *Planta Med.* **2017**, *83*, 1076–1084. [CrossRef] [PubMed]

5. European Medicines Agency Committee on Herbal Medicinal Products (HMPC). Available online: https://www.ema.europa.eu/en/committees/committee-herbal-medicinal-products-hmpc#.Accessed (accessed on 11 January 2019).

6. EDQM Council of Europe. *European Pharmacopoeia*, 9th ed.; Deutscher Apotheker Verlag: Stuttgart, Germany, 2016.

7. Ditz, R.; Gerard, D.; Hagels, H.; Igl, N.; Schäffler, M.; Schulz, H.; Stürtz, M.; Tegtmeier, M.; Treutwein, J.; Chemat, F.; et al. *Phytoextracts: Proposal towards a new comprehensive Research Focus*; DECHEMA Gesellschaft für Chemische Technik und Biotechnologie e.V.: Frankfurt am Main, Germany, 2017.

8. Sixt, M.; Gudi, G.; Schulz, H.; Strube, J. In-line Raman spectroscopy and advanced process control for the extraction of anethole and fenchone from fennel (Foeniculum vulgare L. MILL.). *Comptes Rendus Chim.* **2018**, *21*, 97–103. [CrossRef]

9. Gavan, A.; Colobatiu, L.; Mocan, A.; Toiu, A.; Tomuta, I. Development of a NIR Method for the In-Line Quantification of the Total Polyphenolic Content: A Study Applied on Ajuga genevensis L. Dry Extract Obtained in a Fluid Bed Process. *Molecules* **2018**, *23*, 2152. [CrossRef] [PubMed]

10. Wang, P.; Zhang, H.; Yang, H.; Nie, L.; Zang, H. Rapid determination of major bioactive isoflavonoid compounds during the extraction process of kudzu (Pueraria lobata) by near-infrared transmission spectroscopy. *Spectrochim. Acta Part A Mol. Biomol. Spectrosc.* **2015**, *137*, 1403–1408. [CrossRef]

11. Wu, Z.; Sui, C.; Xu, B.; Ai, L.; Ma, Q.; Shi, X.; Qiao, Y. Multivariate detection limits of on-line NIR model for extraction process of chlorogenic acid from Lonicera japonica. *J. Pharm. Biomed. Anal.* **2013**, *77*, 16–20. [CrossRef] [PubMed]

12. Wu, Y.; Jin, Y.; Ding, H.; Luan, L.; Chen, Y.; Liu, X. In-line monitoring of extraction process of scutellarein from Erigeron breviscapus (vant.) Hand-Mazz based on qualitative and quantitative uses of near-infrared spectroscopy. *Spectrochim. Acta Part A Mol. Biomol. Spectrosc.* **2011**, *79*, 934–939. [CrossRef] [PubMed]

13. Chen, X.; Li, Y.; Chen, Y.; Wang, L.; Sun, C.; Liu, X. Study on fast quality control in extracting process of Paeonia lactiflora using near infrared spectroscopy. *China J. Chin. Mater. Med.* **2009**, *34*, 1355–1358.

14. Hu, T.; Li, T.; Nie, L.; Zang, L.; Zang, H.; Zeng, Y. Rapid monitoring the water extraction process of Radix Paeoniae Alba using near infrared spectroscopy. *J. Innov. Opt. Health Sci.* **2017**, *10*, 1750002. [CrossRef]

15. Siesler, H.W.; Ozaki, Y.; Kawata, S.; Heise, H.M. *Near-infrared spectroscopy: Principles, Instruments, Applications*; 2nd reprint ed.; Wiley-VCH Verlag: Weinheim, Germany, 2002.

16. Aeschbach, R.; Löliger, J.; Scott, B.C.; Mucia, A.; Butler, J.; Halliwell, B.; Aruoma, O.I. Antioxidant actions of thymol, carvacrol, 6-gingerol, zingerone and hydroxytyrosol. *Food Chem. Toxicol.* **1994**, *32*, 31–36. [CrossRef]

17. Schönbichler, S.A.; Falser, G.F.J.; Hussain, S.; Bittner, L.K.; Abel, G.; Popp, M.; Bonn, G.K.; Huck, C.W. Comparison of NIR and ATR-IR spectroscopy for determination of the capacity of Primulae flox cum calycibus. *Anal. Methods* **2014**, *16*, 6343–6351. [CrossRef]

18. Rover, M.R.; Brown, R.C. Quantification of total phenols in bio-oil using the Folin–Ciocalteu method. *J. Anal. Appl. Pyrolysis* **2013**, *104*, 366–371. [CrossRef]

19. Colowick, S.P.; Kaplan, N.O.; Abelson, J.N.; Simon, M.I.; Packer, L. *Methods in Enzymology*; Academic Press: San Diego, CA, USA, 1999.

20. Waterhouse, A.L. Determination of Total Phenolics. *Curr. Protoc. Food Anal. Chem.* **2002**, *6*, I1.1.1–I1.1.8.

21. Kirchler, C.G.; Pezzei, C.K.; Bec, K.B.; Mayr, S.; Ishigaki, M.; Ozaki, Y.; Huck, C.W. Critical evaluation of spectral information of benchtop vs. portable near-infrared spectrometers: Quantum chemistry and two-dimensional correlation spectroscopy for a better understanding of PLS regression models of the rosmarinic acid content in Rosmarini folium. *Analyst* **2017**, *142*, 455–464.

22. Guidance for Industry: Q2B Validation of Analytical Procedures: Methodology. Available online: https://www.fda.gov/downloads/Drugs/GuidanceComplianceRegulatoryInformation/Guidances/UCM073384.pdf (accessed on 24 May 2018).

23. Shabir, G.A. A practical approach to validation of HPLC methods under current good manufacturing practices. *J. Valid. Technol.* **2004**, *10*, 210–218.

24. Singleton, V.L.; Orthofer, R.; Lamuela-Raventós, R.M. [14] Analysis of total phenols and other oxidation substrates and antioxidants by means of folin-ciocalteu reagent. *Methods Enzym.* **1999**, *299*, 152–178.

25. Barnes, R.J.; Dhanoa, M.S.; Lister, S.J. Standard Normal Variate Transformation and De-trending of Near-Infrared Diffuse Reflectance Spectra. *Appl. Spectrosc.* **1989**, *43*, 772–777. [CrossRef]

26. Geladi, P.; MacDougall, D.; Martens, H. Linearization and Scatter-Correction for Near-Infrared Reflectance Spectra of Meat. *Appl. Spectrosc.* **1985**, *39*, 491–500. [CrossRef]

27. Savitzky, A.; Golay, M.J.E. Smoothing and Differentiation of Data by Simplified Least Squares Procedures. *Anal. Chem.* **1964**, *36*, 1627–1639. [CrossRef]

28. Haaland, D.M.; Thomas, E.V. Partial least-squares methods for spectral analyses. 1. Relation to other quantitative calibration methods and the extraction of qualitative information. *Anal. Chem.* **1988**, *60*, 1193–1202. [CrossRef]

Sample Availability: Samples of the compounds are not available from the authors.

molecules

MDPI

Article

Counterfeit and Substandard Test of the Antimalarial Tablet Riamet® by Means of Raman Hyperspectral Multicomponent Analysis

Timea Frosch [1,†], Elisabeth Wyrwich [1,†], Di Yan [1], Christian Domes [1], Robert Domes [1], Juergen Popp [1,2,3] and Torsten Frosch [1,2,3,*]

[1] Leibniz Institute of Photonic Technology, 07745 Jena, Germany
[2] Friedrich Schiller University, Institute of Physical Chemistry, 07745 Jena, Germany
[3] Friedrich Schiller University, Abbe Centre of Photonics, 07745 Jena, Germany
*   Correspondence: torsten.frosch@uni-jena.de or torsten.frosch@gmx.de
†   These authors contributed equally to this study.

Academic Editors: Christian Huck and Krzysztof B. Bec
Received: 21 July 2019; Accepted: 2 September 2019; Published: 5 September 2019

Abstract: The fight against counterfeit pharmaceuticals is a global issue of utmost importance, as failed medication results in millions of deaths every year. Particularly affected are antimalarial tablets. A very important issue is the identification of substandard tablets that do not contain the nominal amounts of the active pharmaceutical ingredient (API), and the differentiation between genuine products and products without any active ingredient or with a false active ingredient. This work presents a novel approach based on fiber-array based Raman hyperspectral imaging to qualify and quantify the antimalarial APIs lumefantrine and artemether directly and non-invasively in a tablet in a time-efficient way. The investigations were carried out with the antimalarial tablet Riamet® and self-made model tablets, which were used as examples of counterfeits and substandard. Partial least-squares regression modeling and density functional theory calculations were carried out for quantification of lumefantrine and artemether and for spectral band assignment. The most prominent differentiating vibrational signatures of the APIs were presented.

Keywords: Raman spectroscopy; hyperspectral imaging; analytical spectroscopy; counterfeit and substandard pharmaceuticals; DFT calculations; chemometrics; PLSR; API; lumefantrine; artemether; antimalarial tablets

1. Introduction

Confidential reports to the World Health Organization (WHO) in the last few years from 20 countries relating to counterfeit drugs revealed that the three highest incidences of fake products were those without active pharmaceutical ingredients (about 30%), followed by incorrect quantities of active ingredients and products with wrong ingredients (about 20% each) [1]. It is estimated that every 10th pharmaceutical product in low- and middle-income countries is substandard or falsified (SF). Antimalarials are the most frequently falsified medicines, representing about 20% of the overall SF products reported in 2017 [2]. Out of the 12 major antimalarial drugs used in the world today, 8 are regularly counterfeited, and more than a third of antimalarial drugs available in sub-Saharan Africa and southeast Asia are reported to be counterfeit or substandard [3].

A report from 2014 [4] showed that among the over 9000 antimalarials sampled, nearly every third failed chemical or packaging quality tests, from which about 40% were classified as counterfeit or substandard and up to 20 wrong active ingredients were found in falsified antimalarials [4].

In 2012 and 2013, one of the most commonly used first-line antimalarials, Riamet®, with active pharmaceutical ingredients (APIs) lumefantrine and artemether (also commercialized as Coartem®), has been involved in one of the greatest counterfeit scandals of our time. The producing company, Novartis, also officially informed customers of the potential counterfeit "dummy tablets"—without active ingredients—saying "counterfeiting medicines is a serious crime against patients who rely on safe and quality-assured medicines to prevent and cure disease, alleviate pain and save lives" and "reports of adverse reactions [...] could materially affect patient confidence in the authentic product, and harm the business of companies such as ours" [5].

Since developing countries are especially concerned of falsified antimalarials, there is an urgent need for low-cost, low-maintenance, easy-to-use, and rapid analytical methods to combat the counterfeit and substandard problem [2]. The Food and Drug Administration (FDA) developed a handheld device named CD-3 [6], which compares scanned images with a stored image of the original product, picking up minute differences in the packaging, pill color, or shape. Although this method is quick and helps to recognize fake packing, it is not chemically selective and does not detect false APIs or false concentrations. Standard techniques, such as high-performance liquid chromatography (HPLC) and mass spectrometry, are highly accurate and reliable, but these methods are strictly lab-based, expensive, time-consuming, and require trained personal. For a quick check, the pH and crystal morphology of the products can be analyzed [3], or a colorimetric test using sulfuric and acetic acid can be applied [7]. This method is based on a color-coded reaction for qualification coupled with color intensity analysis to determine the concentrations of the APIs [7] but chemical selectivity is not ensured.

Raman spectroscopic methods are based on intrinsic molecular vibrations [8–14] and provide an extremely high chemical selectivity [15–22]. The technique is direct and non-invasive [23–25], can be miniaturized, and is also available for on-site applications [26–28]. Hence, Raman spectroscopy has already paved its way in counterfeit detection [29–33]. Handheld Raman devices are commercially available from Rigaku Raman Technologies [29], Ahura Scientific, Inc. [30], and B&W Tek, Inc. [32], and all use 785-nm lasers for excitation. These systems are applicable for solid dosage forms. Still, they are not fully reliable for substandard medicine detection and are used as semi-quantitative methods [32]. Another approach for solid pharmaceutical analysis is spatial offset Raman spectroscopy (SORS), where an excitation wavelength also in the near-infrared (NIR) range is applied (824 nm), focusing on the suppression of signals from colored tablets and capsules' coating [31]. Recently, a line-scanning Raman imaging technique with an excitation wavelength of 785 nm was also reported for API quantification [33].

In this work, we present a proof-of-principle study using fiber-array based Raman spectroscopy [34] with an excitation wavelength in the visible range (532 nm) for multicomponent concentration analysis and counterfeit testing of the antimalarial tablet Riamet®.

Our method allows us to reliably qualify and quantify the active ingredients lumefantrine and artemether in tablets without dissolving them, as it is done for the standard HPLC analysis. By using an 8×8 fiber array, 64 spectra can be collected simultaneously, thus analyzing a larger area of the tablets is possible with only one measurement in a time-efficient way. This advantage is of great importance, as pharmaceutical samples are often heterogeneous. By illuminating the sample surface with a bigger field-of-view (FOV) instead of a mere spot, variations of the spatial concentration distribution can be visualized. The fiber array imaging setup presented here operates with an excitation wavelength of λ = 532 nm, thus the Raman scattering intensity is enhanced in comparison to excitation wavelengths in the NIR according to Equation (1), where N is the number of scatterers, I_0 is the laser intensity, v_0 is the frequency of the excitation laser, and α is the polarizability of the molecule. This offers the chance to quantify substandard drugs with lower amounts of API.

$$I_{STOKES} \propto N {\cdot} I_0 {\cdot} (v_0 - v_r)^4 {\cdot} |\alpha|^2. \tag{1}$$

2. Results and Discussion

This work reports the simultaneous qualification and quantification of two APIs in a pharmaceutical tablet by means of fiber array-based hyperspectral Raman imaging for the first time. First, the Raman spectra of the pure tablet ingredients, lumefantrine, artemether, and hypromellose, were acquired (Figure 1). The vibrational band assignments of the active ingredients were performed based on density functional theory (DFT) calculations and are summarized in Table 1. A comparison of the calculated Raman spectra with the experimentally acquired FT-Raman spectra confirmed a very good agreement (Figure S1). The characteristic Raman bands of lumefantrine were assigned to the vibrational modes from the benzene ring stretching (L3), C=C stretching (L4), and CH deformational vibrations (L1, L2). The dominant Raman bands of artemether were mostly assigned to different CH vibrations (A1—CH_3 wagging, A2—asymmetric stretching of CH_2 combined with slight CH-stretching, A3—asymmetric CH_2 stretching). The latter ones overlap with the Raman modes of the excipient hypromellose. The quantification of artemether in the presence of hypromellose is therefore challenging. To qualify and quantify the APIs lumefantrine and artemether based on the Raman spectra of the tablets in a reliable way, it is necessary to apply multivariate data analysis approaches. A very robust quantitative chemometric method is partial least squares regression (PLSR).

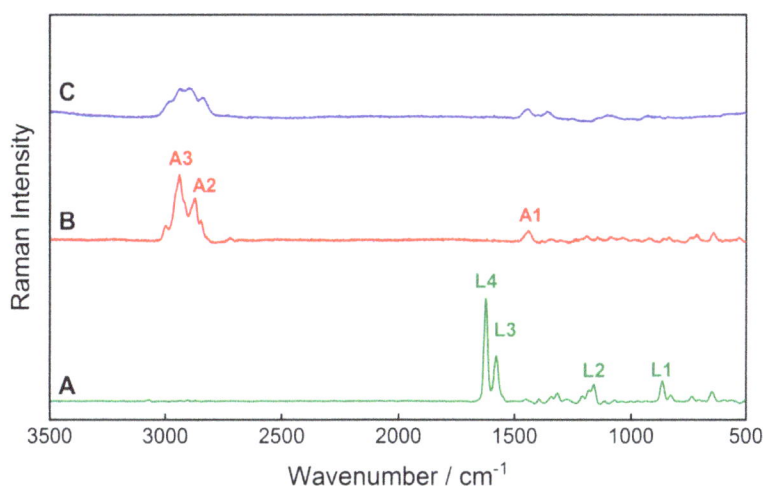

Figure 1. Raman spectra of the active pharmaceutical ingredients lumefantrine (**A**) and artemether (**B**), as well as the excipient hypromellose (**C**), with an excitation wavelength of $\lambda = 532$ nm. The spectra of artemether and hypromellose were scaled with a factor of five for better visibility. The band assignment of the prominent Raman bands A1–A3 and L1–L4 and their spectral positions are listed in Table 1.

Table 1. Band assignment of the prominent Raman peaks of lumefantrine and artemether.

| | Lumefantrine | | | | | Artemether | | | |
| | Peak Position/cm⁻¹ | | | | | Peak Position/cm⁻¹ | | | |
Identification	Measured 532 nm	Measured FT-Raman	Calculated *	Band Assignment	Identification	Measured 532 nm	Measured FT-Raman	Calculated *	Band Assignment
L1	865	876	875	δCH + ωCH	A1	1442	1454	1455	δ_sCH$_3$
L2	1180	1172	1170	δCH	A2	2872	2873	2874	ν_{as}CH$_2$ (+ νCH)
L3	1580	1589	1587	νb + δCH	A3	2940	2937	2937	ν_{as}CH$_2$
L4	1623	1635	1640	νC=C + δCH					

ν_s: symmetric stretching vibration; ν_{as}: asymmetric stretching vibration; ν_b (benzene ring stretching); δ: scissoring; δ_{as} asymmetric bending vibration (twisting); δ_s: symmetric bending vibration (wagging); ω: bending vibration; ν_b (benzene ring stretching); * For the band position assignments based on the DFT calculations different scaling factors were applied: 0.98 for the spectral regions below 2000 cm⁻¹ and 0.95 for the region above 2000 cm⁻¹.

2.1. Development of Partial Least-Squares Regression Model for Ingredient Quantification

Spectral preprocessing is an essential part of modeling to increase the accuracy of the predictions by reducing influences that account to noise-related signal contributions. First, a fiber intensity correction was applied on the hyperspectral image data of the pure substances lumefantrine, artemether, and the model tablets Lu100Ar100, Lu50Ar100, Lu100Ar0, and Lu0Ar100. Afterwards, unit vector normalization was used to correct for Raman intensity variations due to technical effects like different optical path lengths or sample density variations, etc. [35] followed by Savitzky–Golay smoothing. Multiplicative scatter correction (MSC) was section-wise applied for an expanded baseline correction to reduce Raman intensity variations due to different particle sizes [36].

PLSR combines a factorial analysis and a regression method. First, a PLSR calibration model was built, considering simultaneously the responses from the analytes, such that the concentrations exactly summed up to 100% (PLS2 approach). Afterwards, the PLSR calibration model was applied to the hyperspectral images of the model tablets. For validation of the model, external validation is preferred [37]. In case of hyperspectral images, it was possible to use one half of the image for calibration and the other half for validation [38]. However, this approach was not beneficial in the case of the tablets, as they are heterogeneous, and thus the spatial variations of concentrations did not match the input reference values for the model development. Influences caused by outliers and heterogeneities can be reduced by summarizing a single hyperspectral image as a median spectrum. To build up a representative data set for calibration and validation, the Kennard–Stone algorithm was applied in combination with a prior cross validation to remove outliers that would otherwise be taken as extreme samples [39,40]. A good correlation between the predicted and reference data for both the calibration ($R^2 = 0.9829$ for lumefantrine and $R^2 = 0.9989$ for artemether) and for the validation PLSR-model ($R^2 = 0.9827$ for lumefantrine and $R^2 = 0.9982$ for artemether) was achieved. The predictive error for the validation (RMSE) were 5.00 wt% for lumefantrine and 1.59 wt% for artemether.

2.2. Active Ingredient Concentration Prediction and Interpretation of the Spectral Information of the Model

The prediction model was applied to 30 hyperspectral images of each model tablet and for the Riamet® tablet, respectively. The predicted concentrations and the corresponding error ranges are listed in Table 2. The occurrence of outliers was reduced by using median-averaged images.

Table 2. Predictions of the lumefantrine and artemether concentrations in the model tablets and the genuine tablet as follows: Lu100Ar100 (100% nominal lumefantrine and 100% artemether content, corresponding to 60% lumefantrine and 10% artemether in the tablet), Lu50Ar100 (50% nominal lumefantrine and 100% artemether content, corresponding to 30% lumefantrine and 10% artemether in the tablet), Lu100Ar0 (100% nominal lumefantrine and 0% artemether content), and Lu0Ar100 (0% nominal lumefantrine and 100% artemether content) based on the partial least squares regression (PLSR) model.

Tablet	Lumefantrine Concentration/wt%			Artemether Concentration/wt%		
	Expected	Predicted	y_{dev}	Expected	Predicted	y_{dev}
Lu100Ar100	60.0	57.8	4.5	10.0	9.5	1.4
Lu50Ar100	30.0	44.1	6.1	10.0	9.1	1.9
Lu100Ar0	60.0	59.8	7.7	0.0	1.0	2.4
Lu0Ar100	0.0	1.2	6.3	10.0	11.4	2.0
Riamet®	50.0	44.1	14.6	8.3	5.6	4.7

y_{dev} describes the deviation of the concentration prediction.

The predicted mean concentrations for lumefantrine were found around the expected 60 wt%, (4.5–7.7 wt% deviation) (Table 2). For the substandard tablet Lu50Ar100 (containing 50% of the nominal lumefantrine and 100% of the artemether content), the predicted mean concentration was above the expected one, whereas for Riamet® it was 5 wt% below the expected value (Table 2). For artemether, the predicted concentrations fitted very well to the expected ones, deviating only 0.5 to 1.5 wt% in the content of the model tablets and 2.7 wt% in the case of Riamet® (1.4–4.7 wt% deviation) (Table 2). The United States Pharmacopoea requires at least 30 samples for the content uniformity test and allows a maximum range of 25% for deviation from the reference value of a single dosage unit tested [41]. Thus, our observed deviations are well covered in this range. The observed deviations from the expected values are partly caused by the inhomogeneous scattering effects of the surface, combined with limited signal-to-noise ratios, and partly with the uncertainty of the regression model (RMSE of prediction are 5.00 wt% for lumefantrine and 1.59 wt% for artemether). It should also be noted that for the model, the target wt% values in the training group were defined based on the nominal added amounts of the ingredients. This can also lead to some minor errors in the prediction. Lumefantrine is a strong Raman scatterer, and the absolute Raman signal variations of the different concentrations of lumefantrine are much higher than those of artemether. Hence, their simultaneous quantification requires a compromise in the accuracy of the predictions.

For better prediction accuracy for the genuine tablet Riamet®, it would be beneficial to include more excipients in the calibration and validation model. Only hypromellose was used as an excipient, but microcrystalline cellulose, croscarmellose sodium, magnesium stearate, polysorbat 80, and highly dispersed SiO_2 were not considered in the calibration model. As the producing company does not share such detailed information on the exact composition of the tablets, this aspect remains challenging. However, the comparison between the Raman spectrum of the model tablet Lu100Ar100, containing the full content of the APIs lumefantrine and artemether, with the spectrum of the genuine Riamet® tablet show a high similarity (Figure 2) and justifies this approximation.

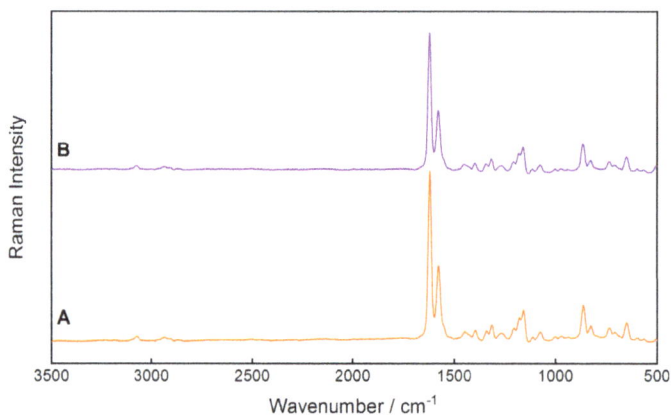

Figure 2. Comparison of the Raman spectra of (**A**) the genuine Riamet® tablet and (**B**) the model tablet Lu100Ar100 with the nominal 100% content of the active ingredients.

The most-representative Raman bands of the active ingredients correlate well to the large regression coefficients (Figure 3A), which account for a high influence of the respective Raman signal in the prediction. The prominent Raman bands of both lumefantrine and artemether correlate with high positive coefficients of their own prediction factors (especially L3 and L4 and A2 and A3). This underlines that the model differentiated correctly between the active ingredients based on the respective spectral information.

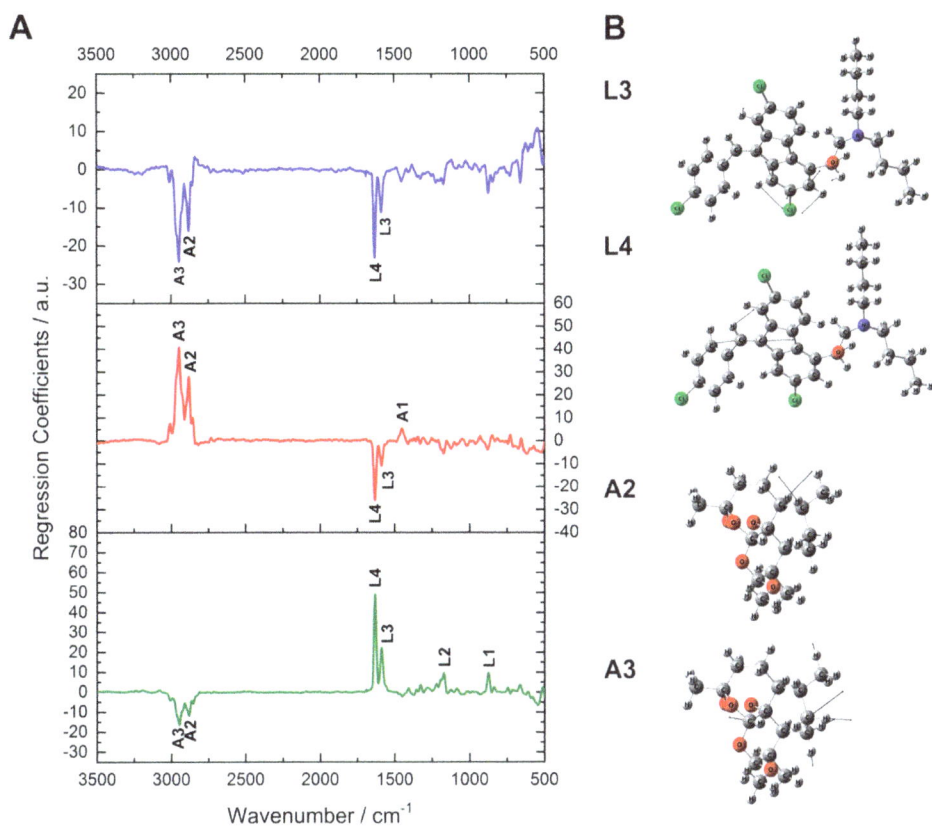

Figure 3. (**A**): Regression coefficients for the prediction of lumefantrine (green, lower part), artemether (red, middle part), and hypromellose (blue, upper part). The coefficients from the first two factors for each analyte correlate perfectly to the characteristic Raman bands of lumefantrine and artemether. Strong contribution for the differentiation is attributed to the peaks L4, L3, A3, and A2. (**B**): Vibrational assignment of the peaks that contribute most to the PLSR model: L3: benzene ring stretching + CH scissoring, L4: C=C stretching vibration + CH scissoring, A2: asymmetric stretching vibration + slight contribution from CH stretching, A4: asymmetric CH_2 stretching vibration.

Hypromellose and artemether have their strongest Raman bands in the same spectral regions between 2800 and 3000 cm^{-1} and some spectral overlap occurs. Nevertheless, the developed model enabled the quantification of artemether in the presence of hypromellose. This is demonstrated by the high negative coefficients for the prediction of hypromellose at the positions of A2 and A3 (Figure 3A). For better visualization of the molecular information underneath the Raman bands, the vibrational assignments of Raman bands L4, L3, A3, and A2 are depicted (Figure 3B). L3 is a combination of a benzene ring stretching and CH scissoring of lumefantrine. L4 is a C=C stretching vibration combined with a less-intensive CH scissoring of lumefantrine. A2 is an asymmetric CH_2 stretching vibration with a slight contribution from CH stretching of artemether, whereas A3 is an asymmetric CH_2 stretching vibration of artemether.

The predicted concentration values and the corresponding uncertainty ranges of Riamet® were presented for 64 random regions from 30 hyperspectral images (Figure 4). Differences of the API concentrations in different parts of the tablets were revealed. For lumefantrine, the local concentrations varied between 21.8 and 54.5 wt% and for artemether between 4.1 and 15.2 wt%, most probably due to an inhomogeneous API distribution. The active ingredients in the model tablets were more homogenously distributed (Figure 5). It is easily obvious that the model tablet with 50% of the nominal lumefantrine and 100% of the artemether content (Figure 5A) has a lower lumefantrine content than the one with a full nominal content (Lu100Ar100) (Figure 5B), as it was expected. This demonstrates the suitability of the presented method to gain information about substandard tablets directly and non-invasively (without dissolution). The concentrations varied on the spot level between 16.1 and 49.6 wt% in the substandard model Lu50Ar100, which corroborates the necessity of acquiring data over numerous areas of pharmaceutical tablets. This can be done in a very time-efficient manner with the presented fiber array-based Raman imaging technique, which allows the simultaneous measurement of 64 sample spots with one measurement. Furthermore, local concentration variations can also be easily visualized (Figure 5), which will be an extremely helpful ability in non-invasive quality control of tablets.

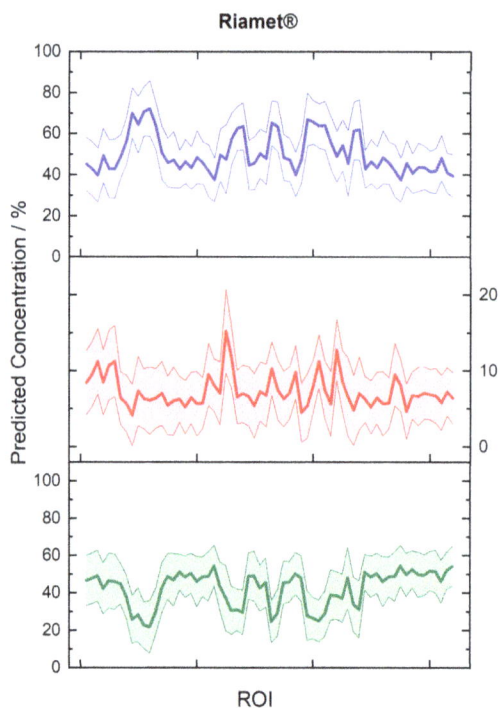

Figure 4. Predicted concentrations for 64 random spots from 30 regions (30 hyperspectral images) of the three constituents in Riamet®: lumefantrine (green line, lower graph), artemether (red line, middle part), and hypromellose (blue line, upper graph). The respective prediction error ranges are shown. Local differences in the distribution of the concentrations of active ingredients in the tablet are revealed. Each region of interest (ROI) indicates the imaged area from a single fiber in the fiber array.

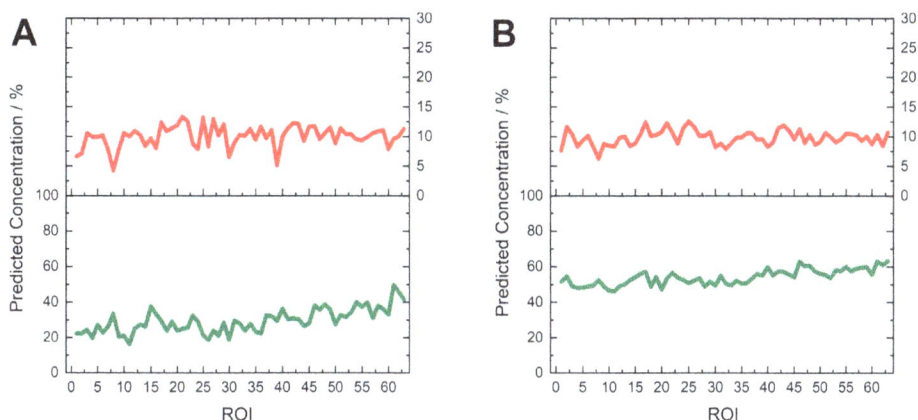

Figure 5. Predicted concentrations of lumefantrine (green, lower line) and artemether (red, upper line) in different spots on the model tablets. Each region of interest (ROI) indicates the imaged area from a single fiber in the fiber array. (**A**) Lu50Ar100: 50% of the nominal lumefantrine and 100% of the nominal artemether content, corresponding to 30 wt% lumefantrine and 10 wt% artemether in the tablet. (**B**) Lu100Ar100: 100% the nominal content of lumefantrine and artemether, corresponding to 60 wt% lumefantrine and 10 wt% artemether in the tablet).

2.3. Potential of Fiber Array-Based Technique for Counterfeit and Substandard Tablet Testing

The fiber array-based Raman hyperspectral imaging technique provides the following advantages, which can be exploited for counterfeit and substandard testing of pharmaceutical tablets: The presented method is non-invasive and non-destructive, without using any aggressive or toxic solvents. Thus, this method is environment-friendly and cost-effective.

Combining Raman measurements with chemometric modeling, both qualitative and quantitative information of several analytes are captured in one single measurement procedure, granting high potential for the efficient investigations of pharmaceutical samples to detect low-quality issues. Using a high magnification objective with a high NA additionally allows visualization of the API distribution in a highly resolved way (e.g., lumefantrine in Figure S2A). Another strong advantage is the time-efficient measurement procedure, as 64 Raman spectra can be acquired simultaneously (Figure S2B). The setup presented in this proof-of-principle study is flexible and can adapt to different experimental settings, as the amount of collected spectra in one shot can be further extended using different fiber array configurations and the dimensions of the FOV at the sample can easily be changed.

3. Materials and Methods

3.1. Chemicals and Tablets

Lumefantrine (Lu), artemether (Ar), and hypromellose were purchased from Sigma Aldrich (Taufkirchen, Germany). Model fake tablets were manufactured, containing the APIs lumefantrine and artemether in different concentration ratios by direct compression. The total weight for each model tablet was 200 mg and the pharmaceutical excipient hypromellose was used to fill up the formulation. The composition of the analyzed tablets is visualized in Figure 6. Riamet® tablets (Novartis) were purchased from a local pharmacy (Jena, Germany) and investigated. The coating of this tablet was removed for better conformity with the model tablets.

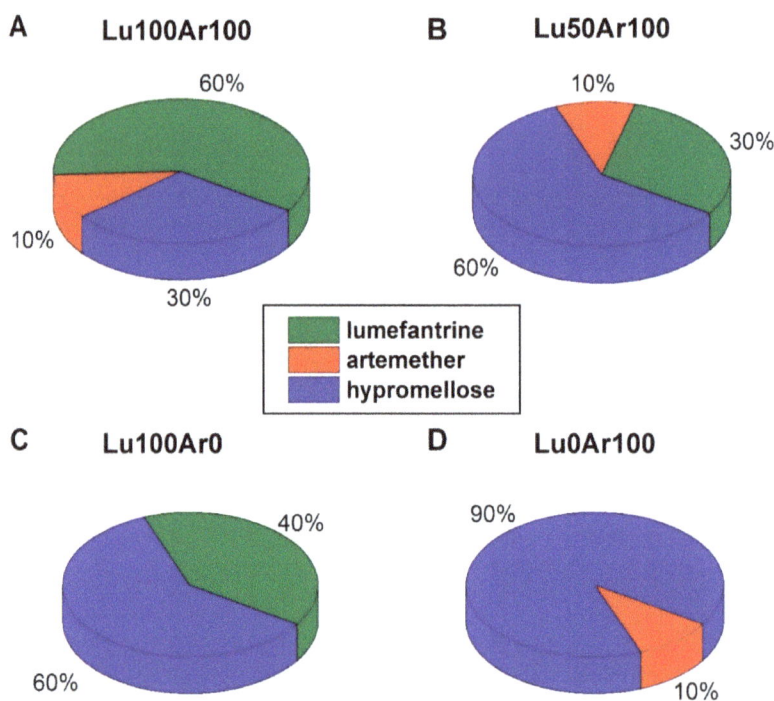

Figure 6. Composition of the anti-malarial model tablets. 100% refers to the nominal content in the original Riamet® tablet, which are 120 mg lumefantrine and 20 mg artemether, corresponding to 60 wt% lumefantrine, 10 wt% artemether, and 30 wt% filling excipient hypromellose in the tablet. The total mass of each tablet is 200 mg. (**A**) Lu100Ar100: Content of nominal 100% lumefantrine and nominal 100% artemether (60 wt% lumefantrine, 10 wt% artemether and 30 wt% filling excipient hypromellose in the tablet). (**B**) Lu50Ar100: Content of nominal 50% lumefantrine and nominal 100% artemether (30 wt% lumefantrine, 10 wt% artemether, and 60 wt% filling excipient hypromellose in the tablet). (**C**) Lu100Ar0: Content of nominal 100% lumefantrine and nominal 0% artemether content (40 wt% lumefantrine and 60 wt% filling excipient hypromellose in the tablet). (**D**) Lu0Ar100: Content of nominal 0% lumefantrine and nominal 100% artemether content (10 wt% artemether and 90 wt% filling excipient hypromellose in the tablet).

3.2. FT-Raman Spectroscopy

The FT-Raman spectra of the active ingredients lumefantrine and artmether were recorded using a Bruker FT-Raman spectrometer (Ram II) (Bruker Optik GmbH, Germany) with an Nd:YAG laser operating at 1064 nm. The spectral resolution was set to 4 cm^{-1}.

3.3. Density Functional Theory Calculation

To better assign and interpret the Raman bands of the active ingredients, the vibrational modes and Raman scattering activities were calculated with the help of density functional theory (DFT) using Gaussian 16 [42]. The hybrid exchange correlation functional with Becke's three-parameter exchange functional (B3) [43] slightly modified by Stephens et al. [44] coupled with the correlation part of the functional from Lee, Yang, and Parr (B3LYP) [45] and Dunning's triple (cc-pVTZ) correlation consistent basis sets of contracted Gaussian functions with polarized and diffuse functions [46] at standard conditions were applied. Separate scaling factors for the lower (<2000 cm^{-1}) and for the higher (>2000 cm^{-1}) wavenumber regions and an intensity correction were estimated [13,47].

3.4. Fiber-Array Based Hyperspectral Imaging

The spectroscopic measurements of the samples (the powder form APIs lumefantrine and artemether, the excipient hypromellose, the model tablets, and Riamet®) were carried out with a hyperspectral imaging setup. The sample area was illuminated with an FOV of 10×10 µm² (Figure 7). The laser power in the sample plane was 600 mW and an exposure time of 10 s was used with three accumulations. A specially designed fiber-array bundle was applied for signal collection (Figure 7). The sample surface was imaged onto the entrance face of the fiber array and the shape of the bundle was transformed from an 8×8 square to a linear array of 64 fibers. The line of fibers was then placed in the plane of the spectrometer slit (IsoPlane, Princeton Instruments) and enabled the simultaneous acquisition of 64 spectra (Figure 7). After the acquisition of the spectra, pre-processing tools, such as baseline correction (rolling-ball algorithm) and spike correction, were applied using LabVIEW. To provide a representative spectrum of the tablets Riamet® and the model tablet Lu100Ar100 (Figure 2), 10 hyperspectral images per tablet were acquired and for each image the median spectrum was calculated. From the 10 median spectra, an average spectrum was calculated, and a second baseline correction was carried out with the SNIP algorithm (2nd order). Each spectrum was assigned to a specific spot in the sample area and hyperspectral images were built based on the desired chemical information.

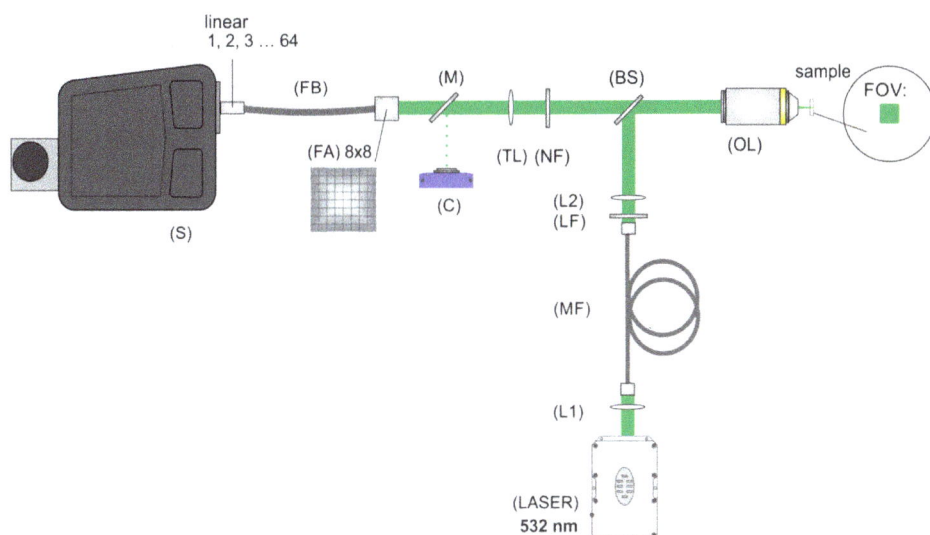

Figure 7. The experimental setup for fiber array-based Raman hyperspectral imaging is divided into an illumination and an imaging part, separated by a beam splitter (BS). The illumination part consists of a laser for excitation (LASER), two lenses (L1 and L2), a step index multimode fiber (MF), a cleanup filter (LF), and an objective lens (OL). Light is scattered back from the sample, collected by the same objective lens (OL), and imaged with the help of a tube lens (TL) onto the entrance face of a fiber array (FA). A suitable sample region can be chosen by directing the light onto a camera (C) with the help of a flip mirror (M). A notch filter (NF) removes the laser excitation wavelength and elastically scattered light. The scattered light is collected by the 8×8 array and is transformed with the help of a specially designed fiber bundle (FB) into a linear fiber array at the distal end and positioned in the slit plane of the spectrometer (S).

3.5. Partial Least-Squares Regression Model for the Ingredients' Quantification

For the spectral analysis and modeling, the chemometrics software 'The Unscrambler® X 10.3' (Camo Software AS., Oslo, Norway) was used.

4. Conclusions

In this work a proof-of-principle study using a novel method to qualify and quantify substances in pharmaceutical tablets that are potentially counterfeit or substandard was presented. Based on a fiber array-based Raman hyperspectral imaging technique combined with PLSR modeling, the concentrations of the APIs lumefantrine and artemether were simultaneously determined in model tablets and in the tablet Riamet®. The analysis was carried out in a non-destructive way, without dissolution, which is an advantage in comparison to conventional methods. In addition, the concentration distribution of active ingredients could also be assessed. Being able to identify and quantify counterfeits (Lu100Ar0, Lu0Ar100) and even substandard (Lu50Ar100) antimalarial tablets fast and directly on the tablet gives us a new tool for the fight against falsification of pharmaceuticals. The analyzed tablet Riamet® is of high importance, since antimalarial tablets are the most frequent targets of counterfeiting in the world, as highlighted by the WHO and the FDA.

In future work, we intend to test "real fake" samples, thus complementing our training model. It would be highly beneficial to apply the presented easily applicable and flexible technique as a first test to detect peculiarities or abnormalities before analyzing the tablets with destructive and more expensive analytical techniques.

In summary, fiber array-based Raman hyperspectral imaging in combination with PLSR analysis enables a fast and chemically selective, noninvasive, and spatially resolved determination of multicomponent API concentrations in pharmaceutical tablets, showing high potential as a future "anti-fake and substandard tool".

Supplementary Materials: The following are available online at http://www.mdpi.com/1420-3049/24/18/3229/s1, Figure S1: Comparison of the calculated Raman spectra (DFT) with the experimentally acquired FT-Raman spectra of the active ingredients lumefantrine and artemether. Figure S2: Exemplary visualization of the spatial distribution of the lumefantrine concentration along one hyperspectral image in the model tablet Lu100Ar100.

Author Contributions: Conceptualization, T.F.; Investigation, T.F., E.W., D.Y., C.D., R.D. and T.F.; Writing—original draft, T.F., E.W. and T.F.; Writing—review & editing, C.D., R.D. and J.P.

Funding: Funding from the federal state of Thuringia (FKZ: 2015 FE 9012 and 2017 FGI 0026), the European Union (EFRE) is gratefully acknowledged.

Conflicts of Interest: The authors declare no conflict of interest.

References

1. WHO. *Substandard and Counterfeit Medicines*; Fact Sheet 275; WHO Media Centre: Geneva, Switzerland, 2003.
2. WHO. *WHO Global Surveillance and Monitoring System for Substandard and Falsified Medical Products*; WHO: Geneva, Switzerland, 2017; ISBN 978-92-4-151342-5.
3. Newton, P.N.; Green, M.D.; Fernández, F.M.; Day, N.P.; White, N.J. Counterfeit anti-infective drugs. *Lancet Infect. Dis.* **2006**, *6*, 602–613. [CrossRef]
4. Tabernero, P.; Fernández, F.M.; Green, M.; Guerin, P.J.; Newton, P.N. Mind the gaps-the epidemiology of poor-quality anti-malarials in the malarious world-analysis of the WorldWide Antimalarial Resistance Network database. *Malar. J.* **2014**, *13*, 139. [CrossRef] [PubMed]
5. Faucon Benoit, M.C.; Whalen, J. Africa's Malaria Battle: Fake Drug Pipeline Undercuts Progress. *Wall Str. J.* **2013**. Available online: https://www.wsj.com/articles/thieves-hijacking-malaria-drugs-in-africa-1384216610 (accessed on 3 September 2019).
6. FDA. *FDA Facts: FDA's Counterfeit Detection Device CD-3*; FDA: Rockville, MD, USA, 2013.
7. Green, M.D.; Hostetler, D.M.; Nettey, H.; Swamidoss, I.; Ranieri, N.; Newton, P.N. Integration of Novel Low-Cost Colorimetric, Laser Photometric, and Visual Fluorescent Techniques for Rapid Identification of Falsified Medicines in Resource-Poor Areas: Application to Artemether–Lumefantrine. *Am. J. Trop. Med. Hyg.* **2015**, *92* (Suppl. S6), 8–16. [CrossRef] [PubMed]
8. Knebl, A.; Domes, R.; Yan, D.; Popp, J.; Trumbore, S.; Frosch, T. Fiber-Enhanced Raman Gas Spectroscopy for 18O–13C-Labeling Experiments. *Anal. Chem.* **2019**, *91*, 7562–7569. [CrossRef] [PubMed]

9. Yan, D.; Frosch, T.; Kobelke, J.; Bierlich, J.; Popp, J.; Pletz, M.W.; Frosch, T. Fiber-Enhanced Raman Sensing of Cefuroxime in Human Urine. *Anal. Chem.* **2018**, *90*, 13243–13248. [CrossRef] [PubMed]

10. Yan, D.; Popp, J.; Pletz, M.W.; Frosch, T. Highly Sensitive Broadband Raman Sensing of Antibiotics in Step-Index Hollow-Core Photonic Crystal Fibers. *ACS Photonics* **2017**, *4*, 138–145. [CrossRef]

11. Keiner, R.; Herrmann, M.; Kusel, K.; Popp, J.; Frosch, T. Rapid monitoring of intermediate states and mass balance of nitrogen during denitrification by means of cavity enhanced Raman multi-gas sensing. *Anal. Chim. Acta* **2015**, *864*, 39–47. [CrossRef] [PubMed]

12. Jochum, T.; Michalzik, B.; Bachmann, A.; Popp, J.; Frosch, T. Microbial respiration and natural attenuation of benzene contaminated soils investigated by cavity enhanced Raman multi-gas spectroscopy. *Analyst* **2015**, *140*, 3143–3149. [CrossRef]

13. Frosch, T.; Popp, J. Structural analysis of the antimalarial drug halofantrine by means of Raman spectroscopy and density functional theory calculations. *J. Biomed. Opt.* **2010**, *15*, 041516. [CrossRef]

14. Frosch, T.; Chan, K.L.; Wong, H.C.; Cabral, J.T.; Kazarian, S.G. Nondestructive three-dimensional analysis of layered polymer structures with chemical imaging. *Langmuir* **2010**, *26*, 19027–19032. [CrossRef] [PubMed]

15. Hanf, S.; Fischer, S.; Hartmann, H.; Keiner, R.; Trumbore, S.; Popp, J.; Frosch, T. Online investigation of respiratory quotients in Pinus sylvestris and Picea abies during drought and shading by means of cavity-enhanced Raman multi-gas spectrometry. *Analyst* **2015**, *140*, 4473–4481. [CrossRef] [PubMed]

16. Jochum, T.; Fastnacht, A.; Trumbore, S.E.; Popp, J.; Frosch, T. Direct Raman Spectroscopic Measurements of Biological Nitrogen Fixation under Natural Conditions: An Analytical Approach for Studying Nitrogenase Activity. *Anal. Chem.* **2017**, *89*, 1117–1122. [CrossRef] [PubMed]

17. Jochum, T.; Rahal, L.; Suckert, R.J.; Popp, J.; Frosch, T. All-in-one: A versatile gas sensor based on fiber enhanced Raman spectroscopy for monitoring postharvest fruit conservation and ripening. *Analyst* **2016**, *141*, 2023–2029. [CrossRef] [PubMed]

18. Bogozi, T.; Popp, J.; Frosch, T. Fiber-enhanced Raman multi-gas spectroscopy: What is the potential of its application to breath analysis? *Future Sci. Bioanal.* **2015**, *7*, 281–284. [CrossRef]

19. Yan, D.; Popp, J.; Frosch, T. Analysis of Fiber-Enhanced Raman Gas Sensing Based on Raman Chemical Imaging. *Anal. Chem.* **2017**, *89*, 12269–12275. [CrossRef] [PubMed]

20. Domes, C.; Domes, R.; Popp, J.; Pletz, M.W.; Frosch, T. Ultrasensitive Detection of Antiseptic Antibiotics in Aqueous Media and Human Urine Using Deep UV Resonance Raman Spectroscopy. *Anal. Chem.* **2017**, *89*, 9997–10003. [CrossRef]

21. Yan, D.; Domes, C.; Domes, R.; Frosch, T.; Popp, J.; Pletz, M.W.; Frosch, T. Fiber enhanced Raman spectroscopic analysis as a novel method for diagnosis and monitoring of diseases related to hyperbilirubinemia and hyperbiliverdinemia. *Analyst* **2016**, *141*, 6104–6115. [CrossRef]

22. Yan, D.; Popp, J.; Pletz, M.W.; Frosch, T. Fiber enhanced Raman sensing of levofloxacin by PCF bandgap-shifting into the visible range. *Anal. Methods-UK* **2018**, *10*, 586–592. [CrossRef]

23. Knebl, A.; Yan, D.; Popp, J.; Frosch, T. Fiber enhanced Raman gas spectroscopy. *Trac-Trend Anal. Chem.* **2018**, *103*, 230–238. [CrossRef]

24. Jochum, T.; von Fischer, J.C.; Trumbore, S.; Popp, J.; Frosch, T. Multigas Leakage Correction in Static Environmental Chambers Using Sulfur Hexafluoride and Raman Spectroscopy. *Anal. Chem.* **2015**, *87*, 11137–11142. [CrossRef] [PubMed]

25. Domes, R.; Domes, C.; Albert, C.R.; Bringmann, G.; Popp, J.; Frosch, T. Vibrational spectroscopic characterization of arylisoquinolines by means of Raman spectroscopy and density functional theory calculations. *Phys. Chem. Chem. Phys.* **2017**, *19*, 29918–29926. [CrossRef] [PubMed]

26. Sieburg, A.; Jochum, T.; Trumbore, S.E.; Popp, J.; Frosch, T. Onsite cavity enhanced Raman spectrometry for the investigation of gas exchange processes in the Earth's critical zone. *Analyst* **2017**, *142*, 3360–3369. [CrossRef] [PubMed]

27. Sieburg, A.; Schneider, S.; Yan, D.; Popp, J.; Frosch, T. Monitoring of gas composition in a laboratory biogas plant using cavity enhanced Raman spectroscopy. *Analyst* **2018**, *143*, 1358–1366. [CrossRef] [PubMed]

28. Keiner, R.; Gruselle, M.C.; Michalzik, B.; Popp, J.; Frosch, T. Raman spectroscopic investigation of 13CO2 labeling and leaf dark respiration of Fagus sylvatica L. (European beech). *Anal. Bioanal. Chem.* **2015**, *407*, 1813–1817. [CrossRef] [PubMed]

29. Mustapha Hajjou, P.L. Potential use of handheld Raman devices as tools for screening medicines for quality. *BioPharma Asia* **2014**, *2014*, 15–21.

30. Ricci, C.; Nyadong, L.; Yang, F.; Fernandez, F.M.; Brown, C.D.; Newton, P.N.; Kazarian, S.G. Assessment of hand-held Raman instrumentation for in situ screening for potentially counterfeit artesunate antimalarial tablets by FT-Raman spectroscopy and direct ionization mass spectrometry. *Anal. Chim. Acta* **2008**, *623*, 178–186. [CrossRef] [PubMed]

31. Eliasson, C.; Matousek, P. Noninvasive authentication of pharmaceutical products through packaging using spatially offset Raman spectroscopy. *Anal. Chem.* **2007**, *79*, 1696–1701. [CrossRef] [PubMed]

32. Visser, B.J.; de Vries, S.G.; Bache, E.B.; Meerveld-Gerrits, J.; Kroon, D.; Boersma, J.; Agnandji, S.T.; van Vugt, M.; Grobusch, M.P. The diagnostic accuracy of the hand-held Raman spectrometer for the identification of anti-malarial drugs. *Malar. J.* **2016**, *15*, 160. [CrossRef] [PubMed]

33. Kandpal, L.M.; Cho, B.-K.; Tewari, J.; Gopinathan, N. Raman spectral imaging technique for API detection in pharmaceutical microtablets. *Sens. Actuators B Chem.* **2018**, *260*, 213–222. [CrossRef]

34. Brueckner, M.; Becker, K.; Popp, J.; Frosch, T. Fiber array based hyperspectral Raman imaging for chemical selective analysis of malaria-infected red blood cells. *Anal. Chim. Acta* **2015**, *894*, 76–84. [CrossRef] [PubMed]

35. Brauchle, E.; Thude, S. Cell death stages in single apoptotic and necrotic cells monitored by Raman microspectroscopy. *Sci. Rep.* **2014**, *4*, 4698. [CrossRef] [PubMed]

36. Zhang, J.; Liu, S.-Z.; Yang, J.; Song, M.; Song, J.; Du, H.-L.; Chen, Z.-P. Quantitative spectroscopic analysis of heterogeneous systems: Chemometric methods for the correction of multiplicative light scattering effects. *Rev. Anal. Chem.* **2013**, *32*, 113–125. [CrossRef]

37. Kessler, W. *Multivariate Datenanalyse für die Pharma-, Bio- und Prozessanalytik*; Wiley VCH: Weinheim, Germany, 2007.

38. Salzer, R.; Siesler, H.W. *Infrared and Raman Spectroscopic Imaging*; Wiley-VCH: Weinheim, Germany, 2014; Volume 2.

39. Galvao, R.K.H.; Araujo, M.C.U.; José, G.E.; Pontes, M.J.C.; Silva, E.C.; Saldanha, T.C.B. A method for calibration and validation subset partitioning. *Talanta* **2005**, *67*, 736–740. [CrossRef] [PubMed]

40. De Groot, P.J.; Postma, G.J.; Melssen, W.J.; Buydens, L.M.C. Selecting a representative training set for the classification of demolition waste using remote NIR sensing. *Anal. Chim. Acta* **1999**, *392*, 67–75. [CrossRef]

41. USP. *<905> Uniformity of Dosage Units*; United States Pharmacopeia Convention: Rockville, MD, USA, 2015.

42. Frisch, M.; Trucks, G.; Schlegel, H.; Scuseria, G.; Robb, M.; Cheeseman, J.; Scalmani, G.; Barone, V.; Mennucci, B.; Petersson, G. *GAUSSIAN09*; Revision E. 01; Gaussian Inc.: Wallingford, CT, USA, 2009.

43. Becke, A. Density-functional thermochemistry. II. The effect of the Perdew–Wang generalized-gradient correlation correction. *J. Chem. Phys.* **1992**, *97*, 9173–9177. [CrossRef]

44. Stephens, P.; Devlin, F.; Chabalowski, C.; Frisch, M.J. Ab initio calculation of vibrational absorption and circular dichroism spectra using density functional force fields. *J. Phys. Chem.* **1994**, *98*, 11623–11627. [CrossRef]

45. Lee, C.; Yang, W.; Parr, R. Development of the Colle-Salvetti correlation-energy formula into a functional of the electron density. *Phys. Rev. B* **1988**, *37*, 785. [CrossRef]

46. Dunning, T.H. A road map for the calculation of molecular binding energies. *J. Phys. Chem. A* **2000**, *104*, 9062–9080. [CrossRef]

47. Polavarapu, P. Ab initio vibrational Raman and Raman optical activity spectra. *J. Phys. Chem.* **1990**, *94*, 8106–8112. [CrossRef]

Sample Availability: Samples of the compounds are not available.

molecules

MDPI

Article

Osteopathic Manipulation Treatment Improves Cerebro–splanchnic Oximetry in Late Preterm Infants

Benedetta Marinelli [1], Francesca Pluchinotta [2], Vincenzo Cozzolino [1], Gina Barlafante [1], Maria Chiara Strozzi [3], Eleonora Marinelli [1], Simone Franchini [4] and Diego Gazzolo [3,*]

[1]   Department, Italian Academy of Traditional Osteopathy (AIOT), 65100 Pescara, Italy
[2]   Laboratory Research, Department of Paediatric Cardiovascular Surgery, San Donato Milanese University Hospital, 20097 San Donato, Italy
[3]   Department of Maternal, Fetal and Neonatal Medicine, C. Arrigo Children's Hospital, 15121 Alessandria, Italy
[4]   Neonatal Intensive Care Unit G. D'Annunzio University of Chieti, 66100 Chieti, Italy
*   Correspondence: dgazzolo@hotmail.com

Academic Editors: Christian Huck and Krzysztof B. Bec
Received: 25 June 2019; Accepted: 27 August 2019; Published: 4 September 2019

Abstract: Background: To evaluate the effectiveness/side-effects of osteopathic manipulation treatment (OMT) performed on the 7th post-natal day, on cerebro–splanchnic oximetry, tissue activation and hemodynamic redistribution in late preterm (LP) infants by using near infrared spectroscopy (NIRS). **Methods**: Observational pretest-test study consisting in a cohort of 18 LPs who received OMT on the 7th post-natal day. NIRS monitoring was performed at three different time-points: 30 min before (T0), (30 min during (T1) and 30 min after OMT (T2). We evaluated the effects of OMT on the following NIRS parameters: cerebral (c), splanchnic (s) regional oximetry (rSO2), cerebro–splanchnic fractional tissue oxygen extraction (FTOE) and hemodynamic redistribution (CSOR). **Results**: crSO2 and cFTOE significantly ($P < 0.001$) improved at T0-T2; srSO2 significantly ($P < 0.001$) decreased and sFTOE increased at T0-T1. Furthermore, srSO2 and sFTOE significantly improved at T1-T2. Finally, CSOR significantly ($P < 0.05$) increased at T0-T2. **Conclusions**: The present data show that OMT enhances cerebro–splanchnic oximetry, tissue activation and hemodynamic redistribution in the absence of any adverse clinical or laboratory pattern. The results indicate the usefulness of further randomized studies in wider populations comparing the effectiveness of OMT with standard rehabilitation programs.

Keywords: NIRS; osteopathy; late preterm; brain; splanchnic

---

1. Introduction

Despite recent advances in perinatal therapeutic strategies, prematurity still constitutes one of the major causes of neonatal mortality and morbidity [1]. Preterm infants are at higher risk for a variety of complications, of long stays in hospital and of long-term neurodevelopmental disabilities, often associated with higher economic and social costs [2]. In this regard, it has been reported that average days of hospitalization can range from 4 to 135 days [3] and the social costs for preterm births are estimated to be more than US$ 26.2 billion with average first-year medical costs of about US$ 32,325 [4].

In the last decade there has been increasing interest in late preterm infants (LP), who account for about 70% of total preterm births. Epidemiological data show that the risks for LP of adverse neonatal outcome are seven times more than those for term infants [5].

The post-critical neurological management of LP infants focuses on complementary and alternative treatments and on an early rehabilitation program [6]. Osteopathy (OP) is a drug-free treatment that uses a manual approach to diagnose and treat so-called somatic dysfunctions (SD). SD are commonly considered as bodily areas which manifest an altered tissue texture, a restricted range of motion, tenderness and asymmetry. These areas are characterized by a pro-inflammatory state as well as altered autonomic control [7].

OP is widely practiced in adults, especially in connection with muscular–skeletal problems and, more recently, osteopathic clinical trials have been conducted to investigate the role and impact of OP treatment in the care of preterm infants, showing a decrease in the length of hospital stays and in social costs [7]. The use of OP techniques in a perinatal setting is still a matter of debate. The term "indirect technique" refers to a gentle manipulative touch (OMT) rather than a passive touch, which has been shown to improve neonatal behavior in both animal and human models [7,8]. It has been suggested that OMT modulates autonomic nervous system functions and reduces pro-inflammatory cytokines [9–12]. However, despite these encouraging findings, data on the possible positive or side effects on neonatal brain oximetry and function are still lacking.

The purpose of the present study was to investigate whether OMT could improve or affect brain–splanchnic oximetry and function in LP infants, using near infrared spectroscopy (NIRS) monitoring before, during and after the OMT procedure.

2. Results

2.1. Main Perinatal Outcomes

Perinatal characteristics, main neonatal outcomes and standard monitoring parameters of the procedures applied to the 18 recruited infants are reported in Table 1. Gestational age (GA) and birthweight (BW) were within the 10–90° centiles for our population standards, emergency caesarean section (CS) was necessary in four out of 18 pregnant women, 11 out of 18 newborns were males and all LP admitted to the study were inborn. Maternal age at delivery was within the reference standard for our population, none of the mothers had a history of chorioamnionitis or pre-eclampsia (EPH), seven out of 18 pregnant women received antenatal steroid prophylaxis and premature rupture of membranes (PROM) occurred in three cases.

Table 1. Perinatal characteristics, main outcome measures and standard monitoring parameters recorded in late preterm infants. Data are given as mean ± SD.

	Late Preterm (N = 18)
Maternal Characteristics	
Maternal age (y)	32 ± 2
Chorioamnionitis (n°/tot)	0/18
Glucocorticoids (n°/tot)	7/18
PROM (n°/tot)	3/18
EPH (n°/tot)	0/10
Neonatal Characteristics	
GA (wks)	35 ± 1
BW (g)	1762 ± 111
CS (n°/tot)	4/18
Gender (M/F)	11/7
Outborn/Inborn (n°/tot)	0/18
Apgar 1'	8 ± 1
Apgar 5'	8 ± 1
Main Outcomes	
RDS (n°/tot)	5/18
Surfactant administration (n°/tot)	5/18
MV (n°/tot)	5/18
PDA (n°/tot)	0/18
IVH (n°/tot)	0/18
EOS (n°/tot)	2/18
NEC (n°/tot)	0/18
ROP (n°/tot)	0/18
LOS (n°/tot)	2/18
BPD (n°/tot)	0/10

Abbreviations: premature rupture of membranes, PROM; EPH, preeclampsia; Gestational age, GA; birthweight, BW; caesarean section, CS; respiratory distress syndrome, RDS; mechanical ventilation, MV; patent ductus arteriosus, PDA; intraventricular hemorrhage, IVH; early-onset sepsis, EOS; broncho-pulmonary dysplasia, BPD.

At birth all LP had an Apgar score >7 and respiratory distress syndrome (RDS) requiring surfactant administration and mechanical ventilation occurred in five out 18 LP. No LP required drug treatment for patent ductus arteriosus persistence (PDA); none of them developed intraventricular hemorrhage (IVH), early onset sepsis (EOS), necrotizing enterocolitis (NEC), retinopathy of prematurity (ROP) or bronchopulmonary dysplasia (BPD), whilst two out of 18 LP needed antibiotic treatment for late onset sepsis (LOS).

2.2. Monitoring Parameters

At T0 (before OMT), laboratory parameters such as venous blood pH, bilirubinemia, hemoglobin levels and hematocrit rate were within reference values. Standard monitoring parameters (heart rate, HR; respiratory rate, RR; pulsed arterial oxygen saturation, SaO_2) at monitoring time-points were within reference values and no significant differences ($P > 0.05$, for all) were found at T0-T2.

Neurological examination and cerebral ultrasound were normal in all LP admitted to the study (Table 2).

Table 2. Standard laboratory and monitoring parameters recorded before osteopathic manipulation treatment and near infrared spectroscopy performance in late preterm infants. Data are given as mean ± SD.

	Late Preterm (N = 18)
Monitoring Parameters	
GA (wks)	36 ± 1
BW (g)	1846 ± 265
pH	7.35 ± 0.02
pCO_2 (mmHg)	43.9 ± 4.7
pO_2 (mmHg)	40.1 ± 2.3
Base excess	0.9 ± 1.1
Bilirubinemia (mg/dL)	4.3 ± 1.5
Hb (g/dL)	13.9 ± 1.3
Hematocrit rate (%)	40.1 ± 2.1
HR (bpm)	146 ± 12
RR (breath pm)	56 ± 9
SaO_2 at T0	98 ± 1
SaO_2 at T1	98 ± 1
SaO_2 at T2	99 ± 1
Neurological Examination	
Normal/suspect/abnormal	18/0/0
Cerebral Ultrasound	
Normal/abnormal	18/0/0

Abbreviations: gestational age, GA; weeks, wks; birthweight, BW; grams g; venous carbon dioxide partial pressure, pCO_2; millimeter of mercury, mmHg; venous oxygen partial pressure, pO_2; hemoglobin, Hb; heart rate, HR; respiratory rate, RR; arterial oxygen saturation, SaO_2.

2.3. NIRS Parameters

NIRS parameters in cerebral district were measurable in all LP infants recruited to the study (Figure 1).

Cerebral regional oxygen saturation ($crSO_2$) values started to increase from T0 to T2 ($P < 0.001$, for all). Significant ($P < 0.001$) differences in $crSO_2$ values were found between T1 and T2 (Figure 2).

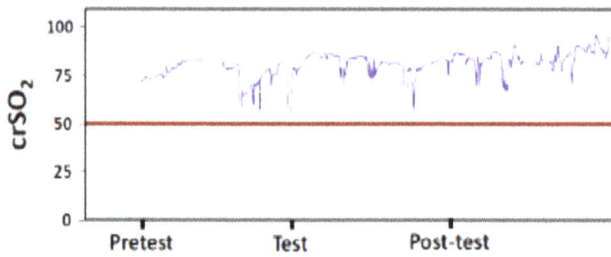

Figure 1. Cerebral regional oxygen saturation (crSO$_2$) pattern recorded before, during and after osteopathic manipulation treatment.

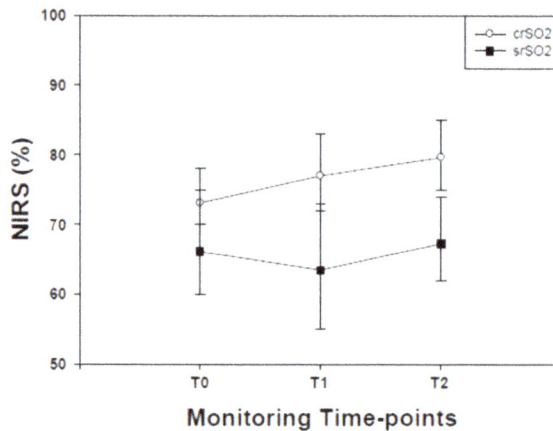

Figure 2. Cerebral (c) and splanchnic (s) regional oximetry (rSO$_2$) values recorded in late preterm infants before (T0) during (T1) and after (T2) osteopathic manipulation treatment. Data are given as median and 25–75° centiles. crSo2 values at T0 were significantly (P <0.001, for both) lower than those recorded at T1 and T2. srSO2 values at T0 were significantly (P <0.001, for both) higher than those recorded at T1 whilst no differences (P <0.05) were found between T0 and T2. srSo2 values at T1 were significantly (P <0.001) lower than T2.

Identically, after normalizing all NIRS results with T0 values set to 0, crSO$_2$ values started to increase from T0 to T2 (T0-T1 median: 3.00; 25/75° centile: 2.00/4.00; T0-T2 median: 2.00; 25/75° centile 2.00/3.00) (P <0.001, for all). Significant (P <0.001) differences in crSO$_2$ values were found between T1 and T2 (T1-T2 median: 0.00; 25/75° centile −2.00/1.00).

Cerebral fractional tissue oxygen extraction (cFTOE) values started to decrease from T0 to T1, reaching their lowest point at T2 (P <0.001, for all). Significant (P <0.001) differences in cFTOE values were found between T1 and T2 (Figure 3).

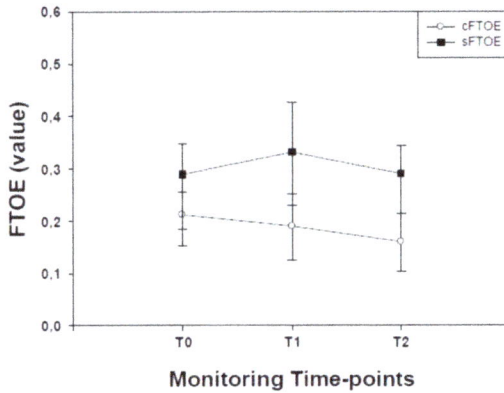

Figure 3. Cerebral (c) and splanchnic (s) fractional tissue oxygen extraction (FTOE) values recorded in late preterm infants before (T0) during (T1) and after (T2) osteopathic manipulation treatment. Data are given as median and 25–75° centiles. cFTOE values at T0 were significantly (*P* <0.001, for both) higher than those recorded at T1 and T2. sFTOE values at T0 were significantly (*P* <0.001, for both) lower than those recorded at T1 whilst no differences (*P* <0.05) were found between T0 and T2 time-points. sFTOE values at T1 were significantly (*P* <0.001) higher than T2.

NIRS parameters in splanchnic district were measurable in all LP infants recruited to the study (Figure 4).

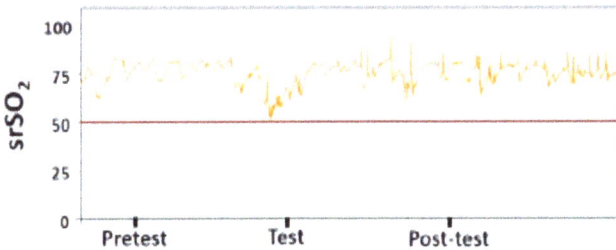

Figure 4. Splanchnic regional oxygen saturation (srSO$_2$) pattern recorded before, during and after osteopathic manipulation treatment.

The pattern of splanchnic regional oxygen saturation (srSO$_2$) values showed a significant (*P* < 0.001) decrease in splanchnic oximetry from T0, reaching their lowest point at the end of the OMT procedure (T1).From T1 to T2, srSO$_2$ values started to increase, being significantly (*P* <0.001) higher than T1 and superimposable (*P* >0.05) at T0 (Figure 2).

Identically, after normalizing all NIRS results with T0 values set to 0, srSO2 values started to decrease from T0 to T1 (T0-T1 median: −3.00; 25/75° centile: −14.00/6.00; T0-T2 median: −1.00; 25/75° centile: −10.00/8.00) (*P* <0.001, for all). Significant (*P* <0.001) differences in srSO2 values were found between T1 and T2 (T1-T2 median: 1.00; 25/75° centile: −8.00/11.00).

Splanchnic fractional tissue oxygen extraction (sFTOE) values showed a significant (*P* <0.001) increase in sFTOE from T0 reaching their peak at the end of T1. From T1 to T2 sFTOE values started to decrease being significantly (*P* <0.001) lower than T1. No differences (*P* >0.05) in sFTOE values were found between T0 and T2 (Figure 3).

Cerebral–splanchnic ratio (CSOR) values at T0 were significantly lower (*P* <0.001, for both) than those recorded at T1 and T2. Lower (*P* <0.05) CSOR values were found between T1 and T2 (Figure 5).

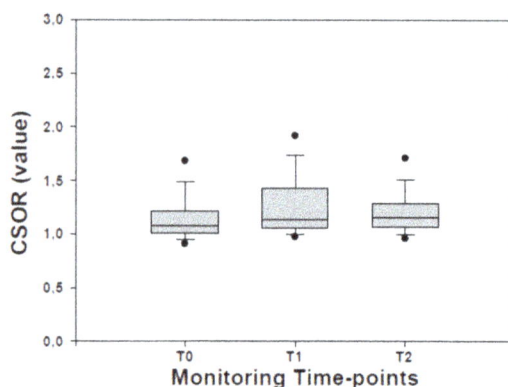

Figure 5. Cerebral–splanchnic ratio (CSOR) values recorded in late preterm infants before (T0) during (T1) and after (T2) osteopathic manipulation treatment. Data are given as median and 25–75° centiles.

3. Discussion

Nowadays, thanks to highly advanced medical technology in NICUs, the mortality rate for preterm infants is significantly reduced. Conversely, the incidence of disability and neurodevelopmental problems among survivors still remains high and problematic [2,5]. Based on the flat trend of morbidity rates, it has been suggested that brain injury should be considered as a complex amalgam of diseases (i.e., damage-related, maturational and trophic disturbances) rather than due to a single agent [13,14]. Alterations in neurological development/damage can occur independently of gestational age and often in the absence of evident signs of injury [15]. The less mature, healthy or sick preterm newborn may be unable or only partially able to manage environmental inputs, demonstrating over-reactive responses and poor tolerance of even minimal input. In addition to standard neuro-therapeutic strategies performed in neonatal intensive care units (NICU), an important role is played by the early introduction of individualized rehabilitation treatments [6–8,16]. In this regard, further progresses and new rehabilitation programs are eagerly awaited.

In the present study we showed, in a cohort of late preterm infants, that cerebral and splanchnic oximetry and tissue activation levels, evaluated by the fractional tissue oxygen extraction ratio, significantly changed during and after osteopathic manipulation treatment. Furthermore, significant hemodynamic changes in the cerebro–splanchnic regions were found during and after OMT.

To the best of our knowledge the present study constitutes the first observation in which brain oximetry and tissue activation levels were longitudinally monitored during osteopathic treatment. There are only a few observations on the potential positive effects of OMT on selected outcomes such as newborns' hospital stays [7].

The findings of improved cerebral oximetry and tissue activation levels in LP warrant further consideration. In particular, $crSO_2$ increased soon after the start of OMT and reached its highest peak 30 min after its conclusion; in parallel, cFTOE significantly decreased. This finding is noteworthy, since $crSO_2$ expresses an adequate cerebral oxygenation status while cFTOE is a reliable indicator of tissue oxygen extraction, reflecting the balance between delivery and central nervous system (CNS) tissue consumption [17–20]. Thus, these findings suggest that the "over-oxygenation rate" due to the OMT procedure increased tissue metabolic activity in the CNS. The issue is of relevance bearing in mind that at this stage the major metabolic activity of the CNS concerns its growth. In an animal model and in some patients, it has been shown that the late preterm period is crucial for the development of the CNS, which reaches 65% of its total weight, and of the cerebral cortex, which reaches 53% of its total volume, while components of structural and functional brain development such as synaptogenesis and dendritic arborization reach up to 35% of their total growth [21–24]. These findings are also corroborated by a significant increase in biological fluids of the concentrations of well-established neurobiomarkers at trophic action [25–27]. Altogether, it is reasonable to argue that OMT improves oximetry and tissue activation levels in the CNS of newborns throughout the monitoring period and may have a beneficial effect on CNS development. The issue is also highlighted by the significant increase in CSOR values, expressing a stable preferential hemodynamic redistribution in favor of the brain. Of course, additional studies in a wider population are needed to investigate further the possible short/long term effects of OMT on CNS development and function.

In the present study, we also found that splanchnic NIRS parameters changed during and after OMT. Briefly, srSO$_2$ decreased and sFTOE increased during the course of the treatment, and this was followed by a significant increase in srSO$_2$ and a decrease in sFTOE from the end of treatment to the recovery period.

These findings merit further consideration. In particular, i) the significant changes in NIRS parameters during OMT seem to suggest a reduction in splanchnic oximetry with a significant change (increase in sFTOE) in the balance between delivery and splanchnic tissue consumption in favor of the former. The explanation may lie in the hemodynamic redistribution from the splanchnic to the brain area following OMT, ii) the strong recovery of NIRS parameters characterized by an increase in splanchnic oximetry and tissue activation levels (decrease in sFTOE) in the post-OMT period are suggestive, as shown by the stable improvement in NIRS parameters in the CNS, of later positive effects of OMT. This finding is also corroborated by increased CSOR ratio values (>1), as expression of a hemodynamic redistribution from the splanchnic to the brain region. Another explanation may lie in the delivery of extra blood volume by the liver, as occurs in other intrauterine conditions such as growth retardation [20,28]. Altogether, it is possible to argue that OMT may reasonably be responsible for a redistribution from the splanchnic to the brain region followed by a significantly improved splanchnic oximetry and function in the presence of a stable increased CNS oximetry and tissue activation levels. In other words, OMT seems to exert beneficial effects both on brain (early) and splanchnic (late) oximetry and function. Further studies in wider populations are needed to confirm our observation bearing in mind the variability in NIRS parameters output due to different devices and last but not least the potential disturbances in splanchnic NIRS oximetry patterns due to stool or transitional meconium that can affect recordings availability [29,30].

We recognize that the present study has several limitations. In particular: i) the small sample size, although we were able to record a considerable number of NIRS values (median: 10,500 values), thereby reducing the possibility of potential bias, ii) external light interference, which can result in photon scattering, and (iii) movement artifacts as a consequence of problems associated with the fixing of sensors [31,32].

Lastly, it should be noted that OMT and NIRS recordings were performed at least 1 h before or after feeding in order to avoid potential bias effects due to feeding regimens [33]. Yet, this possibility was avoided since each case acted as his or her own control.

In conclusion, the present results on improved CNS and splanchnic oximetry and function in late preterm infants given OMT offer additional support for the need for new individually tailored rehabilitation programs. These findings pave the way to further studies in wider populations aimed at investigating the effectiveness of OMT and standard individualized rehabilitation programs on short/long term neurological outcome.

4. Materials and Methods

All subjects gave their informed consent for inclusion before they participated in the study. The study was conducted in accordance with the Declaration of Helsinki, and the protocol was approved by the local Ethics Committee (n°ASO.Neonat.12.01; 2630/71).

From December 2017 to December 2018 we conducted an observational pretest–test design involving 18 LP consecutively admitted to our third-level referral centers for NICU, where they acted as their own controls. GA was determined by clinical data and by longitudinal ultrasound scan monitoring according to the nomograms of Campbell and Thoms [34] and by postnatal confirmation in agreement with Villar J. et al. [35].

The exclusion criteria were: congenital heart diseases, congenital malformations, gastrointestinal anomalies and cutaneous diseases impeding the placement of probes.

The perinatal data, neonatal characteristics and main outcomes are reported in Table 1.

4.1. NIRS Monitoring

Hemodynamic and oxygenation changes in the cerebral district/region were monitored using the Sen Smart X-100 NIRS device (Nonin Medical, Plymouth, MN, USA). Self-adhesive transducers that contain the light-emitted diodes and two distant Equanox Advance sensors (Nonin Medical, Plymouth, MN, USA) were fixed on the central region of the neonatal skull. $crSO_2$ and $srSO_2$ and SaO_2 were calculated by the in-built software. Fractional tissue oxygen extraction values in the cerebral and splanchnic districts were assessed according to the following formula: $(SaO_2 - c(s)rSO_2)/SaO_2$ [17,18].

We also calculated the CSOR, which is the ratio of cerebral versus splanchnic district oximetry [19], according to the following formula: $crSO_2/srSO_2$. This ratio has been found to be a valuable index of hemodynamic redistribution in chronic hypoxic infants [20].

All OMT infants were monitored on the 7th day of age at three time-points: 30 min before, 30 min during and 30 min after completion of the OMT. No differences were observed in the duration of OMT procedures ($P > 0.05$).

4.2. Standard Monitoring Parameters and Main Outcomes

HR and RR rates and SaO_2 monitoring were continuously recorded by MX700 monitors (Philips, Eindhoven, The Netherlands) at 12" intervals (Figure 6).We also recorded the following main outcome measures: maternal age; the incidence of chorioamnionitis, PROM or EPH; the need of antenatal steroid prophylaxis; GA and BW; the incidence of CS, RDS, PDA, IVH, EOS, NEC, ROP, LOS or BPD.

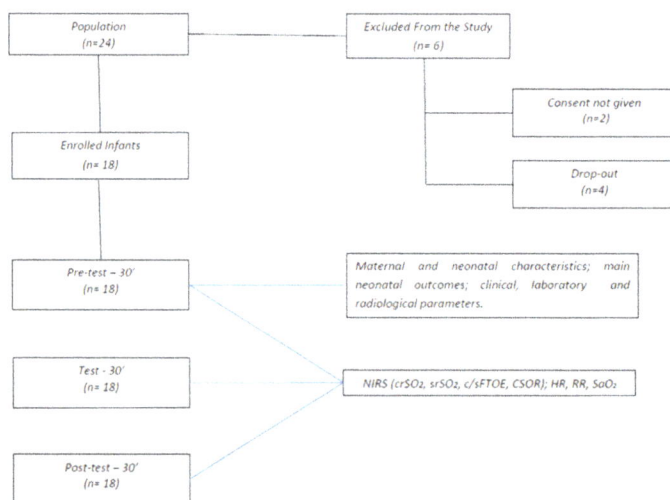

Figure 6. Flowchart of data collection and measurement scenarios. Abbreviations: near infrared spectroscopy, NIRS; regional oxygen saturation, rSO_2; cerebral, c; splanchnic, s; fractional tissue oxygen extraction, FTOE; cerebro–splanchnic oxygenation ratio, CSOR; heart rate, HR; respiratory rate, RR; pulsed arterial oxygen saturation, SaO_2.

4.3. Cranial Assessment

Standard cerebral ultrasonography was performed by a real-time ultrasound machine (Acuson 128SP5, Mountain View, CA, USA) using a transducer frequency emission of 3.5 MHz. Cerebral ultrasound patterns were evaluated at 24 h from birth, at 7 days and before discharge from hospital.

4.4. Neurological Examination

Neurological examination was performed daily according to the method of Prechtl [15]. Each infant was assigned to one of three diagnostic groups: normal, suspect, abnormal. An infant was considered to be abnormal when one or more of the following neurological syndromes were unequivocally present: (a) increased or decreased excitability (hyperexcitability syndrome, convulsion, apathy syndrome, coma); (b) increased or decreased motility (hyperkinesia, hypokinesia); (c) increased or decreased tonus (hypertonia, hypotonia); (d) asymmetries (peripheral or central); (e) defects of the central nervous system; (f) any combination of the above. When indications of the presence of a syndrome were inconclusive or if only isolated symptoms were present, e.g., mild hypotonia or only a slight tremor, the infants were classified as suspect.

4.5. Osteopathic Procedure

Osteopathic procedures were performed by osteopaths with experience in the neonatology field. The procedures included a structural evaluation followed by treatment. The structural evaluation was performed with the infant lying down in the open crib or incubator and was addressed to diagnose somatic dysfunctions [16]. It included rigorous and precise manual assessments of the skull, spine, pelvis, abdomen, and upper and lower limbs to locate bodily areas with an alteration in tissue, asymmetry, range of motion, and tenderness criteria [7]. The findings of that diagnostic procedure formed the basis of treatment that included the application of a selected range of manipulative techniques aimed at relieving the somatic dysfunctions. Techniques used were in line with the benchmarks on osteopathic treatment available in the medical literature and were limited to indirect techniques such as: myofascial release and balanced ligamentous/membranous tension. The whole OMT procedure on infants lasted 30 min, ten minutes for evaluation and 20 min for treatment and re-evaluation.

4.6. Statistical Analysis

For the calculation of sample size, we used $crSO_2$ change as the main parameter. As no basic data are available for the studied population, we assumed a decrease of 0.5 SD in $crSO_2$ to be clinically significant. Indeed, considering an $\alpha = 0.05$ and using a two-sided test, we estimated a power of 0.80 recruiting 16 preterm infants. We therefore added 2 cases to allow for any dropout.

The sample size was calculated using nQuery Advisor (Statistical Solutions, Saugus, MA, USA), version 5.0. Main outcome measures are summarized by mean ± SD. NIRS parameters were summarized by median and interquartile ranges. Comparisons among different groups were analyzed for statistically significant differences by one way ANOVA followed by Dunn's test. Categorical data were analyzed by means of Fisher's test. A value of $P < 0.05$ was considered statistically significant.

5. Conclusions

The present data offer additional support to the need of an additional and individualized rehabilitation program in the neurological care of high risk infants. Further randomized controlled trials comparing OMT with rehabilitation procedures that are considered as standard of care are, therefore, suggested.

Author Contributions: B.M. participated in designing the study, patient recruitment, data analysis and writing the manuscript. F.P. participated in designing the study, data collection and drafting the manuscript. V.C. participated in designing the study, data collection and drafting the manuscript. G.B. participated in patient recruitment, data collection and drafting the manuscript. M.C.S. participated in the design of the project, in patient recruitment and in drafting the manuscript. E.M. participated in in designing the study, in patient recruitment and in data collection and analysis. S.F. participated in designing the study, in patient recruitment and in data collection and analysis. D.G. project leader, participated in data analysis and in writing the manuscript. All authors agreed to be accountable for all aspects of the work in ensuring that questions related to the accuracy or integrity of any part of the work were appropriately investigated and resolved and approved the final manuscript as submitted.

Funding: This research received no external funding.

Acknowledgments: This study is part of the I.O. PhD International Program, under the auspices of the Italian Society of Neonatology, and was partially supported by grants to Diego Gazzolo from I Colori della Vita Foundation, Italy.

Conflicts of Interest: The authors declare no conflict of interest.

References

1. WHO. March of Dimes, PMNCH, Save the Children. In *Born Too Soon: The Global Action Report on Preterm Birth*; World Health Organization: Geneva, Switzerland, 2012. [CrossRef]
2. Beck, S.; Wojdyla, D.; Say, L.; Betran, A.P.; Merialdi, M.; Requejo, J.H.; Rubens, C.; Menon, R.; Van Look, P.F.A. The worldwide incidence of preterm birth: A systematic review of maternal mortality and morbidity. *Bull. World Health Organ.* **2010**, *88*, 31–38. [CrossRef] [PubMed]
3. Italian Ministry of Health. *Annual Report on Hospitals Activity*; Ministry of Health: Rome, Italy, 2012.
4. National Academy Press. *Preterm Birth: Causes, Consequences, and Prevention*; National Academy Press: Cambridge, MA, USA, 2007.
5. Boyle, E.M.; Poulsen, G.; Field, D.J.; Kurinczuk, J.J.; Wolke, D.; Alfirevic, Z.; Quigley, M.A. Effects of gestational age at birth on health outcomes at 3 and 5 years of age: Population based cohort study. *BMJ* **2012**, *344*, e896. [CrossRef] [PubMed]
6. Pineda, R.; Guth, R.; Herring, A.; Reynolds, L.; Oberle, S.; Smith, J. Enhancing sensory experiences for very preterm infants in the NICU: An integrative review. *J. Perinatol.* **2017**, *37*, 323–332. [CrossRef] [PubMed]
7. Cerritelli, F.; Pizzolorusso, G.; Renzetti, C.; Cozzolino, V.; D'Orazio, M.; Lupacchini, M.; Marinelli, B.; Accorsi, A.; Lucci, C.; Lancellotti, J.; et al. A multicenter, randomized, controlled trial of osteopathic manipulative treatment on preterms. *PLoS ONE* **2015**, *10*, e0127370. [CrossRef] [PubMed]
8. Pizzolorusso, G.; Turi, P.; Barlafante, G.; Cerritelli, F.; Renzetti, C.; Cozzolino, V.; D'Orazio, M.; Fusilli, P.; Carinci, F.; D'Incecco, C. Effect of osteopathic manipulative treatment on gastrointestinal function and length of stay of preterm infants: An exploratory study. *Chiropr. Man. Ther.* **2011**, *19*, 15. [CrossRef]
9. Henley, C.; Ivins, D.; Mills, M.; Wen, F.; Benjamin, B. Osteopathic manipulative treatment and its relationship to autonomic nervous system activity as demonstrated by heart rate variability: A repeated measures study. *Osteopath Med. Prim. Care* **2008**, *2*, 7. [CrossRef]
10. Giles, P.; Hensel, K.; Pacchia, C.; Smith, M. Suboccipital decompression enhances heart rare variability indices of cardiac control in healthy subjects. *J. Altern. Complement. Med.* **2013**, *19*, 92–96. [CrossRef]
11. Ruffini, N.; D'Alessandro, G.; Mariani, N.; Pollastrelli, A.; Cardinali, L.; Cerritelli, F. Variations of high frequency parameter of heart rate variability following osteopathic manipulative treatment in healthy subjects compared to control group and sham therapy: Randomized controlled trial. *Front. Neurosci.* **2015**, *9*, 272. [CrossRef]
12. McGlone, F.; Cerritelli, F.; Walker, S.; Esteves, J. The role of gentle touch in perinatal osteopathic manual therapy. *Neurosci. Biobehav. Rev.* **2017**, *72*, 1–9. [CrossRef]
13. Volpe, J.J. Brain injury in premature infants: A complex amalgam of destructive and developmental disturbances. *Lancet Neurol.* **2009**, *8*, 110–124. [CrossRef]
14. Arpino, C.; Argenzio, L.; Ticconi, C.; Di Paolo, A.; Stellin, V.; Lopez, L.; Curatolo, P. Brain damage in preterm infants: Etiological pathways. *Ann. Ist. Super. Sanità* **2005**, *41*, 229–237.
15. Prechtl, H.F.R. Assessment methods for the newborn infant: A critical evaluation. In *Psychobiology of Human Newborn*; Stratton, P., Ed.; John Wiley & Sons: Hoboken, NJ, USA, 1982; pp. 21–52.
16. Cerritelli, F.; Martelli, M.; Renzetti, C.; Pizzolorusso, G.; Cozzolino, V.; Barlafante, G. Introducing an osteopathic approach into neonatology ward: The NE-O model. *Chiropr. Man Ther.* **2014**, *22*, 18. [CrossRef]

17. Naulaers, G.; Meyns, B.; Miserez, M.; Leunens, V.; Van Huffel, S.; Casaer, P.; Weindling, M.; Devlieger, H. Use of tissue oxygenation index and fractional tissue oxygen extraction as non-invasive parameters for cerebral oxygenation: A validation study in piglets. *Neonatology* **2007**, *92*, 120–126. [CrossRef]
18. Wardle, S.P.; Yoxall, C.W.; Weindling, A.M. Determinants of cerebral fractional oxygen extraction using near infrared spectroscopy in preterm neonates. *J. Cereb. Blood Flow Metab.* **2000**, *20*, 272–279. [CrossRef]
19. Dani, C.; Pratesi, S.; Fontanelli, G.; Barp, J.; Bertini, G. Blood transfusions increase, cerebral, splanchnic and renal oxygenation in anemic preterm infants. *Transfusion* **2010**, *50*, 1220–1226. [CrossRef]
20. Bozzetti, V.; Paterlini, G.; Van Bel, F.; Visser, G.H.A.; Tosetti, L.; Gazzolo, D.; Tagliabue, P.E. Cerebral and somatic NIRS-determined oxygenation in IUGR and non-IUGR preterm infants during transition. *J. Matern. Fetal Neonatal Med.* **2015**, *25*, 1–4.
21. Kinney, H.C. The near-term (late preterm) human brain and risk for periventricular leukomalacia: A review. *Semin. Perinatol.* **2006**, *30*, 81–88. [CrossRef]
22. Haynes, R.L.; Borenstein, N.S.; Desilva, T.M.; Folkerth, R.D.; Liu, L.G.; Volpe, J.J.; Kinney, H.C. Axonal development in the cerebral white matter of the human fetus and infant. *J. Comp. Neurol.* **2005**, *484*, 156–167. [CrossRef]
23. Back, S.A.; Luo, N.L.; Borenstein, N.S.; Levine, J.M.; Volpe, J.J.; Kinney, H.C. Late oligodendrocyte progenitors coincide with the developmental window of vulnerability for human perinatal white matter injury. *J. Neurosci.* **2001**, *21*, 1302–1312. [CrossRef]
24. Huppi, P.S.; Warfield, S.; Kikinis, R.; Barnes, P.D.; Zientara, G.P.; Jolesz, F.A.; Tsuji, M.K.; Volpe, J.J. Quantitative magnetic resonance imaging of brain development in premature and mature newborns. *Ann. Neurol.* **1998**, *43*, 224–235. [CrossRef]
25. Sannia, A.; Zimmermann, L.J.; Gavilanes, A.W.; Vles, H.J.; Serpero, L.; Frulio, R.; Michetti, F.; Gazzolo, D. S100B protein maternal and fetal bloodstreams gradient in healthy and small for gestational age pregnancies. *Clin. Chim. Acta* **2014**, *12*, 1337–1340. [CrossRef]
26. Sannia, A.; Risso, F.M.; Zimmermann, L.J.; Gavilanes, A.W.; Vles, H.J.; Gazzolo, D. S100B urine concentrations in late preterm infants are gestational age and gender dependent. *Clin. Chim. Acta* **2013**, *417*, 31–34. [CrossRef]
27. Tina, L.G.; Frigiola, A.; Abella, R.; Artale, B.; Puleo, G.; D'Angelo, S.; Musmarra, C.; Tagliabue, P.; Li Volti, G.; Florio, P.; et al. Near infrared spectroscopy in healthy preterm and term newborns: Correlation with gestational age and standard monitoring parameters. *Curr. Neurovascular Res.* **2009**, *6*, 148–154. [CrossRef]
28. Gazzolo, D.; Pluchinotta, F.; Lapergola, G.; Franchini, S. The Ca2+-Binding S100B Protein: An Important Diagnostic and Prognostic Neurobiomarker in Pediatric Laboratory Medicine. *Methods Mol. Biol.* **2019**, *1929*, 701–728. [CrossRef]
29. Kleiser, S.; Ostojic, D.; Andresen, B.; Nasseri, N.; Isler, H.; Scholkmann, F.; Karen, T.; Greisen, G.; Wolf, M. Comparison of tissue oximeters on a liquid phantom with adjustable optical properties: An extension. *Biomed. Opt. Express* **2018**, *9*, 86–101. [CrossRef]
30. Isler, H.; Shenk, D.; Bernhard, J.; Kleiser, S.; Scholkmann, F.; Ostojic, D.; Kalyanov, A.; Ahnen, L.; Wolf, M.; Karen, T. Absorption spectra of early stool from preterm infants need to be considered in abdominal NIRS oximetry. *Biomed. Opt. Express* **2019**, *10*, 2784–2791. [CrossRef]
31. Naulaers, G.; Morren, G.; Van Huffel, S.; Casaer, P.; Devlieger, H. Cerebral tissue oxygenation index in very premature infants. *Arch. Dis. Child. Fetal Neonatal Ed.* **2002**, *87*, F189–F192. [CrossRef]
32. Toet, M.C.; Lemmers, P.M.; van Schelven, L.J.; Van Bel, F. Cerebral oxygenation and electrical activity after birth asphyxia: Their relation to outcome. *Pediatrics* **2006**, *117*, 333–339. [CrossRef]
33. Grometto, A.; Pizzo, B.; Strozzi, M.C.; Gazzolo, F.; Gazzolo, D. Near-infrared spectroscopy is a promising noninvasive technique for monitoring the effects of feeding regimens on the cerebral and splanchnic regions. *Acta Paediatr.* **2018**, *107*, 234–239. [CrossRef]

34. Campbell, S.; Thoms, A. Ultrasound measurement of the fetal head to abdomen circumference ratio in the assessment of growth retardation. *Br. J. Obstet. Gynaecol.* **1977**, *84*, 165–174. [CrossRef]
35. Villar, J.; Cheikh Ismail, L.; Victora, C.G.; Ohuma, E.O.; Bertino, E.; Altman, D.G.; Lambert, A.; Papageorghiou, A.T.; Carvalho, M.; Jaffer, Y.A.; et al. International standards for newborn weight, length, and head cir- cumference by gestational age and sex: The newborn Cross-Sectional Study of the INTERGROWTH-21st Project. *Lancet* **2014**, *384*, 857–868. [CrossRef]

Sample Availability: Not Available.

      MDPI

Article

Predicting Ewing Sarcoma Treatment Outcome Using Infrared Spectroscopy and Machine Learning

Radosław Chaber [1,*], Christopher J. Arthur [2], Kornelia Łach [1], Anna Raciborska [3], Elżbieta Michalak [4], Katarzyna Bilska [3], Katarzyna Drabko [5], Joanna Depciuch [6], Ewa Kaznowska [7,8] and Józef Cebulski [9]

[1] Clinic of Paediatric Oncology and Haematology, Faculty of Medicine, University of Rzeszow, ul. Kopisto 2a, 35-310 Rzeszow, Poland; kornelia_lach@wp.pl

[2] School of Chemistry, University of Bristol, Bristol BS8 1TS, UK; Chris.Arthur@bristol.ac.uk

[3] Department of Surgical Oncology for Children and Youth, Institute of Mother and Child, 01-211 Warsaw, Poland; anna.raciborska@hoga.pl (A.R.); katarzyna.bilska@gmail.com (K.B.)

[4] Department of Pathology, Institute of Mother and Child, 01-211 Warsaw, Poland; elzbieta.michalak@gmail.com

[5] Department of Pediatric Hematology, Oncology and Bone Marrow Transplant, Medical University of Lublin, 20-081 Lublin, Poland; katarzynadrabko@umlub.pl

[6] Institute of Nuclear Physics, Polish Academy of Sciences, 31-342 Krakow, Poland; joannadepciuch@gmail.com

[7] Laboratory of Molecular Biology, Centre for Innovative Research in Medical and Natural Sciences, Faculty of Medicine, University of Rzeszow, 35-959 Rzeszow, Poland; e.kaznowska@op.pl

[8] Department of Human Histology, Chair of Morphological Sciences, Faculty of Medicine, University of Rzeszow, 35-959 Rzeszow, Poland

[9] Center for Innovation and Transfer of Natural Sciences and Engineering Knowledge, University of Rzeszow, 35-959 Rzeszow, Poland; cebulski@ur.edu.pl

[*] Correspondence: radoslaw.chaber@gmail.com or pracowniabnd@gmail.com; Tel.: +48-17-8664-589; Fax: +48-17-8664-588

Academic Editor: Christian Huck
Received: 1 February 2019; Accepted: 14 March 2019; Published: 19 March 2019

Abstract: Background: Improved outcome prediction is vital for the delivery of risk-adjusted, appropriate and effective care to paediatric patients with Ewing sarcoma—the second most common paediatric malignant bone tumour. Fourier transform infrared (FTIR) spectroscopy of tissues allows the bulk biochemical content of a biological sample to be probed and makes possible the study and diagnosis of disease. Methods: In this retrospective study, FTIR spectra of sections of biopsy-obtained bone tissue were recorded. Twenty-seven patients (between 5 and 20 years of age) with newly diagnosed Ewing sarcoma of bone were included in this study. The prognostic value of FTIR spectra obtained from Ewing sarcoma (ES) tumours before and after neoadjuvant chemotherapy were analysed in combination with various data-reduction and machine learning approaches. Results: Random forest and linear discriminant analysis supervised learning models were able to correctly predict patient mortality in 92% of cases using leave-one-out cross-validation. The best performing model for predicting patient relapse was a linear Support Vector Machine trained on the observed spectral changes as a result of chemotherapy treatment, which achieved 92% accuracy. Conclusion: FTIR spectra of tumour biopsy samples may predict treatment outcome in paediatric Ewing sarcoma patients with greater than 92% accuracy.

Keywords: Ewing sarcoma; Fourier transform infrared spectroscopy; FTIR; chemotherapy; bone cancer

1. Introduction

Improved patient outcomes need not only better therapeutic approaches, but also the reduction of treatment-related complications. Risk-adapted therapeutic approaches have, therefore, been key to recent improvements in paediatric oncology [1–3]. Central to this are the discovery and application of prognostic factors for the risk allocation of patients. Consequently, a risk-adapted approach can be taken whereby treatment may be intensified amongst patients in the high-risk cohort or de-escalated in those considered to be low-risk, minimising toxicity and late sequelae without compromising survival [4].

Ewing sarcoma (ES) is the second most common paediatric malignant bone tumour and comprises 3% of all paediatric malignancies [5]. However, this is a rare neoplasm and affects about 2.9 people per million annually [6]. Overall survival rates for patients with localised disease approach 69%. Treatment of patients with metastatic, refractory or relapsed ES is more challenging though, with only 42% surviving five years [7]. Methods that improve the stratification of patients with ES could, therefore, result in improved therapeutic outcomes whilst reducing toxicity.

Fourier transform infrared (FTIR) spectroscopy is a physicochemical, non-invasive method that provides information about the bulk chemical composition of a biological sample [8]. The frequency range of absorption by molecules is correlated with their structure making it amenable for the study of all classes of biomolecules [9]. Consequently, FTIR spectroscopy can potentially detect changes in the biochemical composition of tissues that mark the progression from healthy to cancerous tissue. Driven by applications such as the identification of cancer, endoscopy and spectral histopathology, FTIR has been applied to many different tumour types; for example, breast cancer [10,11], lung cancer [12,13], ovarian cancer [14], brain tumours [15], cervical cancer [16], gastric cancer [17], colon cancer [18], prostate cancer [19] and melanoma [20].

We have previously shown that the peak absorbance maxima in the bio-fingerprinting region (1000–1100 cm^{-1}) of FTIR spectra of Ewing sarcoma bone sections can be predictive of patient outcome. [21,22] Following on from this, we considered whether treatment outcome might be better predicted by the analysis of the whole FTIR spectra. Herein we report a small-cohort retrospective study into the prognostic value of the whole spectrum obtained from ES tumours before and after neoadjuvant chemotherapy administration in combination with data-reduction and machine learning approaches.

2. Materials and Methods

2.1. Patients

Twenty-seven patients between 5 and 20 years of age with newly diagnosed Ewing sarcoma of bone were included in this study. Each patient was treated according to the Euro-EWING protocols during 2010–2016 and their clinical characteristics are presented in Table 1.

In each case, identical induction neoadjuvant chemotherapy (neoCTX) consisting of six VIDE cycles (vincristine, ifosfamide, doxorubicin, etoposide) were administrated. Surgery was undertaken at the Department of Surgical Oncology, Institute of Mother and Child in Warsaw, Poland. Microscopically complete resection was possible in 26 cases. Each histopathological sample was verified centrally at the same institution and their response to chemotherapy was measured by the percentage of viable tumour cells remaining after neoCTX completion. A good response was defined as greater than or equal to 90% of necrosis. Post-operative treatment was conducted according to Euro-EWING protocols and depended on the clinical status of the patient. Patients were thus treated uniformly and relatively contemporaneously. There were no deaths observed for reasons other than cancer progression. Informed consent was obtained from all patients, or their guardians, before treatment. This retrospective study was conducted in compliance with international regulations for the protection of human research subjects and was authorised by the Institutional Review Board of the University of Rzeszow on June 2014 (Protocol No. KBET/6/06/2014).

Table 1. The clinical characteristic of patients (n = 27).

Gender Male/Female	12/15
Age (years) range, median	5–20 years 14 years
Localised/disseminated	11/16
Tumour resection complete/incomplete	1/26
Necrosis $\geq$ 90% vs. <90%	20/7
Local radiotherapy	17
Auto HSCT	9
Relapses (progressions)/deaths	14/10
Follow up time (months) range (median)	14–74 (34)

HSCT—hematopoietic stem cell transplantation.

2.2. Sample Preparation

Twenty-seven samples consisted of formalin-fixed paraffin-embedded (FFPE) tissues collected during a diagnostic biopsy prior to neoCTX and after completion of the sixth chemotherapy cycle (VIDE). All samples were prepared and verified by pathologists experienced in ES.

FFPE bone tissue blocks were sectioned to a thickness of 10 μm using a rotary microtome and applied to calcium fluoride slides. Sections were placed on the surface of a tub holding warm water to allow them to flatten and were then gently pulled onto a slide. Samples were then dewaxed by washing twice in xylene and rehydrated by rinsing in an alcohol series ranging from absolute alcohol (99.8%) to 96%, 80% and 70% alcohol. Finally, the samples were rinsed with distilled water and dried.

2.3. FTIR Spectroscopy

FTIR spectra were recorded using a Bruker Vertex 70v FTIR spectrometer. Tissue specimens were applied to the attenuated total reflection (ATR) plate and mid-infrared radiation was passed through the sample using a single-reflection snap ATR crystal diamond (recorded at 1 cm^{-1} of spectral resolution, 32 scans). Spectra were recorded in the range of 800–3500 cm^{-1}. As the samples were dewaxed, the air was measured as the background. All measurements were recorded in triplicate. Initial data analysis and baseline corrections were performed using OPUS 7.0 from Bruker Optik GmbH 2011.

2.4. Statistical Analysis

All models and methods used in the experiments were implemented in Python 3.7 and R 3.5.0. Specifically, the Python package Pandas v0.23.0 was used to manipulate the data and Scikit-Learn 0.20 was used to implement the machine learning techniques. Prior to analysis spectra were normalised using Scikit-Learn's StandardScaler, which removes the means and scales to unit variance. Spectra were smoothed with Savitzky Golay filter (SciPy). Survival analysis was performed with the R package 'Survival' version 2.42-6. Two-sided p values of <0.05 were considered statistically significant. All data reduction methods were used as implemented in Scikit-Learn.

3. Results

3.1. Exploratory Data Analysis

Spectra were mean-centred, scaled to unit variance, smoothed using a Savitzky-Golay filter and a linear detrend was applied. As expected, peaks corresponding to functional groups within nucleic acids, phospholipids, polysaccharides, proteins and remaining lipids are observable in the FTIR spectra of bone tissues (Figure 1).

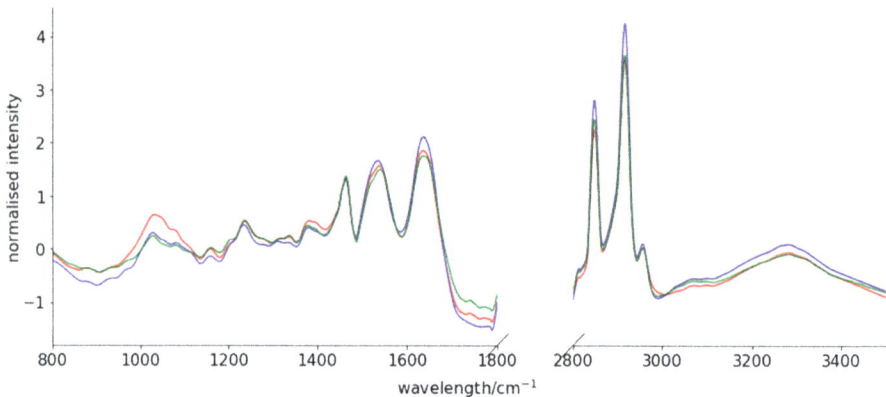

Figure 1. Fourier transform infrared (FTIR) spectrum of normal bone tissue collected outside the area of Ewing sarcoma (ES) infiltration (red line), ES tumour tissue before chemotherapy (blue line) and ES tumour tissue after induction chemotherapy (green line). Measuring range: 800–3500 cm^{-1}.

We first considered whether it was possible to differentiate the treatment outcomes for the ES patients and turned in the first instance to unsupervised dimensionality reduction. Dimensionality reduction approaches are broadly based on the selection of the informative features, or the generation of variables, that retain the information present in the original dataset. In principal component analysis (PCA), this dimensionality reduction is achieved by finding the linear combination of a set of variables that have maximum variance. When the spectral dataset was analysed by PCA (Figure 2a), we saw some clustering of the patients who lived vs. those who died due to tumour progression, although with considerable overlap between these groups in the first two principal components.

Other methods for dimensionality reduction, including non-linear methods, have emerged and we applied a suite of these, including both matrix deconvolution and manifold learning methods as implemented in the python library Scikit-Learn (Figure 2 shows the first two components of each method tested). Many of these methods afforded greater visual separation of the classes than PCA, although all tested methods resulted in some degree of overlap. Of note were the kernel PCA methods (Figure 2f,g), which resulted in a clear clustering of the spectra based on patient outcome. Kernel PCA methods differ from the more commonly used linear PCA in that they use an arbitrary function as opposed to a linear function [23]. Scikit-Learn implements several functions for use in Kernel PCA including a sigmoid, polynomial (poly), cosine and a radial basis function (RBF). This analysis was repeated for the necrosis and relapse responses; however, none of the applied methods resulted in the visual separation of the classes (please see Supplementary Material).

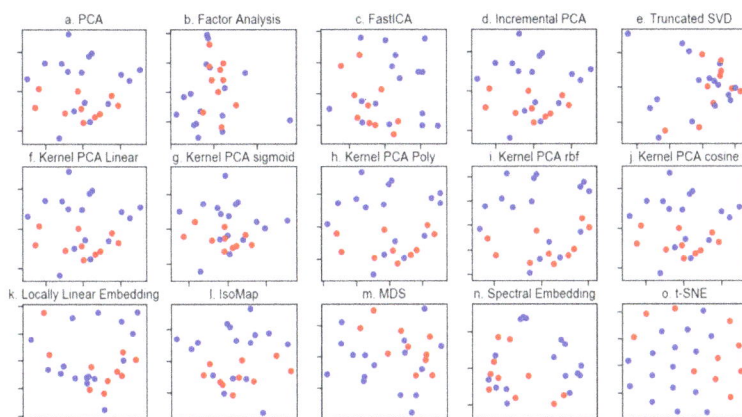

Figure 2. Dimensionality reduction methods applied to the FTIR dataset. Throughout, patients who survived are coloured blue whilst those who did not are coloured red. Matrix decomposition methods: (**a**). PCA (**b**). Factor analysis (**c**). Fast Independent Components Analysis (FastICA) (**d**). Incremental PCA (**e**). Truncated singular value decomposition (SVD) (**f**). Kernel PCA using a linear kernel (**g**). Kernel PCA using a sigmoid kernel (**h**). Kernel PCA using a polynomial kernel (**i**). Kernel PCA using a radial basis function kernel and (**j**). Kernel PCA using a cosine kernel. Manifold learning methods: (**k**). Locally linear embedding (**l**). Isomap m. Multidimensional scaling (MDS) (**m**). Spectral embedding and (**n**). t-distributed stochastic neighbour embedding (t-SNE).

3.2. Spectral Changes after Neo-CTX

It is expected that treatment with neo-CTX would result in changes in the biochemical composition of the tumour tissue. Moreover, we hypothesised that patients who responded positively to neo-CTX treatment would show a different spectral change than those who did not. To test this hypothesis, we subtracted the mean standardised-spectra pre-neoadjuvant chemotherapy (preCTX) treatment and post-neoadjuvant chemotherapy (postCTX) for patients who lived and those who died. As can be seen in Figure 3, consistent with our hypothesis, differences in the mean spectral changes were observed between the patient groups. Particularly, these changes included an increase in lipid and nucleic acid spectral intensity in those who lived as opposed to those who died and an increase in protein amide signal in those who died, with little change in those who lived.

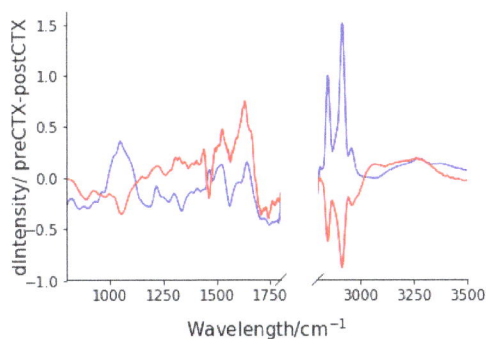

Figure 3. Spectral changes after neoadjuvant chemotherapy (neoCTX) for patients who lived (blue) and for those who died (red). Spectral changes were calculated by subtracting the mean, standardised and normalised spectra after post-neoadjuvant chemotherapy (postCTX) from that of the pre-neoadjuvant chemotherapy (preCTX) for each group.

3.3. Generating a Predictive Model for Prognosis

Prognostic and predictive models are particularly important in the clinical decision-making process. We, therefore, sought to develop a model for the prediction of Ewing sarcoma treatment outcome. Given the limited number of samples in this study (due to the low incidence of Ewing Sarcoma) the aim of this study was not to generate a clinically applicable predictive model (which would require a larger clinical trial) but rather to validate and test the potential of FTIR for the prediction of Ewing Sarcoma prognosis.

Furthermore, taking into consideration the limited number of patients in our study group we were mindful of the risk of over-fitting the data (where an overly complex machine learning model effectively memorises the data and does not generalise to unseen new data). The sample numbers herein have, however, impeded the use of an external test set for validation and, consequently, we have throughout used a leave-one-out cross-validation approach to assess model accuracy.

3.4. Feature Generation

Data such as tissue FTIR spectra consists of many variables, with each constituting the absorbance at different wavenumbers. Given this multivariate nature, it is desirable to simplify or dimensionally reduce the data [7] prior to the generation of a predictive model or clustering. Any dimensionality reduction should, however, result in minimal loss of information.

We report four approaches to predicting prognosis. In the first, we use the pre-neo-CTX spectra alone (henceforth preCTX). In the second we use the post-neo-CTX spectra (postCTX) alone. In the third, we explore the spectral changes between the preCTX-postCTX spectra changes as predictive features and, finally, we combine the dimensionally reduced representations of the preCTX and postCTX data.

Given the multivariate nature of FTIR spectra, and based on our earlier graphical analysis, the normalised and standardised spectra were dimensionally reduced using both PCA and kernel PCA (cosine function) (taking the first 15 and 10 principal components respectively). This therefore mapped the preCTX, postCTX and preCTX-postCTX to a 25-dimensional feature space and mapped the preCTX and postCTX data to a 50-dimensional feature space. This approach has several advantages. First, it reduces the contribution of noise to the spectral data. Second, the inclusion of hundreds of potential variables (from the raw FTIR spectra) into a classification model would likely lead to over-fitting and reduce the predictive performance against new samples.

3.5. Supervised Learning

The goal of supervised learning is to find a model that will correctly associate the inputs with the outputs. In the case of Ewing sarcoma diagnosis, it would be clinically useful to be able to determine the outcome of treatment.

It is not possible to tell a priori which machine learning method will be most suitable for a predictive task. Linear Support Vector Machine (SVM), Random Forest (RF) Decision Tree, Linear Discriminant Analysis (LDA) and Gradient Boosted Classifier (GDM) models were, therefore, trained on the reduced spectral feature sets. All classifiers were used with their default settings as implemented in SciKit-Learn. Table 2 lists the leave-one-out cross-validation accuracy for these models.

We began by attempting to develop a predictor for patient death based on the reduced spectral dataset. Of the tested classifiers a Linear Discriminant Analysis (LDA) classifier was able to predict patient death with an 81% accuracy using only feature sets based on preCTX spectra. Greater success was seen when the analysis was repeated on the postCTX spectra only with the random forest classifier able to predict patient death with better than 92% accuracy. Subtracting the preCTX and postCTX did not typically offer good predictive accuracy, except in the case of the k-nearest neighbours (KNN) classifier. We observed a general improvement in classification accuracy when both the preCTX and postCTX spectra are concatenated with one another. This resulted in an LDA model that was able to classify 92% of the patients correctly. This corresponded to the misclassification of only two patients.

Table 2. Leave-one-out cross-validation accuracy for models calculated on reduced spectral feature sets.

	preCTX	postCTX	preCTX-postCTX	preCTX+postCTX
Death				
KNN	0.692	0.692	0.808	0.692
Linear SVM	0.615	0.885	0.538	0.846
Random Forest	0.769	**0.923**	0.577	0.808
LDA	**0.808**	0.538	0.692	**0.923**
GaussianBoosted	0.769	0.769	0.654	0.692
Relapse				
KNN	0.615	0.5	0.769	0.615
Linear SVM	**0.923**	0.808	0.577	0.769
Random Forest	0.692	0.769	0.462	0.808
LDA	0.692	0.654	0.577	0.692
GaussianBoosted	0.615	0.769	0.5	0.654
Necrosis > 90%				
KNN	0.769	0.769	0.769	0.731
Linear SVM	0.769	0.538	**0.846**	0.692
Random Forest	0.654	0.769	0.692	0.654
LDA	0.654	0.808	0.615	0.769
GaussianBoosted	0.577	0.769	0.538	0.538

PreCTX—pre-neoadjuvant chemotherapy; postCTX—post-neoadjuvant chemotherapy; KNN—k-nearest neighbours; SVM—Support Vector Machine; RF—Random Forest; LDA—Linear Discriminant Analysis. Models highlighted in bold are those which show the best performance for each task.

Prediction of patient relapse was more challenging with the postCTX reduced spectral dataset generally being more useful predictively. A Linear SVM classifier trained on the preCTX spectral data set was, however, found to have the highest accuracy for relapse prediction (92%). Again, this corresponded to the misclassification of two patients. Concatenating or subtracting the preCTX and postCTX spectra did not improve classification accuracy as it did for mortality prediction.

Finally, the prediction of patients whose tumours showed greater than 90% necrosis was more difficult. Specifically, KNN and linear SVM models were able to predict patient outcome in only 77% of cases from preCTX spectra alone. Accuracy could be improved, however, to 84.6% by training a linear SVM model on the spectral changes due to CTX treatment. This is consistent with the work of Bergner et al. who found that, when using FTIR for spectral imaging, tumorous and necrotic tissue were difficult to distinguish and could be distinguished with a sensitivity of 75.3%. Overall linear SVM classifiers offered the most consistent performance in prediction accuracy across all classification tasks.

3.6. Survival Analysis

Kaplan Meier curves were used to assess the survival of the patients over three years after the initial biopsy results. There were no non-cancer related deaths during the assessment period. The median follow-up for the analysed patients was 29 months (14–74 months). The three-year progression-free survival rate was 41.36%, and the three-year overall survival (OS) rate was 56.66%. The Kaplan-Meier plots were calculated for progression-free survival and overall survival according to leave-one-out cross-validation prediction (i.e., where the patient is not included in the training set). Figure 4 shows the calculated Kaplan–Meier plots which use the LDA (concatenated preCTX and postCTX spectra) and RF (postCTX spectra only) for mortality prediction and the Linear SVM (using preCTX-postCTX difference spectra) model for relapse prediction. For the longitudinal analysis, hazard ratios (HR) and their 95% confidence intervals (CI) for death were estimated using Cox proportional-hazard ratios. Log-rank tests were used to estimate survival difference based on the model predictions and were calculated using the R package Survival. This analysis indicated that the survival times are indeed different with each predictive model. Specifically, the preCTX+postCTX LDA mortality model has a *p*-value of 0.00023, the postCTX RF mortality model has a *p*-value of 0.0002 and the linear SVM relapse model has a *p*-value of 0.0004. In all except one case, patients who were predicted to die by either the LDA or RF mortality prediction model did so.

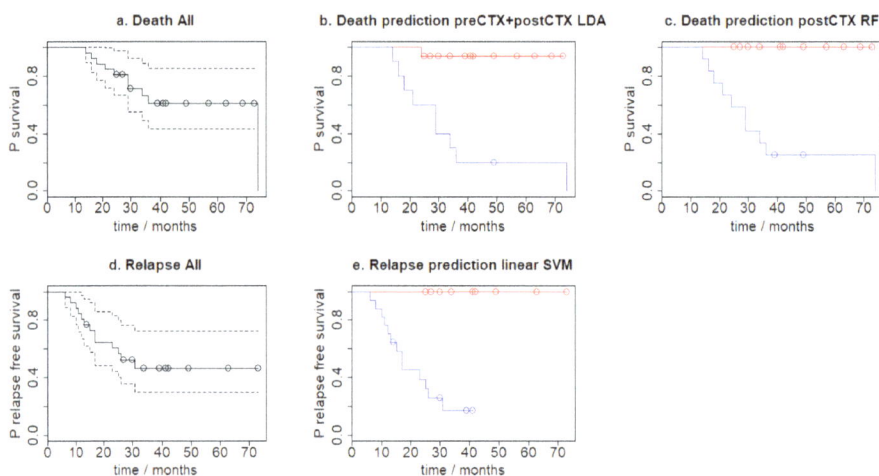

Figure 4. Kaplan–Meier plots for (**a**) overall survival, overall survival according to leave-one-out cross-validation prediction using the (**b**) preCTX+postCTX LDA model (*p* = 0.00023) (**c**) postCTX RF model (*p* = 0.000197), (**d**) progression-free survival and (**e**) according to leave-one-out cross-validation prediction using the linear SVM model (*p* = 0.000387). In (**b**) and (**c**) red indicates those predicted to die, whilst blue are those predicted to live. In (**e**) red indicates those patients who are predicted to relapse, whilst blue are those predicted to not relapse.

4. Discussion

The ability to predict treatment outcome offers significant clinical benefits including the prescription of effective treatments whilst avoiding costly and hazardous regimens that do not benefit the patient. In this retrospective study, we described how machine learning methods applied to the FTIR spectroscopy of bone tissue sections offers a non-destructive, label-free and rapid method for the prediction of treatment outcome in Ewing sarcoma.

Several prognostic factors have been reported previously for ES, such as patient age, tumour size and localisation, stage of disease and degree of tumour necrosis after neoCTX [24,25]. Of these, the extent of necrosis in the resected tumour after neoCTX is a significant predictor of treatment outcome [26]. Indeed, Albergo et al. [27] have suggested that only patients with 100% necrosis after chemotherapy should be classified as having a good response as they have significantly better survival rates compared to those with viable tumour in the surgical specimen.

The effect of CTX on neoplasm depends on many factors including tumour specific biology and the patient's particular pharmacogenomics. [28] The chemical composition of a tumour and its change in response to CTX can be associated with the tumour response to cytostatic drugs [29–31]. FTIR spectra of tissue sections, which reflect this compositional change, may potentially discriminate tumours susceptible to CTX.

By applying machine learning methods to the interpretation of FTIR spectra of ES tissue we were able to predict patient outcomes such as relapse, death and the degree of neoadjuvant CTX response (as a percentage of necrosis) with high accuracy. The risk of death can be predicted with up to 92% accuracy from post-CTX spectra alone using a Random Forest classifier. There were no patient deaths due to toxicity or side effects, with deaths only arising due to tumour progression with chemotherapy resistance. Patient relapse can be predicted with up to 92% accuracy using preCTX treatment spectra alone. High tumour necrosis (>90%) after neoadjuvant CTX may be predicted with 76.9% accuracy (using a KNN or linear SVM model) at the time of diagnosis; the more valuable are spectra from post-CTX specimens and the difference between pre- and post-CTX spectra which achieve an 84.6% accuracy.

We considered whether the deparaffinisation process may have resulted in changes to the lipid composition, which may result in a decrease in classification accuracy. Consequently, we reanalysed the data considering only the absorbances <1800 cm^{-1}. As expected, this resulted in some changes to the classification accuracy of the individual models. Overall, however, average change across all models was less than 1%, suggesting that the inclusion or exclusion of the lipid region does not significantly affect the classification accuracy. The best predictor of patient death using only the absorbances <1800 cm^{-1} is a linear SVM classifier that achieved 96.2% accuracy (corresponding to only one patient misclassified). A linear SVM model was also the best predictor of patient relapse (88.5%) (see Supplementary Table S3). This was a decrease, however, compared to the predictor which had access to features calculated from the full collected spectral width (92.3% accuracy) corresponding to a further misclassified patient.

ES typically presents with systemic metastases, also as invisible foci in standard images (micrometastases). [32] Consequently, curing them all by local measures alone is highly unlikely. Curative therapy for Ewing sarcoma thus requires the combination of effective systemic therapy and local control of all macroscopic tumours [33]. The second aim is achieved by appropriate surgery and irradiation, which is dependent on tumour and macroscopic metastases localisation and their proximity to important structures and organs. Effective systemic therapy of ES is therefore crucial for full recovery except in these cases where a tumour seems to initially form only one visible focus [34].

Our earlier work [21,22] focused on applying statistical approaches to study the prognostic ability of FTIR. These statistical models are based on assumptions drawn from the underlying problem and have concentrated on specific changes to peak intensities or wavelength. If these underlying assumptions are wrong, however, then the predictive power of the model will be poor. In contrast, machine learning models make no assumptions on the problem itself.

In this study, the spectra obtained from Ewing sarcoma samples were reanalysed without searching for any specific differences in tissue chemical. We focused on the potential application of the FTIR spectroscopy as a tool for the rapid and effective selection of patients with good and poor prognosis before starting chemotherapy. This would be a significant step in personalised therapy. Every patient in our cohort was treated with neoadjuvant chemotherapy based on VIDE cycles, so our results should be limited to this scheme. It is also important to note that this study was restricted in sample size due to the low number of ES cases that arise annually and the practical limitation that only patients with full clinical data, where specimens were sampled during biopsy and total surgery, were included. Confirmation of our findings will consequently require a larger sample size and we are actively following this work, including collecting more samples when new cases arise.

Due to the limited size of our study group, the results presented in this paper should be interpreted carefully. We believe, though, that spectral changes are not random and reflect prognostically-meaningful compositional differences in the tissues. While it is not possible for us to determine the potential levels of specificity and sensitivity when using FTIR to discriminate prognostic outcome, our results strongly suggest that it is worthy of further research.

5. Conclusions

The FTIR spectra from ES patients treated initially with VIDE-based CTX were analysed using a variety of machine learning approaches. This analysis demonstrates the possibility that such spectra may predict patient death or relapse with greater than 92% accuracy. Moreover, some of these data can be collected before beginning CTX, offering clinically useful information that might aid patient

Molecules **2019**, 24, 1075

treatment. The significance of these results is limited by the small size of the study group, but we believe that these results point to the potential development of spectroscopic methods for the diagnosis, prognostication and treatment planning of paediatric Ewing Sarcoma.

Supplementary Materials: Figure S1. Dimensionality reduction methods applied to the FTIR dataset. Patients who suffer necrosis are coloured red whilst those who do not are coloured blue. Matrix decomposition methods: (a). PCA (b). Factor analysis (c). Fast Independent Components Analysis (FastICA) (d). Incremental PCA (e). Truncated singular value decomposition (SVD) (f). Kernel PCA using a linear kernel (g). Kernel PCA using a sigmoid kernel (h). Kernel PCA using a polynomial kernel (i). Kernel PCA using a radial basis function kernel and (j). Kernel PCA using a cosine kernel. Manifold Learning methods: (k). Locally linear embedding. (l). Isomap m. Multidimensional scaling (MDS) (m). Spectral embedding and (n). t-distributed stochastic neighbour embedding (t-SNE). Figure S2. Dimensionality reduction methods applied to the FTIR dataset. Throughout, patients who did not relapse are colored blue whilst those who did are coloured red. Matrix decomposition methods: (a). PCA (b). Factor analysis (c). Fast Independent Components Analysis (FastICA) (d). Incremental PCA (e). Truncated singular value decomposition (SVD) (f). Kernel PCA using a linear kernel (g). Kernel PCA using a sigmoid kernel (h). Kernel PCA using a polynomial kernel (i). Kernel PCA using a radial basis function kernel (j). Kernel PCA using a cosine kernel. Manifold Learning methods: (k). Locally linear embedding (l). Isomap m. Multidimensional scaling (MDS) (m). Spectral embedding and (n). t-distributed stochastic neighbour embedding (t-SNE). Supplementary Table S3. The comparison of accuracy for models calculated on all obtained spectral data and truncated (<1800 cm^{-1}) data.

Author Contributions: Conceptualisation, R.C., J.C., methodology, R.C., J.C., writing—review and editing, R.C., C.J.A., A.R., E.M., J.C., supervision, R.C., J.C., data curation, K.Ł., C.J.A., K.B., K.D., E.K., J.D. formal analysis, K.Ł. writing—original draft preparation, C.J.A., A.R., E.M., investigation, K.B., K.D., E.K., resources, K.B., K.D., E.K., J.D.

Funding: This research received no external funding.

Conflicts of Interest: The authors declare no competing interests.

Compliance with Ethical Standards: Competing financial interests: None. The authors of this manuscript declare no relationships with any companies whose products or services may be related to the subject matter of the article. All procedures performed in studies involving human participants were in accordance with the ethical standards of the institutional and/or national research committee and with the 1964 Helsinki declaration and its later amendments or comparable ethical standards.

Abbreviations

FTIR	Fourier Transform Infrared
ES	Ewing sarcoma
SVM	Support Vector Machine
neoCTX	neoadjuvant chemotherapy
preCTX	pre-neoadjuvant chemotherapy
postCTX	post-neoadjuvant chemotherapy
VIDE	vincristine, ifosfamide, doxorubicin, etoposide
FFPE	formalin-fixed paraffin-embedded
ATR	attenuated total reflection
PCA	principal component analysis
poly	polynomial
rbf	radial basis function
RF	Random Forest
LDA	Linear Discriminant Analysis
KNN	k-nearest neighbours
GDM	Gradient Boosted Classifier
CI	confidence interval
HR	hazard ratio

References

1. Gaspar, N.; Rey, A.; Bérard, P.M.; Michon, J.; Gentet, J.C.; Tabone, M.D.; Roché, H.; Defachelles, A.S.; Lejars, O.; Plouvier, E.; et al. Risk adapted chemotherapy for localised Ewing's sarcoma of bone: The French EW93 study. *Eur. J. Cancer* **2012**, *48*, 1376–1385. [CrossRef] [PubMed]

2. Belgaumi, A.F.; Al-Seraihy, A.; Siddiqui, K.S.; Ayas, M.; Bukhari, A.; Al-Musa, A.; Al-Ahmari, A.; El-Solh, H. Outcome of risk adapted therapy for relapsed/refractory acute lymphoblastic leukemia in children. *Leuk. Lymphoma* **2013**, *54*, 547–554. [CrossRef] [PubMed]

3. Ali, A.; Sayed, H.; Farrag, A.; El-Sayed, M. Risk-based combined-modality therapy of pediatric Hodgkin's lymphoma: A retrospective study. *Leuk. Res.* **2010**, *34*, 1447–1452. [CrossRef] [PubMed]

4. Saletta, F.; Seng, M.S.; Lau, L.M.S. Advances in paediatric cancer treatment. *Transl. Pediatr.* **2014**, *3*, 156–182. [PubMed]

5. Gurney, J.G.; Swensen, A.R.; Bulterys, M. Malignant bone tumors. In *Cancer Incidence and Survival among Children and Adolescents: United States SEER Program 1975–1995*; NIH Pub. No. 99–4649; Ries, L.A.G., Smith, M.A., Gumey, J.G., Eds.; National Cancer Institute, SEER Program: Bethesda, MD, USA, 1999.

6. Toretsky, J.A.; Kim, A. Medscape, Ewing Sarcoma Epidemiology. Available online: http://emedicine. medscape.com/article/990378-overview#a6 (accessed on 11 July 2017).

7. Raciborska, A.; Bilska, K.; Drabko, K.; Chaber, R.; Sobol, G.; Pogorzała, M.; Wyrobek, E.; Połczyńska, K.; Rogowska, E.; Rodriguez-Galindo, C.; et al. Validation of a multi-modal treatment protocol for Ewing sarcoma a report from the polish pediatric oncology group. *Pediatr. Blood Cancer* **2014**, *61*, 2170–2174. [CrossRef] [PubMed]

8. Baker, M.J.; Trevisan, J.; Bassan, P.; Bhargava, R.; Butler, H.J.; Dorling, K.M.; Fielden, P.R.; Fogarty, S.W.; Fullwood, N.J.; Heys, K.A.; et al. Using Fourier transform IR spectroscopy to analyze biological materials. *Nature Protoc.* **2014**, *9*, 1771–1791. [CrossRef] [PubMed]

9. Bellisola, G.; Sorio, C. Infrared spectroscopy and microscopy in cancer research and diagnosis. *Am. J. Cancer Res.* **2012**, *2*, 1. [PubMed]

10. Depciuch, J.; Kaznowska, E.; Zawlik, I.; Wojnarowska, R.; Cholewa, M.; Heraud, P.; Cebulski, J. Application of Raman Spectroscopy and Infrared Spectroscopy in the Identification of Breast Cancer. *Appl. Spectrosc.* **2016**, *70*, 251–263. [CrossRef]

11. Zawlik, I.; Kaznowska, E.; Cebulski, J.; Kolodziej, M.; Depciuch, J.; Vongsvivut, J.; Cholewa, M. FPA-FTIR Microspectroscopy for Monitoring Chemotherapy Efficacy in Triple-Negative Breast Cancer. *Sci. Rep.* **2016**, *6*, 37333. [CrossRef] [PubMed]

12. Kaznowska, E.; Depciuch, J.; Łach, K.; Kołodziej, M.; Koziorowska, A.; Vongsvivut, J.; Zawlik, I.; Cholewa, M.; Cebulski, J. The classification of lung cancers and their degree of malignancy by FTIR, PCA-LDA analysis, and a physics-based computational model. *Talanta* **2018**, *186*, 337–345. [CrossRef]

13. Kaznowska, E.; Łach, K.; Depciuch, J.; Chaber, R.; Koziorowska, A.; Slobodian, S.; Kiper, K.; Chlebus, A.; Cebulski, J. Application of infrared spectroscopy for the identification of squamous cell carcinoma (lung cancer). Preliminary study. *Infrared Phys. Technol.* **2018**, *89*, 282–290. [CrossRef]

14. Mehrotra, R.; Tyagi, G.; Jangir, D.K.; Dawar, R.; Gupta, N. Analysis of ovarian tumor pathology by Fourier Transform Infrared Spectroscopy. *J. Ovarian Res.* **2010**, *3*, 27. [CrossRef]

15. Hands, J.R.; Clemens, G.; Stables, R.; Ashton, K.; Brodbelt, A.; Davis, C.; Dawson, T.P.; Jenkinson, M.D.; Lea, R.W.; Walker, C.; et al. Brain tumour differentiation: Rapid stratified serum diagnostics via attenuated total reflection Fourier-transform infrared spectroscopy. *J. Neurooncol.* **2016**, *127*, 463–472. [CrossRef]

16. Lyng, F.M.; Faoláin, E.O.; Conroy, J.; Meade, A.D.; Knief, P.; Duffy, B.; Hunter, M.B.; Byrne, J.M.; Kelehan, P.; Byrne, H.J. Vibrational spectroscopy for cervical cancer pathology, from biochemical analysis to diagnostic tool. *Exp. Mol. Pathol.* **2007**, *82*, 121–129. [CrossRef]

17. Liu, H.; Su, Q.; Sheng, D.; Zheng, W.; Wang, X. Comparison of red blood cells from gastric cancer patients and healthy persons using FTIR spectroscopy. *J. Mol. Struct.* **2017**, *1130*, 33–37. [CrossRef]

18. Khanmohammadi, M.; Garmarudi, A.B.; Ghasemi, K.; Jaliseh, H.K.; Kaviani, A. Diagnosis of colon cancer by attenuated total reflectance-fourier transform infrared microspectroscopy and soft independent modeling of class analogy. *Med. Oncol.* **2009**, *26*, 292–297. [CrossRef]

19. Baker, M.J.; Gazi, E.; Brown, M.D.; Shanks, J.H.; Gardner, P.; Clarke, N.W. FTIR-based spectroscopic analysis in the identification of clinically aggressive prostate cancer. *Br. J. Cancer* **2008**, *99*, 1859–1866. [CrossRef]

20. Wald, N.; Le Corre, Y.; Martin, L.; Mathieu, V.; Goormaghtigh, E. Infrared spectra of primary melanomas can predict response to chemotherapy: The example of dacarbazine. *Biochim. Biophys. Acta* **2016**, *1862*, 174–181. [CrossRef]

21. Chaber, R.; Łach, K.; Szmuc, K.; Michalak, E.; Raciborska, A.; Mazur, D.; Machaczka, M.; Cebulski, J. Application of infrared spectroscopy in the identification of Ewing sarcoma: A preliminary report. *Infrared Phys. Technol.* **2017**, *83*, 200–205. [CrossRef]

22. Chaber, R.; Łach, K.; Arthur, C.J.; Raciborska, A.; Michalak, E.; Ciebiera, K.; Bilska, K.; Drabko, K.; Cebulski, J. Prediction of Ewing Sarcoma treatment outcome using attenuated tissue reflection FTIR tissue spectroscopy. *Sci. Rep.* **2018**, *8*, 12299. [CrossRef]

23. Schoelkopf, B.; Smola, A.J.; Mueller, K.-R. Kernel principal component analysis. In *Advances in Kernel Methods*; MIT Press: Cambridge, MA, USA, 1999; pp. 327–352.

24. Rodríguez-Galindo, C.; Navid, F.; Liu, T.; Billups, C.A.; Rao, B.N.; Krasin, M.J. Prognostic factors for local and distant control in Ewing sarcoma family of tumors. *Ann. Oncol.* **2008**, *19*, 814–820. [CrossRef] [PubMed]

25. Rodríguez-Galindo, C.; Liu, T.; Krasin, M.J.; Wu, J.; Billups, C.A.; Daw, N.C.; Spunt, S.L.; Rao, B.N.; Santana, V.M.; Navid, F. Analysis of prognostic factors in Ewing sarcoma family of tumors: Review of St. Jude Children's Research Hospital studies. *Cancer* **2007**, *110*, 375–384. [CrossRef] [PubMed]

26. Picci, P.; Böhling, T.; Bacci, G.; Ferrari, S.; Sangiorgi, L.; Mercuri, M.; Ruggieri, P.; Manfrini, M.; Ferraro, A.; Casadei, R.; et al. Chemotherapy-induced tumor necrosis as a prognostic factor in localized Ewing's sarcoma of the extremities. *J. Clin. Oncol.* **1997**, *15*, 1553–1559. [CrossRef]

27. Albergo, J.I.; Gaston, C.L.; Laitinen, M.; Darbyshire, A.; Jeys, L.M.; Sumathi, V.; Parry, M.; Peake, D.; Carter, S.R.; Tillman, R.; et al. Ewing's sarcoma only patients with 100% of necrosis after chemotherapy should be classified as having a good response. *Bone Jt. J.* **2016**, *98-B*, 1138–1144. [CrossRef] [PubMed]

28. Hattinger, C.M.; Vella, S.; Tavanti, E.; Fanelli, M.; Picci, P.; Serra, M. Pharmacogenomics of second-line drugs used for treatment of unresponsive or relapsed osteosarcoma patients. *Pharmacogenomics* **2016**, *17*, 2097–2114. [CrossRef]

29. Zambelli, D.; Zuntini, M.; Nardi, F.; Manara, M.C.; Serra, M.; Landuzzi, L.; Lollini, P.; Ferrari, S.; Alberghini, M.; Llombart-Bosch, A.; et al. Biological indicators of prognosis in Ewing's sarcoma: An emerging role for lectin galactoside-binding soluble 3 binding protein (LGALS3BP). *Int. J. Cancer* **2010**, *126*, 41–52. [CrossRef]

30. Kikuta, K.; Tochigi, N.; Shimoda, T.; Yabe, H.; Morioka, H.; Toyama, Y.; Hosono, A.; Beppu, Y.; Kawai, A.; Hirohashi, S.; et al. Nucleophosmin as a Candidate Prognostic Biomarker of Ewing's Sarcoma Revealed by Proteomics. *Clin. Cancer Res.* **2009**, *15*, 2885–2894. [CrossRef]

31. Shukla, N.; Schiffman, J.; Reed, D.; Davis, I.J.; Womer, R.B.; Lessnick, S.L.; Lawlor, E.R. COG Ewing Sarcoma Biology Committee, Biomarkers in Ewing Sarcoma: The Promise and Challenge of Personalized Medicine. A Report from the Children's Oncology Group. *Front. Oncol.* **2013**, *3*, 141. [CrossRef]

32. Vo, K.T.; Edwards, J.V.; Epling, C.L.; Sinclair, E.; Hawkins, D.S.; Grier, H.E.; Janeway, K.A.; Barnette, P.; McIlvaine, E.; Krailo, M.D.; et al. Impact of Two Measures of Micrometastatic Disease on Clinical Outcomes in Patients with Newly Diagnosed Ewing Sarcoma: A Report from the Children's Oncology Group. *Clin. Cancer Res.* **2016**, *22*, 3643–3650. [CrossRef]

33. Meyers, P.A. Systemic therapy for osteosarcoma and Ewing sarcoma. *Am. Soc. Clin. Oncol. Educ. Book* **2015**, e644–e647. [CrossRef]

34. Gaspar, N.; Hawkins, D.S.; Dirksen, U.; Lewis, I.J.; Ferrari, S.; Le Deley, M.C.; Kovar, H.; Grimer, R.; Whelan, J.; Claude, L.; et al. Ewing Sarcoma: Current Management and Future Approaches Through Collaboration. *J. Clin. Oncol.* **2015**, *33*, 3036–3046. [CrossRef] [PubMed]

Sample Availability: The histological samples of tumours are partially available from the authors.

molecules

MDPI

Article

Detection of the BRAF V600E Mutation in Colorectal Cancer by NIR Spectroscopy in Conjunction with Counter Propagation Artificial Neural Network

Xue Zhang [1], Yang Yang [1], Yalan Wang [2] and Qi Fan [1,*]

[1]   School of Pharmacy, Chongqing Medical University, Chongqing 400016, China;
     zhangxue2017@139.com (X.Z.); yyang139@139.com (Y.Y.)
[2]   Department of Pathology, Molecular Medicine and Cancer Research Center, Chongqing Medical University,
     Chongqing 400016, China; wangyalan074@126.com
[*]   Correspondence: fanqi787@cqmu.edu.cn; Tel.: +86-236-848-5161

Academic Editors: Christian Huck and Krzysztof B. Bec
Received: 22 May 2019; Accepted: 13 June 2019; Published: 15 June 2019

Abstract: This paper proposes a sensitive, sample preparation-free, rapid, and low-cost method for the detection of the B-rapidly accelerated fibrosarcoma (BRAF) gene mutation involving a substitution of valine to glutamic acid at codon 600 (V600E) in colorectal cancer (CRC) by near-infrared (NIR) spectroscopy in conjunction with counter propagation artificial neural network (CP-ANN). The NIR spectral data from 104 paraffin-embedded CRC tissue samples consisting of an equal number of the BRAF V600E mutant and wild-type ones calibrated and validated the CP-ANN model. As a result, the CP-ANN model had the classification accuracy of calibration (CAC) 98.0%, cross-validation (CACV) 95.0% and validation (CAV) 94.4%. When used to detect the BRAF V600E mutation in CRC, the model showed a diagnostic sensitivity of 100.0%, a diagnostic specificity of 87.5%, and a diagnostic accuracy of 93.8%. Moreover, this method was proven to distinguish the BRAF V600E mutant from the wild type based on intrinsic differences by using a total of 312 CRC tissue samples paraffin-embedded, deparaffinized, and stained. The novel method can be used for the auxiliary diagnosis of the BRAF V600E mutation in CRC. This work can expand the application of NIR spectroscopy in the auxiliary diagnosis of gene mutation in human cancer.

Keywords: near-infrared spectroscopy; counter propagation artificial neural network; detection; auxiliary diagnosis; BRAF V600E mutation; colorectal cancer; tissue; paraffin-embedded; deparaffinized; stained

1. Introduction

Colorectal cancer (CRC) is one of the human malignant tumors with high incidence and mortality rates [1]. In particular, the mutations in CRC often make the treatment more difficult [2–4]. One of the most common mutations in CRC is the B-rapidly accelerated fibrosarcoma (BRAF) gene mutation, which involves a substitution of valine to glutamic acid at codon 600 (V600E) [5]. Figure 1 gives the structural formulas for valine and glutamic acid. The BRAF V600E mutation in CRC significantly reduces the efficacy of the drugs that are used in the treatment of patients with BRAF V600E wild type in CRC. The drug treatment regimen for patients with BRAF V600E mutant in CRC needs to be redesigned [6,7]. Therefore, it is crucial to detect the BRAF V600E mutation for the targeted therapy in CRC.

Figure 1. The structural formulas for valine (**a**) and glutamic acid (**b**).

The typical methods for the clinical diagnosis of the BRAF V600E mutation in CRC are immunohistochemistry (IHC) in conjunction with microscopy [8], polymerase chain reaction (PCR) [9], and gene sequencing [10]. However, in IHC, the staining for target molecules that are associated with the BRAF V600E mutation is a multistep process. This process is frequently disturbed by many factors, resulting in staining failures, such as all negatives, all positives, too dark background, the positive control stained well but positive samples unstained or heterogeneous. Moreover, the diagnostic accuracy of microscopy is limited by the experience of pathologists. On the other hand, both PCR and gene sequencing are at least time-consuming and high-cost. Consequently, it is imperative to establish a sensitive, sample preparation-free, rapid, and low-cost method for the auxiliary diagnosis of the BRAF V600E mutation in CRC.

Near-infrared (NIR) spectroscopy can be used to characterize the properties of an analyte containing the X-H groups (X = C, N, O, S). Typically, the vibration of one X-H group absorbs NIR light at several overtone frequencies, while the absorption intensity at a certain NIR frequency is the sum of the absorption intensities of a plurality of X-H groups. That is, the NIR absorption bands are seriously overlapping, so that NIR spectra are not directly interpreted and utilized. Thence, it is necessary to extract the information on the analytes from the NIR data for the sample by chemometric techniques [11–13]. NIR spectroscopy, assisted by chemometric techniques, is used to discriminate cancer from benign tumor, such as breast cancer [14], endometrial cancer [15], gastric cancer [16], and colorectal cancer [17], because it is easy-to-use, robust, inherently rapid (measuring a NIR spectrum in seconds), as well as nondestructive and low-cost [18,19].

Therefore, in this work, the feasibility of sensitive, sample preparation-free, rapid, and low-cost detection of the BRAF V600E mutation in CRC was explored with NIR spectroscopy and counter propagation artificial neural network (CP-ANN). The specific objectives are: (1) distinguishing the BRAF V600E mutant from the wild type by a CP-ANN model; (2) exploring the mechanism for NIR detection of the BRAF V600E mutation in CRC. This work can expand the application of NIR spectroscopy in the auxiliary diagnosis of gene mutation in human cancer.

2. Results and Discussion

2.1. Samples

Table 1 lists 312 CRC tissue samples. Therein, the paraffin-embedded (Class 1) CRC sample is the most suitable for auxiliary diagnosis, because it is the most common form of pathological specimen storage. That is, the preparation of Class 1 samples is free. This means that the method of using Class 1 samples is top-priority, rapid, reagent-free, and nondestructive.

The deparaffinized (Class 2) and stained (Class 3) samples were used to explore the mechanism for NIR detection of the BRAF V600E mutation in CRC. However, the preparations of the Class 2 and Class 3 samples are both cumbersome and time-consuming. In addition, the samples of the combination of Class 2 with Class 1 samples (1:1) were named Class 2&1 samples. The Class 2&3 samples were named as similar to the Class 2&1 samples. Both Class 2&1 and Class 2&3 samples were also used to explore the mechanism for NIR detection of the BRAF V600E mutation in CRC.

Table 1. The numbers of models calibrated and validated using 312 colorectal cancer (CRC) tissue samples.

Model Number	Class of Samples	Number of Calibration Samples		Number of Validation Samples	
		Mutant	Wild-type	Mutant	Wild-type
1	Class 1	40	40	12	12
2	Class 2	40	40	12	12
3	Class 3	40	40	12	12
4	Class 2&1	20&20	20&20	NA	NA
5	Class 2&3	20&20	20&20	NA	NA

Note: NA for not available.

The samples in each class consisted of an equal number of the BRAF V600E mutant and wild-type samples. The models calibrated while using an equal number of the BRAF V600E mutant and wild-type samples did not have classification biases that were caused by unequal numbers of samples in two subgroups. The number of validation samples was 30% of the number of calibration samples.

2.2. Spectral Acquisition

The NIR spectra of 312 CRC tissue samples were acquired while using the following means. The transflectance spectra, rather than transmission spectra, for the thin tissue samples were measured to increase the detection sensitivity. The sample signal intensity in the transflectance spectrum is twice that in the transmission spectrum, since the transflectance optical pathlength is twice the transmission one. Each sample was measured at three tissue locations, as the mutation may occur unevenly. The mutant and wild-type samples were alternately measured to avoid systematic errors that are caused by sequential measurement. Both 8 cm^{-1} resolution and 64 co-added scans were selected to obtain a spectrum with sufficient sample information and low noise in about 31.39 s.

Figure 2 shows the mean NIR transflectance spectra for the mutant and wild-type samples of Class 1, Class 2, and Class 3. Red, light red, and dark red represent the mutant samples of Class 1, Class 2, and Class 3, respectively. Blue, light blue, and dark blue represent the wild-type samples of Class 1, Class 2, and Class 3, respectively.

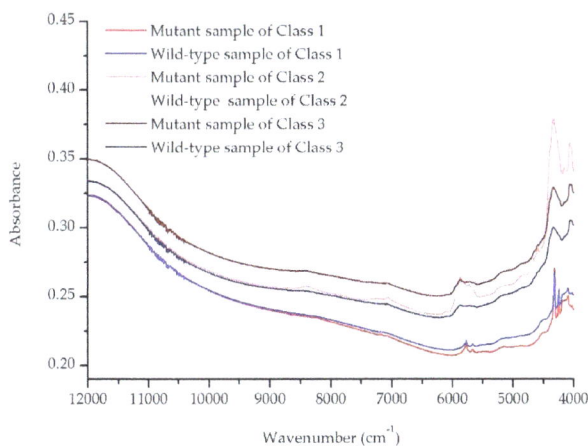

Figure 2. Mean near-infrared (NIR) transflectance spectra for colorectal cancer (CRC) tissue sections. Red, light red, and dark red represent, respectively, the mutant samples of Class 1, Class 2, and Class 3. Blue, light blue, and dark blue represent, respectively, the wild-type samples of Class 1, Class 2, and Class 3.

2.3. Data Processing

2.3.1. Selection of the Spectral Preprocessing Strategy

Table 2 lists the vital preprocessing strategies, spectral subranges, numbers of PCs, numbers of neurons on each side, and corresponding model performances of the CP-ANN models built while using NIR data for the samples. The models, from Model 1 to Model 1.12, were built while using the same Class 1 samples, but changing preprocessing strategy, spectral subrange, number of PCs, and/or number of neurons on each side. Other models are similar to the above. As can be seen from Table 2, the models that were built using only mean centering (MC) have better model performances than those using other preprocessing strategies, respectively, for the models that were built using Class 1, Class 2, and Class 3 samples.

Table 2. Vital preprocessing strategies, spectral subranges, numbers of principal components (PCs), numbers of neurons on each side, and corresponding model performances of the counter propagation artificial neural network (CP-ANN) models built respectively using NIR data for Class 1, Class 2, Class 3, Class 2&1, and Class 2&3 samples.

Model Number	Preprocessing	Spectral Subrange (cm⁻¹)	Number of PCs/ Cumulative Variance Contribution Rate (%)	Number of Neurons on Each Side	Model Performances		
					CAC (%)	CACV (%)	CAV (%)
1	MC	9000–6800, 6500–4000	6/100.0	12	98.0	95.0	94.4
1.1	MSC + MC	9000–6800, 6500–4000	6/99.9	12	97.0	93.0	90.3
1.2	SNV + MC	9000–6800, 6500–4000	6/99.9	12	97.0	94.0	81.9
1.3	FD + MC	9000–6800, 6500–4000	6/98.8	12	93.0	86.0	88.9
1.4	SD + MC	9000–6800, 6500–4000	6/95.7	12	89.0	71.0	73.6
1.5	SGS + MC	9000–6800, 6500–4000	6/100.0	12	98.0	94.0	90.3
1.6	SGS + FD + MC	9000–6800, 6500–4000	6/99.1	12	94.0	88.0	90.3
1.7	NDS + FD + MC	9000–6800, 6500–4000	3/100.0	12	92.0	85.0	87.5
1.8	MSC + SD + MC	9000–6800, 6500–4000	6/ 96.0	12	90.0	74.0	77.8
1.9	SNV + NDS + FD + MC	9000–6800, 6500–4000	6/100.0	12	95.0	88.0	90.3
1.10	MC	9000–4000	6/100.0	12	98.0	94.0	91.7
1.11	MC	9000–6800, 6500–4000	6/100.0	10	97.0	94.0	88.9
1.12	MC	9000–6800, 6500–4000	6/100.0	15	98.0	96.0	88.9
2	MC	9000–6800, 6500–4000	6/100.0	12	97.0	92.0	94.4
2.1	MSC + MC	9000–6800, 6500–4000	6/100.0	12	94.0	85.0	79.2
2.2	SNV + MC	9000–6800, 6500–4000	6/ 99.9	12	89.0	83.0	83.3
2.3	FD + MC	9000–6800, 6500–4000	6/97.2	12	90.0	82.0	86.1
2.4	SD + MC	9000–6800, 6500–4000	20/84.6	12	NA	NA	NA
2.5	SGS + MC	9000–6800, 6500–4000	6/100.0	12	96.0	94.0	90.3
2.6	SGS + FD + MC	9000–6800, 6500–4000	6/97.8	12	92.0	88.0	81.9
2.7	NDS + FD + MC	9000–6800, 6500–4000	2/100.0	12	88.0	80.0	79.2
2.8	MSC + SD + MC	9000–6800, 6500–4000	20/80.6	12	NA	NA	NA
2.9	SNV + NDS + FD + MC	9000–6800, 6500–4000	3/100.0	12	90.0	85.0	87.5
2.10	MC	9000–4000	6/100.0	12	96.0	91.0	93.1
2.11	MC	9000–6800, 6500–4000	6/100.0	10	96.0	90.0	87.5
2.12	MC	9000–6800, 6500–4000	6/100.0	15	97.0	92.0	94.4
3	MC	9000–6800, 6500–4000	5/100.0	12	95.0	88.0	93.1
3.1	MSC + MC	9000–6800, 6500–4000	5/99.9	12	86.0	71.0	66.7
3.2	SNV + MC	9000–6800, 6500–4000	5/99.9	12	85.0	72.0	68.1
3.3	FD + MC	9000–6800, 6500–4000	13/85.5	12	90.0	77.0	79.2
3.4	SD + MC	9000–6800, 6500–4000	20/75.0	12	NA	NA	NA
3.5	SGS + MC	9000–6800, 6500–4000	5/100.0	12	93.0	89.0	90.3
3.6	SGS + FD + MC	9000–6800, 6500–4000	10/ 85.6	12	88.0	79.0	76.4
3.7	NDS + FD + MC	9000–6800, 6500–4000	2/100.0	12	90.0	82.0	77.8
3.8	MSC + SD + MC	9000–6800, 6500–4000	20/74.5	12	NA	NA	NA
3.9	SNV + NDS + FD + MC	9000–6800, 6500–4000	4/100.0	12	87.0	65.0	72.2
3.10	MC	9000–4000	5/100.0	12	95.0	88.0	87.5
3.11	MC	9000–6800, 6500–4000	5/100.0	10	93.0	89.0	86.1
3.12	MC	9000–6800, 6500–4000	5/100.0	15	95.0	89.0	88.9
4	MC	9000–6800, 6500–4000	5/100.0	12	97.0	97.0	NA
5	MC	9000–6800, 6500–4000	6/100.0	12	95.0	90.0	NA

Notes: MC for mean centering; MSC for multiplicative scatter correction; SNV for standard normal variate; FD for first derivative; SD for second derivative; SGS for Savitzky-Golay smoothing; NDS for Norris derivative smoothing; PC for principal component; CAC, CACV and CAV respectively for the classification accuracy of calibration, cross-validation and validation; NA for not available.

2.3.2. Selection of the Spectral Subrange for Modeling

Figure 3 indicates the differences between the mean spectra for the mutant and wild-type samples. The full, long dashed, and short dashed lines represent Class 1, Class 2, and Class 3 samples, respectively. On the full, long dashed, and short dashed lines, we can see significant changes in the two subranges 9000–6800 cm^{-1} and 6500–4000 cm^{-1}.

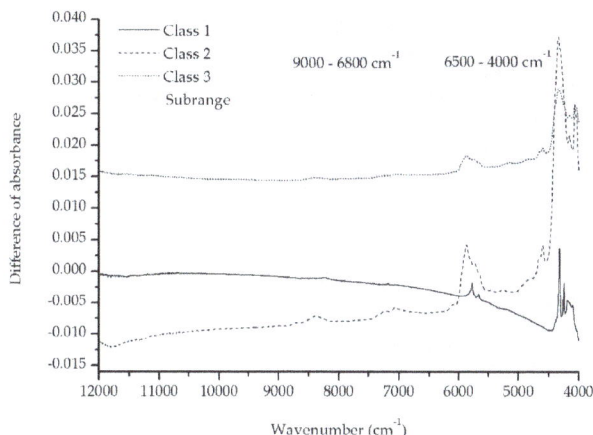

Figure 3. The differences between the mean spectra for the mutant and wild-type samples. The full, long dashed, and short dashed lines represent respectively Class 1, Class 2, and Class 3 samples.

The differences between the mutant and wild-type samples, in fact, are caused by the substitution of valine to glutamic acid. Figure 1 indicates that the largest structural difference between valine and glutamic acid is the difference between $(CH_3)_2CH$- in valine and $-(CH_2)_2COOH$ in glutamic acid. Consequently, the spectral subranges 9000–6800 cm^{-1} and 6500–4000 cm^{-1} can be mainly attributed to the following overtones: the second overtones of CH_3 and CH_2 near 8696–8264 cm^{-1}, CH near 8163 cm^{-1}; the first overtones of CH_3 near 5905 and 5872 cm^{-1}, CH_2 near 5680 cm^{-1}, CH near 5882–5555 cm^{-1}; the combination bands of CH_3 near 7355, 7263, 4545–4500 and 4395 cm^{-1}, CH_2 near 7186 and 7080 cm^{-1}, CH near 6944 cm^{-1}; the combination bands of O-H in COOH near 4500–4000 cm^{-1} [20].

Table 2 shows the models that were built while using various spectral subranges. Model 1, Model 2, and Model 3 were built while using two spectral subranges 9000–6800 cm^{-1} and 6500–4000 cm^{-1}. Model 1.10, Model 2.10, and Model 3.10 were built while only using one spectral subrange 9000–4000 cm^{-1}. The two spectral subranges 9000–6800 cm^{-1} and 6500–4000 cm^{-1} were selected to build the detection model since Model 1, Model 2, and Model 3 had better model performances separately than Model 1.10, Model 2.10, and Model 3.10.

2.3.3. Calibration and Validation of the CP-ANN Model

Principal component analysis (PCA) was used to reduce the redundant dimensionalities of the spectral data for the samples. The scores of the principal components (PCs, cumulative variance contribution rate exceeding 85.0%), as selected from both 9000–6800 cm^{-1} and 6500–4000 cm^{-1}, were used as the inputs to the CP-ANN model. CP-ANN has the advantages of artificial neural network (ANN), such as nonlinearity, self-learning, self-organization, and self-adaptation [21]. Table 2 shows that the optimal structure of the CP-ANN model is 12 × 12, because the performances of the 12 × 12 model are better than the 10 × 10 one and nearly equal to the 15 × 15 one.

In Table 2, Model 1, Model 2, and Model 3 are optimal, respectively, for Class 1, Class 2, and Class 3 samples, because of the highest classification accuracies of calibration (CAC) and validation

(CAV). Furthermore, Model 1, Model 2, and Model 3 have successively the best, medium, and worst classification accuracies. Figure 4a–c illustrate that the mutant and wild-type samples are assigned to the gray and white regions, respectively, by Model 1, Model 2, and Model 3, not only in the calibration (uppercase letter), but also in the validation (lowercase letter), although a few samples that are near the boundary are not correctly assigned.

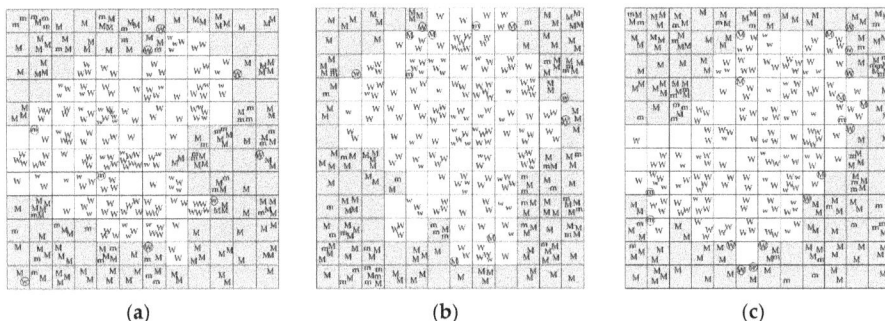

(a)   (b)   (c)

Figure 4. Projection maps for the 12 × 12 CP-ANN models: (**a**) Model 1; (**b**) Model 2; and (**c**) Model 3. The uppercase letter "M" and the lowercase letter "m" for the mutant samples, respectively, in calibration and validation; "W" and "w" for the wild-type samples, respectively, in calibration and validation; "○" for the samples assigned incorrectly; the gray region for mutant; the white region for wild type.

2.3.4. Diagnostic Performances of the CP-ANN Model

Table 3 gives the diagnostic performances of five CP-ANN models that were sequentially built using an equal number of Class 1, Class 2, Class 3, Class 2&1, and Class 2&3 samples.

Table 3. The diagnostic performances of five CP-ANN models built sequentially using an equal number of Class 1, Class 2, Class 3, Class 2&1, and Class 2&3 samples.

Model Number	Diagnostic Performances		
	Sensitivity (%)	Specificity (%)	Accuracy (%)
1	100.0	87.5	93.8
2	100.0	95.0	97.5
3	100.0	82.5	91.3
4	100.0	92.5	96.3
5	100.0	85.0	92.5

As can be seen from Table 3, each model shows a sensitivity of 100.0%. It can be inferred that the sample information in the acquired NIR transflectance spectra is sufficient for detecting the BRAF V600E mutation in CRC. That is, the structural differences between valine and glutamic acid on C-H, N-H, and O-H groups were characterized by NIR spectroscopy. In particular, a sensitivity of 100.0% is critical for auxiliary diagnosis, because it avoids missing the mutant.

In Table 3, Model 1, Model 2, and Model 3 have, respectively, medium, the best, and the worst specificities and accuracies. The probable cause is that the NIR spectra for the Class 1 samples are disturbed by the NIR absorption of paraffin; the NIR spectra for the Class 3 samples are disturbed by the NIR absorption of hematoxylin and eosin (HE). Moreover, the interference from paraffin is weaker than HE. However, the NIR spectra for the Class 2 samples are not disturbed by the NIR absorption of paraffin or HE. These inferences are supported by the following evidences. Model 2 is superior to Model 4 (built using Class 2&1 samples) and Model 5 (built using Class 2&3 samples) regarding the specificity and the accuracy; Model 4 and Model 5 are separately superior to Model 1 and Model 3. In addition, Model 4 is superior to Model 5.

On the other hand, HE is used to increase the color difference between the cancer and non-cancer tissues in pathological diagnosis. it is demonstrated that HE increases the absorbance difference between the mutant and wild-type samples since the color on the stained mutant tissue is darker than the color on the stained wild-type tissue, as shown in Figure 3. However, Model 3 (built using the HE-stained samples) has the worst diagnostic performances. A possible explanation is that HE interferes with the NIR detection and it does not increase the fundamental difference between the mutant and wild-type samples, that is, between valine and glutamic acid.

There are two kinds of differences in the calibration samples used in Model 4, as shown in Table 1. The first is the difference between the mutant and wild-type samples, i.e., the difference between valine and glutamic acid. The second is the difference between the deparaffinized and paraffin-embedded samples, i.e., the difference between no paraffin and paraffin. In fact, Model 4 distinguishes 80 calibration samples into two subgroups that are based on the difference between the mutant and wild-type, rather than between deparaffinized and paraffin-embedded. In other words, Model 4 detects the BRAF V600E mutation in CRC based on the difference between valine and glutamic acid in the deparaffinized and paraffin-embedded samples, rather than between no paraffin and paraffin in the mutant and wild-type samples. Similar results are obtained using the calibration samples in Model 5.

These findings suggest that the CP-ANN models built by the NIR data can detect the BRAF V600E mutation in CRC based directly on the fundamental difference between mutant and wild type, i.e., the difference between valine and glutamic acid, rather than among paraffin, HE, and nothing.

3. Materials and Methods

3.1. Samples

312 CRC tissue sections of BRAF V600E mutant or wild type and their reference information were obtained from the Department of Clinical Pathology and the Molecular Medical Testing Center at Chongqing Medical University. The Ethics Committee of our university approved the collection and use of these specimens for current research. Informed consent was obtained from these patients.

These CRC tissue samples include three classes, as shown in Table 1. Class 1 is the paraffin-embedded sample on a glass slide, which is the most common form of pathological specimen storage; Class 2 is the deparaffinized sample between a glass slide and a coverslip; Class 3 is the HE-stained sample between a glass slide and a coverslip. Each class consisted of an equal number of the BRAF V600E mutant and wild-type samples.

The reference information on the BRAF V600E mutation in the CRC tissue sample was detected by real-time fluorescent quantitative PCR (RT-qPCR). The detection was performed on a Roche LightCycler 480 II system (Roche, Basel, Switzerland) while using the Human BRAF Gene V600E Mutation Detection Kit (Wuhan YZY, China). The detection involved not only the PCR reaction, but also the PCR reaction for quality control (QC).

3.2. Instrument and Spectral Acquisition

The NIR spectra were measured while using a Nicolet iS50 FT-IR analyzer (Thermo Fisher Scientific, Waltham, MA, USA) that was equipped with an indium gallium arsenide detector and an integrating sphere. The instrument was controlled by OMNIC 9.2 software (Thermo Fisher Scientific, Waltham, MA, USA).

A sample (glass slide up) was placed on the detection window of the integrating sphere and was covered by a lid with a gold inner top. The transflectance spectra for the samples were measured in the range of $12,000-4000$ cm^{-1} while using the selected resolution and the selected number of co-added scans. The resolution was selected from 2, 4, 8, and 16 cm^{-1} to obtain sufficient sample information in a shorter time; the number of co-added scans was selected from 16, 32, 64, and 128 to reduce the noise in a shorter time. Each sample was measured at three tissue locations. The mutant and wild-type samples were alternatively measured. The background spectrum was measured, prior to the sample spectra, under the same conditions to eliminate any ambient interferences on the sample spectra.

3.3. Data Processing

In the detection, the sample spectra were preprocessed by a preprocessing strategy that was selected from MC for subtracting the calculated mean of a variable from the spectral data, multiplicative scatter correction (MSC) or standard normal variate (SNV) for eliminating the interferences from granularity and compactness, derivative for deducting the background and separating overlapping signals, smoothing for denoising, and the combinations of various preprocessing techniques, as shown in Table 2.

The spectral subrange for modeling was selected from 12,000–4000 cm^{-1} based primarily on the differences in characteristic absorptions between the mutant and wild-type samples.

Subsequently, the CP-ANN model was calibrated using the reference value of the calibration sample and the scores of the selected PCs in the spectral subrange of spectral data for the calibration sample. The three spectra per sample were all used in modeling. As shown in Table 1, Model 1, Model 2, Model 3, Model 4, and Model 5 were sequentially calibrated by 40 mutant samples and 40 wild-type samples of Class 1, Class 2, Class 3, Class 2&1, and Class 2&3. Next, Model 1, Model 2, and Model 3 were sequentially validated by 12 mutant samples and 12 wild-type samples of Class 1, Class 2, and Class 3. The structure of the CP-ANN model was selected from 10 × 10, 12 × 12, and 15 × 15 based on the model performances CAC, classification accuracy of cross-validation (CACV), and CAV.

The diagnostic performances of the CP-ANN model were evaluated with sensitivity, specificity, and accuracy. Sensitivity is defined as the ratio of TP/(TP + FN), where TP and FN are, respectively, the number of true positive (mutant) and false negative diagnostic results; specificity the ratio of TN/(TN + FP), TN and FP the number of true negative (wild type) and false positive diagnostic results; accuracy the ratio of (TP + TN)/(TP + FP + TN + FN). In the calculation of sensitivity, specificity, and accuracy, the final diagnostic result for the sample was calculated as a wild-type sample when the three prediction results for three spectra per sample were all wild type; otherwise, as a mutant sample. In other words, the final diagnostic result for the sample was calculated as a mutant sample when at least one of three prediction results for three spectra per sample was mutant.

TQ Analyst 8.0 software (Thermo Fisher Scientific, Waltham, MA, USA) was used for spectral preprocessing, selection of the spectral subrange for modeling, and PCA. Matlab 8.0 software (The Math Works, Natick, MA, USA) was used for the calibration and validation of the CP-ANN model.

4. Conclusions

The NIR strategy on the basis of the principle different from the clinical diagnostic methods can be used for the auxiliary diagnosis of the BRAF V600E mutation in CRC. The NIR detection is directly based on the molecular differences between the BRAF V600E mutant and wild type, so that it is undisturbed by the factors affecting sample staining in IHC. When compared to the time-consuming and high-cost PCR and gene sequencing, the NIR detection is sensitive, sample preparation-free, inherently rapid, and low-cost. This research expanded the application of NIR spectroscopy in the auxiliary diagnosis of gene mutation in human cancer. In addition, when combined with our previous work, i.e., the NIR spectroscopy for the auxiliary diagnosis of CRC while using the paraffin-embedded samples [22], it is expected to simultaneously diagnose CRC and the BRAF V600E mutation using the NIR spectra for colorectal tissue.

Author Contributions: Conceptualization, X.Z., Y.W. and Q.F.; Investigation, X.Z. and Y.Y.; Methodology, X.Z.; Project administration, Q.F.; Resources, Y.W. and Q.F.; Supervision, Y.W. and Q.F.; Validation, X.Z. and Y.Y.; Visualization, X.Z. and Q.F.; Writing—original draft, X.Z.; Writing—review & editing, X.Z., Y.Y. and Q.F.

Funding: This research received no external funding.

Acknowledgments: We are grateful for the support of samples and their reference information from the Department of Clinical Pathology and the Molecular Medical Testing Center at Chongqing Medical University.

Conflicts of Interest: The authors declare no conflict of interest.

References

1. Siegel, R.L.; Miller, K.D.; Jemal, A. Cancer statistics, 2018. *CA Cancer J. Clin.* **2018**, *68*, 7–30. [CrossRef] [PubMed]

2. Bouchahda, M.; Karaboué, A.; Saffroy, R.; Innominato, P.; Gorden, L.; Guettier, C.; Adam, R.; Lévi, F. Acquired KRAS mutations during progression of colorectal cancer metastases: Possible implications for therapy and prognosis. *Cancer Chemother. Pharmacol.* **2010**, *66*, 605–609. [CrossRef] [PubMed]

3. Bahrami, A.; Hesari, A.R.; Khazaei, M.; Hassanian, S.M.; Ferns, G.; Avan, A. The therapeutic potential of targeting the BRAF in patients with colorectal cancer. *J. Cell Physiol.* **2017**, *9999*, 1–8. [CrossRef] [PubMed]

4. Wang, Q.; Shi, Y.L.; Zhou, K.; Wang, L.L.; Yan, Z.X.; Liu, Y.L.; Xu, L.L.; Zhao, S.W.; Chu, H.L.; Shi, T.T.; et al. PIK3CA mutations confer resistance to first-line chemotherapy in colorectal cancer. *Cell Death Dis.* **2018**, *9*, 739–749. [CrossRef] [PubMed]

5. Davies, H.; Bignell, G.R.; Cox, C.; Stephens, P.; Edkins, S.; Clegg, S.; Teague, J.; Woffendin, H.; Garnett, M.J.; Bottomley, W.; et al. Mutations of the BRAF gene in human cancer. *Nature* **2002**, *417*, 949–954. [CrossRef] [PubMed]

6. Di Nicolantonio, F.; Martini, M.; Molinari, F.; Sartore-Bianchi, A.; Arena, S.; Saletti, P.; De Dosso, S.; Mazzucchelli, L.; Frattini, M.; Siena, S.; et al. Wild-type BRAF is required for response to panitumumab or cetuximab in metastatic colorectal cancer. *J. Clin. Oncol.* **2008**, *26*, 5705–5712. [CrossRef] [PubMed]

7. Cappuzzo, F.; Varella-Garcia, M.; Finocchiaro, G.; Skokan, M.; Gajapathy, S.; Carnaghi, C.; Rimassa, L.; Rossi, E.; Ligorio, C.; Tommaso, L.D. Primary resistance to cetuximab therapy in EGFR FISH-positive colorectal cancer patients. *Br. J. Cancer* **2008**, *99*, 83–89. [CrossRef] [PubMed]

8. Affolter, K.; Samowitz, W.; Tripp, S.; Bronner, M.P. BRAF V600E mutation detection by immunohistochemistry in colorectal carcinoma. *Genes Chromosomes Cancer* **2013**, *52*, 748–752. [CrossRef] [PubMed]

9. Benlloch, S.; Payá, A.; Alenda, C.; Bessa, X.; Andreu, M.; Jover, R.; Castells, A.; Llor, X.; Aranda, F.L. Detection of BRAF V600E mutation in colorectal cancer: Comparison of automatic sequencing and real-time chemistry methodology. *J. Mol. Diagn.* **2006**, *8*, 540–543. [CrossRef] [PubMed]

10. Tan, Y.H.; Liu, Y.; Eu, K.W.; Ang, P.W.; Li, W.Q.; Salto-Tellez, M.; Iacopetta, B.; Soong, R. Detection of BRAF V600E mutation by pyrosequencing. *Pathology* **2008**, *40*, 295–298. [CrossRef] [PubMed]

11. Lavine, B.; Workman, J. Chemometrics. *Anal. Chem.* **2010**, *82*, 4699–4711. [CrossRef] [PubMed]

12. Toledo-Martín, E.M.; García-García, M.D.C.; Font, R.; Moreno-Rojas, J.M.; Salinas-Navarro, M.; Gómez, P.; Río-Celestino, M.D. Quantification of total phenolic and carotenoid content in blackberries (*Rubus fructicosus* L.) using near infrared spectroscopy (NIRS) and multivariate analysis. *Molecules* **2018**, *23*, 3191. [CrossRef] [PubMed]

13. Xia, F.; Li, C.; Zhao, N.; Li, H.; Chang, Q.; Liu, X.; Liao, Y.; Pan, R. Rapid determination of active compounds and antioxidant activity of okra seeds using fourier transform near infrared (FT-NIR) spectroscopy. *Molecules* **2018**, *23*, 550. [CrossRef]

14. Nioka, S.; Chance, B. NIR spectroscopic detection of breast cancer. *Technol. Cancer Res. Treat.* **2005**, *4*, 497–512. [CrossRef] [PubMed]

15. Yang, F.; Tian, J.; Xiang, Y.; Zhang, Z.; Harrington, P.D.B. Near infrared spectroscopy combined with least squares support vector machines and fuzzy rule-building expert system applied to diagnosis of endometrial carcinoma. *Cancer Epidemiol.* **2012**, *36*, 317–323. [CrossRef] [PubMed]

16. Yi, W.; Cui, D.; Li, Z.; Wu, L.; Shen, A.; Hu, J. Gastric cancer differentiation using fourier transform near-infrared spectroscopy with unsupervised pattern recognition. *Spectrochim. Acta Part A: Mol. Biomol. Spectrosc.* **2013**, *101*, 127–131. [CrossRef] [PubMed]

17. Chen, H.; Lin, Z.; Wu, H.; Wang, L.; Wu, T.; Tan, C. Diagnosis of colorectal cancer by near-infrared optical fiber spectroscopy and random forest. *Spectrochim. Acta A Mol. Biomol. Spectrosc.* **2015**, *135*, 185–191. [CrossRef] [PubMed]

18. McClure, W.F. 204 years of near infrared technology: 1800–2003. *J. Near Infrared Spectrosc.* **2003**, *11*, 487–518. [CrossRef]

Molecules **2019**, *24*, 2238

19. Pasquini, C. Near infrared spectroscopy: A mature analytical technique with new perspectives—A review. *Anal. Chim. Acta* **2018**, *1026*, 8–36. [CrossRef] [PubMed]

20. Workman, J.; Weyer, L. *Practical Guide to Interpretive Nearinfrared Spectroscopy*; CRC Press: Boca Raton, FL, USA, 2008.

21. Lu, W.Z. *Modern near Infrared Spectroscopy Analytical Technology*, 2nd ed.; China Petrochemical Press: Beijing, China, 2006; pp. 46–47.

22. Fan, Q.; Cao, L.Y.; Wang, Y.L.; Chen, Y.; Dong, Y.H. A Fast Identification Method of Human Colorectal Cancer Tissue by near Infrared Diffuse Reflectance Spectroscopy: 201410353552.0. 2018-03-30. Available online: http://epub.cnipa.gov.cn/patentoutline.action (accessed on 9 May 2019).

Sample Availability: Not available.

molecules

MDPI

Article

Detecting Zn(II) Ions in Live Cells with Near-Infrared Fluorescent Probes

Mingxi Fang [1], Shuai Xia [1], Jianheng Bi [1], Travis P. Wigstrom [1], Loredana Valenzano [1], Jianbo Wang [1,2,*], Marina Tanasova [1], Rudy L. Luck [1,*] and Haiying Liu [1,*]

[1] Department of Chemistry, Michigan Technological University, 1400 Townsend Drive, Houghton, MI 49931, USA; mfang@mtu.edu (M.F.); shuaix@mtu.edu (S.X.); jbi1@mtu.edu (J.B.); tpwigstr@mtu.edu (T.P.W.); lvalenza@mtu.edu (L.V.); mtanasov@mtu.edu (M.T.)

[2] College of Biological, Chemical Sciences and Engineering, Jiaxing University, Jiaxing 314001, China

* Correspondence: wjb4207@mail.ustc.edu.cn (J.W.); rluck@mtu.edu (R.L.L.); hyliu@mtu.edu (H.L.)

Received: 25 March 2019; Accepted: 8 April 2019; Published: 22 April 2019

Abstract: Two near-infrared fluorescent probes (**A** and **B**) containing hemicyanine structures appended to dipicolylamine (DPA), and a dipicolylamine derivative where one pyridine was substituted with pyrazine, respectively, were synthesized and tested for the identification of Zn(II) ions in live cells. In both probes, an acetyl group is attached to the phenolic oxygen atom of the hemicyanine platform to decrease the probe fluorescence background. Probe **A** displays sensitive fluorescence responses and binds preferentially to Zn(II) ions over other metal ions such as Cd^{2+} ions with a low detection limit of 0.45 nM. In contrast, the emission spectra of probe **B** is not significantly affected if Zn(II) ions are added. Probe **A** possesses excellent membrane permeability and low cytotoxicity, allowing for sensitive imaging of both exogenously supplemented Zn(II) ions in live cells, and endogenously releases Zn(II) ions in cells after treatment of 2,2-dithiodipyridine.

Keywords: near-infrared fluorescence; fluorescent probes; Zn(II); di-(2-picolyl)amine; living cells; cellular imaging

---

1. Introduction

After iron, zinc is the most abundant metal ion in the human body. Zn occurs strongly bonded within metalloproteins providing structural support and accomplishing various catalytic functions [1–7]. Zn(II) is essential for biological processes including signal transmission, gene expression, cellular metabolism, brain function, apoptosis, metalloenzyme regulation, neurotransmission, and mammalian reproduction [1–7]. Malfunctioning Zn(II) homeostasis can result in various diseases in the human body ranging from Alzheimer's disease to cancers and infantile diarrhea [1,8,9]. Therefore, accurate quantification and imaging of Zn(II) ions are vital for understanding their function in biological, physiological, and pathological processes [1]. The detection of Zn(II) ions at 10^{-4} M concentration in some vesicles and 10^{-10} M concentration in the cytoplasm has been reported [1,10–14]. Unfortunately, many of the reported probes suffer from interferences such as the water Raman peak, metal ions such as Cd^{2+} ions, autofluorescence from biological samples, and photo-damage to cells and tissues because the excitation wavelength is usually less than 600 nm. Fluorescent probes have been devised to address most of these issues and some offer deep tissue penetration in vivo imaging application. However, only a few near-infrared (NIR) fluorescent probes were reported to achieve sensitive visualization of Zn(II) in live cells [15–17].

In this study, we report on the syntheses and properties of fluorescent probes for sensitive identification of Zn(II), see Scheme 1. These probes emit in the near-infrared region and they also possess excellent photostability and high molar absorptivity [18,19]. We outline the syntheses of probes **A** and **B**, see Schemes 1 and 2, and utilize these to determine Zn(II) concentrations in live cells by coordinating different Zn(II)-binding ligands on the hemicyanine platform. We find a significant reduction in the fluorescence background of the probes if an acetyl group is attached to the hydroxyl group on the hemicyanine platform. This may be due to less delocalization of the lone pair from the oxygen atom directly bonded to the acyl/hemicyanine fluorophore resulting in less charge delocalization via resonance as if there was only a hydroxyl group. The acetyl group is very stable to endogenous esterases and acetylated probes display an extremely weak and imperceptible fluorescence background in live cells that do not contain free Zn(II) ions [20]. However when present, Zn(II) ions coordinate with probe **A** and promote acetyl hydrolysis that yields a significant increase in the fluorescence of the hemicyanine fluorophore in both solution and live cells. We demonstrate that probe **A** can reversibly monitor Zn(II) ion concentrations in live cells and allows for sensitive detection of free endogenous Zn(II) released after 2,2-dithiodipyridine is added to cells. In contrast, probe **B** displays a much smaller fluorescence increase with the addition of Zn (II) ions.

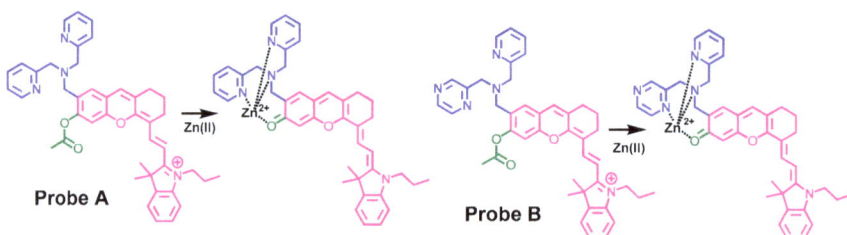

Scheme 1. Drawings of the fluorescent probes and their interactions with Zn(II) ions.

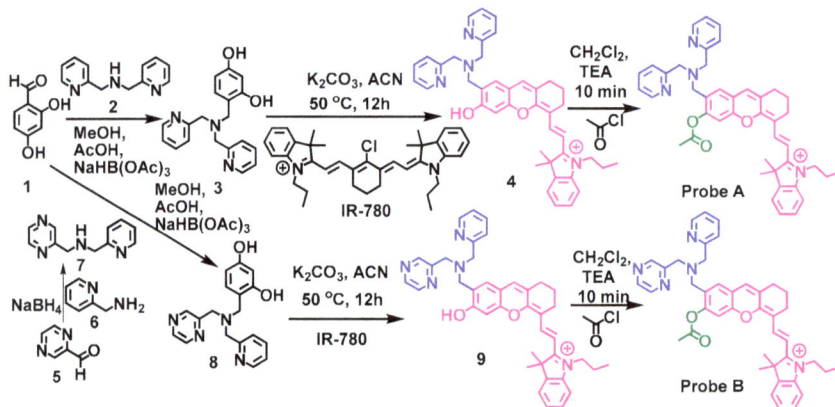

Scheme 2. Synthetic approaches to prepare fluorescent probes **A** and **B**.

2. Experimental Section

2.1. Instrumentation

Details of equipment utilized for NMR and mass spectrometry are as previously reported [21]. Standard 1 cm path length quartz fluorescence cuvettes were used to collect all absorption and emission spectra at room temperature. Absorbance spectra were obtained on a Perkin Elmer Lambda 35 UV/Vis Spectrometer (PerkinElmer, Inc., Maltham, MA, USA). The fluorescence spectra were obtained on a Horiba (Jobin Yvon, Edison, NJ, USA) Fluoromax-4 Spectrofluorometer equipped with a 150 W CW ozone-free xenon arc lamp. A buffer of 10 mM HEPES (pH = 7.0) was used to measure the sensing performance of Zn(II) and for the selectivity measurements of the probe. $ZnCl_2$ was used as the source of Zn(II) for all optical measurements. For each experiment, zinc chloride stock solutions (10 mM) were prepared freshly. In the selectivity measurements, individual solutions of different metal ions were added to solutions containing 5 μM of either probe A or B in an EtOH/buffer (1/99, *v/v*) co-solvent system. Subsequently, a different volume of the zinc stock solution was added to each solution and then thoroughly mixed. The absorption and fluorescence spectra were obtained 10 min after the addition of Zn(II) ions. Both the excitation and emission slit widths were set to 5 nm and fluorescence spectra were obtained at excitation of 640 nm.

In order to explain and confirm the excitation patterns observed experimentally for intermediates **4** and **9**, see Scheme 2, probes **A** and **B** with and without the presence of a [Zn (OH)]⁻ moiety, the density functional theory (DFT) was employed as implemented in the Gaussian09 program package [22]. The solvated environment of aqueous buffer was mimicked via the CPCM continuum model. Structures were first allowed to geometrically affirm at the PBE/6-311G(d,p) level of theory which allowed the HOMO/LUMO energy gaps to be calculated at the PBE/6-311++G(d,p) level of theory. The time-dependent density functional theory (TDDFT) was employed at the level of theory mentioned above. For the initial calculations, three excited states were explored for all the above molecular models investigated. In addition, we conducted additional modeling with $[Zn(OH_2)]^{2+}$ and $[Zn(OH_2)_2]^{2+}$ moieties attached to probes **A** and **B**. These were conducted with the APFD functional [23] and the 6-311+G(2d,p) basis set for the TDDFT calculation using the Gaussian16 program package [24]. Full details of the calculations are presented in the Supplementary Materials.

2.2. Cell Culture and Fluorescence Imaging

HeLa cells were obtained and prepared for confocal imaging as previously reported [25,26]. This pertains to their initial incubation, addition of either probe, addition of fresh serum free media, and final washings with PBS [15]. The live cell images were taken by a confocal fluorescence microscope (Olympus IX 81). The fluorescence images were obtained at 60× magnification and the laser energies were kept constant for each image series. The red channel fluorescence of the probes was excited at a wavelength of 635 nm and fluorescence was collected from 650–750 nm.

For the Zn(II) chelate test, after adding the corresponding concentration of Zn(II) plus sodium pyrithione (Pyr), 100 μM of TPEN (*N,N,N′,N′*-tetrakis(2-pyridylmethyl)ethylenediamine) was added to each confocal dish, and the cells were incubated for 10 min at room temperature. The intracellular level of Zn(II) ions in HeLa cells was evaluated using 2,2′-dithiodipyridine (DTDP). The cells were serum starved for 3 h at 37 °C with 5% CO_2 and then incubated with 1 μM of probe **A** for 30 min. Subsequently, 100 μM DTDP was added and cells were further incubated for 30 min at 37 °C with 5% CO_2. Then, 100 μM TPEN (*N,N,N,N*-tetrakis(2-pyridinylmethyl)-1,2-ethanediamine) was added and the mixture incubated for 10 min at room temperature. Cell images were taken by a confocal fluorescence microscope (Olympus IX 81, Olympus America Inc., Nelville, NY, USA). The excitation wavelength was 635 nm and the images were collected at 675–725 nm. The cytotoxicity of the probes was determined using an MTS assay as previously reported [25,26].

2.3. Materials

Unless indicated, all reagents and solvents were obtained from commercial suppliers and used without further purification.

Synthesis of compound **3**: 2,4-Dihydroxybenzaldehyde (1.086 g, 7.9 mmol) and di-(2-picolyl)amine (1.57g, 7.9 mmol) were dissolved in 30 mL methanol in a 100-mL round-bottom flask. Acetic acid (5 mL) and sodium triacetoxyborohydride (1.5 g, 7 mmol) were added and the reaction mixture stirred at room temperature for 3 days. After the starting materials were depleted (verified by TLC monitoring), dilute HCl solution was added to quench the unreacted sodium salt, and methanol was removed under vacuum. The pH of the reaction mixture was adjusted to 9.0. The resulting crude was extracted with ethyl acetate and washed with water three times, followed by drying over Na_2SO_4. The residue was collected by removing the solvent under vacuum and purified by recrystallization with diethyl ether to afford the product **3**, Scheme 2, in the form of a yellow flaky solid (1.2 g, 47%).

Synthesis of compound **4**: Compound **3**, Scheme 2, (302 mg, 0.94 mmol) and K_2CO_3 (130 mg, 0.94 mmol) were dissolved in acetonitrile (10 mL) in a 50 mL round-bottom flask. After the mixture was stirred at room temperature under a N_2 atmosphere for 15 min, a solution of IR-780 iodide (314 mg, 0.47 mmol) in CH_3CN (2.5 mL) was added to the mixture via a syringe, and the mixture was heated to 50 °C for 5 h. When the solvent was removed under reduced pressure, the crude product was purified by silica gel chromatography using CH_2Cl_2/MeOH (15:1, *v/v*) as eluent, affording compound **4** as an aquamarine solid (160 mg, 55%). ^{1}H NMR (400 MHz, $CDCl_3$): δ 8.57–8.50 (m, 2H), 7.64–7.61 (m, 1H), 7.42–7.38 (m, 2H), 7.31–7.27 (m, 3H), 7.17–7.14 (m, 2H), 7.10–7.03 (m, 3H), 6.86 (s, 1H), 6.67–6.63 (m, 2H), 6.34–6.29 (m, 1H), 4.26 (s, 2H), 3.87–3.67 (m, 6H), 2.65 (s, 4H), 1.89–1.56 (m, 10H), 1.04–0.97 (s, 3H); ^{13}C NMR (100 MHz, $CDCl_3$) δ 176.1, 163.4, 162.9, 157.9, 155.0, 149.4, 148.6, 145.0, 142.0, 141.5, 137.6, 136.1, 130.0, 129.3, 126.8, 126.3, 123.8, 123.4, 122.8, 122.5, 115.2, 114.7, 112.4, 103.5, 102.7, 58.6, 55.9, 53.7, 50.3, 47.2, 29.2, 28.6, 28.0, 24.7, 21.4, 20.6, 11.9. HRMS (ESI): calculated for $C_{41}H_{43}N_4O_2^+$ [M]$^+$: 623.3381; found: 623.3629.

Synthesis of fluorescent probe **A**: Compound **4**, Scheme 2, (100 mg, 0.13 mmol) was added to a solution of acetyl chloride (0.05 mL, 0.26 mmol) and triethylamine (0.1 mL) in anhydrous dichloromethane (10 mL) and stirred for 15 min under a nitrogen atmosphere. The solvent was removed under reduced pressure yielding a blue crude product which was washed with water and brine thrice. The product was dissolved in dichloromethane and dried over Na_2SO_4 and subsequently filtered. The filtrate was collected and the solvent removed affording probe **A**. ^{1}H NMR (400 MHz, $CDCl_3$): δ 8.59–8.52 (m, 3H), 7.72–7.63 (m, 2H), 7.51–7.45 (m, 3H), 7.31 (s, 1H), 7.22–7.15 (m, 3H), 7.00–6.93 (m, 3H), 6.82 (s, 1H), 6.68–6.64 (m, 1H), 4.74–4.53 (m, 2H), 3.91–3.63 (m, 6H), 2.71 (s, 4H), 2.31–2.20(m, 3H), 1.90–1.33 (m, 10H), 1.22 (s, 3H); ^{13}C NMR (100 MHz, CDCl3) δ 178.6, 169.2, 160.3, 152.2, 151.3, 150.1, 148.9, 148.7, 146.4, 142.2, 141.6, 137.2, 137.1, 132.3, 132.2, 130.4, 129.6, 129.5, 129.2, 128.7, 128.4, 128.1, 123.6, 122.8, 122.6, 120.2, 115.7, 113.6, 110.3, 106.6, 60.0, 54.1, 52.3, 51.2, 48.2, 32.1, 29.9, 28.5, 24.7, 22.9, 21.8, 21.3, 11.9. HRMS (ESI): calculated for $C_{43}H_{45}N_4O_3^+$ [M]$^+$: 665.3486; found: 665.3474.

Synthesis of compound **7**: Pyrazine-2-carbaldehyde (1 g, 9.25 mmol) was dissolved in methanol (90 mL) and yridine-2-methylamine (0.96 mL, 9.25 mmol) was added by a syringe. This mixture was stirred at 25 °C for 30 min followed by rapid addition of $NaBH_4$ (1.06 g, 27.76 mmol), and then stirred at 25 °C for an additional 2 h. The solvent was evaporated under reduced pressure, water (20 mL) was added to the residue and the solution pH was adjusted to approximately 10 by adding Na_2CO_3 solution. The mixture was extracted with CH_2Cl_2 (3×30 mL) and purified by column chromatography with CH_2Cl_2/MeOH (30:1, *v/v*) to give the product **7**, Scheme 2, as a pale-yellow oil (1.6 g, 90% yield).

Synthesis of compound **8**: Compound **8**, Scheme 2, was prepared using compound **7** (0.9 g, 4.5 mmol) and 2,4-dihydroxybenzaldehyde (0.68 g, 5 mmol) according to the method used for compound **3**, affording the product as a yellow flaky solid (0.7 g, 44%). ^{1}H NMR (400 MHz, $CDCl_3$): δ 8.54 (d, J = 1.6 Hz, 1H), 8.50–8.49 (m, 1H), 8.46–8.45 (m, 1H), 8.36 (d, J = 2.4 Hz, 1H), 7.60–7.56 (m, 1H), 7.27 (d, J = 7.6 Hz, 1H), 7.14–7.11 (m, 1H), 6.81 (d, J = 8.4 Hz, 1H), 6.40 (d, J = 2.4 Hz, 1H), 6.27–6.24 (m, 1H), 3.87 (s, 4H), 3.66 (s, 2H); ^{13}C NMR (100 MHz, CDCl3) δ 158.4, 158.3, 157.8, 154.5, 148.8, 145.3, 144.0, 143.2, 137.5, 131.1, 123.6, 122.8, 114.2, 106.9, 104.2, 58.9, 56.8. HRMS (ESI): calculated for $C_{18}H_{19}N_4O_2^+$ [M + H]$^+$: 323.1530; found: 323.1495.

Synthesis of compound **9**: Compound **9**, Scheme 2, was carried out using compound **8** (160 mg, 0.25 mmol) and IR-780 (330 mg, 0.49 mmol) in a similar manner to that used for compound **4**, affording 130 mg product (46% yield). ^{1}H NMR (400 MHz, $CDCl_3$): δ 8.62–8.42 (m, 7H), 7.69 (s, 1H), 7.43–7.29 (m, 6H), 7.04 (s, 1H), 6.44 (s, 1H), 4.38 (s, 2H), 3.98–3.90 (m, 6H), 2.75 (s, 4H), 1.92–1.76 (m, 10H), 1.23 (s, 3H); ^{13}C NMR (100 MHz, CDCl3) δ 176.5, 158.7, 157.8, 157.1, 155.0, 154.4, 154.1, 149.1, 148.7, 145.5, 143.9, 143.3, 142.0, 141.6, 137.8, 137.2, 135.5, 131.1, 129.8, 129.3, 126.9, 123.4, 122.7, 115.4, 114.6, 112.6, 106.6, 104.1, 58.9, 56.8, 56.1, 50.5, 47.5, 29.3, 28.6, 24.9, 21.8, 21.4, 20.6, 11.9. HRMS (ESI): calculated for $C_{40}H_{42}N_5O_2^+$ [M]$^+$: 624.3333; found: 624.3314.

Synthesis of fluorescent probe **B**: Probe **B**, Scheme 2, was carried out by reacting compound **9** with acetyl chloride in a similar manner to that for probe **A**. ^{1}H NMR (400 MHz, CDCl3): δ 8.71–8.45 (m, 7H),

7.68–7.67 (m, 1H), 7.51–7.41 (m, 6H), 7.11 (s, 1H), 6.97 (s, 1H), 4.73 (s, 2H), 3.87–3.71 (m, 6H), 2.86–2.70 (m, 4H), 2.30 (s, 3H), 2.02–1.77 (m, 10H), 1.23 (s, 3H); ^{13}C NMR (100 MHz, CDCl3) δ 178.7, 169.1, 168.8, 152.3, 151.2, 150.8, 146.5, 145.5, 145.4, 144.2, 144.1, 143.6, 142.3, 141.6, 141.5, 132.7, 131.1, 130.6, 129.5, 129.0, 128.1, 122.7, 122.5, 120.4, 120.3, 115.9, 113.5, 110.3, 68.4, 57.9, 52.7, 51.2, 47.9, 32.1, 30.6, 29.9, 28.4, 24.7, 21.3, 20.5, 11.8. HRMS (ESI): calculated for $C_{42}H_{44}N_5O_3^+$ [M]$^+$: 666.3439; found: 666.3436.

3. Results and Discussion

3.1. Probe Design and Synthesis

The synthetic route to prepare probes **A** and **B** is outlined in Scheme 2. In order to bind dipicolylamine (DPA) (a ligand known to bind Zn(II) ions) to the hemicyanine dye, we prepared 4-((bis(yridine-2-ylmethyl)amino)methyl)benzene-1,3-diol (**3**) by first reacting 2,4-dihydroxybenzaldehyde with DPA and then reducing the enamine intermediate with sodium triacetoxyborohydride. Compound **3** was reacted with cyanine dye IR-780 (a chloro-substituted tricarbocyanine dye) under basic conditions in acetonitrile, to afford the hemicyanine dye bearing the dipicolylamine residue (**4**). An acetyl group was then bonded to the hydroxyl moiety on the hemicyanine fluorophore to significantly reduce the probe fluorescence yielding fluorescent probe **A**. Probe **B** was prepared in a similar manner to that of probe **A** except that the compound 1-(pyrazin-2-yl)-N-(yridine-2-ylmethyl)methanamine, **7**, prepared by reacting compounds **5** with **6**, see Scheme 2, was utilized. This should result in weaker binding for the Zn(II) ions due to reduced basicity and potentially a reduction in probe fluorescence.

3.2. Optical Properties of Fluorescent Probes A and B in Different Solvents

The absorption spectra of intermediates **4** and **9**, and probes **A** and **B** were investigated in ethanol, tetrahydrofuran (THF), and buffer (pH 7.0) containing 1% ethanol, see Table 1. Intermediate **4** displays absorption peaks at 708 nm, 705 nm and 687 nm, and fluorescence peaks at 725 nm, 718 nm and 703 nm in ethanol, tetrahydrofuran (THF), and buffer (pH 7.0) containing 1% ethanol at 635 nm excitation, respectively. Increasing solution polarity results in blue shifts of absorption and fluorescence peaks for intermediate **4** and causes fluorescence quenching as intermediate **4** has the lowest fluorescence quantum yield of 1.4% in buffer (pH 7.0) containing 1% ethanol. This may be due to fluorescence quenching photo-induced electron transfer from the tertiary amine of the Zn(II)-binding dipicolylamine residue to the hemicyanine fluorophore. The fluorescent probe A displays lower fluorescence quantum yields than intermediate **4** and this may be due to the presence of the acetyl group attached to the phenolic oxygen atom of the hemicyanine platform in probe **A**. This significantly reduces the electron donating ability of the phenolic oxygen atom and decreases fluorescence. Similar results to intermediate **4** and probe **A** were observed with intermediate **9** and probe **B**.

3.3. Absorbance Responses of Intermediates and the Probes to Zn(II) Ions

The absorption responses of intermediates **4** and **9**, and fluorescent probes **A** and **B** to Zn(II) ions were evaluated in aqueous HEPES buffer (pH 7.0) containing 1% ethanol (Figure 1). Intermediate 4 contains a main absorption peak at 691 nm and a shoulder peak at 636 nm (Figure 1). The addition of Zn(II) ions from 0.1 µM to 2.0 µM results in gradual increases of the main absorption peak at 691 nm (Figure 1). Probe **A** at the same concentration shows a lower absorbance than intermediate **4** due to the presence of the acetyl group and for reasons discussed above. However, the addition of Zn(II) ions from 0.1 µM to 2.0 µM significantly increases the main absorption peak at 691 nm with a more moderate increase in the shoulder peak at 636 nm (Figure 1). This is because Zn(II) facilitates hydrolysis of the acetyl group and affords rigidity to the overall structure facilitating the absorption. Interestingly, both intermediate **9** and probe **B** are not as sensitive to increasing concentrations of Zn(II) ions, as addition of up to 20 µM concentration only results in slight increases in both the main absorption and the shoulder peaks at 691 nm and 636 nm, respectively (Figure 2).

Table 1. Optical properties of intermediates and fluorescent probes **A** and **B**.

	Solvent	λ_{abs} (nm)	λ_{em} (nm)	ε_{max} (10^4 M^{-1}cm^{-1})	Φ_f (%)
	Buffer (pH7.0)	687	703	3.2×10^4	1.4
Compound 4	Ethanol	705	718	5.5×10^4	6.7
	THF	708	725	5.6×10^4	7.0
	Buffer (pH7.0)	695	701	3.0×10^4	0.6
Probe A	Ethanol	707	717	4.6×10^4	6.0
	THF	710	723	3.1×10^4	6.8
	Buffer (pH7.0)	685	712	3.5×10^4	1.6
Compound 9	Ethanol	705	717	6.1×10^4	5.2
	THF	706	721	4.2×10^4	7.5
	Buffer (pH7.0)	688	709	4.1×10^4	0.8
Probe B	Ethanol	705	714	6.0×10^4	4.1
	THF	717	719	5.1×10^4	5.0

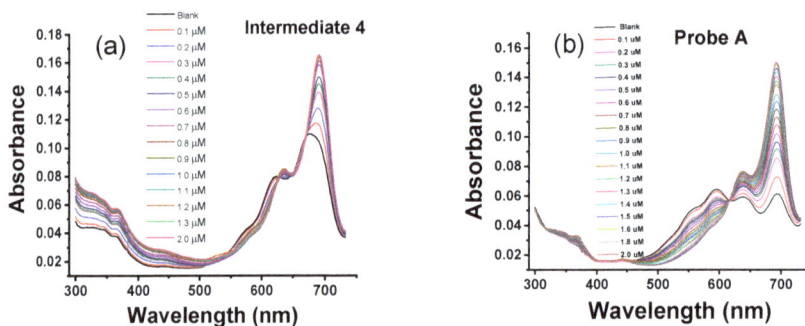

Figure 1. (**a**) UV-Vis absorption spectra of 1.0 µM intermediate **4** and (**b**) UV-Vis absorption spectra of probe **A** upon gradual addition of Zn(II) from 0.1 µM to 2.0 µM to 10 mM HEPES buffer solutions (pH 7.0) containing 1.0 µM probe **A**.

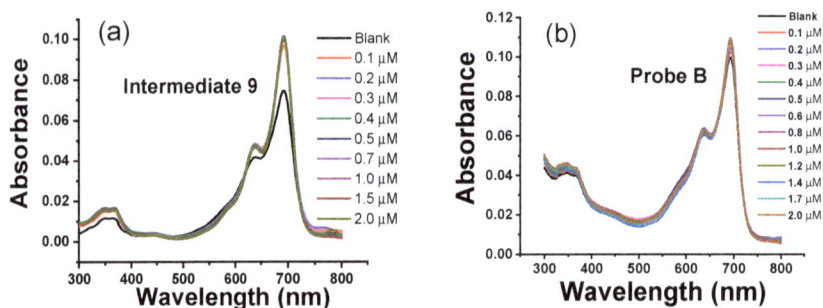

Figure 2. (**a**) UV-Vis absorption spectra of 1.0 µM intermediate **9** and (**b**) UV-Vis absorption spectra of 1.0 µM probe **B** upon gradual addition of Zn(II) from 0.1 µM to 2.0 µM to 10 mM HEPES buffer solutions (pH 7.0) containing 1.0 µM) probe **B**.

3.4. Fluorescence Response of the Intermediates and Probes to Zn(II) Ions

Intermediate **4** displays strong fluorescence in the absence of Zn(II) ions, indicating that the photo-induced electron transfer from the tertiary amine of the dipicolylamine residue to the hemicyanine fluorophores is unable to completely quench the fluorescence of the hemicyanine platform (Figure 2). The addition of Zn(II) ions, up to 2.0 µM, cause a moderate increase in the fluorescence peak at 708 nm, see Figure 3. However, fluorescent probe **A** exhibits a much weaker fluorescence in the absence of

Zn(II) ions and its fluorescence dramatically increases upon the addition of Zn(II) ions from 0.1 μM to 2.0 μM (Figure 2). This sensitive response of probe **A** to Zn(II) ions arises from the quick hydrolysis of the acetyl group via binding of Zn(II) to the dipicolylamine residue. Probe **A** displays a linear fluorescence response to Zn(II) from 0.1 μM to 1.5 μM with a detection limit of 0.45 nM, see Figure S8. The titration curve fitted well up to a 1:1 stoichiometry and the Hill plot displays a linear relationship with a slope of 1.0 and the Job's plot contains a maximum point at a mole fraction of 0.50, see Figure S11. Probe **A** shows high affinity for Zn(II) ions with a lower dissociation constant of 9.5×10^5 M^{-1} as determined by a fluorescence titration curve, see Figure S7. In contrast, intermediate **9** and probe **B** display slight increases in fluorescence if up to 10 times higher concentration of Zn(II) ions is added to their solutions (Figure 4). This difference may be ascribed to the weaker binding of Zn(II) ions as a result of the substitution of one pyridine by pyrazine.

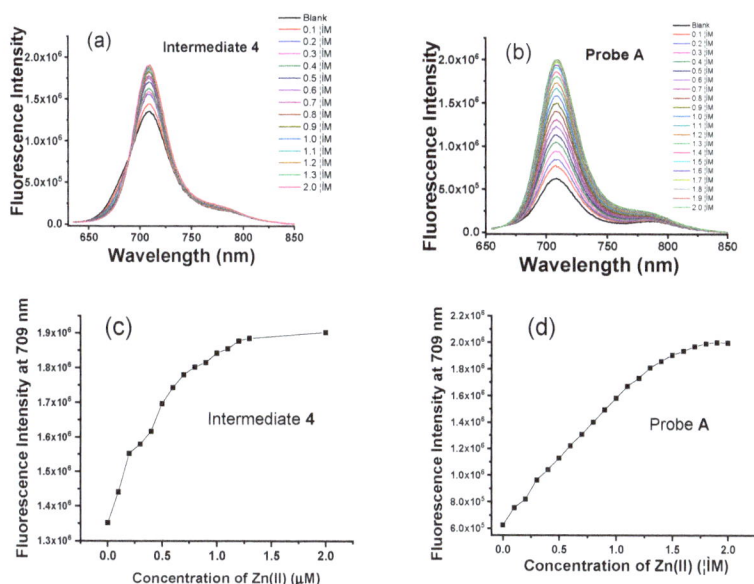

Figure 3. (**a**) Fluorescence spectra of intermediate **4**, and (**b**) probe **A** at 635 nm excitation upon gradual addition of Zn(II) from 0.1 μM to 2.0 μM to 10 mM HEPES buffer solutions (pH 7.0) containing 1.0 μM probe **A**. (**c**) Intensity dependence on zinc ion concentration of 1.0 μM intermediate **4** and (**d**) probe **A** upon gradual addition of Zn(II) from 0.1 μM to 2.0 μM to 10 mM HEPES buffer solutions (pH 7.0) containing 1.0 μM probe **A**.

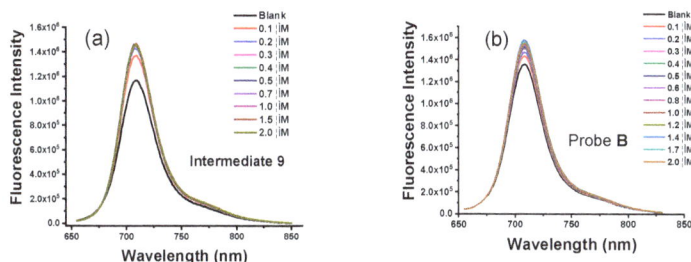

Figure 4. (**a**) Fluorescence spectra of 1.0 μM intermediate **9** and (**b**) probe **B** at 635 nm excitation upon gradual addition of Zn(II) from 0.1 μM to 2.0 μM to 10 mM HEPES buffer solutions (pH 7.0) containing 1.0 μM probe **B**.

The accuracy of the fluorescence response for probe **A** was assessed as greater than 92%. Concentrations of Zn^{2+} were accurately determined using a calibration curve based on atomic absorption, Figure S30, and compared to fluorescence measurements using probe **A**, Figure S32.

3.5. Selectivity Studies

The selectivity of fluorescent probes **A** and **B** to Zn(II) over other metal ions was evaluated in a 10 mM HEPES buffer (pH 7.0). No change in fluorescence was observed with probe **A** if up to 20 μM alkali and alkaline-earth metal ions, specifically Na^+, K^+, Ca^{2+}, and Mg^{2+} ions, and some transition metal ions, i.e., Mn^{2+}, Fe^{2+}, Ni^{2+}, Cu^{2+}, Cd^{2+}, and Hg^{2+} ions were added (Figure S9(a)). In contrast probe **B** did not display selectivity for Zn(II) in a similar comparison (Figure S9(b)). Probe **A** shows high selectivity to Zn(II) ions over other metal ions in contrast to the other reported fluorescent probes for detection of Zn(II) ions which show significant interference from Cd^{2+} ions.

3.6. Photostability of the Probes

The photostability of the probes was evaluated by comparative assessment of fluorescent intensity every 10 min under continuous excitation at 635 nm using a 150 W xenon arc lamp. Probe **A** displays excellent photostability as its fluorescence intensity decreases by only 3.0% after one hour of excitation and by 7.6% after three hours excitation (Figure S10). Probe **B** exhibits similar photostability to probe **A**. Fluorescent intensity losses of 4.2% after one hour of excitation and 7.0% under three hours of excitation are obtained (Figure S10).

3.7. Theoretical Modeling Results

Analysis of the molecular charges allowed for appreciating changes in charge distribution with the PBE functional calculations. The distal nitrogen atom in the pyrazine moiety (i.e., that which binds to the Zn atom) has a notable charge change (+0.049|e|) when comparing pyrazine in probe **B** to the N atom in pyridine in probe **A**, which is in agreement with the known lesser basicity for pyrazine. This change in charge results in an increase in the bond length of 0.046 Å in the N-Zn bond (now 2.0995 Å), indicating a decrease in bond strength. This weaker coordination of the Zn(II) ion may be responsible for the subdued increase in fluorescence when pyrazine is present (probe **B**) in place of pyridine (probe **A**).

The time dependent excitation patterns for all the compounds investigated is reported in Table S9. The results for the probes, including the presence of an additional ligand (i.e., OH^-), are summarized in Table S12. The absorption peaks computed for intermediate **4**, intermediate **9**, and probes **A** and **B** are 699 nm, 701 nm, 693nm, and 699 nm, respectively, and they are in excellent agreement with experimental evidence (see Table S10). When Zn(II) is added to probe **A**, a slight shift in the absorption peak (~+40 nm) is observed. Interestingly, visual inspection of the molecular orbitals indicates that such a value corresponds to an electron excitation from the HOMO to the LUMO, instead of the HOMO-1/LUMO transition observed for the other molecules calculated (see Table S11). The overestimation of the excitation energy may be related to a deficiency in the PBE function in properly handling the delocalization of the electronic charge of the nitrogen-rich molecular moiety when forming the Zn complex.

However, the result computed for the binding of probe **B** to the Zn atom is in excellent agreement with experimental results with a calculated absorption peak of 702 nm. It is also ascribed to the excitation that occurs from the HOMO-1 to the LUMO. The lobes for the HOMO orbital for probe A-Zn(OH) indicate a substantial contribution from the lone pairs on the tertiary amine on the dipicolylamine moiety in contrast to a lack of such overlap in the HOMO-1 orbital for probe B-Zn(OH). Additionally, while Zn complex **4**, see Table S11, has a HOMO/LUMO transition, delocalization into the tertiary N atom is not observed, indicating that completing the coordination sphere around Zn is required for this to occur. An exploration of the geometrical features of the Zn first coordination sphere reveals that the Zn atom coordinates with the available nitrogen and oxygen atoms of the probes. While the average atomic distances are not affected, some individual values do change, in particular, the Zn-N atom distance for the pyridine and pyrazine rings shows an increase of approximately 0.05 Å from probe **A** to probe **B**. Such a structural change is reflected by a decrease in the negative charge of the nitrogen of approximately 0.07 |e|. This observation seems to confirm a limitation in the chosen DFT function in dealing with the Zn complex arising from probe **A**. It is interesting to note that results obtained in the presence of the counter anion, confirm the observations reported above with absorption peaks for Probe **A-Zn** and Probe **B-Zn** of 765 nm and 711 nm, respectively.

The coordination sphere bond distance results from the additional calculations, i.e., with the APFD/6-311+G(2d,p) functional/basis set combination, which consisted of the $[Zn(OH_2)]^{2+}$ and $[Zn(OH_2)_2]^{2+}$ moieties attached to probes **A** and **B**, are illustrated in Figure S29, (see supporting information pages 11–35. As is evident, trigonal bipyramidal and octahedral geometries are obtained with these various entities and there is crystallographic evidence of the former mode of coordination. The fact that pyrazole is a weaker ligand is again confirmed by the longer Zn-N (pz) distances obtained with both probes, i.e., 2.075 $Zn(OH_2)^{2+}$ and 2.120 $[Zn(OH_2)_2]^{2+}$ compared to 2.049 and 2.091 Å for probes **B** and **A**, respectively. In these structures the phenolic C-O distances were equivalent, leading to the conclusion that the conjugation and fluorescence intensities for the hemicyanine platform should be equivalent. This is also evident in the calculated excited state transitions which consisted of the HOMO to LUMO transition in all cases, see Table S12. It is noteworthy that this calculation strategy, i.e., APFD/6-311+G(2d,p), resulted in values for the transitions that were ~140 nm less than those observed experimentally (and at 0.426 and 0.412 eV for probes **A** and **B**, respectively, and were greater than the expected error range of 0.20–0.25 eV [23]) in contrast to the excellent agreement noted above for PBE/6-311++G(d,p).

3.8. Cytotoxicity of the Probes

The cytotoxicity of probes **A** and **B** was determined by carrying out the MTS assay (Figure 5). Incubation of the HeLa cells with the probes did not have any significant impact on cell viability even at 10 µM concentration. Overall, the probe showed insignificant cytotoxicity over 48 h at the test concentrations, with cell viability of 95%, indicating that the probes have good biocompatibility and can serve as an excellent staining reagent for live cell.

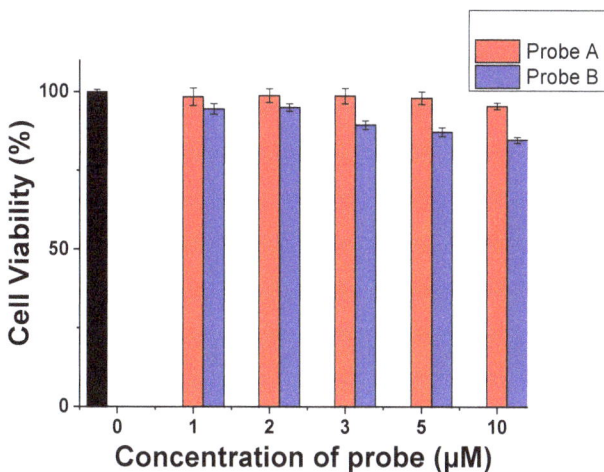

Figure 5. Cytotoxicity and cell proliferation of probes **A** and **B** conducted by MTS assay. HeLa cells were incubated with 1, 2, 3, 5, and 10 µM of probes for 48 h, and cell viability was measured by adding MTS reagent and measuring at 490 nm at the average of three times. The black bar represents a blank control.

3.9. Live Cell Imaging of Fluorescent Probes

In light of the aforementioned fluorescence analysis, we investigated only probe **A** to detect Zn(II) ions in live cells by incubating HeLa cells with different concentrations of probe **A** (0.1 μM, 0.5 μM, 1.0 μM) for 30 min at 37 °C prior to imaging. There was an extremely weak and barely perceptible fluorescence due to the low level of endogenous auto-fluorescence from the cells in the near-infrared region and the low fluorescence quantum yield of probe **A**, see Figure 6. This indicated that the ester bond of the probe was very stable inside live cells. The addition of 10 μM exogenous Zn(II) to the cells without the ionophore pyrithione did not cause a noticeable change in the fluorescence background. However, intracellular fluorescence increased significantly in response to exogenous Zn(II) in the presence of pyrithione. The cellular fluorescence intensity increased with an increase of the probe concentration from 0.1 μM to 1.0 μM.

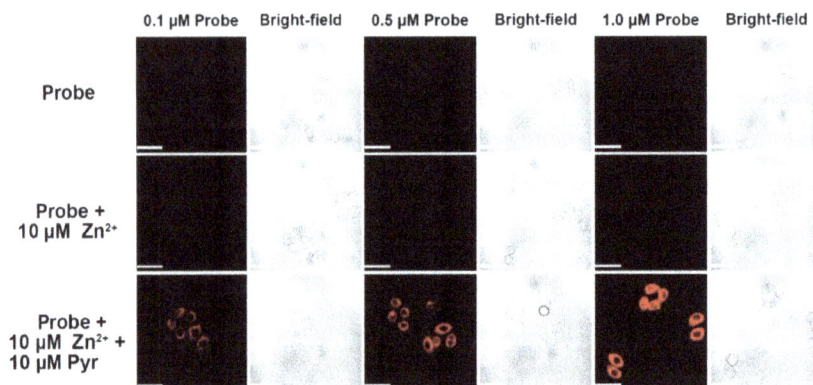

Figure 6. Fluorescence images of fluorescent probe **A** with concentration at 0.1 μM, 0.5 μM, and 1.0 μM in HeLa cells. Cells were incubated with probe **A** with specific concentration for 30 min. Cells were then supplemented with either 10 μM of zinc(II) chloride or 10 μM each of zinc(II) chloride plus sodium pyrithione (Pyr) for 30 min before acquiring images. Scale bar: 50 μm. λ_{ex}: 635 nm.

The sensitivity of the probe to the intracellular Zn(II) level was also investigated by using a probe concentration of 1.0 μM, followed by stepwise increases in the concentration of the Zn(II)/ionophore pyrithione from 0.1 μM, 0.5 μM, 1.0 μM, 5 μM, and 10 μM, see Figure 7. The intracellular fluorescence increased with an increase of Zn(II) ion concentration from 0.1 to 5 μM, and it did not change significantly with further increases in the Zn(II) ion concentration. Clearly, probe **A** can sensitively detect intracellular Zn(II) ion with at least 0.1 μM concentration. The intracellular fluorescence decreased significantly after a cell-permeable Zn(II) ion chelator N,N,N',N'-tetrakis(2-pyridylmethyl)ethylenediamine (TPEN) was added to the cells, [27] indicating that the probe responded to intracellular Zn(II) ions reversibly since TPEN effectively removed Zn(II) ions from the probe because of its much stronger binding affinity.

In addition, we investigated the fluorescent probes to determine if they could be used to detect Zn(II) ions endogenously released from intracellular metalloproteins after the treatment with 2,2-dithiodipyridine (DTDP), see Figure 8. The DTDP treatment alone was reported to give modest increases of intracellular Zn(II) ions (nM) in cells. The probe fluorescence has imperceptible background in cells without DTDP treatment. However, a very strong fluorescence intensity from probe **A** (1.0 μA) was observed after DTDP treatment of HeLa cells without the addition of external ionophores, indicating the accumulation of endogenously released Zn(II) ions. The intracellular fluorescence was considerably and immediately reduced by TPEN treatment. These results clearly proved that fluorescent probe **A** sensitively detected endogenous Zn(II) ions in live cells.

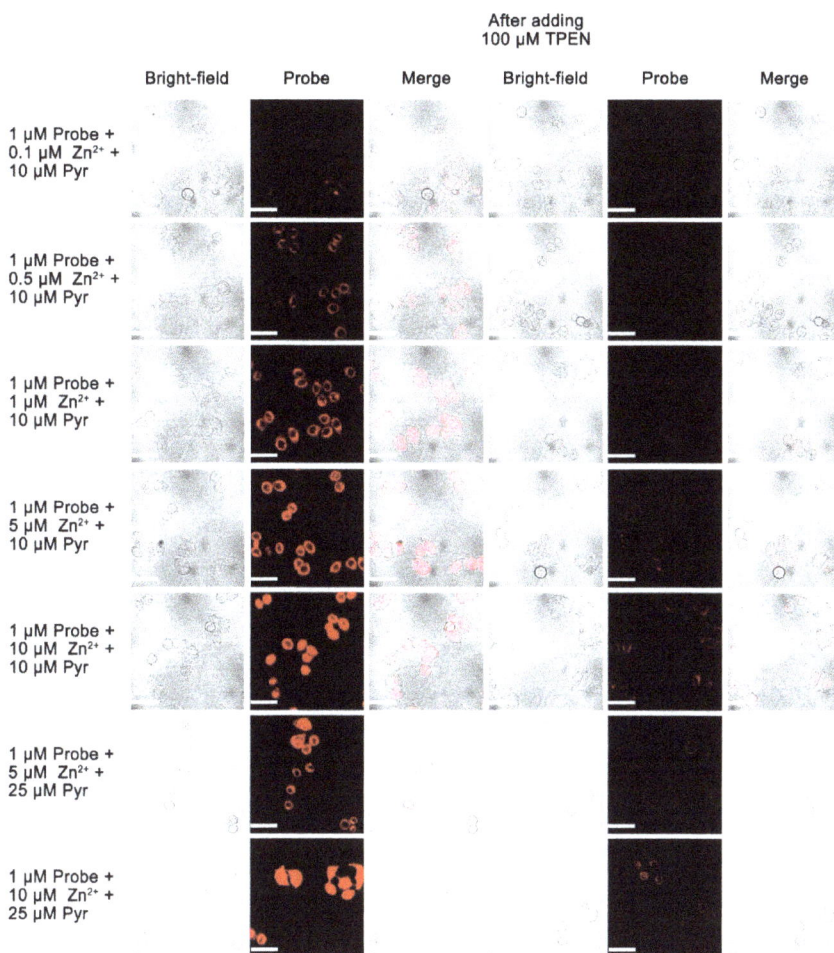

Figure 7. Fluorescence images of 1.0 μM fluorescent probe **A** in HeLa cells. Cells were incubated with probe **A** with specific concentration for 30 min. Cells were then supplemented with either zinc(II) chloride with different concentrations or zinc II) chloride with different concentrations plus sodium pyrithione (Pyr) for 30 mins before acquiring images. Then 100 μM TPEN (zinc chelator) was added for 10 min before acquiring the second set of images. Scale bar: 50 μm. λ_{ex}: 635 nm.

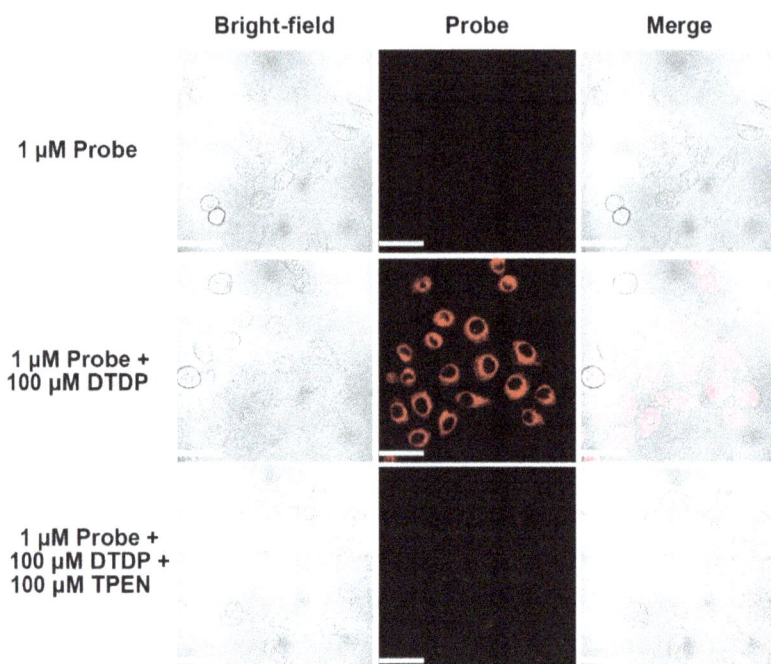

Figure 8. Fluorescence images of HeLa cells incubated with probe **A**. Cells were first incubated with 1.0 μM of probe **A** for 1 h and then were supplemented with 100 μM of 2,2′-dithiodipyridine (DTDP) for 30 mins before images were taken. Scale bar: 50 μm. λ_{ex}: 635 nm.

4. Conclusions

We reported on the rational design, syntheses, and characterization of two novel near-infrared fluorescent probes for sensing Zn(II) ions based on the hemicyanine fluorophore. The fluorescent probe **A** containing the dipicolylamine residue is highly selective and sensitive to Zn(II) ions over other metal ions. The strong binding of Zn(II) ions to probe **A** facilitates the cleavage of the acetyl group and effectively increases the fluorescence of the fluorophore. The fluorescent probe **A** offers a way to live cell imaging of both exogenously supplemented Zn(II) ions and free endogenous Zn(II) ions released from intracellular metalloproteins. In contrast, probe **B** exhibits selectivity but only moderate sensitivity to Zn(II) ions, and our theoretical calculations did not provide an answer for this difference.

Supplementary Materials: The following are available online at http://www.mdpi.com/1420-3049/24/8/1592/s1, The supplementary materials include high-resolution mass spectra of probes **A** and **B** in the absence and presence of Zn(II) ions, Zn(II) binding constants with the probes, probe selectivity over other metal ions, Job's plot, fluorescence quantum yields of the probes, and theoretical calculation results.

Author Contributions: Synthesis and compound characterization, M.F. and J.W. cell imaging, S.X.; optical measurement, writing, and original draft preparation, J.B.; theoretical calculations, T.P.W., L.V., and R.L.L.; writing, H.Y.L. and M.F.; review and editing, R.L.L., M.T. and L.V.; visualization, S.X.; project administration and funding acquisition, H.Y.L.

Funding: Funding from the National Institute of General Medical Sciences of the National Institutes of Health under Award Number R15GM114751 (to H.Y. Liu) is gratefully acknowledged. A high-performance computing infrastructure at Michigan Technological University was used for the calculations.

Conflicts of Interest: The authors declare no conflict of interest.

References

1. Pluth, M.D.; Tomat, E.; Lippard, S.J. Biochemistry of Mobile Zinc and Nitric Oxide Revealed by Fluorescent Sensors. *Annu. Rev. Biochem.* **2011**, *80*, 333–355. [CrossRef] [PubMed]
2. Kikuchi, K. Design, synthesis and biological application of chemical probes for bio-imaging. *Chem. Soc. Rev.* **2010**, *39*, 2048–2053. [CrossRef] [PubMed]
3. Truong-Tran, A.Q.; Ho, L.H.; Chai, F.; Zalewskiet, P.D. Cellular zinc fluxes and the regulation of apoptosis/gene-directed cell death. *J. Nutr.* **2000**, *130*, 1459S–1466S. [CrossRef] [PubMed]
4. Truong-Tran, A.Q.; Carter, J.; Ruffin, R.; Zalewski, P.D. New insights into the role of zinc in the respiratory epithelium. *Imunol. Cell Biol.* **2001**, *79*, 170–177. [CrossRef] [PubMed]
5. Chai, F.G.; Truong-Tran, A.Q.; Ho, L.H.; Zalewskiet, P.D. Regulation of caspase activation and apoptosis by cellular zinc fluxes and zinc deprivation: A review. *Imunol. Cell Biol.* **1999**, *77*, 272–278. [CrossRef]
6. Sensi, S.L.; Paoletti, P.; Bush, A.I.; Sekler, I. Zinc in the physiology and pathology of the CNS. *Nat. Rev. Neurosci.* **2009**, *10*, 780. [CrossRef]
7. Kay, A.R.; Toth, K. Is Zinc a Neuromodulator? *Sci. Signal.* **2008**, *1*, 3. [CrossRef]
8. Frederickson, C.J.; Koh, J.Y.; Bush, A.I. The neurobiology of zinc in health and diseas. *Nat. Rev. Neurosci.* **2005**, *6*, 449–462. [CrossRef]
9. Takeda, A.; Tamano, H. Insight into zinc signaling from dietary zinc deficienc. *Brain Res. Rev.* **2009**, *62*, 33–44. [CrossRef]
10. Wong, B.A.; Friedle, S.; Lippard, S.J. Solution and Fluorescence Properties of Symmetric Dipicolylamine-Containing Dichlorofluorescein-Based Zn^{2+} Sensors. *JACS* **2009**, *131*, 7142–7152. [CrossRef] [PubMed]
11. Wong, B.A.; Friedle, S.; Lippard, S.J. Subtle Modification of 2,2-Dipicolylamine Lowers the Affinity and Improves the Turn-On of Zn(II)-Selective Fluorescent Sensors. *Inorg. Chem.* **2009**, *48*, 7009–7011. [CrossRef]
12. Nolan, E.M.; Jaworski, J.; Racine, M.E.; Sheng, M.; Lippard, S.J. Midrange affinity fluorescent Zn(II) sensors of the Zinpyr family: Syntheses, characterization, and biological imaging applications. *Inorg. Chem.* **2006**, *45*, 9748–9757. [CrossRef]
13. Nolan, E.M.; Lippard, S.J. The zinspy family of fluorescent zinc sensors: Syntheses and spectroscopic investigations. *Inorg. Chem.* **2004**, *43*, 8310–8317. [CrossRef] [PubMed]
14. Nolan, E.M.; Ryu, J.W.; Jaworski, J.; Feazell, R.P.; Sheng, M.; Lippard, S.J. Zinspy sensors with enhanced dynamic range for imaging neuronal cell zinc uptake and mobilization. *JACS* **2006**, *128*, 15517–15528. [CrossRef]
15. Zhang, S.W.; Adhikari, R.; Fang, M.; Dorh, N.; Li, C.; Jaishi, M.; Zhang, J.; Tiwari, A.; Pati, R.; Luo, F.-T.; et al. Near-Infrared Fluorescent Probes with Large Stokes Shifts for Sensing Zn(II) Ions in Living Cells. *ACS Sens.* **2016**, *1*, 1408–1415. [CrossRef]
16. Tian, X.H.; Zhang, Q.; Zhang, M.; Uvdal, K.; Wang, Q.; Chen, J.; Du, W.; Huang, B.; Wu, J.; Tian, Y. Probe for simultaneous membrane and nucleus labeling in living cells and in vivo bioimaging using a two-photon absorption water-soluble Zn(II) terpyridine complex with a reduced pi-conjugation system. *Chem. Sci.* **2017**, *8*, 142–149. [CrossRef]
17. Huang, Y.W.; Lin, Q.; Wu, J.; Fu, N. Design and synthesis of a squaraine based near-infrared fluorescent probe for the ratiometric detection of Zn^{2+} ions. *Dyes Pigm.* **2013**, *99*, 699–704. [CrossRef]
18. Zhang, J.T.; Li, C.; Dutta, C.; Fang, M.; Zhang, S.; Tiwari, A.; Werner, T.; Luo, F.-T.; Liu, H. A novel near-infrared fluorescent probe for sensitive detection of beta-galactosidase in living cells. *Anal. Chim. Acta.* **2017**, *968*, 97–104. [CrossRef] [PubMed]
19. Yuan, L.; Lin, W.; Zhao, S.; Gao, W.; Chen, B.; He, L.; Zhu, S. A Unique Approach to Development of Near-Infrared Fluorescent Sensors for in Vivo Imaging. *JACS* **2012**, *134*, 13510–13523. [CrossRef] [PubMed]
20. Zastrow, M.L.; Radford, R.J.; Chyan, W.; Anderson, C.T.; Zhang, D.Y.; Laos, A.; Tzounopoulos, T.; Lippard, S.J. Reaction-Based Probes for Imaging Mobile Zinc in Live Cells and Tissues. *ACS Sens.* **2016**, *1*, 32–39. [CrossRef] [PubMed]
21. Zhang, S.W.; Chen, T.-H.; Lee, H.-M.; Bi, J.; Ghosh, A.; Fang, M.; Gian, Z.; Xie, F.; Ainsley, J.; Christov, C.; et al. Luminescent Probes for Sensitive Detection of pH Changes in Live Cells through Two Near-Infrared Luminescence Channels. *ACS Sens.* **2017**, *2*, 924–931. [CrossRef]

22. Frisch, M.J.; Trucks, G.M.; Schlegel, H.B.; Scuseria, G.E.; Robb, M.A.; Cheeseman, J.R.; Scalmani, G.; Barone, V.; Mennucci, B.; Petersson, G.A.; et al. *Gaussian 09*, Revision B. 01. 2009; Gaussian, Inc.: Wallingford, CT, USA, 2009.

23. Austin, A.; Petersson, G.A.; Frisch, M.J.; Dobek, F.J.; Scalmani, G.; Throssell, K. A Density Functional with Spherical Atom Dispersion Terms. *JCTC* **2012**, *8*, 4989–5007. [CrossRef]

24. Frisch, M.J.; Trucks, G.W.; Schlegel, H.B.; Scuseria, G.E.; Robb, M.A.; Cheeseman, J.R.; Scalmani, G.; Barone, V.; Petersson, G.A.; Nakatsuji, H.; et al. *Gaussian 16*, Revision A.03. 2016; Gaussian, Inc.: Wallingford, CT, USA, 2016.

25. Xia, S.; Wang, J.; Bi, J.; Wang, X.; Fang, M.; Phillips, T.; May, A.; Conner, N.; Tanasova, M.; Luo, F.-T.; et al. Fluorescent probes based on pi-conjugation modulation between hemicyanine and coumarin moieties for ratiometric detection of pH changes in live cells with visible and near-infrared channels. *Sens. Actuator B-Chem.* **2018**, *265*, 699–708. [CrossRef]

26. Wang, J.B.; Xia, S.; Bi, J.; Fang, M.; Mazi, W.; Zhang, Y.; Conner, N.; Luo, F.-T.; Lu, H.P.; Liu, H. Ratiometric Near-Infrared Fluorescent Probes Based On Through Bond Energy Transfer and pi-Conjugation Modulation between Tetraphenylethene and Hemicyanine Moieties for Sensitive Detection of pH Changes in Live Cells. *Bioconjug. Chem.* **2018**, *29*, 1406–1418. [CrossRef]

27. Zhu, S.L.; Zhang, J.; Janjanam, J.; Vegesna, G.; Luo, F.T.; Tiwari, A.; Liu, H. Highly water-soluble BODIPY-based fluorescent probes for sensitive fluorescent sensing of zinc(II). *J. Mater. Chem. B* **2013**, *1*, 1722–1728. [CrossRef]

Sample Availability: Probes A and B are available from the authors.

molecules

MDPI

Article

Insight into Rapid DNA-Specific Identification of Animal Origin Based on FTIR Analysis: A Case Study

Yahong Han, Lin Jian, Yumei Yao, Xinlei Wang, Lujia Han and Xian Liu *

College of Engineering, China Agricultural University, Beijing 100083, China; hyhmelody@cau.edu.cn (Y.H.); jianlin@cau.edu.cn (L.J.); Yaoao@cau.edu.cn (Y.Y.); wangxinlei917@cau.edu.cn (X.W.); hanlj@cau.edu.cn (L.H.)
* Correspondence: lx@cau.edu.cn; Tel.: +86-10-6273-6778

Received: 11 October 2018; Accepted: 29 October 2018; Published: 1 November 2018

Abstract: In this study, a methodology has been proposed to identify the origin of animal DNA, employing high throughput extension accessory Fourier transform infrared (HT-FTIR) spectroscopy coupled with chemometrics. Important discriminatory characteristics were identified in the FTIR spectral peaks of 51 standard DNA samples (25 from bovine and 26 from fish origins), including 1710, 1659, 1608, 1531, 1404, 1375, 1248, 1091, 1060, and 966 cm^{-1}. In particular, the bands at 1708 and 1668 cm^{-1} were higher in fish DNA than in bovine DNA, while the reverse was true for the band at 1530 cm^{-1} was shown the opposite result. It was also found that the PO_2^- V_{as}/V_s ratio (1238/1094 cm^{-1}) was significantly higher ($p < 0.05$) in bovine DNA than in fish DNA. These discriminatory characteristics were further revealed to be closely related to the base content and base sequences of different samples. Multivariate analyses, such as principal component analysis (PCA) and partial least squares-discriminant analysis (PLS-DA) were conducted, and both the sensitivity and specificity values of PLS-DA model were one. This methodology has been further validated by 20 meat tissue samples (4 from bovine, 5 from ovine, 5 from porcine, and 6 from fish origins), and these were successfully differentiated. This case study demonstrated that FTIR spectroscopy coupled with PLS-DA discriminant model could provide a rapid, sensitive, and reliable approach for the identification of DNA of animal origin. This methodology could be widely applied in food, feed, forensic science, and archaeology studies.

Keywords: DNA; FTIR spectroscopy; rapid identification; PLS-DA; animal origin

1. Introduction

Deoxyribonucleic acid (DNA) is the genetic material in all living organisms. Many studies on DNA-specific identification continue to be developed for a range of applications, for example, differentiating murine from mutton meat [1]; identifying poisonous mushrooms [2]; identifying fish species in cooked products [3]; detecting meat and poultry species in ground meats, deli meats, canned meats; and dried meats [4] to ensure food safety. In addition, DNA analysis has also been applied in identifying pork, beef, chicken and mutton origins in food products [5] and feed products [6,7] to control adulteration. Besides, it has contributed to the identification of parasite for the rapid detection of infectious disease [8]. In general, methods for the identification of DNA, such as polymerase chain reaction (PCR), DNA metabarcoding, and polymerase chain reaction-restriction fragment length polymorphism (PCR-RFLP) have been commonly used in these studies. These methods are generally based on specific DNA fragments, and can obtain high accurate rates. However, these methods are time-consuming, expensive, and require skilled labor [9].

Compared with traditional molecular technologies, Fourier transform infrared (FTIR) spectroscopy has the advantage of great simplicity, rapidity and cheap. It is widely employed in the fingerprint identification of molecular composition and structure. Therefore, FTIR spectroscopy could provide a new insight into rapid identification based on the difference in DNA molecular composition and structure. To date, only a few studies have focused on rapid identification using FTIR. For example, it was used for differentiating *japonica* from *indica* rice varieties based on DNA structural differences [10]. Subsequently, this method has been extended to differentiate other species, such as varieties of Chinese cabbage [11], and *Camellia reticulata* Lindl in the Chuxiong population [12]. All these studies deal with plant species. Furthermore, it was used in microorganism identification, such as enterococci [13], and invertebrate animals, such as nematode [14]. However, investigations on the rapid identification of vertebrate animal DNA by FTIR spectroscopy can also be helpful in a range of applications, such as the detection of bovine meat and bone meal in fishmeal, due to that bovine meat and bone meal were prohibited from feeding animals to control bovine spongiform encephalopathy crises in Europe [15], differentiating horse meat from beef to control the "European horsemeat crisis" [5], and the identification of wildlife forensic animal to avoid illegal mammalian wildlife trafficking [16,17]. To the best of our knowledge, the mechanism and model of animal DNA discrimination based on FTIR has not been reported previously.

Based on our previous work [18], high throughput extension accessory FTIR (HT-FTIR) spectroscopy was used in the present study for identification of animal DNA. Bovine and fish standard DNA samples were involved in a case study. The discrimination analysis was developed by a combination of principal component analysis (PCA) and partial least squares-discriminant analysis (PLS-DA). Furthermore, the mechanism of the discrimination was investigated. The methodology reported herein was also validated by use on 20 market meat tissue samples, with emphasis on the analysis of discriminatory characteristics.

2. Results and Discussion

2.1. Infrared Spectral Characteristics of Bovine and Fish Standard DNA Samples

Figure 1 shows the FTIR spectra of bovine and fish standard DNA samples, which shared similar spectral characteristics. FTIR spectra of DNA samples showed major peaks in three regions, including $1800–1500 \text{ cm}^{-1}$, $1500–1250 \text{ cm}^{-1}$, and $1250–800 \text{ cm}^{-1}$. Similar IR spectral peaks have been identified in related studies [19–24]. The assignment of specific absorption bands in three regions are summarized in Table 1. Absorption bands in the region between 1800 and 1500 cm^{-1} were mainly derived from the vibration of double bonds, such as C=N, C=O and C=C [21,25]. These vibrations might be easily sensitive to the effects of base stacking and pairing [21,22]. Absorption intensities in the region from 1500 to 1250 cm^{-1} were mainly attributed to pyrimidine and purine ring modes [19,21,23]. The vibrations of peaks in this region were easily affected by the sugar puckering modes, glycosidic bond rotation, and backbone conformation [21]. Finally, absorption bands in the region ($1250–800 \text{ cm}^{-1}$) could mostly be attributed to the vibration of PO_2^- groups and deoxyribose stretching, which was extremely sensitive to DNA backbone conformation [11,21,26].

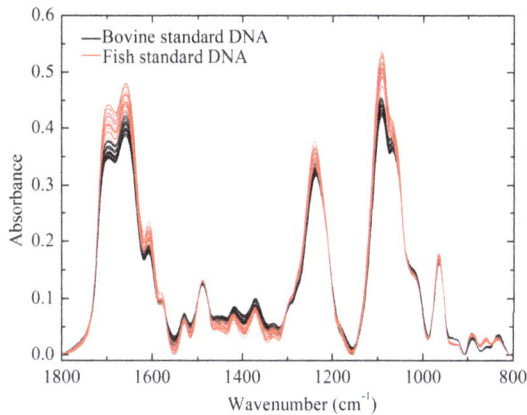

Figure 1. FTIR spectra of calf and salmon DNA.

Table 1. Assignment of characteristic peaks of bovine and fish standard DNA samples.

Assignment	Comment	Frequencies		References
		Bovine Standard DNA	Fish Standard DNA	
Thymine	C=O stretching	1712	1711	[19,21,23,24]
Thymine	C=O stretching	1659	1659	[19,20,22]
Adenine	Base/in-plane vibration	1608	1608	[20,22]
Cytosine, Guanine	Base/in-plane vibration	1529	1529	[19]
Adenine, Guanine	Ring vibration, C=N	1489	1489	[21]
Adenine, Guanine	Base/in-plane vibration	1420	1420	[19]
Adenine, Guanine	dA, dG anti	1371	1371	[23]
Backbone-A form	$V_{as} PO_2^-$	1238	1240	[22]
Backbone	$V_s PO_2^-$	1094	1092	[20]
Backbone	O-P-O bending	964	964	[19]
Deoxyribose	Deoxyribose ring vibration	889	889	[21]
Deoxyribose	Deoxyribose-phosphate, B-marker	831	833	[19]

2.2. Multivariate Analysis of FTIR Spectral Data

Multivariate analysis was used for further identifying DNA of different animal origins. Initially, PCA was used to assess the separation ability of FTIR spectra of bovine and fish standard DNA samples. The PCA score plot was built using the first two principal components (PCs), as illustrated in Figure 2. PC1 and PC2 could explain most of the variances in the sample clustering, which accounted for 99.84% and 0.14% of the total variation, respectively. As presented in Figure 2, bovine standard DNA showed positive values on PC2, while fish standard DNA samples showed negative scores on PC2. Bovine and fish standard DNA samples were clearly distinguished from each other. It was concluded from PCA result that bovine and fish standard DNA samples were discriminated explicitly without false prediction. Furthermore, a PLS-DA discriminant model was developed based on the DNA characteristics. The calibration model was chosen with leave-one-out cross validation, independent validation was used to assess model accuracy. The PLS-DA model result is shown in Table 2. For all data sets, including the calibration model set, the cross validation set and the independent validation set, both sensitivity and specificity values were 1.00, and the classification error was 0.00. These results indicate that this PLS-DA model was valid and reliable. From the combined results of PCA and PLS-DA, it was concluded that FTIR spectroscopy coupled with chemometrics can be successfully used to differentiate bovine from fish standard DNA samples.

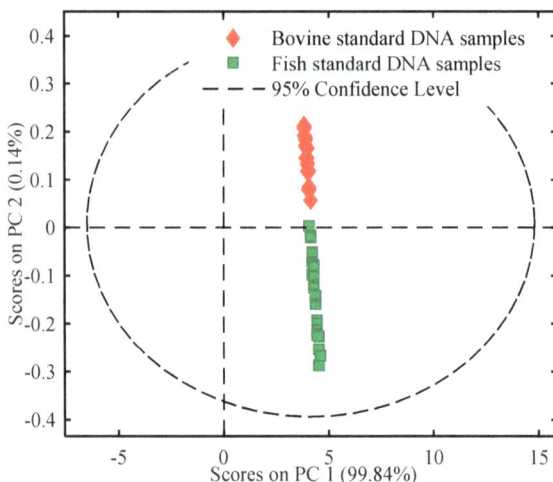

Figure 2. PCA score plot of FTIR spectral characteristics of DNA.

Table 2. Result of PLS-DA discriminant analysis.

	Bovine DNA Standard Samples	Fish DNA Standard Samples
Sensitivity (Cal)	1.00	1.00
Specificity (Cal)	1.00	1.00
Classification error (Cal)	0.00	0.00
Sensitivity (CV)	1.00	1.00
Specificity (CV)	1.00	1.00
Classification error (CV)	0.00	0.00
Sensitivity (Val)	1.00	1.00
Specificity (Val)	1.00	1.00
Classification error (Val)	0.00	0.00

The Cal refers a calibration set; The CV refers a cross-validation set; The Val refers an independent-validation set.

2.3. Discussion on the Mechanism of the Discrimination Model

2.3.1. The Average-Difference Profile Analysis

In order to explore the difference between bovine and fish standard DNA samples, the average-difference profile was constructed in Figure 3. The absorption intensities in the region between 1740 and 1600 cm^{-1} were below zero, while the band at around 1530 cm^{-1} was above zero. Both bands at 1708 and 1668 cm^{-1} in the region (1700–1600 cm^{-1}) could be due to C=O stretching vibration of thymine, while the band at 1530 cm^{-1} could be associated with the vibration of cytosine and guanine. A similar result was obtained by Mello and Vidal [27], who suggested that FTIR spectra of bovine DNA was higher than that of fish DNA at the band at 1661 cm^{-1}, while it showed the opposite result in the 1600–1500 cm^{-1}. An important cause may be that bovine DNA has higher GC content, and thus lower AT content, which could have an effect on the vibration of the base pairing, as well as hydrogen bonds [28]. Besides these characteristic peaks, there were other obvious spectral differences at 1404 and 1091 cm^{-1}, which may be ascribed to vibration of all bases (adenine, thymine, cytosine, and guanine) and PO$_2$$^-$ groups [19,24,29].

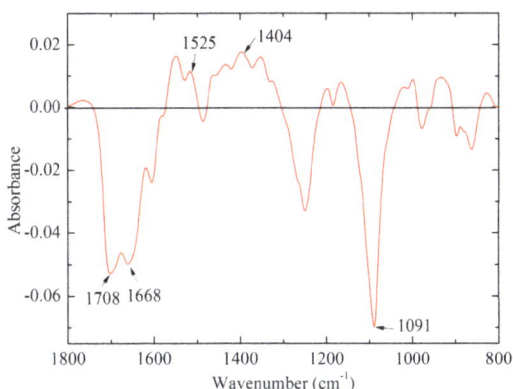

Figure 3. A difference profile between bovine (minuend) and fish (subtrahend) DNA standard samples.

2.3.2. Comparison of Spectral Characteristics of DNA of Two Groups

Besides the analysis of mean spectra of bovine and fish standard DNA samples, spectral distances, including the intergroup and intragroup distances, were analyzed via Euclidian distance. For each standard DNA sample, the intragroup distance was the average of 21 Euclidian distances, which were calculated between one sample and the other within the same type. The intergroup distance was the average of 22 Euclidian distances, between one sample and all the samples in the different group. The intergroup and intragroup distances of each standard DNA sample are illustrated in Figure 4. Average intragroup distance was 0.18 ± 0.06, while intergroup distance between bovine and fish standard DNA samples was 0.48 ± 0.13. More importantly, a nonparametric test of intergroup and intragroup distances was conducted. This revealed that there was a significant difference ($p < 0.05$) between intragroup and intergroup distances, which may explain why bovine and fish standard DNA samples could be differentiated.

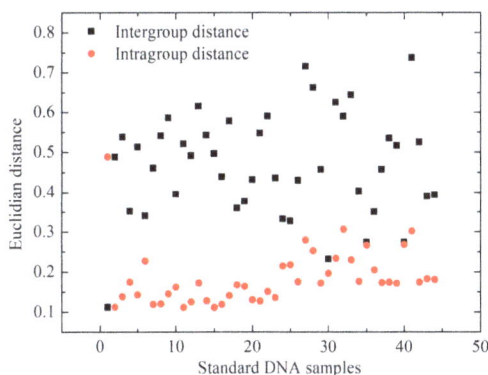

Figure 4. Spectral distance of standard DNA samples including intergroup and intragroup distances.

2.3.3. The Contribution of IR Spectral Characteristic Peaks of DNA

As mentioned above, bovine and fish standard DNA samples could be readily separated by the PCA score plot, especially by PC2 (Figure 2). The PCA loading plot could be used to reveal the most discriminative peaks [30]. Therefore, the loading plot of PC2 was further analyzed for exploring the highest contribution in the discrimination model. As illustrated in Figure 5A, several infrared absorption bands, including 1710, 1531, 1404, 1375, 1248, 1091, and 1060 cm^{-1}, could be considered as the highest contribution to the distinction between bovine and fish standard DNA samples. It should be noted that these peaks at 1710, 1531, 1404, and 1091 cm^{-1} were discussed previously (seen in Section 2.3.1). Apart from these bands, the peak at 1375 cm^{-1} was attributed to anti vibration of adenine and guanine [23]. It is consistent with a previous study by Qiu [12], who demonstrated that the band at 1388 cm^{-1} contributed the most to the differentiation among ten varieties of *Camellia reticulata* Lindl. from Chuxiong population. The peak at 1248 cm^{-1} was ascribed to phosphate antisymmetric stretching. Furthermore, it was found that the PO_2^- V_{as}/V_s ratio of bovine standard DNA samples was 0.75 ± 0.0046, while it was 0.70 ± 0.0048 for fish standard DNA samples. It was further confirmed that the PO_2^- V_{as}/V_s ratio was significantly higher ($p < 0.05$) in bovine than fish standard DNA samples, which is in close agreement with a previous finding of Mello and Vidal [27]. It could be speculated that the PO_2^- group may play an important role in differentiating two types of DNA. It was also seen that this ratio was different from that reported by Mello and Vidal [27], who suggested that the PO_2^- V_{as}/V_s ratio of bovine and fish DNA obtained by using ARO objective were 0.92 and 0.87, respectively; while they were 0.67 and 0.61, respectively, when obtained by using ATR objective. The difference may be due to the different sampling techniques used in each case.

To explore the specific contribution of more characteristic peaks in the discriminant model, an independent samples test-analysis of spectral intensities of bovine and fish standard DNA samples was conducted at each spectral wavenumber. Moreover, adjusting *p* values were calculated by false discovery rates to make multiple testing corrections, as shown in Figure 5B. It was found that there were significant differences ($p < 0.05$) in 469 sites of 520 wavenumbers, particularly at 452 sites ($p < 0.01$). This indicated the ability of FTIR spectral analysis for differentiating bovine from fish DNA. It may be speculated that, together, these spectral peaks have an effect on distinguishing bovine from fish DNA samples.

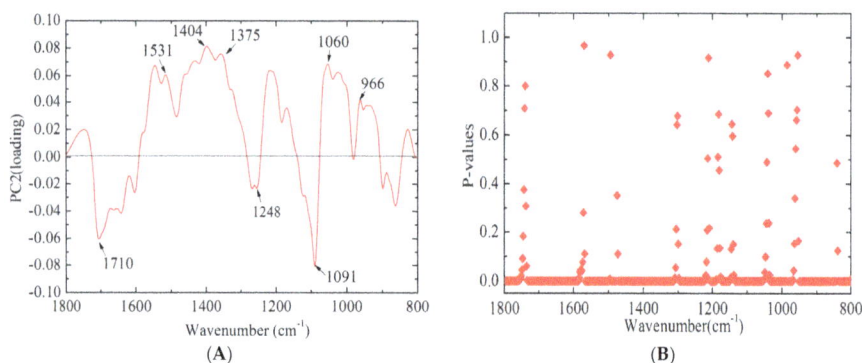

Figure 5. Analysis of the mechanism of the discriminant model ((**A**) loading plot for second principal component (PC2) loading; and (**B**) statistical significance of DNA standard samples in different wavenumbers, based on a *t*-test).

More importantly, these bands were sensitive to base stacking and base pairing, while the alteration of base stacking and pairing could be associated with different base sequences of DNA from different animal origins. Base stacking interactions between pyrimidine and purine bases in the following trend have been demonstrated: Pyrimidine–pyrimidine < purine–pyrimidine < purine–purine [31–33]. These interactions could have an influence on absorption intensities of the DNA bands. It has been found that the degree of propeller twist is sensitive to the base sequence of DNA [34].

2.4. Methodology Validation of Market Meat Samples

In order to verify the practical applicability of the method developed, genomic DNA of meat tissue samples were used for HT-FTIR spectroscopic analysis.

2.4.1. Bovine and Fish Samples

Ten meat tissues (four from bovine and six from fish origins) were used to build a PCA score plot to verify this developed method. PC2 and PC4 were used, which could account for 1.76% and 0.14% of the total variation, respectively. As illustrated in Figure 6A, only bovine DNA had negative values onPC2 and PC4, which indicated that bovine and fish DNA samples were successfully differentiated. Furthermore, a PLS-DA model with leave-one-out cross-validation has been developed. Results showed that the sensitivity, specificity and classification error were one, one, and zero, respectively, which indicated that bovine and fish DNA could be successfully separated. In addition, the PO_2^- V_{as}/V_s ratio was found to be significantly higher ($p < 0.05$) in bovine DNA samples than that in fish DNA samples, which is consistent with results in Section 2.3.3. Thus, it may be concluded that the PO_2^- V_{as}/Vs ratio could be an important biomarker for differentiation bovine from fish DNA samples.

In order to explore the contribution of characteristic peaks, the difference profile and the loading plot were analyzed. The difference profile between bovine (minuend) and fish (subtrahend) DNA samples is shown in Figure 6B. Comparison with the results of standard DNA samples (Figure 3), there are several similarities. The peaks at around 1676 cm^{-1} (thymine) and 1604 cm^{-1} (adenine) were below zero, while the band at 1530 cm^{-1} (cytosine and guanine) was above zero (Figure 6B). This phenomenon was consistently observed in Figure 3. The loading plot of PC2 and PC4 is presented in Figure 6C. Several high marked bands, including 1699, 1531, 1404, 1375, 1250, 1106, 1070, and 960 cm^{-1} were identified. All marked bands were in good agreement with Figure 5A. However, compared with Figure 5A, there were several peak shifts, including 1699, 1250, and 1106 cm^{-1}. One possible explanation could be that genomic DNA extracted from meat samples was not as pure as standard DNA. Overall, high marked bands, including 1710, 1659, 1608, 1531, 1404, 1375, 1248, 1091, 1060, and 966 cm^{-1}, were confirmed in DNA spectra of market meat samples.

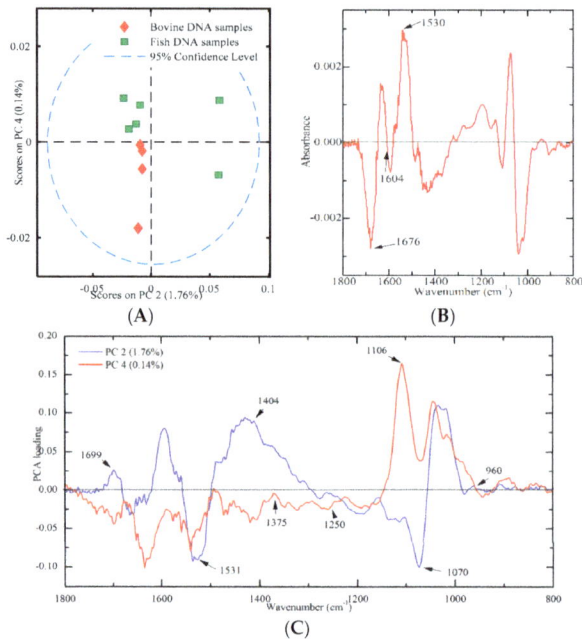

Figure 6. Multivariate analysis of FT-IR spectral data on genomic DNA from meat tissues ((**A**) a PCA plot; (**B**) a difference profile between bovine (minuend) and fish (subtrahend) DNA samples; and (**C**) a PCA loading plot).

2.4.2. Bovine and Porcine Samples

Nine meat tissues (four from bovine and five from porcine origins) were used to build a PCA score plot to verify this proposed method. A PCA score plot was established by using PC1 and PC2, which could account for 91.80% and 7.00% of the total variation, respectively. As presented in Figure 7A, porcine DNA had positive values, while bovine DNA samples had negative values on PC2, which indicated that bovine and porcine DNA samples could be clearly separated. In order to explore the contribution of characteristic peaks, the difference profile and the loading plot were also analyzed. The difference profile between bovine (minuend) and porcine (subtrahend) DNA samples and the loading plot of PC2 are presented in Figure 7B,C. Several characteristic peaks at 1670, 1659, 1608, 1110, 1070, and 1042 cm^{-1} were identified.

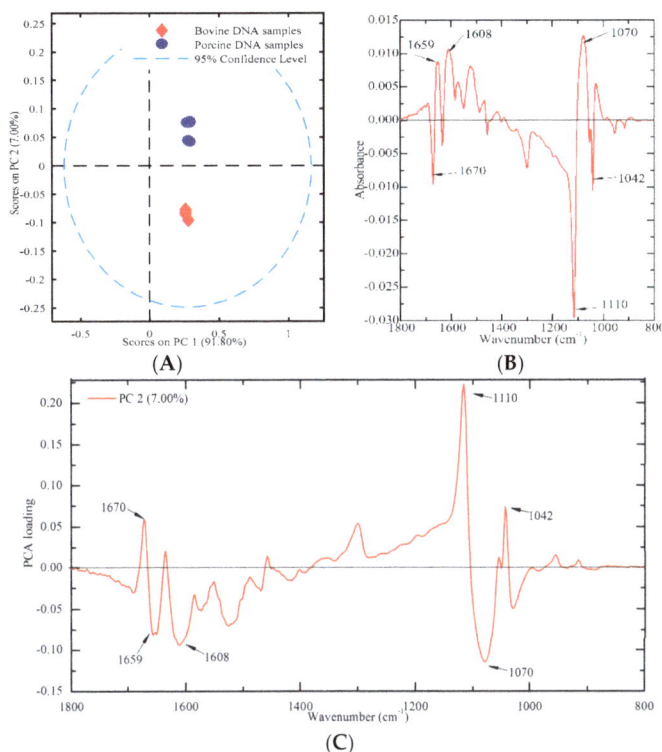

Figure 7. Multivariate analysis of FT-IR spectral data on genomic DNA from meat tissues ((**A**) a PCA plot; (**B**) difference profile between bovine (minuend) and porcine (subtrahend) DNA samples; and (**C**) a PCA loading plot).

2.4.3. Bovine and Ovine Samples

Nine meat tissues (four from bovine and five from ovine origins) were used to build a PCA score plot to verify this established method. A PCA score plot was built using PC1 and PC2, which could account for 92.09% and 7.41% of the total variation, respectively. As presented in Figure 8A, ovine DNA had negative values, while bovine DNA samples had positive values on PC2, which indicated that bovine and ovine DNA samples could be clearly separated. In order to explore the contribution of characteristic peaks, the difference profile and the loading plot were also analyzed. The difference profile between bovine (minuend) and ovine (subtrahend) DNA samples are presented in Figure 8B. The most significant spectral variations were a concentrated distribution in the region of 1800–1500 cm^{-1}. It may be due to that the characteristics bands in this region, which are sensitive to DNA base pairing, base stacking, and the propeller twist of the DNA structure. Furthermore, the loading plot of PC2 was analyzed in Figure 8C. Several high marked bands, including 1650, 1608, 1418, and 1070 cm^{-1} were revealed.

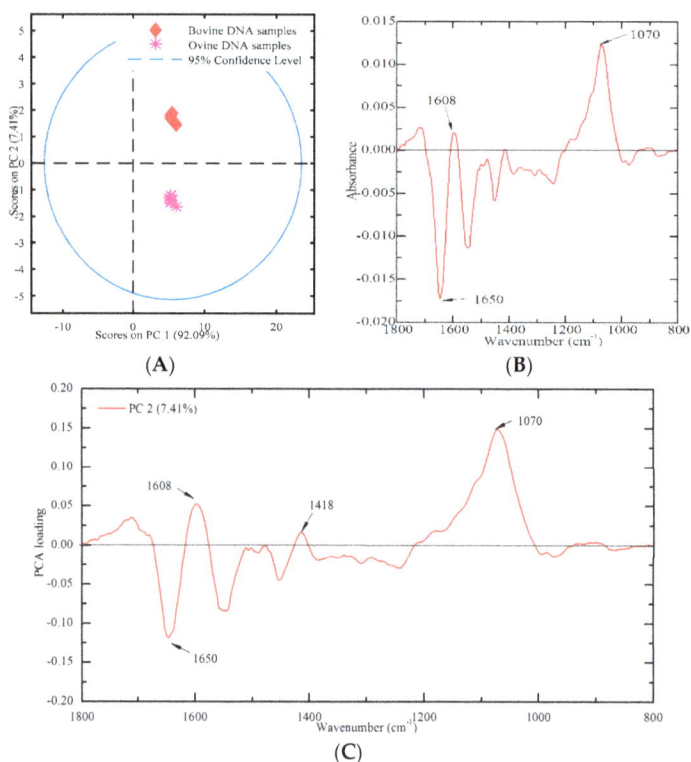

Figure 8. Multivariate analysis of FT-IR spectral data on genomic DNA from meat tissues ((**A**) a PCA plot; (**B**) difference profile between bovine (minuend) and ovine (subtrahend) DNA samples; and (**C**) a PCA loading plot).

3. Material and Methods

3.1. Materials

Double-stranded DNA derived from calf thymus (bovine standard DNA) and salmon testis (fish standard DNA) were purchased from Sigma (Sigma-Aldrich, St. Louis, MO, USA). 20 meat tissue samples (four from bovine, five from ovine, five from porcine, and six from fish origins), were purchased from a local market in Beijing, China. Minced beef tenderloin, porcine leg, ovine leg and fish meat tissues were used in our study. TIANamp genomic DNA kit (DP304, supplied with buffer PW, Rnase A, GD, GB, GA, proteinase K and column CB3) was obtained from Tiangen (Beijing, China). Ethanol (99.9%) was purchased from Beijing Chemical Co., Ltd. (Beijing, China) and nucleasefree water was obtained from Promega (Madison, WI, USA) was used.

3.2. Preparation of DNA Samples

Genomic DNA was extracted from 20 meat tissue samples by TIANamp genomic DNA kit following supplier instructions [35]. Briefly, 20 mg of animal tissue were treated to cells suspension, which were then centrifuged (10,000 rpm) for one min. The supernatant was discarded and 200 µL of GA buffer were added to resuspend the cell pellet. Next, 4 µL of RNase A and 20 µL of Proteinase K were added in succession. The solution was then incubated at 56 °C until the tissue is completely lysed; next, 200 µL of GB buffer were mixed with the solution, and this was incubated at 70 °C for 10 min for homogenizing before adding 200 µL of ethanol, mixing and then pipetting into a CB3 spin column. This column was centrifuged at 12,000 rpm for 30 s and then placed into the collection tube. This procedure was repeated three times, with 500 μL of GD buffer, 600 μL of PW buffer and another 600 μL of PW buffer in succession. After this, the CB3 spin column was centrifuged at 12,000 rpm for two min to dry the membrane completely. Finally, 50 μL of water were pipetted to the membrane of the CB3 spin column, which was then incubated at room temperature for two min, and again centrifuged at 12,000 rpm for two min. The solution was then collected into a new, clean 1.5-mL microcentrifuge tube.

In all, 71 DNA samples were involved in this study, including 44 standard DNA samples (25 from bovine and 26 from fish origins) as a calibration set, 7 standard DNA samples (3 from bovine and 4 from fish origins) as an independent validation set, and 20 DNA samples for market meat-tissue validation. The final concentration of all DNA samples was adjusted 100 ng/μL.

DNA sample purity was evaluated by calculating the absorption ratio at 260/280 nm using a micro-UV spectrophotometer (Nanodrop, Thermo Fisher Scientific, San Jose, CA, USA) for further confirmation [36–38].

3.3. HT-FTIR Measurements

The sampling technique and pretreatment temperature in FTIR spectroscopic analysis of DNA were according to our previous report [18]. Briefly, prior to FTIR measurements, 15 μL of DNA solution were dried on a 96-well silicon plate (Bruker, Rheinstetten, Germany) at 30 °C for 30 min. A Tensor 27 FTIR spectrometer model coupled with a high throughput extension (HTS-XT) accessory (HT-FTIR, Bruker Inc., Germany) was used in all of the measurements. All spectra were collected between 4000 and 400 cm^{-1} with a spectral resolution of 4 cm^{-1}. 64 scans were co-added to improve the signal-to-noise ratio [39]. The IR spectra of each sample were run in triplicate.

3.4. Data Processing and Multivariate Analysis

The FTIR spectral data in the region (1800−800 cm^{-1}) was analyzed using the Matlab version R2015 (Mathworks, Natick, MA, USA). Multiplicative scattering correction (MSC) was used prior to the multivariate analysis, including PCA and PLS-DA [40–42]. Preprocessing and multivariate analysis were performed with the PLS Toolbox 8.0 (Eigenvector Research, Wenatchee, WA, USA). Moreover, sensitivity (percentage of correct positive results), specificity (percentage of correct negative results), and classification error (percentage of false results) were used to evaluate the discrimination model [43].

In addition, the mechanism of differentiation between bovine and fish DNA samples was explored by a combination of difference profile analysis, the nonparametric test of spectral distances and the independent samples test analysis of infrared intensities at each wavenumber. The nonparametric test and the independent samples tests were analyzed by software package SPSS 20.0 for Windows (SPSS Inc., Chicago, IL, USA).

4. Conclusions

In conclusion, HT-FTIR spectroscopy was demonstrated to be a simple, rapid and sensitivity method to identify genomic DNA from different animal origins. Both DNA standard samples and DNA from meat tissues samples were correctly differentiated. Important discriminatory peaks for bovine/fish model were identified. These peaks were sensitive to base pairing, base stacking, and glycosidic bond rotation, which were closely associated with the base sequence and GC contents. These results, combined with literature analysis, allow us to further speculate that HT-FTIR spectroscopy coupled with PLS-DA discriminant model could identify the DNA of animal origin within different subspecies. This methodology may be used in a wide array of applications, including food adulteration, archaeology, and forensic authentication.

Author Contributions: Y.H., L.H., and X.L. designed the research; Y.H. and X.L. wrote the manuscript; L.J., Y.Y. and X.W. performed the statistical analysis; L.J. and Y.Y. helped map the figures and revise the manuscript. All authors discussed, edited and approved the final version.

Funding: This research was supported by National Key R&D Program of China (2017YFGH001504) and China Agriculture Research System (CARS-36) and Innovative Research Team in University of Ministry of Education of China (IRT-17R105).

Acknowledgments: Special thanks go to Shihong Ma and Meiqin Yang (Microbiology division of the National Institutes for Food and Drug Control) for providing access to the HT-FTIR spectrophotometer (HTS-XT extension, Tensor 27, Bruker Inc., Germany).

Conflicts of Interest: The authors declare no conflicts of interest.

References

1. Fang, X.; Zhang, C. Detection of adulterated murine components in meat products by TaqMan© real-time PCR. *Food Chem.* **2016**, *192*, 485–490. [CrossRef] [PubMed]
2. Jansson, D.; Wolterink, A.; Bergwerff, L.; Hough, P.; Geukens, K.; Åstot, C. Source attribution profiling of five species of Amanita mushrooms from four European countries by high resolution liquid chromatography-mass spectrometry combined with multivariate statistical analysis and DNA-barcoding. *Talanta* **2018**, *186*, 636–644. [CrossRef] [PubMed]
3. Pollack, S.J.; Kawalek, M.D.; Williams-Hill, D.M.; Hellberg, R.S. Evaluation of DNA barcoding methodologies for the identification of fish species in cooked products. *Food Control* **2018**, *84*, 297–304. [CrossRef]
4. Hellberg, R.S.; Hernandez, B.C.; Hernandez, E.L. Identification of meat and poultry species in food products using DNA barcoding. *Food Control* **2017**, *80*, 23–28. [CrossRef]
5. Dai, Z.; Qiao, J.; Yang, S.; Hu, S.; Zuo, J.; Zhu, W.; Huang, C. Species authentication of common meat based on PCR analysis of the mitochondrial COI gene. *Appl. Biochem. Biotechnol.* **2015**, *176*, 1770–1780. [CrossRef] [PubMed]
6. Okuma, T.A.; Hellberg, R.S. Identification of meat species in pet foods using a real-time polymerase chain reaction (PCR) assay. *Food Control* **2015**, *50*, 9–17. [CrossRef]
7. Lahiff, S.; Glennon, M.; Lyng, J.; Smith, T.; Maher, M.; Shilton, N. Species-specific PCR for the identification of ovine, porcine and chicken species in meat and bone meal (MBM). *Mol. Cell. Probes* **2001**, *15*, 27–35. [CrossRef] [PubMed]
8. Luque-González, M.A.; Tabraue-Chávez, M.; López-Longarela, B.; Sánchez-Martín, R.M.; Ortiz-González, M.; Soriano-Rodríguez, M.; García-Salcedo, J.A.; Pernagallo, S.; Díaz-Mochón, J.J. Identification of Trypanosomatids by detecting Single Nucleotide Fingerprints using DNA analysis by dynamic chemistry with MALDI-ToF. *Talanta* **2018**, *176*, 299–307. [CrossRef] [PubMed]
9. Lecrenier, M.; Ledoux, Q.; Berben, G.; Fumière, O.; Saegerman, C.; Baeten, V.; Veys, P. Determination of the ruminant origin of bone particles using fluorescence in situ hybridization (FISH). *Sci. Rep.* **2014**, *4*, 5730. [CrossRef] [PubMed]
10. Emura, K.; Yamanaka, S.; Isoda, H.; Watanabe, K.N. Estimation for different genotypes of plants based on DNA analysis using near-infrared (NIR) and fourier-transform infrared (FT-IR) spectroscopy. *Breed. Sci.* **2006**, *56*, 399–403. [CrossRef]
11. Song, S.Y.; Jie, E.Y.; Ahn, M.S.; Lee, I.H.; Nou, I.; Min, B.W.; Kim, S.W. Fourier transform infrared (FT-IR) spectroscopy of genomic DNA to discriminate F1 progenies from their paternal lineage of Chinese cabbage (*Brassica rapa* subsp. *pekinensis*). *Mol. Breed.* **2014**, *33*, 453–464.
12. Qiu, L.; Wang, Z.; Liu, P.; Liu, R.; Cai, C.; Fan, S. Fourier Transform Infrared Spectroscopy of the DNA of the Chuxiong Population of Camellia reticulata Lindl. of China. *Spectrosc. Lett.* **2015**, *48*, 120–127. [CrossRef]
13. Kirschner, C.; Maquelin, K.; Pina, P.; Ngo Thi, N.A.; Choo-Smith, L.P.; Sockalingum, G.D.; Sandt, C.; Ami, D.; Orsini, F.; Doglia, S.M.; et al. Classification and Identification of Enterococci: A Comparative Phenotypic, Genotypic, and Vibrational Spectroscopic Study. *J. Clin. Microbiol.* **2001**, *39*, 1763–1770. [CrossRef] [PubMed]
14. Ami, D.; Natalello, A.; Zullini, A.; Doglia, S.M. Fourier transform infrared microspectroscopy as a new tool for nematode studies. *FEBS Lett.* **2004**, *576*, 297–300. [CrossRef] [PubMed]
15. Regulation (EC). No 999/2001 of the European Parliament and of the Council laying down rules for the prevention, control and eradication of certain transmissible spongiform encephalopathies. *Off. J. Eur. Union* **2001**, *31*, L147.

16. Staats, M.; Arulandhu, A.J.; Gravendeel, B.; Holst-Jensen, A.; Scholtens, I.; Peelen, T.; Prins, T.W.; Kok, E. Advances in DNA metabarcoding for food and wildlife forensic species identification. *Anal. Bioanal. Chem.* **2016**, *408*, 4615–4630. [CrossRef] [PubMed]

17. Kitpipit, T.; Chotigeat, W.; Linacre, A.; Thanakiatkrai, P. Forensic animal DNA analysis using economical two-step direct PCR. *Forensic Sci. Med. Pathol.* **2014**, *10*, 29–38. [CrossRef] [PubMed]

18. Han, Y.; Han, L.; Yao, Y.; Li, Y.; Liu, X. Key factors in FTIR spectroscopic analysis of DNA: The sampling technique, pretreatment temperature and sample concentration. *Anal. Methods* **2018**, *10*, 2436–2443. [CrossRef]

19. Mello, M.L.S.; Vidal, B.C. Changes in the infrared microspectroscopic characteristics of DNA caused by cationic elements, different base richness and single-stranded form. *PLoS ONE* **2012**, *7*, e43169. [CrossRef] [PubMed]

20. Jangir, D.K.; Tyagi, G.; Mehrotra, R.; Kundu, S. Carboplatin interaction with calf-thymus DNA: A FTIR spectroscopic approach. *J. Mol. Struct.* **2010**, *969*, 126–129. [CrossRef]

21. Banyay, M.; Sarkar, M.; Gräslund, A. A library of IR bands of nucleic acids in solution. *Biophys. Chem.* **2003**, *104*, 477–488. [CrossRef]

22. Ahmad, R.; Naoui, M.; Neault, J.F.; Diamantoglou, S.; Tajmir-Riahi, H.A. An FTIR spectroscopic study of calf-thymus DNA complexation with Al (III) and Ga (III) cations. *J. Biomol. Struct. Dyn.* **1996**, *13*, 795–802. [CrossRef] [PubMed]

23. Taillandier, E.; Liquier, J. Infrared spectroscopy of DNA. *Methods Enzymol.* **1991**, *211*, 307–335.

24. Alex, S.; Dupuis, P. FT-IR and Raman investigation of cadmium binding by DNA. *Inorg. Chim. Acta* **1989**, *157*, 271–281. [CrossRef]

25. Brewer, S.H.; Anthireya, S.J.; Lappi, S.E.; Drapcho, D.L.; Franzen, S. Detection of DNA hybridization on gold surfaces by polarization modulation infrared reflection absorption spectroscopy. *Langmuir* **2002**, *18*, 4460–4464. [CrossRef]

26. Blout, E.R.; Lenormant, H. Changes in the infrared spectra of solutions of deoxypentose nucleic acid in relation to its structure. *Biochim. Biophys. Acta* **1955**, *17*, 325–331. [CrossRef]

27. Mello, M.L.S.; Vidal, B.C. Analysis of the DNA Fourier transform-infrared microspectroscopic signature using an all-reflecting objective. *Micron* **2014**, *61*, 49–52. [CrossRef] [PubMed]

28. Adams, R.L. *The Biochemistry of the Nucleic Acids*; Springer Science & Business Media: New York, NY, USA, 2012.

29. Hembram, K.P.S.S.; Rao, G.M. Studies on CNTs/DNA composite. *Mater. Sci. Eng. C* **2009**, *29*, 1093–1097. [CrossRef]

30. Sivertsen, A.H.; Kimiya, T.; Heia, K. Automatic freshness assessment of cod (*Gadus morhua*) fillets by Vis/Nir spectroscopy. *J. Food Eng.* **2011**, *103*, 317–323. [CrossRef]

31. Guckian, K.M.; Schweitzer, B.A.; Ren, R.X.; Sheils, C.J.; Tahmassebi, D.C.; Kool, E.T. Factors contributing to aromatic stacking in water: Evaluation in the context of DNA. *J. Am. Chem. Soc.* **2000**, *122*, 2213–2222. [CrossRef] [PubMed]

32. Friedman, R.A.; Honig, B. A free energy analysis of nucleic acid base stacking in aqueous solution. *Biophys. J.* **1995**, *69*, 1528. [CrossRef]

33. Egli, M.; Saenger, W. *Principles of Nucleic Acid Structure*; Springer Science & Business Media: New York, NY, USA, 2013.

34. Nelson, H.C.; Finch, J.T.; Luisi, B.F.; Klug, A. The structure of an oligo (dA)·oligo (dT) tract and its biological implications. *Nature* **1987**, *330*, 221–226. [CrossRef] [PubMed]

35. Cheng, X.; He, W.; Huang, F.; Huang, M.; Zhou, G. Multiplex real-time PCR for the identification and quantification of DNA from duck, pig and chicken in Chinese blood curds. *Food Res. Int.* **2014**, *60*, 30–37. [CrossRef]

36. Yalçınkaya, B.; Yumbul, E.; Mozioğlu, E.; Akgoz, M. Comparison of DNA extraction methods for meat analysis. *Food Chem.* **2017**, *221*, 1253–1257. [CrossRef] [PubMed]

37. Steward, K.F.; Robinson, C.; Waller, A.S. Transcriptional changes are involved in phenotype switching in Streptococcus equi subspecies equi. *Mol. Biosyst.* **2016**, *12*, 1194–1200. [CrossRef] [PubMed]

38. Zhao, H.; Wei, J.; Xiang, L.; Cai, Z. Mass spectrometry investigation of DNA adduct formation from bisphenol A quinone metabolite and MCF-7 cell DNA. *Talanta* **2018**, *182*, 583–589. [CrossRef] [PubMed]

39. Kondepati, V.R.; Heise, H.M.; Oszinda, T.; Mueller, R.; Keese, M.; Backhaus, J. Detection of structural disorders in colorectal cancer DNA with Fourier-transform infrared spectroscopy. *Vib. Spectrosc.* **2008**, *46*, 150–157. [CrossRef]

40. Dina, N.E.; Muntean, C.M.; Leopold, N.; Fălămaş, A.; Halmagyi, A.; Coste, A. Structural changes induced in grapevine (*Vitis vinifera* L.) DNA by femtosecond IR laser pulses: A surface-enhanced Raman spectroscopic study. *Nanomaterials* **2016**, *6*, 96. [CrossRef] [PubMed]

41. Abbas, O.; Fernández Pierna, J.A.; Codony, R.; von Holst, C.; Baeten, V. Assessment of the discrimination of animal fat by FT-Raman spectroscopy. *J. Mol. Struct.* **2009**, *924–926*, 294–300. [CrossRef]

42. Correia, R.M.; Tosato, F.; Domingos, E.; Rodrigues, R.R.T.; Aquino, L.F.M.; Filgueiras, P.R.; Lacerda, V.; Romão, W. Portable near infrared spectroscopy applied to quality control of Brazilian coffee. *Talanta* **2018**, *176*, 59–68. [CrossRef] [PubMed]

43. Gao, F.; Xu, L.; Zhang, Y.; Yang, Z.; Han, L.; Liu, X. Analytical Raman spectroscopic study for discriminant analysis of different animal-derived feedstuff: Understanding the high correlation between Raman spectroscopy and lipid characteristics. *Food Chem.* **2018**, *240*, 989–996. [CrossRef] [PubMed]

Sample Availability: Samples of the compounds are not available from the authors.

molecules

MDPI

Article

On-The-Go VIS + SW − NIR Spectroscopy as a Reliable Monitoring Tool for Grape Composition within the Vineyard

Juan Fernández-Novales [1,2,*], Javier Tardáguila [1,2], Salvador Gutiérrez [3] and María Paz Diago [1,2,*]

[1] University of La Rioja, Department of Agriculture and Food Science, 26006 Logroño, Spain
[2] Institute of Grapevine and Wine Sciences (University of La Rioja, Consejo Superior de Investigaciones Científicas, Gobierno de La Rioja), 26007 Logroño, Spain
[3] Department of Computer Science and Engineering, University of Cádiz, Avda, de la Universidad de Cádiz 10, 11519 Puerto Real, Spain
* Correspondence: juan.fernandezn@unirioja.es (J.F.-N.); maria-paz.diago@unirioja.es (M.P.D.);
   Tel.: +34-941-894-980 (J.F.-N.); +34-941-299-731 (M.P.D.)

Received: 10 July 2019; Accepted: 30 July 2019; Published: 31 July 2019

Abstract: Visible-Short Wave Near Infrared (VIS + SW − NIR) spectroscopy is a real alternative to break down the next barrier in precision viticulture allowing a reliable monitoring of grape composition within the vineyard to facilitate the decision-making process dealing with grape quality sorting and harvest scheduling, for example. On-the-go spectral measurements of grape clusters were acquired in the field using a VIS + SW − NIR spectrometer, operating in the 570–990 nm spectral range, from a motorized platform moving at 5 km/h. Spectral measurements were acquired along four dates during grape ripening in 2017 on the east side of the canopy, which had been partially defoliated at cluster closure. Over the whole measuring season, a total of 144 experimental blocks were monitored, sampled and their fruit analyzed for total soluble solids (TSS), anthocyanin and total polyphenols concentrations using standard, wet chemistry reference methods. Partial Least Squares (PLS) regression was used as the algorithm for training the grape composition parameters' prediction models. The best cross-validation and external validation (prediction) models yielded determination coefficients of cross-validation (R^2_{cv}) and prediction (R^2_p) of 0.92 and 0.95 for TSS, $R^2_{cv} = 0.75$, and $R^2_p = 0.79$ for anthocyanins, and $R^2_{cv} = 0.42$ and $R^2_p = 0.43$ for total polyphenols. The vineyard variability maps generated for the different dates using this technology illustrate the capability to monitor the spatiotemporal dynamics and distribution of total soluble solids, anthocyanins and total polyphenols along grape ripening in a commercial vineyard.

Keywords: *Vitis vinifera* L.; proximal sensing; precision viticulture; near infrared; chemometrics; non-destructive sensor

---

1. Introduction

Grape berry ripening is usually described as the accumulation of sugars, and it is measured in terms of total soluble solids (°Brix). However, there are other compositional variables taken into account to determine the optimal maturity for harvest, such as berry acidity, often expressed as pH, titratable acidity and concentrations of tartaric and malic acids, berry weight, as well as the anthocyanin and total phenol concentrations (in red varieties) [1]. In red grape varieties, anthocyanins and other polyphenols are located in berry skins, and their accumulation starts at veraison and continues during ripening [2]. Anthocyanins are red coloured phenolic compounds or pigments, which are responsible for the red wine colour [3], while the other phenolic compounds, such as the flavonols, flavanols, hydroxycinnammic acids, and stilbenes may increase and stabilize the wine color by means of the copigmentation phenomenon [4,5] and contribute to the wine's mouthfeel and taste perception attributes [6]. Both the anthocyanin and total phenolic contents in berries are currently measured using 'wet chemistry' procedures [7,8], which are time-consuming and labour-intensive, while the total soluble solids (TSS) are more easily measured using a refractometer. Both methodologies are destructive and require time-consuming berry sampling and sample preparation (for anthocyanins and total polyphenols content), which hinder their massive application (analysis of a large number of samples) to assess the spatial variability of grape composition within a given vineyard plot at a specific date. The occurrence of spatial variability of grape compositional parameters in many vineyards is nothing new, and most widely used berry sampling procedures [9,10] do attempt to have it into consideration when defining the trajectories and manual sampling protocols within a plot. However, these same procedures end up with a berry sample of 100 to 200 berries or 20 to 40 clusters, regardless the size of the plot, in most cases. Should we consider a 1.0 ha vineyard plot, the estimated number of total berries could be well beyond 4–5 million berries. Therefore, a 200-berry sample, even if it is picked across many rows, barely represents less than ~0.005% of the berry production in this 1.0 ha plot, whose analytical measurements are often used to drive decisions on harvest scheduling or grape quality classification and pricing, in some cases.

Within the context of precision agriculture, the development of new sensors, especially based on spectroscopy, enables high resolution data acquisition that could be used to track crop development and ripening. In this regard, the capability to assess ripening in a fast, non-destructive way would substantially and positively impact the process of harvest scheduling and classification.

Visible and near-infrared (Vis-NIR) spectroscopy is a well-known technique for the non-destructive measurement of quality attributes of fruits and vegetables [11,12]. The Vis-NIR region covers the range of the electromagnetic spectrum between 380 and 2500 nm. Spectroscopy techniques combined with multivariate analysis have been widely used for quantitative determination of several quality properties or chemical compounds in fruit, to determine ripeness, and to measure quality indices [13–20].

In the context of grape composition monitoring during the ripening process using NIR spectroscopy, a variety of works has been published [21–25], but all these were conducted under controlled conditions, such as illumination, temperature, humidity, and sample positioning, among others-i.e. in a laboratory. In-field measurements have deserved less attention, and fewer number of works have addressed the utilization of manual, portable, hand-held spectrometers to assess the composition of grape berries while they are still on the grapevines [26–28].

One step forward has been recently given by Gutiérrez et al. (2019) [29] who reported the quantification of TSS and anthocyanins in grape berries under field conditions, using on-the-go hyperspectral imaging (HSI) between 400 and 900 nm, acquired from a moving platform. HSI is a very powerful technology, capable of recording the whole spectrum within a given spectral range in each specific pixel of a two-dimensional image. Its potential to yield a huge amount of relevant information is well recognized, but the difficulties in analyzing this information are also HSI's main drawback.

On-the-go VIS-NIR spectroscopy has been successfully used to assess the grapevine water status [30,31]. Using this technology, an average spectrum of a circular measuring spot with a diameter of ~1.9 cm is acquired at a rate of acquisition between 15 to 28 measurements per second, which provides a sufficient amount of spectral measurements, whose analysis is less complex than that of HSI.

Therefore, given the good prediction results that manual VIS-NIR spectroscopy demonstrated under control conditions, and the experiences of on-the-go VIS-NIR spectroscopy and HSI in the vineyard, the aim of the present work was to assess and map the grape composition parameters (TSS, anthocyanins content and total polyphenols) along the ripening period using a proximal VIS + SW-NIR sensor from a motorized platform in the vineyard.

2. Results

2.1. Berry Composition

The boxplots for the berry composition analysis in four different dates throughout the field experiment are shown in Figure 1. This type of graphs provides a convenient way of visually displaying the data distribution through their quartiles. The boxplots in Figure 1 are not skewed indicating that the data were normally distributed. Moreover, they illustrate the different ripening rates among grapevines and inherent variability per measurement date within the vineyard for the monitored compositional variables, such as TSS (Figure 1A), total anthocyanins (Figure 1B) and polyphenols (Figure 1C).

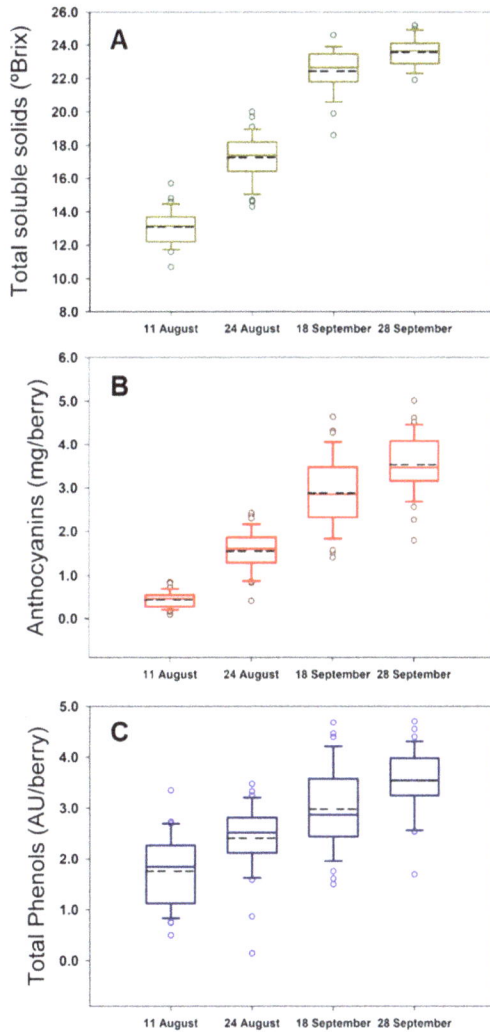

Figure 1. Box plots for total soluble solids (**A**), anthocyanins (**B**), and total polyphenols (**C**) during the four dates of the experimental study. Dashed lines represent mean values.

The studied parameters were well represented with an adequate variability. Likewise, TSS varied between 10.7 °Brix to 25.2 °Brix, while anthocyanins and total polyphenols ranged from 0.09 mg/berry and 0.14 AU/berry to 4.64 mg/berry and 4.70 AU/berry, respectively. The mean values increased throughout the season as illustrated in Figure 1.

2.2. Regression Models and Mapping for Grape Composition Parameters

The performance statistics of the best regression models of calibration, cross-validation, and external validation (prediction) for the prediction of TSS, anthocyanins and total polyphenols in grape clusters under field conditions from on-the-go Vis + SW − NIR spectroscopy are summarized in Table 1.

Table 1. Calibration, cross-validation, and external validation (prediction) of the best models obtained to predict the total soluble solids, anthocyanins and total polyphenols concentrations in grape clusters under field conditions from on-the-go Vis + SW − NIR spectroscopy (570–990 nm).

Parameters	Spectral Treatment	N	SD	Range	PLS Factor	Calibration		Cross-Validation		External Validation	
						RMSEC	R^2_c	RMSECV	R^2_{cv}	RMSEP	R^2_p
Total soluble solids (°Brix)	D1W15	116	4.403	10.70–25.20	7	1.119	0.93	1.248	0.92	1.011	0.95
Anthocyanins (mg/berry)	D1W15	116	1.329	0.09–4.64	6	0.607	0.79	0.664	0.75	0.618	0.79
Total polyphenols (Au/berry)	SNV + DT D1W15	116	0.947	0.14–4.70	7	0.642	0.54	0.728	0.42	0.749	0.43

SNV: standard normal variate. DnWm, Savitzky–Golay filter with n-degree derivative, window size of m. N: number of samples used for calibration and cross-validation models after outlier detection. SD: standard deviation. RMSEC: root mean square error of calibration. R^2_c: determination coefficient of calibration. RMSECV: root mean square error of cross-validation. R^2_{cv}: determination coefficient of cross-validation. RMSEP: root mean square error of prediction. R^2_p: determination coefficient of prediction.

Diverse pre-processing operations were applied. However, the best models involved the implementation of the Savitzky–Golay first derivative and a window size of 15. In the case of total polyphenols, the standard normal variate (SNV) filtering was also applied for spectra pre-processing. No anomalous spectra were identified following the Residuals (Q) and Hotelling values (T^2) and, moreover, the models were developed with low number of latent variables (Table 1), which calls for increased robustness and higher capability of generalization.

The best models for cross and external validations (also called prediction) returned determination coefficient values higher than 0.90 for TSS (R^2_{cv} = 0.92, R^2_p = 0.95), higher than 0.75 for anthocyanins (R^2_{cv} = 0.75, R^2_p = 0.79) and more modest values for total polyphenols (R^2_{cv} = 0.42, R^2_p = 0.43). The accuracy of the models in terms of RMSECV and RMSEP values were below 1.24 °Brix for TSS, 0.664 mg/berry for anthocyanins and 0.749 AU/berry for total polyphenols, respectively (Table 1).

Figure 2 shows the regression plots for the best prediction models for TSS, anthocyanins and total polyphenols in grape clusters under field conditions from on-the-go spectral measurements. A greater scattering of the points around the regression line was observed for the TSS in comparison with the other two parameters. All the samples from the TSS regression model exhibited a very good fit along the correlation line and were also inside of the 95% confidence bands, except three of them (Figure 2A).

A wide data range was covered by the samples from anthocyanins and total polyphenols regression models. The 1:1 line displayed a better fit over the anthocyanins regression line (Figure 2B) than over the total polyphenols regression line (Figure 2C) Additionally, seven out of 144 samples lied outside the anthocyanins prediction bands, keeping 95.14% of the samples (Figure 2B), while for total polyphenols 93.75% of the samples lied within prediction bands (Figure 2C).

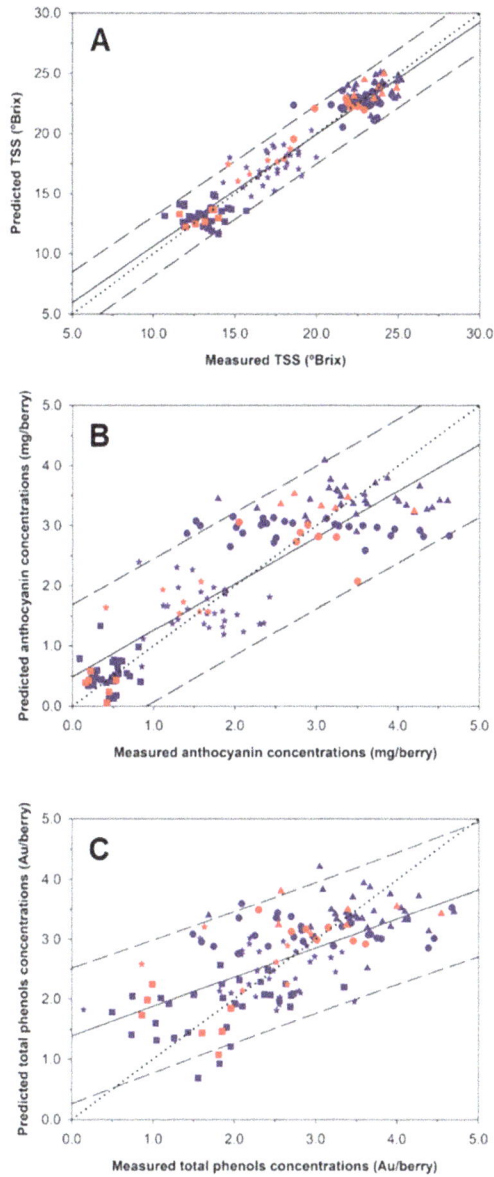

Figure 2. Regression plots for the total soluble solids (**A**), anthocyanins (**B**) and total polyphenols (**C**) using the best Partial Least Squares (PLS) models generated from on-the-go grape clusters spectral measurements. (blue color) 10-fold cross validation; (red color) external validation. (■: 11 August; *: 24 August; ●: 18 September; ▲: 28 September). Solid line represents the regression line and dotted line refers to the 1:1 line. Prediction confidence bands are shown at a 95% level (dashed lines).

To analyze the spatial variability of the grape composition parameters in a commercial vineyard along the different maturity stages (11 August to 28 September), maps for TSS, anthocyanins and total polyphenols were computed and presented in Figure 3. The highest values of TSS for each stage were mainly found at the north and south areas of the vineyard plot, while in the central part of the plot generally lower values were detected.

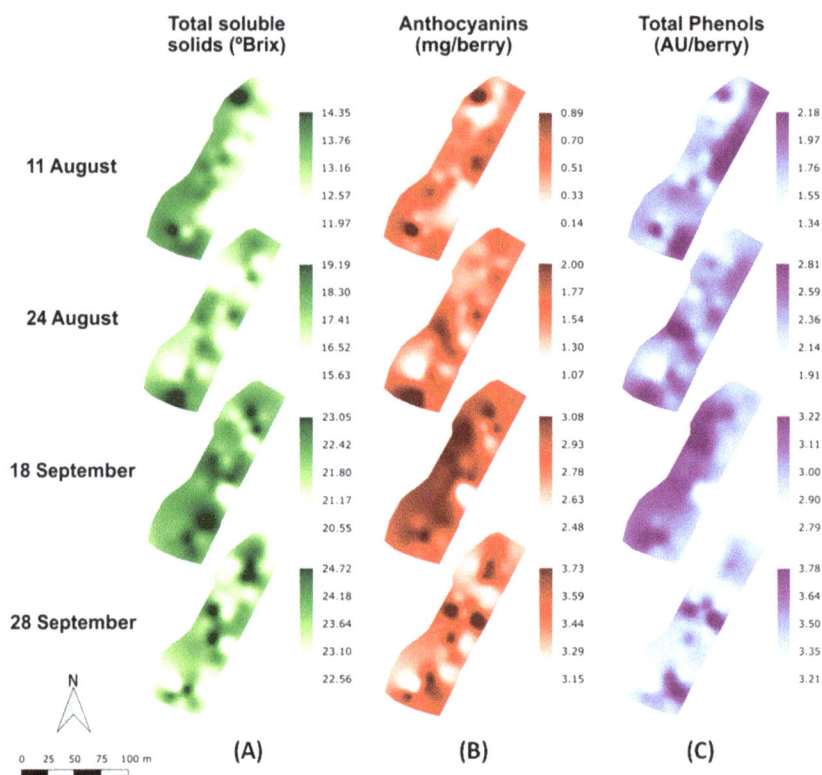

Figure 3. Prediction maps of the spatial variability of anthocyanins (**A**), total soluble solids (**B**), and total polyphenols concentrations (**C**) along the grape ripening period (11 August to 28 September).

The general evolution trend along the vineyard of anthocyanins and total polyphenols concentrations showed a different pattern than that of TSS, but very similar between them. Along the four measuring dates, a monotonous increase of the anthocyanin and total phenol concentration was observed. Since the anthocyanins largely contribute to the phenolic pool of compounds in the berries the similarity between their spatiotemporal evolution from veraison to harvest is coherent.

3. Discussion

The results presented in this work have demonstrated the capability of contactless VIS + SW − NIR reflectance spectroscopy in the range of 570–990 nm acquired on-the-go, from a motorized platform in the field, to estimate key parameters of grape composition. To achieve this, robust and reliable prediction models were generated for three important grape composition indicators from spectra of grape clusters acquired non-destructively from a moving vehicle at a speed (5 km/h) similar to that used for conventional machine operations. Furthermore, on-the-go VIS-SW-NIR spectroscopy has also been successful in characterizing the spatiotemporal dynamics of the accumulation of total soluble solids, anthocyanins and total polyphenols along ripening, within a commercial vineyard.

Many authors have reported that the monitoring of grape quality non-destructively through the ripening process under laboratory conditions is predominantly represented by two vibrational spectroscopy-related technologies: Near infrared spectroscopy [21–25] and hyperspectral imaging [32–35]. Nevertheless, little research has been conducted directly in the field using NIR spectroscopy [26–28]. In these works, a contact portable instrument was used to determine TSS in different red varieties, reporting prediction RMSE values of 1.25, 1.24, and 1.68 °Brix, respectively. In terms of the anthocyanin concentrations, [27] reported values of $R^2_v = 0.624$, RMSEV = 0.302 mg/g using the spectral range between 640 and 1300 nm. These performance numbers are in good agreement with the RMSEP values obtained in the present work, although in this case, spectral acquisition was carried out contactless, on-the-go, from a moving vehicle along the vineyard.

Very recently, the capability of in-field, on-the-go hyperspectral imaging for the assessment of total soluble solids and anthocyanin concentrations in wine grapes in a commercial vineyard has been tested [29]. In this study, the performance of the models obtained with support vector machines for TSS and anthocyanin concentrations returned determination coefficients for external prediction R^2 of 0.92 and 0.83 with RMSEP values of 1.274 and 0.211 mg/g berry, respectively. The accuracy and precision of the prediction statistics were in line with the ones presented in the present work.

In terms of the computational time, the processing of the spectral measurements for each block (five consecutive vines) took approximately 5 minutes. This process, partly automated, divided in four steps (Section 4.4.1), involved two different software packages. The prediction of the unknown spectra using PLS models required less than 1 minute per block. Therefore, the total time needed to process the spectra and to predict these three grape composition parameters (TSS, total anthocyanins, and total polyphenols) per block would be 6 minutes, that is less than 1.5 minutes per vine. Compared to hyperspectral imaging, the contactless spectrometer provided the spectra directly, without the need of computationally expensive processing. This simpler nature of the acquired data accounts for a reduced computational time (around 3.6 hours for 36 blocks of five vines each vs 5.5 hours for 36 hyperspectral images [29]). Moreover, since both a light source and the reference (white and dark) measurements are enclosed in the VIS + SW − NIR system, in-field monitoring and data acquisition becomes less affected by environmental lighting conditions than measurements with a hyperspectral camera.

The potential of VIS + SW − NIR contactless spectroscopy has also been confirmed through the development of prediction maps for grape TSS, anthocyanins and total polyphenols concentrations and their evolution along the four different dates during the ripening process. The mapping of the grape composition parameters in the vineyard plot can be addressed to classify the vineyard plot into different grape composition zones during ripening and to determine the optimal timing of harvest in each delineated zone, enabling the winegrowers with a new monitoring tool towards improved and optimized decision-making. Moreover, the developed system provides spatial information on grape composition at each measuring date, which could not be achieved with the traditional manual sampling protocol of 100 to 200 berries.

One important point to take into account is the need to establish predictive relationships for entire cluster compositional characters based on the composition of visible berries as a first step towards the development of non-destructive methods, such as the one presented in this work to measure grape composition [36]. In this regard, the work of Tang et al. [36] concluded that variation in cluster compactness could contribute to variation in predictive relationships among varieties. Additionally, other factors, such as row orientation, or the level of defoliation at the fruiting zone, could potentially affect in a dissimilar way the grape metabolism of exposed vs non-exposed berries. Therefore, further research should be conducted involving vineyards planted with different varieties, row orientations, trellising, subjected to different climates to ensure the robustness of the predictive models using contactless VIS + SW − NIR spectroscopy on-the-go.

The remarkable outcomes obtained in this study reveal the actual applicability of bringing this non-destructive methodology based on spectroscopy from indoor applications to the field, either embedded in agricultural vehicles (during another viticultural operation, e.g., tilling) or mounted on phenotyping platforms or even robots to monitor agricultural crops directly in the field [37–39].

4. Materials and Methods

The experimental study involved the on-the-go acquisition of grape cluster spectra from the fruiting zone of the grapevine canopy using a spectral device for the estimation of chemical grape composition parameters at several dates during the grape ripening period. The regions of grape clusters scanned by the device were picked for the analysis of TSS, anthocyanins and total polyphenols concentrations. Grape cluster spectral measurements were filtered, averaged and analysed to be modelled using Partial Least Squares (PLS). The last step was the spatiotemporal evaluation of these composition parameters by mapping them during all the measurement dates.

4.1. Experimental Layout

The experiment was performed in a commercial vineyard located in Ábalos, La Rioja, Spain (Lat. 42° 34′ 45.7″, Long. −2° 42′ 27.78″, Alt. 628 m) during four dates from early August to late September 2017, along the grape ripening period. The vineyard was planted in 2010 with grapevines of (*Vitis vinifera* L.) Tempranillo, grafted on rootstock R-110. The vines were trained to a vertically shoot-positioned (VSP) trellis system on a double-cordon Royat with vine spacing of 2.20 m between rows and 1.0 m between vines in a northeast-southwest orientation.

With the purpose of ensuring an appropriate variability of grape composition, three different equally-distanced rows were selected and, within each one of them, 12 blocks with five plants each were chosen for the spectral and grape analyses, making up a total of 36 blocks, that were monitored during four dates from veraison to harvest. The four dates belong to different phenological stages according to the modified Eichhorn and Lorenz system [40]: 11 August, stage 36; 24 August, stage 37; 18 September, stage 38; and 28 September, stage 38.

The spectral measurements (a total of 144 throughout the whole experiment) were carried out on the east side of the canopy. This side was defoliated at the end of July, following a common viticultural practice in the region, to promote air circulation and sun exposure in the cooler, morning hours of the day. Hence, 36 blocks were measured each date, making a total of 144 measurements.

4.2. On-The-Go VIS+SW-NIR Measurements

On-the-go spectral measurements in the vineyard were acquired using a VIS + SW − NIR PSS 1050 spectrometer (Polytec GmbH, Waldbronn, Germany) operating in the 570–990 nm spectral range, at a 2 nm resolution, with 215 datapoints per spectrum.

The spectrometer was an active VIS + SW − NIR optical device with a polychromator as reflection light source selector, and Silicon (Si) detectors. The system includes a sensor head for light emission (by an integrated 20W tungsten lamp) and capturing, and a processing unit, both linked by an optical fiber.

The system was mounted in the front part of an all-terrain-vehicle (ATV) (Trail Boss 330, Polaris Industries, Minnesota, USA), aiming to the left and able to make spectral acquisitions controlled by a physical trigger while the ATV was in motion. The sensor head was placed at a height of 0.80 m from the ground, pointed to the canopy on a lateral point of view at 0.30 m of distance in order to cover the fruiting zone (Figure 4).

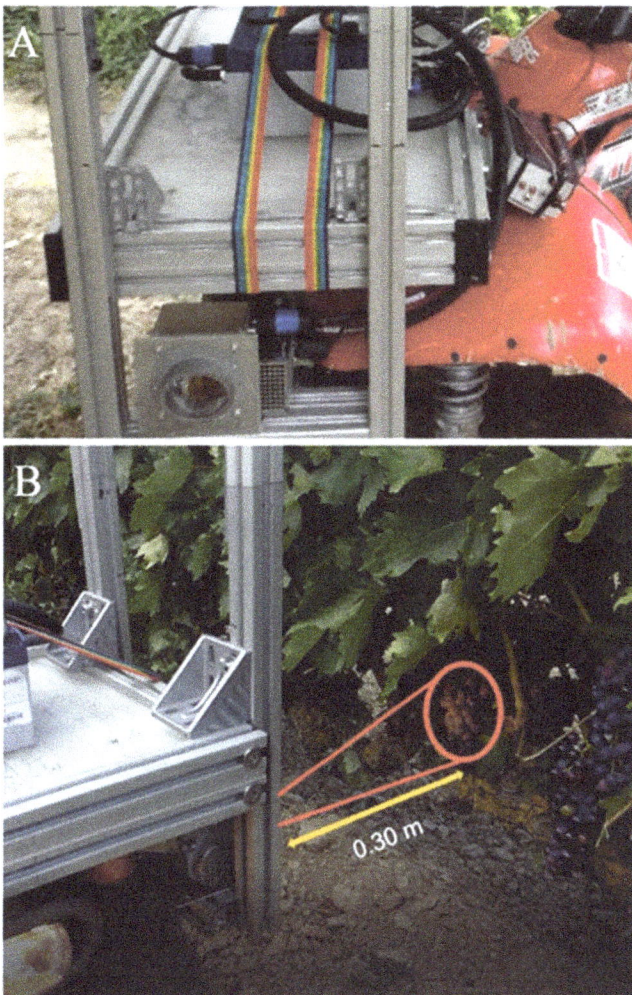

Figure 4. (**A**) Visible-Short Wave Near Infrared (VIS + SW − NIR) spectral acquisition system installed on the all-terrain vehicle (ATV) used for contactless on-the-go grape clusters spectral measurements in the vineyard. (**B**) Head sensor monitoring the grape cluster in motion.

The circular measuring spot area had a diameter of around 1.9 cm (area of 2.83 cm^2). On-the-go spectral measurements were acquired on the east canopy side at a constant speed of 5 km/h and rate of spectral acquisition of 18Hz. Spectral measurements were georeferenced using a GPS receiver Ag Leader 6500 (Ag Leader Technology, Inc., Ames, IA, USA) with RTK correction installed on the ATV.

4.3. Berry Composition Analysis

A total of 200 grape berries from all exposed clusters per block were collected and labeled immediately after the on-the-go VIS+SW-NIR measurements during the four dates.

The samples were transported to the laboratory of the University of La Rioja in portable refrigerators where they were stored in a freezer at −20 °C until chemical analysis. Once defrosted overnight in a cold room at 4 °C, a subsample of 100 berries was hand crushed and filtered. Total soluble solids (TSS), anthocyanins and total polyphenols concentrations were measured from the berries corresponding to each block. The TSS concentration was determined using a temperature compensating digital refractometer Quick-Brix 60 (Mettler Toledo, LLC, Columbus, OH, USA), expressed as °Brix. The remaining berry sub-sample was homogenized using a high-performance disperser T25 Ultra-Turrax (IKA, Staufen, Germany) at high speed (14,000 rpm for 60 s). Subsequently, anthocyanin and total polyphenols were analyzed following the Iland method [8]. Anthocyanin concentrations were expressed as mg/ fresh berry mass, whereas total polyphenols were expressed as absorbance units (AU) at 280 nm/ fresh berry mass.

4.4. Data Analysis

4.4.1. Spectral Processing

The spectral processing followed four essential steps (Figure 5). The first one consisted on the allocation of the acquired spectra to the different blocks of vines within each equally-distanced rows in the field experiment. Within each block the raw on-the-go spectral measurements captured information from leaves, gaps, wood, metal, etc., so a filtering step to retain only the spectral information of grape berries was needed.

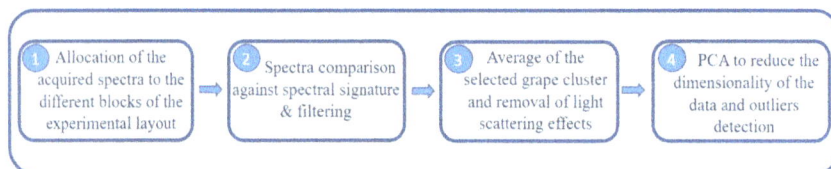

Figure 5. Design of the spectral processing procedure required to analyze on-the-go spectral measurements of grape clusters under field conditions.

In order to retain those spectra corresponding to grape clusters, spectra comparison against manually-taken spectral signatures of grape clusters (Figure 6) was performed using the "Spectra Comparison & Filtering" tool from the SL Utilities software (version 3.1, Polytec GmbH, Waldbronn, Germany). Only spectra which passed the "Spectra Comparison & Filtering" thresholds were considered as valid to be used in calculating the average spectrum per block. The settings Cosine was the method used to adjust the threshold value to determine the required similarity of the raw on-the-go spectra to the defined signature of grape clusters spectrum. A higher threshold value (close to 1) means that greater similarity is required to accept the measured spectrum as a true grape berry spectrum. The threshold value in the field experiment was set to 0.993.

The third step involved the averaging of the filtered grape cluster spectra per block and removal of the effects of light scattering. Different combinations of several spectral pre-processing filters were tested. These filters involved the use of standard normal variate (SNV) [41,42] and the application of the Savitzky–Golay smoothing and derivative process [43], selecting different values for the window size and the degree of the derivative. Both SNV and Savitzky–Golay derivative techniques contributed to the removal of light scattering effects. Figure 7 shows the average absorbance raw (Figure 7A) and processed (after application of Savitzky–Golay first derivative) for grape cluster spectra from one date collected on-the-go.

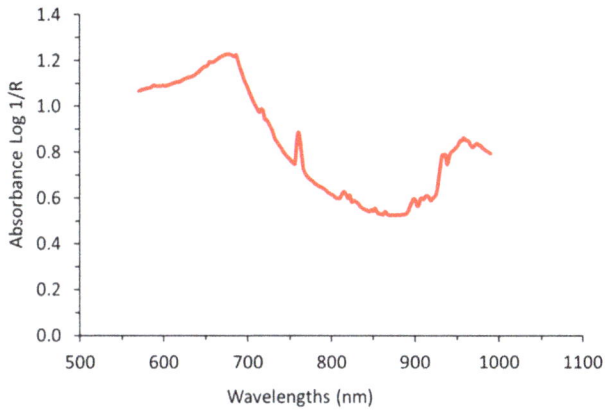

Figure 6. Spectral signature manually taken and averaged previous to on-the-go acquisitions from several grape clusters. This signature was used for filtering the on-the-go spectra and to select only those spectral signals belonging to grape clusters.

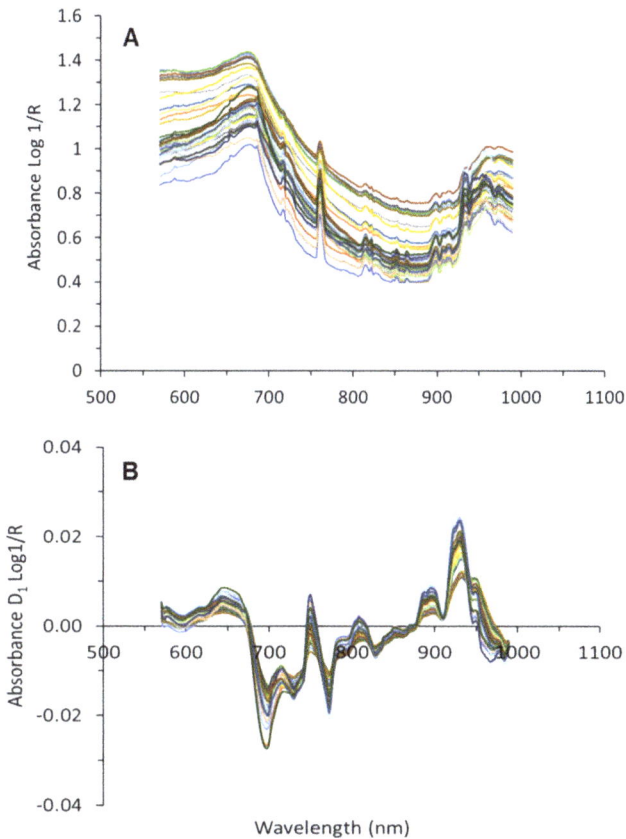

Figure 7. Average raw (**A**), and processed with Savitzky–Golay smoothing filtering (1st derivative, window size 15) (**B**) spectra collected on-the-go from grape clusters.

In the last step principal component analysis (PCA) was used to reduce the dimensionality of the data structure, to visualize the presence of spectra outliers and also to identify the main sources of variability in the spectra [44,45]. Outlier detection was performed based on the Residuals (Q) and Hotelling values (T^2) for the detection of samples with atypical spectra [46].

Spectral data manipulation and calibration models were performed with algorithms programmed in MATLAB (version 8.5.0, The Mathworks Inc., Natick, MA, USA). The partial least squares (PLS) Toolbox (version 8.1, Eigenvector Research, Inc., Manson, WA, USA) was used for principal component analysis (PCA) and partial least square regression (PLS).

4.4.2. Calibration and Prediction Models

Once the grape cluster spectra were processed and the chemical grape compositional parameters were obtained for each block, they were used to build the dataset, in which each spectrum was linked with its corresponding berry composition analysis (TSS, anthocyanins and total polyphenols concentrations). Considering 12 blocks per three rows and four different measurement dates, the dataset comprised a total of 144 samples.

In order to train robust models, capable of predicting totally unknown samples, the original dataset, was split up into two independent randomized datasets: a calibration one (comprising 80% of all data), consisting of 116 samples, and an external validation (prediction) set, which comprised the remaining 28 samples (20%). Samples of both data sets were appropriately distributed and covered the entire range of the grape composition parameters. The calibration dataset was used to train and to perform an internal cross-validation of the model, while the external validation (prediction) set was only utilized for prediction purposes, using the calibration models.

Partial Least Squares (PLS) regression was used as the algorithm for training the grape composition parameters prediction models. This algorithm has proved to be an accurate, robust, and reliable chemometric method [47] to analyze spectral data, as it is capable to deal with a vast amount of data, especially when the number of attributes (wavelengths in this case) largely surpasses the number of samples. The input independent variables X were the 215 wavelengths within the spectral range of 570–1000 nm, while TSS, anthocyanins and total polyphenols concentrations were used as dependent variables Y, each one for the training of three different models.

The calibration dataset was used to train the model, and statistics of calibration and cross validation, using a 10-fold venetian blind approach, were computed to assess the performance of the built models. The optimal number of latent variables (LVs) was selected as the one yielding the lowest root mean square error of cross-validation (RMSECV). The validation dataset was never used in the training process. It was employed only for testing with external samples, also called external validation or prediction. To evaluate the quality of the obtained models, the determination coefficient of calibration (R^2_c), cross-validation (R^2_{cv}), and prediction (R^2_p), the root mean square error of calibration (RMSEC), cross-validation (RMSECV) and prediction (RMSEP) were calculated.

4.5. Mapping

Based on the developed prediction models, prediction maps of TSS, anthocyanins and total phenol concentrations were created to monitor and illustrate the spatial variability of a commercial vineyard's grape composition using VIS + SW − NIR spectroscopy during the ripening period. Multilevel b-spline interpolation [48] with QGIS 2.18 (Free Software Foundation, Boston, MA, USA) was used to carry out the mapping tasks.

5. Conclusions

VIS + SW − NIR technology has proven to be a real alternative to appraise and map the vineyard grape composition variability in VSP vineyards, with a high spatial and temporal resolution, in a fast and non-destructive way. The capability to monitor the spatiotemporal evolution and distribution of total soluble solids, anthocyanins, and total polyphenols along the grape ripening process in a commercial vineyard will greatly enhance the decision-making about differential fruit allocation and harvest according to grape composition and quality.

Author Contributions: J.F.-N., J.T. and M.P.D.: Designed the experimental layout; J.F.-N., S.G. and M.P.D.: Acquired spectral measurements and determined the berry composition analysis; S.G.: Created the algorithm for spectral assignment and built the maps; J.F.-N.: Analyzed the data and built the models. J.F.-N. and M.P.D.: wrote the paper.

Funding: This work has received funding from the European Union under grant agreement number 737669 (Vinescout project).

Acknowledgments: Salvador Gutierrez would like to acknowledge the research founding FPI grant 299/2016 by Universidad de La Rioja, Gobierno de La Rioja. Maria P. Diago is funded by the Spanish Ministry of Science, Innovation and University with a Ramon y Cajal grant RYC-2015-18429).

Conflicts of Interest: The authors declare no conflict of interest.

References

1. Boulton, R.B.; Singleton, V.L.; Bisson, L.F.; Kunkee, R.E. *Principles and Practices of Winemaking*; Springer Science & Business Media: Berlin, Germany, 1996.
2. Boss, P.K.; Davies, C.; Robinson, S.P. Analysis of the expression of anthocyanin pathway genes in developing *Vitis vinifera* L. cv Shiraz grape berries and the implications for pathway regulation. *Plant Physiol.* **1996**, *111*, 1059–1066. [CrossRef] [PubMed]
3. Waterhouse, A.L. Wine phenolics. *Ann. N. Y. Acad. Sci.* **2002**, *957*, 21–36. [CrossRef] [PubMed]
4. Nogales-Bueno, J.; Baca-Bocanegra, B.; Jara-Palacios, M.J.; Hernández-Hierro, J.M.; Heredia, F.J. Evaluation of the influence of white grape seed extracts as copigment sources on the anthocyanin extraction from grape skins previously classified by near infrared hyperspectral tools. *Food Chem.* **2017**, *221*, 1685–1690. [CrossRef] [PubMed]
5. Gordillo, B.; Rodriguez-Pulido, F.J.; González-Miret, M.L.; Quijada-Morin, N.; Rivas-Gonzalo, J.C.; García-Estévez, I.; Heredia, F.J.; Escribano-Bailón, M.T. Application of differential colorimetry to evaluate anthocyanin–flavonol–flavanol ternary copigmentation interactions in model solutions. *J. Agric. Food Chem.* **2015**, *63*, 7645–7653. [CrossRef] [PubMed]
6. Cosme, F. Wine phenolics: Looking for a smooth mouthfeel. *SDRP J. Food Sci. Technol.* **2016**, *1*, 20–28.
7. Glories, Y. La couleur des vins rouges: 2e. Partie: Mesure, origine et interpretation. *Connaiss. la Vigne du Vin* **1984**, *18*, 253–271. [CrossRef]
8. Iland, P. *Chemical Analysis of Grapes and Wine*; Patrick Iland Wine Promotions PTYLTD: Athelstone, SA, Australia, 2004.
9. Nail, W. Collecting Berry Samples to Assess Grape Maturity. USDA National Institute of Food and Agriculture, New Technologies for Ag Extension Project. Available online: http://articles.extension.org/pages/33154/collecting-berry-samples-to-asses-grape-maturity (accessed on 31 July 2017).
10. Dami, I. *Midwest Grape Production Guide*; Ohio State University Extension: Columbus, OH, USA, 2005.
11. Nicolai, B.M.; Beullens, K.; Bobelyn, E.; Peirs, A.; Saeys, W.; Theron, K.I.; Lammertyn, J. Nondestructive measurement of fruit and vegetable quality by means of NIR spectroscopy: A review. *Postharvest Biol. Technol.* **2007**, *46*, 99–118. [CrossRef]
12. Bramley, R.G.V. Understanding variability in winegrape production systems 2. Within vineyard variation in quality over several vintages. *Aust. J. Grape Wine Res.* **2005**, *11*, 33–42. [CrossRef]
13. Li, X.; Jiang, Y.; Bai, X.; He, Y. Non-destructive measurement of SSC of apple using Vis. *Prog. Biomed. Opt. imaging* **2005**, *7*, 37.
14. Christen, D.; Camps, C.; Summermatter, A.; Gabioud Rebeaud, S.; Baumgartner, D. Prediction of the pre-and postharvest apricot quality with different VIS/NIRs devices. In Proceedings of the XV International Symposium on Apricot Breeding and Culture 966, Yerevan, Armenia, 20–24 June 2011; pp. 149–153.
15. Magwaza, L.S.; Landahl, S.; Cronje, P.J.R.; Nieuwoudt, H.H.; Mouazen, A.M.; Nicolai, B.M.; Terry, L.A.; Opara, U.L. The use of Vis/NIRS and chemometric analysis to predict fruit defects and postharvest behaviour of Nules Clementine mandarin fruit. *Food Chem.* **2014**, *163*, 267–274. [CrossRef]
16. Liu, H.J.; Ying, Y.B. Evaluation of sugar content of Huanghua Pear on trees by visible/near infrared spectroscopy. *Guang Pu Xue Yu Guang Pu Fen xi= Guang Pu* **2015**, *35*, 3078–3081. [PubMed]

17. Cen, H.; Bao, Y.; He, Y.; Sun, D.-W. Visible and near infrared spectroscopy for rapid detection of citric and tartaric acids in orange juice. *J. Food Eng.* **2007**, *82*, 253–260. [CrossRef]

18. Davey, M.W.; Saeys, W.; Hof, E.; Ramon, H.; Swennen, R.L.; Keulemans, J. Application of visible and near-infrared reflectance spectroscopy (Vis/NIRS) to determine carotenoid contents in banana (*Musa* spp.) fruit pulp. *J. Agric. Food Chem.* **2009**, *57*, 1742–1751. [CrossRef] [PubMed]

19. Moghimi, A.; Aghkhani, M.H.; Sazgarnia, A.; Sarmad, M. Vis/NIR spectroscopy and chemometrics for the prediction of soluble solids content and acidity (pH) of kiwifruit. *Biosyst. Eng.* **2010**, *106*, 295–302. [CrossRef]

20. Cortés, V.; Blanes, C.; Blasco, J.; Ortiz, C.; Aleixos, N.; Mellado, M.; Cubero, S.; Talens, P. Integration of simultaneous tactile sensing and visible and near-infrared reflectance spectroscopy in a robot gripper for mango quality assessment. *Biosyst. Eng.* **2017**, *162*, 112–123. [CrossRef]

21. Arana, I.; Jarén, C.; Arazuri, S. Maturity, variety and origin determination in white grapes (*Vitis vinifera* L.) using near infrared reflectance technology. *J. Near Infrared Spectrosc.* **2005**, *13*, 349–357. [CrossRef]

22. Beghi, R.; Giovenzana, V.; Marai, S.; Guidetti, R. Rapid monitoring of grape withering using visible near-infrared spectroscopy. *J. Sci. Food Agric.* **2015**, *95*, 3144–3149. [CrossRef]

23. Bellincontro, A.; Cozzolino, D.; Mencarelli, F. Application of NIR-AOTF spectroscopy to monitor Aleatico grape dehydration for Passito wine production. *Am. J. Enol. Vitic.* **2011**, *62*, 256–260. [CrossRef]

24. González-Caballero, V.; Sánchez, M.T.; López, M.I.; Pérez-Marín, D. First steps towards the development of a non-destructive technique for the quality control of wine grapes during on-vine ripening and on arrival at the winery. *J. Food Eng.* **2010**, *101*, 158–165. [CrossRef]

25. Musingarabwi, D.M.; Nieuwoudt, H.H.; Young, P.R.; Eyéghè-Bickong, H.A.; Vivier, M.A. A rapid qualitative and quantitative evaluation of grape berries at various stages of development using Fourier-transform infrared spectroscopy and multivariate data analysis. *Food Chem.* **2016**, *190*, 253–262. [CrossRef]

26. Herrera, J.; Guesalaga, A.; Agosin, E. Shortwave-near infrared spectroscopy for non-destructive determination of maturity of wine grapes. *Meas. Sci. Technol.* **2003**, *14*, 689–697. [CrossRef]

27. Larrain, M.; Guesalaga, A.R.; Agosin, E. A multipurpose portable instrument for determining ripeness in wine grapes using NIR spectroscopy. *IEEE Trans. Instrum. Meas.* **2008**, *57*, 294–302. [CrossRef]

28. Urraca, R.; Sanz-Garcia, A.; Tardaguila, J.; Diago, M.P. Estimation of total soluble solids in grape berries using a hand-held NIR spectrometer under field conditions. *J. Sci. Food Agric.* **2016**, *96*, 3007–3016. [CrossRef] [PubMed]

29. Gutiérrez, S.; Tardaguila, J.; Fernández-Novales, J.; Diago, M.P. On-the-go hyperspectral imaging for the in-field estimation of grape berry soluble solids and anthocyanin concentration. *Aust. J. Grape Wine Res.* **2019**, *25*, 127–133. [CrossRef]

30. Diago, M.P.; Fernández-Novales, J.; Gutiérrez, S.; Marañón, M.; Tardaguila, J. Development and validation of a new methodology to assess the vineyard water status by on-the-go near infrared spectroscopy. *Front. Plant. Sci.* **2018**, *9*, 59. [CrossRef]

31. Fernández-Novales, J.; Tardaguila, J.; Gutiérrez, S.; Marañón, M.; Diago, M.P. In field quantification and discrimination of different vineyard water regimes by on-the-go NIR spectroscopy. *Biosyst. Eng.* **2018**, *165*, 47–58. [CrossRef]

32. Diago, M.P.; Fernández-Novales, J.; Fernandes, A.M.; Melo-Pinto, P.; Tardaguila, J. Use of Visible and Short-Wave Near-Infrared Hyperspectral Imaging to Fingerprint Anthocyanins in Intact Grape Berries. *J. Agric. Food Chem.* **2016**, *64*, 7658–7666. [CrossRef]

33. Gomes, V.M.; Fernandes, A.M.; Faia, A.; Melo-Pinto, P. Comparison of different approaches for the prediction of sugar content in new vintages of whole Port wine grape berries using hyperspectral imaging. *Comput. Electron. Agric.* **2017**, *140*, 244–254. [CrossRef]

34. Fernandes, A.M.; Franco, C.; Mendes-Ferreira, A.; Mendes-Faia, A.; da Costa, P.L.; Melo-Pinto, P. Brix, pH and anthocyanin content determination in whole Port wine grape berries by hyperspectral imaging and neural networks. *Comput. Electron. Agric.* **2015**, *115*, 88–96. [CrossRef]

35. Piazzolla, F.; Amodio, M.L.; Colelli, G. The use of hyperspectral imaging in the visible and near infrared region to discriminate between table grapes harvested at different times. *J. Agric. Eng.* **2013**, *44*, e7. [CrossRef]

36. Tang, J.; Petrie, P.R.; Whitty, M. Modelling relationships between visible winegrape berries and bunch maturity. *Aust. J. Grape Wine Res.* **2019**, *25*, 116–126. [CrossRef]

37. Wendel, A.; Underwood, J. Illumination compensation in ground based hyperspectral imaging. *ISPRS J. Photogramm. Remote Sens.* **2017**, *129*, 162–178. [CrossRef]

38. Underwood, J.; Wendel, A.; Schofield, B.; McMurray, L.; Kimber, R. Efficient in-field plant phenomics for row-crops with an autonomous ground vehicle. *J. F. Robot.* **2017**, *34*, 1061–1083. [CrossRef]

39. Deery, D.; Jimenez-Berni, J.; Jones, H.; Sirault, X.; Furbank, R. Proximal Remote Sensing Buggies and Potential Applications for Field-Based Phenotyping. *Agronomy* **2014**, *4*, 349–379. [CrossRef]

40. Coombe, B.G. Growth stages of the grapevine: Adoption of a system for identifying grapevine growth stages. *Aust. J. Grape Wine Res.* **1995**, *1*, 104–110. [CrossRef]

41. Barnes, R.J.; Dhanoa, M.S.; Lister, S.J. Standard Normal Variate Transformation and De-trending of Near-Infrared Diffuse Reflectance Spectra. *Appl. Spectrosc.* **1989**, *43*, 772–777. [CrossRef]

42. Dhanoa, M.S.; Lister, S.J.; Barnes, R.J. On the scales associated with near-infrared reflectance difference spectra. *Appl. Spectrosc.* **1995**, *49*, 765–772. [CrossRef]

43. Savitzky, A.; Golay, M.J.E. Smoothing and Differentiation of Data by Simplified Least Squares Procedures. *Anal. Chem.* **1964**, *36*, 1627–1639. [CrossRef]

44. Næs, T.; Isaksson, T.; Fearn, T.; Davies, T. *A User Friendly Guide to Multivariate Calibration and Classification*; NIR Publications: Chichester, UK, 2002.

45. Massart, D.L.; Vandeginste, B.G.M.; Deming, S.N.; Michotte, Y.; Kaufman, L. *Data Handling in Science and Technology: Chemometrics a Textbook*; Elsevier: Amsterdam, The Netherlands, 1988.

46. Brereton, R.G. *Chemometrics: Data Analysis for the Laboratory and Chemical Plant*; John Wiley & Sons: Hoboken, NJ, USA, 2003.

47. Wold, S.; Sjöström, M.; Eriksson, L. PLS-regression: A basic tool of chemometrics. *Chemom. Intell. Lab. Syst.* **2001**, *58*, 109–130. [CrossRef]

48. Lee, S.; Wolberg, G.; Shin, S.Y. Scattered data interpolation with multilevel B-splines. *IEEE Trans. Vis. Comput. Graph.* **1997**, *3*, 228–244. [CrossRef]

molecules

MDPI

Article

Rapid and Nondestructive Measurement of Rice Seed Vitality of Different Years Using Near-Infrared Hyperspectral Imaging

Xiantao He [1,2,†], Xuping Feng [1,2,†], Dawei Sun [1,2], Fei Liu [1,2], Yidan Bao [1,2] and Yong He [1,2,*]

[1] College of Biosystems Engineering and Food Science, Zhejiang University, Hangzhou 310058, China; hxt@zju.edu.cn (X.H.); pimmmx@163.com (X.F.); DZS0015@zju.edu.cn (D.S.); fliu@zju.edu.cn (F.L.); ydbao@zju.edu.cn (Y.B.)

[2] Key Laboratory of Spectroscopy Sensing, Ministry of Agriculture and Rural Affairs, Zhejiang University, Hangzhou 310058, China

* Correspondence: yhe@zju.edu.cn; Tel.: +86-571-8898-2143

† These authors contributed equally to this work.

Academic Editor: Christian Huck
Received: 31 May 2019; Accepted: 13 June 2019; Published: 14 June 2019

Abstract: Seed vitality is one of the primary determinants of high yield that directly affects the performance of seedling emergence and plant growth. However, seed vitality may be lost during storage because of unfavorable conditions, such as high moisture content and temperatures. It is therefore vital for seed companies as well as farmers to test and determine seed vitality to avoid losses of any kind before sowing. In this study, near-infrared hyperspectral imaging (NIR-HSI) combined with multiple data preprocessing methods and classification models was applied to identify the vitality of rice seeds. A total of 2400 seeds of three different years: 2015, 2016 and 2017, were evaluated. The experimental results show that the NIR-HSI technique has great potential for identifying vitality and vigor of rice seeds. When detecting the seed vitality of the three different years, the extreme learning machine model with Savitzky–Golay preprocessing could achieve a high classification accuracy of 93.67% by spectral data from only eight wavebands (992, 1012, 1119, 1167, 1305, 1402, 1629 and 1649 nm), which could be developed for a fast and cost-effective seed-sorting system for industrial online application. When identifying non-viable seeds from viable seeds of different years, the least squares support vector machine model coupled with raw data and selected wavelengths of 968, 988, 1204, 1301, 1409, 1463, 1629, 1646 and 1659 nm achieved better classification performance (94.38% accuracy), and could be adopted as an optimal combination to identify non-viable seeds from viable seeds.

Keywords: seeds vitality; rice seeds; near-infrared spectroscopy; hyperspectral image; discriminant analysis

1. Introduction

Rice (*Oryza sativa* L.) is one of the three most important crops in the world, with a harvested area of 167 million ha and 769 million tons of total yield in 2017 [1]. However, the world population is increasing rapidly, and the total population will grow up to nearly 7.7 billion in 2019, compared with 6.9 billion in 2010, which will affect food security greatly and may lead to a food crisis around the world [2]. Numerous efforts have been made to satisfy this demand, such as optimizing the agronomic process, improving post-harvest technologies and biotechnology improvements in seeds and breeding mechanisms [3]. As an optimization means of agronomic processes, ensuring seed vitality and vigor is one of the most effective methods to increase crop production, which is particularly important for direct seeding, as it can not only enhance crop establishment but also increase the plant's ability to compete against weeds [4].

Seed vitality and vigor directly affect the performance of seedling emergence and stand establishment [5]. Usually, any physical or biochemical damage to seeds can cause reduced or complete loss of vitality. More specifically, any changes in field conditions (e.g., humidity, temperature, pests, diseases) and post-harvest processes (e.g., drying, storage) can lead to seed damage, and thus cause retardation or complete vitality loss if not carefully controlled. These factors are, however, difficult to control. Therefore, the knowledge of whether a seed is viable or not before sowing is important both to seed companies and farmers. For seed companies, knowing seed vitality in advance helps them to determine the quality of their products, while for farmers it plays an important role in yield increase and prediction [6]. Determination of seed vitality is therefore necessary, and relevant studies should be conducted to build such a detection system for seed vitality.

Traditional detection methods of seed vitality, such as immunoassay tests, polymerase chain reaction tests and germination tests, could obtain the seed vigor intuitively, but they are expensive, time-consuming and destructive, which results in their low application in seed vigor detection [3]. Many research works have been conducted to construct potential rapid and non-destructive methods to measure seed vigor. Four non-destructive approaches with different techniques or principles, i.e., nuclear magnetic resonance spectroscopy [7], X-ray [8,9], laser speckle technique [10] and the measuring technology of seed conductivity [11] were investigated, however, they have not been widely used because of the low efficiency and complicated operation. Fortunately, recent studies show that molecular spectroscopic techniques, such as point-based and image-based hyperspectral techniques, have great potentials in the detection of seed ingredients with the advantages of high detection speed, non-destructive nature and low cost [12].

Point-based spectroscopic techniques, such as Raman, mid infrared, and Fourier transform-near infrared spectroscopies, acquire chemical information in a fixed-point area of the sample, and provide a large number of spectral details, but do not offer the spatial information that is important for seed detection application [13]. Hyperspectral imaging (HSI) is one of the most feasible methods for rapidly and non-destructively detecting the substances of agricultural products. It combines the technologies of spectroscopy and digital imaging, and is able to obtain spectral and spatial information simultaneously from testing samples in the form of a hypercube with two spatial dimensions and one spectral dimension [14]. Based on the spatial data, the HSI technique has the ability to collect hyperspectral information from samples of different sizes and shapes [15]. In addition, the detection speed of HSI is faster than that of point-based techniques, as many samples can be scanned and analyzed at the same time by using an HSI camera.

The HSI technique coupled with visible (vis) and/or near infrared (NIR) spectroscopy is generally used to identify or inspect different substances of seed by recognizing the molecular bonds in the sample. Many studies have been conducted to detect the vitality of seeds for different species. The corn with a large grain size and flat shape has been paid more attention for seed vitality detecting. Collins et al. measured corn seed vitality using short wave infrared line-scan hyperspectral imaging, and the results indicated that hyperspectral imaging can be used to accurately classify corn based on vitality [3]. Ashabahebwa et al. assessed the performance of testing corn seed vitality by applying the Fourier transform near-infrared spectroscope [16]. In addition, the detections of vitality and vigor for seeds of oat [17], muskmelon [18], soybean [19,20] and watermelon [21] were developed with the HSI technique. Previous studies have shown the potential of using HSI coupled with multivariate data analysis for the detection of internal conditions of rice seeds, such as origin [22], variety [23–25], nitrogen content [26], moisture content [27] and heavy metal concentration [28]. To the best of our knowledge, many studies were conducted only for vitality detection of artificially aged seed, and, so far, no study has been carried out to detect the vitality of rice seeds under natural ageing conditions by using HSI, even though the results obtained from natural ageing seeds were more consistent with the actual situation.

This study was conducted to determine the optimal spectral wavebands and multivariable classification model to acquire or detect the vigor of rice seeds stored for different years based on the near-infrared hyperspectral imaging (NIR-HSI) technique, and attempt to build a model to identify non-viable seeds from viable seeds of different years, and ultimately provide an alternative approach of rapidly and non-destructively measuring the rice seed vitality for industrial or large-scale application.

2. Results and Discussion

2.1. Spectral Interpretation

A raw spectral data plot and mean raw spectral data plot from selected regions of interest (ROI) are shown in Figure 1a,b, respectively. The change trends of the spectral reflectance curves of all rice kernels showed clear similarities. As shown in Figure 1b, the seed spectral curves of three different years had large differences in the reflectance of wavebands, while the differences were negligible after all three year seeds were artificially aged to lose vigor. The germination tests on the representative samples showed a high vitality, with a germination rate of 95%, 92.86% and 80.71%, and vitality index of 261.26, 225.6 and 154.15 for rice seeds of the years 2017, 2016, and 2015, respectively, as shown in Table 1. It is obvious that the germination rate and vitality index reduced as the year of preservation increased, which was consistent with the spectral change of rice seeds, and could be used as a basic principle for classifying rice seeds of different years. All germination rate values were higher than the factory labelled 80% germination rate, indicating the seeds stored within three years still have enough vitality to be used in rice production. The seeds that were subjected to microwave heat treatment were similar to the non-viable seeds, and their germination rate and vitality index were both tested to be zero, which resulted in a higher spectral reflectance of artificial aging seeds. Moreover, the spectral reflectance of aged seeds with non-vitality had high similarity, no matter the year of seeds. Therefore, it is difficult to identify the year of aged non-viable seeds using hyperspectral imaging; however, it is highly possible to identify non-viable seeds from common viable seeds of three different years.

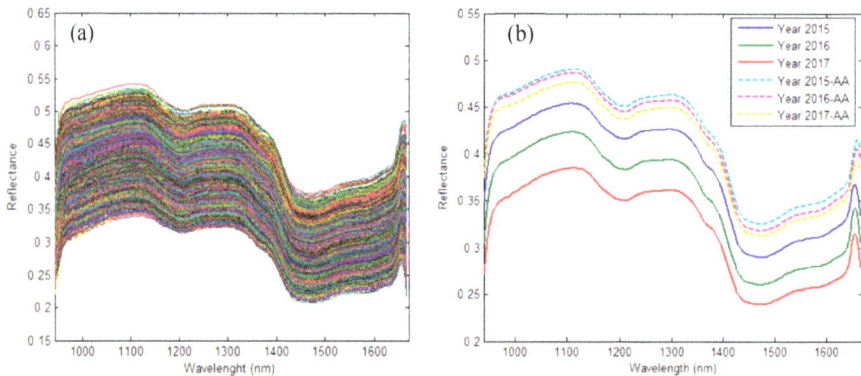

Figure 1. (**a**) Raw spectra of all rice simples and (**b**) mean spectra for rice seeds.

Table 1. Germination rate and vitality index of all sets of seeds as determined by germination test.

Years of Seed	Treatment	Germination Number	Non-Germination Number	Germination Rate (GR)	Vitality Index (VI)
2015	–	113	27	80.71%	154.15
	AA	0	140	0	0
2016	–	130	10	92.86%	225.6
	AA	0	140	0	0
2017	–	133	7	95%	261.26
	AA	0	140	0	0

AA: artificial ageing.

2.2. The Results of Principal Component Analysis

Principal component analysis (PCA) is one of the most popular multivariate statistical techniques in almost all scientific disciplines. It uses an orthogonal transformation to convert a set of observations of possibly correlated variables into a set of values of linearly uncorrelated variables called principal components [29]. PCA was used in the study for data exploration and classification feasibility analysis. Figure 2 shows PCA results for raw data based on the spectral data of all groups of seeds. The analysis of PCA results shows that the first two principal components (PCs) were found in up to 99.58% of all the variability—PC1 and PC2 had 97.94% and 1.64% variance, respectively. That is to say, these two PCs showed the most significant variation among samples, and could explain 99.58% of all the variability. As illustrated in Figure 2, the PCA data of non-viable seeds of three different years in this plane projection were more concentrated, while an obvious difference occurred for the viable seeds of three different years. As a result, the viable seeds of different years were more likely to be classified with each other, while the seeds were difficult to differentiate after the three kinds of seeds were artificially aged to lose vigor because of the high overlap between the groups shown in Figure 2.

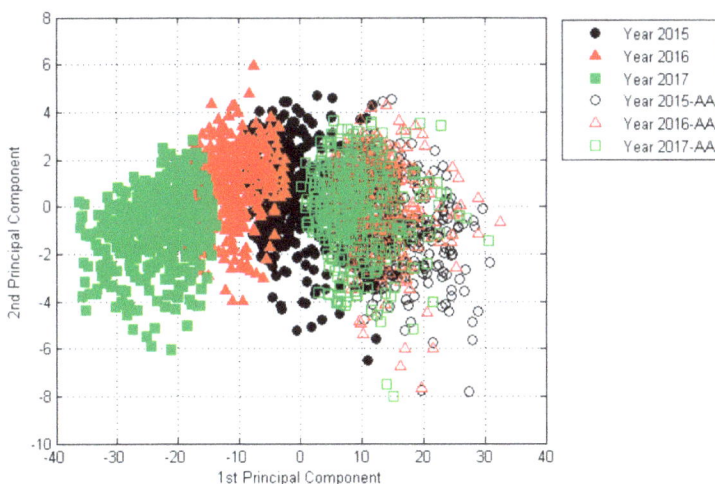

Figure 2. Principal component analysis (PCA) results for raw data based on the spectral data of all six seed groups. AA: artificial ageing.

The PCA technique was utilized to analyze the spectral data of viable seeds of three different years at the three different preprocessing methods, and the results are illustrated in Figure 3. The PCA results showed that differences among three samples have better data clustering performance using Savitzky–Golay (SG) preprocessing algorithms compared with other models (Figure 3b). However, PCA results for Savitzky–Golay first derivative (SG-D1) and multiplicative scatter correction (MSC) showed preprocessed data generated much less distinctive clustering results (Figure 3c,d), which was worse than the raw data (Figure 3a). It may have been that the noise was overamplified when spectral data was preprocessed by SG-D1 and MSC methods, thus resulting in a lower signal-to-noise ratio and less distinctive clustering for the three groups. The raw data obtained a better clustering performance due to the data being calculated and obtained based on the mean spectral data of the region of one rice seed, which could remove spectral noises in the seed to some extent.

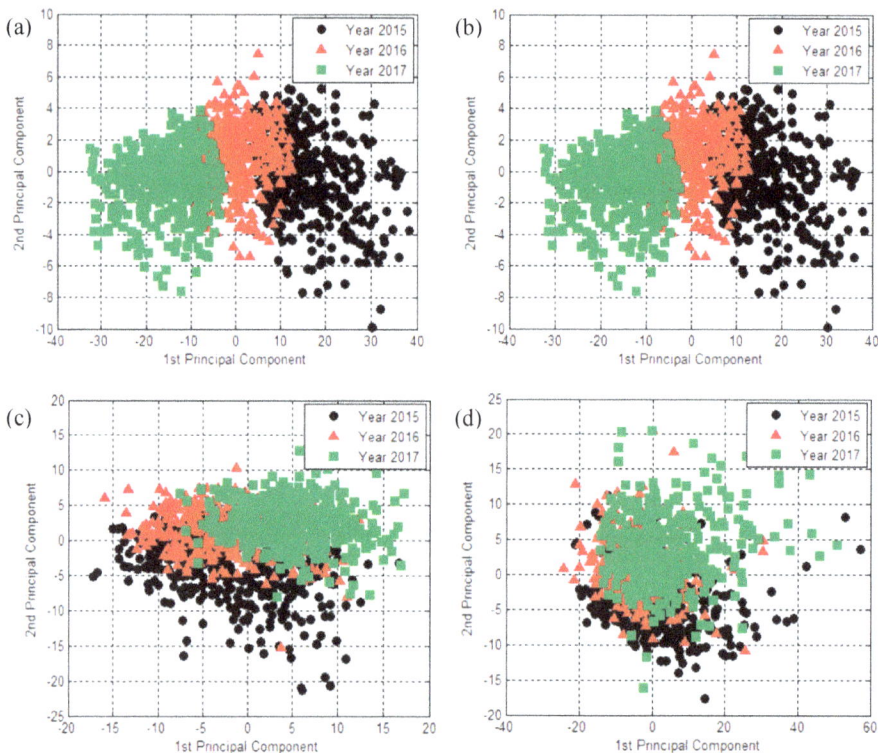

Figure 3. PCA results for (**a**) raw data and preprocessed data of (**b**) Savitzky–Golay (SG), (**c**) Savitzky–Golay first derivative (SG-D1) and (**d**) multiplicative scatter correction (MSC), based on the spectral data of rice seeds of different years.

2.3. Optimal Wavelengths Selection

A classification model established by applying a number of highly correlated variables would increase the computational complexity for predicting. Thus, selecting important and irrelevant wavelengths from hyperspectral data is necessary before establishing the discriminant model. In this study, the successive projections algorithm (SPA) was proposed to determine the optimal wavelengths for predicting rice seed vitality based on SG, SG-D1 and MSC preprocessed data and the raw data. The numbers of wavelengths selected by SPA were decreased to 4.2, 3.7, 5.1 and 2.8% of all 216 wavelengths. Then, the selected wavelengths were used to build multivariate classification models for the determination of rice vigor, including the partial least square-discriminant analysis (PLS-DA), the least squares support vector machines (LS-SVM) and the extreme learning machine (ELM).

In general, spectral absorptions at the optimum wavelengths had a notable correlation with the molecular structures of chemical components. Some important wavelengths (988, 1409, 1629 and 1659 nm) were shared by data of raw, SG, SG-D1 and MSC (Figure 4), and may have been responsible for the germination ability of the rice seed. The absorption band near 988 nm may be assigned to the second overtone of the O–H vibration bond overtone of water [18,30]. The wavelength band near 1409 was primarily attributed to the O–H first overtone, which are common in starch and lipids [31]. The wavelengths near 1629 and/or 1659 nm were assigned to the first overtone of O–H stretching, C–H from the methylene group and the N–H stretch first overtone, which refer to the CONH representing the protein content [31]. Lipid peroxidation, loss of protein function and hydrolysis of starch have been suggested as causes for loss of seed vitality [32]. Thus, the selected wavelengths related to starch, lipids and protein structures were the foundation for discrimination between the three groups. In addition, wavelengths selected from SG preprocessed data had roughly the same distribution as that of raw data, and the common wavebands of 1204 nm and 1301 nm were connected to the second overtone of C–H harmonic stretching [33]. As for the SG-D1 and MSC preprocessing methods, most wavelengths were located in the range of 1392–1514 nm, which mainly corresponded to the first overtone of C–H stretching and deformation of CH_2 and CH_3 groups [33].

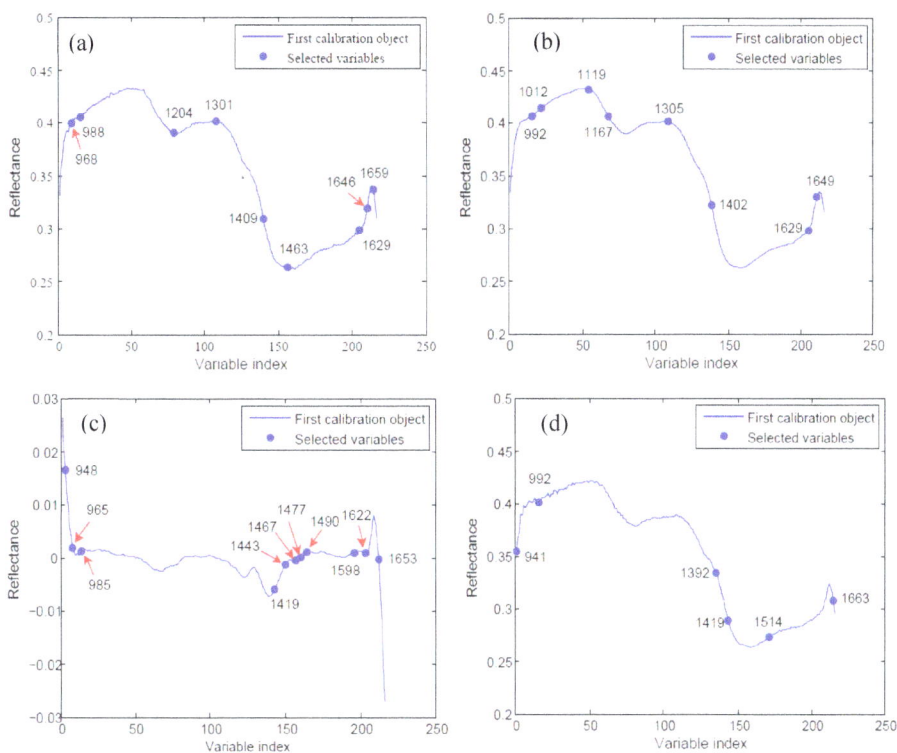

Figure 4. Selection of optimal wavelengths by successive projections algorithm (SPA). Distributions of important variables (marked with 'filled circle') for (**a**) raw data and preprocessed data of (**b**) SG, (**c**) SG-D1 and (**d**) MSC.

2.4. Classification Model Results

After optimal wavelength selection, the whole spectral data set was reduced to a matrix of dimensions m × n, where m represents the number of samples (m = 2400) and n was the number of selected wavelengths including 9, 8, 11 and 6 for raw data, SG, SG-D1 and MSC preprocessed data, respectively. To determine the suitability of optimal variables selected by SPA, the optimal wavelengths were used to build multivariate models, including PLS-DA, LS-SVM and ELM for classifying the samples.

2.4.1. Assessment of Seed Vitality of Three Different Years

The seed vitality was different in the seeds of three different years, and the seeds stored in later years could obtain a higher vigor, which was consistent with the change trend of spectral reflectance of the seeds (i.e., the reflectance of seeds stored in earlier years was generally higher than that of later years). Based on this principle, three models of PLS-DA, LS-SVM and ELM were built to identify the vitality of seed samples of different years. The classification accuracy of calibration set varied from 64.67% to 97.5%, and the accuracy range of prediction set was 67.5–95.67%. The lowest 64.67% accuracy of the calibration set and 67.5% accuracy of the prediction set were obtained by using the PLS-DA model with MSC preprocessing and the SPA method, while the highest values of 97.5% and 95.67% for the calibration and prediction set, respectively, were achieved when using the LS-SVM mode with SG preprocessing and full-wave bands. As for the classification results of the prediction set (Figure 5), the PLS-DA model with selected wavelengths had the lowest classification accuracy in the three classification models (87.83, 87.5, 75 and 67.5% for raw, SG, SG-D1 and MSC, respectively). Applying preprocessing and wavelength selection methods before model application had no improvement in classification accuracy. The LS-SVM model gave the highest accuracy of the three models with/without data preprocessing procedures—with up to 95.67% accuracy using the data of SG preprocessing in the full-wave bands—and could reach the high accuracy of 93.33% by applying the reduced wavelengths selected by SPA. The good performance of the LS-SVM model is probably because its decision boundaries can become much clearer after transforming the data into higher dimensions, and as a result it classified different groups more accurately. However, the PLS-DA models establish decision boundaries based on the thresholds under low dimensions, and thus this results in misclassifications due to outliers [34]. The ELM model, a simple tuning-free three-step algorithm with a fast learning speed, achieved a result of accuracy of 93.67% based on the reduced wavelengths of SPA, along with SG preprocessing, which was even a little higher than the 93.33% accuracy of LS-SVM under the same condition. Though the accuracy of 93.67% was lower than the 95.67% accuracy of the LS-SVM model with SG preprocessed data in the full-wave bands, its data processing load with only eight wave bands (992, 1012, 1119, 1167, 1305, 1402, 1629 and 1649 nm) decreased to 3.7% of the classification model of full wavelengths, which is a significant performance improvement for an almost 27-fold increase in data processing speed. Therefore, the ELM model coupled with the variable-selection method of SPA and the preprocessing method of SG could be adopted as an optimal combination to classify the seed of different years for a fast and cost-effective seed-sorting system for industrial online application.

2.4.2. Identifying Non-Viable Seeds from Viable Seeds of Different Years.

The seed samples, no matter whether they were stored in year 2015, 2016 or 2017, all lost vitality completely, with a germination rate of 0% and vitality index of 0 after they underwent artificial aging. The spectral reflectance of aged seeds increased greatly and differed from that of the seeds of three years (Figure 1b), which provides a possibility to pick out non-viable seeds from normal viable seeds stored in different years. Furthermore, 133, 133 and 134 seeds were selected randomly from aged seeds of the years 2015, 2016 and 2017, respectively. In total, 400 aged seeds were obtained and then used as a non-viable group with other three viable groups of different years (i.e., 2015, 2016 and 2017) to build classification models for evaluating the performance of identifying non-viable seeds. The results are shown in Table 2. The classification accuracy of the calibration set varied from 48.75% to 96.38% and the accuracy range of the prediction set was 46.63–95.57%. The classification accuracy of the calibration set was generally higher than the accuracy of the prediction set at the same conditions. As for the classification results of the prediction set (Figure 5b), PLS-DA with less than 70% accuracy was the model of lowest classification accuracy, which was even lower than the accuracy of the PLS-DA model used for classifying seed vitality of merely three different years. The LS-SVM model gave the highest accuracy of the three models with/without data preprocessing procedures, with up to 95.57% accuracy using the raw data in the full-wave bands, and could reach the high accuracy of 94.38%, applying the reduced wavelengths selected by SPA. The ELM model achieved a result of accuracy of 93.75% based on the reduced wavelengths of SPA with raw data, which was slightly lower than the accuracy of 94.38% of LS-SVM under the same condition. Though the accuracy of 94.38% of the LS-SVM model was lower than the 95.57% accuracy of LS-SVM with raw data in the full-wave bands, its data processing load with nine wave bands (968, 988, 1204, 1301, 1409, 1463, 1629, 1646 and 1659 nm) decreased to 4.2% of the classification model of full wavelengths, which is a significant performance improvement for an almost 23.8-fold increase in data processing speed. Therefore, the LS-SVM model coupled with the variable-selection method of SPA and raw data could be adopted as an optimal combination to identify non-viable seeds from viable seeds.

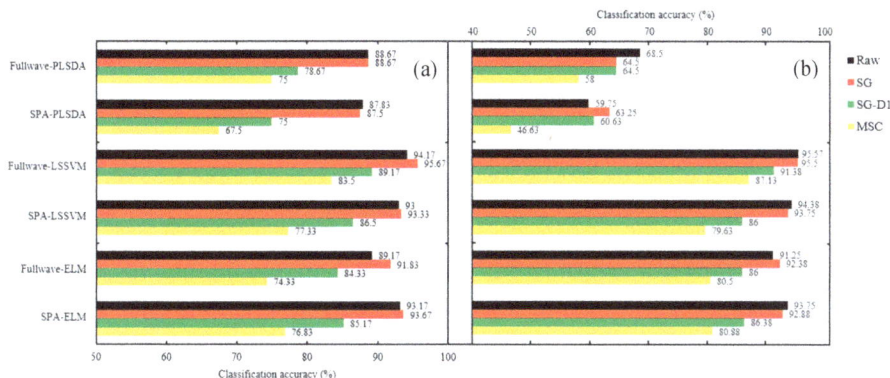

Figure 5. The prediction results of classification models for identifying (**a**) seed vitality of three different years and (**b**) non-viable seeds from viable seeds of three different seeds.

Table 2. The results of classification models established by full and selected wavelengths with different preprocessing methods.

| | | IVY | | | | | | INV | | | | | |
| | | PLS-DA | | LS-SVM | | ELM | | PLS-DA | | LS-SVM | | ELM | |
		Full.	Sel.	Full.	Sel.	Full.	Sel.	Full.	Sel.	Full.	Sel.	Full.	Sel.
Raw	Cal.	92.17	86.83	96.67	95.83	95.5	93.5	69.75	58.38	96	95.13	94.75	94.13
	Pre.	88.67	87.83	94.17	93	89.17	93.17	68.5	59.75	95.57	94.38	91.25	93.75
SG	Cal.	87.75	87	97.5	94.33	95.67	94.17	62.63	62.13	96.38	93.5	95.25	93.13
	Pre.	88.67	87.5	95.67	93.33	91.83	93.67	64.5	63.25	95.5	93.75	92.38	92.88
SG-D1	Cal.	79.17	73.67	94.67	86.17	90.17	85.5	66.25	61.13	95.75	87.13	91	86.38
	Pre.	78.67	75	89.17	86.5	84.33	85.17	64.5	60.63	91.38	86	86	86.38
MSC	Cal.	78.83	64.67	87.33	78	82.83	79	61.25	48.75	94.25	77.88	86.25	80.63
	Pre.	75	67.5	83.5	77.33	74.33	76.83	58	46.63	87.13	79.63	80.5	80.88

Cal.: calibration; Pre.: prediction; Raw: raw data; IVY: identification of the seed vitality of three different years; INV: identifying non-viable seeds from viable seeds; Full.: full wavelengths; Sel.: selected wavelengths by SPA; PLS-DA: partial least square-discriminant analysis; LS-SVM: least squares support vector machines; ELM: extreme learning machine.

3. Materials and Methods

3.1. Samples and Sample Preparation

In this study, the rice seeds of ShenLiangYou862 from three different years, including 2015, 2016 and 2017, were selected to be investigated, which were kindly provided by a commercial company (Jiangsu Tomorrow Seed Technology LLC, Nanjing, China). The seeds were cleaned first, and damaged seeds were removed. When acquiring hyperspectral images, three samples in different years could not be differentiated by the naked eye. For each category, 800 kernels were acquired with 400 seeds used as the different-year sample and the other 400 seeds used as the aged sample for comparison. Artificial aging of seeds was induced in the rice simples using microwave heat treatment at 700 W input power and 60 s exposure time, which was optimized in advance for this experiment in accordance with the study by Ambrose et al. [35].

3.2. Hyperspectral Image Collection

A line-scan NIR-HSI system was used to acquire the hyperspectral images of rice seeds, as shown in Figure 6. The system comprised an imaging spectrograph (ImSpector N17E; Spectral Imaging Ltd., Oulu, Finland) that covered the spectral range of 874–1734 nm with a spectral resolution at 3.36 nm, a charge coupled device camera (Xeva 992; Xenics Infrared Solutions, Leuven, Belgium) with the spatial resolution of 320 × 256 pixels, two line light sources (Fiber-Lite DC950, Dolan Jenner Industries Inc., Boxborough, MA, USA), a transmission platform (IRCP0076, Isuzu Optics Crop, Taiwan), a dark box and a computer. In order to acquire clear and non-deformable hyperspectral images, the moving speed of the transmission platform, the exposure time and the work distance between samples and the camera were adjusted to 19 mm/s, 3.5 ms and 23.4 cm, respectively. Rice seeds were placed on a dark-background sampling plate irrespective of whether the germinal side of the kernel was facing the camera, then the sampling plate was transferred to the transmission platform for scanning seeds line by line. Spatial and spectral data were obtained from the sample when it was moved into the range of the camera filed. After scanning the samples for hyperspectral data, the hyperspectral images were calibrated by the following equation:

$$I_{cal} = (I_{raw} - I_{dark}) / (I_{ref} - I_{dark}),$$ (1)

where I_{cal}, I_{raw}, I_{dark} and I_{ref} are the corrected images, original images, dark current and reference images, respectively. I_{ref} was measured using a white Teflon tile with the reflectance close to 99%, and I_{dark} was collected by covering the camera lens completely with the cap provided by the manufacturer. The calibrated HSI image was ultimately obtained to analyze the spectral data in every single seed (Figure 7). Spectral data before 941 nm and after 1666 nm were omitted because of low signal-to-noise ratio, which was mainly caused by bad pixels on the camera detector, lighting characteristics and the movement of the transmission platform.

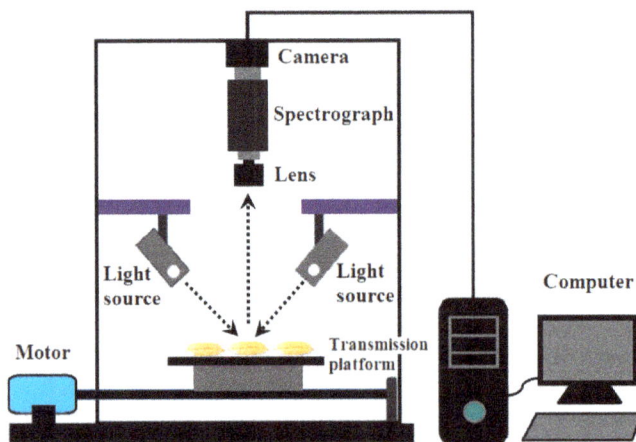

Figure 6. Schematic of line-scan near-infrared hyperspectral imaging (NIR-HSI) system and scanning of seed samples.

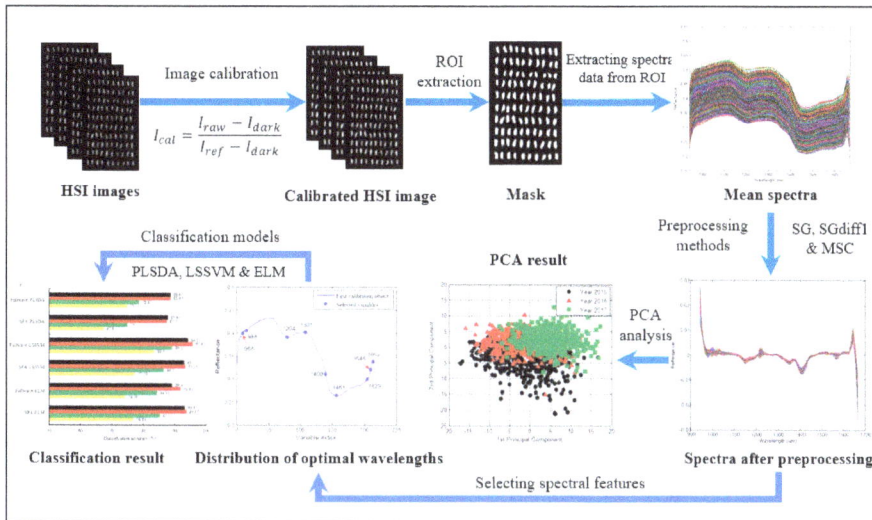

Figure 7. Schematic overview of the analytical procedure for identifying the vitality of different years. ROI: regions of interest.

3.3. Data Extraction and Preprocessing

A threshold value of 0.15 was used to segment calibrated hyperspectral images to remove the effect of the background and to obtain only seed pixels. The regions of interest (ROI) were selected by applying the 1301 nm band image, and then spectral information of the respective rice sample in the HSI images was extracted relying on the ROI. The spectra of each pixel in the ROI were averaged for each seed, and, in total, 2400 average spectra representing 2400 scanned seeds were calculated and saved for further analysis.

Three preprocessing methods were used in this paper to correct the spectral data, including the Savitzky–Golay smoothing (SG), the Savitzky–Golay first derivative (SG-D1) and multiplicative scatter correction (MSC).

The SG method is a digital filter that can be applied to a set of digital data points for the purpose of smoothing the data, which can effectively keep useful information and reduce high-frequency noise in a hyperspectral image. The polynomial order and number of points in the SG method are two computation parameters, which were adjusted to 3 and 15, respectively, for a good effect in spectrum smoothness.

The SG-D1 method is the first derivative form of the SG method. By deriving SG data, it has the advantages of emphasizing the spectral features of the data and removing the additive baseline; however, it inevitably amplifies the noise at the same time, which may have a large impact on the classification results. The polynomial order and number of points was also set to 3 and 15, respectively, when the SG process was executed.

The MSC method was used to remove physical effects, such as particle size and surface blaze, from the spectra, which do not carry any chemical or physical information. This method is capable of correcting differences in the baseline and has an advantage of the transformed spectra being similar to the original spectra, and optical interpretation is therefore more easily accessible [36].

3.4. Spectral Feature Selection

Hyperspectral images could provide a large amount of spectral and spatial information related to the vitality properties of the rice seeds; nevertheless, they also contain overlapping and redundant information. It is necessary to apply a feature selection algorithm to obtain representative and important wavelengths for reducing irrelevant information and improving computation speed.

The successive projections algorithm (SPA) is a variable-selection technique that has attracted increasing interest in the analytical-chemistry community in the past 10 years. In SPA, the selection of variables is cast in the form of a combinatorial optimization problem with constraints, and projection operations in a vector space are used to choose subsets of variables with a small degree of multi-collinearity in order to minimize redundancy and ill-conditioning problems [37]. The algorithm SPA was applied in this study to select the optimal wavelengths. The selected wavelengths with the minimum collinearity have the maximum projection value on the orthogonal subspace.

3.5. Construction and Analysis of Classification Models

In this paper, three discriminant models were built and analyzed, including the partial least square-discriminant analysis (PLS-DA), the least squares support vector machine (LS-SVM) and the extreme learning machine (ELM).

The partial least squares (PLS) algorithm was first induced for regression tasks and then evolved into a classification method that is well known as PLS-DA. This method is a popular chemometrics technique used to optimize the separation between different groups of samples, which is accomplished by linking raw data and class membership [38], as described in Equation (2):

$$Y = X \cdot B + F, \tag{2}$$

where Y is the n × 1 vector of the response variables that relates to the measured sample categories, B is the regression coefficients matrix for the spectral variables, F is the n × 1 error vector of residuals, X is the n × j data matrix of the spectral variables for each measured sample category, n is the number of samples and j is the number of variables. During the model development and updating stages, the number of main components was optimized by 10-fold cross validation and ultimately 10 main components were determined.

Known as the least square form of the support vector machine (SVM) approach, LS-SVM applies an equality constraint instead of an inequality constraint that has been used in SVM to obtain a linear set of equations. As a result, it simplifies the complex calculation and is easy to train. It has been reported that the LS-SVM could present a remarkable performance, as it maps the data input space into a high-dimensional feature space through a kernel function (the radial basis function (RBF) kernel function was applied in this paper). The two main parameters of the SVM method, including the penalty factor and the radial width of the kernel function, are optimized using a grid-search algorithm coupled with 10-fold cross validation during the model development and updating stages [39].

ELM has shown the advantages of fast learning speed and excellent generalization performance compared to traditional feedforward network learning algorithms such as back-propagation (BP). In most cases, ELM is used as a simple learning algorithm for single-hidden layer feedforward neural network (SLFN). Due to its different learning algorithm implementations for regression, classification or clustering, ELM has also been used to form multi hidden layer networks, deep learning or hierarchical networks [40]. The hidden node in ELM is a computational element, which is considered as a classical neuron, and its number was tuned to 100 for high accuracy.

Based on the spectral data with different preprocessing methods—i.e., SG, SG-D1 and MSC—the performances of the three models above were analyzed and evaluated to classify the vitality of seeds stored for different years. For 400 samples of each category, 200 seeds were used as the training sample and the other 200 seeds were used as the testing sample.

3.6. Germination Test

After hyperspectral images of all seeds were collected, 140 seeds were randomly selected from each group for the germination test following the International Seed Testing Association (ISTA) guidelines [41]. Seeds for germination were placed between two wet germination papers and incubated in a germination chamber for 7 days. The germination chamber was set as day-night mode at 30 °C, 80% RH and 10,000 Lx during the day (16 h), and 20 °C, 80% RH and 0 Lx during the night (8 h). Germination results of seeds were recorded daily and seeds with a 1 cm germ length were counted as germinated according to ISTA standards. The germination rate (GR, %) was calculated by Equation (3). The seeds high in vigor generally provided early and uniform stands, indicating that the seeds had the potential to produce vigorous seedlings under favorable conditions. Therefore, in this study, germination days were considered as a standard for seed vigor, and used as a factor to determine the vitality index (VI), as shown in Equation (4):

$$GR = GN/SN \cdot 100\%,\qquad(3)$$

$$VI = S \cdot \sum \frac{Gt}{Dt},\qquad(4)$$

where GN and SN are the numbers of germinated and non-germinated rice seeds, respectively, which were recorded on the last day of the germination test, S is the average value of germ length (cm), Dt is the number of the day t and Gt is the germination number recorded on the day of Dt.

4. Conclusions

The NIR-HSI technique, combined with multiple preprocessing methods and classification models, was used to identify the vitality of rice seeds. Spectral data was extracted from the ROI of the hyperspectral image and three preprocessing methods, including SG, SG-D1 and MSC, were applied to reduce the effect of irregularities in the spectral data caused by factors such as random noise, light scattering and sample texture. The SPA algorithm was adopted to obtain optimal wavelengths for the vitality of seeds, and to reduce computational cost. The numbers of selected wavelengths were 9, 8, 11 and 6 for raw data, SG, SG-D1 and MSC preprocessed data, respectively, which could decrease data processing load greatly compared to the classification model of full wavelengths. Then, these optimal wavelengths, as well as full wavelengths, were used to build multivariate models, including PLS-DA, LS-SVM and ELM, for determinate seed vitality of three different years and non-viable seeds from viable seeds of three different seeds. As for the detection of seed vitality of the three different years, better performance could be achieved by using pretreatment SG compared with the other two preprocessing methods. The classification accuracies for the seed vitality of three different years obtained using PLS-DA, LS-SVM and ELM with selected wavelengths and SG preprocessing were 87.5%, 93.33% and 93.67%, respectively. The ELM-SG method with spectral data from only eight wavebands (992, 1012, 1119, 1167, 1305, 1402, 1629 and 1649 nm) had better and faster classification performance, and could be developed to a fast and cost-effective seed-sorting system for industrial online application. As for identifying non-viable seeds from viable seeds of different years, the LS-SVM model coupled with raw data and selected wavelengths of 968, 988, 1204, 1301, 1409, 1463, 1629, 1646 and 1659 nm, achieved a classification accuracy of 94.38%, which decreased the data processing load to 4.2% of the classification model of full wavelengths and could be adopted as an optimal combination to identify non-viable seeds from viable seeds.

Author Contributions: Conceptualization, X.H., X.F. and Y.H.; data curation, X.F., D.S., Y.B. and Y.H.; formal analysis, F.L., Y.B. and Y.H.; funding acquisition, X.F. and Y.H.; investigation, X.H., X.F., D.S. and F.L.; methodology, X.H., X.F., D.S., F.L., Y.B. and Y.H.; project administration, Y.H.; resources, X.H. and X.F.; software, X.H. and D.S.; supervision, F.L., Y.B. and Y.H.; validation, X.H. and D.S.; visualization, X.H.; writing—original draft, X.H.; writing—review and editing, X.H., X.F. and D.S.

Funding: This research was funded by National key R&D program of China (Grant No. 2018YFD0101002) and National Natural Science Foundation of China (Grant No. 31801257).

Conflicts of Interest: The authors declare no conflict of interest.

References

1. Database of Crops from Food and Agriculture Organization of the United Nations. Available online: http://www.fao.org/faostat/en/#data/QC (accessed on 20 April 2019).
2. World Population (2019 and Historical) from Worldometers. Available online: http://www.worldometers.info/world-population/#table-historical (accessed on 20 April 2019).
3. Wakholi, C.; Kandpal, L.M.; Lee, H.; Bae, H.; Park, E.; Kim, M.S.; Mo, C.; Lee, W.; Cho, B. Rapid assessment of corn seed viability using short wave infrared line-scan hyperspectral imaging and chemometrics. *Sens. Actuators B Chem.* **2018**, *255*, 498–507. [CrossRef]
4. Ramesh, K.; Rao, A.N.; Chauhan, B.S. Role of crop competition in managing weeds in rice, wheat, and maize in India: A review. *Crop Prot.* **2017**, *95*, 14–21. [CrossRef]
5. Angélica, B.R.; Julio, M.F. Onion seed vigor in relation to plant growth and yield. *Hortic. Bras.* **2003**, *21*, 220–226.
6. Finch-Savage, W.E.; Bassel, G.W. Seed vigour and crop establishment: Extending performance beyond adaptation. *J. Exp. Bot.* **2016**, *67*, 567–591. [CrossRef] [PubMed]
7. Krishnan, P.; Joshi, D.K.; Nagarajan, S.; Moharir, A.V. Characterization of germinating and non-viable soybean seeds by nuclear magnetic resonance (NMR) spectroscopy. *Seed Sci. Res.* **2004**, *14*, 355–362. [CrossRef]
8. Al-Turki, T.A.; Baskin, C.C. Determination of seed viability of eight wild Saudi Arabian species by germination and X-ray tests. *Saudi J. Biol. Sci.* **2017**, *24*, 822–829. [CrossRef]
9. Al-Hammad, B.A.; Al-Ammari, B.S. Seed viability of five wild Saudi Arabian species by germination and X-ray tests. *Saudi J. Biol. Sci.* **2017**, *24*, 1424–1429. [CrossRef]
10. Braga, R.A.; Dal Fabbro, I.M.; Borem, F.M.; Rabelo, G.; Arizaga, R.; Rabal, H.J.; Trivi, M. Assessment of Seed Viability by Laser Speckle Techniques. *Biosyst. Eng.* **2003**, *86*, 287–294. [CrossRef]
11. Fatonah, K.; Suliansyah, I.; Rozen, N. Electrical conductivity for seed vigor test in sorghum (Sorghum bicolor). *Cell Biol. Dev.* **2017**, *1*, 6–12. [CrossRef]
12. Feng, X.; Yu, C.; Chen, Y.; Peng, J.; Ye, L.; Shen, T.; Wen, H.; He, Y. Non-Destructive Determination of Shikimic Acid Concentration in Transgenic Maize Exhibiting Glyphosate Tolerance Using Chlorophyll Fluorescence and Hyperspectral Imaging. *Front. Plant Sci.* **2018**, *9*, 468. [CrossRef]
13. Doherty, B.; Daveri, A.; Clementi, C.; Romani, A.; Bioletti, S.; Brunetti, B.; Sgamellotti, A.; Miliani, C. The Book of Kells: A non-invasive MOLAB investigation by complementary spectroscopic techniques. *Spectrochim. Acta Part A Mol. Biomol. Spectrosc.* **2013**, *115*, 330–336. [CrossRef] [PubMed]
14. Caporaso, N.; Whitworth, M.B.; Fisk, I.D. Near-Infrared spectroscopy and hyperspectral imaging for non-destructive quality assessment of cereal grains. *Appl. Spectrosc. Rev.* **2018**, *53*, 667–687. [CrossRef]
15. Park, B. Future Trends in Hyperspectral Imaging. *NIR News* **2016**, *27*, 35–38. [CrossRef]
16. Ambrose, A.; Lohumi, S.; Lee, W.; Cho, B.K. Comparative nondestructive measurement of corn seed viability using Fourier transform near-infrared (FT-NIR) and Raman spectroscopy. *Sens. Actuators B Chem.* **2016**, *224*, 500–506. [CrossRef]
17. Han, L.; Mao, P.; Wang, X.; Wang, Y. Study on vigour test of oat seeds with near infrared reflectance spectroscopy. *J. Infrared Millim. Waves* **2008**, *27*, 86–90. [CrossRef]
18. Kandpal, L.M.; Lohumi, S.; Kim, M.S.; Kang, J.; Cho, B. Near-infrared hyperspectral imaging system coupled with multivariate methods to predict viability and vigor in muskmelon seeds. *Sens. Actuators B Chem.* **2016**, *229*, 534–544. [CrossRef]
19. Al-Amery, M.; Geneve, R.L.; Sanches, M.F.; Armstrong, P.R.; Maghirang, E.B.; Lee, C.; Vieira, R.D.; Hildebrand, D.F. Near-infrared spectroscopy used to predict soybean seed germination and vigour. *Seed Sci. Res.* **2018**, *28*, 245–252. [CrossRef]
20. Qi, X.W.; Li, W.K.; Li, W.; Li, H. Study on the Vigour Testing of Soybean Seed Based on Near Infrared Spectroscopy Technology. *Appl. Mech. Mater.* **2011**, *58*, 458–462. [CrossRef]
21. Lohumi, S.; Mo, C.; Kang, J.; Hong, S.; Cho, B. Nondestructive Evaluation for the Viability of Watermelon (Citrullus lanatus) Seeds Using Fourier Transform Near Infrared Spectroscopy. *J. Biosyst. Eng.* **2013**, *38*, 312–317. [CrossRef]
22. Sun, J.; Lu, X.; Mao, H.; Jin, X.; Wu, X. A method for rapid identification of rice origin by hyperspectral imaging technology. *J. Food Process Eng.* **2015**, *40*, e12297. [CrossRef]

23. Liu, W.; Liu, C.; Ma, F.; Lu, X.; Yang, J.; Zheng, L. Online Variety Discrimination of Rice Seeds Using Multispectral Imaging and Chemometric Methods. *J. Appl. Spectrosc.* **2016**, *82*, 993–999. [CrossRef]
24. Wang, L.; Liu, D.; Pu, H.; Sun, D.; Gao, W.; Xiong, Z. Use of Hyperspectral Imaging to Discriminate the Variety and Quality of Rice. *Food Anal. Methods* **2015**, *8*, 515–523. [CrossRef]
25. Qiu, Z.; Chen, J.; Zhao, Y.; Zhu, S.; He, Y.; Zhang, C. Variety Identification of Single Rice Seed Using Hyperspectral Imaging Combined with Convolutional Neural Network. *Appl. Sci.* **2018**, *8*, 212. [CrossRef]
26. Chen, S.; Huang, C.; Huang, C.; Yang, C.; Wu, T.; Tsai, Y.; Miao, P. Determination of Nitrogen Content in Rice Crop Using Multi-Spectral Imaging. In Proceedings of the 2003 ASAE Annual Meeting, Las Vegas, NV, USA, 27–30 July 2003; Available online: https://elibrary.asabe.org/azdez.asp?JID=5&AID=13741&CID=lnv2003&T=2 (accessed on 21 April 2019).
27. Lin, L.H.; Lu, F.M.; Chang, Y.C. Development of a Near-Infrared Imaging System for Determination of Rice Moisture. *Cereal Chem.* **2006**, *83*, 498–504. [CrossRef]
28. Liu, M.; Liu, X.; Wu, M.; Li, L.; Xiu, L. Integrating spectral indices with environmental parameters for estimating heavy metal concentrations in rice using a dynamic fuzzy neural-network model. *Comput. Geosci.* **2011**, *37*, 1642–1652. [CrossRef]
29. Abdi, H.; Williams, L.J. Principal component analysis. *WIREs Comput. Stat.* **2010**, *2*, 433–459. [CrossRef]
30. Tsenkova, R.; Munćan, J.; Pollner, B.; Kovacs, Z. Essentials of Aquaphotomics and Its Chemometrics Approaches. *Front. Chem.* **2018**, *6*, 363. [CrossRef]
31. Aenugu, H.P.R.; Kumar, D.S.; Srisudharson, N.P.; Ghosh, S.; Banji, D. Near infra red spectroscopy—An overview. *Int. J. ChemTech Res.* **2011**, *3*, 825–836.
32. Lee, J.; Welti, R.; Roth, M.; Schapaugh, W.T.; Li, J.; Trick, H.N. Enhanced seed viability and lipid compositional changes during natural ageing by suppressing phospholipase Dα in soybean seed. *Plant Biotechnol. J.* **2012**, *10*, 164–173. [CrossRef]
33. Eldin, A.B. *Near Infra Red Spectroscopy*; INTECH Open Access Publisher: London, UK, 2011; Available online: https://www.intechopen.com/books/wide-spectra-of-quality-control/near-infra-red-spectroscopy (accessed on 19 April 2019).
34. Uarrota, V.G.; Moresco, R.; Coelho, B.; Nunes, E.C.; Peruch, L.A.M.; Neubert, E.O.; Rocha, M.; Maraschin, M. Metabolomics combined with chemometric tools (PCA, HCA, PLS-DA and SVM) for screening cassava (Manihot esculenta Crantz) roots during postharvest physiological deterioration. *Food Chem.* **2014**, *161*, 67–78. [CrossRef]
35. Ambrose, A.; Lee, W.H.; Cho, B.K. Effect of microwave heat treatment on inhibition of corn seed germination. *J. Biosyst. Eng.* **2015**, *40*, 224–231. [CrossRef]
36. Maleki, M.R.; Mouazen, A.M.; Ramon, H.; Baerdemaeker, J. Multiplicative scatter correction during on-line measurement with near infrared spectroscopy. *Biosyst. Eng.* **2007**, *96*, 427–433. [CrossRef]
37. Araújo, M.C.U.; Saldanha, T.C.B.; Galvão, R.K.H.; Yoneyama, T.; Chame, H.C.; Visani, V. The successive projections algorithm for variable selection in spectroscopic multicomponent analysis. *Chemom. Intell. Lab. Syst.* **2001**, *57*, 65–73. [CrossRef]
38. Zhang, J.; Dai, L.; Cheng, F. Classification of Frozen Corn Seeds Using Hyperspectral VIS/NIR Reflectence Imaging. *Molecules* **2019**, *24*, 149. [CrossRef] [PubMed]
39. Khosravani, H.S.; Nabipour, M. Application of LSSVM algorithm as a novel tool for prediction of density of bitumen and heavy n-alkane mixture. *Pet. Sci. Technol.* **2018**, *36*, 1137–1142. [CrossRef]
40. Huang, G.B.; Zhu, Q.Y.; Siew, C.K. Extreme learning machine: Theory and applications. *Neurocomputing* **2006**, *70*, 489–501. [CrossRef]
41. International Seed Testing Association. *International Rules for Seed Testing 2018*; International Seed Testing Association: Bassersdorf, Switzerland, 2018.

Sample Availability: Samples of the compounds are not available from the authors.

molecules

MDPI

Article

Discrimination of Trichosanthis Fructus from Different Geographical Origins Using Near Infrared Spectroscopy Coupled with Chemometric Techniques

Liang Xu [1,2], Wen Sun [1], Cui Wu [1,2], Yucui Ma [1,2] and Zhimao Chao [1,2,*]

[1] Institute of Chinese Materia Medica, China Academy of Chinese Medical Sciences, Beijing 100700, China; xuliang9988@126.com (L.X.); sun.w@outlook.com (W.S.); wucuidalian@163.com (C.W.); yucuim@gmail.com (Y.M.)
[2] Storage & Packaging Position, Chinese Materia Medica, China Agriculture Research System, Beijing 100700, China
*   Correspondence: chaozhimao@163.com; Tel.: +86-135-2270-5161

Academic Editors: Christian Huck and Krzysztof B. Bec
Received: 26 March 2019; Accepted: 17 April 2019; Published: 19 April 2019

Abstract: Near infrared (NIR) spectroscopy with chemometric techniques was applied to discriminate the geographical origins of crude drugs (i.e., dried ripe fruits of *Trichosanthes kirilowii*) and prepared slices of Trichosanthis Fructus in this work. The crude drug samples (120 batches) from four growing regions (i.e., Shandong, Shanxi, Hebei, and Henan Provinces) were collected, dried, and used and the prepared slice samples (30 batches) were purchased from different drug stores. The raw NIR spectra were acquired and preprocessed with multiplicative scatter correction (MSC). Principal component analysis (PCA) was used to extract relevant information from the spectral data and gave visible cluster trends. Four different classification models, namely *K*-nearest neighbor (KNN), soft independent modeling of class analogy (SIMCA), partial least squares-discriminant analysis (PLS-DA), and support vector machine-discriminant analysis (SVM-DA), were constructed and their performances were compared. The corresponding classification model parameters were optimized by cross-validation (CV). Among the four classification models, SVM-DA model was superior over the other models with a classification accuracy up to 100% for both the calibration set and the prediction set. The optimal SVM-DA model was achieved when C =100, γ = 0.00316, and the number of principal components (PCs) = 6. While PLS-DA model had the classification accuracy of 95% for the calibration set and 98% for the prediction set. The KNN model had a classification accuracy of 92% for the calibration set and 94% for prediction set. The non-linear classification method was superior to the linear ones. Generally, the results demonstrated that the crude drugs from different geographical origins and the crude drugs and prepared slices of Trichosanthis Fructus could be distinguished by NIR spectroscopy coupled with SVM-DA model rapidly, nondestructively, and reliably.

Keywords: near infrared spectroscopy; Trichosanthis Fructus; geographical origin; chemometric techniques; crude drugs; prepared slices; support vector machine-discriminant analysis

---

1. Introduction

Trichosanthis Fructus, the dried ripe fruits of *Trichosanthes kirilowii* Maxim. or *T. rosthornii* Harms (Fam. Cucurbitaceae), has been commonly used in Traditional Chinese Medicine (TCM) for the treatment of cough with lung heat, sticky phlegm, constipation, thoracic obstruction, and cardiodynia [1]. In modern clinical practice, Trichosanthis Fructus and its TCM prescriptions played a very important role in treating cardiovascular diseases including angina, cardiac failure, myocardial cinfarction, arrhythmia during acute myocardial infarction reperfusion, pulmonary heart disease, and cerebral ischaemic disease [2–4]. Because of its high medicinal value and good economic benefit, *T. kirilowii* has been cultivated widely in China [5], such as Shandong, Shanxi, Henan, and Hebei Provinces [6]. Specially, Trichosanthis Fructus produced from Shandong Province, was considered to be genuine since it showed the highest curative effect in traditional clinical use and active constituent content [6–8]. However, it was not easy to discriminate the geographical origin by visual inspection. There were a few studies on the discrimination of Trichosanthis Fructus from different cultivars or geographical origins using analytical methods including high pressure liquid chromatography (HPLC) fingerprint [9], seed protein electrophoresis [10], scanning electron microscope [11], and random amplified polymorphic DNA (RAPD), internal transcribed spacer (ITS), and sequence-related amplified polymorphism (SRAP) molecular markers [12,13]. However, these methods were time-consuming, costly, and destructive. Therefore, a fast, accurate, and non-destructive analytical method was established to discriminate the geographical origins of Trichosanthis Fructus in this work.

Near infrared (NIR) spectroscopy is a fast, accurate, and nondestructive technique requiring minimal sample processing before analysis. Coupled with chemometric techniques, it appears to be an effective and powerful analytical tool widely used in different fields, such as agricultural food [14,15], petrochemical [16], pharmaceutical [17], environment [18], metabolomic profiling [19], etc. The NIR region spans the wavelength range between 780 and 2500 nm. The absorption bands in this region correspond mainly to combinations and overtones of the fundamental vibrations of O-H, C-H, S-H, and N-H bonds, which are the primary structural components of organic chemical constituents [20]. As the environment factors including light, climate, water, soil, planting methods, etc. have great influences on the growth quality of the medicinal plants, some chemical constituents of the same crude drug from different geographical origins vary in content. Therefore, NIR spectroscopy has been also used to determine the geographical origins of TCM, such as Radix Pseudostellariae [21], Herba Epimedii [22], and Gastrodiae Rhizoma [23] in recent years. However, there has not been any reports until now on the use of NIR spectroscopy for the discrimination of Trichosanthis Fructus from different geographical origins, and the discrimination between crude drugs and prepared slices. It has been found that the concentrations of total saponins, amino acids and total flavonoids were different in different geographic origins [7,9,24], and the concentrations of 5-hydroxymethylfurfural, vanillic acid, quercetin, luteolin, and sugar in prepared slices showed significant changes compared with crude drugs, especially the concentration of 5-hydroxymethylfurfural increased to nearly 26 times as much as crude drugs of Trichosanthis Fructus ($p < 0.05$ or $p < 0.01$) [25–27]. All of the above laid the foundation for the feasibility of the following experiment.

In this study, four chemometric techniques including *K*-nearest neighbors (KNN), soft independent modeling of class analogy (SIMCA), partial least squares discriminant analysis (PLS-DA), and support vector machine discrimination analysis (SVM-DA) were attempted to discriminate Trichosanthis Fructus from different geographical origins. Among them, KNN, SIMCA, and PLS-DA were three linear methods, while SVM-DA was a non-linear method. Principal component analysis (PCA) was conducted on the NIR data to extract some principal components (PCs) as the inputs of the supervised pattern classification models. The number of PCs was optimized by cross-validation.

2. Results and Discussion

2.1. Spectra Investigation

The raw NIR spectra of 150 Trichosanthis Fructus samples were shown in Figure 1a. It can be seen that the raw spectra of Trichosanthis Fructus samples from different geographical origins and drug stores were similar. The variation of baseline shifts in spectra was wide, which was attributed to noise, packing density, and particle-size effect. It was difficult to determine specific bands in the raw spectra based on geographical origin because of the high degree of band overlapping. Moreover, the background information and noises contained in the raw spectra could weaken the model performance. Hence, in order to reduce the systematic noise and achieve a reliable model, mathematical spectral preprocessing before calibration was necessary. The multiplicative scatter correction (MSC), as a mathematical transformation method for spectra, was used to remove slope variation and correct scatter effects on the basis of different particle sizes, and correct for additive and multiplicative effects in the spectra. The Savitzky–Golay (SG) filter algorithm could be used to avoid the augmentation of noise which came from the derivatization. Therefore, MSC spectral preprocessing method with SG smoothing was applied in this research and the preprocessing spectra was presented in Figure 1b.

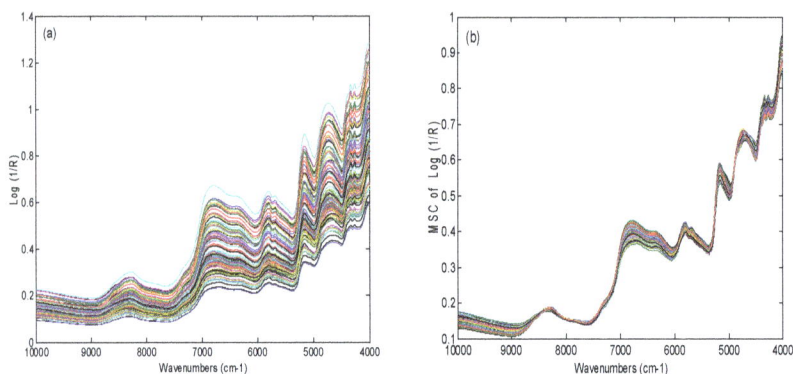

Figure 1. Spectra of Trichosanthis Fructus (**a**) raw data and (**b**) with multiplicative scatter correction (MSC) pretreatment.

As shown in Figure 1b, ten absorption bands in the spectra could be clearly observed. There was a water absorption band around 5155 cm^{-1} corresponding to the vibration of O-H stretching. There were strong absorption bands belonged to the vibration of C-H stretching (4261 and 4333 cm^{-1}), C=O stretching (4673 cm^{-1}), -CH$_2$ (5797cm^{-1}), and N-H stretching (6369 and 6798 cm^{-1}). These vibrations were caused by the chemical constituents such as lipids, alkaloids, polysaccharides, free amino acids, proteins, volatile compounds, and so on in Trichosanthis Fructus [28–30].

2.2. Principal Component Analysis

Principal component analysis (PCA) was a widely used technique for exploring and modeling multivariate data by reducing the dimension of the data matrix and compressing the information into a smaller number of uncorrelated variables called principle components (PCs), which were linear combinations of the original variables [31]. The first principal component, PC1, covered the maximum of the total variance; the second, PC2, was orthogonal to the first one and covered as much of the remaining variation as possible, and so on, until the total variance was accounted for. By plotting the PCs, one could view the interrelationships between different variables and interpret sample patterns, groupings, similarities, and differences. PCA was applied to examine the natural grouping of samples and develop the SIMCA models [32].

To visualize the data trends, a score plot was obtained by using the top three principal components (PC1, PC2, and PC3). Figure 2 showed the outcome of the principal component analysis and there were separations in the geographical origins. Figure 2 showed a three-dimensional (3D) space of Trichosanthis Fructus samples represented by PC1, PC2, and PC3. The variance interpreted by PC1, PC2, and PC3 were 64.87%, 15.77%, and 12.69%, respectively. In other words, the top three PCs could load almost the whole NIR spectral information of the samples. There was a clear separation between prepared slices and crude drugs. The crude drugs from different origins were separated roughly. The results indicated that there were differences in the chemical compositions between prepared slices and crude drugs of Trichosanthis Fructus. Though the PCA analysis gave the cluster trend in the 3D space, it could not separate the samples completely. Therefore, effective multivariate classification models were utilized and optimized in the following studies.

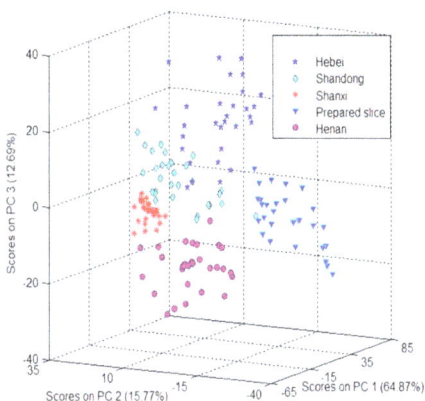

Figure 2. Score cluster plot of the top three principal components (PCs) of Trichosanthis Fructus for the data set.

2.3. Optimation of Models

2.3.1. The Establishment of the KNN Model

The KNN method was based on the Euclidean distance between neighbors. Parameter K, the number of neighbors which significantly influenced the model performance, could be determined by the classification accuracy (%) for each class. The prediction ability of the model for a given set of K values was evaluated by cross-validation, and the K value which gave the highest prediction rate was selected as the optimal one. Figure 3 showed the classification accuracy of the KNN model according to different K values. The optimal K value gave the highest classification accuracy by cross-validation. As shown in Figure 3, eight K values ($K = 1, 2, \ldots, 8$) were tested simultaneously for building model and the optimal KNN model was obtained when $K = 3$. The classification accuracy was 92% in the calibration set and 94% in the prediction set, respectively.

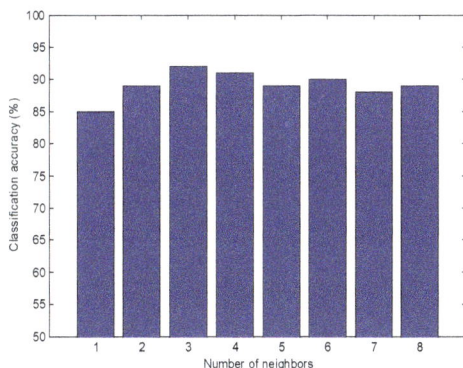

Figure 3. Cross-validation classification accuracy of the K-nearest neighbor (KNN) model according to varying parameter K.

2.3.2. The Establishment of SIMCA Model

In SIMCA model, after the initial cluster trends by PCA, the samples were divided into the calibration set and prediction set. Twenty samples were randomly selected from drug stores and each geographical origin as the calibration set, and five sub-models were established. The remaining 50 samples were used for the prediction set to test reliability and stability of each sub-model. The optimal number of the principal components (PCs) for each sub-model was selected based on the root mean square error of cross-validation (RMSECV) by cross-validation. Figure 4 showed an example for the selection of optimal number of PCs in sub-model 1 construction. As shown in Figure 4, most of the improvement in error was achieved before six PCs and the addition of another PC did not greatly lower the RMSECV. These results suggested six PCs for the final sub-model 1. The optimal number of PCs for another four sub-models was determined based on the same criterion. Table 1 listed the optimization and results of SIMCA model. The optimal number of PCs selected for the five sub-models was 6 (Hebei), 4 (Shanxi), 4 (Shandong), 5 (Henan), and 5 (Prepared slices). In the calibration set, the classification accuracy was all above 90% except Shandong and Henan. In the prediction set, the classification accuracies were all 100%, except for Hebei (90%) and Shandong (90%). The average classification accuracy was 93% in the calibration set and 96% in the prediction set. Therefore, NIR spectra combined with SIMCA model had certain feasibility in the discrimination of Trichosanthis Fructus.

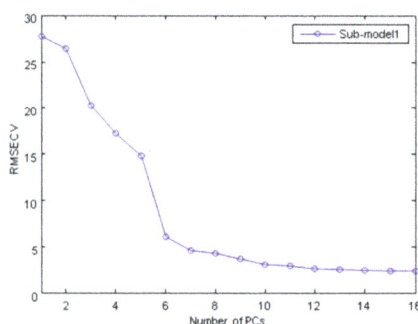

Figure 4. Root mean square error of cross-validation (RMSECV) values according to different number of PCs in sub-model 1.

Table 1. Optimization and results of the soft independent modeling of class analogy (SIMCA) model of Trichosanthis Fructus samples.

Sub-Models	Labels	PCs	Calibration Set		Prediction Set	
			N_{right}/N_0	CA%	N_{right}/N_0	CA%
1	Hebei	6	19/20	95	9/10	90
2	Shanxi	4	19/20	95	10/10	100
3	Shandong	4	17/20	85	9/10	90
4	Henan	5	18/20	90	10/10	100
5	Prepared slice	5	20/20	100	10/10	100

2.3.3. The Establishment of the PLS-DA Model

Figure 5 showed the classification accuracy according to different number of PCs. The optimal model was obtained when the number of PCs equaled 8 by cross-validation. Its classification accuracy was 95% in the calibration set and 98% in the prediction set.

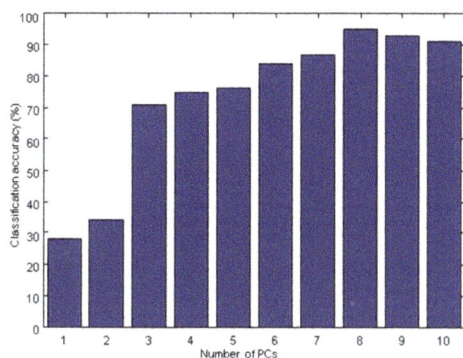

Figure 5. Classification accuracy of the partial least squares-discriminant analysis (PLS-DA) model according to different number of PCs.

2.3.4. The Establishment of SVM-DA Model

To obtain the best performance of SVM-DA model, parameters C and γ were optimized by cross-validation. The term C was the penal parameter, which determined the tradeoff between minimizing training error and minimizing model complexity. The γ term was the RBF kernel parameter. In this study, 15 γ values from 10^{-6} to 10 and 11 C values from 10^{-3} to 100 spaced uniformly in log (Log_{10} (γ) = $-6.0, -5.5, -5.0, -4.5, \ldots, 1$; Log_{10} (C) = $-3.0, -2.5, -2.0, -1.5, \ldots, 2$) were tested simultaneously for searching the optimal parameter. Figure 6 showed the classification accuracy of the SVM-DA model influenced by values of Log_{10} C and Log_{10} γ. The optimal SVM-DA model was obtained according to the highest classification accuracy by cross-validation. It could be found that the optimal model was achieved when C = 100 and γ = 0.00316 (i.e., Log_{10} (C) = 2, Log_{10} (γ) = -2.5). After parameters C and γ were determined, the optimal number of PCs was obtained according to the highest classification accuracy by 5-fold cross-validation. Figure 7 showed that the highest classification accuracy was achieved when PCs = 6. The classification accuracy in the calibration set and prediction set were both 100%.

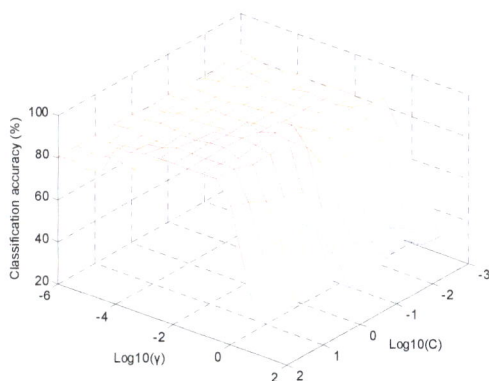

Figure 6. Classification accuracy of the support vector machine-discriminant analysis (SVM-DA) model according to different Log_{10} (γ) and Log_{10} (C) values.

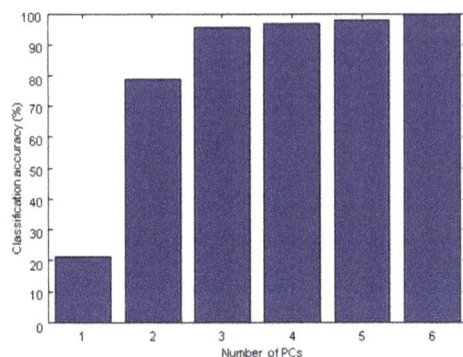

Figure 7. Classification accuracy of the SVM-DA model according to number of PCs.

2.4. Comparison of Four Models

To highlight the good performance in discrimination of Trichosanthis Fructus from different geographical origins, we attempted to compare the performance of 4 classification models of KNN, SIMCA, PLS-DA, and SVM-DA. Table 2 showed the optimal parameters and classification results from the 4 classification models. As shown in Table 2, the classification accuracies of the KNN and SIMCA models did not show better performance than those of PLS-DA model, presumably due to that they put the emphasis more on the similarity within a class. The classification accuracies of SVM-DA gave the best performance compared with KNN, SIMCA, and PLS-DA. The superiority of SVM-DA suggested its superb ability in solving the nonlinear problem in dataset.

Table 2. Comparison of the classification accuracy of the *K*-nearest neighbor (KNN), soft independent modeling of class analogy (SIMCA), partial least squares-discriminant analysis (PLS-DA), and support vector machine-discriminant analysis (SVM-DA) models.

Classification Models	Optimal Parameters	Classification Accuracy	
		Calibration Set (%)	Prediction Set (%)
KNN	$K = 3$	92	94
SIMCA	PCs = 6, 4, 4, 5, 5	93	96
PLS-DA	PCs = 8	95	98
SVM-DA	C = 100, γ = 0.00316, PCs = 6	100	100

3. Materials and Methods

3.1. Sample Preparation

In this experiment, the fresh ripe fruits of *T. kirilowii* Maxim. were picked up from four geographical origins including Shandong, Shanxi, Hebei, and Henan Provinces (30 samples from each) in October, 2017. They were strung together with their vines, and hung in a cool and drafty room for 6 months. The geographic location of the samples from Shandong Province was Zhuangke Village, Mashan Town, Changqing District, which was acknowledged as the traditional genuine producing area of Trichosanthis Fructus [33]. The crude drugs were obtained after cutting their vines and carpopodium, broke, and smashed. Additionally, 30 batches of prepared slices of uncertain geographical origins were purchased from multiple drug stores from January to March in 2018. The plant materials were identified by Prof. Zhimao Chao (Institute of Chinese Materia Medica, China Academy of Chinese Medical Sciences) as the fruits of *T. kirilowii* Maxim. Voucher specimens were deposited at the 1016 room of Institute of Chinese Materia Medica, China Academy of Chinese Medical Sciences.

Before data acquisition, all samples were dried in a DHG-9053A electric thermostatic drying oven from Shanghai Yiheng Scientific Instrument Co. Ltd. (Shanghai, China) at 60 °C for 4 h to remove moisture. Considering the heterogeneities of the samples, all the samples were crushed into powder by a FW-100 high speed universal grinder from Tianjin Taiste Instrument Co. Ltd. (Tianjin, China). The powder was then screened through a 60-mesh sieve and stored in a glass desiccator for further analysis. After MSC pretreatment, the 150 spectra data were partitioned into a calibration set and validation set, respectively. The former was used to build the calibration model, and the latter was used to test the robustness of the model constructed. About two thirds of the samples were randomly selected to make up the calibration set, and the rest were used as the prediction set. Table 3 shows the details of the tested samples.

Table 3. A summary of Trichosanthis Fructus samples.

Sample No.	Sample Type	Geographic Origins	Geographic Location *	Harvesting Time
1–30	Crude drug	Jinan, Shandong	36°19′ N, 116°19′ E, 127–131 m	Oct 12, 2017
31–60	Crude drug	Anyang, Henan	36°03′ N, 114°23′ E, 68–70 m	Oct 17, 2017
61–90	Crude drug	Anguo, Hebei	38°21′ N, 115°16′ E, 32–33 m	Oct 1, 2017
91–120	Crude drug	Houma, Shanxi	35°19′ N, 111°04′ E, 461–466 m	Oct 4, 2017
121–150	Prepared slices	Uncertain	Uncertain	Jan-Mar, 2018

* Geographic location is marked in the order of latitude, longitude, and altitude.

3.2. Spectral Measurement

The NIR spectra were collected in the diffuse reflectance mode using an MPA multi-purpose FT-NIR spectrometer (Bruker Optics, Ettlingen, Germany) with a Pbs detector and an internal gold background as the reference. Before the sample measurement, the spectrometer needed to be preheated for about 30 min. The spectral data were recorded as the logarithm of the reciprocal reflectance, i.e., log (1/R). Each spectrum was collected by an average of 32 scans performed at 3.857 cm^{-1} interval over the wavelength range of 10,000–4000 cm^{-1}. About 5 g of the dried sample powders were densely packed into the sample cup with the loading height of 2 cm. Each sample was collected 3 times in the standard procedure. The average of the 3 spectra collected from the same sample was used in the further analysis. The temperature was controlled at 23 ± 1 °C and the relative humidity at ambient level in the laboratory.

3.3. Data Analysis

An OPUS 7.0 from Bruker was used for instrumental and measurement control of the NIR spectrometer as well as for data analysis. The software Solo, version 6.7.1 (Eigenvector Research Inc., Wenatchee, WA, USA) was used for classification methods realization. The NIR spectra (files in MATLAB format, collected by OPUS 7.0) could be recognized by the software for further calculation.

3.4. Chemometrics Study

3.4.1. KNN

KNN is a linear and non-parametric supervised pattern recognition method which was first introduced by Fix and Hodges [34]. In this method, distance between the unknown object and each of the objects of the calibration set is determined. The unknown object is classified in the group which the majority of K objects belongs [35]. It is of great importance to select the optimal parameter *K*, which has a great influence on the classification accuracy of the KNN model. The *K* value is optimized by comparing the prediction ability with several *K* values and the one which gives the highest classification accuracy is chosen.

3.4.2. SIMCA

SIMCA is a supervised data classification method based on PCA which was first raised by Wold [36]. In SIMCA, a PCA is performed on each class in the data set and optimal number of PCs is retained to account for most of the variation within each class [37]. Hence, a PCA model is used to represent each class in the data set. The number of PCs retained for each class is important, as retention of too few components can distort the signal or information content contained in the model, whereas retention of too many PCs diminishes the signal-to-noise. RMSECV based on cross-validation is used to find the optimal number of PCs. To perform cross-validation, segments of data are predicted and compared to the actual values, using one, two, three, etc., PCs. The optimal number of PCs is selected when the addition of another PC does not greatly improve the performance of the model.

3.4.3. PLS-DA

PLS-DA is performed in order to find models that allow the maximum separation among classes of objects [38] by hopefully rotating PCA components and to understand which variables carry the class separating information. PLS-DA consists of a classical PLS regression where the response variable is a categorical one (replaced by the set of dummy variables describing the categories) expressing the class membership of the statistical units. Therefore, PLS-DA does not allow for other response variables than the one for defining the groups of individuals. As a consequence, all measured variables play the same role with respect to the class assignment. PLS-DA simultaneously decomposes spectral matrix and class matrix, and extracts the spectral information most related to the classes, which can lead to the establishment of a more accurate classification model [39].

3.4.4. SVM-DA

SVM-DA is a chemometric technique that is originated from binary classification but supports classification of multiple classes [40]. Each classification model is achieved by creating a hyperplane that allows linear separation in the higher dimension feature space, unless the linear boundary in lower dimension input space would accomplish a proper classification. In SVM-DA, this transformation into higher dimensional space is achieved through a kernel function. There exist three classical kernel functions: polynomial kernel function, Gaussian kernel function, and sigmoid kernel function. Selection of kernel function is of great importance on the performance of SVM-DA.

In this work, the popularly used Gaussian kernel function was applied. Its structure was the radial basic function (RBF), also called RBF kernel function. RBF kernel took the form as Equation (1):

$$K\left(x_i, x_j\right) = \exp\left(-\gamma \|x_i - x_j\|^2\right) \tag{1}$$

In order to obtain a good performance of SVM-DA model, two parameters including the penalty parameter (C) and kernel width (γ) in Gaussian kernel function should be optimized. The optimization was achieved by a combination of grid-search approach and 5-fold cross-validation (CV). The optimal C and γ were selected when the highest classification accuracy achieved. After the selection of parameters C and γ, the number of PCs was also optimized based on the highest classification accuracy by CV.

3.4.5. Model Efficiency Estimation

To evaluate the classification performances of the different classification models, the classification accuracy (%) by cross-validation was used as N_{right}/N_0. N_{right} and N_0 referred to the number of rightly classified and total number of samples in data set, respectively.

Five-fold CV was used to evaluate the efficiency of classification models. The calibration set was first divided into five subsets of equal size. Sequentially, one subset was tested using the classification model trained on the remaining four subsets. Thus, each instance of the whole calibration set was predicted once so the CV accuracy was the percentage of data.

For SIMCA model construction, RMSECV was used to evaluate the model efficiency based on CV. The RMSECV was defined as Equation (2):

$$RMSECV = \sqrt{\frac{\sum_{i=1}^{n} (y_i - \hat{y}_i)^2}{n}} \tag{2}$$

4. Conclusions

This study sufficiently demonstrated that NIR spectroscopy coupled with chemometric techniques had high potential to distinguish the crude drugs of Trichosanthis Fructus from different geographical origins and to discriminate the crude drugs and prepared slices in an accurate and non-destructive way. The successful discrimination using chemometric analysis was based on their differences in NIR spectra, which mainly correlated with the differences in their chemical compositions. The differences of crude drugs from different geographic origins might be caused by soil, climate, light, planting methods, and other factors. Light affected the synthesis and accumulation of carbohydrates and nitrogen metabolism of plants, soil affected the absorption of mineral elements in plants, and climate affected the growth cycle of plants resulting in the inconsistency of fruit maturity. All these led to changes in the types and contents of the constituents in Trichosanthis Fructus of different geographic origins. The variations between crude drugs and prepared slices might lie in that the former was only dried from the fresh fruits in the air, while the latter also needed to be steamed through, pressed, shredded, and sun-cured after dried fruits.

Four chemometric techniques (KNN, SIMCA, PLS-DA, and SVM-DA) were applied comparatively to construct the classification models. Among the four classification models, SVM-DA as a non-linear classification method showed superior performance over the linear ones of KNN, SIMCA, and PLS-DA after preprocessing with MSC. The classification accuracy of the calibration set and prediction set were both 100% when C = 100, γ = 0.00316, and PCs = 6. Generally, the non-linear model performed better than the linear models.

The genuineness of herbal medicine depends mostly on its geographical origins. It can be concluded that NIR spectroscopy coupled with chemometric techniques will have more application on the discrimination of TCM according to different geographical origins similarly, which is essential for quality control and traceability management. It also has a promising future to distinguish crude drugs and prepared slices, authenticity, adulteration, and storage period of TCM, all of which cannot be easily recognized by simple visual inspection. Therefore, more representative TCM samples need to be collected and experimented to develop more robust models for prediction in further studies.

Author Contributions: Z.C. conceived of and designed research. L.X., W.S., C.W., and Y.M. collected the fresh ripe fruits of *T. kirilowii* and prepared slices of Trichosanthis Fructus. L.X. and W.S. carried out the drying procedures from these fresh fruits. L.X., W.S., and C.W. performed the experiment and analyzed the spectral data. L.X. contributed significantly to the result interpretation and the manuscript preparation. All authors have read, discussed, and approved the final manuscript.

Funding: This work was supported by China Agriculture Research System (Grant No. CARS-21).

Acknowledgments: Our thanks will go to some planting bases for providing us the fresh ripe fruits of *T. kirilowii* in different geographical origins. We are also grateful to our academic colleagues for constructive discussion in this work.

Conflicts of Interest: The authors declare no conflict of interest.

References

1. The Pharmacopoeia Commission of PRC. *Pharmacopoeia of the People's Republic of China (Part 1)*; China Medical Science and Technology Press: Beijing, China, 2015; pp. 112–113.
2. Lu, P.; Shi, W.; Wang, Z.; Song, J. Clinical application and mechanism of Trichosanthis Pericarpium. *J. Tradit. Chin. Med.* **2013**, *54*, 1428–1431.

3. Chu, D.H.; Zhang, Z.Q. Trichosanthis Pericarpium aqueous extract protects H9c2 cardiomyocytes from hypoxia/roxygenation injury by regulating PI3K/Akt/NO pathway. *Molecules* **2018**, *23*, 2409. [CrossRef] [PubMed]

4. Cheng, J.; Huang, L.; Kong, C.C.; He, F. Gualou Xiebai Baijiu decoction alleviates myocardial ischemia reperfusion injury via Akt/GSK-3β signaling pathway. *Acad. J. Shanghai Univ. Tradit. Chin. Med.* **2018**, *32*, 82–85, 97.

5. Liu, J.N.; Xie, X.L.; Yang, T.X.; Liu, M.; Jia, D.S.; Wen, C.X. Study on resources and cultivation progress of Trichosanthis Fructus. *Food Res. Dev.* **2014**, *35*, 125–127.

6. Wu, C.; Chao, Z.M. Process control system about Trichosanthis Fructus quality. *Chin. J. Exp. Tradit. Med. Form.* **2016**, *22*, 230–234.

7. Xin, J.; Zhang, R.C.; Guo, Q.M.; Zhang, Y.Q. Effect of germplasms difference on the content of total saponins and adenosine in Trichosanthis Fructus. *Lishizhen Med. Mater. Med. Res.* **2015**, *26*, 2236–2237.

8. Hao, B.; Pan, L.L.; Yuan, S.X.; Yang, S.; Li, T.T.; Xu, H.F.; Li, X.R. HPLC determination of arginine in *Trichosanthes Pericarpium* with OPA by pre-column derivatization. *Acta Chin. med. Pharm.* **2014**, *42*, 17–19.

9. Hao, B.; Pan, L.L.; Yuan, S.X.; Yang, S.; Li, T.T.; Xu, H.F.; Li, X.R. Fingerprints of amino acid of Trichosanthis Pericarpium from different varieties and habitats by HPLC. *Acta Chin. med. Pharm.* **2015**, *43*, 14–18.

10. Sun, Z.Y.; Zhou, F.Q.; Guo, Q.M. Analysis of seed protein electrophoresis of farm cultivars of Fructus Trichosanthis from Shandong province. *Lishizhen Med. Mater. Med. Res.* **2005**, *16*, 1224–1225.

11. Guo, Q.M.; Zhou, F.Q.; Yang, J.L.; Gao, H. Study on micro-morphological characters of fruit coats in cultivated Fructus Trichosanthis in Shandong province. *Chin. J. Chin. Mater. Med.* **2005**, *30*, 1580–1582.

12. Sun, Z.Y.; Zhou, F.Q. RAPD analysis on farm cultivars of fruits of *Trichosanthes kirilowii* from Shandong province. *Chin. Tradit. Herb. Drugs* **2006**, *37*, 426–429.

13. Cao, L.; Huang, Y.N.; Xie, J.; Peng, S.W.; Xu, R.; Zhu, X.Q. Genetic diversity of Trichosanthis Fructus for seeds cultivating. *Chin. Tradit. Herb. Drugs* **2017**, *48*, 4316–4322.

14. Teye, E.; Huang, X.Y.; Dai, H.; Chen, Q.S. Rapid differentiation of *Ghana* cocoa beans by FT-NIR spectroscopy coupled with multivariate classification. *Spectrochim. Acta A Mol. Biomol. Spectrosc.* **2013**, *114*, 183–189. [CrossRef] [PubMed]

15. Zhao, Y.Y.; Zhang, C.; Zhu, S.S.; Gao, P.; Feng, L.; He, Y. Non-destructive and rapid variety discrimination and visualization of single grape seed using near-infrared hyperspectral imaging technique and multivariate analysis. *Molecules* **2018**, *22*, 1352. [CrossRef] [PubMed]

16. Reboucas, M.V.; Dos Santos, J.B.; Domingos, D.; Massa, A.R.C.G. Near-infrared spectroscopic prediction of chemical composition of a series of petrochemical process streams for aromatics production. *Vib. Spectrosc.* **2010**, *52*, 97–102. [CrossRef]

17. Jamrógiewicz, M. Application of the near-infrared spectroscopy in the pharmaceutical technology. *J. Pharm. Biomed. Anal.* **2012**, *66*, 1–10. [CrossRef]

18. Samiei Fard, R.; Matinfar, H.R. Capability of vis-NIR spectroscopy and Landsat 8 spectral data to predict soil heavy metals in polluted agricultural land (Iran). *Arab. J. Geosci.* **2016**, *9*, 745. [CrossRef]

19. Vergouw, C.G.; Botros, L.L.; Roos, P.; Lens, J.W.; Schats, R.; Hompes, P.G.A.; Burns, D.H.; Lambalk, C.B. Metabolomic profiling by near-infrared spectroscopy as a tool to assess embryo viability: a novel, non-invasive method for embryo selection. *Hum. Reprod.* **2008**, *23*, 1499–1504. [CrossRef] [PubMed]

20. Williams, P.; Norris, K. *Near-infrared Technology in the Agricultural and Food Industries*, 2nd ed.; American Association of Cereal Chemist: St. Paul, MN, USA, 2002.

21. Lin, H.; Zhao, J.W.; Chen, Q.S.; Zhou, F.; Sun, L. Discrimination of Radix Pseudostellariae according to geographical origins using NIR spectroscopy and support vector data description. *Spectrochim. Acta A Mol. Biomol. Spectrosc.* **2011**, *79*, 1381–1385. [CrossRef] [PubMed]

22. Yang, Y.; Wu, Y.J.; Li, W.; Liu, X.S.; Zheng, J.Y.; Zhang, W.T.; Chen, Y. Determination of geographical origin and icariin content of Herba Epimedii using near infrared spectroscopy and chemometrics. *Spectrochim Acta Part A Mol. Biomol. Spectrosc.* **2018**, *191*, 233–240. [CrossRef]

23. Zuo, Y.M.; Deng, X.H.; Wu, Q. Discrimination of *Gastrodia elata* from different geographical origin for quality evaluation using newly-build near infrared spectrum coupled with multivariate analysis. *Molecules* **2018**, *23*, 1088. [CrossRef] [PubMed]

24. Wang, Z.Z.; Wang, X.H.; Zhu, Y.; Wang, C.Q. Determination of content of quercetin and total flavonoids in Trichosanthis Pericarpium from different areas. *Chin. J. Exp. Tradit. Med. Form.* **2014**, *20*, 86–89.

25. Zou, C.C.; Zong, Q.N.; Yan, H.Y.; Xie, S.X.; Zhang, M.J.; Zhang, Y.M.; Li, N. HPLC fingerprint and quantitative analysis of 6 components of Fructus Trichosanthis and its steamed products. *Chin. Pharm. J.* **2017**, *52*, 597–601.

26. Sun, W.; Chao, Z.M.; Wang, C.; Wu, X.Y.; Tan, Z.G. Determination of 5-hydroxmethylfurfural in commercial Trichosanthis Fructus by HPLC. *Chin. J. Exp. Tradit. Med. Form.* **2012**, *18*, 73–76.

27. Yan, H.Y.; Zong, Q.N.; Zou, C.C.; Zhang, N. A comparative study on reducing sugar and total sugar in Trichosanthis Fructus and its steamed products. *J. Dali Univ.* **2018**, *3*, 71–74.

28. Chao, Z.M.; He, B.; Zhang, Y.; Akihisa, T. Studies on the chemical constituents of unsaponifiable lipids from the seeds of *Trichosanthes kirilowii*. *Chin. Pharm. J.* **2000**, *35*, 733–736.

29. Li, X.; Tang, L.Y.; Xu, J.; Yu, X.K.; Yang, H.J.; Li, D.F.; Zhang, Y.; Fan, J.W.; Su, R.Q.; Wu, H.W.; et al. Analysis and identification of chemical components in Trichosanthis Fructus by UPLC-LTQ-Orbitrap-MS. *Chin. J. Exp. Tradit. Med. Form.* **2019**, *25*, 201–210.

30. Ye, X.; Ng, C.C.; Ng, T.B.; Chan, G.H.; Guan, S.; Sha, O. Ribosome-inactivating proteins from root tubers and seeds of *Trichosanthes kirilowii* and other Trichosanthes species. *Protein Pept. Lett.* **2016**, *23*, 699–706. [CrossRef] [PubMed]

31. Luna, A.S.; Silva, A.P.; Pinho, J.S.A.; Ferré, J.; Boqué, R. Rapid characterization of transgenic and non-transgenic soybean oils by chemometric methods using NIR spectroscopy. *Spectrochim Acta Part A Mol. Biomol. Spectrosc.* **2013**, *100*, 115–119. [CrossRef] [PubMed]

32. Chen, Q.S.; Zhao, J.W.; Zhang, H.D.; Liu, M.H. Application of near infrared reflectance spectroscopy to the identification of tea using SIMCA pattern recognition method. *Food Sci.* **2006**, *27*, 186–189.

33. Lou, Z.Q.; Qin, B. *Species Systematization and Quality Evaluation of Commonly Used Chinese Traditional Drugs (Volume 3)*; Associated Press of Beijing Medical University and Peking Union Medical College: Beijing, China, 1996; pp. 579–680.

34. Fix, E.; Hodges, J.L. Discriminatory analysis. Nonparametric discrimination: consistency properties. *Int. Stat. Rev.* **1989**, *57*, 238–247. [CrossRef]

35. Coomans, D.; Massart, D.L. Alternative k-nearest neighbour rules in supervised pattern recognition: Part 2. Probabilistic classification on the basis of the kNN method modified for direct density estimation. *Anal. Chim. Acta.* **1982**, *138*, 153–165. [CrossRef]

36. Wold, S. Pattern recognition by means of disjoint principal components models. *Pattern Recogn.* **1976**, *8*, 127–139. [CrossRef]

37. Sjöström, M.; Kowalski, B.R. A comparison of five pattern recognition methods based on the classification results from six real data bases. *Anal. Chim. Acta.* **1979**, *112*, 11–30. [CrossRef]

38. Wold, S.; Sjöström, M.; Eriksson, L. PLS in chemistry. In *The Encyclopedia of Computational Chemistry*; Schleyer, P.V.R., Allinger, N.L., Clark, T., Gasteiger, J., Kollman, P.A., Schaefer, H.F., Schreiner, P.R., Eds.; Wiley: Chichester, UK, 1998; pp. 2006–2020.

39. Dong, G.; Guo, J.; Wang, C.; Chen, Z.L.; Zheng, L.; Zhu, D.Z. The classification of wheat varieties based on near infrared hyperspectral imaging and information fusion. *Spectrosc. Spect. Anal.* **2015**, *35*, 3369–3374.

40. Chen, Q.S.; Zhao, J.W.; Lin, H. Study on discrimination of roast green tea (*Camellia sinensis* L.) according to geographical origin by FT-NIR spectroscopy and supervised pattern recognition. *Spectrochim. Acta Part A Mol. Biomol. Spectrosc.* **2009**, *72*, 845–850. [CrossRef]

Sample Availability: Samples of the compounds 5-hydroxymethylfurfural, vanillic acid, and quercetin are available from the authors.

molecules

MDPI

Article

Quantification of Total Phenolic and Carotenoid Content in Blackberries (*Rubus Fructicosus* L.) Using Near Infrared Spectroscopy (NIRS) and Multivariate Analysis

Eva María Toledo-Martín [1], María del Carmen García-García [2], Rafael Font [3], José Manuel Moreno-Rojas [4], María Salinas-Navarro [5], Pedro Gómez [1] and Mercedes del Río-Celestino [1,*]

[1]   Department of Genomics and Biotecnology, IFAPA Center La Mojonera, Camino San Nicolás 1, La Mojonera, 04745 Almería, Spain; ortiztoledo@hotmail.com (E.M.T.-M.); pedro.gomez.j@juntadeandalucia.es (P.G.)
[2]   Department Agrifood Engineering and Technology, IFAPA Center La Mojonera, Camino San Nicolás 1, La Mojonera, 04745 Almería, Spain; mariac.garcia.g@juntadeandalucia.es
[3]   Department of Food Science and Health, IFAPA Center La Mojonera, Camino San Nicolás 1, La Mojonera, 04745 Almería, Spain; rafaelm.font@juntadeandalucia.es
[4]   Department of Food Science and Health, IFAPA Center Alameda del Obispo, 14080 Córdoba, Spain; josem.moreno.rojas@juntadeandalucia.es
[5]   Department of Applied Biology (Genetic), University of Almería, Edificio CITE II-B, Ctra. Sacramento s/n, La Cañada de San Urbano, 04120 Almería, Spain; msalinas@ual.es
*   Correspondence: mercedes.rio.celestino@juntadeandalucia.es; Tel.: +34-671-532-238

Received: 14 October 2018; Accepted: 30 November 2018; Published: 4 December 2018

Abstract: A rapid method to quantify the total phenolic content (TPC) and total carotenoid content (TCC) in blackberries using near infrared spectroscopy (NIRS) was carried out aiming to provide reductions in analysis time and cost for the food industry. A total of 106 samples were analysed using the Folin-Ciocalteu method for TPC and a method based on Ultraviolet-Visible Spectrometer for TCC. The average contents found for TPC and TCC were 24.27 mg·g^{-1} dw and 8.30 µg·g^{-1} dw, respectively. Modified partial least squares (MPLS) regression was used for obtaining the calibration models of these compounds. The RPD (ratio of the standard deviation of the reference data to the standard error of prediction (SEP)) values from external validation for both TPC and TCC were between 1.5 < RPDp < 2.5 and RER values (ratio of the range in the reference data to SEP) were 5.92 for TPC and 8.63 for TCC. These values showed that both equations were suitable for screening purposes. MPLS loading plots showed a high contribution of sugars, chlorophyll, lipids and cellulose in the modelling of prediction equations.

Keywords: blackberries; *Rubus fructicosus*; phenolics; carotenoids; bioanalytical applications; near infrared; chemometrics

1. Introduction

Consumers have high awareness of the health benefits of increased fruit and vegetable consumption, especially those rich in phytochemicals with nutraceutical properties. Vegetables and fresh fruit are reported to decrease the risk of cardiovascular diseases, certain forms of cancer and to prevent degenerative diseases [1,2]. This protection has been attributed to the fact that these foods may contain an optimal content of phytochemicals, such as antioxidants, fibre and other bioactive compounds [3]. These phytochemicals are in higher concentrations in small fruit, such as berries (blueberries, blackberries and strawberries) and this has motivated a large demand for fresh fruit.

Berries are widely grown in Spain especially in Huelva (South-eastern Spain) where the cultivated area has increased in the last years to approximately 11,145 hectares in 2018, being more than 95% of national volume [4]. With reference to commercialization and sales prospects, the blackberry is the most promising reference [5,6].

Previous studies have reported that berry fruits contain a high total phenolic content (1.97–3.62 mg gallic acid equivalent GAE g^{-1} fresh weight (fw) and 16.94–31.13 mg GAE g^{-1} dry weight (dw) representing a rich source of antioxidants [5,7–10].

The total carotenoid content was also high in blackberry fruits with values from 0.86 µg·g^{-1} fw (7.4 µg·g^{-1} dw) to 21.40 µg·g^{-1} dw in blueberry fruits [11,12]. However, other authors also reported lower carotenoid contents ranging from 0.162 µg·g^{-1} fw (1.39 µg·g^{-1} dw) [13] to 1.84 µg·g^{-1} dw [14].

The demonstrated antioxidant capacity of blackberry fruits suggests that can play an important role against oxygen-free radical in the organism [15,16] and therefore for use in the development of functional food or nutraceuticals [17]. Due to the recognized importance of these antioxidant compounds, it is essential to characterize their content of them in the fruits.

Nowadays, the measuring of phenolic compounds and carotenoids is carried out using methods such as high performance liquid chromatography (HPLC) [18,19], gas chromatography (GC), or combinations of these methods with different systems of detection such as UV-Vis or mass spectrometry (MS) [20]. These methodologies are efficient for a rapid separation and quantification of these compounds. Although their use is common, these methods require sophisticated and expensive equipment, skilled labour and a variety of reagents which contain pollutants. Another relevant method includes spectrophotometry since it represents a relatively simple method for measuring phenolic compounds and carotenoids. As an alternative to these methods of analysis, NIRS (Near Infrared Reflectance Spectroscopy) technique offers several advantages such as high response, non-sample destruction, non-polluting and low analytical cost that does not require sophisticated sample preparation [21]. This methodology measures the interaction of the material with the light, which is in turn determined by the vibration of the chemical bonds of the sample constituents [22].

With regards to berry fruits, the studies with NIRS have been focused on determining the total phenolic content and antioxidant activity in intact berries (multispecies calibration) [23], in quality control and identification of food product adulteration of wild berry fruit extracts during storage [24], in evaluation of quality and nutraceutical compounds such as anthocyanin, polyphenol and flavonoid content of blueberries (*Vaccinium corymbosum* L.) [25] and also for detecting of an underground insect named *Eurhizococcus colombianus* (Hemiptera: *Margarodidae*) in blackberry leaves [26].

Since Andalusia (Southern Spain) is an important exporter of blackberries, there is interest in developing methodologies for the rapid analysis of antioxidant compounds as is demanded in both, food industries and in germplasm-screening programs. Nutritional quality improvement has been initiated in blackberry breeding programs; thus, rapid techniques such as those based on NIRS are needed for quick screening of lines with higher quality in early generations.

Therefore, the objectives of the present work were: (i) to study the potential of the NIRS technology for predicting the total phenolic and total carotenoid contents in blackberries, being these compounds constituting some of the main responsible molecules of the antioxidant properties in this fruit; (ii) to provide some knowledge about the mechanism used by NIRS for successfully determining these compounds in the fruits of this species.

2. Results and Discussion

2.1. Reference Analysis of Total Phenolic and Carotenoid Contents in the Samples

Figure 1a,b showed frequency distribution plots of total phenolic and carotenoid contents for the samples (*n* = 106) used in this work, respectively. Such Total phenolic content (TPC) as Total carotenoid content (TCC) exhibited normal distributions in their intervals.

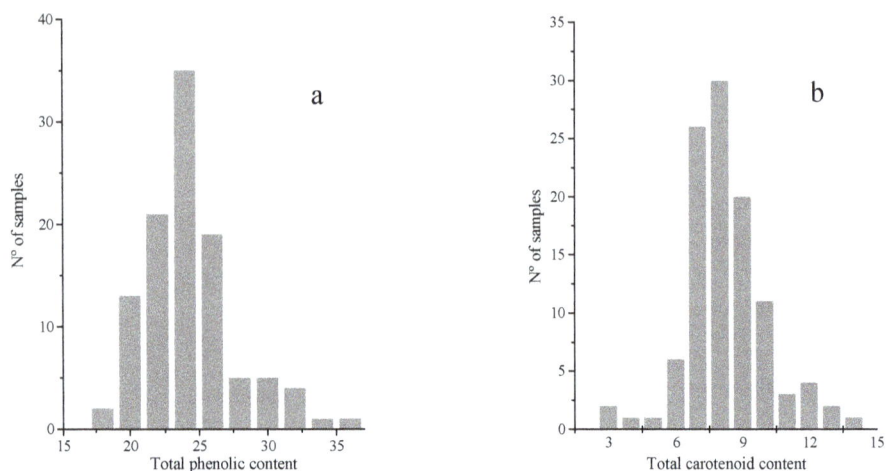

Figure 1. Frequency distribution plots for total phenolic (**a**) and total carotenoid content (**b**) by reference analysis.

TPC values ranged from 17.36 to 35.67 mg·g^{-1} dw with mean and variation coefficient values of 24.27 and 13.67, respectively. These values were similar to those contents reported previously in studies carried out on blackberry fruits by Souza et al. [11] and Contessa et al. [10] with 34.53 and 36.78 mg·g^{-1} dw, respectively. Previous works have reported higher TPC concentrations (43.29–99.47 mg·g^{-1} dw) [23,27] and lower findings (5.58 mg·g^{-1} dw) [13,17] than those found in this study. The qualitative and quantitative differences found among fruits for the phenolic compounds could be due to factors such as environmental conditions, genotype, storage conditions and agro-techniques as observed by Aaby et al. [28] in berry fruits.

Regarding TCC, the values varied from 2.84 to 13.73 µg·g^{-1} dw with mean and variation coefficients of 8.30 and 21.92, respectively. Souza et al. [11] obtained similar results with 12.14 µg·g^{-1} dw of TCC in blackberries. Higher TCC contents (21.40 µg·g^{-1} dw) have been described in previous studies by Rutz et al. [13] and Lashmanova et al. [12].

2.2. Spectral Data Pre-Treatments and Equation Performances. Second Derivative Spectra of Blueberry Fruit

Figure 2 shows the peaks and troughs corresponding to the points of maximum curvature in the raw spectrum.

The bands in the visible region at 558 nm are due to electric transitions in the green, the band at 614 nm corresponding to electric transitions in the orange and 678 nm to electronic transitions in the red. The absorption band at 674 nm is assigned to absorption by chlorophyll [29].

The characteristic bands for phenolics can be observed in the NIR regions from 1415 nm to 1512 nm and from 1955 to 2035 nm [30]. The wavelength regions of the spectra in the ranges 1100–1250, 1300–1350 and 1650–1700 nm correspond to the 3rd overtone, the combination bands and the 1st overtone, respectively, of the C–H bonds of carotenoids [31]. In addition to these bands, the main absorption bands in the NIR segment of the spectrum were displayed at 1404 nm related to O-H stretch 1st overtone; at 1436 nm, which is characteristic of sugars [32] and related to combination O-H stretch/HOH deformation (O-H bend 2nd overtone; the band at 1724 nm related to lipid-specific 1st overtone [33]; at 1924 nm assigned to O-H stretch first overtone; the band at 2278 nm was assigned to the CH- stretch of cellulose [32]. The band at 2350 nm is related to C-H stretching first overtone of lipids, the peak at 2388 nm is associated with the C-H functional group present in hemicellulose and cellulose. Other absorptions were associated to the O-H 1st overtone (1364 nm) [33], the O-H group hydroxyl (1514 nm), C-O stretch of phenols (2056 nm), C-H stretch first overtone (1762 nm, 2142 nm and 2170 nm) [32].

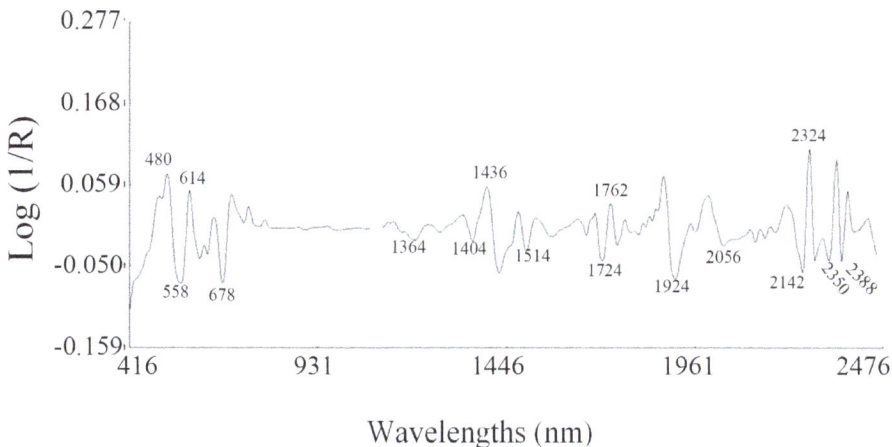

Figure 2. Second-derivative NIR spectra of blueberry samples.

2.3. Calibration Development for TPC and TCC

For calibration purposes, the wavelength ranges between 400–2500 nm were used. Table 1 summarizes the statistics of calibration and prediction models after the application of spectral pre-treatment. For the development of NIRS calibrations, four derivative mathematical treatments were tested: 1,4,4,1; 1,10,10,1; 2,5,5,2 and 2,20,20,2 (where the first digit is the number of the derivative, the second is the gap over which the derivative is calculated, the third is the number of data points in a running average or smoothing and the fourth is the second smoothing) [34]. The use of the second derivative to the raw spectra resulted in an increased complexity of spectra and assisted in a clear separation between peaks.

Table 1. Calibration and cross-validation statistics of total phenolic content (TPC expressed as $mg \cdot g^{-1}$ dw) and total carotenoid content (TCC expressed as $\mu g \cdot g^{-1}$ dw) for blackberry fruit measured by FNS-6500 with different treatments.

Trait	Range	SD [a]	R^2 [b]	SEC [c]	R^2_{CV} [d]	SECV [e]	RPDcv [f]	Treatment	Factor [g]
TPC	17.36–35.67	3.06	0.86	1.14	0.69	1.69	1.81	2,5,5,2	8
	17.36–35.67	3.06	0.71	1.66	0.59	1.95	1.58	1,4,4,1	8
	17.36–35.67	3.06	0.70	1.68	0.59	1.97	1.57	1,10,10,1	8
	17.36–35.67	3.06	0.76	1.49	0.67	1.75	1.75	2,20,20,2	8
TCC	2.84–13.73	1.82	0.92	0.52	0.76	0.95	1.91	2,5,5,2	8
	2.84–13.73	1.82	0.85	0.72	0.71	1.03	1.83	1,4,4,1	8
	2.84–13.73	1.82	0.83	0.75	0.71	0.99	1.82	1,10,10,1	8
	2.84–13.73	1.82	0.84	0.75	0.70	1.05	1.80	2,20,20,2	8

[a] SD: standard deviation; [b] R^2: coefficient of determination in calibration, [c] SEC: standard error in calibration, [d] R^2_{CV}: coefficient of determination in cross-validation, [e] SECV: standard error of cross-validation, [f] RPDcv: ratio of the standard deviation to standard error of cross-validation; [g] Factor: number of latent variables.

The coefficient of determination for cross-validation (R^2_{CV}) for TPC was 0.70 in this study. The second derivative resulted in a better prediction in the cross-validation. This Modified partial least squares (MPLS) model (2,5,5,2) reached the best prediction precision. The number of latent variables was determined by cross validation of MPLS procedure and it was 8 for all models.

Figure 3 shows the plots of Standard error of cross-validation (SECV) versus the different number of factors included in the cross validation of MPLS for TPC and TCC models (2,5,5,2; standard normal variate and de-trending transformations (SNV + DT)). The number of factor of 8 were optimum for both parameters as it resulted in the minimum SECV of 1.69 and 0.95 for TPC and TCC models, respectively. This SECV value was close to the value of SEC which then shows that the calibration carried out was feasible.

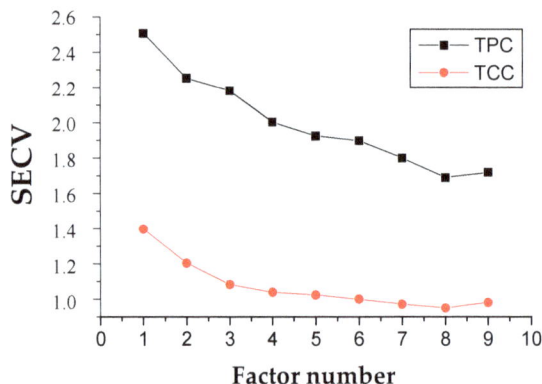

Figure 3. Plot of standard error of cross-validation (SECV) vs. the different number of factors included in the cross-validation of the modified partial least model for Total phenolic content (TPC) and total carotenoid content (TCC).

Other authors have shown the ability of NIRS to predict the content of phenolic compounds in blueberries (*Vaccinium corymbosum* L.) [25], in methanolic extracts of berry fruits (wild blueberries, blackberries, raspberries, strawberries and red currants) reporting high R^2 coefficients (ranging from 0.864 to 0.975).

R^2_{CV} and RPD_{CV} coefficients for the cross-validation (treatment 2,5,5,2) were 0.76 and 1.91, respectively for total carotenoid content in blackberry fruits.

2.4. External Validation

Table 2 shows the external validation statistics (SEP, Q^2, RPDp and RER) for both compounds measured in blackberries.

Table 2. Reference values and external validation statistics of the NIRS calibrations for total phenolic content (TPC expressed as mg·g^{-1} dw) and total carotenoid content (TCC expressed as µg·g^{-1} dw) in blackberry fruit.

Parameters	Reference Values (*n* = 30)				External Validation		
	Range	Mean	SD [a]	Q^2 [b]	SEP [c]	RDPp [d]	RER [e]
TPC	20.77–27.97	23.41	1.85	0.65	1.22	1.52	5.92
TCC	5.02–11.66	8.21	1.40	0.71	0.77	1.82	8.63

[a] SD: standard deviation; [b] Q^2: coefficient of determination in external validation; [c] SEP: standard error of prediction corrected for bias; [d] RPDp: ratio of the standard deviation to standard error of prediction (performance); [e] RER: ratio of the range to standard error of prediction (performance).

The SEP values in the external validation were lower than their respective standard deviation, which point that NIRS is able to determine these traits in blackberry fruits.

The Q^2 values give an indication of the percentage variation in the Y variable that is accounted for by the X variable. Therefore, Q^2 values above 0.50 indicate that over 50% of the variation in Y is attributable to variation in X and this allows discrimination. In our study, external validation resulted in Q^2 of 0.65 and 0.71 for TPC and TCC respectively (Table 2) which indicated that 65% and 71% of the variability in the data was explained by the respective calibration model.

According to Williams and Norris [35], the values for Q^2 obtained from the external validations in this work, indicated that the models for both TPC and TCC can be classified as models that can be used for rough predictions of samples (Table 2).

The Q^2 statistic obtained in external validation for TPC was lower than those reported in previous works. Sinelli et al. [25] indicated a Q^2 value of 0.87 for TPC in blueberry fruits. Other authors found Q^2 values of 0.98 for different berry species [23]. According to the guidelines for interpretation of RPDp from external validation [35], if this ratio is between 1.5 < RPDp < 2.5 this characterizes the equations as suitable for screening purposes, which was obtained for TPC (1.52) and TCC (1.82). Similar results were obtained in blueberry [25] with RPDp = 2.05. However, a higher RPDp value (RPDp = 3.05) has been reported for different berry species than those found in this work [23].

The results were corroborated by the Figure 4 of predicted values versus reference values obtained using the MPLS model (second derivative) for the TPC and TCC validation sets.

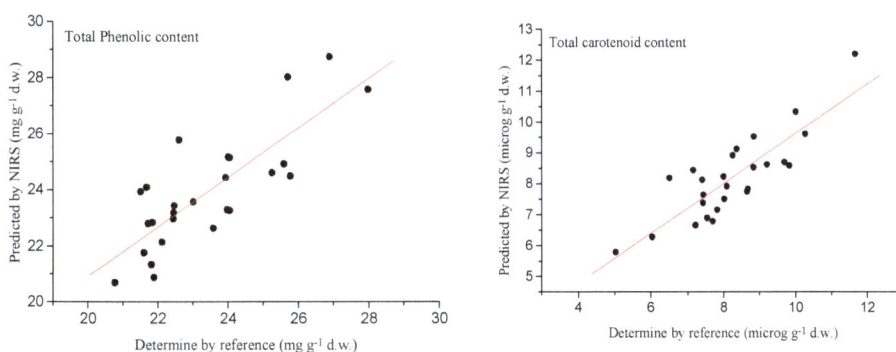

Figure 4. External validation scatter plot for near infrared predicted values versus reference values for total phenolic content (TPC) and total carotenoid content (TCC) in blackberries.

The calibration equation obtained in the present work for TPC showed an RPDp value lower than that reported by Gajoš [23], which developed a multicalibration equation for wild blueberries, blackberries, raspberries, strawberries and red currants. This could be due to the narrower range of TPC content found in the samples of our study which varied from 20.77 to 27.97 mg·g^{-1} dw (Table 2). As the RPDp value is highly dependent on the range of the sample population for a determined parameter [36], a wider range implies a higher RPD value.

In terms of RER coefficients, predictive ability of the prediction models in this work ranged from 5.92 to 8.63. For TPC and TCC, the validation yielded RER (5.92 and 8.63, respectively) values which indicated models that can be used for screening purposes, thus being very useful in quality control and as a selection tool in blackberry breeding programmes [36].

2.5. Modified Partial Least Square Loadings for Total Phenolic Content

MPLS regression was used to obtain the spectral information and predict the sample composition.

Figure 5 shows the equation corresponding to TPC. Some of the spectral regions used by the TPC models for calibrating these compounds have been previously reported by other authors [32]. The first MPLS term was influenced by absorption bands characteristic of electronic vibrations at 632 nm, it was also influenced by absorption bands at 1412 and 1668 nm. Vibration differences in the range 1399–1699 nm have been identified for fruit products such as wine, grape juice and orange fruit [21,37,38] presenting the vibration range of the C-H and O-H bonds, corresponding to water and phenolic absorbance [21]. There was also a peak at 1908 nm influenced by absorption assigned to the first OH stretch. At 1980 nm it corresponded to C-H aromatic 2nd overtone, it also relate to one or more aromatic rings and hydroxyl groups, mainly related to combination bands of the -OH functional group, symmetric and anti-symmetric stretching. The absorption vibrations at 2236, 2300 and 2388 nm were due to N-H bend [32]. The second and third terms were influenced by absorption bands characteristic to electronic vibrations at 640 nm and 672 nm, respectively.

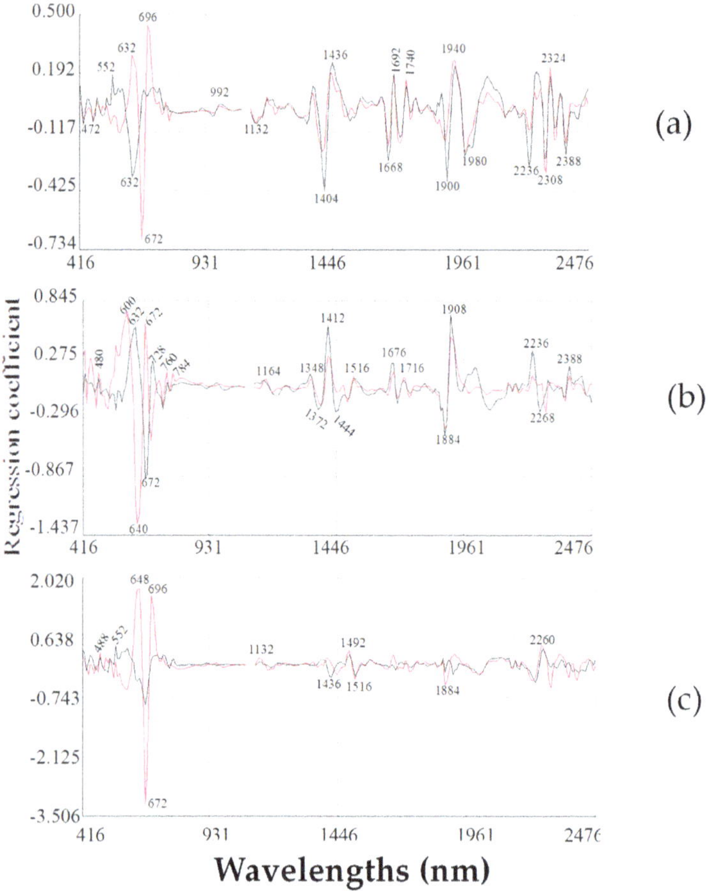

Figure 5. MPLS loading plots for TPC (black line) and TCC (red line) using near-infrared reflectance spectroscopy. (**a**) First loading; (**b**) Second loading; (**c**) Third loading.

2.6. Modified Partial Least Square Loadings for Total Carotenoid Content

MPLS loading plots of the TCC equation are shown in Figure 5. The first term (Figure 5a) was influenced by bands which corresponding to electronic vibration assigned to chlorophyll at 672 nm [39], second C=O stretch at 1900 nm. Absorptions at 1980 nm and 2300 nm correspond to vibrations in N-H stretch bending and C-H combination tones by lipids [32].

Those wavelengths corresponding to absorptions by electronic vibration assigned to chlorophyll (672 nm) and stretch groups: C=O and O-H (1444 nm), C-O (1692 nm), N-H with C-O (2068 nm) and OH cellulose stretch (2268 nm) highly influenced the second factor of the equation (Figure 5b). The third term (Figure 5c) of the equation was modelled with those wavelengths corresponding to electronic vibrations (672 nm) with the following stretches: C=O (1420 nm), N-H (1516 nm) and O-H (1908 nm) [32].

3. Materials and Methods

3.1. Plant Material and Greenhouse Experiments

Blackberry (*Rubus fructicosus* L.) cv. Tupy was chosen for the field trials.

The plant transplant took place on November 29th 2013 in a greenhouse of 600 m^2 in the IFAPA Centre La Mojonera, Almería (36°47′19″ N, 02°42′11″ W; 142 m a.s.l.), following standard cultural practices for disease control, insect pest and plant nutrition.

The blackberry plants were transplanted on polypropylene containers of 15 l capacity using coconut fibre substrate. The irrigation water conditions were pH 8.1 and 1.26 mS·cm^{-1} conductivity and the nutrient solution had pH 5.8 and 2.50 mS·cm^{-1} electrical conductivity.

The trial (Figure 6) was designed as a randomized complete block with 3 replicates and 20 plants per repetition. Thirty fruits were collected per each plant and stored at −80 °C until lyophilization, then were lyophilized (Telstar LyoQuest, Terrassa, Spain) and ground in a mill (Janke & Kunkel, model A10, IKA®-Labortechnik). The samples were lyophilized to remove the strong absorbance of water in the infrared region, which overlaps with important bands of nutritional parameters present in low concentration [40].

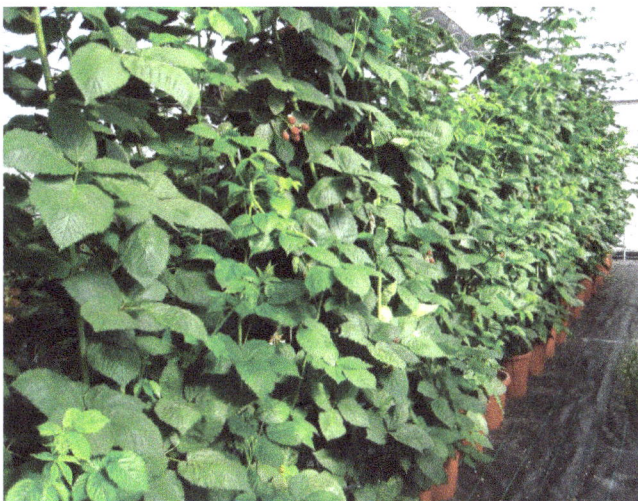

Figure 6. Trial of Tupy blackberry variety growing in greenhouse.

Two samplings were performed at the time of maximum production (21 April 2014 and 20 May 2014).

The fruits harvested were classified according to their colour with a colorimeter to avoid fruit-to-fruit variation in ripeness, thus these were considered to be ripe when the CIE L*a*b (CIELAB) values were L: 21.11; a: 0.835; b: 0.073 y C: 1.27.

3.2. Determination of the Total Phenolic Fraction

Five grams of each sample (fresh weight) were homogenized in 20 mL of ethanol (99.7%) and stored at −20 °C for 2 weeks. An aliquot of 60 µL supernatant was taken previous to centrifugation of the extracts and then prepared according to the modified method by Dewanto et al. [41]. After 75 min, the absorbance was measured at 765 nm using a Thermo Spectronic UV–visible Spectrometer (Thermo Fisher Scientific, Waltham, MA, USA). The external standard gallic acid, 3,4,5-trihydroxybenzoic acid (Sigma–Aldrich, Steinheim, Germany) was used for quantifying. The results were expressed in mg GAE (gallic acid equivalent) g^{-1} dry weight.

3.3. Determination of the Total Carotenoid Content

Analysis of total carotenoid content was carried out by the method described by Rutz et al. [13].

Five grams of each sample and 2 g of celite were added to 20 mL of cold acetone and the mixture was shaken for 10 min. The material was filtered with a Buchner funnel with filter paper, washing the sample with acetone until the extract was colourless. The filtrate was transferred to a separatory funnel, to which 30 mL of petroleum ether and 30 mL of distilled water were added. The lower phase was discarded, then distilled water was added; this procedure was repeated four times to achieve total removal of the acetone. The upper phase was transferred to a 50-mL volumetric flask and the volume was completed with petroleum ether. The absorbance was measured with a Thermo Spectronic UV–visible Spectrometer (Thermo Fisher Scientific, Delaware, USA) at 450 nm, using petroleum ether as a blank. The total carotenoid content was determined by Equation (1) and the results were expressed in mg of total carotenoids per g dry weight.

3.4. NIRS Analysis Calibration and Validation Development

One hundred and twenty freeze-dried blackberry samples were analysed by NIRS (90 calibration, 30 calibration). An spectrometer (Model 6500 Foss-NIRSystems, Inc., Silver Spring, MD, USA) was used for registrating the spectra in the range from 400–2500 nm each 2 nm in reflectance mode.

Freeze-dried, ground samples of the blackberries were placed in the sample holder (3 cm diameter, 10 mL volume approximately) until it was full (sample weight: 3.50 g) and then were scanned. Their spectra were acquired at 2 nm wavelength resolution as log $1/R$ (R is reflectance) over a wavelength range from 400 to 2500 nm (visible and near-infrared regions).

The spectral variability and structure of the sample population was performed using the CENTER algorithm; samples with a statistical value >3 were considered anomalous spectra or outliers [42].

Calibration equations for total phenolic content and total carotenoid content were developed on the whole set (n = 90) using the application GLOBAL v. 1.50 (WINISI II, Infrasoft International, LLC, Port Matilda, PA, USA). Calibration equations were computed using different mathematical treatments although only those that displayed the higher predictive capacity were showed: [(1,4,4,1); (1,10,10,1); (2,5,5,2); (2,20,20,2)] where the meaning of each term is the derivative order of the log $1/R$ data (being R the reflectance), segment of the derivative, first smooth and second smooth). Additionally to the use of derivatives, standard normal variate and de-trending (SNV-DT) transformations [43] were used, which are algorithms used to correct baseline offset due to scattering effects (differences in particle size and path length among samples) and improve the accuracy of the calibration.

Modified partial least squares (PLSm) was used as a regression method to correlate the spectral information (raw optical data or derived spectra) of the samples and TPC and TCC contents determined by the reference method, using different number of wavelengths from 400 to 2500 nm for the calculation. The objective was to perform a linear regression in a new coordinate system with a lower dimensionality than the original space of the independent variables. The PLS loading factors (latent variables) were determined by the maximum variance of the independent (spectral data) variables and by a maximum correlation with the dependent (chemical) variables. The model obtained used only the most important factors, the "noise" being encapsulated in the less important factors.

Cross-validation was performed on the calibration set to determine both, the ability to predict on unknown samples and the best number of terms to use in the equation [44]. The number of principal component terms used in the equation to explain the analyte variance was also taken into account before selecting the equation for use. The cross validation process used in the software should prevent over fitting of the equation to the calibration set as the optimum number of terms are selected when the SECV is at its lowest and R^2_{CV} is at its highest. Addition of more terms than necessary will increase the prediction error and over fit the equation to its calibration set resulting in poor predictive performance on samples outside the calibration set. Usually a medium sized model is preferred. An external validation in 30 independent samples was carried out to evaluate the accuracy and precision of the calibration equations for total phenolic and carotenoid content following the protocol outline by Shenk et al. [44]. The 30 samples of the validation set were selected by taking one of every 5 samples in the 120 samples set; finally, the calibration set was constituted of the 90 remaining samples. The standard error (SE) and coefficient of determination were calculated for cross-validation (R^2_{CV}) and external validation (Q_2). The predictive ability of the equations was assessed in the external validation from the Q^2 coefficient, the RPD (the ratio of the standard deviation for the samples of the validation to the SEP (standard error of prediction (performance) and the RER (the ratio of the range in the reference data (validation set) to the SEP). NIR models can be classified depending the Q^2 from the external validation [36] as: if $0.26 < Q^2 < 0.49$, the models show a low correlation;); if $0.50 < Q^2 < 0.64$) models can be used for rough predictions of samples; if $0.65 < Q^2 < 0.81$) the models can be used to discriminate between low and high values of the samples; (if $0.82 < Q^2 < 0.90$ are models with good prediction; if $Q^2 > 0.90$ the models show excellent precision. RPD values > 3 are desirable for excellent calibration equations, however equations with an RPD < 1.5 are unusable [35]. The RER (ratio of the range to standard error of prediction (performance), it should be at least 10 [36].

The mathematical expressions of these statistics are as follows:

$$\text{RPD} = SD \langle [(\textstyle\sum_{i=1}^{n} (y_i - \hat{y}_i)^2)(N - K - 1)^{-1}]^{1/2} \rangle^{-1}$$

where y_i = laboratory reference value for the ith sample; $\hat{y}$ = NIR value; K = number of wavelengths used in an equation; N = number of samples; SD = standard deviation.

$$\text{RER} = range \langle [(\textstyle\sum_{i=1}^{n} (y_i - \hat{y}_i)^2)(N - K - 1)^{-1}]^{1/2} \rangle^{-1}$$

where y_i = laboratory reference value for the ith sample; $\hat{y}$ = NIR value; K = number of wavelengths used in an equation; N = number of samples.

4. Conclusions

The NIRS technique has the potential to reduce the cost and time in analysing the total phenolic and carotenoid content in blackberries for both agri-food applications and research. Approximately each 1.5 min we can analyse a sample for both quality components by using the NIR spectroscopy. From the different mathematical treatments tested the second derivative produced the better results for predicting, however the models reported here are usable for routine screening of a large number of samples in breeding programs.

The spectral regions corresponding to absorbance by cellulose, lipids, chlorophyll and sugars were used by MPLS for modelling the prediction equations for total phenolic and carotenoid content in blackberries.

Author Contributions: M.d.R.-C. and R.F. conceived and designed the experiments; M.d.C.G.-G. performed the field trials; P.G. and M.S.-N. determined the phenolic and carotenoid content; J.M.M.-R. registered the NIRS spectra; E.M.T.-M. developed the NIRS calibrations and wrote the paper.

Funding: The authors wish to express their thanks to the Project entitled Innovación sostenible en horticultura protegida (PP.AVA.AVA201601.7), FEDER y FSE (Programa Operativo FSE de Andalucia 2007–2013 "Andalucía se mueve con Europa") for the funding of this research.

Acknowledgments: We thank Nicholas Davies for his help in the grammatical revision of the manuscript.

Conflicts of Interest: The authors declare no conflict of interest.

References

1. Joshipura, K.J.; Hu, F.B.; Manson, J.E.; Stampfer, M.J.; Rimm, E.B.; Speizer, F.E.; Colditz, G.; Ascherio, A.; Rosner, B.; Spiegelman, D.; et al. The effect of fruit and vegetable intake on risk for coronary heart disease. *Ann. Intern. Med.* **2001**, *134*, 1106–1114. [CrossRef] [PubMed]

2.   Maynard, M.; Gunnell, D.; Emmett, P.; Frankel, S.; Davey Smith, G. Fruit, vegetables, and antioxidants in childhood and risk of adult cancer: The Boyd Orrcohort. *J. Epidemiol. Commun. Health* **2003**, *57*, 218–225. [CrossRef]

3.   Ames, B.N.; Shigenaga, M.K.; Hagen, T.M. Oxidants, antioxidants, and the degenerative diseases of aging. *Proc. Natl. Acad. Sci. USA* **1993**, *90*, 7915–7922. [CrossRef] [PubMed]

4.   Freshfruitportal. 2018. Available online: http://www.freshfruitportal.com/news/2018/02/01/spain-sees-strong-growth-berry-plantings/ (accessed on 20 July 2017).

5.   Prior, R.; Cão, G.H.; Martin, A.; Sofic, E.; Mcewen, J.; O'brien, C.; Lischner, N.; Ehlenfeldt, M.; Kalt, W.; Krewer, G.; et al. Antioxidant capacity as influenced by Total phenolic and anthocyanin content, maturity, and variety of *Vaccinium* species. *J. Agric. Food Chem.* **1998**, *46*, 2686–2693. [CrossRef]

6.   Castrejón, A.; Eichholz, I.; Rohn, S.; Kroh, L.W.; Huyskens-Keil, S. Phenolic profile and antioxidant activity of highbush blueberry (*Vaccinium corymbosum* L.) during fruit maturation and ripening. *Food Chem.* **2008**, *109*, 564–572. [CrossRef]

7.   Cardeñosa, V.; Medrano, E.; Lorenzo, P.; Sánchez-Guerrero, M.C.; Cuevas, F.; Pradas, I.; Moreno-Rojas, J.M. Effects of salinity and nitrogen supply on the quality and health-related compounds of strawberry fruits (*Fragaria* × *ananassa* cv. Primoris). *J. Sci. Food Agric.* **2015**, *95*, 2924–2930. [CrossRef]

8.   Cardeñosa, V.; Girones-Vilaplana, A.; Muriel, J.L.; Moreno, D.A.; Moreno-Rojas, J.M. Influence of genotype, cultivation system and irrigation regime on antioxidant capacity and selected phenolics of blueberries (*Vaccinium corymbosum* L.). *Food Chem.* **2016**, *202*, 276–283. [CrossRef]

9.   Ruiz, D.; Egea, J.; Gil, M.I.; Tomas-Barberan, F.A. Characterization and quantitation of phenolic compounds in new apricot (*Prunus armeniaca* L.) varieties. *J. Agric. Food Chem.* **2005**, *53*, 9544–9552. [CrossRef]

10.   Contessa, C.; Mellano, M.G.; Beccaro, G.L.; Giusiano, A. Total antioxidant capacity and total phenolic and anthocyanin contents in fruit species grown in Northwest Italy. *Sci. Hortic.* **2013**, *160*, 351–357. [CrossRef]

11.   Souza, D.; Vera de Rosso, V.; Zerlotti, A. Compostos bioactivos presentes en amora-preta (*Rubus* spp.). *Rev. Bras. Frutic.* **2010**, *32*, 664–674.

12.   Lashmanova, K.; Kuzivanova, O.; Dymova, O. Northern berries as a source of carotenoids. *ABP* **2012**, *59*, 133–134.

13.   Rutz, J.K.; Voss, G.B.; Zambiazi, R.C. Influence of the degree of maturation on the bioactive compounds in blackberry (*Rubus* spp.) cv. Tupy. *Food Nutr. Sci.* **2012**, *3*, 1453–1460.

14.   Perkins-Veazie, P.; Fernández, G.E. Screening of Raspberry Fruit for Carotenoids: Impact on Flavor and Color. Available online: http://www.raspberryblackberry.com/wp-content/uploads/2011-Screening-of-Raspberry-Fruit-for-Carotenoids-Impact-on-Flavor-and-Color.pdf (accessed on 10 December 2012).

15.   Reyes, J.; Yousef, G.; Martínez, R.; Lila, M. Antioxidant capacity of fruit extracts of blackberry (*Rubus* sp.) produced in different climatic regions. *J. Food Sci.* **2005**, *70*, 497–503. [CrossRef]

16.   Heinonen, M.; Lehtonen, P.J.; Hopia, A.I. Antioxidant activity of berry and fruit wines and liquors. *J. Agric. Food Chem.* **1998**, *46*, 25–31. [CrossRef] [PubMed]

17.   Huang, W.; Zhang, H.; Liu, W.; Li, C. Survey of antioxidant capacity and phenolic composition of blueberry, blackberry, and strawberry in Nanjing. *J. Zhejiang Univ. Sci. B* **2012**, *13*, 94–102. [CrossRef] [PubMed]

18.   Berté, K.A.S.; Beurx, M.R.; Spada, P.K.D.S.; Slavador, M.; Hoffmann-Ribani, R. Chemical composition and antioxidant activity of yerba-mate (*Ilex paraguariensis* A.St.-Hil., *Aquifoliaceae*) extract as obtained by spray drying. *J. Agric. Food Chem.* **2011**, *59*, 5523–5527. [CrossRef] [PubMed]

19.   Cardozo, J.R.; Ferrarese-Filho, O.; Filho, L.C.; LucioFerrarese, M.D.; Donaduzzi, C.M.; Sturion, J.A. Methylxanthines and phenolic compounds in mate (*Ilex paraguariensis* St. Hil.) progenies grown in Brazil. *J. Food Compos. Anal.* **2007**, *20*, 553–558. [CrossRef]

20.   Jacques, R.A.; Santos, J.G.; Dariva, C.; Oliveira, J.V.; Caramarao, E.B. GC/MS ~ characterization of mate tea leaves extracts obtained from high-pressure CO_2 extraction. *J. Supercrit. Fluids* **2007**, *40*, 354–359. [CrossRef]

21.   Nicolaï, B.M.; Beullens, K.; Bobelyn, E.; Peirs, A.; Saeys, W.; Theron, K.I. Nondestructive measurement of fruit and vegetable quality by means of NIR spectroscopy: A review. *Postharvest Biol. Technol.* **2007**, *46*, 99–118. [CrossRef]

22.   Pasquini, C. Near Infrared Spectroscopy: Fundamentals, practical aspects and analytical applications. *J. Braz. Chem. Soc.* **2003**, *14*, 198–219. [CrossRef]

23. Gajdoš Kljusurić, J.; Mihalev, K.; Bečić, I.; Polović, I.; Georgieva, M.; Djaković, S.; Kurtanjek, Z. Near-infrared spectroscopic analysis of total phenolic content and antioxidant activity of berry fruits. *Food Technol. Biotechnol.* **2016**, *54*, 236–242. [CrossRef] [PubMed]

24. Georgieva, M.; Nebojan, I.; Mihalev, K.; Yoncheva, N.; Gajdoš Kljusurić, J.; Kurtanjek, Ž. Application of NIR spectroscopy and chemometrics in quality control of wild berry fruit extracts during storage. *Croat. J. Food Technol. Biotechnol. Nutr.* **2013**, *8*, 67–73.

25. Sinelli, N.; Spinardi, A.; Di Egidio, V.; Mignani, I.; Casiraghi, E. Evaluation of quality and nutraceutical content of blueberries (*Vaccinium corymbosum* L.) by near and mid-infrared spectroscopy. *Postharvest Biol. Technol.* **2008**, *50*, 1–36. [CrossRef]

26. Meneses, E.; Arango, G.; Correa, G.; Ruíz, O.; Vargas, L.; Pérez, J. Detection of *Eurhizococcus colombianus* (*Hemiptera*: *Margarodidae*) in blackberry plants by near-infrared spectroscopy. *Acta Agron.* **2015**, *64*, 280–288. [CrossRef]

27. Ali, L.; Svensson, B.; Alsanius, B.W.; Olsson, M.E. Late season harvest and storage of Rubus berries—Major antioxidant and sugar levels. *Sci. Hortic.* **2011**, *129*, 376–381. [CrossRef]

28. Aaby, K.; Ekeberg, D.; Skrede, G. Characterization of phenolic compounds in strawberry (*Fragaria × ananassa*) fruits by different HPLC detectors and contribution of individual compounds to total antioxidant capacity. *J. Agric. Food Chem.* **2007**, *55*, 4395–4406. [CrossRef]

29. Naes, T.; Isaksson, T.; Fearn, T.; Davies, T.A. *2 User-Friendly Guide to Multivariate Calibration and Classification*; NIR: Chichester, UK, 2002.

30. Miller, C.E. Chemical principles of near-infrared technology. In *Near-Infrared Technology in the Agricultural and Food Industries*, 2nd ed.; Williams, P.C., Norris, K.H., Eds.; American Association of Cereal Chemists: St. Paul, MN, USA, 2001.

31. Tamburini, E.; Costa, S.; Rugiero, I.; Pedrini, P.; Marchetti, M.G. Quantification of lycopene, β-carotene, and Total soluble solids in intact red-flesh watermelon (*Citrullus lanatus*) using on-line Near-Infrared Spectroscopy. *Sensors* **2017**, *17*, 746. [CrossRef]

32. Osborne, B.G.; Fearn, T.; Hindle, P. *Practical NIR Spectroscopy with Applications in Food and Beverage Analysis*; Longman Scientific and Technical: London, UK, 1993.

33. Murray, I.; Williams, P.C. Chemical principles of near-infrared technology. In *Near-Infrared Technology in the Agricultural and Food Industries*; Williams, P., Norris, K., Eds.; American Association of Cereal Chemists, Inc.: St. Paul, MN, USA, 1987; pp. 17–34.

34. Shenk, J.S.; Westerhaus, M.O. The application of near infrared reflectance spectroscopy (NIRS) to forage analysis. In *Forage Quality, Evaluation and Utilization*; Fahey, G.C., Jr., Collins, M., Mertens, D.R., Moser, L.E., Madison, W.I., Eds.; Crop Science Society of America, Soil Science Society of America: Madison, WI, USA, 1994; pp. 406–450.

35. Williams, P.; Norris, K. *Near-Infrared Technology in the Agricultural and Food Industries*; Minnesota American Association: St. Paul, MN, USA, 2001.

36. Williams, P.C.; Sobering, D.C. How do we do it: A brief summary of the methods we use in developing near infrared calibrations. In *Near Infrared Spectroscopy: The Future Waves*; Davies, A.M.C., Williams, P.C., Eds.; NIR Publications: Chichester, UK, 1996; pp. 185–188.

37. Xie, L.; Ye, X.; Liu, D.; Ying, Y. Quantification of glucose, fructose and sucrose in bayberry juice by NIR and PLS. *Food Chem.* **2009**, *114*, 1135–1140. [CrossRef]

38. Budić-Leto, I.; Gajdos Kljusurić, J.; Zdunić, G.; Tomić-Potrebuješ, I.; Banović, M.; Kurtanjek, Ž.; Lovrić, T. Usefulness of near infrared spectroscopy and chemometrics in screening of the quality of dessert wine Prošek. *Croat. J. Food Sci. Technol.* **2011**, *3*, 9–15.

39. Tkachuk, R.; Kuzina, F.D. Chlorophyll analysis of whole rape-seed kernels by near infrared reflectance. *Can. J. Plant Sci.* **1982**, *62*, 875–884. [CrossRef]

40. Venyaminov, S.Y.; Prendergast, F.G. Water (H_2O and D_2O) molar absorptivity in the 1000–4000 cm^{-1} range and quantitative infrared spectroscopy of aqueous solutions. *Anal. Biochem.* **1997**, *248*, 234–245. [CrossRef] [PubMed]

41. Dewanto, V.; Wu, X.; Liu, R.H. Processed sweet corn has higher antioxidant activity. *J. Agric. Food Chem.* **2002**, *50*, 4959–4964. [CrossRef] [PubMed]

42. Shenk, J.S.; Westerhaus, M.O. Population structuring of near infrared spectra and modified partial least squares regression. *Crop Sci.* **1991**, *31*, 1548–1555. [CrossRef]

43. Barnes, R.J.; Dhanoa, M.S.; Lister, S.J. Standard normal variate transformation and de-trending of near infrared diffuse reflectance spectra. *Appl. Spectrosc.* **1989**, *43*, 772–777. [CrossRef]
44. Shenk, J.S.; Workman, J.; Westerhaus, M.O. Application of NIR spectroscopy to agricultural products. In *Marcel Dekker Handbook of Near Infrared Analysis*; Burns, D.A., Ciurczac, E.W., Eds.; Taylor Francis Group: New York, NY, USA, 2001; pp. 419–474.

Sample Availability: Samples of the compounds are not available from the authors.

Article

Classification of Frozen Corn Seeds Using Hyperspectral VIS/NIR Reflectance Imaging

Jun Zhang, Limin Dai and Fang Cheng *

College of Biosystems Enginaeering and Food Science, Zhejiang University, Hangzhou 310058, China;
11613006@zju.edu.cn (J.Z.); lmdai@zju.edu.cn (L.D.)
* Correspondence: fcheng@zju.edu.cn; Tel.: +86-571-8898-2713

Received: 3 December 2018; Accepted: 27 December 2018; Published: 2 January 2019

Abstract: A VIS/NIR hyperspectral imaging system was used to classify three different degrees of freeze-damage in corn seeds. Using image processing methods, the hyperspectral image of the corn seed embryo was obtained first. To find a relatively better method for later imaging visualization, four different pretreatment methods (no pretreatment, multiplicative scatter correction (MSC), standard normal variation (SNV) and 5 points and 3 times smoothing (5-3 smoothing)), four wavelength selection algorithms (successive projection algorithm (SPA), principal component analysis (PCA), X-loading and full-band method) and three different classification modeling methods (partial least squares-discriminant analysis (PLS-DA), K-nearest neighbor (KNN) and support vector machine (SVM)) were applied to make a comparison. Next, the visualization images according to a mean spectrum to mean spectrum (M2M) and a mean spectrum to pixel spectrum (M2P) were compared in order to better represent the freeze damage to the seed embryos. It was concluded that the 5-3 smoothing method and SPA wavelength selection method applied to the modeling can improve the signal-to-noise ratio, classification accuracy of the model (more than 90%). The final classification results of the method M2P were better than the method M2M, which had fewer numbers of misclassified corn seed samples and the samples could be visualized well.

Keywords: VIS/NIR hyperspectral imaging; corn seed; classification; freeze-damaged; image processing; imaging visualization

1. Introduction

Corn (*Zea mays* L.), one of the world's three major food crops, is currently one of the most grown food crops in several parts of the world [1]. As the world's second-largest corn producing and consuming country, China has most of its corn growing areas located in the north. In these regions, the corn seed is often damaged due to low temperatures and high seed moisture content before harvest or dehydration, which is an agricultural disaster.

Seed embryo is the most important part of the seed which contains a large number of nutrients. If damage takes place in this part, it must have great impact on subsequent growth. After suffering from low-temperature freeze damage, the seed quality declines, and it is easy for mildew to grow when the seeds are stored at later stage. The internal components of the seed will change, which results in a great impact on the subsequent germination, root growth and development. To investigate the vigor change in seed and how it changes, the International Seed Testing Association (ISTA) recommended two kinds of seed viability measurement methods in 1995 [2], the electrical conductivity test could be conveniently conducted due to its simplicity and low cost [3,4].

Therefore, a key factor in current study is how to quickly and accurately determine the characteristic changes in the freeze-damage seeds and identify the freeze condition (especially, the slightly freeze-damaged seeds), which will provide guidance for the seed agricultural production. In particular, it is of more specific significance to study the frost damage status of the embryo.

In recent studies, through the use of near-infrared spectroscopy technology, the vitality [5], internal essential constituents such as lipids [6], starches [7,8] and the toxin-infected pests [9–11] to corn seed batches have been studied, all of which have a rapid non-destructive advantage, but this technology only processes the corn seed in batches, and it is hard to determine the characteristics of individual corn seeds.

At present, the application of hyperspectral imaging technology in non-destructive testing of agricultural products has become more extensive. As a technology that combines the advantages of traditional image technology and spectral technology, it can obtain the spectral information of every pixel on the collected image, which can be used to effectively analyze the chemical composition index of each part of the seed and avoid the instability of experimental results. Another advantage of this technology is that it can test the seeds individually. Many research studies such as those on water content [12], hardness [13,14], internal component testing [15], variety classification [16–19], vitality [1,20], different storage periods [21–23], fungi and toxin detection [24–32] have been reported. Huang et al. (2015) used hyperspectral imaging techniques to predict the consistency of seed moisture content with correlation coefficient of prediction set of 0.848 [33]; Williams et al. (2016) used NIR hyperspectral imaging to classify maize kernels into three hardness categories, where pixel-wise and object-wise methods were compared and they had similar results [34]; Zhao et al. (2018) used hyperspectral imaging techniques to study a total of 12,900 maize seeds of 3 different varieties, first to determine the optimal calibration set of each variety, and then the performance of the back propagation neutral network and support vector machine models were compared to obtain the best model, the overall results indicated that hyperspectral imaging was a potential technique for varietal classification of maize seeds [35].

According to the current research, no study has been conducted on the freeze damage of corn seed, although there are some studies using hyperspectral imaging techniques to study frozen grown crops [36–39] or fruit [40], but these studies could not be used for the freeze damage identification of the seeds due to different technical methods and objectives.

To summarize, the objectives of this study were to: (1) conduct an electrical conductivity test on corn seeds to consider if the seed is damaged or not; (2) obtain the corn seed embryo hyperspectral image, and assess the potential of applying hyperspectral imaging technology for the classification of different degrees of freeze damage to the corn seeds; (3) evaluate the models established by different spectral pretreatment methods, wavelength selection methods and modeling methods, and then compare and identify the optimal model among them; (4) visualize the classification results of two different methods (M2M and M2P), and to identify the optimal method.

2. Results and Discussions

2.1. Results of Conductivity Test

Figure 1 shows the results of the conductivity of three varieties of corn seeds after soaking for 24 h, it can be found that the highest conductivity is at the frost condition of $-20\,°C$, for 10 h, the second is at $-10\,°C$, for 5 h while the lowest is at the normal condition. It indicates that during the freeze-damage process, the membrane integrity of the corn seed deteriorated, and then the leakage of the cell contents was serious after the seeds absorbing water, thus a higher conductivity was obtained.

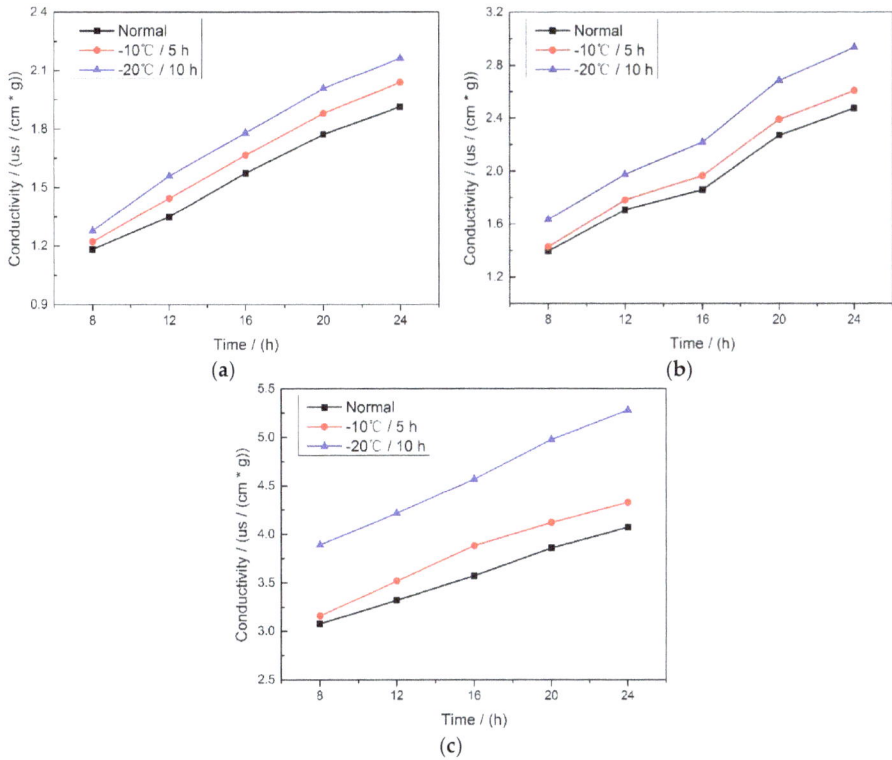

Figure 1. The conductivity of (**a**) Haoyu21, (**b**) Haihe78 and (**c**) Jindan10 soaking for 24 h.

2.2. The Analysis of Spectral Features

According to the experiment description, Figure 2 shows the average spectrums of three different corn seeds varieties.

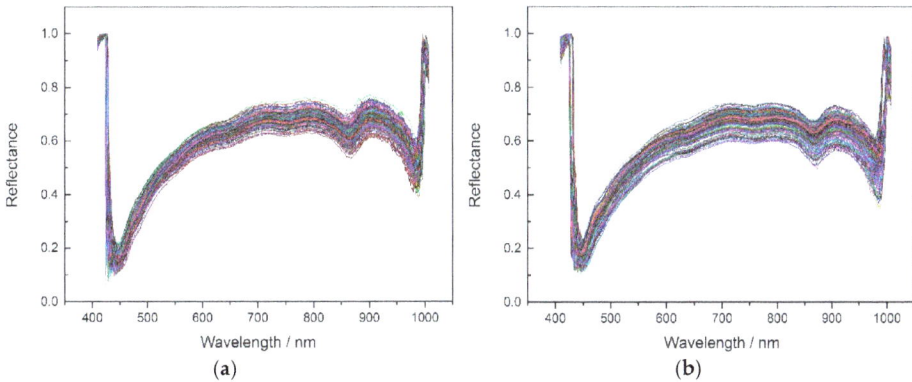

Figure 2. *Cont.*

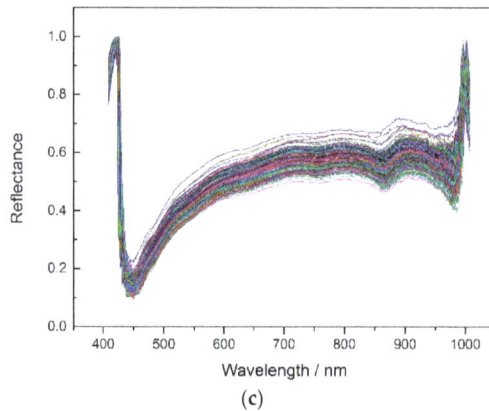

(c)

Figure 2. The average spectrums of (**a**) Haoyu21, (**b**) Haihe78 and (**c**) Jindan10.

In order to get better classification results, the noise wavelengths (before 444.23 nm and after 985.37 nm) were excluded and the remaining 430 wavelengths were used for later modeling. The SPA, PCA and X-loading methods were applied to select feature wavelengths after applying the pretreatment methods (no pretreatment, MSC, SNV and 5-3 smoothing) to the spectrums. The feature wavelength results for the three varieties are shown in appendixes Tables A1–A3.

From Tables A1–A3, the original input of 430 wavelengths was dramatically reduced to several or no more than 20 inputs, thus the calculating time was greatly shortened. To take the selected wavelengths in the condition of 5-3 smoothing pretreatment method and SPA method as an example (Figure 3). Most of the selected wavelengths were in the range of 450–700 nm, they are possibly related to the change in Chlorophyll, β-carotene or other components related to the embryo [41]. In the range of 850–950 nm, it is mainly related to the 3rd overtune vibrations of the hydrocarbon C-H bond [42].

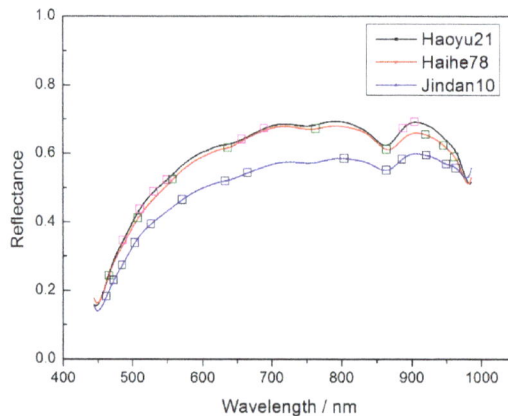

Figure 3. The description of the selected wavelengths by 5-3 smoothing pretreatment and the SPA method for the three varieties.

2.3. The Results of Established Classification Models

When the feature wavelengths were selected, the classification models were established. The accuracy results of the three varieties of corn seed are shown as Figures 4–6.

From Figure 4, an 80% pink line for both calibration sets and validation sets was firstly set. At full-band treatment, the classification accuracy results of 5-3 smoothing and no pretreatment method were better than those of the MSC and SNV pretreatment methods, because the results of the validation sets with the MSC and SNC pretreatment methods were all lower than 80%. With the same pretreatment method, the classification accuracies of the full-band, SPA, PCA and X-loading methods were sorted: The classification accuracies for the full-band method were higher than those of SPA, and those of SPA were higher than those of the PCA and X-loading methods. With the same pretreatment method and the same wavelength selection algorithm, both the PLS-DA and SVM modeling methods had higher accuracy results than the KNN method. The >80% classification accuracy results for the calibration set and validation set with the KNN method only appeared in no pretreatment and 5-3 smoothing pretreatment method.

By counting the number of >80% classification accuracy results, the 5-3 smoothing and no pretreatment method had similar classification result, and in the meantime, the PLS-DA and SVM modeling methods had similar classification result.

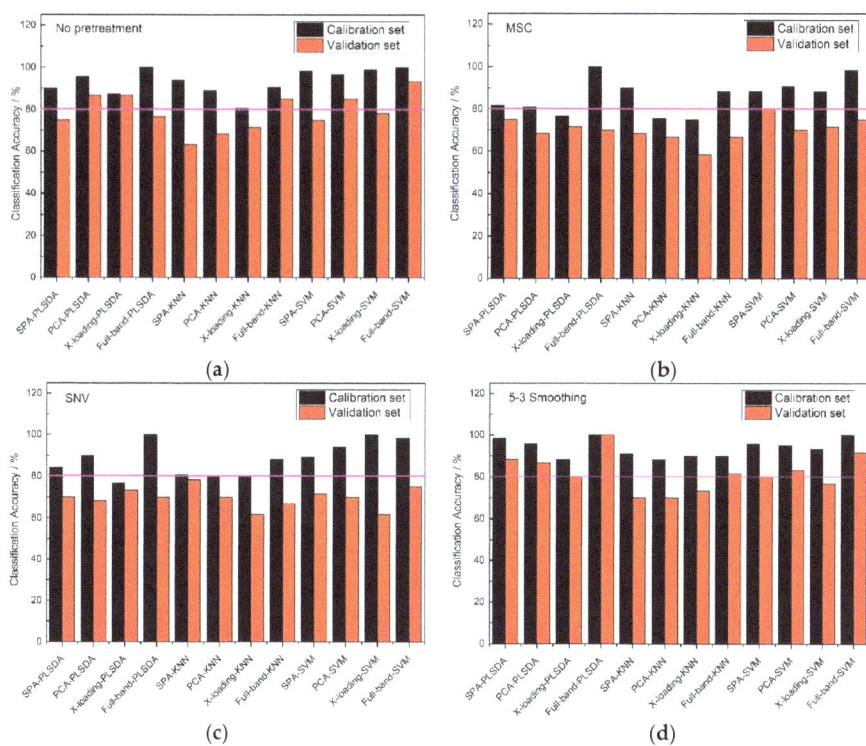

Figure 4. The classification accuracy results of Haoyu21 with (**a**) no pretreatment method, (**b**) the MSC pretreatment method, (**c**) the SNV pretreatment method and (**d**) the 5-3 smoothing pretreatment method.

From Figure 5, the classification accuracy results of the no pretreatment and three pretreatment methods at full band, were very high, and most of them could reach an accuracy of 100%, which perhaps shown that a wonderful classification model could be established on the premise of it containing all the reflectance spectral information of the samples. Though there were good accuracy results among each pretreatment method for full-band spectrums, irrelevant information for the sample is still existed, so it was necessary to find several wavelengths to represent the 430 wavelengths to reduce the calculation time. With the same pretreatment methods, the classification accuracies of the full-band, SPA, PCA and X-loading methods were sorted: the classification accuracies for the full-band method were higher than those of the SPA, and those for SPA and PCA methods were higher than those of the X-loading method. In some respects, it could be found that many of the classification accuracies for PCA were slightly higher than those for SPA, but the number of > 80% classification accuracies for SPA was one more than that for PCA. Though the full-band method had the best classification results among the four pretreatment methods, the accuracies of SPA were also much higher, and most of the accuracies of the validation sets were almost more than 90%. So it could also be used for the classification of frozen corn seeds. Moreover, with the same pretreatment method and the same wavelength selection algorithm, the PLS-DA and SVM modeling methods had higher accuracy results than the KNN method.

By counting the number of >80% classification accuracies, all of the pretreatment methods had similar classification result, and in the meantime, the number of >80% classification accuracy results for the calibration set and validation set of the KNN method were fewer than for the other two classification modeling methods.

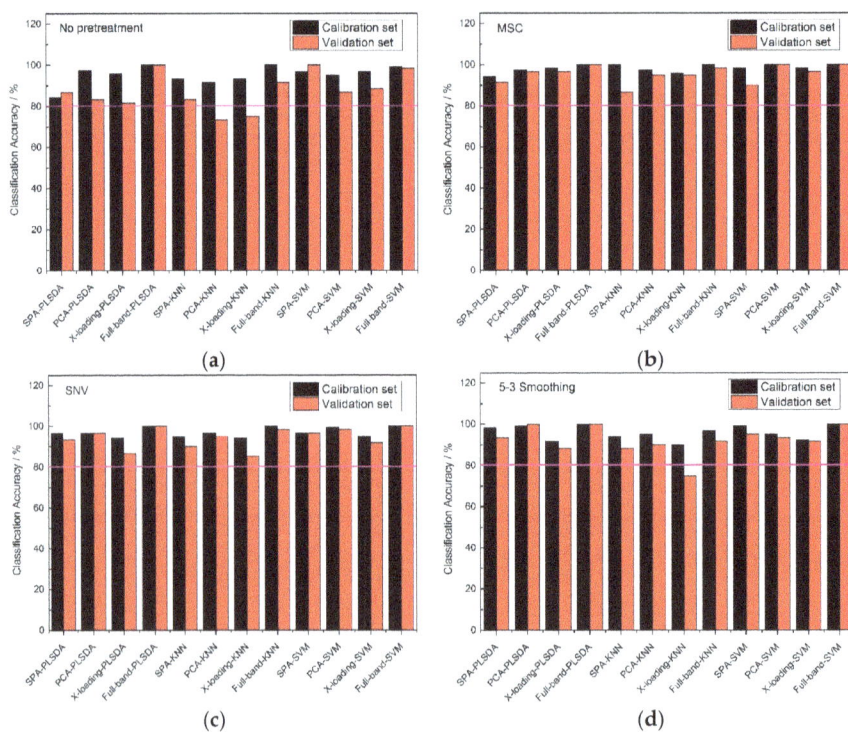

Figure 5. The classification accuracy results of Haihe78 with (**a**) no pretreatment method, (**b**) the MSC pretreatment method, (**c**) the SNV pretreatment method and (**d**) the 5-3 smoothing pretreatment method.

From Figure 6, the 5-3 smoothing pretreatment method had a greater number of >80% classification accuracy results than the no pretreatment and the other two pretreatment methods at full band. Almost all of the calibration sets could reach an accuracy of 85% or higher, while most of the accuracies of the validation sets were lower than 80%. With the same pretreatment method, the classification accuracies of the full-band, SPA, PCA and X-loading methods were sorted: the classification accuracy results of the full-band than 80%. With the same pretreatment method and the same wavelength selection algorithm, the PLS-DA modeling method had higher accuracy results than the SVM and

KNN modeling methods. There was no >80% classification accuracy result for the calibration set and validation set in KNN modeling method.

By counting the number of >80% classification accuracies, the conclusion was drawn that the classification accuracy results for the PLS-DA modeling method with the 5-3 smoothing pretreatment method had better results than any other pretreatment methods.

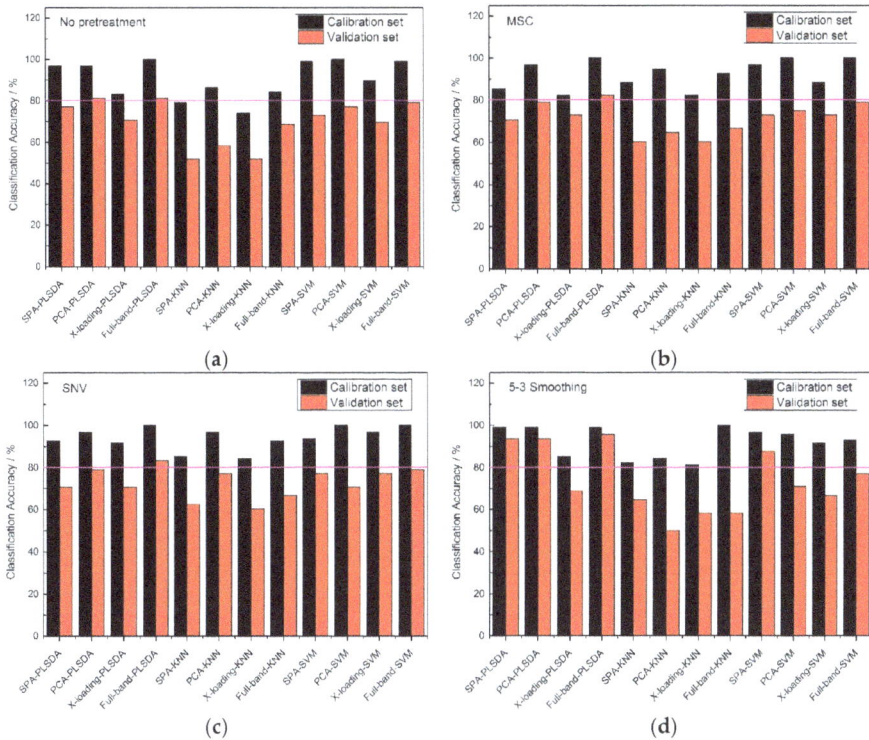

(a)

(b)

(c)

(d)

Figure 6. The classification accuracy results of Jindan10 with (**a**) no pretreatment method, (**b**) the MSC pretreatment method, (**c**) the SNV pretreatment method and (**d**) the 5-3 smoothing pretreatment method.

2.4. The Visualization Images of the Classification Results

A better model was found when using the 5-3 smoothing pretreatment method and the PLS-DA classification modeling method. To more clearly understand the classification results of the corn seed samples, the spectrum of each pixel in the embryo image was classified to realize the visualization of the corn embryo images.

2.4.1. The Visualization Images of Haoyu21

At first, the six images (the first two are of normal corn seeds, the middle two are of slightly freeze-damaged corn seeds and the last two are of severely freeze-damaged corn seeds) of three different degrees of freeze-damage in corn seed were merged, Figure 7 shows the two different classified images. Figure 7a,b, were the results images obtained by method M2M and method M2P, respectively.

From Figure 7a, the visualization image with the SPA and PLS-DA model was almost matched with the above figure results. As for Figure 7b, each pixels had a classification value and each corn seed image was obtained to form a whole image with different color gradients (from light blue to yellow and then to deep red) although not all of the pixels were the same color in one corn seed.

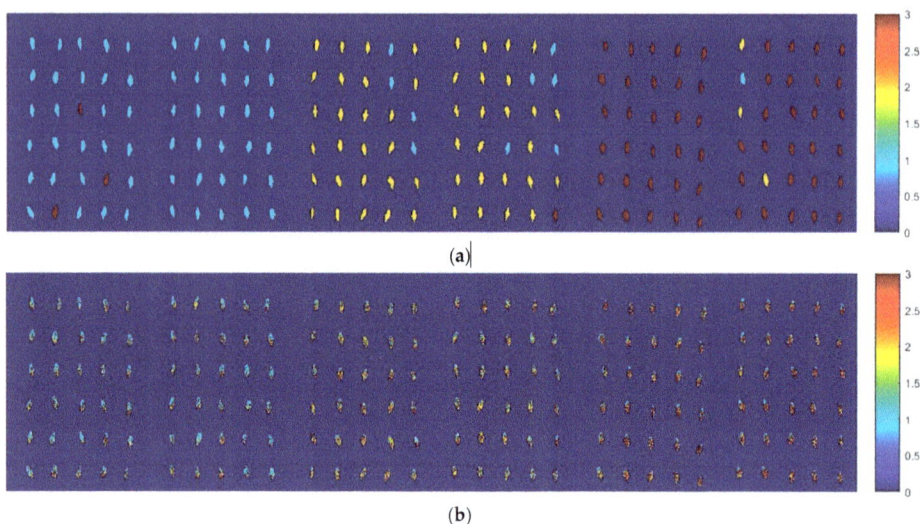

Figure 7. The visualization images of (**a**) method M2M and (**b**) method M2P for Haoyu21 with the SPA and PLS-DA model.

Now, how to identify the final category for the corn seeds was next step. The percentage of each category of each corn seed was calculated, and the results are shown in Figure 8.

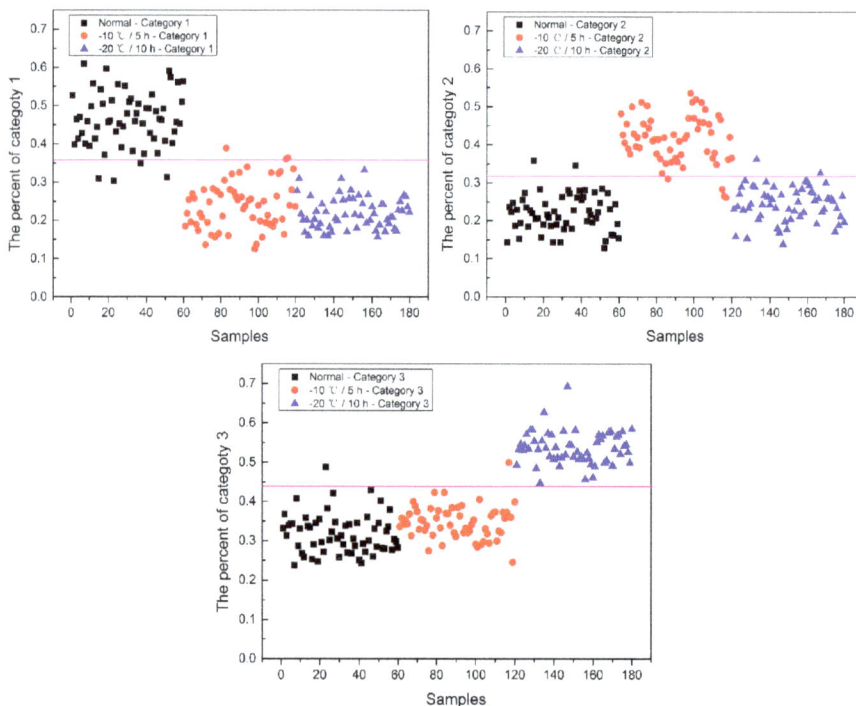

Figure 8. The percentage of each category of Haoyu21 with the SPA and PLS-DA model.

From Figure 8, each type of corn seed had its own concentrated percentage distribution area. For example, the first 60 corn seed samples had a larger percentage of category 1 than those of the other corn seed samples because they were the normal corn seeds; With a threshold of 0.37, the number 83 corn seed sample was misclassified as category 1. The middle 60 corn seed samples had a larger percentage of category 2 than those of the other corn seed samples because they were the slightly freeze-damaged corn seeds; With a threshold of 0.32, the number 15, 37 and 167 corn seed samples were misclassified as category 2. The last 60 corn seed samples had a larger percentage of category 3 than those of the other corn seed samples because they were the severely freeze-damaged corn seed; With a threshold of 0.44, and all of the category 3 corn seeds were classified correctly, the number 23 and 117 corn seed samples were misclassified as category 3.

Among numbers 15, 23, 37, 83, 117 and 167, It was found that numbers 83 and 167 were classified into two categories (number 83 was classified as categories 1 and 2, and number 167 was classified as categories 2 and 3), shown in Table 1. One sample should only have one category. Thus, a method to classify them into one category need to be found. In this study, the percentages of two categories were compared, and the bigger one was the final category and the final category was obtained with the smallest value in the deep black color. In the end, the number 167 was classified correctly.

Table 1. The percentages of the two categories of Haoyu21 corn seed samples.

The Number (the Original Category) of the Sample	The Percentage—Threshold of Category 1	The Percentage—Threshold of Category 2	The Percentage—Threshold of Category 3
83 (2)	**0.38833–0.37**	0.32435–0.32	0.28732–0.44
167 (3)	0.17004–0.37	0.32591–0.32	**0.50405–0.44**

It was also found that the category 1 percentage of number 15, 23, 37 and 51 corn seeds were lower than 0.37, and they were not included in category 1; the category 2 percentage of number 86, 115, 116 and 117 corn seeds were lower than 0.32, and they were not included in category 2. From the above study the number 15, 23, 37 and 117 corn seeds were classified, but the 51, 86, 115 and 116 corn seeds did not have their own category, so the percentage of each category was compared shown in Table 2. After subtracting the percentage from the threshold, the final category was obtained with the smallest value in the deep black color. At last, the number 86 corn seed was classified as category 2 correctly, while the number 51 (should be category 1) was classified as category 2 and numbers 115 and 116 (should be category 2) were classified as category 1. The final classification results of method M2P are shown in Figure 9.

To compare the above results, the number of misclassified corn seed samples were counted. There were 17 corn seed samples misclassified by method M2M while eight corn seed samples were misclassified by method M2P. Meanwhile, none of the severely freeze-damaged samples were misclassified by method M2P. In some respects, it could be drawn that method M2P shown better results than method M2M.

Table 2. The percentage and subtraction of each uncertain Haoyu21 corn seed sample.

The Number (the Original Category) of the Sample	The Percentage—Threshold (Subtraction) of Category 1	The Percentage—Threshold (Subtraction) of Category 2	The Percentage—Threshold (Subtraction) of Category 3
51 (1)	0.31313–0.37 (0.0587)	**0.28535–0.32 (0.03465)**	0.40152–0.44 (0.03848)
86 (2)	0.321–0.37 (0.049)	**0.31026–0.32 (0.00974)**	0.36874–0.44 (0.07126)
115 (2)	**0.36018–0.37 (0.00982)**	0.28308–0.32 (0.03692)	0.35673–0.44 (0.08327)
116 (2)	**0.36364–0.37 (0.00636)**	0.26477–0.32 (0.05523)	0.37159–0.44 (0.06841)

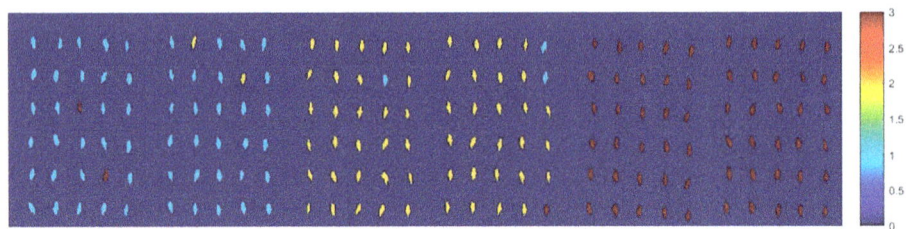

Figure 9. The final visualization images of Haoyu21 with the SPA and PLS-DA model with method M2P.

2.4.2. The Visualization Images of Haihe78

From Figure 10, the top visualization image (Figure 10a) with the SPA and PLS-DA model was almost matched with the above figure results. The 6 misclassified corn seed samples of Haihe78 were fewer than the 17 of Haoyu21.

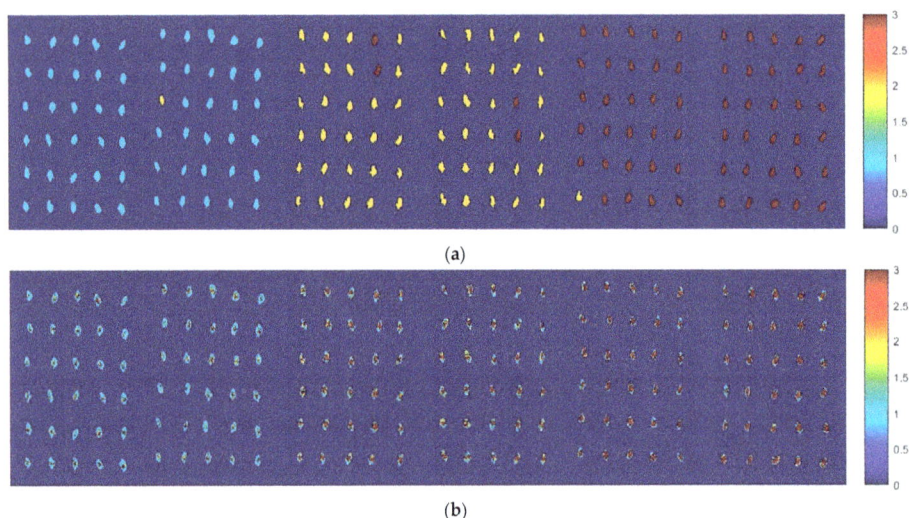

(a)

(b)

Figure 10. The visualization images (**a**) method M2M and (**b**) method M2P for Haihe78 with the SPA and PLS-DA model.

Next, the percentage of each category of each corn seed was calculated, and the results are shown in Figure 11.

From Figure 11, the first 60 corn seed samples had a higher percentage of category 1 with a threshold of 0.55, and all of the category 1 corn seeds were classified correctly and no other category corn seeds were classified to category 1 in this situation. The middle 60 corn seed samples had a larger percentage of category 2 with a threshold of 0.239, and the number 121, 122 and 155 corn seed samples were misclassified as category 2. The last 60 corn seed samples had a larger percentage of category 3 with a threshold of 0.45, and the number 113 corn seed sample was misclassified as category 3.

Among numbers 113, 121, 122 and 155, the number 122 was classified as categories 2 and 3, shown in Table 3. Similar to the above study, the percentages of two categories were compared, and the bigger one was the final category and the final category was shown with the smallest value in the deep black color. In the end, number 122 was classified correctly.

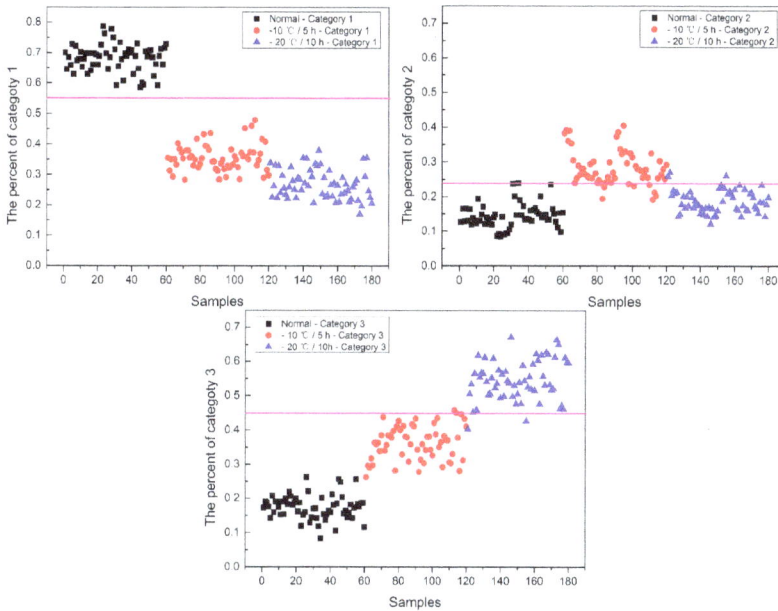

Figure 11. The percentage of each category of Haihe78 with the SPA and PLS-DA model.

Table 3. The percentage of the two categories of Haihe78 corn seed samples.

The Number (the Original Category) of The Sample	The Percentage—Threshold of Category 1	The Percentage—Threshold of Category 2	The Percentage—Threshold of Category 3
122 (3)	0.22491–0.55	0.26965–0.239	**0.50544–0.45**

It was also found that the category 2 percentage of numbers 83, 84, 98, 101 112, 113 and 114 corn seeds were lower than 0.239, and they were not included in category 2. The category 3 percentage of numbers 121 and 155 corn seeds were lower than 0.45, and they were not included in category 3. Numbers 113, 121 and 155 corn seeds were classified, but the 83, 84, 98, 101, 112 and 114 corn seeds did not have their own category, so the percentage of each category were compared and shown in Table 4. After subtracting the percentage from the threshold, the final category was shown with the smallest value in the deep black color. At last, the numbers 84, 98, 101, 112 corn seeds were classified as category 2 correctly, while number 83 (should be category 2) was classified as category 1 and number 114 (should be category 2) was classified as category 3. The final classification results for method M2P are shown in Figure 12.

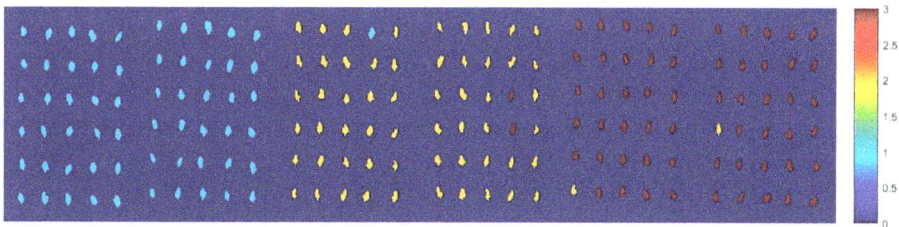

Figure 12. The final visualization images of Haihe78 with the SPA and PLS-DA model with method M2P.

To compare the above results, the number of misclassified corn seed samples are counted. There are six corn seed samples misclassified by method M2M while five corn seed samples are misclassified by method M2P. In some respects, it can be drawn that the effect of method M2P is similar to that of method M2M.

Table 4. The percentage and subtraction of each uncertain Haihe78 corn seed sample.

The Number (the Original Category) of the Sample	The Percentage—Threshold (Subtraction) of Category 1	The Percentage—Threshold (Subtraction) of Category 2	The Percentage—Threshold (Subtraction) of Category 3
83 (2)	**0.39402–0.55 (0.15598)**	0.19421–0.239 (0.04479)	0.41176–0.45 (0.03824)
84 (2)	0.39179–0.55 (0.15821)	**0.22754–0.239 (0.01146)**	0.38067–0.45 (0.06933)
98 (2)	0.38179–0.55 (0.16821)	**0.23726–0.239 (0.00174)**	0.38095–0.45 (0.06905)
101 (2)	0.34878–0.55 (0.20122)	**0.23059–0.239 (0.00841)**	0.42063–0.45 (0.02937)
112 (2)	0.47799–0.55 (0.07201)	**0.19227–0.239 (0.04673)**	0.32974–0.45 (0.12026)
114 (2)	0.34722–0.55 (0.20278)	0.20313–0.239 (0.03587)	**0.44965–0.45 (0.00035)**

2.4.3. The Visualization Images of Jindan10

From Figure 13, the top visualization image (Figure 13a) with the SPA and PLS-DA model was almost matched with the above figure results. The 4 misclassified corn seed samples of Jindan10 was less than that the 17 of Haoyu21 and the 6 of Haihe78.

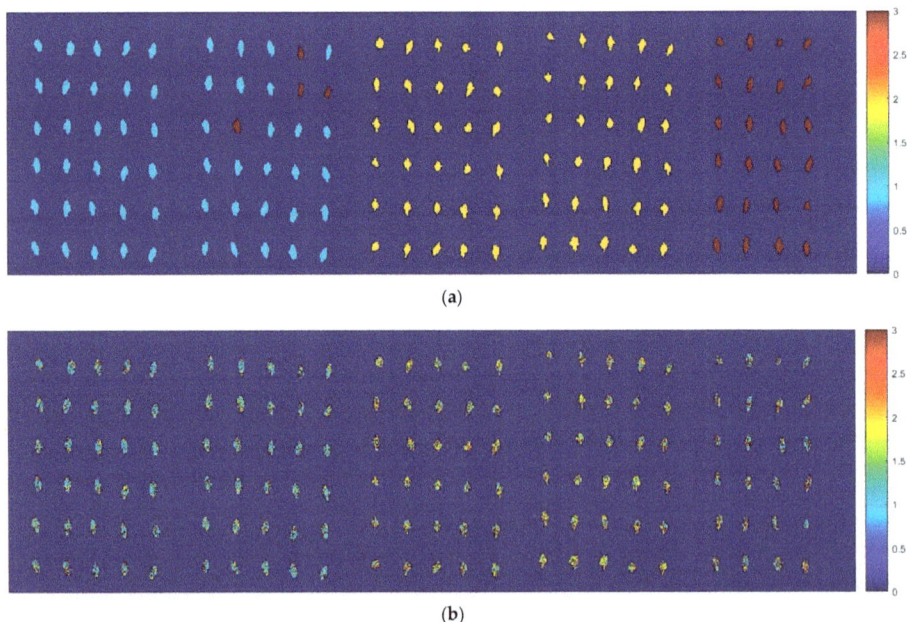

(a)

(b)

Figure 13. The visualization images (**a**) method M2M and (**b**) method M2P for Jindan10 with the SPA and PLS-DA model.

Next, the percentage of each category of each corn seed was calculated, and the results are shown in Figure 14.

From Figure 14, the first 60 corn seed samples had a larger percentage of category 1 with a threshold of 0.38, and the number 135 corn seed sample was misclassified as category 1. The middle 60 corn seed samples had a larger percentage of category 2 with a threshold of 0.385, no other category corn seeds were classified as category 2 in this situation. The last 24 corn seed samples had a larger percentage of category 3 with a threshold of 0.34, and the number 39 and 59 corn seed samples were misclassified as category 3.

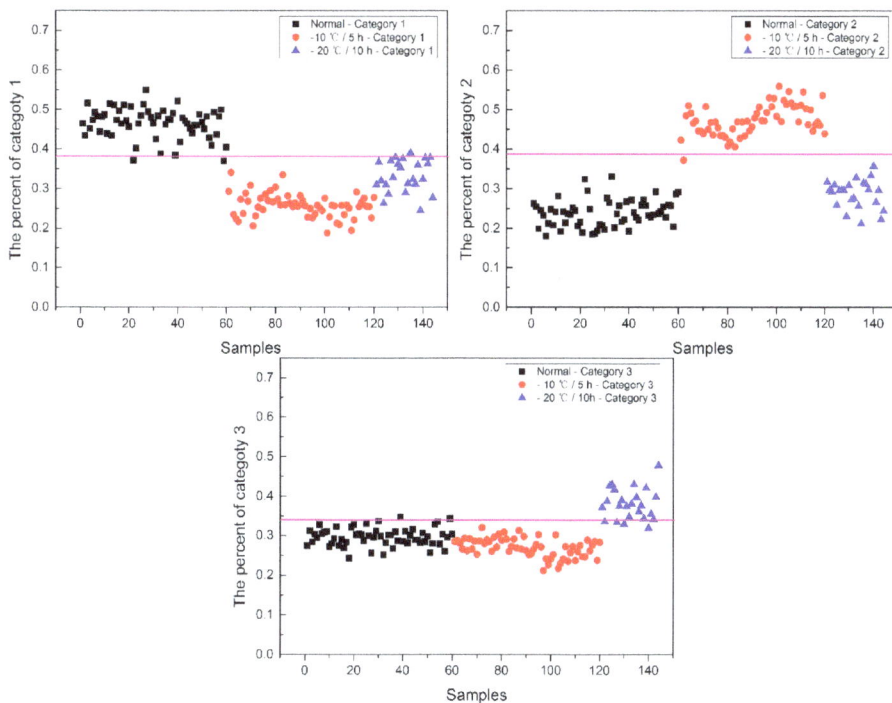

Figure 14. The percentage of each category of Jindan10 with the SPA and PLS-DA model.

Among numbers 39, 59 and 135, numbers 39 and 135 were classified to two categories (number 39 was classified to categories 1 and 3, and number 135 was classified to categories 1 and 3), shown in Table 5. Similar to the above study, the percentage of two categories was compared, the bigger one was the final category and the final category was shown with the smallest value in the deep black color. In the end, both numbers 39 and 135 were classified correctly.

Table 5. The percentage of the two categories of Jindan10 corn seed samples.

The Number (the Original Category) of the Sample	The Percentage—Threshold of Category 1	The Percentage—Threshold of Category 2	The Percentage—Threshold of Category 3
39 (1)	**0.38468–0.38**	0.26813–0.385	0.34719–0.34
135 (3)	0.38983–0.38	0.21243–0.385	**0.39774–0.34**

It was also found that the category 1 percentage of numbers 22 and 59 corn seeds was lower than 0.38, and they were not included in category 1. The category 2 percentage of number 62 was lower than 0.385; the category 3 percentage of numbers 122, 127, 130 and 140 corn seeds were lower than 0.34, and they were not included in category 3. The number 59 corn seed was classified from the above study, but the 22, 62, 122, 127, 130 and 140 corn seeds did not have their own category. The percentage of each category was compared and shown in Table 6. After subtracting the percentage from the threshold, the final category was obtained with the smallest value in the deep black color. At last, the number

22, 62, 122, 127, 130 and 140 corn seeds were classified correctly and the final classification results of method M2P are shown in Figure 15.

Table 6. The percentage and subtraction of each uncertain Jindan10 corn seed sample.

The Number (the Original Category) of the Sample	The Percentage—Threshold (Subtraction) of Category 1	The Percentage—Threshold (Subtraction) of Category 2	The Percentage—Threshold (Subtraction) of Category 3
22 (1)	**0.37094–0.38 (0.00906)**	0.32401–0.385 (0.06099)	0.30505–0.34 (0.03495)
62 (2)	0.34061–0.38 (0.03939)	**0.37212–0.385 (0.01288)**	0.28727–0.34 (0.05273)
122 (3)	0.36697–0.38 (0.01303)	0.29702–0.385 (0.08798)	**0.33601–0.34 (0.00399)**
127 (3)	0.36983–0.38 (0.01017)	0.29603–0.385 (0.08897)	**0.33414–0.34 (0.00586)**
130 (3)	0.36148–0.38 (0.01852)	0.30871–0.385 (0.07629)	**0.32982–0.34 (0.01018)**
140 (3)	0.324–0.38 (0.056)	0.35688–0.385 (0.02812)	**0.31912–0.34 (0.02088)**

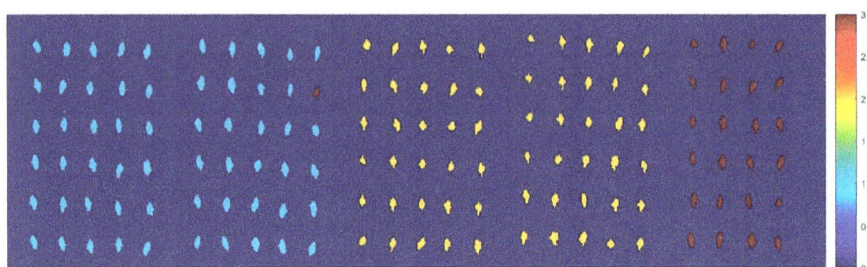

Figure 15. The final visualization images of Jindan10 with the SPA and PLS-DA model with method M2P.

To compare the above results, the number of misclassified corn seed samples were counted. There were four corn seed samples misclassified by method M2M while one corn seed sample was misclassified by method M2P. Meanwhile, none of the severely freeze-damaged samples were misclassified by method M2P. In some respects, it could be drawn that method M2P had better results than method M2M.

To summarize, the visualization results of three corn varieties using method M2M and method M2P were compared. By setting several category thresholds, and comparing the percentage value or subtracting the percentage from the threshold, method M2P could get fewer numbers of misclassified corn seed samples than in method M2M. In some respects, the method M2P had better results than the method M2M.

3. Materials and Methods

3.1. Sample Preparation

Three fresh corn seed varieties (Haoyu21, Haihe78 and Jindan10) were collected from the Jiuquan Julong Tengfei Seed Industry Co., Ltd. in Gansu province, China, with a moisture content of about 30% before harvesting in 2017. In order to obtain different freezing-damage corn seeds, the seeds were dried in an oven until the moisture content was 18% and were put in plastic bags for later experiment. Then the seeds were placed in different freezing temperatures for different duration and eventually were divided into three categories (normal samples, slightly freeze-damaged samples, severely freeze-damaged samples). The frozen environment is shown in Table 7.

180 corn samples (60 normal samples, 60 slightly freeze-damaged samples, 60 severely freeze-damaged samples) of Haoyu21; 180 corn samples (60 normal samples, 60 slightly freeze-damaged samples, 60 severely freeze-damaged samples) of Haihe78; and 144 corn samples (60 normal samples,

60 slightly freeze-damaged samples, 24 severely freeze-damaged samples) of Jindan10. The three varieties of corn samples were randomly divided into calibration sets and validation sets (calibration sets:validation sets = 2:1). Next, the seeds were stored in a refrigerator with a temperature of 4 °C to prevent moisture absorption. Before collecting the hyperspectral images, the moisture content was tested again to ensure all the samples had nearly the same moisture content. The final moisture content was controlled at 13%.

Table 7. The frozen condition (freezing temperature and freezing duration) of corn seeds.

Frozen Condition	Normal	Slight Freeze-Damage	Severely Freeze-Damage
Freezing temperature	Room temperature	−10 °C	−20 °C
Freezing duration	/	5 h	10 h

3.2. Conductivity Test

Before the collection of hyperspectral images, an electrical conductivity test of corn seeds was conducted to check for the effects of the freeze damage on the corn seeds. Fifty corn seed samples for each treatment condition were weighed and placed in a 500 mL beaker containing 250 mL of deionized water. The conductivity was measured after the corn seed soaked for 8, 12, 16, 20 and 24 h.

3.3. Hyperspectral Image Acquisition

3.3.1. Hyperspectral Imaging System

The hyperspectral data of the corn seeds were collected using a VIS/NIR hyperspectral imaging system (Dualix Spectral Imaging, Inc., Sichuan, China). The system consisted of a CCD camera (C8484-05G, Hamamatsu Photonics, Shizuoka, Japan), an imaging spectrometer (Impressor V10E-QE, Spectral Imaging Ltd., Oulu, Finland), a lens (V23-f/2.4 030603, Specim Ltd., Oulu, Finland), line light sources (P/N9130, Illumination Technologies, Inc., East Syracuse, NY, USA) and its controller (2900ER, Illumination Technologies, Inc., East Syracuse, NY, USA), sample stage (GZ02DS20, Guangzheng Instruments Co., Ltd., Beijing, China) and a moving stage controller (PSA200-11-X, Zolix Instruments Co., Ltd., Beijing, China).

The CCD camera, VIS/NIR imaging spectrometer, lens, line light source, sample stage, moving stage and 0.5 mm extension tube were all placed in a dark box. The moving stage was installed in the bottom of the dark box and connected with the moving stage controller. The sample stage was installed on the top of the moving stage and the corn seed was placed on the sample stage; above the sample stage, two line tungsten halogen light sources were installed on both side walls of the dark box, which provided illumination for the corn seeds. The two line light sources were symmetrically at 60° and installed at a height of 26 cm above the moving stage to provide a stable and uniform diffuse reflection light for the corn seeds. The CCD camera, VIS/NIR imaging spectrometer and lens were installed on the top surface of the dark box and were connected to each other vertically. Above the lens, there was a 0.5 mm extension tube between the lens and the VIS/NIR infrared spectrometer. After adding the extension tube, more clear corn seed hyperspectral images were obtained.

3.3.2. Hyperspectral Image Acquisition and Correction

The hyperspectral image acquisition software, Spectracube 2.75b (Spectral Imaging Ltd., Oulu, Finland), was used for image acquisition and correction. At last, the corn seed embryo was placed upside down on black sample stage with an object distance of 28.5 cm, with an exposure time of 1 ms and the moving stage speed of 2.6 mm/s. The hyperspectral images were acquired with a spectral range of 400–1000 nm which had a total of 477 wavelengths.

In order to overcome the inhomogeneity of the light source intensity at each wavelength and the influence of the dark current of the acquisition sensor, the collected hyperspectral images required black and white correction according to Equation (1).

$$I_c = (I_0 - I_b)/(I_w - I_b) \tag{1}$$

where I_c is the relative reflectance intensity of each wavelength, I_0 is the original reflectance intensity of the hyperspectral image, I_b is the intensity of the dark current, which was obtained by turning off the light source and completely covering the lens with its cap, and I_w is the reflectance intensity of the Teflon white surface (Spectralon, Labsphere Inc., North Sutton, NH, USA), which was obtained under the same conditions as the raw image.

The hyperspectral images were analyzed using the ENVI 4.6.1 software (ITT Visual Information Solutions, Boulder, CO, USA) and MATLAB R2017b (MathWorks, Natick, MA, USA).

3.4. Data Analysis

3.4.1. Image Segmentation and Processing

After the hyperspectral images were obtained, the feature image, i.e., the embryo image was extracted for later processing. At first, the average spectrums of 10 pixel × 10 pixel regions of interest (ROI) in the endosperm and embryo were compared, and the gray image in the wavelength of 500 nm was selected as a segment image to separate the embryo (the brighter region) from the whole corn seed (in this method, image enhancement and the Otsu threshold segmentation method was used). Finally, the final embryo-binary image was obtained.

3.4.2. The Spectral Pretreatment Methods

After the embryo-binary image was obtained, it was masked with the original hyperspectral image and the average spectrum of corn seed embryo hyperspectral image was pretreated for spectral feature extraction. Since the spectrometer had low response and noise at the edge of the spectral region, the middle 31 to 460 wavelength spectrum was selected for analysis.

In this paper, multiplicative scatter correction (MSC), standard normal variation (SNV) and 5 points and 3 times smoothing (5-3 smoothing) pretreatment methods were applied [43,44].

The MSC method is used to correct the scattering of each sample's spectrum, get the desired spectrum, and remove the undesirable scatter effect which can improve the signal-to-noise ratio [44].

The SNV method is commonly used to eliminate spectral errors among samples due to different solid particle sizes, scattering, or measurement path lengths. It converts the data mean to 0 and the standard deviation to 1 and is generally used for scatter correction and to remove the slope variation from the spectra [43].

The smoothing method can effectively remove high-frequency noise which may be produced by instrument noise, random errors, etc., which can keep the original useful signal information and improve the signal-to-noise ratio at the same time [44]. In this study, 5-3 smoothing was applied which is based on 5 points and 3 times polynomial fitting, and the smoothing time was 2000.

3.4.3. Spectral Feature Extraction

It is well known that a hyperspectral image has a large amount of spectral data and the whole spectral wavelength contains much noise and irrelevant information. According to the introduction, it is necessary to apply some algorithms (in this study, the successive projection algorithm (SPA), principal component analysis (PCA) and other methods were applied) which have faster speed in data computation and higher accuracy to obtain important wavelengths and establish the classification model [42].

SPA is a forward selection algorithm that looks for variable combinations with minimal redundancy information in the variable matrix to minimize collinearity among the variables. Therefore, the SPA is used to extract the characteristic wavelengths during spectral data analysis processing [42].

PCA is a commonly used dimension reduction mapping method, which maps original features with strong original correlation into a set of new features. Each new feature constructed is a linear combination of the original features. Each new feature is not related to each other [45].

X-loading method uses spectral variables for partial least-squares modeling, and the wavelengths are extracted based on the absolute value of the regression coefficients for each wavelength. As the absolute value of the regression coefficient gets larger, the greater the influence of these wavelengths on the modeling, that is, the final extracted feature wavelength [42].

3.4.4. The Classification Methods

In this paper, the partial least squares-discriminant analysis (PLS-DA), K-nearest neighbor (KNN) and support vector machine (SVM) models were established for classification of different frozen corn seeds.

The PLS-DA is based on the partial least squares (PLS) technique that is a commonly used in multivariate statistical analysis methods [46]. The main principle of the PLS-DA model is briefly described as Equation (2):

$$Y = X_{n*p} \, B + E \tag{2}$$

where Y is the matrix of the response variables that relates to the measured sample categories; X is the $n*p$ matrix of the spectral variables for each measured sample category; n is the number of samples, p is the number of variables; B is the matrix of regression coefficients for the spectral variables and E is the matrix of residuals. The number of main components is optimized by using ten-fold cross validation during the model development and updating stages.

KNN classifies by measuring the distance (the distance can be Euclidean distance or Manhattan distance) between different feature values [47]. If a sample is in the feature space, most of the k most similar samples belong to a certain category, and the sample also belongs to this category. The main principles of KNN are as follows:

(1) Calculating the distance d (this paper Euclidean distance was applied as d) between the test data and each set of training data; (2) Sorting the distance increasingly; (3) Selecting the K points with the smallest distance; (4) Calculating the frequency f of each category of the K points; and (5) The category with the highest frequency among the K points is the predictive category for the test data. K is optimized by using ten-fold cross validation during the model development and updating stages.

SVM is an important classification method and has many unique advantages in solving small sample sets, nonlinear and high-dimensional pattern recognition problems [48]. It establishes the model from the limited training samples and obtains small errors for the independent test set. It tries to improve the generalization ability of the learning machine. The data input space is mapped into a high-dimensional feature space through a kernel function (in this paper, the radial basis function (RBF) kernel function was applied); and c (the penalty factor) and g (the radial width of the kernel function) are the two main parameters of the SVM method, which are optimized using a grid-search algorithm coupled with ten-fold cross validation during the model development and updating stages.

To analyze the classification accuracy results, there were three aspects can be considered: (1) the results of different pretreatment methods at full-band treatment; (2) the results of different wavelength selection algorithms with the same pretreatment method; and (3) the results of different classification methods with the same pretreatment method and the same wavelength selection algorithm. The >80% classification accuracy results of both the calibration sets and the validation sets could be observed clearly by a pink 80% line among the three corn varieties.

3.4.5. The Visualization Images of the Classification Results

When the PLS-DA, KNN, and SVM methods were applied to establish the classification models, the robust classification model was selected to achieve the visualization of classification results of the corn seeds.

The average spectrum and the pixel spectrum are explained. When the ROI of a hyperspectral image is extracted, it contains many pixels, and each pixel corresponds to a spectrum; in this paper, the average spectrum refers to the average spectrum of the seed embryo ROI, the pixel spectrum refers to the spectrum of each pixel in the seed embryo ROI.

In this paper, the visualization images were obtained in Method M2M and Method M2P. Method M2M: the classification results were obtained from the above study and each corn seed was corresponding with the classification value. Method M2P: the spectrum of each pixel of the embryo images was classified, and each pixel of the corn seed was corresponding with the classification value. Mean-spectrum to mean-spectrum (M2M) means that both of the calibration and validation sets were the mean spectrums of the ROI; Mean-spectrum to pixel-spectrum (M2P) means that the calibration sets were the mean spectrums of the ROI while the validation sets were the spectrum of each pixel in the seed embryo ROI.

It is known that the surface of the corn seed sample is not very flat where the height of the embryo edge is higher than the height of the embryo inside, thus, in this paper, 1~3 thresholds for each category were set to classify the different degrees of freeze-damage in corn seed after the percentage of each category was calculated.

4. Conclusions

Conductivity test is a general method to distinguish different degrees of freeze-damage in corn seed. The more serious the freeze damage is, the higher its conductivity. In this feasibility attempt to classify different degrees of freeze-damage in corn seed with hyperspectral imaging technology, four different pretreatment methods (no pretreatment, SNV, MSC and 5-3 smoothing), four wavelength selection algorithms (SPA, PCA, X-loading and full-band methods) and three different classification modeling methods (PLS-DA, KNN and SVM) were applied to find a relatively better method for the three different corn seed varieties. In order to better represent the freeze damage of the seed embryos, comparisons were made between method M2M and method M2P.

The following conclusions are drawn from this study:

(1) By using related image preprocessing methods on the gray image of corn seeds at 500 nm wavelength, the final embryo hyperspectral images can be clearly obtained to achieve the following classification of different degrees of freeze-damaged corn seed.

(2) The 5-3 smoothing pretreatment method has higher classification accuracy than the other pretreatment methods from the results of different pretreatment methods at full-band treatment. The classification accuracy almost reaches 90%, maybe because the pretreatment has the advantages of keeping the signal from the original spectrums and improving the signal-to-noise ratio.

(3) The classification accuracies of the full-band, SPA, PCA and X-loading algorithms can be sorted as follows: the classification accuracies of the full-band method are higher than those of the SPA and PCA methods, and the X-loading method has the lowest classification accuracy from the results of different wavelength selection algorithms with the same preprocessing method. Maybe this is because a great deal of information about the corn seeds can be got in the full band situation, but in some way, it necessary to find several wavelengths to reduce the modeling time and improve the efficiency, so the SPA algorithm could be a good choice.

(4) The classification accuracies of the PLS-DA, KNN and SVM methods can be sorted as follows: the PLS-DA modeling method has the best classification accuracy compared to the SVM and KNN modeling methods, while the KNN modeling method has the lowest classification accuracy.

Molecules **2019**, *24*, 149

(5) By setting several category thresholds, and comparing the percentage value or subtracting with method M2P, fewer numbers of misclassified corn seed samples can be obtained. In some respects, method M2P has better results than method M2M.

Based on the above several conclusions, it is feasible that the hyperspectral imaging technology used to establish classification models for the embryos of corn seeds with different degrees of freeze damage. The smoothing method and wavelength selection method can be applied to modeling to improve the signal-to-noise ratio, classification efficiency and result accuracy of the model, although good modeling results can be obtained in the full-band case. The method M2P allowed visualization of the classification result of each embryo pixel and the final classification result is better than the method M2M.

Author Contributions: J.Z. and F.C. conceived of and designed the experiment. J.Z. performed the experiment. J.Z. and L.D. analyzed the spectral data and contributed to the result interpretation and writing. J.Z. and F.C. reviewed and revised the paper. All authors have read, revised and approved the final manuscript.

Funding: This research was funded by the National Natural Science Foundation of China (Grant No. 61873231 and No. 61573309).

Acknowledgments: The authors were grateful to the Jiuquan Julong Tengfei Co., Ltd. in Gansu province (China) for providing experimental samples.

Conflicts of Interest: The authors declare no conflict of interest.

Appendix A

Table A1. The selected wavelengths using different methods under different pretreatments of Haoyu21.

No Pretreatment	The Number of Wavelengths	Wavelength/nm
SPA	5	445.44 484.16 645.23 977.66 984.09
PCA	6	465.96 482.94 535.58 608.84 667.92 878.6
X-loading	7	461.12 637.68 861.89 881.17 918.48 968.65 976.37
Full-band	430	444.24~985.27

MSC	The number of wavelengths	Wavelength/nm
SPA	5	470.8 474.44 492.68 824.66 906.9
PCA	11	459.91 465.96 469.59 470.80 535.58 684.36 832.3500 847.7600 878.6 919.76 968.65
X-loading	7	472.01 854.1 863.18 865.75 964.79 976.37 977.66
Full-band	430	444.24~985.27

SNV	The number of wavelengths	Wavelength/nm
SPA	6	449.05 455.08 470.8 784.94 906.9 927.48
PCA	11	459.91 465.96 469.59 470.8 535.58 684.36 832.35 847.76 878.6 919.76 968.65
X-loading	7	472.01 854.18 863.18 865.75 964.79 968.65 977.66
Full-band	430	444.24~985.27

5-3 smoothing	The number of wavelengths	Wavelength/nm
SPA	8	486.59 511.01 530.66 549.16 656.57 689.42 887.6 904.33
PCA	9	449.05 467.17 484.16 641.46 671.71 861.89 874.74 913.33 981.51
X-loading	7	449.05 451.46 463.54 467.17 647.75 864.46 976.37
Full-band	430	444.24~985.27

Table A2. The selected wavelengths using different methods under different pretreatments of Haihe78.

No Pretreatment	The Number of Wavelengths	Wavelength/nm
SPA	3	465.96 523.28 637.68
PCA	10	451.46 458.7 478.08 540.51 551.63 605.09 756.83 864.46 870.89 876.03
X-loading	7	464.75 478.08 864.46 868.32 870.89 891.46 969.94
Full-band	430	444.24~985.27

MSC	The number of wavelengths	Wavelength/nm
SPA	6	461.12 533.12 735.15 840.06 843.91 921.05
PCA	14	457.49 462.33 464.75 482.94 509.79 525.74 531.89 643.97 756.83 814.4 842.62 851.61 877.32 924.91
X-loading	6	457.49 531.89 863.18 869.6 969.94 976.37
Full-band	430	444.24~985.27

SNV	The number of wavelengths	Wavelength/nm
SPA	6	465.96 508.56 557.82 636.43 763.21 864.46 919.76 945.5 960.94
PCA	12	457.49 464.75 482.94 509.79 525.74 531.89 643.97 756.83 814.4 842.62 851.61 924.91
X-loading	6	457.49 484.16 863.18 869.6 969.94 976.37
Full-band	430	444.24~985.27

5-3 smoothing	The number of wavelengths	Wavelength/nm
SPA	9	465.96 508.56 557.82 636.43 763.21 864.46 919.76 945.5 960.94
PCA	10	456.29 472.01 540.51 602.59 662.88 781.1 870.89 912.04 966.08 972.51
X-loading	7	449.05 462.33 475.65 865.75 969.94 971.23 976.37
Full-band	430	444.24~985.27

Table A3. The selected wavelengths using different methods under different pretreatments of Jindan10.

No Pretreatment	The Number of Wavelengths	Wavelength/nm
SPA	9	450.26 451.46 452.67 465.96 491.46 520.82 964.79 978.94 985.37
PCA	8	449.05 484.16 490.24 525.74 554.1 665.4 868.32 919.76
X-loading	7	450.26 465.96 491.46 877.32 964.79 967.37 976.37
Full-band	430	444.24~985.27
MSC	**The number of wavelengths**	**Wavelength/nm**
SPA	6	472.01 551.63 722.43 904.33 915.9 985.37
PCA	16	462.33 472.01 479.3 498.78 506.11 523.28 539.28 550.39 655.31 833.64 842.62 845.19 859.32 883.75 888.89 919.76
X-loading	7	457.49 459.91 463.54 484.16 865.75 879.89 984.09
Full-band	430	444.24~985.27
SNV	**The number of wavelengths**	**Wavelength/nm**
SPA	9	472.01 551.63 722.43 904.33 915.9 922.34 923.62 951.93 985.37
PCA	18	456.29 462.33 472.01 479.3 484.16 498.78 506.11 523.28 550.39 655.31 758.1 833.64 842.62 845.19 859.32 883.75 888.89 919.76
X-loading	7	457.49 459.91 463.54 484.16 865.75 873.46 984.09
Full-band	430	444.24~985.27
5-3 smoothing	**The number of wavelengths**	**Wavelength/nm**
SPA	14	462.33 473.23 485.37 503.67 526.97 571.46 632.66 665.4 804.14 864.46 886.32 921.05 949.36 963.51
PCA	13	446.64 450.26 469.59 491.46 523.28 526.97 554.1 661.61 872.17 918.48 939.06 962.22 980.23
X-loading	7	450.26 485.37 878.6 922.34 960.94 968.65 977.66
Full-band	430	444.24~985.27

References

1. Ambrose, A.; Kandpal, L.M.; Kim, M.S.; Lee, W.; Cho, B. High speed measurement of corn seed viability using hyperspectral imaging. *Infrared Phys. Technol.* **2016**, *75*, 173–179. [CrossRef]
2. Hampton, J.G.; Tekrony, D.M. *Handbook of Vigour Test Methods*; ISTA: Zurich, Switzerland, 1995.
3. Ferreira, L.B.D.S.; Fernandes, N.A.; Aquino, L.C.D.; Silva, A.R.D.; Nascimento, W.M.; Leão-Araújo, É.F. Temperature and seed moisture content affect electrical conductivity test in pea seeds. *J. Seed Sci.* **2017**, *39*, 410–416. [CrossRef]
4. Fessel, S.A.; Vieira, R.D.; Cruz, M.C.P.D.; Paula, R.C.D.; Panobianco, M. Electrical conductivity testing of corn seeds as influenced by temperature and period of storage. *Pesquisa Agropecuária Brasileira* **2006**, *41*, 1551–1559. [CrossRef]
5. Qiu, G.; Lü, E.; Lu, H.; Xu, S.; Zeng, F.; Shui, Q. Single-Kernel FT-NIR Spectroscopy for Detecting Supersweet Corn (*Zea mays* L. Saccharata Sturt) Seed Viability with Multivariate Data Analysis. *Sensors* **2018**, *18*, 1010. [CrossRef] [PubMed]
6. Egesel, C.Ö.; Kahrıman, F.; Ekinci, N.; Kavdır, İ.; Büyükcan, M.B. Analysis of Fatty Acids in Kernel, Flour, and Oil Samples of Maize by NIR Spectroscopy Using Conventional Regression Methods. *Cereal Chem.* **2016**, *93*, 487–492. [CrossRef]
7. Zhong, J.; Qin, X. Rapid Quantitative Analysis of Corn Starch Adulteration in Konjac Glucomannan by Chemometrics-Assisted FT-NIR Spectroscopy. *Food Anal. Methods* **2016**, *9*, 61–67. [CrossRef]
8. Plumier, B.M.; Danao, M.C.; Singh, V.; Rausch, K.D. Analysis and Prediction of Unreacted Starch Content in Corn Using FT-NIR Spectroscopy. *Trans. ASABE* **2013**, *56*, 1877–1884. [CrossRef]
9. Falade, T.D.O.; Sultanbawa, Y.; Fletcher, M.; Fox, G. Near Infrared Spectrometry for Rapid Non-Invasive Modelling of Aspergillus-Contaminated Maturing Kernels of Maize (*Zea mays* L.). *Agriculture* **2017**, *7*, 77. [CrossRef]
10. Levasseur-Garcia, C.; Bailly, S.; Kleiber, D.; Bailly, J. Assessing Risk of Fumonisin Contamination in Maize Using Near-Infrared Spectroscopy. *J. Chem.* **2015**. [CrossRef]
11. Tallada, J.G.; Wicklow, D.T.; Pearson, T.C.; Armstrong, P.R. Detection of fungus-infected corn kernels using near-infrared reflectance spectroscopy and colorimaging. *Trans. ASABE* **2011**, *54*, 1151–1158. [CrossRef]
12. Sun, Y.; Chen, S.S.; Ning, J.F.; Han, W.T.; Weckler, P.R. Prediction of Moisture Content in Corn Leaves Based on Hyperspectral Imaging and Chemometric Analysis. *Trans. ASABE* **2015**, *58*, 531–537. [CrossRef]
13. McGoverin, C.M.; Manley, M. Classification of maize kernel hardness using near infrared hyperspectral imaging. *J. Near Infrared Spectrosc.* **2012**, *20*, 529. [CrossRef]
14. Williams, P.; Geladi, P.; Fox, G.; Manley, M. Maize kernel hardness classification by near infrared (NIR) hyperspectral imaging and multivariate data analysis. *Anal. Chim. Acta* **2009**, *653*, 121–130. [CrossRef] [PubMed]
15. Weinstock, B.A.; Janni, J.; Hagen, L.; Wright, S. Prediction of oil and oleic acid concentrations in individual corn (*Zea mays* L.) kernels using near-infrared reflectance hyperspectral imaging and multivariate analysis. *Appl. Spectrosc.* **2006**, *60*, 9–16. [CrossRef] [PubMed]
16. Feng, X.; Zhao, Y.; Zhang, C.; Cheng, P.; He, Y. Discrimination of Transgenic Maize Kernel Using NIR Hyperspectral Imaging and Multivariate Data Analysis. *Sensors* **2017**, *17*, 1894. [CrossRef] [PubMed]
17. Wang, L.; Sun, D.; Pu, H.; Zhu, Z. Application of Hyperspectral Imaging to Discriminate the Variety of Maize Seeds. *Food Anal. Methods* **2016**, *9*, 225–234. [CrossRef]
18. Yang, X.; Hong, H.; You, Z.; Cheng, F. Spectral and Image Integrated Analysis of Hyperspectral Data for Waxy Corn Seed Variety Classification. *Sensors* **2015**, *15*, 15578–15594. [CrossRef]
19. Zhang, X.; Liu, F.; He, Y.; Li, X. Application of Hyperspectral Imaging and Chemometric Calibrations for Variety Discrimination of Maize Seeds. *Sensors* **2012**, *12*, 17234–17246. [CrossRef]
20. Wakholi, C.; Kandpal, L.M.; Lee, H.; Bae, H.; Park, E.; Kim, M.S.; Mo, C.; Lee, W.; Cho, B. Rapid assessment of corn seed viability using short wave infrared line-scan hyperspectral imaging and chemometrics. *Sens. Actuators B Chem.* **2018**, *255*, 498–507. [CrossRef]
21. Guo, D.; Zhu, Q.; Huang, M.; Guo, Y.; Qin, J. Model updating for the classification of different varieties of maize seeds from different years by hyperspectral imaging coupled with a pre-labeling method. *Comput. Electron. Agric.* **2017**, *142*, 1–8. [CrossRef]

22. He, C.; Zhu, Q.; Huang, M.; Mendoza, F. Model Updating of Hyperspectral Imaging Data for Variety Discrimination of Maize Seeds Harvested in Different Years by Clustering Algorithm. *Trans. ASABE* **2016**, *59*, 1529–1537. [CrossRef]
23. Huang, M.; Tang, J.; Yang, B.; Zhu, Q. Classification of maize seeds of different years based on hyperspectral imaging and model updating. *Comput. Electron. Agric.* **2017**, *122*, 139–145. [CrossRef]
24. Kimuli, D.; Wang, W.; Lawrence, K.C.; Yoon, S.; Ni, X.; Heitschmidt, G.W. Utilisation of visible/near-infrared hyperspectral images to classify aflatoxin B1 contaminated maize kernels. *Biosyst. Eng.* **2018**, *166*, 150–160. [CrossRef]
25. Kimuli, D.; Wang, W.; Wang, W.; Jiang, H.; Zhao, X.; Chu, X. Application of SWIR hyperspectral imaging and chemometrics for identification of aflatoxin B1 contaminated maize kernels. *Infrared Phys. Technol.* **2018**, *89*, 351–362. [CrossRef]
26. Chu, X.; Wang, W.; Yoon, S.; Ni, X.; Heitschmidt, G.W. Detection of aflatoxin B1 (AFB 1) in individual maize kernels using short wave infrared (SWIR) hyperspectral imaging. *Biosyst. Eng.* **2017**, *157*, 13–23. [CrossRef]
27. Zhao, X.; Wang, W.; Chu, X.; Li, C.; Kimuli, D. Early Detection of Aspergillus parasiticus Infection in Maize Kernels Using Near-Infrared Hyperspectral Imaging and Multivariate Data Analysis. *Appl. Sci.* **2017**, *7*, 90. [CrossRef]
28. Zhu, F.; Yao, H.; Hruska, Z.; Kincaid, R.; Brown, R.; Bhatnagar, D.; Cleveland, T. Integration of Fluorescence and Reflectance Visible Near-Infrared (VNIR) Hyperspectral Images for Detection of Aflatoxins in Corn Kernels. *Trans. ASABE* **2016**, *59*, 785–794. [CrossRef]
29. Yao, H.; Hruska, Z.; Kincaid, R.; Brown, R.L.; Bhatnagar, D.; Cleveland, T.E. Hyperspectral Image Classification and Development of Fluorescence Index for Single Corn Kernels Infected with Aspergillus flavus. *Trans. ASABE* **2013**, *56*, 1977–1988. [CrossRef]
30. Williams, P.J.; Geladi, P.; Britz, T.J.; Manley, M. Investigation of fungal development in maize kernels using NIR hyperspectral imaging and multivariate data analysis. *J. Cereal Sci.* **2012**, *55*, 272–278. [CrossRef]
31. Del Fiore, A.; Reverberi, M.; Ricelli, A.; Pinzari, F.; Serranti, S.; Fabbri, A.A.; Bonifazi, G.; Fanelli, C. Early detection of toxigenic fungi on maize by hyperspectral imaging analysis. *Int. J. Food Microbiol* **2010**, *144*, 64–71. [CrossRef]
32. Firrao, G.; Torelli, E.; Gobbi, E.; Raranciuc, S.; Bianchi, G.; Locci, R. Prediction of milled maize fumonisin contamination by multispectral image analysis. *J. Cereal Sci.* **2010**, *52*, 327–330. [CrossRef]
33. Huang, M.; Zhao, W.; Wang, Q.; Zhang, M.; Zhu, Q. Prediction of moisture content uniformity using hyperspectral imaging technology during the drying of maize kernel. *Int. Agrophys.* **2015**, *29*, 39–46. [CrossRef]
34. Williams, P.J.; Kucheryavskiy, S. Classification of maize kernels using NIR hyperspectral imaging. *Food Chem.* **2016**, *209*, 131–138. [CrossRef] [PubMed]
35. Zhao, Y.; Zhu, S.; Zhang, C.; Feng, X.; Feng, L.; He, Y. Application of hyperspectral imaging and chemometrics for variety classification of maize seeds. *RSC Adv.* **2018**, *8*, 1337–1345. [CrossRef]
36. Feng, M.; Guo, X.; Wang, C.; Yang, W.; Shi, C.; Ding, G.; Zhang, X.; Xiao, L.; Zhang, M.; Song, X. Monitoring and evaluation in freeze stress of winter wheat (*Triticum aestivum* L.) through canopy hyperspectrum reflectance and multiple statistical analysis. *Ecol. Indic.* **2018**, *84*, 290–297. [CrossRef]
37. Perry, E.M.; Nuttall, J.G.; Wallace, A.J.; Fitzgerald, G.J. In-field methods for rapid detection of frost damage in Australian dryland wheat during the reproductive and grain-filling phase. *Crop Pasture Sci.* **2017**, *68*, 516. [CrossRef]
38. Wang, H.; Huo, Z.; Zhou, G.; Liao, Q.; Feng, H.; Wu, L. Estimating leaf SPAD values of freeze-damaged winter wheat using continuous wavelet analysis. *Plant Physiol. Biochem.* **2016**, *98*, 39–45. [CrossRef] [PubMed]
39. Wu, Q.; Zhu, D.; Wang, C.; Ma, Z.; Wang, J. Diagnosis of freezing stress in wheat seedlings using hyperspectral imaging. *Biosyst. Eng.* **2012**, *112*, 253–260. [CrossRef]
40. Sun, Y.; Gu, X.; Sun, K.; Hu, H.; Xu, M.; Wang, Z.; Tu, K.; Pan, L. Hyperspectral reflectance imaging combined with chemometrics and successive projections algorithm for chilling injury classification in peaches. *LWT Food Sci.Technol.* **2017**, *75*, 557–564. [CrossRef]
41. Mo, C.; Kim, G.; Lee, K.; Kim, M.; Cho, B.; Lim, J.; Kang, S. Non-Destructive Quality Evaluation of Pepper (Capsicum annuum L.) Seeds Using LED-Induced Hyperspectral Reflectance Imaging. *Sensors* **2014**, *14*, 7489–7504. [CrossRef] [PubMed]

42. Liu, D.; Sun, D.; Zeng, X. Recent advances in wavelength selection techniques for hyperspectral image processing in the food industry. *Food Bioprocess Technol.* **2013**, *7*, 307–323. [CrossRef]

43. Candolfi, A.; Maesschalck, R.D.; Jouan-Rimbaud, P.A.; Massart, D.L. The influence of data pre-processing in the pattern recognition of excipients near-infrared spectra. *J. Pharm. Biomed. Anal.* **1999**, *21*, 115–132. [CrossRef]

44. Chen, H.; Song, Q.; Tang, G.; Feng, Q.; Lin, L. The combined optimization of Savitzky-Golay smoothing and multiplicative scatter correction for FT-NIR PLS models. *ISRN Spectrosc.* **2013**, *2013*, 642190. [CrossRef]

45. Wold, S.; Esbensen, K.; Geladi, P. Principal component analysis. *Chemometr. Intell. Lab. Syst.* **1987**, *2*, 37–52. [CrossRef]

46. Ståhle, L.; Wold, S. Partial least squares analysis with cross-validation for the two-class problem: A Monte Carlo study. *J. Chemometr.* **1987**, *1*, 185–196. [CrossRef]

47. Ruiz, J.R.R.; Parello, T.C.; Gomez, R.C. Comparative study of multivariate methods to identify paper finishes using infrared spectroscopy. *IEEE Trans. Instrum. Meas.* **2012**, *61*, 1029–1036. [CrossRef]

48. Mavroforakis, M.; Theodoridis, S. A geometric approach to support vector machine (SVM) classification. *IEEE Trans. Neural Netw.* **2006**, *17*, 671–682. [CrossRef]

Sample Availability: Not available.

Article

Identification of Maize Kernel Vigor under Different Accelerated Aging Times Using Hyperspectral Imaging

Lei Feng [1,2], Susu Zhu [1,2], Chu Zhang [1,2], Yidan Bao [1,2], Xuping Feng [1,2] and Yong He [1,2,3,*]

[1] College of Biosystems Engineering and Food Science, Zhejiang University, Hangzhou 310058, China; lfeng@zju.edu.cn (L.F.); sszhu@zju.edu.cn (S.Z.); chuzh@zju.edu.cn (C.Z.); ydbao@zju.edu.cn (Y.B.); pimmmx@163.com (X.F.)
[2] Key Laboratory of Spectroscopy Sensing, Ministry of Agriculture and Rural Affairs, Hangzhou 310058, China
[3] State Key Laboratory of Modern Optical Instrumentation, Zhejiang University, Hangzhou 310058, China
* Correspondence: yhe@zju.edu.cn; Tel.: +86-571-8898-2143

Received: 6 November 2018; Accepted: 23 November 2018; Published: 25 November 2018

Abstract: Seed aging during storage is irreversible, and a rapid, accurate detection method for seed vigor detection during seed aging is of great importance for seed companies and farmers. In this study, an artificial accelerated aging treatment was used to simulate the maize kernel aging process, and hyperspectral imaging at the spectral range of 874–1734 nm was applied as a rapid and accurate technique to identify seed vigor under different accelerated aging time regimes. Hyperspectral images of two varieties of maize processed with eight different aging duration times (0, 12, 24, 36, 48, 72, 96 and 120 h) were acquired. Principal component analysis (PCA) was used to conduct a qualitative analysis on maize kernels under different accelerated aging time conditions. Second-order derivatization was applied to select characteristic wavelengths. Classification models (support vector machine−SVM) based on full spectra and optimal wavelengths were built. The results showed that misclassification in unprocessed maize kernels was rare, while some misclassification occurred in maize kernels after the short aging times of 12 and 24 h. On the whole, classification accuracies of maize kernels after relatively short aging times (0, 12 and 24 h) were higher, ranging from 61% to 100%. Maize kernels with longer aging time (36, 48, 72, 96, 120 h) had lower classification accuracies. According to the results of confusion matrixes of SVM models, the eight categories of each maize variety could be divided into three groups: Group 1 (0 h), Group 2 (12 and 24 h) and Group 3 (36, 48, 72, 96, 120 h). Maize kernels from different categories within one group were more likely to be misclassified with each other, and maize kernels within different groups had fewer misclassified samples. Germination test was conducted to verify the classification models, the results showed that the significant differences of maize kernel vigor revealed by standard germination tests generally matched with the classification accuracies of the SVM models. Hyperspectral imaging analysis for two varieties of maize kernels showed similar results, indicating the possibility of using hyperspectral imaging technique combined with chemometric methods to evaluate seed vigor and seed aging degree.

Keywords: maize kernel; hyperspectral imaging technology; accelerated aging; principal component analysis; support vector machine model; standard germination tests

---

1. Introduction

Seeds enter an aging process after natural maturity. During this process, the vitality of the seed gradually decreases, which is a common phenomenon during the seed storage period. Seed vigor is an important indicator synthesizing seed germination, seedling rate, seedling growth potential, plant stress resistance and production potential [1,2]. For farmers, seeds with low viability will have low germination rates, which will increase their costs. Compared with seeds with low viability, seeds with high vigor which can save time, labor and material resources have obvious advantages [3]. Thus, an appropriate seed vigor detection method can help farmers engage in agricultural production activities in a better way. For seed companies, the seeds should be dried, processed and stored after harvest. If certain conditions are not suitable for seeds during these processes, it is possible to cause damage to the seeds, therefore, a rapid, non-destructive and high-accuracy method for seed vigor detection is of great help to them too.

The aging process of maize kernels can be influenced by maize varieties and environment factors such as temperature and humidity [4]. Generally, the natural aging of seeds is a long-duration procedure, which increases the cost of sampling for research purposes. In order to facilitate the research process, artificial accelerated aging tests are applied as a common method to simulate the seed aging procedure in a short time compared with natural aging. Studies have shown that artificial aging tests are an effective method to study seed vigor instead of natural aging. Han et al. identified quantitative trait loci (QTLs) for four maize seed vigor-related traits under artificial aging treatment [5]. Gelmond et al. applied accelerated aging to obtain six different levels of vigor of sorghum seeds from an identical lot [6]. Souza et al. also adopted an accelerated aging test during their study of the physiological quality of quinoa seeds under different storage conditions [7].

Most of the current research methods for seed aging determination are traditional physical and chemical detection methods. Mcdonough et al. studied the effects of accelerated aging on the vigor of maize, sorghum and sorghum powder. They detected both physical and chemical attributes that reveal the vitality of seeds. The density of maize and sorghum physical attribute was tested using a gas comparison pycnometer and tangential abrasive huller. The chemical attribute content of soluble protein in aged maize and sorghum was detected by gel chromatography with reagents [8]. Among all the seed vigor test methods, the standard germination test is the most widely used method for seed vigor detection, but it needs a complete sprouting procedure with the manual measurement of shoot length, root length and germination, which will take a long time. The disadvantages of traditional physical and chemical methods lies in that they are destructive, inefficient, time-consuming and usually involve complex operating procedures, thus a rapid, non-destructive method is needed for seed vigor detection.

Hyperspectral imaging technology is a new non-destructive test method which combines imaging information and spectral information [9–13]. Hyperspectral imaging can obtain the chemical information of heterogeneous samples and the spatial distribution of chemical components [14–20]. The hyperspectral imaging can be used to study the quality of seeds. Wei et al. used a visible/near-infrared hyperspectral imaging technique to detect the spatial distribution of aflatoxin B1 in kernels [21]. Wang et al. used hyperspectral imaging to predict the texture of maize seeds after different storage periods. The established quadrature signal correction-continuous new algorithm-piece partial least squares regression model (OSC-SPA-PLSR) had good prediction results of corn hardness and elasticity [22]. Williams et al. used near infrared (NIR) hyperspectral imaging to distinguish hard, intermediate and soft maize kernels from inbred lines. They used a Spectral Dimensions MatrixNIR camera and a short wave infrared (SWIR) hyperspectral imaging system to acquire the images of whole maize kernels. The authors used principal component analysis (PCA) to remove background, bad pixels, shading and found histological classes including glassy (hard) and floury (soft) endosperm on the cleaned images. They used PCA to discriminate endosperm from different kinds of maize kernels. Then PLS-DA was applied in classifying two kinds of maize. The result verified the effectiveness of the proposed method [10].

Hyperspectral imaging technology can also be used to detect the changes in seeds which underwent artificial accelerated aging test. Mcgoverin et al. investigated the viability of barley, wheat and sorghum grains using NIR hyperspectral imaging [11]. Nansen et al. adopted hyperspectral imaging to detect the germination rate of two native Australian tree species. During the process, hyperspectral images were acquired of individual seeds after 0, 1, 2, 5, 10, 20, 30 and 50 days of standard accelerated aging, and they found the loss of germination was associated with a significant change in seed coat spectral reflectance profiles [12]. Kandpal et al. predicted the viability of muskmelon seeds using NIR hyperspectral imaging system. After image collection, all seeds underwent a germination test to confirm their viability and vigor. The muskmelon seeds used in the study were vacuum-packed in plastic bags and stored in 45 °C hot water to age for 2, 4 and 6 days, while another set of seeds did not undergo artificial aging and were kept as the control (0 h). They found the spectral reflectance intensity decreases when there was an increment of seed viability, and this could reveal the changes in the chemical components in the seed as the artificial aging time increasing [13].

The main objective of this study was to explore the feasibility of using hyperspectral imaging to identify maize kernels vigor undergoing different accelerated aging time. The specific objectives were to: (1) conduct qualitative analysis of differences among maize kernels under different aging time by PCA; (2) build classification models and select optimal wavelengths to identify maize kernels undergoing different accelerated aging time; and (3) validate the results of hyperspectral imaging by standard germination tests.

2. Results and Discussion

2.1. Spectral Profile

The average spectral reflectance curves of maize kernels of two varieties at eight different aging times are shown in Figure 1. Similarity was observed in the change trends of the spectral reflectance curves of maize kernels which belonged to same variety but underwent different aging processes. The change trends of the reflectance curves of two varieties of maize showed clear similarities. Reflectance curves of maize kernels had differences in the reflectance of broad wavebands, so it was difficult to identify optimal wavelengths to discriminate maize kernels processed for different aging times.

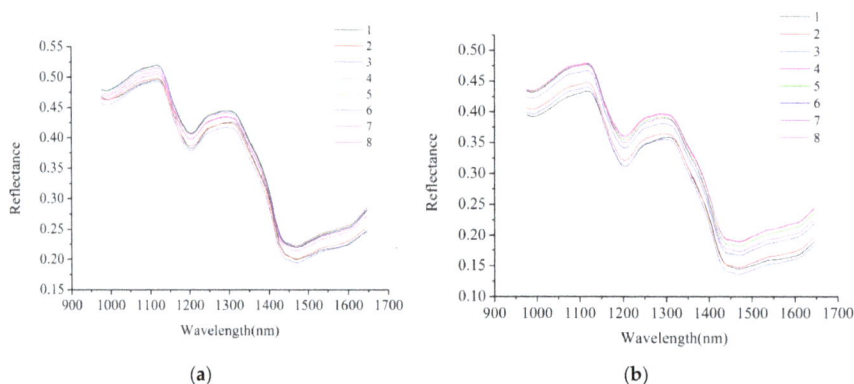

(a) (b)

Figure 1. Average spectra of unprocessed spectra: (a) Maize 1; (b) Maize 2. Average spectra of maize kernels under different aging duration time differs in reflectance value.

The average spectra of two varieties of maize preprocessed by the second-order derivative with three smoothing points are shown in Figure 2. The second-order derivative spectra in Figure 2 show the main changes in the spectral reflectance among maize kernels for eight aging durations. The wavelengths with obvious difference in reflectance data were manually selected as the optimal wavelengths by comparing maize kernels exposed to the eight different aging treatments. The wavelengths showing obvious differences could be easily identified in Figure 2, and the interferences of unimportant wavelengths were greatly suppressed. The second-order derivative spectra showed more obvious differences of maize kernels among different aging time than unpreprocessed spectra. Moreover, it could be found in Figure 2 that the second-order derivative spectra of maize kernels under different aging time between two varieties of maize showed similar trends in their spectral curves.

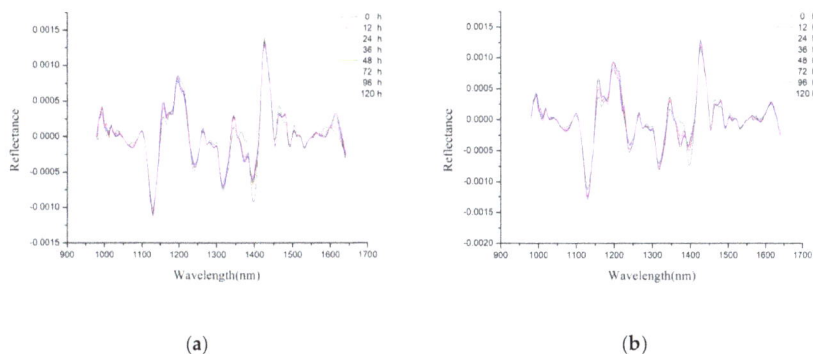

(a)   (b)

Figure 2. Average spectra preprocessed by second-order derivative: (**a**) Maize 1; (**b**) Maize 2. Spectral differences of maize kernels under different aging duration time at certain wavelengths could be observed.

2.2. PCA Analysis

2.2.1. Pixel-Wise PCA Scores Visualization

One hyperspectral image under each aging time of each variety was randomly selected to conduct PCA analysis. The PCA score images of PC1, PC2 and PC3 of Maize 1 and Maize 2 were shown in Figures S1 and S2 of the Supplementary Materials, respectively. Using the original images as references, it can be seen in Figures S1 and S2 that warm colours (yellow-red) were related to soft endosperm, while cold colours (green-blue) were associated with hard endosperm. The compositional structure of unprocessed maize kernels was shown more clearly in score images, while the structure outline inside maize kernels after accelerated aging treatment were fuzzier. That might be because the accelerated aging treatment altered the physical and chemical attributes of material inside the seeds, causing the hardness to change to varying degrees in different parts of maize kernels.

2.2.2. Object-Wise PCA Scores Scatter Plots Analysis

PCA analysis was conducted on average spectra of maize kernels to explore the scores scatter of different PCs. The first three PCs for each kind of maize were used to conduct qualitative analysis because the first three PCs contained the most of information of maize kernel, with 99.98% explained variance for Maize 1 (93.75% for PC1, 6.04% for PC2 and 0.14% for PC3) and 99.84% for Maize 2 (94.98% for PC1, 4.67% for PC2 and 0.14% for PC3). According to Figure 3, maize kernels after different aging processing treatments were grouped together depending on different features of their own spectral characteristics though there were some overlapping among the eight clusters. Maize kernels without aging processing were partly separated on Figure 3b–e, which revealed that the differences among maize kernels without aging treatment and maize kernels under seven different aging treatments had more differences in hyperspectral imaging. In order to obtain satisfactory classification results, further processing should be conducted.

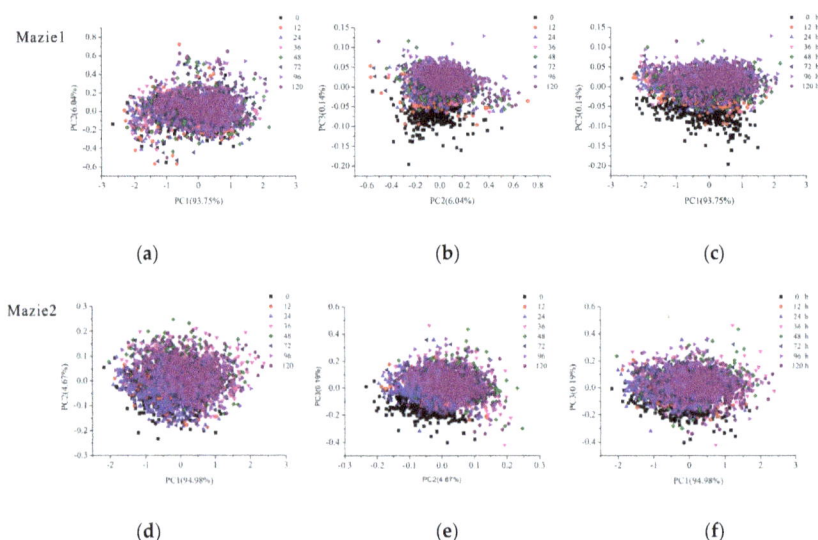

Figure 3. PCA scores scatter plots of maize kernels under different aging duration time: (**a**) PC1 versus PC2 for Maize 1; (**b**) PC2 versus PC3 for Maize 1; (**c**) PC1 versus PC3 for Maize 1; (**d**) PC1 versus PC2 for Maize 2; (**e**) PC2 versus PC3 for Maize 2; (**f**) PC1 versus PC3 for Maize 2. Clusters show the differences of maize kernels under different aging duration time.

2.3. Classification Models Based on Full Spectra

PCA analysis indicated that differences existed among maize kernels after different aging times. SVM models were built to evaluate the differences among maize kernels exposed to different aging times. To build SVM models, the maize kernels of each category were randomly split into a calibration set and a prediction set at the ratio of 2:1 (400 maize kernels in the calibration set and 200 maize kernels in the prediction set).

The overall classification results of Maize 1 and Maize 2 are shown in Table 1. A SVM model using the spectra of the combination of Maize 1 and Maize 2 was also built, and the overall classification results also presented in Table 1. As shown in Table 1, the overall classification accuracy of the calibration sets for Maize 1 and Maize 2 was approximately 80%, but the prediction accuracy of Maize 1 was a little higher than that of Maize 2, with Maize 1 reaching 70% and Maize 2 only 60%. The SVM model using the combined dataset showed close classification results to the SVM models using Maize 1 and Maize 2.

Table 1. The classification accuracy of SVM models using full spectra.

Sample Variety	C [1]	G [2]	Cal. [3] (%)	Pre. [4] (%)	Cv. [5]
Maize 1	256.00	1.74	81.53	68.15	58.13
Maize 2	256.00	3.03	78.47	60.16	63.84
Maize Mixed	256.00	5.28	73.43	59.90	57.23

[1] The regularization parameter of SVM; [2] The kernel function parameter of SVM; [3] Calibration set; [4] Prediction set; [5] Five-fold cross-validation.

To explore the details of the classification results of maize kernels under different aging times, Table 2 shows the confusion matrix of maize kernels of each category obtained according to the results of the SVM models using the full spectra. From Table 2, it could be seen that for Maize 1, Maize 2 and the combination of Maize 1 and Maize 2, maize kernels could be divided into three groups. The first group (Group 1) contained the maize kernels aged for 0 h, with nearly no misclassification with other categories. Because maize kernels aged for 12 and 24 h were more likely to be misclassified with each other, these two categories were sorted as Group 2, but the categories from Group 2, still had a high classification accuracy in both the calibration set and prediction set. The third group (Group 3) contained the maize kernels aged for 36, 48, 72, 96, 120 h. The maize kernels aged for each duration time in Group 3 gave lower classification accuracies, and they were grouped together because they were misclassified with each other more often.

Table 2. Confusion matrix of SVM models using full spectra.

Sample Variety		Sample Number	Pre.								Accuracy (%)
			1 [1]	2	3	4	5	6	7	8	
Maize 1	Cal.	1 (400)	400	0	0	0	0	0	0	0	100.00
		2 (400)	0	384	16	0	0	0	0	0	96.00
		3 (400)	0	14	356	26	3	0	1	0	89.00
		4 (400)	0	3	34	306	40	11	5	1	76.50
		5 (400)	0	0	16	74	228	68	2	12	57.00
		6 (400)	0	0	4	25	76	261	8	26	65.30
		7 (400)	0	0	0	1	5	18	327	49	81.80
		8 (400)	0	0	0	5	9	20	19	347	86.80
	Pre.	1 (200)	199	1	0	0	0	0	0	0	99.50
		2 (200)	1	150	47	0	0	2	0	0	75.00
		3 (200)	0	19	158	17	6	0	0	0	79.00
		4 (200)	0	4	26	114	39	16	0	1	57.00
		5 (200)	0	0	2	11	92	87	0	8	46.00
		6 (199)	0	1	2	22	66	100	3	5	50.30
		7 (200)	0	0	0	1	10	12	117	60	58.50
		8 (199)	0	0	0	1	6	17	16	159	79.90
Maize 2	Cal.	1 (400)	400	0	0	0	0	0	0	0	100.00
		2 (400)	0	374	24	0	0	0	2	0	93.50
		3 (400)	0	16	384	0	0	0	0	0	96.00
		4 (400)	0	0	0	279	37	27	10	47	69.80
		5 (400)	0	0	0	38	322	6	0	34	80.50
		6 (400)	0	1	0	30	2	256	95	16	64.00
		7 (400)	0	1	0	17	1	105	259	17	64.80
		8 (400)	0	1	0	79	53	22	8	237	59.30
	Pre.	1 (200)	196	0	3	0	0	0	0	1	98.00
		2 (200)	1	156	36	0	0	1	6	0	78.00
		3 (199)	1	21	177	0	0	0	0	0	88.90
		4 (200)	0	2	0	90	36	23	10	39	45.00
		5 (200)	1	0	0	47	109	4	2	37	54.50
		6 (200)	0	3	0	12	3	80	92	10	40.00
		7 (200)	0	6	1	19	0	71	93	10	46.50
		8 (200)	0	1	0	71	41	16	10	61	30.50
Maize mixed	Cal.	1 (800)	800	0	0	0	0	0	0	0	100.00
		2 (800)	0	654	134	0	0	8	0	4	81.80
		3 (800)	0	123	657	17	1	2	0	0	82.10
		4 (800)	0	5	30	499	126	35	40	65	62.40
		5 (800)	0	4	19	156	422	129	22	48	52.80
		6 (800)	0	7	4	52	70	480	110	77	60.00
		7 (800)	0	0	0	57	51	125	409	158	51.00
		8 (800)	0	2	1	40	64	75	87	531	66.40
	Pre.	1 (400)	394	2	3	0	0	1	0	0	98.50
		2 (400)	4	246	130	0	2	11	3	4	61.50
		3 (399)	2	94	287	9	1	4	0	2	72.20
		4 (400)	0	9	16	205	90	26	25	29	51.30
		5 (400)	0	3	3	55	136	130	20	53	34.00
		6 (399)	0	3	5	36	64	187	74	30	46.90
		7 (400)	0	3	4	36	59	167	77	54	19.30
		8 (399)	0	3	0	40	59	51	58	188	47.10

[1] 1, 2, 3, 4, 5, 6, 7 and 8 are assigned respectively as the category value of the maize kernels processed under different aging duration (12, 24, 36, 48, 72, 96 and 120 h).

From Table 2, although maize kernels within one group would be misclassified with each other, maize kernels were not easily misclassified with categories from the other groups. Table 3 showed the classification accuracy of SVM models of three groups (Group 1 (0 h), Group 2 (12 and 24 h) and Group 3 (36, 48, 72, 96, 120 h)) using full spectra. All accuracies were above 90%.

Table 3. The classification accuracy of SVM models of three groups using full spectra.

Sample Variety		Sample Number	Pre.			Accuracy (%)
			Group 1	Group 2	Group 3	
Maize 1	Cal.	Group 1 (400)	400	0	0	100.00
		Group 2 (800)	0	770	30	96.25
		Group 3 (2000)	0	57	1943	97.15
	Pre.	Group 1 (200)	199	1	0	99.50
		Group 2 (400)	1	374	25	93.50
		Group 3 (998)	0	35	963	96.49
Maize 2	Cal.	Group 1 (400)	400	0	0	100.00
		Group 2 (800)	0	798	2	99.75
		Group 3 (2000)	0	3	1997	99.85
	Pre.	Group 1 (200)	196	3	1	98.00
		Group 2 (399)	2	390	7	97.74
		Group 3 (1000)	1	13	986	98.60
Maize mixed	Cal.	Group 1 (800)	800	0	0	100.00
		Group 2 (1600)	0	1568	32	98.00
		Group 3 (4000)	0	72	3928	98.20
	Pre.	Group 1 (400)	394	5	1	98.50
		Group 2 (799)	6	757	36	94.74
		Group 3 (1998)	0	49	1949	97.55

As shown in Tables 1 and 2 the SVM model using the combined dataset showed close classification results to the SVM models using Maize 1 and Maize 2, and the general trend of classification accuracy of each aging duration time of the combined dataset was similar to Maize 1 and Maize 2. The results indicated that it would be possible to build a non-variety specific classification model for maize kernel vigor detection.

2.4. Optimal Wavelengths Selection

In this study, the second-order derivative was adopted to select the optimal wavelengths. As shown in Figure 2, the wavelengths with larger differences among maize kernels aged for different duration were highlighted in the spectra. The peaks and valleys with larger differences were selected as the optimal wavelengths to identify maize kernels at different aging times. The selected optimal wavelengths for Maize 1 and Maize 2 are shown in Table 4, and 19 and 18 optimal wavelengths were obtained finally to reduce the data volume. The optimal wavelengths near 995 nm were related to the second vibration of N–H bonds in proteins or amino acids [23]. The attributes of the secondary stretching vibration of C–H bonds in starch, proteins or lipids were revealed at optimal wavelengths near 1200 nm [24]. The spectral bands near 1463 nm were concerned with the absorption region of water [25]. From Table 4, the differences of the optimal wavelengths selected for Maize1 and Maize2 might be related to the genotypic differences for two varieties of maize.

Table 4. Corresponding optimal wavelengths selected by second-order derivative spectra.

Sample Variety	No.	Optimal Wavelengths (nm)
Maize 1	19	995, 1005, 1035, 1076, 1130, 1156, 1167, 1207, 1241, 1264, 1321, 1375, 1399, 1426, 1463, 1480, 1504, 1585, 1615
Maize 2	18	1005, 1072, 1130, 1156, 1160, 1167, 1197, 1241, 1264, 1318, 1345, 1372, 1396, 1426, 1453, 1463, 1480, 1612

2.5. Classification Models on the Optimal Wavelengths

The overall results of SVM models of Maize 1 and Maize 2 using the optimal wavelengths selected by second-order derivative spectra are shown in Table 5. For Maize 1, the classification accuracy of the calibration set was over 70%, and the classification accuracy of the prediction set was about 60%. For Maize 2, the classification accuracy of the calibration set reached 71%, and the classification accuracy of the prediction set was 62%. A slight decrease could be found between the overall results of the calibration set of SVM models using full spectra and optimal wavelengths, and the results of the prediction set were quite close to each other. The number of wavelengths reduced to 18 and 19 by optimal wavelengths selection, resulting in reduction of spectral data volume to 91% and 90.5%. During this process, some useful information was lost, leading to the accuracy reduction in SVM models based on optimal wavelengths. In the case of small difference between the classification results based on full spectra and optimal wavelengths, it was still meaningful to adopt the classification model based on optimal wavelengths.

To explore the classification results of maize kernels at different aging time using optimal wavelengths, the confusion matrix of maize kernels of each category obtained by SVM models using optimal wavelengths of Maize 1 and Maize 2 were shown in Table 6. Classification results of Maize 1 and Maize 2 using optimal wavelengths could also be divided into the three same groups as the SVM models using full spectra. Group 1 contained maize kernels under aging time of 0 h, Group 2 contained maize kernels under aging time of 12 and 24 h, and Group 3 contained maize kernels under the aging time of 36, 48, 72, 96 and 120 h. Maize kernels aged for different duration time within one group were more likely to be misclassified with each other. And maize kernels within different groups had fewer misclassified samples. Table 7 showed the classification accuracy of SVM models of three groups (Group 1 (0 h), Group 2 (12 and 24 h) and Group 3 (36, 48, 72, 96, 120 h)) using optimal wavelengths selected by second-order derivative spectra. Although the accuracies were slightly lower than models based on full spectra, the accuracies were still above 90%.

Table 5. The classification accuracy of SVM models using the optimal wavelengths selected by second-order derivative spectra.

Sample Variety	c	g	Cal. (%)	Pre. (%)	Cv.
Maize 1	256.00	27.86	70.47	57.45	71.31
Maize 2	256.00	16.00	71.66	62.48	63.81

Table 6. Confusion matrix of SVM models using optimal wavelengths selected by second-order derivative spectra.

Sample Variety		Sample Number	Prediction Value								Accuracy (%)
			1	2	3	4	5	6	7	8	
Maize 1	Cal.	1 (400)	400	0	0	0	0	0	0	0	100.00
		2 (400)	0	372	27	1	0	0	0	0	93.00
		3 (400)	0	33	314	14	23	12	1	3	78.50
		4 (400)	0	5	27	190	57	68	11	42	47.50
		5 (400)	0	3	43	54	196	66	3	35	49.00
		6 (400)	0	0	17	92	82	170	3	36	42.50
		7 (400)	0	0	0	19	7	13	319	42	79.80
		8 (400)	0	1	5	31	29	26	14	294	73.50
	Pre.	1 (200)	199	1	0	0	0	0	0	0	99.50
		2 (200)	2	161	36	0	0	1	0	0	80.50
		3 (200)	0	36	131	6	14	6	0	7	65.50
		4 (200)	0	4	31	66	41	47	1	10	33.00
		5 (200)	0	2	25	37	77	37	2	20	38.50
		6 (199)	0	2	13	62	53	43	4	22	21.60
		7 (200)	0	0	0	11	10	10	121	48	60.50
		8 (199)	0	0	0	21	22	16	20	120	60.30

Table 6. *Cont.*

Sample Variety		Sample Number	Prediction Value								Accuracy (%)
			1	2	3	4	5	6	7	8	
Maize 2	Cal.	1 (400)	400	0	0	0	0	0	0	0	100.00
		2 (400)	0	365	33	0	0	2	0	0	91.30
		3 (400)	0	25	375	0	0	0	0	0	93.80
		4 (400)	0	0	0	246	54	29	10	61	61.50
		5 (400)	0	0	0	57	295	6	0	42	73.80
		6 (400)	0	5	0	38	1	230	113	13	57.50
		7 (400)	0	2	0	21	1	165	196	15	49.00
		8 (400)	0	0	0	117	63	27	7	186	46.50
	Pre.	1 (200)	196	0	3	0	0	0	0	1	98.00
		2 (200)	1	164	30	0	0	2	3	0	82.00
		3 (199)	2	15	182	0	0	0	0	0	91.50
		4 (200)	0	0	0	86	35	11	11	57	43.00
		5 (200)	1	0	0	36	126	3	2	32	63.00
		6 (200)	0	2	0	16	2	93	84	3	46.50
		7 (200)	0	2	0	20	0	75	94	9	47.00
		8 (200)	0	0	0	78	46	12	6	58	29.00

Table 7. The classification accuracy of SVM models of three groups using optimal wavelengths selected by second-order derivative spectra.

Sample Variety		Sample Number	Pre.			Accuracy (%)
			Group 1	Group 2	Group 3	
Maize 1	Cal.	Group 1 (400)	400	0	0	100.00
		Group 2 (800)	0	746	54	93.25
		Group 3 (2000)	0	101	1899	94.95
	Pre.	Group 1 (200)	199	1	0	99.50
		Group 2 (400)	2	364	34	91.00
		Group 3 (998)	0	77	921	92.28
Maize 2	Cal.	Group 1 (400)	400	0	0	100.00
		Group 2 (800)	0	798	2	99.75
		Group 3 (2000)	0	7	1993	99.65
	Pre.	Group 1 (200)	196	3	1	98.00
		Group 2 (399)	3	391	5	97.99
		Group 3 (1000)	1	4	995	99.50

2.6. Germination Tests Analysis

A germination test was carried out to validate the accuracy of hyperspectral imaging in this study. Table 8 showed the germination rate, shoot length and root length of Maize 1 and Maize 2 at different aging duration. It could be seen that Maize 1 and Maize 2 at aging duration time from 0 to 24 h had significant differences in shoot length and root length, but there were no significant difference in germination rate among these three categories. Because the germination rate was calculated by seeds with at least 1 cm germ after 10 days, accelerated aging treatment for a short time (12 h and 24 h) may not affect the germination ability for maize kernels obviously, but the significant differences of root length and shoot length could reveal the vigor differences of maize kernels from these three categories, which consisted with the high classification accuracies of SVM models of Group 1 (0 h) and Group 2 (12 and 24 h). It also could be found in Table 8 that there were small significant differences among maize kernels under aging duration of 36, 48, 72, 96 and 120 h, which consisted with the low accuracies of SVM models of Group 3. The overall results indicated that hyperspectral imaging could be used to detect seed vigor under different aging duration time.

Table 8. Germination rate, shoot and root length of Maize 1 and Maize 2 under different accelerated aging time.

Sample Variety	Accelerating Aging Time (hrs)	Germination Rate (%)	Shoot Length (cm/seedling)	Root Length (cm/seedling)
Maize 1	0	92.00a	11.30a	23.15a
	12	90.67a	12.26b	21.42b
	24	86.00a	9.77c	17.31cd
	36	75.33b	6.35d	18.20c
	48	73.67b	8.95c	16.68d
	72	76.33b	6.17d	13.76e
	96	59.00c	5.60d	12.80e
	120	57.00c	5.99d	12.69e
Maize 2	0	96.33a	13.06a	24.68a
	12	97.67a	10.64b	24.11a
	24	93.00a	10.09bc	19.25b
	36	82.67b	8.78cd	17.08c
	48	79.00bc	6.98e	18.46b
	72	75.33c	7.27de	12.80d
	96	62.00d	5.93e	14.63e
	120	63.67d	6.73e	12.13e

The letters (a, b, c, d, e) in each column indicate the significance of difference among maize kernel processed by different duration of aging time at the confidence level of 5% (Duncan's). Within a column, data followed by different letters are significantly different.

3. Materials and Methods

3.1. Sample Preparation

Two varieties of maize kernels cultivated by a commercial seed company (Jiudingjiusheng Seed Industrial Co., Ltd., Beijing, China) with breed numbers of 106101 and 7879 (in this work the names Maize 1 and Maize 2 were used to refer to maize varieties 106101 and 7879, respectively) instead of their original chemical names, complying with the company rules. The two varieties of maize were sown and harvested in the same experimental field simultaneously in 2016. For each variety, 4800 maize kernels were prepared for artificial accelerated aging. Before accelerating aging treatment, maize kernels were disinfected with 1% hypochlorous acid (HClO) solution for 20 min and then the maize kernels were naturally dried after being rinsed with distilled water. The 4800 maize kernels of each variety were randomly divided into eight categories (600 kernels in each category). One category was selected as control group (0 h) placed at room temperature (20 °C, 60% relative humidity) and the other 7 categorizes were used to conduct aging process under different aging time (12, 24, 36, 48, 72, 96 and 120 h). Then the maize kernels were aged in LH-150S artificial aging box (Ansheng Instrument Ltd., Zhengzhou, Henan, China) with temperature of 45 °C and relative humidity of 99%. After accelerated aging treatment, maize kernels were disinfected, rinsed with distilled water, naturally air-dried, and stored in Kraft paper bags. After the acquisition of hyperspectral images, maize kernels of each category were divided into 30 samples (20 kernels in each sample) for standard germination analysis.

3.2. Hyperspectral Imaging System

The experiment was carried out using a hyperspectral imaging system with the spectral range of 874–1734 nm, the spectral resolution of 5 nm and the spatial resolution of 320 × 256 pixels. The system consisted of an ImSpector N17E imaging spectrograph (Spectral Imaging Ltd., Oulu, Finland), a Xeva 992 camera (Xenics Infrared Solutions, Leuven, Belgium) equipped with an OLES22 lens (Spectral Imaging Ltd., Oulu, Finland), two 150 W tungsten halogen lamps (2900 Lightsource, Illumination Technologies Inc., Elbridge, NY, USA) that were symmetrically placed and served as the light source and a conveyer belt (Isuzu Optics Corp., Taiwan, China). The imaging system was controlled by the software (Xenics N17E, Isuzu Optics Corp.), which can be used to calibrated and analyze the images as well.

3.3. Hyperspectral Image Acquisition and Calibration

The maize kernels were placed on a black plate with a very low reflectivity, so it is easy to isolate maize kernels from the background. During the experiment, the exposure time of the camera was 3500 μs. The distance between the lens and the plate was adjusted to 17.9 cm, and the moving speed of the conveyer belt was set to 13.8 mm/s. The above adjustments were aimed at obtaining a clear image without distortion.

Two reference standards were used to calibrate the raw images (I_{raw}). The dark reference image (I_{dark}) was acquired by covering the lens with lens cap whose reflectivity is about 0%. The white reference image (I_{white}) was collected from a piece of pure white Teflon board whose reflectivity is about 100%. The calibrated image (I_c) could be calculated as Equation (1):

$$I_c = \frac{I_{raw} - I_{dark}}{I_{white} - I_{dark}} \tag{1}$$

3.4. Spectral Reflectance Extraction and Preprocessing

After image calibration, the spectral reflectance of each maize kernel was extracted from the hyperspectral images. Hyperspectral imaging provides spectral reflectance data and grayscale images at each wavelength. Prior to image processing, the maize kernels were separated from the background by using a mask 14, 26, 27. In this study, the mask was built by conducting image binaryzation on the gray-scale image at 1116 nm to set maize kernel area as 1 and the background as 0. The maize kernels were then isolated from background by applying the mask to the gray-scale images at each wavelength. Then, calibrated hyperspectral images were pre-processed to minimize noise 11, 28, 29. The original pixel-wise spectra were denoised by the wavelet transformation with decomposition level 3 using Daubechies 8 (db8) as the wavelet basis function. Then, the pixel-wise spectra of all pixels within a maize kernel were averaged as one spectrum.

3.5. Standard Germination Tests

Standard germination tests for Maize 1 and Maize 2 were conducted on ten kernels of each sample after acquiring hyperspectral images. For each sample, 10 maize kernels were picked randomly for germination tests. To obtain the vigor of maize kernels, the standard germination tests were performed according to the guidelines of the International Seed Testing Association (ISTA)30. Maize kernels were placed in round holes of sponges, and sponges were placed in seedling basins with enough water. Then all the seedling basins were stored in germination cabinet at 25 °C with 99% relative humidity for 10 days. According to ISTA standards, seeds with 1 cm germ after germination were considered to be seeds with viability. After germination, the germination percentage, shoot length and root length were calculated and measured manually.

3.6. Data analysis Methods

3.6.1. Principal Component Analysis

Principal component analysis (PCA) is a multivariate statistical method that studies the correlation between multiple variables. It examines how a few principal components can be used to reveal the internal relationship between multiple variables. PCA derives principal components from the original variables, and the first few principal components (PCs) contained most of the useful information. The PCs are linear combinations of original variables, and they are orthotropic and irrelevant to each other. The scores of the first few PCs can be used to explore the differences between samples [24–26].

For hyperspectral images, there are two approaches to conduct PCA analysis, pixel-wise analysis and objective-wise analysis [15]. Pixel-wise analysis is to form PCA scores visualization image. For this method, PCA is calculated on individual pixels of the images. Scores of each pixel within the hyperspectral image of each PC can be obtained to form a scores visualization image. The differences between samples can be visually observed and explored in colormaps for each PC.

Object-wise analysis is to form PCA scores scatter plots. For this method, average spectra of the depicted objects are used instead of individual pixels. The scores of different PCs of samples are scattered in a two-dimensional space or a three-dimensional space. The differences among samples can be explored more clearly in these spaces [15,27].

3.6.2. Optimal Wavelength Selection

The spectral data obtained by hyperspectral images usually have a large data volume and contain a lot of useless information like redundant and collinear information. The existence of useless information will increase the data processing burden, which is likely to casue instability of the model and thus result in a poor performance. Meanwhile, processing of a large amount of data places a high burden on computer hardware and increases the calculation time. Thus, it is necessary to select optimal wavelengths to reduce the inputs, which can simplify the model and improve the model performance. The second derivative is an efficient preprocessing method in spectral data analysis. It can eliminate the interference of other backgrounds, improve the spectral resolution and highlight useful information in the spectra. Differences in peaks and valleys of spectra preprocessed by second-order derivative indicate the physical and chemical changes of the samples, which has been used as an efficient method to identify optimal wavelengths [28,29]. Peaks and valleys with large differences in second-order derivative spectra can be selected as the optimal wavelengths.

3.6.3. Discriminant Model

Support vector machine (SVM) is a supervised machine learning model used for classification and regression. The main idea of SVM is to create a hyperplane as a decision surface, which maximizes the margins of isolation between different samples. SVM could deal with both linear and nonlinear data efficiently with its good generalization ability. Kernel function is important for conducting SVM, and radial basis function (RBF) is a widely used kernel function. The parameters for SVM models should be determined, including the regularization parameter c and kernel function parameter g. The former determines the tradeoff between minimizing the training error and minimizing model complexity, and the latter defines the non-linear mapping from input space to some high dimensional feature space. The search range for c and g ranged from 2^{-8} to 2^{8} in this study. The optimal combination of c and g was determined by the SVM model with the highest classification accuracy. Grid-search was applied to optimize the two parameters for SVM in this study [30–32].

3.6.4. Significance Test

Duncan's multiple range tests were applied to calculate for comparison of maize vitality index (germination, shoot and root length) at different accelerated aging duration time at a significance level of 0.05 [33].

4. Conclusions

Hyperspectral imaging technology combined with SVM models was used to identify the vigor of maize kernels after different aging times. The results of SVM models based on optimal wavelengths were about 10% lower than that of models based on full spectra. However, it was meaningful to conduct optimal wavelengths selection because of the obvious improvement in modeling speed. Confusion matrixes for maize kernels of each category were built for both SVM models using full spectra and optimal wavelengths to reveal the detail of classification results of maize kernels processed under different aging duration. From the results of confusion matrixes, 8 categories of maize kernels could be divided into three groups. Group 1 contained unprocessed maize kernels, Group 2 contained maize kernels aged for 12 and 24 h and Group 3 contained maize kernels with longer aging times (36, 48, 72, 96, 120 h). Maize kernels aged for different durations within one group were more likely to be misclassified with each other. Maize kernels within different groups had fewer misclassified samples. To verify the results of SVM models, traditional seed vigor testing method, standard germination test was conducted. The results of standard germination tests were generally consistent with those of SVM models. Maize kernels belonging to Group 1 (0 h) and Group 2 (12 h and 24 h) had significant differences for root length and shoot length. Maize kernels belonging to Group 3 (36 h, 48 h, 72 h, 96 h and 120 h) had no significant differences with each other comprehensively considering germination rate, root length and shoot length.

The results of this research demonstrate that it is feasible to detect maize kernel vigor with a hyperspectral imaging system combined with SVM models and the second-order derivative spectra could be used to select optimal wavelengths which do great help in shortening modeling time. Thus, as a rapid, non-destructive method, hyperspectral imaging system has great potential for application in seed vigor detection. To improve model performances, different varieties of maize kernels from different crop years, growth locations and storage conditions will be take into consideration to extend the database in the future researches. Variety specific models and non-variety specific models will also be explored for real-world application.

Supplementary Materials: The following are available online, Figure S1: Score images for the first three principal components of Maize 1: (**a**) Score image for PC1. (**b**) Score image for PC2. (**c**) Score image for PC3. The color bar indicates the score value of each pixel, differences of maize kernels under different accelerating aging duration time could be seen according to the score images. Warm color (positive score values) were related to soft endosperm, while cold color (negative score values) were associated with hard endosperm; Figure S2: Score images for the first three principal components of Maize 2: (**a**) Score image for PC1. (**b**) Score image for PC2. (**c**) Score image for PC3. The color bar indicates the score value of each pixel, differences of maize kernels under different accelerating aging duration time could be seen according to the score images. Warm color (positive score values) were related to soft endosperm, while cold color (negative score values) were associated with hard endosperm.

Author Contributions: Conceptualization, L.F.; Data curation, C.Z.; Formal analysis, S.Z.; Funding acquisition, L.F., Y.B. and Y.H.; Investigation, Y.B. and Y.H.; Methodology, L.F. and S.Z.; Project administration, L.F. and Y.H.; Resources, C.Z. and X.F.; Software, C.Z. and X.F.; Supervision, Y.H.; Visualization, Y.B.; Writing—original draft, L.F. and S.Z.; Writing—review & editing, S.Z. and Y.H.

Funding: This research was funded by National key R & D program of China, grant number 2016YFD0300606; Zhejiang Province Public Technology Research Program, grant number 2015C02008; National Natural Science Foundation of China, grant number 31471417 and China Postdoctoral Science Foundation, grant number 2017M610370.

Conflicts of Interest: The authors declare no conflict of interest.

References

1. Xin, X.; Lin, X.H.; Zhou, Y.C.; Chen, X.L.; Liu, X.; Lu, X.X. Proteome analysis of maize seeds: The effect of artificial ageing. *Physiol. Plant.* **2011**, *143*, 126–138. [CrossRef] [PubMed]
2. Woltz, J.M.; Tekrony, D.M. Accelerated aging test for corn seed. *Seed Technol.* **2001**, *23*, 21–34.
3. Williams, P.; Manley, M.; Fox, G.; Geladi, P. Indirect detection of *Fusarium verticillioides* in maize (*Zea maize* L.) kernels by NIR hyperspectral imaging. *J. Near Infrared Spectrosc.* **2010**, *18*, 49–58. [CrossRef]
4. Bittencourt, S.R.M.D.; Grzybowski, C.R.D.S.; Panobianco, M.; Vieira, R.D. Alternative methodology for the accelerated aging test for corn seeds. *Ciênc. Rural* **2012**, *42*, 1360–1365. [CrossRef]
5. Han, Z.; Ku, L.; Zhang, Z.; Zhang, J.; Guo, S.; Liu, H.; Zhao, R.; Ren, Z.; Zhang, L.; Su, H. QTLs for seed vigor-related traits identified in maize seeds germinated under artificial aging conditions. *PLoS ONE* **2014**, *9*, 92535. [CrossRef] [PubMed]
6. Gelmond, H.; Luria, I.; Woodstock, L.W.; Perl, M. The effect of accelerated aging of sorghum seeds on seedling vigour. *J. Exp. Bot.* **1978**, *29*, 489–495. [CrossRef]
7. De Jesus Souza, F.I.F.; Devilla, I.A.; de Souza, R.T.G.; Teixeira, I.R.; Spehar, C.R. Physiological quality of quinoa seeds submitted to different storage conditions. *Afr. J. Agric. Res.* **2016**, *11*, 1299–1308. [CrossRef]

8. Mcdonough, C.M.; Floyd, C.D.; Waniska, R.D.; Rooney, L.W. Effect of accelerated aging on maize, sorghum, and sorghum meal. *J. Cereal Sci.* **2004**, *39*, 351–361. [CrossRef]

9. Ambrose, A.; Kandpal, L.M.; Kim, M.S.; Lee, W.H.; Cho, B.K. High speed measurement of corn seed viability using hyperspectral imaging. *Infrared Phys. Technol.* **2016**, *75*, 173–179. [CrossRef]

10. Williams, P.; Geladi, P.; Fox, G.; Manley, M. Maize kernel hardness classification by near infrared (NIR) hyperspectral imaging and multivariate data analysis. *Anal. Chim. Acta* **2009**, *653*, 121–130. [CrossRef] [PubMed]

11. Mcgoverin, C.M.; Engelbrecht, P.; Geladi, P.; Manley, M. Characterisation of non-viable whole barley, wheat and sorghum grains using near-infrared hyperspectral data and chemometrics. *Anal. Bioanalyt. Chem.* **2011**, *401*, 2283–2289. [CrossRef] [PubMed]

12. Nansen, C.; Zhao, G.; Dakin, N.; Zhao, C.; Turner, S.R. Using hyperspectral imaging to determine germination of native Australian plant seeds. *J. Photochem. Photobiol. B* **2015**, *145*, 19–24. [CrossRef] [PubMed]

13. Kandpal, L.M.; Lohumi, S.; Kim, M.S.; Kang, J.S.; Cho, B.K. Near-infrared hyperspectral imaging system coupled with multivariate methods to predict viability and vigor in muskmelon seeds. *Sens. Actuators B Chem.* **2016**, *229*, 534–544. [CrossRef]

14. Liu, C.; Wei, L.; Lu, X.; Wei, C.; Yang, J.; Lei, Z. Nondestructive determination of transgenic *Bacillus thuringiensis* rice seeds (*Oryza sativa* L.) using multispectral imaging and chemometric methods. *Food Chem.* **2014**, *153*, 87–93. [CrossRef] [PubMed]

15. Williams, P.J.; Kucheryavskiy, S. Classification of maize kernels using NIR hyperspectral imaging. *Food Chem.* **2016**, *209*, 131–138. [CrossRef] [PubMed]

16. Fiore, A.D.; Reverberi, M.; Ricelli, A.; Pinzari, F.; Serranti, S.; Fabbri, A.A.; Bonifazi, G.; Fanelli, C. Early detection of toxigenic fungi on maize by hyperspectral imaging analysis. *Int. J. Food Microbiol.* **2010**, *144*, 64–71. [CrossRef] [PubMed]

17. Ravikanth, L.; Singh, C.B.; Jayas, D.S.; White, N.D.G. Classification of contaminants from wheat using near-infrared hyperspectral imaging. *Biosyst. Eng.* **2015**, *135*, 73–86. [CrossRef]

18. Weinstock, B.A.; Janni, J.; Hagen, L.; Wright, S. Prediction of oil and oleic acid concentrations in individual corn (*Zea mays* L.) kernels using near-infrared reflectance hyperspectral imaging and multivariate analysis. *Appl. Spectrosc.* **2006**, *60*, 9. [CrossRef] [PubMed]

19. Lin, L.H.; Lu, F.M.; Chang, Y.C. Development of a near-infrared imaging system for determination of rice moisture. *Cereal Chem.* **2006**, *83*, 498–504. [CrossRef]

20. Caporaso, N.; Whitworth, M.B.; Fisk, I.D. Protein content prediction in single wheat kernels using hyperspectral imaging. *Food Chem.* **2017**, *240*, 32–42. [CrossRef] [PubMed]

21. Wei, W.; Heitschmidt, G.W.; Windham, W.R.; Peggy, F.; Xinzhi, N.; Xuan, C. Feasibility of detecting aflatoxin B1 on inoculated maize kernels surface using Vis/NIR hyperspectral imaging. *J. Food Sci.* **2014**, *80*. [CrossRef]

22. Wang, L.; Pu, H.; Sun, D.W.; Liu, D.; Wang, Q.; Xiong, Z. Application of hyperspectral imaging for prediction of textural properties of maize seeds with different storage periods. *Food Anal. Methods* **2015**, *8*, 1535–1545. [CrossRef]

23. Kafle, G.K.; Khot, L.R.; Jarolmasjed, S.; Si, Y.; Lewis, K. Robustness of near infrared spectroscopy based spectral features for non-destructive bitter pit detection in honeycrisp apples. *Postharvest Biol. Technol.* **2016**, *120*, 188–192. [CrossRef]

24. Kamruzzaman, M.; Elmasry, G.; Sun, D.W.; Allen, P. Application of NIR hyperspectral imaging for discrimination of lamb muscles. *J. Food Eng.* **2011**, *104*, 332–340. [CrossRef]

25. Qin, J.; Burks, T.F.; Kim, M.S.; Chao, K.; Ritenour, M.A. Detecting citrus canker by hyperspectral reflectance imaging and PCA-based image classification method. In Proceedings of the SPIE—The International Society for Optical Engineering, Orlando, FL, USA, 15 April 2008.

26. Liu, D.; Ma, J.; Sun, D.-W.; Pu, H.; Gao, W.; Qu, J.; Zeng, X.-A. Prediction of color and pH of salted porcine meats using visible and near-infrared hyperspectral imaging. *Food Bioprocess Technol.* **2014**, *7*, 3100–3108. [CrossRef]

27. Jiang, H.; Chen, Q. Development of electronic nose and near infrared spectroscopy analysis techniques to monitor the critical time in SSF process of feed protein. *Sensors* **2014**, *14*, 19441–19456. [CrossRef] [PubMed]

28. Kamruzzaman, M.; Barbin, D.; Elmasry, G.; Sun, D.W.; Allen, P. Potential of hyperspectral imaging and pattern recognition for categorization and authentication of red meat. *Innov. Food Sci. Emerg. Technol.* **2012**, *16*, 316–325. [CrossRef]

29. Zhang, C.; Feng, X.; Wang, J.; Liu, F.; He, Y.; Zhou, W. Mid-infrared spectroscopy combined with chemometrics to detect Sclerotinia stem rot on oilseed rape (*Brassica napus* L.) leaves. *Plant Methods* **2017**, *13*, 39. [CrossRef] [PubMed]

30. Devos, O.; Ruckebusch, C.; Durand, A.; Duponchel, L.; Huvenne, J.-P. Support vector machines (SVM) in near infrared (NIR) spectroscopy: Focus on parameters optimization and model interpretation. *Chemom. Intell. Lab. Syst.* **2009**, *96*, 27–33. [CrossRef]

31. Zhang, L.D.; Su, S.G.; Wang, L.S.; Li, J.H.; Yang, L.M. Study on application of fourier transformation near-infrared spectroscopy analysis with support vector machine (SVM). *Spectrosc. Spect. Anal.* **2005**, *25*, 33–35. [CrossRef]

32. Campsvalls, G.; Gómezchova, L.; Calpemaravilla, J.; Soriaolivas, E.; Martínguerrero, J.D.; Moreno, J. Support vector machines for crop classification using hyperspectral data. *Lecture Notes Comput. Sci.* **2003**, *2652*, 134–141.

33. Dai, Q.; Cheng, J.H.; Sun, D.W.; Pu, H.; Zeng, X.A.; Xiong, Z. Potential of visible/near-infrared hyperspectral imaging for rapid detection of freshness in unfrozen and frozen prawns. *J. Food Eng.* **2015**, *149*, 97–104. [CrossRef]

Sample Availability: Not available.

![molecules logo]

molecules

MDPI

Article

Variety Identification of Raisins Using Near-Infrared Hyperspectral Imaging

Lei Feng [1,2], Susu Zhu [1,2], Chu Zhang [1,2], Yidan Bao [1,2], Pan Gao [3,*] and Yong He [1,2,4,*]

1   College of Biosystems Engineering and Food Science, Zhejiang University, Hangzhou 310058, China;
    lfeng@zju.edu.cn (L.F.); sszhu@zju.edu.cn (S.Z.); chuzh@zju.edu.cn (C.Z.); ydbao@zju.edu.cn (Y.B.)
2   Key Laboratory of Spectroscopy Sensing, Ministry of Agriculture and Rural Affairs,
    Hangzhou 310058, China
3   College of Information Science and Technology, Shihezi University, Shihezi 832000, China
4   State Key Laboratory of Modern Optical Instrumentation, Zhejiang University, Hangzhou 310058, China
*   Correspondence: gp_inf@shzu.edu.cn (P.G.); yhe@zju.edu.cn (Y.H.); Tel.: +86-0993-205-7997 (P.G.);
    +86-571-8898-2143 (Y.H.)

Academic Editors: Christian Huck and Krzysztof B. Bec
Received: 12 October 2018; Accepted: 7 November 2018; Published: 8 November 2018

Abstract: Different varieties of raisins have different nutritional properties and vary in commercial value. An identification method of raisin varieties using hyperspectral imaging was explored. Hyperspectral images of two different varieties of raisins (Wuhebai and Xiangfei) at spectral range of 874–1734 nm were acquired, and each variety contained three grades. Pixel-wise spectra were extracted and preprocessed by wavelet transform and standard normal variate, and object-wise spectra (sample average spectra) were calculated. Principal component analysis (PCA) and independent component analysis (ICA) of object-wise spectra and pixel-wise spectra were conducted to select effective wavelengths. Pixel-wise PCA scores images indicated differences between two varieties and among different grades. SVM (Support Vector Machine), k-NN (k-nearest Neighbors Algorithm), and RBFNN (Radial Basis Function Neural Network) models were built to discriminate two varieties of raisins. Results indicated that both SVM and RBFNN models based on object-wise spectra using optimal wavelengths selected by PCA could be used for raisin variety identification. The visualization maps verified the effectiveness of using hyperspectral imaging to identify raisin varieties.

Keywords: near-infrared hyperspectral imaging; raisins; support vector machine; pixel-wise; object-wise

1. Introduction

Raisins are generally consumed as snacks, and they are also served as popular ingredients in many other food menus. Raisins are dried grapes which are rich in dietary fiber, carbohydrates with a low glycemic index, and minerals like copper and iron, with a low fat content [1,2]. In addition to their nutritional value, they also have medical value, such as regulating blood pressure for individuals with mild increases in blood pressure [2–4]. In general, raisins are important commercial products for the grape industry.

The commercial value of raisins differs according to the production area. In China, Xinjiang Uygur Autonomous Region is one of the major producing regions of grape, the perfect producing conditions and climates make it quite suitable for grape planting and deep processing. Variety is another important factor which influences the taste and nutritional compositions of raisins. To satisfy the demands of producers and consumers, different varieties of grapes are developed. Variety is one of the important factors in pricing the raisins. Varieties of raisins can be identified by specialist, experienced famers, and laboratory-based chemical analysis methods. To improve the identification efficiency, advanced non-destructive methods have been introduced, among which computer vision, spectroscopy, and spectral imaging techniques have shown great efficiency and potential for large scale detection at industrial level. Ma et al. achieved rapid non-destructive identification of apple varieties with 96.67% accuracy based on hyperspectral imaging [5]. Zhang et al. identified coffee variety using mid-infrared transmittance spectroscopy combined with pattern recognition algorithm [6]. Yang et al. developed a model for maize seed variety identification based on hyperspectral imaging [7].

Hyperspectral imaging is a technique combining computer vision and spectroscopy. Images of the study objects can be acquired for image analysis, and spectral information can be extracted from each pixel within the image for spectral analysis. A combination of image analysis and spectral analysis can also be explored. Hyperspectral imaging has been widely used in food analysis [8,9], and it has showed great potential in the grape industry. Fernandes et al. estimated grape anthocyanin concentration using hyperspectral imaging data. The squared correlation coefficient value was 0.65 compared to the values measured using conventional laboratory techniques [10]. Rodríguez-Pulido et al. found it was possible to assess the maturation stage in grape seeds based on the near-infrared spectra with prediction models and multivariate analysis methods [11]. Zhao et al. used hyperspectral imaging to identify different varieties of grape seeds. The results indicated that the variety of each single grape seed was accurately identified with 94.3% accuracy of the calibration set and 88.7% accuracy of the prediction set [12].

The general application of hyperspectral images is to conduct data analysis on a predefined region of interest (ROI) [12–14]. Spectral information is most widely used in hyperspectral image analysis, due to the advantage that spectral information can be precisely extracted from each pixel within ROIs. In general, pixel-wise spectra are averaged to build calibration models, and some researchers have focused on using pixel-wise spectra to build calibration models [15–18]. In fact, the size and the shape of raisins, which are key factors for the classification of different varieties, also play an important role in raisin grading within one variety. The raisin size can be influenced by the harvesting procedure of fresh grapes and the air-drying procedure, which can also beget irregular shapes of raisins in addition to storage.

The objective of this study is to explore the feasibility of using near-infrared hyperspectral imaging to identify raisin varieties. The specific objectives are: (1) exploring the influence of fruit size and shape in classification accuracy; (2) exploring spectral preprocessing of standard normal variate (SNV) in classification accuracy; (3) comparing performances of objective-wise analysis and pixel-wise analysis of SVM (Support Vector Machine), k-NN (k-nearest Neighbors Algorithm), and RBFNN (Radial Basis Function Neural Network) models.

2. Results and Discussion

2.1. Spectral Profiles

In this research, 200 wavelength variables ranging from 975 to 1646 nm of hyperspectral images were studied. Figure 1 presents average spectra of each grade of raisins of Wuhebai (WHB) and Xiangfei (XF) with standard deviation (SD) at peaks and valleys (1123, 1210, 1308, and 1473 nm). The absorbance bands at 1123, 1210, and 1308 nm are largely attributed to the C–H stretching mode and overtone [19]. The wavelength around 1473 nm is a characteristic water wavelength [20]. It was obvious that a large proportion of overlap exists among eight curves, so it was necessary to conduct further study to make a better distinction between the two varieties of raisins.

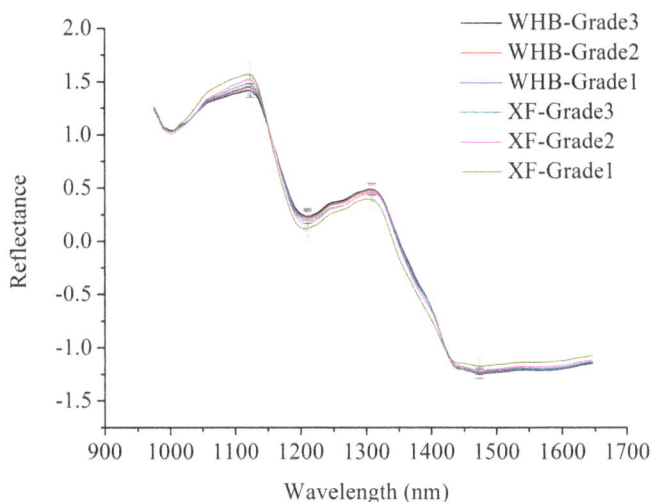

Figure 1. Average spectra with standard deviation (SD) of Wuhebai (WHB) and Xiangfei (XF).

2.2. PCA Scores Image Visualization

Pixel-wise PCA scores could be used to depict the PCA scores image. The first seven PCs explained over 99% of the total variance. Figure 2 shows visualized hyperspectral images of the first seven principal components (PC1–PC7) of two varieties of raisins. As can be seen from Figure 2, the warm color (yellow-red) accounted for the majority in WHB scores image of PC1 and PC2. In contrast to WHB scores image, the cold color (green-blue) was more obvious in XF scores image of PC1 and PC2, which revealed differences between two varieties. PCA scores image of PC5, PC6 and PC7 of XF exhibited obvious difference in color for Grade1 and two other grades, which showed differences among different grades. Although the PCA scores image could be used to distinguish different varieties and grades of raisins to some extent, it was necessary to conduct further study in order to obtain satisfactory classification results.

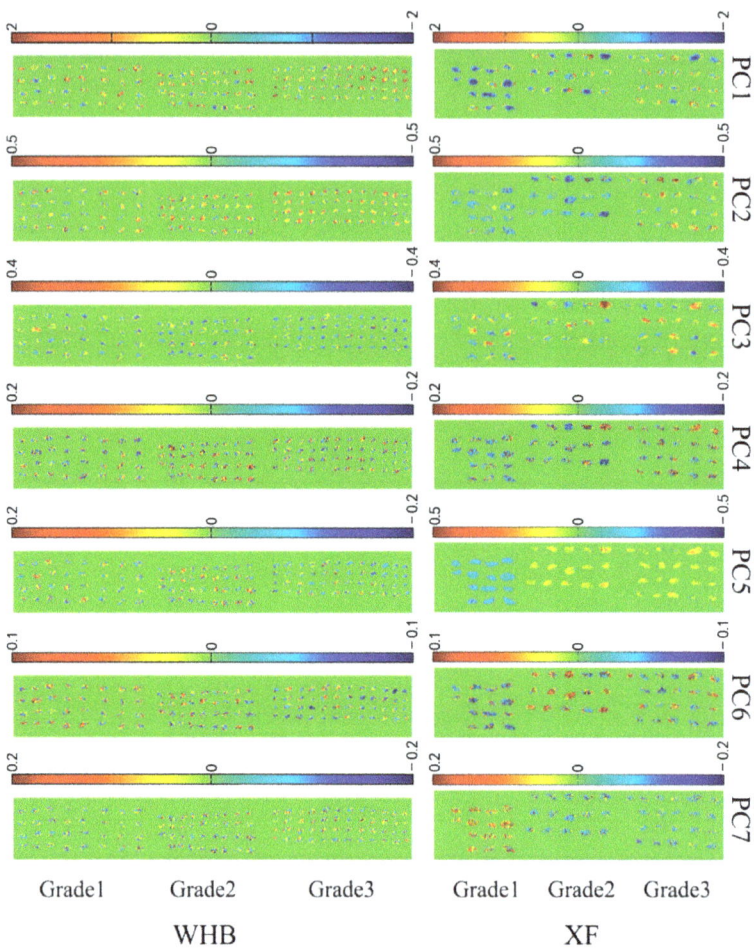

Figure 2. Scores image for the first seven principal components.

2.3. Effective Wavelength Selection

PCA loadings were used to select effective wavelengths for raisin cultivars classification. Since the first seven PCs explained over 99% of total variance, the loadings of these PCs were used. To examine the differences of object-wise analysis (average spectra) and pixel-wise analysis (pixel-wise spectra), PCA was conducted on both object-wise spectra and pixel-wise spectra of two different varieties of raisins. Figure 3 and Table 1 shows 20 or 17 optimal wavelengths selected by PCA based on object-wise analysis or pixel-wise analysis, respectively. PCA loadings plots of object-wise spectra and pixel-wise spectra were quite similar. As shown in Table 1, corresponding optimal wavelengths for object-wise spectra and pixel-wise spectra were nearly the same, with slight differences caused by different varieties and grades.

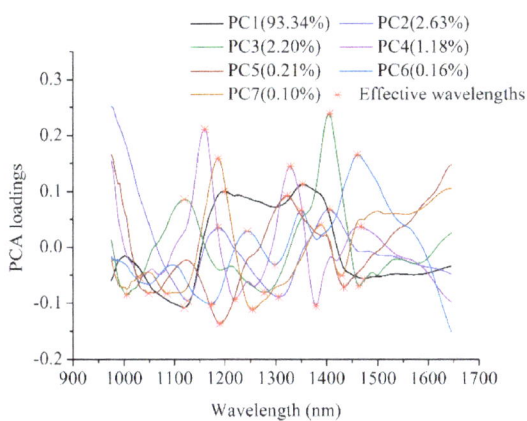

(a)

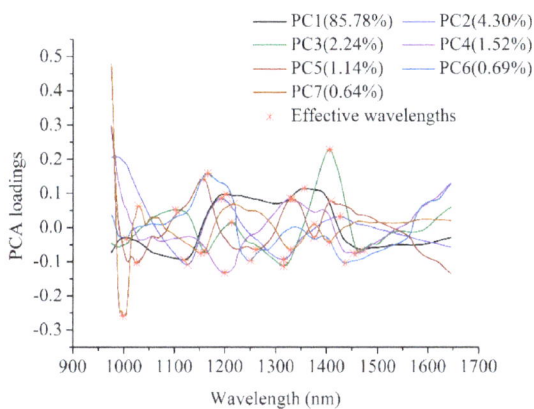

(b)

Figure 3. Corresponding optimal wavelengths selected by principal component analysis (PCA): (**a**) Object-wise analysis. (**b**) Pixel-wise analysis.

Table 1. Corresponding optimal wavelengths selected by PCA.

Type of Analysis	No.	Optimal Wavelengths (nm)
Object-wise	20	1005, 1032, 1049, 1086, 1119, 1160, 1173, 1187, 1200, 1220, 1244, 1254, 1278, 1305, 1328, 1352, 1379, 1406, 1433, 1473
Pixel-wise	17	1005, 1029, 1103, 1119, 1164, 1200, 1214, 1251, 1261, 1315, 1328, 1355, 1375, 1406, 1426, 1436, 1473

ICA was also conducted on object-wise spectra and pixel-wise spectra. To compare with PCA, the same numbers of optimal wavelengths are showed in Table 2.

Table 2. Corresponding optimal wavelengths selected by independent component analysis (ICA).

Type of Analysis	No.	Optimal Wavelengths (nm)
Object-wise	20	982, 985, 995, 999, 1002, 1009, 1012, 1015, 1019, 1022, 1025, 1029, 1032, 1035, 1039, 1042, 1046, 1049, 1052, 1056
Pixel-wise	17	1139, 1143, 1146, 1150, 1153, 1156, 1207, 1210, 1230, 1521, 1527, 1531, 1548, 1554, 1561, 1575, 1582

2.4. Raisin Variety Classification Models Based on Different Grades

The PCA analysis above indicated that there were differences between different varieties of raisins, and there were also differences among different grades of raisins. Thus, grade was an important factor which influenced classification results of two varieties of raisins.

To evaluate the influence of different grades on model performances, samples from the same grade of raisins were formed as calibration set, and the remaining samples were used as prediction set. SVM (Support Vector Machine) models were built using effective wavelengths selected by PCA, and the results are shown in Table 3.

For the calibration model built based on Grade1, classification results of the calibration set were good for both varieties and prediction results of three grades of WHB were good, while prediction results of XF were poor. There were no correctly classified samples for Grade3 of XF. The SVM model built based on Grade2 exhibited better performance compared with SVM model based on Grade1. For the calibration model built based on Grade2, both the calibration set and prediction set obtained satisfactory results, and WHB and XF both obtained good classification results. When the calibration set was built according to Grade3, classification results of calibration set were good. The prediction results of the three grades of WHB were good, and the prediction results of Grade2 and Grade3 of XF were also good. However, the prediction result of Grade1 was poor with classification accuracy lower than 20%.

These results revealed that different grades of raisins had influences on variety classification. As shown in Table 3, classification models based on Grade1 and Grade3 obtained poorer classification results compared with classification models based on Grade2. The reason might be that feature differences between Grade1 and Grade3 of XF were significant (for example the difference of sample size of different grades for same raisin variety was obvious as shown in Figure 4). The calibration set using Grade1 or Grade3 might not cover enough sample features used for PCA scores analysis.

Table 3. Classification models based on different grade using optimal wavelengths selected by PCA.

WHB	XF	C[4]	γ[4]	Cal. Result		Pre. Results		
				WHB	XF	Pre. set	WHB	XF
Grade1 [1]	Grade1	1	3.0	665/665	245/246	Grade3	1382/1382	0/602
						Grade2	930/931	22/453
						Grade1	380/380	99/116
Grade2 [2]	Grade2	256	16	622/622	304/305	Grade3	1371/1382	559/602
						Grade2	305/309	146/148
						Grade1	1040/1045	323/362
Grade3 [3]	Grade3	48.5	9.1	950/950	405/405	Grade3	419/432	197/197
						Grade2	658/931	434/453
						Grade1	1033/1045	51/362

[1] Grade1 represents large size; [2] Grade2 represents medium size; [3] Grade3 represents small size; [4] C and γ are parameters of SVM model.

Table 4 shows the results of SVM models built based on optimal wavelengths selected by ICA. Compared with Table 3, the calibration and prediction accuracies of SVM models based on Grade1 and Grade3 using optimal wavelengths selected by ICA were close to SVM models using optimal wavelengths selected by PCA. The accuracies based on Grade2 using optimal wavelengths selected by ICA were lower than SVM models based on PCA optimal wavelengths selection for both varieties of raisins.

Table 4. Classification models based on different grade using optimal wavelengths selected by ICA.

WHB	XF	C	γ	Cal. Result		Pre. Results		
				WHB	XF	Pre. set	WHB	XF
Grade1	Grade1	147.0	0.3	664/665	242/246	Grade3	1380/1382	0/602
						Grade2	931/931	17/453
						Grade1	379/380	100/116
Grade2	Grade2	147.0	48.5	606/622	255/305	Grade3	1360/1382	267/602
						Grade2	296/309	119/148
						Grade1	1014/1045	306/362
Grade3	Grade3	84.4	3.0	944/950	385/405	Grade3	409/432	197/197
						Grade2	487/931	393/453
						Grade1	899/1045	15/362

2.5. Classification Results of Pixel-Wise and Object-Wise Spectra

According to Zhang et al. (2018) [21], pixel-wise spectra can be used to build classification models, and can achieve good prediction results on sample average spectra. The results of average spectra showed that sample size might be a factor influencing classification results. The advantage of hyperspectral imaging was to obtain spectral information of each pixel within the sample. Previous results have showed that pixel-wise spectra analysis has great value in hyperspectral image analysis [22,23]. For each variety of raisins, pixel-wise spectra were extracted. In all, there were about 300,000 pixels of each grade within the calibration sets of WHB and XF raisins. Establishing calibration models using such a great number of pixels (over 1,800,000 pixels) requires quite heavy computation. Selecting effective wavelengths could reduce the data volume significantly, but the data volume was still large. As for about 300,000 pixels of each grade of raisins, there might be redundant pixels for modelling, so representative pixels should be selected to reduce the data amount.

To select representative pixel-wise spectra, a calibration set selection procedure was proposed by Kang et al. (2004) [24]. Firstly, for pixel-wise spectra of all grades of a variety, the collected pixel-wise spectra were clustered into 3000 groups using the k-means algorithm. Secondly, the Euclidean distance between sample and group centroid was calculated, and the sample with smallest Euclidean distance was selected into the new calibration set.

SVM, k-NN, and RBFNN models were built using selected pixel-wise spectra or object-wise spectra, and prediction set was also formed based on selected pixel-wise spectra or object-wise spectra. The results are shown in Table 5. The value of sensitivity means the classification accuracy of WHB, and the value of sensitivity means the classification accuracy of XF.

From Table 5, the results of SVM and RBFNN models using pixel-wise spectra to predict pixel-wise spectra were acceptable, with classification accuracy of calibration and prediction about 80%–90%. Meanwhile, the results of SVM and RBFNN models using pixel-wise spectra to predict object-wise spectra also obtained good results for calibration set, with 93.62% and 88.40% accuracy, respectively. Compared with SVM and RBFNN models based on pixel-wise spectral, the results of k-NN were slightly poorer, with accuracies varied from 40%–90%. Three models based on object-wise spectra all obtained acceptable results for the calibration set, with accuracies ranging from 87%–99%. SVM, k-NN and RBFNN models using object-wise spectra to predict object-wise spectra obtained better results for the prediction set compared with models using pixel-wise spectra to predict object-wise spectra, with all accuracies above 80%.

The overall results indicated that SVM and RBFNN models using object-wise spectra to predict object-wise spectra could be used to identify raisin varieties. Selection of representative samples was of significance for stable and accurate models, which should be further studied.

Table 5. Classification results for SVM, k-NN, and RBFNN models based on optimal wavelengths selected by PCA.

	Model	Parameter [5]	Calibration Set			Prediction Set		
			Acc. [6] (%)	Sen. [7]	Spe. [8]	Acc. (%)	Sen.	Spe.
Pixel to pixel [1]	SVM	(256, 5.28)	91.83	0.898	0.939	80.10	0.800	0.802
	k-NN	3	78.48	0.700	0.870	78.18	0.642	0.895
	RBFNN	7	88.40	0.842	0.926	80.89	0.797	0.819
Pixel to object [2]	SVM	(256, 5.28)	91.83	0.898	0.939	93.62	0.785	0.998
	k-NN	3	78.48	0.700	0.870	83.82	0.464	0.992
	RBFNN	7	88.40	0.842	0.926	91.40	0.711	0.997
Object to pixel [3]	SVM	(147, 9.12)	99.72	0.994	0.998	71.10	0.817	0.626
	k-NN	5	95.46	0.870	0.991	76.86	0.727	0.803
	RBFNN	3	99.78	0.994	0.999	54.14	0.819	0.317
Object to object [4]	SVM	(147, 9.12)	99.72	0.994	0.998	99.12	0.987	0.993
	k-NN	5	95.46	0.870	0.991	94.06	0.839	0.982
	RBFNN	3	99.78	0.994	0.999	99.30	0.983	0.997

[1] Pixel to pixel means to use models using pixel-wise spectra to predict pixel-wise spectra; [2] Pixel to object means models using pixel-wise spectra to predict object-wise spectra; [3] Object to pixel means to use models using object-wise spectra to predict pixel-wise spectra; [4] Object to object means to use models using object-wise spectra to predict object-wise spectra; [5] Parameters for SVM models are C and γ, parameter for k-NN is number of neighbors (k) and parameter for RBFNN is spread value; [6] Acc. means accuracy; [7] Sen. means sensitivity; [8] Spe. means specificity.

Table 6 shows the classification results for SVM, k-NN, and RBFNN models based on optimal wavelengths selected by ICA. The accuracies based on models using pixel-wise spectra to predict pixel-wise spectra or object-wise spectra were much lower than the same models using optimal wavelengths selected by PCA. The prediction set results of models using object-wise spectra to predict pixel-wise spectra were poor, with accuracies varying from 48%–63%. Models using object-wise spectra to predict object-wise spectra obtained acceptable results, with all accuracies above 90%. However, the calibration results of three models using object-wise spectra as the calibration set using optimal wavelengths selected by ICA were slightly lower than the results of three models using optimal wavelengths selected by PCA.

Table 6. Classification results for SVM, k-NN, and RBFNN models based on optimal wavelengths selected by ICA.

	Model	Parameter [5]	Calibration Set			Prediction Set		
			Acc. [6] (%)	Sen. [7]	Spe. [8]	Acc. (%)	Sen.	Spe.
Pixel to pixel [1]	SVM	(256, 16)	82.15	0.739	0.903	74.9	0.708	0.784
	k-NN	3	85.60	0.791	0.896	71.13	0.618	0.789
	RBFNN	6	78.92	0.695	0.884	76.74	0.797	0.819
Pixel to object [2]	SVM	(256, 9.19)	82.15	0.739	0.903	78.63	0.271	0.998
	k-NN	3	85.60	0.791	0.896	79.58	0.393	0.962
	RBFNN	6	78.92	0.695	0.884	80.47	0.341	0.996
Object to pixel [3]	SVM	(147, 84.45)	94.68	0.879	0.976	54.63	0.870	0.285
	k-NN	5	93.64	0.849	0.974	62.17	0.709	0.551
	RBFNN	3	93.96	0.851	0.977	48.34	0.565	0.417
Object to object [4]	SVM	(147, 84.45)	94.68	0.879	0.976	93.81	0.863	0.969
	k-NN	5	93.64	0.849	0.974	90.58	0.805	0.947
	RBFNN	3	93.96	0.851	0.977	93.30	0.844	0.970

[1] Pixel to pixel means to use models using pixel-wise spectra to predict pixel-wise spectra; [2] Pixel to object means models using pixel-wise spectra to predict object-wise spectra; [3] Object to pixel means to use models using object-wise spectra to predict pixel-wise spectra; [4] Object to object means to use models using object-wise spectra to predict object-wise spectra; [5] Parameters for SVM models are C and γ, parameter for k-NN is number of neighbors (k) and parameter for RBFNN is spread value; [6] Acc. means accuracy; [7] Sen. means sensitivity; [8] Spe. means specificity.

2.6. Prediction Maps of Raisin Variety Detection

Based on the developed models, prediction maps of different raisins varieties could be formed. According to the results in Table 5, we used the pixel-wise SVM model to form prediction maps. Raisin grades of the corresponding pixel were predicted to form classification maps, and prediction maps are shown in Figure 4. Pixel-wise prediction maps indicate that most of the pixels could be correctly classified. The prediction maps show clear a difference between WHB and XF according to different visualized prediction color.

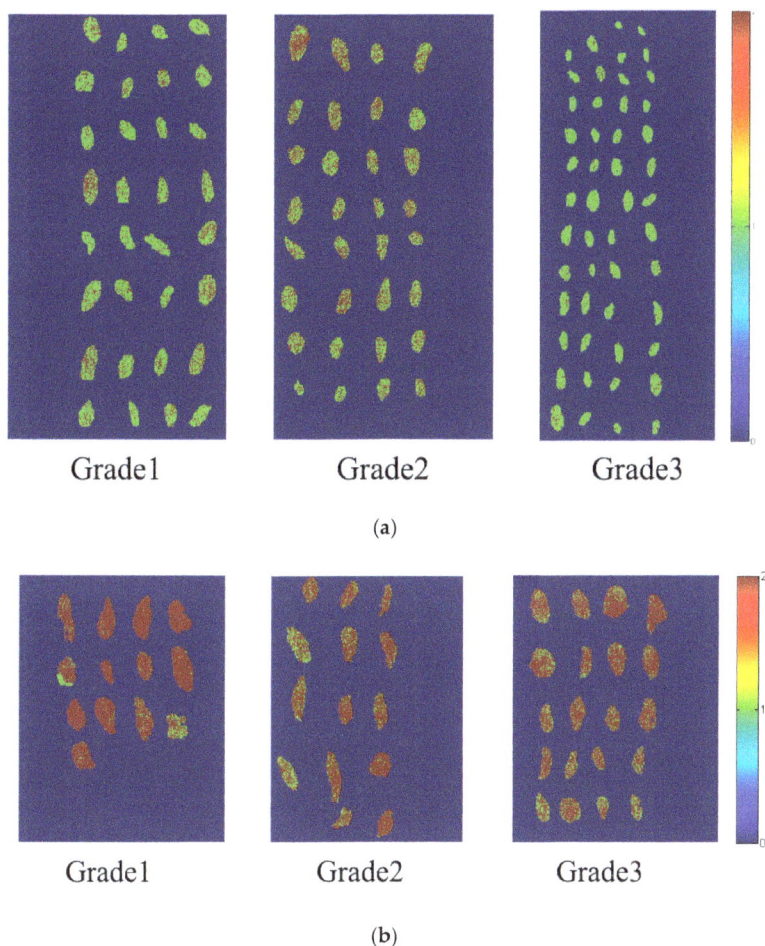

Grade1   Grade2   Grade3

(a)

Grade1   Grade2   Grade3

(b)

Figure 4. Classification results using pixel-wise spectra: (a) WHB; (b) XF.

3. Materials and Methods

3.1. Sample Preparation

Two varieties of raisins, including Wuhebai (WHB) and Xiangfei (XF), were collected from a local market in Shihezi, Xinjiang Uygur Autonomous Region, China. For each variety of raisin, three grades (Grade1-large size, Grade2-medium size, Grade3-small size) of raisins were manually collected according to the raisin size. For each grade, 450 g samples were divided into 30 groups (nearly 15 g per group) to acquire 30 hyperspectral images. Two varieties of raisins were all produced in 2017. RGB images of the two varieties of raisins are shown in Figure 5.

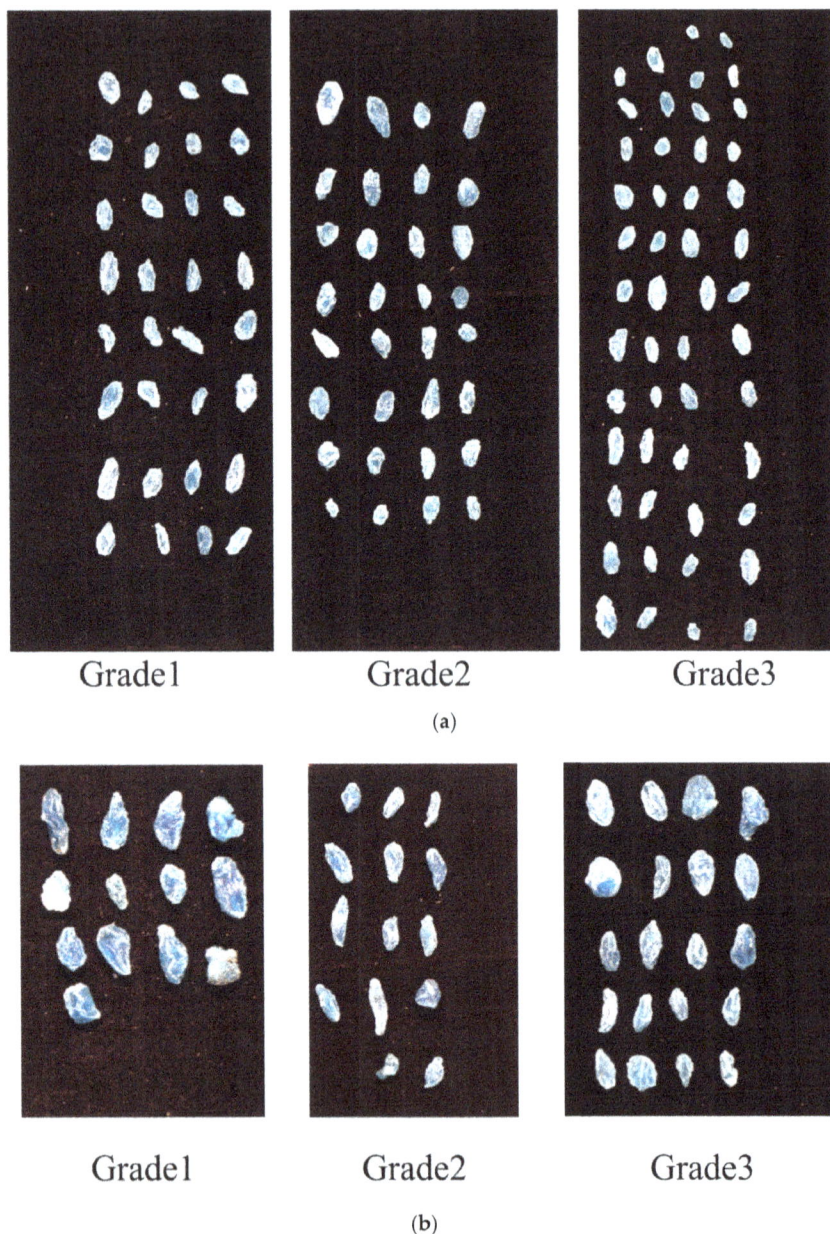

Grade1 Grade2 Grade3

(a)

Grade1 Grade2 Grade3

(b)

Figure 5. RGB images of the two varieties of raisins: (**a**) WHB; (**b**) XF.

3.2. Hyperspectral Imaging System

The hyperspectral image acquisition was carried out using an assembled near-infrared hyperspectral imaging system with the spectral range of 975–1646 nm. The system consisted of an ImSpector N17E imaging spectrograph (Spectral Imaging Ltd., Oulu, Finland), a Xeva 992 camera (Xenics Infrared Solutions, Leuven, Belgium) installed with an OLES22 lens (Spectral Imaging Ltd., Oulu, Finland), two 150 W tungsten halogen lamps (3900 Lightsource, Illumination Technologies Inc., Elbridge, NY, USA) that were symmetrically placed and served as the light source, and a conveyer belt (Isuzu Optics Corp., Taiwan). The imaging system was controlled by the software (Xenics N17E, Isuzu Optics Corp., Taiwan), which can be used to calibrated and analyze the images as well. The sketch of the hyperspectral imaging system is presented in Figure 6.

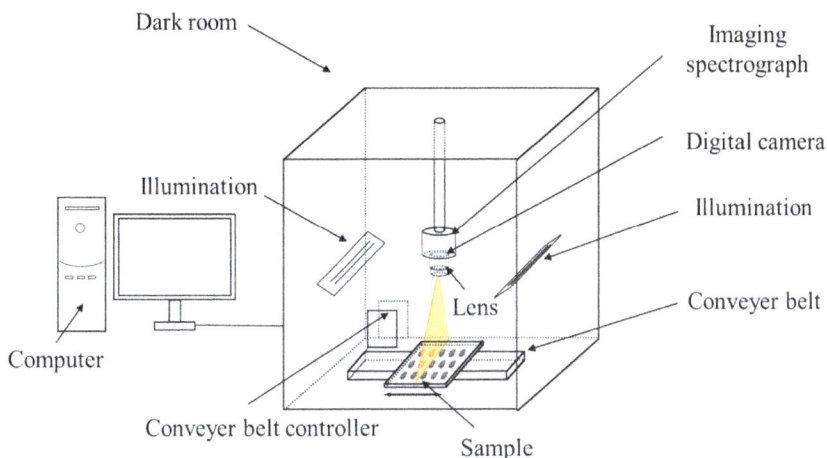

Figure 6. Hyperspectral imaging system.

3.3. Hyperspectral Image Acquisition and Correction

To acquire hyperspectral images, the distance between sample plane and the camera was set to 16 cm, the moving speed of the plate was set to 13.5 mm/s, and the exposure time of the camera was set to 4 ms. After adjustment, white reference image was collected by a white Teflon bar whose reflectivity is approximately 100%, and dark reference image was acquired by turning off the light source and covering the lens with lens cap whose reflectivity is about 0%. The white and dark reference images were used to calibrate the light intensity and reduce the dark current. For each group, one image was acquired. In all, 30 images were acquired for each grade of raisin.

After raw hyperspectral images acquisition, hyperspectral images were then corrected by the following equation:

$$I_c = \frac{I_r - I_d}{I_w - I_d} \tag{1}$$

where I_c is the corrected image, I_r is the raw image, I_w is the white reference image and I_d is the dark reference image.

3.4. Spectral Data Preprocessing and Extraction

The hyperspectral image at 1119 nm was selected for background segmentation since the reflectance difference between sample and background was more obvious. To differentiate background from foreground, we set the segmentation threshold to 0.122 for hyperspectral image binarization.

After image correction, spectral information was extracted from hyperspectral images. Each raisin kernel was defined as the region of interest (ROI). Pixel-wise spectra within the ROI were firstly extracted, and wavelet transform (WT) was used for smoothing. Wavelet function of Daubechies 7 with a decomposition level of 3 was used to reduce random noises. After WT preprocessing, standard normal variate (SNV) was used to reduce the influence of scattering of pixel-wise spectra [25].

Then, average spectra calculated according to pixel-wise spectra within each ROI were used to represent the sample. In this study, pixel-wise spectra and average spectra were both used for analysis.

To extract spectral information, a binary image was obtained using the gray-scale image at 1199 nm, the background was set as 0 and the kernel regions were set as 1. The binary image was applied to the gray-scale images at each wavelength, and the background information was thus removed. After the background removal, pixel-wise spectra were extracted and preprocessed.

3.5. Sample Set Division

For each grade of raisin, 30 hyperspectral images were acquired. Hyperspectral images were randomly split into the calibration set and prediction set at the ratio of 2:1, with 20 hyperspectral images in the calibration set and 10 hyperspectral images in the prediction set for each grade.

3.6. Data Analysis Methods

3.6.1. Principal Component Analysis

Principal component analysis (PCA) was used to explore the qualitative differences among different varieties of raisins [11,26–28]. In hyperspectral images, object-wise analysis and pixel-wise analysis were studied. For object-wise analysis, the average spectrum of each raisin kernel was used to conduct PCA; for pixel-wise analysis, pixel-wise spectra were used to conduct PCA. To explore the differences among raisins, the samples in the calibration set were used to conduct PCA. Then, scores values of each PC were then assigned to each kernel or each pixel to form the PCA scores image.

Hyperspectral imaging suffers from the large volume of data, and effectively reducing the data volume is of significance for data processing. There are also collinearity and redundancy in the spectra, which will affect the data analysis procedure. Variable selection is an effective strategy to reduce the data volume and select informative wavelengths. In this study, PCA loadings were used to select effective wavelengths. Loadings of each principal component (PC) indicate the correlation between the original variables and new feature variables. The higher the loading value is, the more important the variable is. The wavelengths with high absolute loading values can be selected as effective variables.

3.6.2. Independent Component Analysis

Independent component analysis (ICA) is a technique which is widely used in feature selection and feature extraction. It extracts independent source signals which are statistical independent by linear or nonlinear transformation. Independent component (IC) is obtained by a high-order statistic. Given a spectral matrix X, X can be expressed as Equation (2):

$$X = As \tag{2}$$

where s are the independent components (ICs) and A is the mixing matrix. For spectral data matrix X, s is unknown, and the general procedure is to find the estimation of s by the following equation:

$$\hat{s} = WX \tag{3}$$

where $\hat{s}$ is the estimation of s and W is the weight matrix for unmixing.

The procedure to select optimal wavelengths is as follow [29]. The average absolute weight value of each variable in W is calculated, and the variables with larger average absolute weight values are selected as optimal wavelengths. To compare with PCA, the same number of optimal wavelengths was selected by ICA. Fast ICA proposed by Hyvarinen and Oja was used to perform ICA in this study [30].

3.6.3. Discriminant Models

Support vector machine (SVM) is used to build models to classify different varieties of raisins. SVM is a supervised machine learning method, which is efficient to deal with linear and nonlinear data for classification and regression. For classification issues, SVM maps the original data into new feature spaces [31–33]. According to linearly separable data, a simple linear classifier can be constructed. For non-linearly separable data, the original data should be mapped into high-dimensional feature spaces so that the non-linearly separable issue can be transformed to a linearly separable issue. Kernel functions are the key for the mapping. Radial basis function (RBF) is a widely used kernel function with good performances for nonlinear data, and it was used as kernel function in this study. To conduct SVM with RBF kernel function, model penalty coefficient (C) and kernel parameter (γ) were determined by a grid-search procedure. In this study, the range of C and γ was 2^{-8}–2^8.

The k-nearest neighbors algorithm (k-NN) is a type of instance-based learning method used for classification and regression [34,35]. Both for classification and regression, a useful technique can be used to assign weight to the contributions of the neighbors, so that the nearest neighbors contribute more to the average than the more distant ones. The k-NN algorithm is among the simplest of all machine learning algorithms.

Radial basis function neural network (RBFNN) is an efficient feedforward neural network, which has the best approximation performance and global optimal characteristics that outperforms other feedforward networks, and has a simple structure as well as a fast training speed. On the other hand, it is also a neural network model that is widely used in pattern recognition, nonlinear function approximation, and other fields [36,37].

3.6.4. Software and Model Evaluation

The performance of classification models was evaluated by the classification accuracy, specificity, and sensitivity [38]. Hyperspectral images analysis, spectral data extraction, spectral preprocessing, PCA analysis, SVM, K-NN, and RBFNN were conducted on Matlab R2014b (The MathWorks, Natick, MA, USA).

3.6.5. Visualization of Prediction Maps

One of the advantages of hyperspectral imaging is that prediction maps can be formed to visualize the distribution of physical and chemical features. Object-wise or pixel-wise calibration models using spectra extracted from the hyperspectral images can be used to predict object-wise or pixel-wise features, and prediction maps can be formed with the predicted values [6,13,39].

4. Conclusions

Hyperspectral imaging was successfully used to identify different varieties of raisins. Three grades of raisins of Wuhebai and Xiangfei were studied. Object-wise and pixel-wise spectra were extracted. PCA analysis was firstly conducted to form PCA scores images, and scores images of the first seven PCs indicated the differences between different varieties and among different grades. PCA and ICA of object-wise spectra and pixel-wise spectra were conducted to select effective wavelengths. The overall results indicates that SVM models and RBFNN models using object-spectra to predict object-spectra based on optimal wavelengths selected by PCA both obtained acceptable results. The overall results showed that hyperspectral imaging was an effective technique to identify raisin varieties, and that both pixel-wise and object-wise could be used to build classification models. Selection of representative samples was important for building a stable and accurate model, and how to select representative samples should be studied in the future.

Author Contributions: Conceptualization, L.F. and C.Z.; Data curation, Y.H.; Formal analysis, P.G.; Funding acquisition, L.F., Y.B., P.G. and Y.H.; Investigation, S.Z.; Methodology, S.Z. and Y.B.; Project administration, L.F. and P.G.; Resources, P.G.; Software, Y.B., P.G. and Y.H.; Supervision, Y.H.; Validation, L.F., C.Z. and Y.H.; Visualization, L.F.; Writing—original draft, L.F. and S.Z.; Writing—review & editing, C.Z. and Y.H.

Funding: This research was funded by National key R&D program of China, grant number 2018YFD0101002; National Natural Science Foundation of China, grant number 31871526 and Major scientific and technological tackling project of Shihezi University, grant number GXJS2015-ZDGG08.

Acknowledgments: We thank Hongji Lin in College of electrical engineering, Zhejiang University for editing and improving the manuscript.

Conflicts of Interest: The authors declare no conflict of interest.

References

1. Williamson, G.; Carughi, A. Polyphenol content and health benefits of raisins. *Nutr. Res.* **2010**, *30*, 511–519. [CrossRef] [PubMed]
2. Margaret, J.S.; Xinyue, W.; Tiffany, H.; James, E.P. A Comprehensive review of raisins and raisin components and their relationship to human health. *J. Nutr. Health* **2017**, *50*, 203. [CrossRef]
3. Kanellos, P.T.; Kaliora, A.C.; Gioxari, A.; Christopoulou, G.O.; Kalogeropoulos, N.; Karathanos, V.T. Absorption and bioavailability of antioxidant phytochemicals and increase of serum oxidation resistance in healthy subjects following supplementation with raisins. *Plant Food Hum. Nutr.* **2013**, *68*, 411–415. [CrossRef] [PubMed]
4. Bays, H.E.; Schmitz, K.; Christian, A.; Ritchey, M.; Anderson, J. Raisins and Blood Pressure: A Randomized, Controlled Trial. *J. Am. Coll. Cardiol.* **2012**, *59*, E1721. [CrossRef]
5. Huiling, M.; Wang, R.; Cheng, C.; Dong, W. Rapid Identification of Apple Varieties Based on Hyperspectral Imaging. *Trans. CSAE* **2017**, *48*, 305–312. [CrossRef]
6. Zhang, C.; Wang, C.; Liu, F.; He, Y. Mid-Infrared Spectroscopy for Coffee Variety Identification: Comparison of Pattern Recognition Methods. *J. Spectrosc.* **2016**, *2016*, 1–7. [CrossRef]
7. Yang, S.; Zhu, Q.B.; Huang, M.; Qin, J.W. Hyperspectral Image-Based Variety Discrimination of Maize Seeds by Using a Multi-Model Strategy Coupled with Unsupervised Joint Skewness-Based Wavelength Selection Algorithm. *Food Anal. Method* **2017**, *10*, 1–10. [CrossRef]
8. Bao, Y.; Liu, F.; Kong, W.; Sun, D.W.; He, Y.; Qiu, Z. Measurement of Soluble Solid Contents and pH of White Vinegars Using VIS/NIR Spectroscopy and Least Squares Support Vector Machine. *Food Bioprocess Technol.* **2014**, *7*, 54–61. [CrossRef]
9. Wang, H.; Peng, J.; Xie, C.; Bao, Y.; He, Y. Fruit Quality Evaluation Using Spectroscopy Technology: A Review. *Sensors* **2015**, *15*, 11889. [CrossRef] [PubMed]
10. Fernandes, A.M.; Oliveira, P.; Moura, J.P.; Oliveira, A.A.; Falco, V.; Correia, M.J.; Melopinto, P. Determination of anthocyanin concentration in whole grape skins using hyperspectral imaging and adaptive boosting neural networks. *J. Food Eng.* **2011**, *105*, 216–226. [CrossRef]
11. Rodríguez-Pulido, F.J.; Barbin, D.F.; Sun, D.W.; Gordillo, B.; González-Miret, M.L.; Heredia, F.J. Grape seed characterization by NIR hyperspectral imaging. *Postharvest Biol. Technol.* **2013**, *76*, 74–82. [CrossRef]
12. Zhao, Y.; Zhang, C.; Zhu, S.; Gao, P.; Feng, L.; He, Y. Non-Destructive and Rapid Variety Discrimination and Visualization of Single Grape Seed Using Near-Infrared Hyperspectral Imaging Technique and Multivariate Analysis. *Molecules* **2018**, *23*, 1352. [CrossRef] [PubMed]
13. Zhang, C.; Wang, Q.; Liu, F.; He, Y.; Xiao, Y. Rapid and non-destructive measurement of spinach pigments content during storage using hyperspectral imaging with chemometrics. *Measurement* **2016**, *97*, 149–155. [CrossRef]
14. Kong, W.; Zhang, C.; Cao, F.; Liu, F.; Luo, S.; Tang, Y.; He, Y. Detection of Sclerotinia Stem Rot on Oilseed Rape (*Brassica napus* L.) Leaves Using Hyperspectral Imaging. *Sensors* **2018**, *18*, 1764. [CrossRef]
15. Xing, J.; Symons, S.; Shahin, M.; Hatcher, D. Detection of sprout damage in Canada Western Red Spring wheat with multiple wavebands using visible/near-infrared hyperspectral imaging. *Biosyst. Eng.* **2010**, *106*, 188–194. [CrossRef]
16. Williams, P.J.; Kucheryavskiy, S. Classification of maize kernels using NIR hyperspectral imaging. *Food Chem.* **2016**, *209*, 131–138. [CrossRef]
17. Arngren, M.; Hansen, P.W.; Eriksen, B.; Larsen, J.; Larsen, R. Analysis of Pregerminated Barley Using Hyperspectral Image Analysis. *J. Agric. Food Chem.* **2011**, *59*, 11385–11394. [CrossRef] [PubMed]
18. Yao, H.; Hruska, Z.; Kincaid, R.; Brown, R.L.; Bhatnagar, D.; Cleveland, T.E. Hyperspectral image classification and development of fluorescence index for single corn kernels infected with Aspergillus flavus. *Trans. ASABE* **2013**, *56*, 1977–1988. [CrossRef]
19. Khodabux, K.; L'Omelette, M.S.S.; Jhaumeer-Laulloo, S.; Ramasami, P.; Rondeau, P. Chemical and near-infrared determination of moisture, fat and protein in tuna fishes. *Food Chem.* **2007**, *102*, 669–675. [CrossRef]

20. Kinoshita, K.; Miyazaki, M.; Morita, H.; Vassileva, M.; Tang, C.; Li, D.; Ishikawa, O.; Kusunoki, H.; Tsenkova, R. Spectral pattern of urinary water as a biomarker of estrus in the giant panda. *Sci. Rep.* **2012**, *2*, 856. [CrossRef] [PubMed]

21. Zhang, C.; Liu, F.; He, Y. Identification of coffee bean varieties using hyperspectral imaging: Influence of preprocessing methods and pixel-wise spectra analysis. *Sci. Rep.* **2018**, *8*, 2166. [CrossRef] [PubMed]

22. Diezma, B.; Lleó, L.; Roger, J.M.; Herrero-Langreo, A.; Lunadei, L.; Ruiz-Altisent, M. Examination of the quality of spinach leaves using hyperspectral imaging. *Postharvest Biol. Technol.* **2013**, *85*, 8–17. [CrossRef]

23. Lara, M.A.; Lleó, L.; Diezma-Iglesias, B.; Roger, J.M.; Ruiz-Altisent, M. Monitoring spinach shelf-life with hyperspectral image through packaging films. *J. Food Eng.* **2013**, *119*, 353–361. [CrossRef]

24. Kang, J.; Ryu, K.R.; Kwon, H.C. Using Cluster-Based Sampling to Select Initial Training Set for Active Learning in Text Classification. In *Pacific-Asia Conference on Knowledge Discovery and Data Mining*; Springer: Berlin/Heidelberg, Germany, 2004; pp. 384–388.

25. Rinnan, Å.; Berg, F.V.D.; Engelsen, S.B. Review of the most common pre-processing techniques for near-infrared spectra. *Trac-Trend Anal. Chem.* **2009**, *28*, 1201–1222. [CrossRef]

26. Choudhary, R.; Mahesh, S.; Paliwal, J.; Jayas, D.S. Identification of wheat classes using wavelet features from near infrared hyperspectral images of bulk samples. *Biosyst. Eng.* **2009**, *102*, 115–127. [CrossRef]

27. Sun, J.; Jiang, S.; Mao, H.; Wu, X.; Li, Q. Classification of Black Beans Using Visible and Near Infrared Hyperspectral Imaging. *Int. J. Food Sci. Technol.* **2016**, *19*, 1687–1695. [CrossRef]

28. Huang, M.; He, C.; Zhu, Q.; Qin, J. Maize Seed Variety Classification Using the Integration of Spectral and Image Features Combined with Feature Transformation Based on Hyperspectral Imaging. *Appl. Sci.* **2016**, *6*, 183. [CrossRef]

29. Du, H.; Qi, H.; Wang, X.; Ramanath, R. Band selection using independent component analysis for hyperspectral image processing. In Proceedings of the Applied Imagery Pattern Recognition Workshop, Washington, DC, USA, 15–17 October 2003; IEEE: New York, NY, USA, 2003; pp. 93–98.

30. Hyvärinen, A.; Oja, E. Independent component analysis: Algorithms and applications. *Neural Netw.* **2000**, *13*, 411–430. [CrossRef]

31. Feng, X.; Zhao, Y.; Zhang, C.; Cheng, P.; He, Y. Discrimination of Transgenic Maize Kernel Using NIR Hyperspectral Imaging and Multivariate Data Analysis. *Sensors* **2017**, *17*, 1894. [CrossRef] [PubMed]

32. Dumont, J.; Hirvonen, T.; Heikkinen, V.; Mistretta, M.; Granlund, L.; Himanen, K.; Fauch, L.; Porali, I.; Hiltunen, J.; Keski-Saari, S. Thermal and hyperspectral imaging for Norway spruce (*Picea abies*) seeds screening. *Comput. Electron. Agric.* **2015**, *116*, 118–124. [CrossRef]

33. Lee, H.; Kim, M.S.; Song, Y.R.; Oh, C.S.; Lim, H.S.; Lee, W.H.; Kang, J.S.; Cho, B.K. Non-destructive evaluation of bacteria-infected watermelon seeds using visible/near-infrared hyperspectral imaging. *J. Sci. Food Agric.* **2016**, *97*, 1084. [CrossRef] [PubMed]

34. Guo, G.; Wang, H.; Bell, D.; Bi, Y.; Greer, K. KNN Model-Based Approach in Classification. In Proceedings of the Otm Confederated International Conferences on the Move to Meaningful Internet Systems, Catania, Sicily, Italy, 3–7 November 2003; Springer: Berlin, Germany, 2003; pp. 986–996.

35. Hall, P.; Park, B.U.; Samworth, R.J. Choice of Neighbor Order in Nearest-Neighbor Classification. *Ann. Stat.* **2008**, *36*, 2135–2152. [CrossRef]

36. Shcherbakov, V.; Larsson, E. Radial basis function partition of unity methods for pricing vanilla basket options. *Comput. Math. Appl.* **2016**, *71*, 185–200. [CrossRef]

37. Fornberg, B.; Flyer, N.; Hovde, S.; Piret, C. Locality properties of radial basis function expansion coefficients for equispaced interpolation. *IMA J. Numer. Anal.* **2018**, *28*, 121–142. [CrossRef]

38. Baratloo, A.; Hosseini, M.; Negida, A.; El Ashal, G. Part 1: Simple Definition and Calculation of Accuracy, Sensitivity and Specificity. *Emergency* **2015**, *3*, 48–49. [PubMed]

39. He, J.; Zhang, C.; He, Y. Application of Near-Infrared Hyperspectral Imaging to Detect Sulfur Dioxide Residual in the Fritillaria thunbergii Bulbus Treated by Sulfur Fumigation. *Appl. Sci.* **2017**, *7*, 77. [CrossRef]

Sample Availability: Samples of the compounds are available from the authors.

molecules

MDPI

Article

Discrimination of *Chrysanthemum* Varieties Using Hyperspectral Imaging Combined with a Deep Convolutional Neural Network

Na Wu [1,2,3,†], Chu Zhang [1,2,3,†], Xiulin Bai [1,2,3], Xiaoyue Du [1,2,3] and Yong He [1,2,3,*]

[1] College of Biosystems Engineering and Food Science, Zhejiang University, Hangzhou 310058, China; nawu018@zju.edu.cn (N.W.); chuzh@zju.edu.cn (C.Z.); xlbai@zju.edu.cn (X.B.); xydu@zju.edu.cn (X.D.)
[2] Key Laboratory of Spectroscopy Sensing, Ministry of Agriculture and Rural Affairs, Zhejiang University, Hangzhou 310058, China
[3] State Key Laboratory of Modern Optical Instrumentation, Zhejiang University, Hangzhou 310058, China
* Correspondence: yhe@zju.edu.cn; Tel.: +86-571-8898-2143
† These authors contributed equally to this work.

Received: 30 September 2018; Accepted: 25 October 2018; Published: 31 October 2018

Abstract: Rapid and accurate discrimination of *Chrysanthemum* varieties is very important for producers, consumers and market regulators. The feasibility of using hyperspectral imaging combined with deep convolutional neural network (DCNN) algorithm to identify *Chrysanthemum* varieties was studied in this paper. Hyperspectral images in the spectral range of 874–1734 nm were collected for 11,038 samples of seven varieties. Principal component analysis (PCA) was introduced for qualitative analysis. Score images of the first five PCs were used to explore the differences between different varieties. Second derivative (2nd derivative) method was employed to select optimal wavelengths. Support vector machine (SVM), logistic regression (LR), and DCNN were used to construct discriminant models using full wavelengths and optimal wavelengths. The results showed that all models based on full wavelengths achieved better performance than those based on optimal wavelengths. DCNN based on full wavelengths obtained the best results with an accuracy close to 100% on both training set and testing set. This optimal model was utilized to visualize the classification results. The overall results indicated that hyperspectral imaging combined with DCNN was a very powerful tool for rapid and accurate discrimination of *Chrysanthemum* varieties. The proposed method exhibited important potential for developing an online *Chrysanthemum* evaluation system.

Keywords: hyperspectral imaging; variety discrimination; *Chrysanthemum*; deep convolutional neural network

1. Introduction

As one of the most popular flowers throughout the world, *Chrysanthemum* has a long planting history in China. The excellent ornamental, edible and medicinal values make *Chrysanthemum* used in many different forms. *Chrysanthemum* tea is one of the most commonly consumed teas for Chinese consumers. The chemical components such as flavonoids and polysaccharides rich in *Chrysanthemum* tea have antioxidant and antibacterial properties, which can relieve cell damage and improve body immunity [1,2]. The nutritional qualities of *Chrysanthemum* tea are affected by many factors, including climate, soil, water, cultivation management and post-harvest treatment, being the variety a determinant factor. Due to differences in content of chemical compositions, different varieties of *Chrysanthemum* tea have specific effects on human bodies. With the frequent mixing of *Chrysanthemum* from different varieties in the market in recent years, the purity of *Chrysanthemum* is difficult to guarantee. Thus, an appropriate method for discrimination of *Chrysanthemum* varieties is needed. The appearance characteristics such as color, flower diameter, petal shape often serve as the basis to identify *Chrysanthemum* varieties. This visual inspection method is subjective and requires professional knowledge. Some other approaches like high performance liquid chromatography (HPLC) combined with photodiode array detection, employed to determine the quality attributes, are destructive, time consuming, and can only handle very small number of samples [3]. Therefore, a rapid and accurate method would be advantageous when large number of *Chrysanthemum* samples need to be classified.

Near-infrared spectroscopy (NIRS), as a potential technology for rapid measurement, has been widely used in different fields such as geographical origin discrimination of agricultural products [4], quality assessment of agricultural seeds [5], variety identification of Chinese herbal medicines [6]. However, the samples needed to be shattered into powder when using this technology, making extraction of external space information difficult. Moreover, the sample size in these studies was very small which could not cover a broad variation. In contrast to NIRS, hyperspectral imaging (HSI) perfectly integrating visible/near-infrared spectroscopy and optical imaging in one system, can acquire both spectral information and spatial information. The capacity of collection spectra of multiple samples in one scan simultaneously gives HSI the property of batch detection, which makes the practical application possible. In addition, the spectra and the corresponding location of each pixel in image recorded by HSI can be employed to visualize the variety and chemical composition distribution of the samples.

To extract spectral and spatial information of a sample, hyperspectral image contains hundreds of contiguous wavebands for each pixel. Multivariate analysis methods, including spectral and image preprocessing, variable extraction and selection, model building and analysis, are often utilized to process this kind of data [7–9]. Currently, traditional machine learning methods combined with HSI have been widely used in variety identification of agricultural products [10–14], and multiple classification models were utilized, such as multiple logistic regression (MLR) [15], partial least squares discriminant analysis (PLS-DA) [16], support vector machine (SVM) [17], extreme learning machine (ELM) [18].

Deep learning, also known as representational learning, is a research focus in artificial intelligence nowadays. Among a variety of deep learning algorithms, deep convolution neural network (DCNN) aims to automatically extract abstract distributed features layer-by-layer. Various DCNNs has dramatically improved the state-of-the-art results in many vision tasks. In the field of hyperspectral image analysis, DCNN was first introduced in 2015 to classify hyperspectral sensing data [19]. In recent years, researchers have developed different DCNNs according to specific spectral analysis tasks, such as variety identification of rice seeds [20], disease detection of wheat *Fusarium* head blight [21], crop classification from remote sensing images [22]. It is of interest to further investigate if DCNN has the potential to discriminate the *Chrysanthemum* varieties.

The main objective of this study was to explore the feasibility of using HSI technique combined with DCNN for variety discrimination and visualization of *Chrysanthemum*. The specific objectives were to: (1) select important wavelengths that can contribute to identification of *Chrysanthemum* varieties, (2) develop appropriate DCNNS using full wavelengths and optimal wavelengths, (3) compare the results of DCNNs with traditional machine learning methods, including SVM and LR, (4) visualize the identification results of *Chrysanthemum* varieties using the optimal model.

2. Results

2.1. Overview of Spectra

Figure 1 shows the mean spectra with standard deviation (SD) of *Chrysanthemum* samples of seven varieties. The shape of the reflectance curves was similar to that of *Chrysanthemum* in [23]. It can also be seen from the figure that the average spectra of seven *Chrysanthemum* varieties shared the consistent trend with similar peak and valley positions. However, slight differences could be observed from the average spectra of *Chrysanthemum* samples. The different chemical compositions and biochemical characteristics of these seven varieties resulted in these differences in spectral features. The peaks, around 1116 and 1308 nm, and valleys, around 1200 and 1460 nm, in spectral curves could be employed to discriminant the *Chrysanthemum* varieties. Among them, the two peaks and the valley at 1200 nm are attributed to the second overtone of C–H stretching [24,25], while the valley at 1460 nm (around 1450 nm) is attributed to the first overtone of O–H stretching [25]. In addition, it could be clearly observed that the spectral curves of Boju and Hangbaiju are very close and partially overlapping in the range of 975–1200 nm, indicating that the chemical compositions of these two varieties are similar.

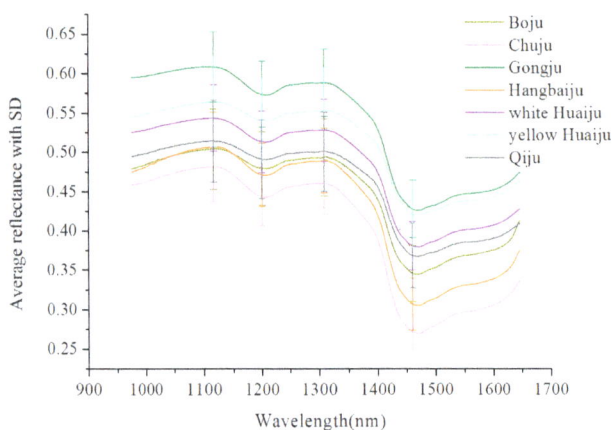

Figure 1. The average spectra of *Chrysanthemum* samples of seven varieties.

2.2. Principal Component Analysis

In the field of spectral analysis, PCA is often used as a method for qualitative analysis. In this study, PCA was employed to explore the differences between seven *Chrysanthemum* varieties. A hyperspectral image for each variety was random selected from the testing set for PCA. The first five PCs reflected 99.95% of information in original spectral data (96.61%, 3.17%, 0.11%, 0.04%, 0.02% for PC1, PC2, PC3, PC4, PC5, respectively). Thus, these five PCs of seven hyperspectral images were extracted. The pixels with PC value in sample region together with pixels with zero value in black background formed the final score images illustrated in Figure 2, from which the scores of *Chrysanthemum* samples of each variety were displayed intuitively, and some varieties could be preliminarily distinguished through combining these five PCs. For example, Boju could be highlighted because of the high scores of most sample pixels in PC4, which caused the samples to appear yellow. Due to the negative scores of most pixels in PC2, it was clear to discriminate Chuju and Hangbaiju from other *Chrysanthemum* varieties. However, the further discrimination between Chuju and Hangbaiju was difficult. In addition, it was easy to distinguish Gongju and yellow Huaiju from other *Chrysanthemum* varieties in PC1 and PC5, since most pixels of these two varieties had high scores in PC1 and negative scores in PC5. And Gongju could be further identified in PC4 for its negative scores. White Huaiju and Qiju having the same clustering pattern as some other varieties could not be identified. To distinguish all *Chrysanthemum* varieties, discriminant models need to be built for quantitative analysis in further study.

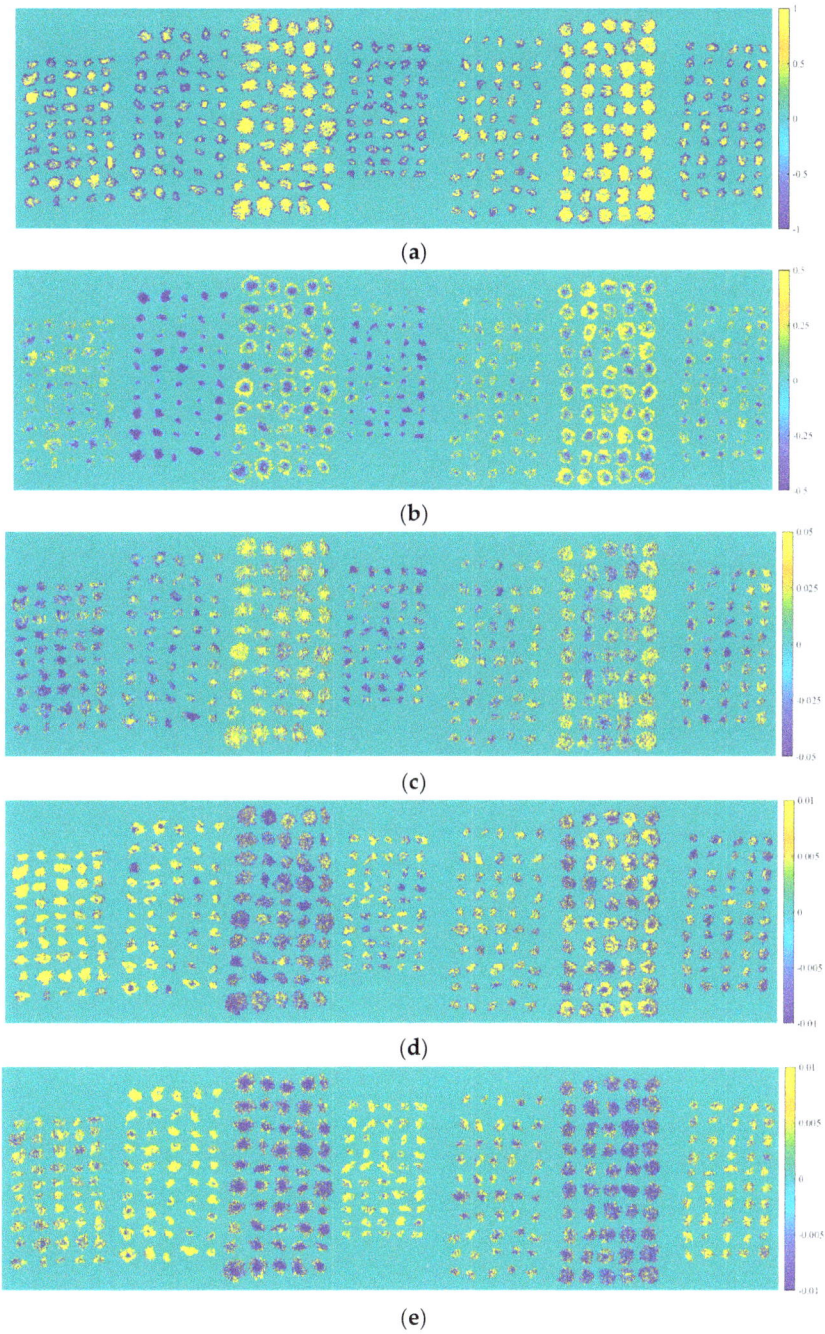

Figure 2. Score images of the first five PCs of seven *Chrysanthemum* varieties (from left to right: Boju, Chuju, Gongju, Hangbaiju, white Huaiju, yellow Huaiju, and Qiju): (**a**) PC1; (**b**) PC2; (**c**) PC3; (**d**) PC4; (**e**) PC5.

2.3. Selection of Optimal Wavelengths

In order to remove the redundancy information contained in hyperspectral images and improve the classification performance of *Chrysanthemum* varieties, second derivative (2nd derivative) method was introduced to select optimal wavelengths from full wavelengths. Figure 3 shows the 2nd derivative spectra of average spectra of seven varieties. There are multiple high peaks and low valleys in the 2nd derivative spectra, and the wavelengths with large differences between *Chrysanthemum* varieties were selected as optimal wavelengths for discrimination. Finally, eighteen optimal wavelengths were selected in total. Among them, the absorption bands at approximately 999, 1005, 1015, 1025 and 1032 nm are attributed to the second overtone of N−H stretching [17]. The wavelengths between 1136 nm and 1311 nm (1136, 1190, 1214, 1244, 1301, 1311 nm) are related to the second overtone of C-H stretching [17,26]. The selected wavelengths of 1321 and 1375 nm are associated with the first overtone of C–H combination bands [27]. The bands at 1406, 1433 and 1456 nm present the first overtone of O–H stretching [27]. The bands at 1470 nm (around 1480 nm) is ascribed to the second overtone of O-H stretching [25,28]. The peak at 1633 nm (around 1630 nm) is attributed to the aromatic C-H bands [29]. These wavelengths carrying the category information are closely related to the constituent differences of chemical composition of different *Chrysanthemum* varieties.

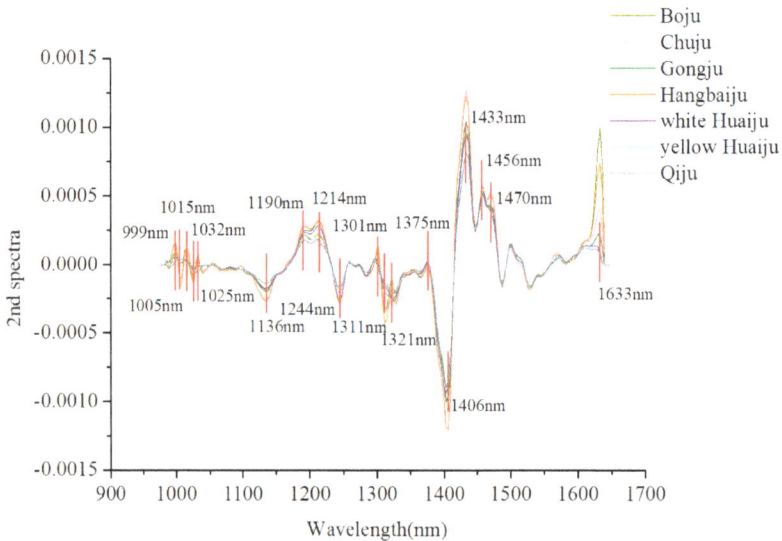

Figure 3. The 2nd derivative spectra and the selected optimal wavelengths.

2.4. Discrimination Results of Different Models

Discriminant models using full and optimal wavelengths were built by SVM, LR, and DCNN for quantitative analysis. The classification accuracies of different models and corresponding parameters were summarized in Table 1. As can be seen in Table 1, SVM, LR and DCNN models all achieved good classification results on both training set and testing set. For full wavelengths, the accuracies of these three models on the training set were greater than 99%, and the accuracies on the testing set were more than 94%. The classification capacity of DCNN was better than those of SVM and LR, showing accuracies of close to 100% on both training set and testing set. Being able to learn deep spectral features automatically, DCNN could provide excellent classification performance.

Since a large amount of redundant information existed in full wavelengths, the optimal wavelengths were often extracted in previous spectral analysis to improve the robustness of the model. In this study, 2nd derivative method was introduced to select the optimal wavelengths. The accuracies of the three models based on optimal wavelengths were slightly lower than those based on full wavelengths. Consistent with the results based on full wavelengths, the best results were still obtained by DCNN model with an accuracy of 98.45% on training set and 94.27% on testing set. LR was most sensitive to wavelength reduction that led to the largest drop of accuracy on testing set. Due to the removal of a part of spectral information, a slight accuracy reduction was understandable. However, the fact that SVM and LR based on optimal wavelengths achieved lower accuracies than DCNN based on full wavelengths further proved that the deep spectral features learnt by DCNN were more distinguishable than the selected feature wavelengths.

In summary, DCNN achieved better classification performance than the traditional machine learning algorithms, including SVM and LR. The overall results indicated that hyperspectral imaging combined with DCNN was feasible to distinguish *Chrysanthemum* varieties. Without any optimal wavelengths extraction, DCNN based on full wavelengths is a very reliable model and is available for identification of more *Chrysanthemum* varieties in future.

Table 1. Discrimination results of *Chrysanthemum* varieties by different models using full wavelengths and optimal wavelengths.

Models	Full Wavelengths			Optimal Wavelengths		
	Parameters [1]	Training	Testing	Parameters	Training	Testing
SVM	$(10^6, 10^{-5})$	99.83%	94.02%	$(10^7, 10^{-4})$	98.26%	90.03%
LR	(L2, 100, liblinear)	99.34%	96.59%	(L2, 100, liblinear)	94.35%	85.75%
DCNN	(4, 32, 93)	99.98%	99.98%	(3, 32, 125)	98.45%	94.27%

[1] The parameters of the discriminant models. (*c*, *g*) for SVM, (*pi*, *c'*, *optimize_algo*) for LR, and (*num_convs*, *num_first_kernels*, *epoch*) for DCNN.

2.5. Visualization of Chrysanthemum Variety Classification

In order to discriminant the *Chrysanthemum* varieties more intuitively, the optimal model, DCNN based on full wavelengths, was used to visualize the classification of *Chrysanthemum* varieties in this study. A hyperspectral image for each variety was randomly selected from testing set. The original grayscale images of seven varieties are shown in Figure 4a. Although some *Chrysanthemum* varieties differed in size from others, it was difficult to identify all varieties according to the external phenotype. The corresponding classification maps were displayed in Figure 4b. The low resolution of hyperspectral images and the application of some morphological operations during image segmentation resulted in some changes of *Chrysanthemums'* shape. However, the main patterns and positions of the *Chrysanthemums* were clearly expressed on the classification maps. It was easy to distinguish different *Chrysanthemum* varieties according to the colors. For these randomly selected hyperspectral images, DCNN based on full wavelengths classified all samples correctly. That is to say, DCNN achieved an accuracy of 100%, which is consistent with the quantitative analysis. The visualization results indicated that hyperspectral imaging combined with DCNN provided a rapid, accurate and intuitive way to distinguish *Chrysanthemum* varieties, which is a potential tool for identifying and locating more *Chrysanthemum* varieties.

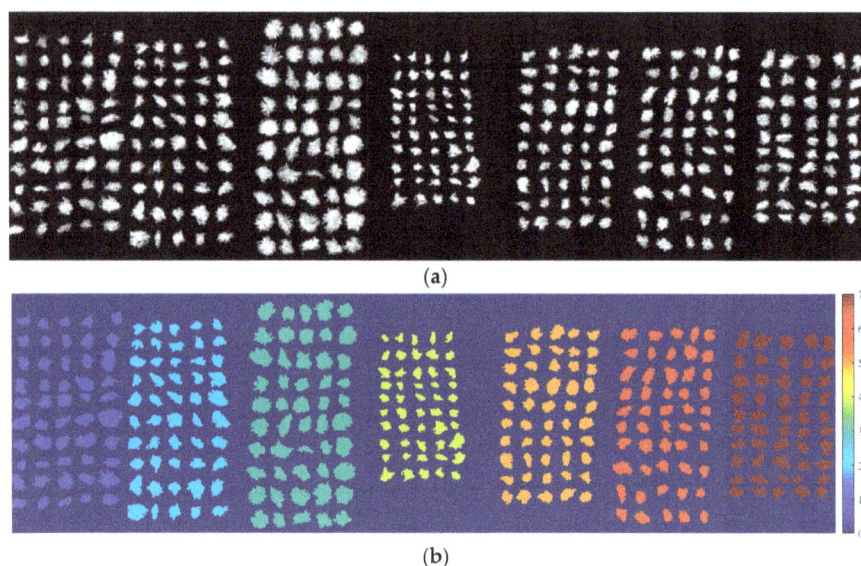

Figure 4. Visualization of *Chrysanthemum* varieties (from left to right: Boju, Chuju, Gongju, Hangbaiju, white Huaiju, yellow Huaiju, and Qiju): (**a**) Original grayscale images; (**b**) The classification maps.

3. Discussion

Influenced by growth environment, cultivation management, picking period and other factors, the chemical compositions of *Chrysanthemums* from same variety may vary greatly. For example, there are significant differences in total polysaccharide content and total flavonoid content between *Chrysanthemum* picked in different periods. As a result of these differences, their pharmacological properties and prices vary widely [23,30]. To include these variations, large-scale samples need to be collected. In previous studies, the classification of *Chrysanthemum* varieties has been reported. A total of 200 samples including five cultivars of *Chrysanthemum* were classified using a multispectral imaging system in [31]. To identify three kinds of white *Chrysanthemum*, a near infrared spectroscopy system was employed to collect the spectra of 139 samples and 92 spectra were selected as calibration set to build the identification model in [32]. In this study, a total of 11,038 samples of seven *Chrysanthemum* varieties were classified using hyperspectral imaging technology. The characteristic of batch detection of hyperspectral imaging makes it possible to acquire large-scale samples, which also provides favorable conditions for the application of deep learning.

As a research focus in machine learning, deep learning has been gradually applied in the field of spectral analysis. DCNN is a typical deep learning algorithm that learns abstract features through multiple convolutional layers. The large-scale samples obtained by hyperspectral imaging technology enable DCNN to fully exploit its advantages and automatically learn the deep spectral features contained in hyperspectral images. In previous studies on spectral analysis, the optimal wavelengths were commonly selected manually and then modeled using traditional machine learning algorithms such as SVM, LR, and KNN [16]. However, deep learning algorithms often achieved good classification results without additional feature selection [33,34]. In this study, DCNN and two traditional machine learning algorithms using full wavelengths and optimal wavelengths were compared. The results showed that DCNN based on full wavelengths achieved the best performance. This further illustrated that DCNN can discriminate *Chrysanthemum* varieties more accurately since it can learn deep spectral features through multiple hidden layers automatically. More *Chrysanthemum* varieties need to be collected to develop a *Chrysanthemum* variety identification instrument. In addition, in order to further evaluate the quality of *Chrysanthemum*, a comprehensive research need to be conducted in future.

Combining the advantages of hyperspectral imaging and DCNN, an on-line detection system of *Chrysanthemum* varieties and quality could be developed.

4. Materials and Methods

4.1. Sample Preparation

Seven varieties of dried *Chrysanthemum,* including Boju, Chuju, Gongju, Hangbaiju, white Huaiju, yellow Huaiju, and Qiju, were collected for our experiment. Among them, Boju, Chuju, and Gongju were bought from the local tea sales companies in Bozhou, Chuzhou and Huangshan, Anhui Province, China, respectively. Hangbaiju were bought from the local market in Hangzhou, Zhejiang Province, China. The two varieties of Huaiju and Qiju were bought from the local tea sales companies in Jiaozuo, Henan Province and Anguo, Hebei Province, China, respectively. All *Chrysanthemums* were harvested in 2017 and had a similar dry state.

In total, 1600, 1500, 1643, 1600, 1500, 1590, 1605 samples were obtained for Boju, Chuju, Gongju, Hangbaiju, white Huaiju, yellow Huaiju, and Qiju, respectively. The dataset of each variety was randomly divided into a training set and a testing set at a ratio of 3:1. Therefore, there were 8280 samples in the training set and 2758 samples in the testing set. All *Chrysanthemum* samples were assigned a category label. Boju, Chuju, Gongju, Hangbaiju, white Huaiju, yellow Huaiju, and Qiju were assigned from 1 to 7, respectively.

4.2. Hyperspectral Image Acquisition and Correction

Hyperspectral images of *Chrysanthemums* were acquired using a near-infrared HSI system. This system consists of a group of devices interacting to each other: an imaging spectrograph (ImSpector N17E; Spectral Imaging Ltd., Oulu, Finland) with a spectral range of 874–1734 nm, a high-performance CCD camera assembled with a camera lens (OLES22; Specim, Spectral Imaging Ltd., Oulu, Finland) having a resolution of 326 × 256 (spatial × spectral) pixels, two 150-W tungsten halogen lamps (3900e Lightsource; Illumination Technologies Inc.; West Elbridge, NY, USA) regarded as the illumination unit, and a conveyer belt controlled by a stepped motor (Isuzu Optics Corp., Zhubei, Taiwan) used for moving samples.

To obtain non-deformable and clear hyperspectral images, dried *Chrysanthemums* were placed on the conveyer belt, and the distance between the camera lens and the conveyer belt, the exposure time of the camera, and the speed of the conveyer belt along X-axis were adjusted to 25 cm, 4 ms and 19.5 mm/s, respectively. The acquired hyperspectral images of *Chrysanthemums* were composed of 256 spectral channels with a spectral resolution of 5 nm.

To reduce the effects of dark current and obtain the reflectivity of samples, raw hyperspectral images I_{raw} should be corrected with the white reference image and black reference image using the following Equation (1):

$$I_c = \frac{I_{raw} - I_{dark}}{I_{white} - I_{dark}} \tag{1}$$

where I_c is the hyperspectral image after corrected, I_{white} is the hyperspectral image of a white Teflon tile with nearly 100% reflectance, I_{dark} is acquired by covering the camera lens with its opaque cap. I_{raw}, I_{white}, I_{dark} are obtained under the same condition during samples collection.

4.3. Spectra Extraction and Pretreatment

Before spectra extraction, the region of interest (ROI), each *Chrysanthemum* sample region, need to be segmented from the black background. A threshold segmentation procedure was conducted on the gray image at 1119 nm where the contrast between the sample regions and the background reached the maximum value, and then the obtained binary mask was applied on the gray images at other wavelengths. After getting ROI of each *Chsrysanthemum* sample, the spectrum of each pixel in each ROI with a spectral range of 874–1734 nm was extracted. Due to the instability of hyperspectral imaging system at the start and end of sample collection, the beginning and the end of the spectral data contained random noise. Thus, the middle 200 wavelengths from 975 nm to 1646 nm were used for analysis. To further reduce the spectral noise and improve the signal-to-noise ratio, wavelet transform (WT) with decomposition scale of 3 and basis function of Daubechies 6 was employed to smooth the pixel-wise spectra. Finally, the preprocessed pixel-wise spectra in each ROI were averaged and the mean spectrum of each *Chrysanthemum* sample was used for discrimination analysis.

4.4. Chemometrics Analysis

PCA is a powerful tool to reduce the dimensionality of high-dimensional data. More importantly, PCA can remove noise and discover patterns inherent data through dimensionality reduction. In spectral analysis, each specific wavelength regarded as a feature variable forms the spectral matrix. PCA is applied to this matrix, and projected the original spectral variables into a new coordinate system by maximizing the sample variance. The variables in the new coordinate system called PCs are a linear transformation of original spectral variables and are orthogonal to each other. The PCs are arranged in descending order of interpreted variance and the first few PCs can reflect most of variance inherent in original matrix. From the score images of PCs, it is possible to identify the pattern difference between different categories of data.

Collinearity and redundancy exist among the contiguous wavelengths in hyperspectral image. Optimal wavelength selection is an efficient way to extract wavelengths that are beneficial for classification. 2nd derivative is a widely-used wavelength selection method, which can highlight spectral change [35]. Subtle changes in original spectra can be projected into the peaks and the valleys in 2nd derivative spectra. The wavelengths corresponding to the peaks and valleys with large difference between spectra could be selected as the optimal wavelengths to discriminant different sample categories.

4.5. Discriminant Methods

To classify the *Chrysanthemum* samples correctly, a DCNN was built as the discriminant model. Traditional machine learning methods, including SVM and LR, were introduced as contrast methods.

4.5.1. Support Vector Machine

SVM is a supervised machine learning approach, widely used in spectral data classification. The basic principle of SVM is to find the optimal hyperplane that maximizes the interval between the positive and negative samples in training set. To solve the nonlinear problem, kernel function is introduced into SVM. The hidden mapping of samples from original feature space into a new high-dimension space using kernel function can make the samples change from the linear indivisible state to a linear separable state [36]. Among the kernel functions, radial basis function (RBF) is efficient to deal with nonlinear classification problem. In this study, RBF was selected as the kernel function of SVM. To obtain a satisfactory classification performance, penalty coefficient c and the kernel parameter g could be determined using a simple grid-search procedure.

4.5.2. Logistic Regression

LR is a commonly-used pattern recognition approach to solve classification problem using regression-like method. Sigmoid function is utilized to map the real value predicted by linear regression model into the value in range 0–1. The output of sigmoid function is treated as the predicted category probability. When solving binary classification problem (labeled by 0 and 1), the sample with a value greater than or equal to 0.5 is classified as category 1, otherwise assigned to category 0. When solving multi-classification problem, multiple one-to-many binary classification models are combined. Structural risk loss is employed as the objective function to be optimized [15]. The penalty item pi can be set to L1 regularization or L2 regularization to reduce the overfitting risk. The inverse of regularization coefficient c' can be adjusted, while small c' causes strong regularization. The optimization algorithms *optimize_algo*, including newton-cg, lbfgs, liblinear, sag, can be selected to optimize the loss function according to the classification performance.

4.5.3. Deep Convolutional Neural Network

A DCNN was further developed to discriminate the *Chrysanthemum* varieties, and its performance was compared with that of SVM and LR. A typical DCNN consists of convolutional layers to extract the local features, pooling layers to reduce the size of parameters and fully-connected layers to output the classification results.

The structure of our designed DCNN for full wavelengths shown in Figure 5 contained four convolutional modules and two full connected layers. Each convolutional module included two convolutional layers followed by a max pooling layer. The number of filters in the first convolutional module was set to 32, and was doubled as the modules going deeper. To process one-dimensional spectral data, the commonly-used two-dimensional convolution kernels were replaced by one-dimensional convolution kernels. The trick of using two consecutive 1×3 kernels instead of a 1×5 kernel was inspired by VGGNet to decrease the number of parameters while increasing the network depth [34]. Each convolution kernel was acted on the local region of the feature maps of the upper layer, and all regions were processed by the same kernel. This mechanism allowed DCNN to quickly learn the local spectral features in parallel. The max pooling layer with a kernel of 1×2 was used to reduce the number of feature maps to the half. The stride and padding of all the filters were set to 1. The two full connected layers were used to combine the features output by the last convolution module.

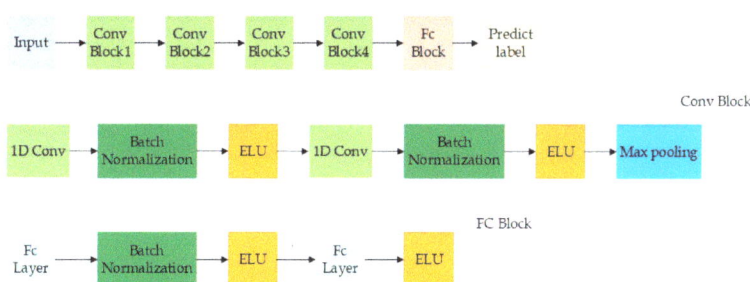

Figure 5. The structure of DCNN based on full wavelengths.

Exponential linear unit (ELU) was selected as the activation function in this study due to its better performance than rectified linear unit (RELU), which was consistent with the results in [20,37]. The right linear part allows ELU to mitigate the gradient disappearance like other activation functions. The left soft-saturated part allows ELU to push the mean of the active unit closer to 0, thereby reducing the offset effect and making ELU more robust to input variations and noise. Combining the advantages of these two parts, ELU can speed up the training process and improve the classification accuracy. The expression of ELU is as shown in Equation (2):

$$ f(x) = \begin{cases} x & x \geq 0 \\ \alpha(\exp(x) - 1) & x < 0 \end{cases} \tag{2} $$

As an important achievement of deep learning in recent years, Batch Normalization has been widely proved to be effective and important [38]. For each neuron in hidden layers, Batch Normalization forces the input distribution closing to the saturation region back to the standard normal distribution to reduce the offset effect like ELU. The consistent scale of data in each layer and each dimension makes parameter adjustment efficient. This accelerates the convergence process, reduces the possibility of overfitting and improves the classification performance. In this study, Batch Normalization was inserted before each ELU (except the last fully connected layer).

At the end of DCNN, a softmax function was introduced to transform the output of last fully-connected layer to the value in range 0–1, which represents the relative probability between different categories. Then, the cross-entropy loss was chosen as the objective function to evaluate the difference between the output of DCNN and the ground-truth in training phase. The cross-entropy loss function can be defined by Equation (3):

$$Loss = -\sum_x p(x) \log q(x) \tag{3}$$

where x is the input of DCNN, $p(x)$ is the probability value of expected output, $q(x)$ is the probability value predicted by DCNN. A Stochastic Gradient Descent (SGD) optimizer with a learning rate of 0.001 and a momentum of 0.9, was used to minimize the cross-entropy loss function during training. And the batch size was set to 256. The network structure for optimal wavelengths was similar. The number of convolution modules *num_convs*, the number of convolution kernels in the first convolution module *num_first_kernels*, and the iterations of network training *epoch* should be adjusted according to the classification performance.

4.6. Chrysanthemum Varieties Visualization

Visualization of *Chrysanthemum* varieties facilitates intuitive and rapid inspection of *Chrysanthemum* varieties by industrial producers and market regulators. The advantage of hyperspectral imaging to obtain spatial and spectral information simultaneously makes visualization of *Chrysanthemum* varieties possible. To build the classification maps, the average spectrum of each sample in hyperspectral image was input into the classification model, and the obtained label was mapped back to each pixel of the corresponding sample in hyperspectral image. In this study, the optimal discriminant model based on hyperspectral imaging was selected to visualize the spatial distribution of *Chrysanthemum* varieties. Different *Chrysanthemum* varieties were assigned to different colors on the chemical imaging maps, which is beneficial for identifying the specific *Chrysanthemums* whose varieties are different from that of most *Chrysanthemums*.

4.7. Software

ENVI 4.6 (ITT Visual Information Solutions, Boulder, CO, USA) was used to crop the *Chrysanthemum* samples from the irrelevant background in hyperspectral images. MATLAB R2018a (The MathWorks, Natick, MA, USA) was used to extract and preprocess the spectral data from hyperspectral images. PCA for pattern recognition between different varieties was also implemented with MATLAB R2018a. Unscrambler 10.1 (CAMO AS, Oslo, Norway) was used to extract the optimal wavelengths by 2nd derivative method. Discriminant models including SVM, LR and DCNN were implemented using python language with Spyder3.2.6 (Anaconda, Austin, TX, USA). The famous machine learning library sklearn (http://scikit-learn.org/stable/) and convenient deep learning framework Pytorch (Facebook, Menlo Park, CA, USA) were used during programming. All software tools were carried out on the software platform of win10 64-bit operating system and the hardware platform of a computer with Inter(R) Core (TM) i5-8500 3.00 HZ CPU and 8 G memory.

5. Conclusions

Hyperspectral imaging combined with DCNN was used to distinguish *Chrysanthemum* varieties. The qualitative analysis of PCA showed that different *Chrysanthemum* varieties could be preliminarily distinguished according to the score images. The optimal wavelengths with certain distinguishing ability were selected by 2nd derivative method. The performance of SVM, LR, and DCNN models using full wavelengths and optimal wavelengths were compared, and the performance of models based on full wavelengths were superior to those based on optimal wavelengths. DCNN based on full wavelengths obtained the best classification results, indicating that the deep spectral features automatically learned by DCNN were more beneficial for discrimination than the artificially selected optimal wavelengths. The classification maps of *Chrysanthemum* varieties formed by DCNN made the spatial distribution of *Chrysanthemum* varieties to be displayed in an intuitive manner, showing great potential of rapid detection of large-scale samples in industrial production. The overall results indicated that the characteristics of non-destructive and batch detection of hyperspectral imaging and the ability of automatically learning deep features of DCNN were the key factors for rapid and accurate discrimination of *Chrysanthemum* varieties. This study provides a new idea for identification of *Chrysanthemum* varieties.

Author Contributions: Conceptualization, N.W., C.Z. and Y.H.; Data curation, N.W., C.Z., X.B. and X.D.; Formal analysis, X.B. and X.D.; Funding acquisition, Y.H.; Investigation, N.W., C.Z. and Y.H.; Methodology, N.W., C.Z. and Y.H.; Project administration, Y.H.; Resources, N.W.; Software, C.Z. and Y.H.; Supervision, Y.H.; Validation, C.Z. and X.B.; Visualization, N.W. and C.Z.; Writing-original draft, N.W.; Writing-review & editing, N.W., C.Z. and Y.H.

Funding: This research was funded by National Key R&D Program of China (2018YFD0101002).

Conflicts of Interest: The authors declare no conflict of interest.

References

1. Zheng, C.; Dong, Q.; Chen, H.; Cong, Q.; Ding, K. Structural characterization of a polysaccharide from *Chrysanthemum morifolium* flowers and its antioxidant activity. *Carbohydr. Polym.* **2015**, *130*, 113–121. [CrossRef] [PubMed]

2. Ke, L.; Chen, H. Enzymatic-assisted microwave extraction of total flavonoids from dud of *Chrysanthemum indicum* L. and evaluation of biological activities. *Int. J. Food Eng.* **2016**, *12*, 607–613. [CrossRef]

3. Yang, Y.; Sun, X.; Liu, J.; Kang, L.; Chen, S.; Ma, B.; Guo, B. Quantitative and qualitative analysis of flavonoids and phenolic acids in snow chrysanthemum (*Coreopsis tinctoria* Nutt.) by HPLC-DAD and UPLC-ESI-QTOF-MS. *Molecules* **2016**, *21*, 1307. [CrossRef] [PubMed]

4. Li, Y.; Zou, X.; Shen, T.; Shi, J.; Zhao, J.; Holmes, M. Determination of geographical origin and anthocyanin content of black goji berry (*Lycium ruthenicum* Murr.) using near-infrared spectroscopy and chemometrics. *Food Anal. Methods* **2017**, *10*, 1034–1044.

5. Shrestha, S.; Knapič, M.; Žibrat, U.; Deleuran, L.; Gislum, R. Single seed near-infrared hyperspectral imaging in determining tomato (*Solanum lycopersicum* L.) seed quality in association with multivariate data analysis. *Sens. Actuators B* **2016**, *237*, 1027–1034. [CrossRef]

6. Han, B.; Yan, H.; Chen, C.; Yao, H.; Dai, J.; Chen, N. A rapid identification of four medicinal chrysanthemum varieties with near infrared spectroscopy. *Pharmacogn. Mag.* **2014**, *10*, 353–358. [CrossRef] [PubMed]

7. Manley, M. Near-infrared spectroscopy and hyperspectral imaging: Non-destructive analysis of biological materials. *Chem. Soc. Rev.* **2014**, *43*, 8200–8214. [CrossRef] [PubMed]

8. Liu, D.; Sun, D.; Zeng, X. Recent advances in wavelength selection techniques for hyperspectral image processing in the food industry. *Food Bioprocess Technol.* **2013**, *7*, 307–323. [CrossRef]

9. Sanz, J.; Fernandes, A.; Barrenechea, E.; Silva, S.; Santos, V.; Gonçalves, N.; Paternain, D.; Jurio, A.; Melo-Pinto, P. Lamb muscle discrimination using hyperspectral imaging: Comparison of various machine learning algorithms. *J. Food Eng.* **2016**, *174*, 92–100. [CrossRef]

10. Huang, M.; Tang, J.; Yang, B.; Zhu, Q. Classification of maize seeds of different years based on hyperspectral imaging and model updating. *Comput. Electron. Agric.* **2016**, *122*, 139–145. [CrossRef]

11. Soares, S.; Medeiros, E.; Pasquini, C.; Morello, C.; Galvão, R.; Araújo, M. Classification of individual cotton seeds with respect to variety using near-infrared hyperspectral imaging. *Anal. Methods* **2016**, *8*, 8498–8505. [CrossRef]

12. William, P.; Geladi, P.; Fox, G.; Manley, M. Maize kernel hardness classification by near infrared (NIR) hyperspectral imaging and multivariate data analysis. *Anal. Chim. Acta* **2009**, *653*, 121–130. [CrossRef] [PubMed]

13. Rodríguez-Pulido, F.; Barbin, D.; Sun, D.; Gordillo, B.; González-Miret, M.; Heredia, F. Grape seed characterization by NIR hyperspectral imaging. *Postharvest Biol. Technol.* **2013**, *76*, 74–82. [CrossRef]

14. Williams, P.; Kucheryavskiy, S. Classification of maize kernels using NIR hyperspectral imaging. *Food Chem.* **2016**, *209*, 131–138. [CrossRef] [PubMed]

15. Wu, Z.; Wang, Q.; Plaza, A.; Li, J.; Sun, L.; Wei, Z. Real-time implementation of the sparse multinomial logistic regression for hyperspectral image classification on GPUs. *IEEE Geosci. Remote Sens. Lett.* **2015**, *12*, 1456–1460.

16. Kong, W.; Zhang, C.; Liu, F.; Nie, P.; He, Y. Rice seed cultivar identification using near-infrared hyperspectral imaging and multivariate data analysis. *Sensors* **2013**, *13*, 8916–8927. [CrossRef] [PubMed]

17. Zhang, C.; Liu, F.; He, Y. Identification of coffee bean varieties using hyperspectral imaging: Influence of preprocessing methods and pixel-wise spectra analysis. *Sci. Rep.* **2018**, *8*, 2166. [CrossRef] [PubMed]

18. Zhou, Y.; Peng, J.; Chen, C. Extreme learning machine with composite kernels for hyperspectral image classification. *IEEE J. Sel. Top. Appl. Earth Obs. Remote Sens.* **2015**, *8*, 2351–2360. [CrossRef]

19. Hu, W.; Huang, Y.; Wei, L.; Zhang, F.; Li, H. Deep convolutional neural networks for hyperspectral image classification. *J. Sens.* **2015**, *2015*. [CrossRef]

20. Qiu, Z.; Chen, J.; Zhao, Y.; Zhu, S.; He, Y.; Zhang, C. Variety identification of single rice seed using hyperspectral imaging combined with convolutional neural network. *Appl. Sci.* **2018**, *8*, 212. [CrossRef]

21. Jin, X.; Jie, L.; Wang, S.; Qi, H.; Li, S. Classifying wheat hyperspectral pixels of healthy heads and Fusarium head blight disease using a deep neural network in the wild field. *Remote Sens.* **2018**, *10*, 395. [CrossRef]

22. Ji, S.; Zhang, C.; Xu, A.; Shi, Y.; Duan, Y. 3D convolutional neural networks for crop classification with multi-temporal remote sensing images. *Remote Sens.* **2018**, *10*, 75. [CrossRef]

23. He, J.; Chen, L.; Chu, B.; Zhang, C. Determination of total polysaccharides and total flavonoids in *Chrysanthemum morifolium* using near-infrared hyperspectral imaging and multivariate analysis. *Molecules* **2018**, *23*, 2395. [CrossRef] [PubMed]

24. Feng, X.; Peng, C.; Chen, Y.; Liu, X.; Feng, X.; He, Y. Discrimination of CRISPR/Cas9-induced mutants of rice seeds using near-infrared hyperspectral imaging. *Sci. Rep.* **2017**, *7*, 15934. [CrossRef] [PubMed]

25. Serranti, S.; Cesare, D.; Marini, F.; Bonifazi, G. Classification of oat and groat kernels using NIR hyperspectral imaging. *Talanta* **2013**, *103*, 276–284. [CrossRef] [PubMed]

26. Zhang, C.; Jiang, H.; Liu, F.; He, Y. Application of near-infrared hyperspectral imaging with variable selection methods to determine and visualize caffeine content of coffee beans. *Food Bioprocess Technol.* **2017**, *10*, 213–221. [CrossRef]

27. Ribeiro, J.; Ferreira, M.; Salva, T. Chemometric models for the quantitative descriptive sensory analysis of Arabica coffee beverages using near infrared spectroscopy. *Talanta* **2011**, *83*, 1352–1358. [CrossRef] [PubMed]

28. Restaino, E.; Fassio, A.; Cozzolino, D. Discrimination of meat patés according to the animal species by means of near infrared spectroscopy and chemometrics Discriminación de muestras de paté de carne según tipo de especie mediante el uso de la espectroscopia en el infrarrojo cercano y la quimiometria. *J. Food* **2011**, *9*, 210–213.

29. Chung, H.; Choi, H.; Ku, M. Rapid identification of petroleum products by near-infrared spectroscopy. *Bull. Korean Chem. Soc.* **1999**, *20*, 1021–1025.

30. Ding, X.; Ni, Y.; Kokot, S. Analysis of different Flos Chrysanthemum tea samples with the use of two-dimensional chromatographic fingerprints, which were interpreted by different multivariate methods. *Anal. Methods* **2015**, *7*, 961. [CrossRef]

31. Shui, S.; Liu, W.; Liu, C.; Yan, L.; Hao, G.; Zhang, Y.; Zheng, L. Discrimination of cultivars and determination of luteolin content of *Chrysanthemum morifolium* Ramat. using multispectral imaging system. *Anal. Methods* **2018**, *10*, 1640. [CrossRef]

32. Chen, C.; Yan, H.; Han, B. Rapid identification of three varieties of *Chrysanthemum* with near infrared spectroscopy. *Rev. Bras. Farmacogn.* **2014**, *24*, 33–37. [CrossRef]

33. Krizhevsky, A.; Sutskever, I.; Hinton, G. ImageNet classification with deep convolutional neural networks. In Proceedings of the Conference on Neural Information Processing Systems, Lake Tahoe, NV, USA, 3–6 December 2012.

34. Simonyan, K.; Zisserman, A. Very deep convolutional networks for large-scale image recognition. *arXiv* **2014**, arXiv:1409.1556.

35. Zhang, C.; Feng, X.; Wang, J.; Liu, F.; He, Y.; Zhou, W. Mid-infrared spectroscopy combined with chemometrics to detect Sclerotinia stem rot on oilseed rape (*Brassica napus* L.) leaves. *Plant Methods* **2017**, *13*, 39. [CrossRef] [PubMed]

36. Mavroforakis, M.; Theodoridis, S. A geometric approach to support vector machine (SVM) classification. *IEEE Trans. Neural Netw.* **2006**, *17*, 671–682. [CrossRef] [PubMed]

37. Clevert, D.; Unterthiner, T.; Hochreiter, S. Fast and accurate deep network learning by exponential linear units (ELUs). *arXiv* **2015**, arXiv:1511.07289.

38. Ioffe, S.; Szegedy, C. Batch normalization accelerating deep network training by reducing internal covariate shift. In Proceedings of the International Conference on Machine Learning, Lille, France, 6–11 July 2015.

Sample Availability: Samples of the compounds are not available from the authors.

molecules

MDPI

Article

Optimized Prediction of Reducing Sugars and Dry Matter of Potato Frying by FT-NIR Spectroscopy on Peeled Tubers

Cédric Camps * and Zo-Norosoa Camps

Institute for Plant Production Sciences IPS, Agroscope, CH-1964 Conthey, Switzerland; zonorosoa@gmail.com
* Correspondence: cedric.camps@agroscope.admin.ch

Academic Editors: Christian Huck and Krzysztof B. Bec
Received: 23 January 2019; Accepted: 4 March 2019; Published: 9 March 2019

Abstract: Dry matter content (DMC) and reducing sugars (glucose, fructose) contents of three potato varieties for frying (Innovator, Lady Claire, and Markies) were determined by applying Fourier-transform near-infrared spectrometry (FT-NIR), with paying particular attention to tubers preparation (unpeeled, peeled, and transversally cut tubers) before spectral acquisitions. Potatoes were subjected to normal storage temperature as it is processed in the industry (8 °C) and lower temperature inducing sugar accumulations (5 °C) for 195 and 48 days, respectively. Prediction of DMC has been successfully modeled for all varieties. A common model to the three varieties reached R^2, root mean square error (RMSEP), and ratio performance to deviation (RPD) values of 0.84, 1.2, and 2.49. Prediction accuracy of reducing sugars was variety dependent. Reducing sugars were accurately predicted for Innovator ($R^2 = 0.84$, RMSEP = 0.097, and RPD = 2.86) and Markies ($R^2 = 0.78$, RMSEP = 0.033, and RPD = 2.15) and slightly less accurate for Lady Claire ($R^2 = 0.63$, RMSEP = 0.036, and RPD = 1.64). The lack of accuracy obtained with the Lady Claire variety is mainly due to the tight variability in sugar content measured over the storage. Finally, the best preparation of the tuber from the point of view of the accuracy of the prediction models was to use the whole peeled potato. Such preparation allowed for the improvement in RPD values by 15% to 38% the RPD values depending on reducing sugars and 35% for DMC.

Keywords: Fourier-transform near-infrared spectroscopy; glucose; fructose; dry matter; partial least square regression

---

1. Introduction

The potato is the fifth most produced agricultural product with 388 million tons, behind sugar cane (1.84 trillion tons), corn (1.13 trillion tons), wheat (771 million tons), and rice (769 million tons) [1]. It can be consumed fresh, dehydrated, or fried as a snack.

Around the world, the potato crisp is a particularly popular snack. At the industrial level, the preparation of chips requires particular attention to the formation of acrylamide. This chemical compound is suspected to of being carcinogenic to humans, and its evolution in industrial products is closely monitored. During cooking, reducing sugars and asparagine interact can lead to browning on the outline and sometimes the whole of the chips, as a result of the reaction of Maillard [2]. The more the potato is rich in reducing sugars (glucose, fructose, and sucrose), the more the browning is visible. Whether for health or commercial reasons, the industry seeks to limit this browning. To limit the formation of acrylamide in food products, FoodDrinkEurope [3] has published tools in the form of "toolboxes" that are intended for the food industry. Among the first recommendations is the need to use potato varieties with low levels of reducing sugars. Some studies have already shown that there is a strong link between the reducing sugar content and the final level of acrylamide [4,5]. From both sanitary and aesthetic points of view, it is important for the industry to find an effective and accurate way to quantify the reducing sugars in their raw material, the fresh potato.

Indeed, the choice of a variety with low content of reducing sugars is a first big step, but there remains a consideration for an intra-varietal variability of reducing sugar content point of view. As a result, the method to be developed by the industry must be able to scan the largest number of individual tubers along the supply and processing chain.

During the last twenty years, near-infrared spectroscopy has been developed as a fast, precise, and mostly non-destructive method for the quality control of the agri-food sector. This method developed in the laboratory has the potential to be implemented in the industry for a chain analysis. Studies described the ability to quantify sugar content in fresh apples [6,7]. Subsequently, this method of measuring sugars has been extended to other fruits [8,9]. Studies have been conducted on potatoes to test the possibility of using near-infrared spectroscopy to measure the sugar or dry matter content of potatoes [10–14]. These studies are difficult to compare because some focused on whole tubers, whole and peeled, in cross-sections, or crushed in the form of puree. However, these studies have shown the potential of this method for application on the potato industry.

In general, the Swiss potato industry only accepts tubers with a reducing sugar content of less than 0.1% (*w/w*) for the manufacture of potato chips. The aim of the present study will be to develop predictive models for reducing sugars and dry matter content of fresh potato tubers based on Fourier-transform near-infrared spectroscopy (FT-NIRs). The novelty of this work is to use the advantages of Fourier-transform spectroscopy compared to spectroscopy from sequential instruments using monochromators or filters and multichannel instruments using diode arrays. The advantages of FT-NIR spectroscopy are significant, e.g., the usage of the interferometer saves time in acquiring spectra (Fellgett advantage) and allows higher throughput by passing through a larger amount of the NIR radiation, which can then be emitted or reflected by the sample (Jaquinot advantage) [15]. From the point of view of the potential future application in industry, it is important to note that FT-NIR spectroscopy has a lower sensitivity to stray light compared to monochromator and diode array devices. Such an advantage is significant since, in the industry, the environmental conditions can be more difficult to control compared to those at the lab [16]. Finally, these advantages are not detrimental to precision because the FT-NIR spectroscopy allows the highest precision of wavenumber (Connes' advantage). FT-NIR spectrometers present the best wavelength precision, accuracy, high signal-to-noise, and scan speed [17]. In this way, the FT-NIR-based technique is fully capable of being transferred from a lab to the potato industry. To date, studies aiming at predicting dry matter and reducing sugars in potatoes have been carried out using monochromators or diode array-based instruments and often present a reduced range of NIR spectral absorbance (i.e., short-wave NIR wavelengths or higher than 1100 nm). The present study aims to highlight the potential of FT-NIR spectroscopy to predict dry matter and reducing the sugar content of potato tubers.

In order to optimize the method, three types of fresh tuber preparations were tested to decide the optimal preparation for maximum accuracy of the prediction models. Finally, the study worked on three varieties adapted to the potato industry, and the models were calibrated and validated with potato provided by industrials gathering tubers from different producers and different production sites.

2. Results and Discussion

2.1. Prediction of Dry Matter Content

Dry matter content (DMC) of individual potatoes was monitored during storage at 5 °C and 8 °C for the three varieties. Before storage, Lady Claire presented significantly ($p < 0.0001$) higher DMC values (26%) compared to Markies (23%) and Innovator (24%). This range of values is consistent with those found in several studies and in particular that of Elmore et al. [18] which compared the DMC of 20 UK-grown varieties intended for frying. The authors determined average values of DMC in a range of 17% to 28% depending on the variety, with Lady Claire being around 26%, and Innovator and Markies around 24%.

During storage at 5 °C, DMC remained steady while higher values were measured after 62 days at 8 °C for Lady Claire ($p < 0.0001$) and after 195 days at 8 °C for Markies ($p < 0.0001$) and Innovator ($p < 0.0001$). Finally, DMC values ranged from 18% to 40% depending on the variety and the storage modalities.

Then, DMC values were used to elaborate prediction models based on FT-NIR spectral data (Table 1). More than 400 and 100 spectra were used to calibrate and validate the models, respectively. Three models were attempted, with the first one based on spectra acquired on entire and unpeeled potatoes (PDTE), the second one with spectra of entire and peeled potatoes (PDTP), and the third one with spectra acquired on transversal cuts of potatoes (PDTC). The best result was obtained with spectra based on PDTP. This model presents the highest R^2 value (0.84), and the lowest root mean square error (RMSE) value (1.23%).

Table 1. Partial least square values of dry matter content prediction. PDTE: entire and unpeeled potatoes, PDTP: entire and peeled potatoes; PDTC: potatoes cut transversally.

PLS Parameters		PDTE	PDTP	PDTC
Spectra (n)	CAL/VAL	420/105	417/104	417/103
Averaged spectra (n)	CAL/VAL	140 (140) [1]/35	139 (140) [1]/35	140 (140) [1]/35
Wavenumber range (cm^{-1})		9403.8–7498.4 6102.1–5446.3	9403.8–7498.4 6102.1–5446.3	9403.8–7498.4 6102.1–5774.2 4601.6–4246.8
LV	CAL/VAL	7/7	7/7	7/7
R^2	CAL/VAL	0.76/0.70	0.84/0.83	0.83/0.78
RMSE (%)	CAL/VAL	1.59/1.66	1.27/1.23	1.34/1.40
RPD	CAL/VAL	2.05/1.85	2.55/2.49	2.45/2.17
Data preprocessing	**CAL/VAL**	**LOS**	**D1 + SNV**	**SNV**
DMC values (min–max) (g$_{DW}$/100 g$_{FW}$)	CAL/VAL	18–41/18–37	18–41/18–37	18–40/18–37
DMC Standard error (%)	CAL/VAL	3.22/3.07	3.22/3.07	3.22/3.07

[1] number of averaged spectra (tubers) before elimination of outlayers. CAL: calibration; VAL: validation; n: number of spectra or potato samples; LV: the number of latent variables; R^2: determination coefficient; RMSE: root mean square error; RPD: ratio performance to deviation; DMC (%): range of dry matter content values used in the models; DW: dry weight; FW: fresh weight. LOS: Linear offset subtraction; SNV: Standard Normal Variate; D1: first derivative.

Furthermore, the model used only five latent variables while seven have been required for the PDTE- and PDTC-based models. In the present modeling, all potato varieties were successfully gathered for elaborating a "trans-varietal" model (Figure 1A). Modeling with separating the varieties allowed us to slightly improve the model performances of the Lady Claire variety in terms of ratio performance to deviation (RPD) value (3.05) (Figure 1C). Concerning the Innovator and Markies varieties, models were less accurate (Figure 1B,D).

The performance of the model gathering all varieties is comparable to that obtained by Helgerud et al. [19] who reached R^2 and RMSE values of 0.8–0.9 and 0.9–1.7, respectively. However, the model developed with peeled potatoes only used five latent variables while the model developed by Helgerud et al. [19] used between five and nine latent variables. When the models were separately built per each variety, the accuracy increased until reaching R^2 and RMSE values of 0.89 and 0.9%, respectively, for the Lady Claire variety. The performances were not improved for the two other varieties since the ranges of DMC values were very tight. Hartmann and Büning-Pfaue [20] were one of the first to study the prediction of DMC of potatoes using NIR spectral data acquired on peeled potato. They concluded that the accuracy of models was cultivar-dependent. In the present study, the feasibility of a model gathering the three cultivars shows that it is possible to predict DMC without developing a cultivar-dependent model. Subedi and Walsh [21] reported accurate predictions of DMC in potatoes using short-wave NIR spectral data with R^2 values that ranged between 0.80 and 0.95 and RMSECV values that ranged from 0.5% to 1.52% depending on the cultivar. The authors used a batch of potatoes with a range of DMC values of 17% to 25%. The results obtained in the present study using FT-NIR spectroscopy allowed for obtaining similar accuracy for a range of DMC values of 18% to 40%.

DMC is crucial for the quality of chips or different forms of potato-based frites. Indeed, several studies showed the tight relationship existing between DMC and starch content [22,23]. Monitoring the DMC of fresh potatoes with FT-NIR spectroscopy could be an indirect indicator of starch content-related quality.

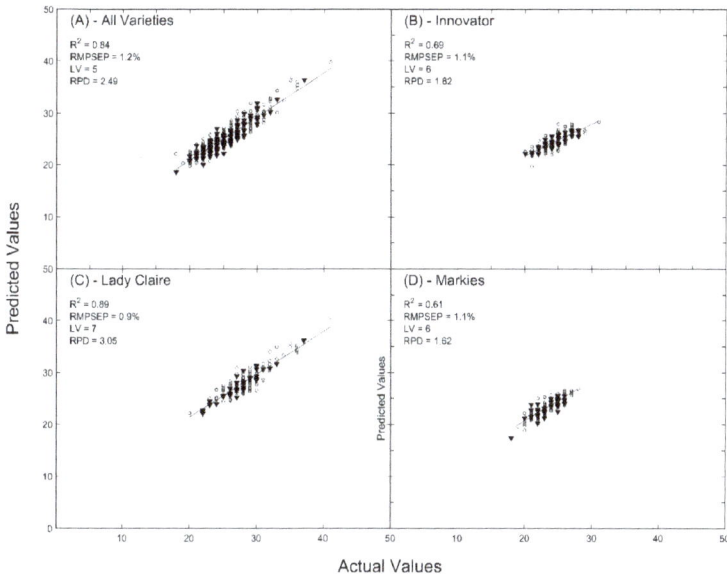

Figure 1. Actual vs. predicted values of dry matter content (DMC) (g of dry weight/100 g fresh weight). Calibration (○), validation (▲). (**A**) Entire and peeled tubers of the three tested varieties, (**B**) Entire and peeled tubers of Innovator, (**C**) Entire and peeled tubers of Lady Claire, (**D**) Entire and peeled tubers of Markies.

Figure 2 shows the beta-coefficients of the first latent variable of the Partial least square models predicting the DMC of peeled, entire, and cut potatoes. Peeled potato model is based on two wavelength bands (1060–1330 nm and 1640–1830 nm). Such bands are mainly related to the second and first C–H overtones, respectively. Subedi and Walsh [21] identified the short-wave NIR region (750–950 nm) and particularly high importance of absorbance at 910 nm as significant to predict DMC in potatoes. This region is essentially related to absorption bands of the third overtone of CH and NH. In their study, the authors used a spectrometer whose wavelengths did not exceed 1100 nm, as they did not have access to the first overtones and combinations of CH and OH molecular bonds. In the present study, the two bands identified as significant in predicting DMC corresponded also to NH and CH molecular bonds, but these are the first and second overtones. The absorbance in the short-wave NIR range did not appear as significant. Finally, it can be considered that the main difference between the study of Subedi and Walsh [21] and the present study rely on the wavelength range availability due to the spectrometer. Hartmann and Büning-Pfaue [20] predicted DMC using a spectrometer with a wavelength range comprised between 1100 nm and 2500 nm. The accuracy of the models was correct, but no information about beta-coefficients was provided to determine the significance of wavelength absorbencies.

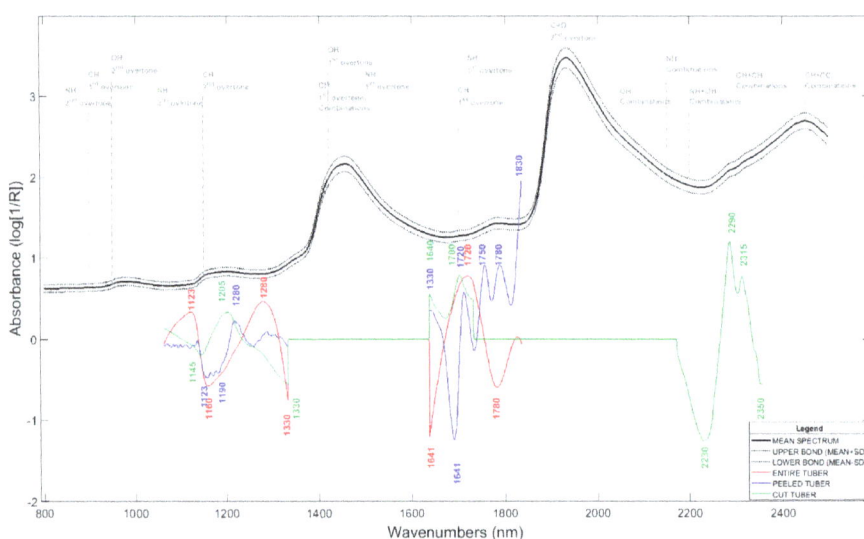

Figure 2. Mean spectra and beta-coefficients of the first PLS models latent variable to predict the DMC based on spectral data acquired on entire tubers (red line), peeled tubers (blue line), and cut tubers (green line).

2.2. Prediction of Reducing Sugars

Reducing sugars were measured in individual potatoes during storage of the three varieties (Table 2). Sugar levels remained low (under 0.1%) over the 195 days of storage at 8 °C for all varieties. During storage at 5 °C, the levels of sugars were a function of the variety. Sugar contents increased slightly for Markies (0.14% fresh weight, FW) and Lady Claire (0.21% FW), and more strongly for the Innovator (0.70% FW) variety (Table 2).

Table 2. Sugar values of individual potatoes. Sugar values were statistically analyzed by the non-parametric test of Kruskal–Wallis ($p = 0.05$); mean values comparisons have been processed by the Dun test. RS: Reducing sugars, FW: fresh weight.

Duration (Days)	Temp. (°C)	Glucose (% FW)			Fructose (% FW)			RS (Glucose + Fructose) (% FW)		
		IN	LC	MA	IN	LC	MA	IN	LC	MA
0	-	0.07 a,b	0.03 c	0.02 b	0.05 b,c	0.01 b,c	0.01 b	0.11 a,b	0.04 b	0.04 b
6	5	0.11 b	0.03 c	0.02 b	0.09 c	0.03 c,d	0.02 b	0.20 b	0.06 b	0.04 b
24	5	0.30 c	0.05 c,d	0.08 c	0.27 d	0.05 d,e	0.10 c	0.57 c	0.10 b,c	0.18 c
48	5	0.37 c	0.07 d	0.10 c	0.33 d	0.07 e	0.11 c	0.70 c	0.14 c	0.21 c
6	8	0.08 a	0.02 b,c	0.02 b	0.06 b,c	0.01 c,d	0.01 b	0.14 a,b	0.03 b	0.03 b
62	8	0.03 a	0.01 a,b	0.01 a,b	0.02 a,b	0.00 a,b	0.01 a,b	0.06 a	0.01 a	0.01 a,b
195	8	0.04 a	0.00 a	0.00 a	0.01 a	0.00 a	0.00 a	0.05 a	0.01 a	0.00 a
p-value		<0.0001	<0.0001	<0.0001	<0.0001	<0.0001	<0.0001	<0.0001	<0.0001	<0.0001

Means followed by different letters within one column differ significantly at $p = 0.05$.

Predictions of sugars contents were attempted for each potato variety. Prediction values of the Innovator, Lady Claire, and Markies varieties have been gathered in Tables 3–5. Predictions of glucose, fructose, and reducing sugars (glucose + fructose) were performed using FT-NIR data acquired on entire and unpeeled potatoes (PDTE), entire and peeled potatoes (PDTP), and transversally cut potatoes (PDTC).

Potatoes preparation for FT-NIR spectra acquisition affected the overall performance of models (Figure 3).

Table 3. PLS values of sugar content prediction of the potato variety Innovator.

PLS Score		Glucose			Fructose			Fructose + Glucose		
		PDTE	PDTP	PDTC	PDTE	PDTP	PDTC	PDTE	PDTP	PDTC
Spectra	CAL	420	405	405	402	402	405	405	405	408
	VAL	105	102	99	99	105	99	102	105	99
Averaged spectra	CAL	140 (140) [1]	135 (140)	135 (140)	134 (140)	134 (140)	135 (140)	135 (140)	135 (140)	136 (140)
	VAL	35 (35)	34 (35)	33 (35)	33 (35)	35 (35)	33 (35)	34 (35)	35 (35)	33 (35)
WL (cm^{-1})		9403.8–7498.4; 6102.1–5446.3	9403.8–7498.4; 5774.2–5446.3	9403.8–7498.4	8451.1–7498.4; 5774.2–5446.3	9403.8–7498.4	9403.8–7498.4	9403.8–7498.4; 6102.1–5446.3	9403.8–7498.4	9403.8–7498.4
LV		7	6	7	8	6	6	7	6	6
R^2	CAL	0.56	0.71	0.82	0.72	0.71	0.77	0.68	0.69	0.76
	VAL	0.70	0.82	0.79	0.84	0.82	0.85	0.77	0.84	0.80
RMSE (%)	CAL	0.108	0.076	0.061	0.065	0.071	0.061	0.149	0.151	0.129
	VAL	0.084	0.059	0.066	0.050	0.050	0.049	0.120	0.097	0.113
RPD	CAL	1.51	1.87	2.35	1.9	1.87	2.08	1.75	1.8	2.05
	VAL	1.89	2.34	2.25	2.5	2.43	2.59	2.07	2.86	2.23
Data Preprocessing		SNV	SNV	LOS	MMN	None	LOS	MMN	None	LOS
Sugar values (% FW)	CAL	0–0.709	0–0.596	0–0.613	0–0.417	0–0.539	0–0.564	0–0.969	0–1.135	0–1.178
	VAL	0–0.494	0–0.476	0–0.493	0–0.39	0–0.39	0–0.39	0–0.779	0–0.779	0–0.779

[1] number of averaged spectra (tubers) before elimination of outlayers. CAL: calibration; VAL: validation; spectra: number of spectra or potato samples; LV: the number of latent variables; R^2: determination coefficient; RMSE: root mean square error; RPD: ratio performance to deviation; SNV: Standard Normal Variate; LOS: Linear offset subtraction; MMN: Min–Max Normalization.

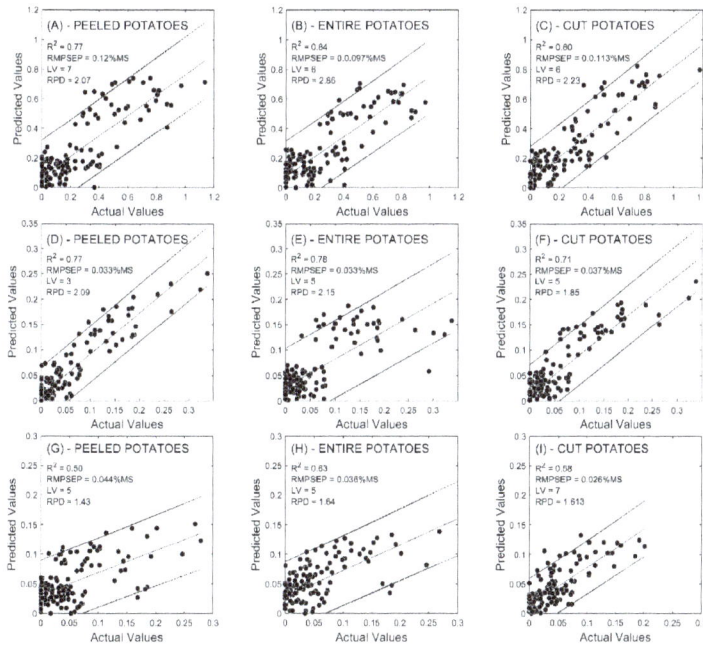

Figure 3. Actual vs. predicted values of Innovator's (**A–C**), Markies' (**D–F**) and Lady Claire's (**G–I**) reducing sugars contents. Predictions were performed based on "Entire and peeled" (**A,D,G**) tubers, "Entire and unpeeled" tubers (**B,E,H**), and tubers "cut transversally" (**C,F,I**).

PDTP (peeled potatoes) configuration allowed us to reach the most accurate models. Reducing sugar (fructose + glucose) were predicted with RPD values of 2.07 (PDTE), 2.86 (PDTP), and 2.23 (PDTC) for Innovator, 2.09 (PDTE), 2.15 (PDTP), and 2.1.85 (PDTC) for Markies, and 1.43 (PDTE), 1.64 (PDTP), and 1.61 (PDTC) for Lady Claire. In the same way, R^2 and RMSE values were generally favored by the PDTP configuration (Tables 3–5).

Table 4. PLS values of sugar content prediction of the potato variety Markies.

PLS Score		Glucose			Fructose			Fructose + Glucose		
		PDTE	PDTP	PDTC	PDTE	PDTP	PDTC	PDTE	PDTP	PDTC
Spectra	CAL	402	396	405	405	405	405	411	405	405
	VAL	102	102	102	102	99	99	102	99	102
Averaged spectra	CAL	134 (140)[1]	132 (140)	135 (140)	135 (140)	135 (140)	135 (140)	137 (140)	135 (140)	135 (140)
	VAL	34 (35)	34 (35)	34 (35)	34 (35)	33 (35)	33 (35)	34 (35)	33 (35)	34 (35)
WL (cm⁻¹)		8451.1–7498.4	9403.8–8451.1; 6102.1–5774.2	9403.8–8451.1; 6102.1–5446.3	8451.1–7498.4	9403.8–8451.1; 6102.1–5774.2	9403.8–7498.4; 5774.2–5446.3	8451.1–7498.4	9403.8–7498.4; 6102.1–5446.3	9403.8–7498.4; 5774.2–5446.3
LV		5	9	4	3	9	9	3	5	5
R²	CAL	0.71	0.83	0.72	0.61	0.83	0.88	0.55	0.81	0.78
	VAL	0.80	0.75	0.70	0.75	0.81	0.77	0.77	0.78	0.71
RMSE (%)	CAL	0.019	0.013	0.019	0.022	0.016	0.013	0.051	0.031	0.034
	VAL	0.015	0.017	0.019	0.018	0.016	0.017	0.033	0.033	0.037
RPD	CAL	1.86	2.4	1.89	1.6	2.39	2.94	1.5	2.31	2.14
	VAL	2.29	2.14	1.83	2.01	2.4	2.11	2.09	2.15	1.85
Data Preprocessing		SLS	None	MSC	1st der.	None	None	1st der.	D1 + MSC	MSC
Sugar values (% FW)	CAL	0–0.165	0–0.128	0–0.165	0–0.173	0–0.173	0–0.173	0–0.338	0–0.338	0–0.338
	VAL	0–0.115	0–0.115	0–0.115	0–0.119	0–0.119	0–0.119	0–0.235	0–0.235	0–0.235

[1] number of averaged spectra (tubers) before elimination of outlayers. CAL: calibration, VAL: validation; spectra: number of spectra or potato samples; LV: the number of latent variables; R²: determination coefficient; RMSE: root mean square error; RPD: ratio performance to deviation; MSC: Multiplicative scatter correction; 1st der.: first derivative; SLS: Straight line subtraction.

Table 5. PLS values of sugar content prediction of the potato variety Lady Claire.

PLS Score		Glucose			Fructose			Fructose + Glucose		
		PDTE	PDTP	PDTC	PDTE	PDTP	PDTC	PDTE	PDTP	PDTC
Spectra	CAL	396	393	408	408	411	411	414	411	393
	VAL	96	105	102	99	105	102	102	105	96
Averaged spectra	CAL	132 (140)[1]	131 (140)	136 (140)	136 (140)	137 (140)	137 (140)	138 (140)	137 (140)	131 (140)
	VAL	32 (35)	35 (35)	34 (35)	33 (35)	35 (35)	34 (35)	34 (35)	35 (35)	32 (35)
WL (cm⁻¹)		7502.2–6098.2	9403.8–7498.4	9403.8–7498.4; 6102.1–5446.3	9403.8–7498.4; 6098.2	9403.8–7498.4; 4601.6–4246.8	9403.8–7498.4; 6102.1–5446.3	9403.8–7498.4	6102.1–5446.3	9403.8–7498.4; 6102.1–5446.3
LV		6	5	6	9	14	6	5	5	7
R²	CAL	0.55	0.57	0.61	0.57	0.64	0.59	0.45	0.38	0.64
	VAL	0.57	0.52	0.61	0.67	0.63	0.50	0.50	0.63	0.58
RMSE (%)	CAL	0.015	0.018	0.018	0.020	0.019	0.018	0.045	0.047	0.028
	VAL	0.016	0.014	0.018	0.016	0.017	0.018	0.044	0.036	0.026
RPD	CAL	1.49	1.52	1.6	1.74	1.68	1.56	1.35	1.27	1.66
	VAL	1.56	1.47	1.62	1.52	1.83	1.66	1.43	1.64	1.61
Data Preprocessing		D1 + SNV	LOS	SLS	MMN	D1	SNV	SNV	None	LOS
Sugar values (% FW)	CAL	0–0.103	0–0.131	0–0.131	0–0.138	0–0.147	0–0.138	0–0.300	0–0.279	0–0.202
	VAL	0–0.102	0–0.077	0–0.117	0–0.122	0–0.122	0–0.122	0–0.239	0–0.239	0–0.144

[1] number of averaged spectra or potatoes before elimination of outlayers. CAL: calibration, VAL: validation; spectra: number of spectra or potato samples; LV: the number of latent variables; R²: determination coefficient; RMSE: root mean square error; RPD: ratio performance to deviation; SNV: Vector normalization; LOS: Linear offset subtraction; MMN: Min–Max Normalization; D1: first derivative; SLS: Straight line subtraction.

In a study, Rady and Guyer [14] worked on the prediction of sugar by near-infrared spectroscopy using whole or sliced potatoes. They showed few differences between the two potato preparations from a model accuracy point of view. This result was similar in our study since the results obtained with PDTE and PDTC were quite similar and less accurate than those obtained with the PDTP. A reason why the PDTP was the most efficient method could be due to the fact that the reducing sugars are concentrated at the periphery of the tuber. On whole tuber, Chen et al. [24] predicted sugar contents with RMSE of 0.26 mg/g (0.026%) for fructose and 0.46 mg/g (0.046%) for glucose. For this, they used a range of NIR wavelength of 400–1100 nm. These results are comparable to ours, although the latter is strongly related to the variety and the total reducing sugar content of potatoes. Furthermore, Chen et al. [24] used particularly drastic conditions of storage, such as 25 °C storage temperature during several months. Such a storage parameter is not representative of the real storage rules for potatoes and induced very high sugar levels in tubers. Consequently, the tested storage conditions would be a simulation and consequence of at-home consumers' storage rather than professional storage conditions. The very large range of sugar content allowed Chen et al. (2010) to develop a correct model in terms of accuracy. Rady and Guyer [14] obtained results showing the variability of model accuracy as a function of the used variety. They obtained R^2 values from 0.55 to 0.88 and RPD values from 1.49 to 2.73. However, it benefited from varieties with a greater range of sugar concentration than the varieties used in our study. Our lower sugar range gives our results a lot of room for improvement and the performance of our models could be increased by adding varieties with other ranges of sugar values.

In the present study, we considered that two of the three tested varieties (Innovator and Markies) were suitable for building a prediction model using near-infrared spectral data, but the last variety was inadequate (Lady Claire). The latter proved to be very insensitive to storage conditions and, as a result, the variations in sugar content were very small. However, this variety is not the right candidate to build our models, but it remains a good candidate for the French fries potato industry.

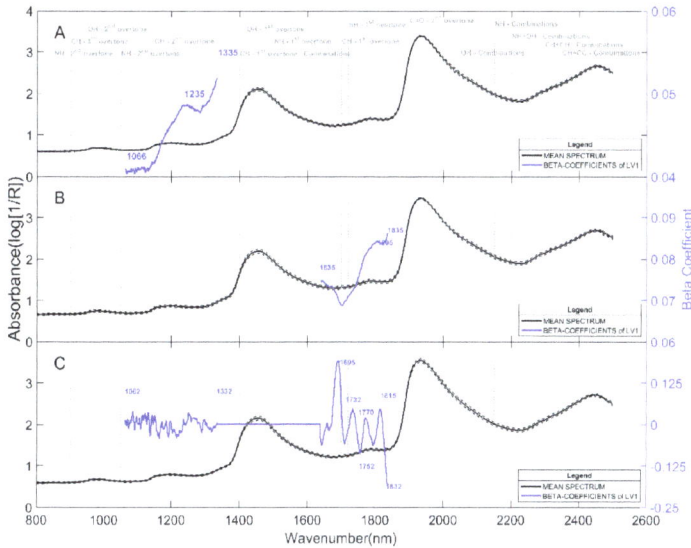

Figure 4. Mean spectra and beta-coefficients of the first PLS models latent variable to predict the reducing sugars of the three potato varieties: Innovator (**A**), Lady Claire (**B**), and Markies (**C**).

Figure 4 shows the beta-coefficients of the first latent variable of the PLS models predicting the reducing sugars of peeled tubers for the three varieties. Prediction of reducing sugars is based on wavelength bands between 1065 nm and 1335 nm, 1635 nm and 1835 nm, and both bands for Innovator, Lady Claire, and Markies varieties, respectively. Thus, models relied on second and first overtones of CH molecular bonds. Chen et al. [24], who used only a short-wave NIR range, identified relevant absorbencies in the vicinity of 710 nm (fourth overtone CH) and 888 nm (third overtone CH), but also bands around 950–960 nm (water band, second overtone OH) and 1020 nm (second overtone NH). The two first bands are consistent with the results obtained in the present study. The water bands were removed from the range used in our models (around 950 nm, 1450 nm, and 1900–1950 nm). The Relevance of wavelength absorbance at 910 nm to predict sugars in potatoes has been assigned by previous studies and is confirmed by the present FT-NIR approach [25]. The use of the full NIR range allowed for assigning other overtones of CH bonds to reduce sugar of potatoes. Finally, the FT-NIR approach suggests a possible transfer from the laboratory to the industry in the medium term. However, the models will still have to be enriched with other varieties of potatoes with different sugar content variations during storage to make the models more robust and confirm the relevant wavelengths already assigned.

3. Materials and Methods

3.1. Potatoes

Three potato varieties for frying were used in the present study: Lady Claire, Innovator, and Markies. In order to increase the variability of quality due to the varieties, the potatoes were stored into two different conditions. The first batch was stored in classical conditions at 8 °C and a second one at 5 °C to increase the sugar contents over time. A first batch of potatoes was analyzed some days after harvest. Then, potato batches stored at 5 °C were analyzed after 7, 21, and 49 days while potatoes stored at 8 °C were analyzed after 7, 62, and 195 days. A given batch was constituted of 25 potatoes per variety. A total of 525 tubers were analyzed.

3.2. Dry Matter Content

A sample of each fresh potato (a cube of about 2 cm on each side) was weighed to obtain the fresh weight (FW). Then, the sample was placed in a dryer at 70 °C for 6 days and weighed again to obtain the dry weight (DW). The relative content of dry matter (DMC) was calculated according to Equation (1).

$$\text{DMC}(\%) = \left[1 - \left(\frac{\text{FW(g)} - \text{DW(g)}}{\text{FW(g)}}\right)\right] \times 100 \tag{1}$$

3.3. Reducing Sugars Analyses

Sugar analyses were performed using enzymatic tests (Enzytec™ Fluid D-Glucose, r-Biopharm, Darmstadt, Germany). The limit of detection (LoD) and quantification (LoQ) calculated according to the method DIN 32645:2008-11 were 4.0 mg/L and 10 mg/L, respectively (Enzytec™ Fluid D-Glucose, r-Biopharm, Germany). A repeatability calculation was performed on 19 potato samples and duplicated. Relative standard errors of 1.6% and 1.7% were calculated for the glucose and fructose, respectively.

3.3.1. Extraction

Approximately 1 g of the potato powder was weighed into a 50 mL centrifuge tube. A total of 10 mL of ethanol at 40% was added. The mixture was then extracted for 1 min using an Ultra-Turrax (Polytron PT3100/Polytron PT 10 20 3500, Kinematica AG, Luzern, Switzerland) on the highest rpm setting. After that, the tubes were centrifuged for 5 min at 4000 rpm. The extract was filled into a 50 mL measuring flask. This extraction was repeated twice, and extracts were pooled together. Tubes were then filled to the calibration mark with ethanol at 40%.

3.3.2. Quantification of Free Sugars

A total of 10 mL of the ethanol extract was pipetted into a 25 mL pointed flask. The solvent was then evaporated using a Rotavap at 60 °C. The residue was solved in 2 mL of distilled water. The results were expressed in milligrams of sugar contained in 100 g of potato fresh weight. An aliquot of the sample was filtered through a 0.45 μm syringe filter (nylon). Glucose/fructose contents of this filtered sample were quantified by photometric methods using the Konelab Arena 20XT (Thermo Fisher Scientific OY, Vantaa, Finland).

3.4. FT-NIR Spectroscopy

FT-NIR measurements (MPA, Bruker, Fällanden, Switzerland) were carried out according to 3 configurations, all using an optical fiber. Spectral acquisitions were performed on (1) entire and unpeeled potatoes (PDTE), (2) entire and peeled potatoes (PDTP), and (3) potatoes cut transversally (PDTC). Spectra were acquired in diffuse reflexion using an optic fiber. A total of 3 spectra were recorded per potato. A given spectra was the average of 16 scans in a wavenumber range comprised between 12,500 cm^{-1} and 4000 cm^{-1}. A total of 4725 spectra was collected in the present study (4725 = (25 potatoes) $\times$ (3 varieties) $\times$ (7 batches of storage) $\times$ (3 spectra per potato) $\times$ (3 configurations of spectra measurements)). All spectra were collected with the OPUS software (Bruker, Germany).

3.5. Chemometric

Partial Least Square Regression

Averaged spectra of each tuber were gathered in a matrix X(n,p) where 'n' is the number of spectra and 'p' is the number of wavenumber steps. The reference values (sugars) were gathered in a column vector y(n,1). Potato batches were separated in a calibration set (3n/4) and a test set (n/4). The accuracy and goodness of models were evaluated according to several indicators: the coefficient of determination (R^2), root mean square errors (RMSE), and the ratio performance to deviation (RPD) [15]. All data analyses were performed with OPUS software and Matlab R2016a (The MathWorks, Inc., Natick, MA, USA).

4. Conclusions

The potato industry needs to precisely monitor the quality of the potatoes along the supply chain to ensure the optimal quality of the final product both in terms of aesthetics and sanitation. The Fourier-transform near-infrared spectroscopy is a possibility since this technology is already implemented in various food industries and presents significant advantages compared to monochromator and diode array-based NIR instruments. The study presented in this paper aimed to evaluate the possibility of determining the dry matter and reducing sugars contents in fresh potatoes which are the primary matters of the industry. The results obtained are promising in terms of accuracy. In addition, sample preparation is important when working with NIR spectroscopy. The present study tested three different tuber preparations in order to optimize the configuration of spectral acquisitions. The peeled but not necessarily crushed potato was determined to be the most interesting. This preparation has made it possible to obtain more precise models for sugar and dry matter contents. In particular, it has improved the RPD values from 15% to 38% for reducing sugars and 35% for DMC. Finally, since the robustness of the models is closely linked to the variability introduced, additional potato varieties adapted to the potato frying industry should be added to the present models. In addition, different storage conditions may be important to make the models more robust.

Author Contributions: C.C. contributed to the conceptualization of the project, supervision of the different tasks of the study, implementation of data analyses methodology, and partially writing the original draft. Z.-N.C. contributed to methodology, software, draft preparation, and data modeling. She mainly performed the practical manipulations such as spectral analyses and data management.

Funding: This research received no external funding.

References

1. FAOSTAT Food and Agriculture Organization of the United Nations. Available online: http://www.fao.org/faostat/fr/#data/QC (accessed on 7 January 2019).
2. Maillard, L.C. Action des acides aminés sur les sucres: formation des mélanoïdines par voie méthodique. *C. R. Hebd. Séances Acad. Sci.* **1912**, *154*, 66–68.
3. FoodDrinkEurope. A "Toolbox" for the Reduction of Acrylamide in Fried Potato Crisps. Available online: https://www.fooddrinkeurope.eu/uploads/publications_documents/crisps-EN-final.pdf (accessed on 7 January 2019).
4. Matthäus, B.; Haase, N.U.; Vosmann, K. Factors affecting the concentration of acrylamide during deep-fat frying of potatoes. *Eur. J. Lipid Sci. Technol.* **2004**, *106*, 793–801. [CrossRef]
5. Williams, J.S.E. Influence of variety and processing conditions on acrylamide levels in fried potato crisps. *Food Chem.* **2005**, *90*, 875–881. [CrossRef]

6. McGlone, V.A.; Jordan, R.B.; Seelye, R.; Clark, C.J. Dry-matter-a better predictor of the post-storage soluble solids in apples? *Postharv. Biol. Technol.* **2003**, *28*, 431–435. [CrossRef]

7. Peirs, A.S.N.; Touchant, K.; Nicolaï, B.M. Comparison of Fourier transform and dispersive near-infrared reflectance spectroscopy for apple quality measurements. *Biosys. Engineer.* **2002**, *81*, 305–311. [CrossRef]

8. Garcia-Jares, C.M.; Medina, B. Application of multivariate calibration to the simultaneous routine determination of ethanol, glycerol, fructose, glucose and total residual sugars in botrytized-grape sweet wines by means of near-infrared reflectance spectroscopy. *Fresen. J. Anal. Chem.* **1997**, *357*, 86–91. [CrossRef]

9. Camps, C.; Christen, D. Non-destructive assessment of apricot fruit quality by portable visible-near infrared spectroscopy. *Lwt-Food Sci. Technol.* **2009**, *42*, 1125–1131. [CrossRef]

10. Mehrübeoğlu, M.; Coté, G.L. Determination of total reducing sugars inpotato samples using near-infrared spectroscopy. *Cereal Food. World* **1997**, *42*, 409–413.

11. Van Dijk, C.; Fischer, M.; Holm, J.; Beekhuizen, J.G.; Stolle-Smits, T.; Boeriu, C. Texture of cooked potatoes (*Solanum tuberosum*). 1. Relationships between dry matter content, sensory-perceived texture, and near-infrared spectroscopy. *J. Agric. Food Chem.* **2002**, *50*, 5082–5088. [CrossRef]

12. Chen, J.Y.; Miao, Y.; Zhang, H.; Matsunaga, R. Non-destructive determination of carbohydrate content in potatoes using near infrared spectroscopy. *J. Near Infrared Spec.* **2004**, *12*, 311–314. [CrossRef]

13. Haase, N.U. Prediction of Potato Processing Quality by near Infrared Reflectance Spectroscopy of Ground Raw Tubers. *J. Near Infrared Spec.* **2011**, *19*, 37–45. [CrossRef]

14. Rady, A.M.; Guyer, D.E. Evaluation of sugar content in potatoes using NIR reflectance and wavelength selection techniques. *Postharvest Biol. Tec.* **2015**, *103*, 17–26. [CrossRef]

15. Jacquinot, P. How the search for a throughput advantage led to Fourier transform spectroscopy. *Infrared Phys.* **1984**, *24*, 99–101. [CrossRef]

16. White, R. *Chromatography/Fourier Transform Infrared Spectroscopy and Its Applications*; CRC Press: New York, NY, USA, 1989; Volume 10.

17. Fernández Pierna, J.A.; Manley, M.; Dardenne, P.; Downey, G.; Baeten, V. Spectroscopic Technique: Fourier Transform (FT) Near-Infrared Spectroscopy (NIR) and Microscopy (NIRM). In *Modern Techniques for Food Authentication*, 2nd ed.; Sun, D.-W., Ed.; Academic Press: London, UK, 2018; pp. 103–138.

18. Elmore, J.S.; Briddon, A.; Dodson, A.T.; Muttucumaru, N.; Halford, N.G.; Mottram, D.S. Acrylamide in potato crisps prepared from 20 UK-grown varieties: Effects of variety and tuber storage time. *Food Chem.* **2015**, *182*, 1–8. [CrossRef] [PubMed]

19. Helgerud, T.; Wold, J.P.; Pedersen, M.B.; Liland, K.H.; Ballance, S.; Knutsen, S.H.; Rukke, E.O.; Afseth, N.K. Towards on-line prediction of dry matter content in whole unpeeled potatoes using near-infrared spectroscopy. *Talanta* **2015**, *143*, 138–144. [CrossRef] [PubMed]

20. Hartmann, R.; Büning-Pfaue, H. NIR determination of potato constituents. *Potato Res.* **1998**, *41*, 327–334. [CrossRef]

21. Subedi, P.P.; Walsh, K.B. Assessment of Potato Dry Matter Concentration Using Short-Wave Near-Infrared Spectroscopy. *Potato Res.* **2009**, *52*, 67–77. [CrossRef]

22. Stark, J.C.; Love, S.L. Tuber quality. *Potato Prod. Sys.* **2003**, 329–343.

23. Storey, R.M.J. The harvested crop. In *Potato Biology and Biotechnology Advances and Perspectives*; Vreugdenhil, D., Ed.; Elsvier: London, UK, 2007; pp. 441–450.

24. Chen, J.Y.; Zhang, H.; Miao, Y.; Asakura, M. Nondestructive determination of sugar content in potato tubers using visible and near infrared spectroscopy. *Jpn. J. Food Eng.* **2010**, *11*, 59–64.

25. Yaptenco, K.F.; Suzuki, T.; Kawakami, S.; Sato, H.; Takano, K.; Kozima, T.T. Nondestructive determination of sugar content in 'Danshaku' potato (*Solanum tuberosum* L.) by near infrared spectroscopy. *J. Agric. Sci.* **2000**, *44*, 284–294.

Sample Availability: Samples are not available from the authors.

Article

Determination of Adulteration Content in Extra Virgin Olive Oil Using FT-NIR Spectroscopy Combined with the BOSS–PLS Algorithm

Hui Jiang [1],* and Quansheng Chen [2],*

[1] School of Electrical and Information Engineering, Jiangsu University, Zhenjiang 212013, China
[2] School of Food and Biological Engineering, Jiangsu University, Zhenjiang 212013, China
* Correspondence: h.v.jiang@ujs.edu.cn (H.J.); qschen@ujs.edu.cn (Q.C.)

Academic Editors: Christian Huck and Krzysztof B. Bec
Received: 15 May 2019; Accepted: 3 June 2019; Published: 6 June 2019

Abstract: This work applied the FT-NIR spectroscopy technique with the aid of chemometrics algorithms to determine the adulteration content of extra virgin olive oil (EVOO). Informative spectral wavenumbers were obtained by the use of a novel variable selection algorithm of bootstrapping soft shrinkage (BOSS) during partial least-squares (PLS) modeling. Then, a PLS model was finally constructed using the best variable subset obtained by the BOSS algorithm to quantitative determine doping concentrations in EVOO. The results showed that the optimal variable subset including 15 wavenumbers was selected by the BOSS algorithm in the full-spectrum region according to the first local lowest value of the root-mean-square error of cross validation (RMSECV), which was 1.4487 % v/v. Compared with the optimal models of full-spectrum PLS, competitive adaptive reweighted sampling PLS (CARS–PLS), Monte Carlo uninformative variable elimination PLS (MCUVE–PLS), and iteratively retaining informative variables PLS (IRIV–PLS), the BOSS–PLS model achieved better results, with the coefficient of determination (R^2) of prediction being 0.9922, and the root-mean-square error of prediction (RMSEP) being 1.4889 % v/v in the prediction process. The results obtained indicated that the FT-NIR spectroscopy technique has the potential to perform a rapid quantitative analysis of the adulteration content of EVOO, and the BOSS algorithm showed its superiority in informative wavenumbers selection.

Keywords: bootstrapping soft shrinkage; partial least squares; extra virgin olive oil; adulteration; FT-NIR spectroscopy

---

1. Introduction

With the rising prices of cooking oil, greedy traders and suppliers may resort to unethical practices, such as mixing low-value cooking oil with high-value cooking oil [1]. The consumers cannot detect these low-value, inexpensive ingredients in cooking oils, so they pay more for them. Extra virgin olive oil (EVOO) is native to the Mediterranean area, is known as "the gold of liquids", "the queen of plant oils", and "the Mediterranean nectar", and is an established Chinese consumer favorite [2]. The consumption of the EVOO has increased in recent years. However, the production of EVOO is not enough to cope with the growing consumer demand in China because of the demanding production conditions of EVOO. Therefore, EVOO adulteration has spread in the Chinese market. Adulteration not only causes confusion in the edible oil market but also violates the rights of consumers. Therefore, a fast and effective analytical method of EVOO adulteration is required to assist government's regulations.

Fourier transform near-infrared (FT-NIR) molecular spectroscopy is a technique widely applied in food quality analysis [3–6] that can provide abundant information about the chemical composition and molecular structure of various food substances. In addition, this technology also has the advantages of being non-destructive, fast, low-cost, with good reproducibility and broad application prospects. Recently, the FT-NIR spectroscopy technique has been extensively used in quality and safety analysis of EVOO [7–9]. In addition, other molecular spectroscopy techniques, such as fluorescence spectroscopy [10–12], infrared spectroscopy [13–15], Raman spectroscopy [16–18], and nuclear magnetic resonance spectroscopy [19,20], have good applications in the analysis of EVOO adulteration. With the technological developments, the amount of spectral data acquired is increasingly large because of the improvement of instrument resolution. Therefore, the selection of spectral characteristic wavenumbers plays an important role in spectral model development. Moreover, more and more researchers have proved that the selection of characteristic wavenumbers in the multivariable model calibration can not only improve the prediction performance of the chemometrics model but also enhance the interpretability of the model [21–24].

Partial least-square (PLS) regression is a statistical method related to principal component regression (PCR), which is to search a linear regression model by projecting predicted variables and observed variables into a new state space [25]. Because of the advantages of variable selection, many PLS-based feature variable selection algorithms have been developed [26], for example, the variable importance in projection (VIP) score [27], the successive projections algorithm (SPA) [28], the uninformative variable elimination (UVE) algorithm [29], and the selectivity ratio (SR) [30]. These methods were developed on the basis of the criteria of variable weights or regression coefficients. Additionally, some other feature wavenumber selection methods based on model population analysis (MPA) strategies have been developed [31], for instance, the iteratively retaining informative variables (IRIV) [32], the variable iterative space shrinkage approach (VISSA) [33,34], the variable combination population analysis (VCPA) [35], and the bootstrapping soft shrinkage (BOSS) [36]. Compared with IRIV, VISSA, and VCPA, an important feature of the BOSS algorithm is the introduction of weighted bootstrap sampling (WBS) criteria that the other three algorithms do not consider. Furthermore, different from other bootstrap-based algorithms, the BOSS algorithm performs the bootstrap criteria in the variable space, while other algorithms perform the criteria in the sample space. Thus, in this study, the BOSS algorithm was applied for the wavenumber selection of spectral data of EVOO doped samples.

The aim of this study was to verify the feasibility of establishing an improved and reliable reduced spectral model which can directly and quantitatively determine the doping content of EVOOs by their spectra. The feature wavenumbers were first selected by the BOSS algorithm, and a detection model based on the PLS regression using the selected wavenumbers by the BOSS algorithm was built. Finally, the performance of the reduced BOSS–PLS model was compared with the performances of the other three commonly used reduced models (i.e., competitive adaptive reweighted sampling PLS (CARS–PLS), Monte Carlo uninformative variable elimination PLS (MCUVE–PLS), and iteratively retaining informative variables PLS (IRIV–PLS)).

2. Results

2.1. Variable Selection by the BOSS Algorithm

In this study, the informative wavenumbers were firstly selected by using the BOSS algorithm during PLS modeling. A five-fold cross validation was used for the optimization of relevant parameters, and the optimal variables were finally determined according to the first local lowest root-mean-square error of cross validation (RMSECV) value. Before running the BOSS algorithm, the number of bootstrap sampling was set to 1000, and the maximum number of principal components (PCs) was set to 15. In this study, in order to verify the repeatability and stability of the algorithm, the approach was conducted repeatedly 10 times, and the best results were recorded.

Figure 1 shows the evolution of the variables and the value of RMSECV in each iteration of sub-models during the run of the BOSS algorithm. The number of wavenumbers selected decreased smoothly with iteration of the BOSS algorithm. The initial number of wavenumbers obtained was

1557 from the full spectrum. As can be seen in Figure 1a, the number of variables selected gradually decreased and became 1 after 14 iterations. Meanwhile, as can be seen in Figure 1b, the values of RMSECV in the sub-models decreased with the increase of the iteration number, reached the minimum value at the eighth iteration, and then started to rise slowly. The best variable subset was finally achieved in the eighth iteration, and the optimal number of wavenumbers selected was 15 at the eighth iteration, according to the first local lowest RMSECV, which was 1.4487 % v/v.

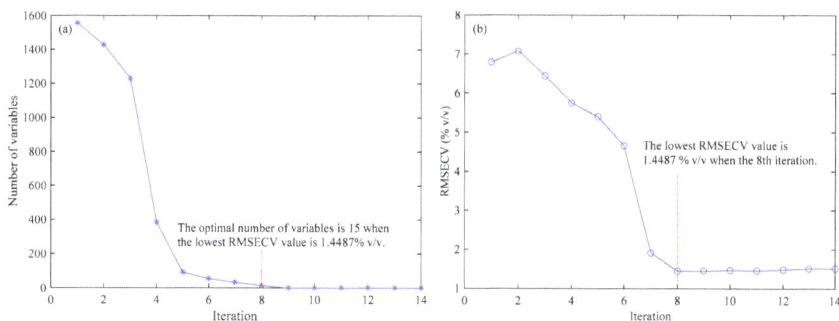

Figure 1. Evolution of the number of variables (**a**) and root-mean-square error of cross validation (RMSECV) (**b**) in each iteration of the sub-models using the bootstrapping soft shrinkage (BOSS) algorithm.

Figure 2 shows the weights and the wavenumbers distribution in the full spectrum of the 15 variables selected at the eighth iteration of the sub-models; it shows the 15 variables selected with their respective weights and the variable with the largest weight and highest importance. By investigating the results in Figure 2, the most informative wavenumbers were finally obtained at around 5900 cm^{-1}. Thus, the 15 variables selected by the BOSS algorithm constituted the best variable subsets for building the final PLS model.

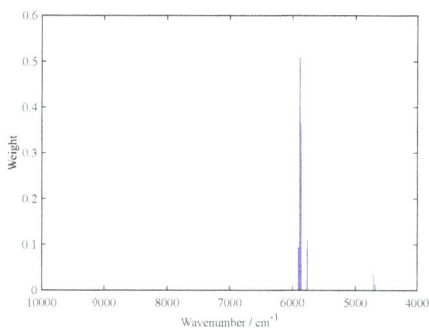

Figure 2. The weights of the variables in the optimal sub-model at the eighth iteration using the BOSS algorithm.

2.2. Results of the PLS Model

The optimal PLS model was built using the 15 wavenumbers selected by the BOSS algorithm when three PLS factors were included. The value of RMSECV was 1.4487 % v/v, and the R^2 was 0.9908 in the calibration set. The predictive accuracy and generalization performance of the constructed model were evaluated using the independent samples from the validation set. The result of the root-mean-square error of prediction (RMSEP) was 1.4889 % v/v, and the R^2 was 0.9922 in the validation set which, as shown in Figure 3.

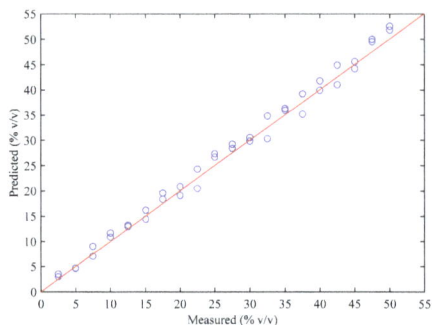

Figure 3. Reference-measured versus FT-NIR-predicted doping concentration of extra virgin olive oil (EVOO) in the validation set.

3. Discussion

In order to show the advantages of the BOSS algorithm in terms of wavenumber selection, it was compared with other three high-performance approaches for wavenumber selection, i.e., CARS, MCUVE, and IRIV. The best results of PLS models based on variables selected from different variable selection algorithms are shown in Table 1. The results in Table 1 show that the prediction accuracy of the PLS model could be improved by the four wavenumber selection algorithms with respect to the full-spectrum PLS model. Moreover, compared with the CARS–PLS model, the MCUVE–PLS model, and the IRIV–PLS model, the BOSS–PLS model achieved better results not only in the calibration process but also in the validation process. The main reason is that, quite likely, the BOSS algorithm combines the strategies of soft shrinkage, MPA, and WBS and makes full use of the regression coefficient information.

Also, the BOSS algorithm adopts the soft shrinkage strategy to select informative variables. Compared with the method of variable selection based on the hard shrinkage strategy, such as CARS and MCUVE, which delete less informative wavenumbers directly, the soft shrink strategy allocates smaller weights to wavenumbers with less information. However, these wavenumbers can still participate in the sub-models' construction for further evaluation considerations in the next iteration. Thus, the advantage of the soft shrink strategy is that it is able to reduce the risk of removing characteristic variables during the iteration and to choose the optimal variable subsets with better prediction ability.

The best variable subset is finally obtained by the BOSS algorithm on the basis of the criteria of the MPA combined with those of the WBS. Concretely, the sub-models are obtained in terms of the weight of each variable by the BOSS algorithm. The weight of each wavenumber is determined according to the value of the regression coefficients of multiple PLS sub-models by using the MPA strategy, rather than by using a single full-spectrum model. Then, the WBS strategy is used to stepwise update the weight of the wavenumbers selected so that the variable space can be compressed better. Thus, the BOSS algorithm considers all possible combinations of the selected wavenumbers, which is reasonable because the best number of variable subsets obtained is unknown before and during wavenumbers selection.

Table 1. Results of different partial least-square (PLS) models for the prediction of doping concentrations in EVOO. CARS: competitive adaptive reweighted sampling; MCUVE: Monte Carlo uninformative variable elimination; IRIV: iteratively retaining informative variables.

Models	Selected Wavenumbers (cm^{-1})	Number of Variables	PLS Factors	Calibration Set		Validation Set	
				R^2	RMSECV	R^2	RMSEP
PLS	9999.10-3999.64	1557	6	0.9421	3.4618	0.9599	3.2520
CARS-PLS	4192.49; 4242.63; 4261.92; 4578.18; 4593.61; 4655.32; 4659.18; 4666.89; 4670.75; 4674.60; 4682.32; 4690.03; 5746.83; 5754.55; 5758.40; 5766.12; 5858.68; 5862.54; 5870.25; 5874.11; 5877.97; 5881.82; 5885.68; 5889.54; 5897.25; 5901.11; 5912.68; 5920.39; 5935.82; 8234.55	30	4	0.9617	2.9647	0.9683	2.7664
MCUVE-PLS	4373.76; 4412.33; 4566.61; 4593.61 4612.89; 4632.18; 4647.61 4670.75; 4690.03; 4709.32; 5750.69; 5762.26; 5777.69; 5866.40; 5885.68; 5904.97; 5924.25; 5939.68; 6001.39; 6028.39; 8238.41; 8253.84; 8261.55; 8265.41	24	3	0.9694	2.6828	0.9778	2.3232
IRIV-PLS	4373.76; 4412.33; 5750.69; 5754.55; 5758.40; 5762.26; 5769.97; 5773.83; 5777.69; 5854.83; 5858.68; 5862.54; 5866.40; 5874.11	14	2	0.9901	1.4877	0.9887	1.8471
BOSS-PLS	4373.76; 4678.46; 4705.46; 5758.40; 5762.26; 5766.12; 5777.69; 5858.68; 5862.54; 5866.40; 5870.25; 5877.97; 5881.82; 5885.68; 5904.97	15	3	0.9908	1.4487	0.9922	1.4889

4. Materials and Methods

4.1. Sample Preparation and Division

In this study, extra virgin olive oil, peanut oil, sunflower seed oil, soybean oil, sesame oil, and maize oil were purchased in local supermarkets. In the experiments, peanut oil, sunflower seed oil, soybean oil, sesame oil, and corn oil were used as adulterating oils, which would be added separately to the EVOO to prepare the samples to be tested. That is to say, the adulterated oil samples were prepared including only two kinds of edible oil, namely, the EVOO and one of adulterating oils. The specific preparation process is reported below.

The doped oil samples were prepared using the EVOO and one of the adulterating oils. The volume fraction of each adulterated oil ranged from 2.5 to 50% v/v, increasing by 2.5% v/v volume fraction. Thus, 100 samples could be obtained in the experiment process.

In this study, the 100 samples were divided into two subsets. One was the calibration set, which was adopted to construct the prediction model, the other was the validation set, which was applied to verify the accuracy and generalization performance of the model. In order to meet the statistical requirements, three samples at the same doping concentration were randomly selected and put into the calibration set during sample division. Thus, there were 60 samples in the calibration set and 40 samples in the validation set. Because the adulterated samples obtained in this study only contained two kinds of edible oils, the calibration model established in this study can only be used to quantitatively detect one adulterated oil mixed with EVOO.

4.2. FT-NIR Spectra Acquisition

In this study, the NIR spectra of the doped samples were collected in transmission mode by means of an Antaris II NIR spectrophotometer (Thermo Scientific Co., Waltham, MA, USA). The number of spectral scanning was set to 32, and the spectral resolution was set to 4 cm^{-1}. The range of spectral scanning was set from 10,000 cm^{-1} to 4000 cm^{-1}. Thus, the original spectrum of each doped sample contained 1557 wavenumbers (i.e., 1557 wavelength variables). The absorbance data were stored as Log (1/T), T being the transmittance.

In spectral collection, each doped sample was first placed in a cuvette with a diameter of 6.0 mm, and then in the sampling chamber of the spectrometer for original spectral collection. The spectra of each doped sample were collected three times, and the mean values of the three measured spectra were taken as the original NIR spectra of the sample. When the spectra were collected, the laboratory temperature maintained at 25 °C.

4.3. Spectra Preprocessing

Figure 4a shows the raw FT-NIR spectra of all collected samples. As can be seen from Figure 4a, the spectra obtained contained not only useful sample information but also certain noise information, even overflow occurred in some wavenumbers. In order to eliminate the influence of these adverse factors, it was necessary to adopt appropriate methods to preprocess the spectra obtained before multivariable model calibration. Standard normal variate (SNV) transformation, which can be used to eliminate not only the baseline drift of diffuse reflectance spectrum but also the overflow phenomenon of diffuse reflectance spectrum, is mainly used to eliminate the influence of surface scattering and optical path change on diffuse reflectance spectra. Therefore, in this study, the SNV method was adopted to pretreat the spectra obtained, and the FT-NIR spectra after SNV preprocessing are presented in Figure 4b.

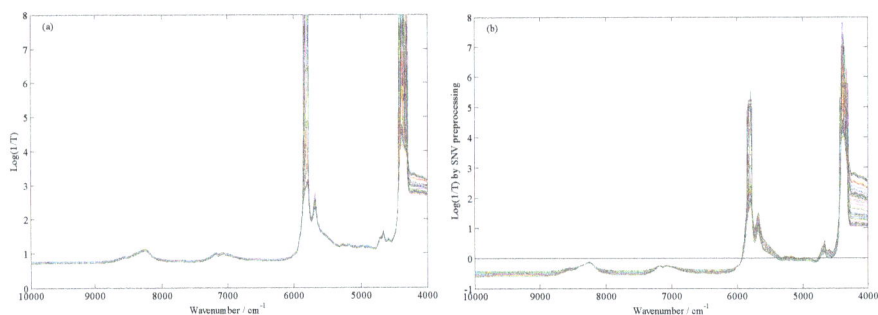

Figure 4. The original FT-NIR spectra (**a**) and the standard normal variate (SNV) preprocessing FT-NIR spectra (**b**) of all adulterated EVOO samples.

4.4. Data Analyses Methods

The BOSS algorithm applied here, which can be used to select the characteristic variables in the presence of collinearity, was described by Deng et al. [36]. The BOSS algorithm is based on a favorable criterion of shrinkage and utilizes the information of regression coefficients instead of the traditional hard shrinkage strategy. The BOSS algorithm, which is based on the bootstrap sampling (BSS) [37] and WBS [38] techniques, was used to determine the random combination wavenumbers and to establish the sub-models. The MPA was applied to extract informative variable subsets from the sub-models developed on the basis of PLS regression. The specific process of the BOSS algorithm was as follows:

In the process of spectral data analysis, suppose the spectral data matrix is X, of size $N \times P$, which includes N samples and P wavenumbers, and a vector Y, of size $N \times 1$, which represents the reference measurements.

Step 1, *K* subsets were generated in a variable space by the BSS. In each subset, one of many redundant variables remained by the BSS to extract characteristic variables. In the step, all wavenumbers were treated equally so that they had the same probability of being selected into the variable subset. That is to say, each variable had the same weights (*w*)

Step 2, the *K* sub-model of PLS were first developed by the data from the subsets selected. Then, the cross-validation RMSECV of each sub-model was calculated, each sub-model was sorted from smallest to largest, according to the RMSECV value, and the sub-model ranked in the top 10% was extracted.

Step 3, the regression coefficients of each sub-model extracted was calculated. By normalizing each regression vector, all elements in the regression vector were transformed into the absolute value of unit length. The new weights of the variable selected were then obtained according to the following summation formula:

$$w_i = \sum_{i=1}^{K} b_{i,k} \tag{1}$$

where *K* represents the number of sub-models that are extracted, and $b_{i,k}$ is the absolute value of the normalized regression coefficients for the *i*th wavenumber in the *k*th sub-model.

Step 4, the WBS was used to generate some new subsets based on the new weight of each variable selected, and the number of substitution wavenumbers in the WBS was obtained according to the average number of wavenumbers selected in the last step.

Step 5, steps 2 to 4 were repeatedly conducted until the number of wavenumbers selected in the renewed variable subset equaled one, and the variable subset was finally selected according to the lowest value of the RMSECV during the iterations as the best variable subset.

4.5. Model Evaluation

The prediction and generalization performances of the models were examined by a five-fold cross validation and an independent validation set. The values of the RMSECV, RMSEP, and coefficient of determination (R^2) were used as measures for model performance evaluation. RMSECV, RMSEP, and R^2 are given by the expressions

$$\text{RMSECV} = \sqrt{\frac{\sum_{i=1}^{n}\left(\hat{y}_{\setminus i} - y_i\right)^2}{n}} \tag{2}$$

$$\text{RMSEP} = \sqrt{\frac{\sum_{i=1}^{n}(y_i - \hat{y}_i)^2}{n}} \tag{3}$$

$$R^2 = 1 - \frac{\sum_{i=1}^{n}(y_i - \hat{y}_i)^2}{\sum_{i=1}^{n}(y_i - \overline{y}_i)^2} \tag{4}$$

For RMSECV, *n* is the number of samples in the calibration set, y_i is the reference measurement value from the *i*th sample, and $\hat{y}_{\setminus i}$ is the estimated value of the *i*th sample, when the model is constructed with the removed *i*th sample. For RMSEP, *n* is the number of samples in validation set, y_i is the reference measurement value of the *i*th sample in the validation set, and $\hat{y}_i$ is the estimated value of the *i*th sample in the validation set. For R^2, *n* is the number of samples, y_i is the reference measurement value from the *i*th sample, $\hat{y}_i$ is the estimated value of the *i*th sample, and $\overline{y}_i$ is the mean of all samples.

4.6. Software

All algorithms were implemented in Matlab R2018a (Mathworks, Natick, MA, USA) under Windows 10. The Matlab codes for implementing BOSS are freely available on the website: http://www.mathworks.com/matlabcentral/fileexchange/52770-boss.

5. Conclusions

The results obtained in this study show the potentials of FT-NIR spectroscopy in the detection of adulterations in EVOO. The BOSS algorithm combines the strategies of soft shrinkage, MPA, and WBS and could be used to extract the informative wavenumbers from the full-spectrum. The BOSS–PLS model revealed its superiority with respect to the full-spectrum PLS, CARS–PLS, MCUVE–PLS, and IVIR–PLS models. It can be concluded that the FT-NIR spectroscopy technique is an effective tool for the determination of EVOO adulteration and has a good guiding significance for the evaluation of EVOO quality. Moreover, the BOSS algorithm is a promising wavenumbers selection algorithm in chemometrics analysis, which can improve the prediction performance of calibration models.

Author Contributions: Conceptualization, H.J.; methodology, Q.C.; software, H.J.; validation, H.J., and Q.C.; formal analysis, H.J.; data curation, H.J.; writing—original draft preparation, H.J.; writing—review and editing, Q.C.; project administration, H.J.; funding acquisition, H.J.

Funding: This research was funded by the National Key Research and Development Program of China (Grant number 2017YFC1600600).

Conflicts of Interest: The authors declare no conflict of interest. This article does not contain any studies with human or animal subjects. The funders had no role in the design of the study; in the collection, analyses, or interpretation of data; in the writing of the manuscript, or in the decision to publish the results. All the authors have been involved with the work agree to submit this paper to Molecules, and all authors claim that none of the material in the paper has been published or is under consideration for publication elsewhere.

References

1. Chen, H.; Lin, Z.; Tan, C. Fast quantitative detection of sesame oil adulteration by near-infrared spectroscopy and chemometric models. *Vib. Spectrosc.* **2018**, *99*, 178–183. [CrossRef]
2. Xu, Y.; Li, H.; Chen, Q.; Zhao, J.; Ouyang, Q. Rapid detection of adulteration in extra-virgin olive oil using three-dimensional fluorescence spectra technology with selected multivariate calibrations. *Int. J. Food Prop.* **2015**, *18*, 2085–2098. [CrossRef]
3. Chen, Q.; Chen, M.; Liu, Y.; Wu, J.; Wang, X.; Ouyang, Q.; Chen, X.H. Application of FT-NIR spectroscopy for simultaneous estimation of taste quality and taste-related compounds content of black tea. *J. Food Sci. Tech. Mys.* **2018**, *55*, 4363–4368. [CrossRef] [PubMed]
4. Hu, W.; He, R.; Hou, F.; Ouyang, Q.; Chen, Q. Real-time monitoring of alcalase hydrolysis of egg white protein using near infrared spectroscopy technique combined with efficient modeling algorithm. *Int. J. Food Prop.* **2017**, *20*, 1488–1499. [CrossRef]
5. Guo, Z.; Huang, W.; Peng, Y.; Chen, Q.; Ouyang, Q.; Zhao, J. Color compensation and comparison of shortwave near infrared and long wave near infrared spectroscopy for determination of soluble solids content of 'Fuji' apple. *Postharvest Biol. Technol.* **2016**, *115*, 81–90. [CrossRef]
6. Zhang, H.; Jiang, H.; Liu, G.; Mei, C.; Huang, Y. Identification of Radix puerariae starch from different geographical origins by FT-NIR spectroscopy. *Int. J. Food Prop.* **2017**, *20*, 1567–1577. [CrossRef]
7. Azizian, H.; Mossoba, M.M.; Fardin-Kia, A.R.; Karunathilaka, S.R.; Kramer, J.K.G. Developing FT-NIR and PLS1 methodology for predicting adulteration in representative varieties/blends of extra virgin olive oils. *Lipids* **2016**, *51*, 1309–1321. [CrossRef]
8. Mossoba, M.M.; Azizian, H.; Fardin-Kia, A.R.; Karunathilaka, S.R.; Kramer, J.K.G. First application of newly developed FT-NIR spectroscopic methodology to predict authenticity of extra virgin olive oil retail products in the USA. *Lipids* **2017**, *52*, 443–455. [CrossRef]
9. Ozdemir, I.S.; Dag, C.; Ozinanc, G.; Sucsoran, O.; Ertas, E.; Bekiroglu, S. Quantification of sterols and fatty acids of extra virgin olive oils by FT-NIR spectroscopy and multivariate statistical analyses. *Lwt-Food Sci. Technol.* **2018**, *91*, 125–132. [CrossRef]
10. Mabood, F.; Boque, R.; Folcarelli, R.; Busto, O.; Jabeen, F.; Al-Harrasi, A.; Hussain, J. The effect of thermal treatment on the enhancement of detection of adulteration in extra virgin olive oils by synchronous fluorescence spectroscopy and chemometric analysis. *Spectrochim. Acta A* **2016**, *161*, 83–87. [CrossRef]

11. Tan, J.; Li, R.; Jiang, Z.T.; Shi, M.; Xiao, Y.Q.; Jia, B.; Lu, T.X.; Wang, H. Detection of extra virgin olive oil adulteration with edible oils using front-face fluorescence and visible spectroscopies. *J. Am. Oil Chem. Soc.* **2018**, *95*, 535–546. [CrossRef]

12. Tavares Melo Milanez, K.D.; Araujo Nobrega, T.C.; Nascimento, D.S.; Insausti, M.; Fernandez Band, B.S.; Coelho Pontes, M.J. Multivariate modeling for detecting adulteration of extra virgin olive oil with soybean oil using fluorescence and UV-Vis spectroscopies: A preliminary approach. *Lwt-Food Sci. Technol.* **2017**, *85*, 9–15. [CrossRef]

13. Poiana, M.A.; Alexa, E.; Munteanu, M.F.; Gligor, R.; Moigradean, D.; Mateescu, C. Use of ATR-FTIR spectroscopy to detect the changes in extra virgin olive oil by adulteration with soybean oil and high temperature heat treatment. *Open Chem.* **2015**, *13*, 689–698. [CrossRef]

14. Sun, X.; Lin, W.; Li, X.; Shen, Q.; Luo, H. Detection and quantification of extra virgin olive oil adulteration with edible oils by FT-IR spectroscopy and chemometrics. *Anal. Methods* **2015**, *7*, 3939–3945. [CrossRef]

15. Xu, Y.; Hassan, M.M.; Kutsanedzie, F.Y.H.; Li, H.H.; Chen, Q.S. Evaluation of extra-virgin olive oil adulteration using FTIR spectroscopy combined with multivariate algorithms. *Qual. Assur. Saf. Crops Foods* **2018**, *10*, 411–421. [CrossRef]

16. Dong, W.; Zhang, Y.; Zhang, B.; Wang, X. Quantitative analysis of adulteration of extra virgin olive oil using Raman spectroscopy improved by Bayesian framework least squares support vector machines. *Anal. Methods* **2012**, *4*, 2772–2777. [CrossRef]

17. Philippidis, A.; Poulakis, E.; Papadaki, A.; Velegrakis, M. Comparative study using Raman and visible spectroscopy of cretan extra virgin olive oil adulteration with sunflower oil. *Anal. Lett.* **2017**, *5*, 1182–1195. [CrossRef]

18. Tiryaki, G.Y.; Ayvaz, H. Quantification of soybean oil adulteration in extra virgin olive oil using portable raman spectroscopy. *J. Food Meas. Charact.* **2017**, *11*, 523–529. [CrossRef]

19. Fragaki, G.; Spyros, A.; Siragakis, G.; Salivaras, E.; Dais, P. Detection of extra virgin olive oil adulteration with lampante olive oil and refined olive oil using nuclear magnetic resonance spectroscopy and multivariate statistical analysis. *J. Agric. Food Chem.* **2005**, *53*, 2810–2816. [CrossRef]

20. Jiang, X.Y.; Li, C.; Chen, Q.Q.; Weng, X.C. Comparison of F-19 and H-1 NMR spectroscopy with conventional methods for the detection of extra virgin olive oil adulteration. *Grasas Aceites* **2018**, *69*. [CrossRef]

21. Zhu, J.; Agyekum, A.A.; Kutsanedzie, F.Y.H.; Li, H.; Chen, Q.; Ouyang, Q.; Jiang, H. Qualitative and quantitative analysis of chlorpyrifos residues in tea by surface-enhanced Raman spectroscopy (SERS) combined with chemometric models. *Lwt-Food Sci. Techol.* **2018**, *97*, 760–769. [CrossRef]

22. Ouyang, Q.; Chen, Q.; Zhao, J.; Lin, H. Determination of amino acid nitrogen in soy sauce using near infrared spectroscopy combined with characteristic variables selection and extreme learning machine. *Food Bioprocess Technol.* **2013**, *6*, 2486–2493. [CrossRef]

23. Jiang, H.; Mei, C.; Li, K.; Huang, Y.; Chen, Q. Monitoring alcohol concentration and residual glucose in solid state fermentation of ethanol using FT-NIR spectroscopy and L1-PLS regression. *Spectrochim. Acta A* **2018**, *204*, 73–80. [CrossRef] [PubMed]

24. Liu, G.H.; Jiang, H.; Xiao, X.H.; Zhang, D.J.; Mei, C.L.; Ding, Y.H. Determination of process variable pH in solid-state fermentation by FT-NIR spectroscopy and extreme learning machine (ELM). *Spectrosc. Spect. Anal.* **2012**, *32*, 970–973.

25. Wold, S.; Sjöström, M.; Eriksson, L. PLS-regression: A basic tool of chemometrics. *Chemom. Intell. Lab. Syst.* **2001**, *58*, 109–130. [CrossRef]

26. Lin, Y.W.; Deng, B.C.; Wang, L.L.; Xu, Q.S.; Liu, L.; Liang, Y.Z. Fisher optimal subspace shrinkage for block variable selection with applications to NIR spectroscopic analysis. *Chemom. Intell. Lab. Syst.* **2016**, *159*, 196–204. [CrossRef]

27. Farrés, M.; Platikanov, S.; Tsakovski, S.; Tauler, R. Comparison of the variable importance in projection (VIP) and of the selectivity ratio (SR) methods for variable selection and interpretation. *J. Chemom.* **2015**, *29*, 528–536. [CrossRef]

28. Li, H.D.; Zeng, M.M.; Tan, B.B.; Liang, Y.Z.; Xu, Q.S.; Cao, D.S. Recipe for revealing informative metabolites based on model population analysis. *Metabolomics* **2010**, *6*, 353–361. [CrossRef]

29. Cai, W.; Li, Y.; Shao, X.A. Variable selection method based on uninformative variable elimination for multivariate calibration of near-infrared spectra. *Chemom. Intell. Lab. Syst.* **2008**, *90*, 188–194. [CrossRef]

30. Rajalahti, T.; Arneberg, R.; Kroksveen, A.C.; Berle, M.; Myhr, K.M.; Kvalheim, O.M. Discriminating variable test and selectivity ratio plot: Quantitative tools for interpretation and variable (biomarker) selection in complex spectral or chromatographic profiles. *Anal. Chem.* **2009**, *81*, 2581–2590. [CrossRef]

31. Deng, B.C.; Yun, Y.H.; Liang, Y.Z. Model population analysis in chemometrics. *Chemom. Intell. Lab. Syst.* **2015**, *149*, 166–176. [CrossRef]

32. Yun, Y.H.; Wang, W.T.; Tan, M.L.; Liang, Y.Z.; Li, H.D.; Cao, D.S. A strategy that iteratively retains informative variables for selecting optimal variable subset in multivariate calibration. *Anal. Chim. Acta* **2014**, *807*, 36–43. [CrossRef] [PubMed]

33. Deng, B.C.; Yun, Y.H.; Liang, Y.Z.; Yi, L.Z. A novel variable selection approach that iteratively optimizes variable space using weighted binary matrix sampling. *Analyst* **2014**, *139*, 4836–4845. [CrossRef] [PubMed]

34. Deng, B.C.; Yun, Y.H.; Ma, P.; Lin, C.C.; Ren, D.B.; Liang, Y.Z. A new method for wavelength interval selection that intelligently optimizes the locations, widths and combinations of the intervals. *Analyst* **2015**, *140*, 1876–1885. [CrossRef] [PubMed]

35. Yun, Y.H.; Wang, W.T.; Deng, B.C.; Lai, G.B.; Liu, X.B.; Ren, D.B. Using variable combination population analysis for variable selection in multivariate calibration. *Anal. Chim. Acta* **2015**, *862*, 14–23. [CrossRef] [PubMed]

36. Deng, B.C.; Yun, Y.H.; Cao, D.S.; Yin, Y.L.; Wang, W.T.; Lu, H.M. A bootstrapping soft shrinkage approach for variable selection in chemical modeling. *Anal. Chim. Acta* **2016**, *908*, 63–74. [CrossRef]

37. Linden, A.; Adams, J.L.; Roberts, N. Evaluating disease management program effectiveness—An introduction to the bootstrap technique. *Dis. Manag. Health Out.* **2005**, *13*, 159–167. [CrossRef]

38. Ma, S.G.; Kosorok, M.R. Robust serniparametric M-estimation and the weighted bootstrap. *J. Multivariate Anal.* **2005**, *96*, 190–217. [CrossRef]

Sample Availability: Samples of the compounds are available from the authors.

molecules

MDPI

Review

Aquaphotomics—From Innovative Knowledge to Integrative Platform in Science and Technology

Jelena Muncan [1,2] **and Roumiana Tsenkova** [2,*]

1 Biomedical Engineering Department, Faculty of Mechanical Engineering, University of Belgrade, 11000 Belgrade, Serbia
2 Biomeasurement Technology Laboratory, Graduate School of Agricultural Science, Kobe University, Hyogo 657-8501, Japan
* Correspondence: jmuncan@people.kobe-u.ac.jp; Tel.: +81 90-5652-3639

Academic Editor: Christian Huck
Received: 30 June 2019; Accepted: 26 July 2019; Published: 28 July 2019

Abstract: Aquaphotomics is a young scientific discipline based on innovative knowledge of water molecular network, which as an intrinsic part of every aqueous system is being shaped by all of its components and the properties of the environment. With a high capacity for hydrogen bonding, water molecules are extremely sensitive to any changes the system undergoes. In highly aqueous systems—especially biological—water is the most abundant molecule. Minute changes in system elements or surroundings affect multitude of water molecules, causing rearrangements of water molecular network. Using light of various frequencies as a probe, the specifics of water structure can be extracted from the water spectrum, indirectly providing information about all the internal and external elements influencing the system. The water spectral pattern hence becomes an integrative descriptor of the system state. Aquaphotomics and the new knowledge of water originated from the field of near infrared spectroscopy. This technique resulted in significant findings about water structure-function relationships in various systems contributing to a better understanding of basic life phenomena. From this foundation, aquaphotomics started integration with other disciplines into systematized science from which a variety of applications ensued. This review will present the basics of this emerging science and its technological potential.

Keywords: aquaphotomics; water; light; near infrared spectroscopy; water-mirror approach; perturbation; biomeasurements; biodiagnosis; biomonitoring

1. Introduction to Aquaphotomics

Aquaphotomics is a young scientific discipline introduced by Professor Dr Roumiana Tsenkova at Kobe University in Japan in 2005 [1–5]. The establishment of a new science came in response to the recognized need in the current state of art for a common platform that can provide integration of knowledge about the water structure and functionality coming from various disciplines and most spectroscopy fields.

Water is the simplest compound and is made of two most common reactive elements. It covers more than 70% of the Earth's surface, comprises almost 2/3 of human body and is the most abundant molecule of all living cells. From nano to micro, meso, and up to the level of galaxies—water is everywhere. Wherever it is found, there are many phenomena involving it for which the mainstream science still does not have an explanation for. In everyday lives water is the first association to the word "liquid", and yet liquid water is such an atypical liquid—with behaviors so different from other liquids—that its properties are called "anomalies". This behavior stems from the capacity of water molecules for hydrogen bonding; if it did not exist, water would be a rather uninteresting material and our world would most likely look profoundly different. The hydrogen bonds connect the water molecules into a dynamic network; a water molecular network, or in other words—into a very complex water molecular system.

The past two decades have seen much progress in water science. Due to the significant role it plays in biological systems, water has received considerable attention. Many interesting phenomena where water is a key player stimulated research across disciplines, revealing the significance of water structure and consequently its functionality in properties of materials or processes such as wettability [6], biocompatibility [7–9], cell communication and carcinogenesis [10], DNA structure [11], molecular recognition and communication [12], protein stability [10,13], membrane stability and survival in desiccated state [14], mechanical properties such as kernel hardness [15] or mechanical behaviors such as curling of the plant stem [16]—to list a few. Spectroscopy methods such as X-ray, infrared spectroscopy (IR), THz spectroscopy, near infrared (NIR) spectroscopy and others, using light as a probe, proved to be especially valuable tools for water studies and have contributed immensely to elucidation of various aspects of water systems. In general, water-light interaction over the entire electromagnetic spectrum, significantly contributed to a better understanding of water molecular systems [5].

Water molecules absorb radiation over the entire range of the electromagnetic spectrum (Figure 1). In contrast to mid- and far-infrared, where water strongly absorbs, allowing analysis of only very thin samples, in the NIR part of the spectrum, water absorption is much weaker, therefore offering the possibility of analyzing thicker samples and objects rapidly, in a completely non-destructive and non-invasive manner, and with none or little sample preparation. Using light of the NIR range, it is very easy to acquire spectral data of various aqueous and biological systems in real time without disruption of their state and dynamics. Near infrared spectroscopy thus offers a unique window of opportunities to observe the water molecular network as a scaffold—a matrix of every system of which it is an intrinsic part of—in relation to all other contributing elements and factors shaping the system structure, state and resulting dynamics—without any disruptions.

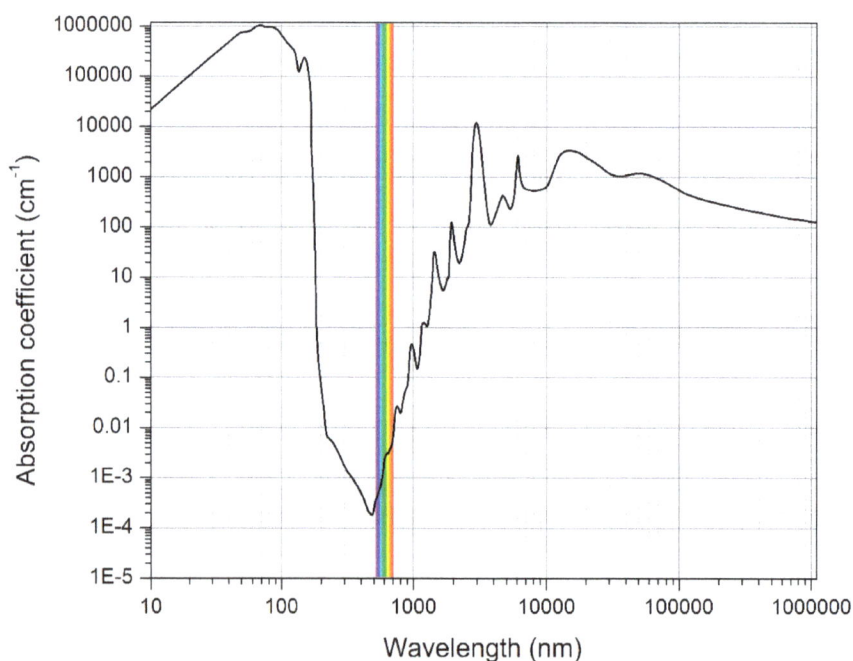

Figure 1. Water spectrum (double logarithmic plot), based on data from Segelstain [17].

Aquaphotomics as a science was laid on a foundation provided by near infrared spectroscopy [5]. The breakthrough knowledge regarding the importance of water stemmed from the observation that NIR spectral data for milk of healthy and dairy animals with mammary gland inflammation (mastitis) differed at water absorbing bands (1440 and 1912 nm) [18–21]. The presence of disease in an organism influenced many biomolecules (fat, lactose, proteins, etc.); these changes were subtle and sometimes not even visible in the spectra at absorbance bands related to those compounds. However, all these components exerted an influence on the water structure, and this cumulative effect was observable and measurable at multiple water absorbance bands corresponding to different water molecular species. In other words, the water molecular network changed when the composition of the aqueous system was altered, and this was reflected in water spectral pattern.

This innovative knowledge changed the approach in spectral analysis and paved way for the development of aquaphotomics. Changes in the water spectrum accurately and sensitively reflect the changes of water molecular species, hydrogen bonding and charges of the solvated and solvent molecules. In liquid water, each water molecule forms bonds with neighboring molecules, and can also establish dipoles and induce dipole interactions with other molecules, which gives the water molecular systems a heterogeneous character responsive to physical and environmental conditions [10]. Specific water species such as free water molecules, dimers, solvation shells and others contribute to the water spectrum in a very distinctive manner [5]. The water on a molecular level behaves as a collective mirror—its spectrum depicts changes as a response to all internal and environmental perturbations [5,22,23]. Rich experience acquired during many years resulted in a big database of spectra acquired under various perturbations, which revealed information regarding the water molecular system dynamics and the functionality of water in bio-aqueous systems [5], supporting the recognition of water as an active molecule and a central player in living processes [10,24].

The aim in establishing aquaphotomics as a science on its own, came in response to the recognized need in the current state of art in "omics" disciplines. Despite huge contributions of genomics, proteomics, metabolomics, transcriptomics and etc. to the comprehensive understanding of the principles underlying basic living functions, they all have an approach focused on single molecules and involve extraction procedures with the sample disruption. Biological systems can be studied using a non-destructive and integrative approach based on aquaphotomics, i.e., the interaction between water and biomolecules in which spectroscopic techniques combined with multivariate analysis represent a powerful tool.

Therefore, aquaphotomics aims at integrating and systematizing the knowledge about water-light interaction into a complementary, novel "omics" discipline whose objective is the large-scale, comprehensive study of water, its structure and related functionality. The first step towards this goal is identification of all the absorbance bands corresponding to specific water species. In this way, by knowing what each frequency means in the terms of the water structure, the absorbance bands become like "letters" that could be used to describe the features of aqueous systems. Relating the observed spectral patterns (combinations of water absorbance bands and the intensities of absorbance at these bands) with the observed characteristics or behaviors of aqueous systems, will clarify the functionality of certain water species and allow for future descriptions of the system states and dynamics solely in terms of the water structure. For various systems under various perturbations, aquaphotomics aims to build an aquaphotome, a comprehensive database of water bands and spectral patterns which describe the system and can therefore be used for future evaluations.

Since its establishment in 2005, aquaphotomics has showed steady progress (Figure 2). From only eight articles published in the first year after it was first introduced, the influence of the general idea and change over the years in the approach of how water is seen and treated in spectroscopy field can be seen. If the current trend continues, the estimate is that in the year 2025, 500 research articles per year can be expected. Through fundamental research, aquaphotomics provided novel insights and a better understanding of the basic phenomena and the role of water. It stimulated research and development of novel signal processing and chemometrics methods for data analysis, and provided a novel common measurement platform for a variety of applications, which led to the development of novel sensing devices and instruments based on water-light interaction.

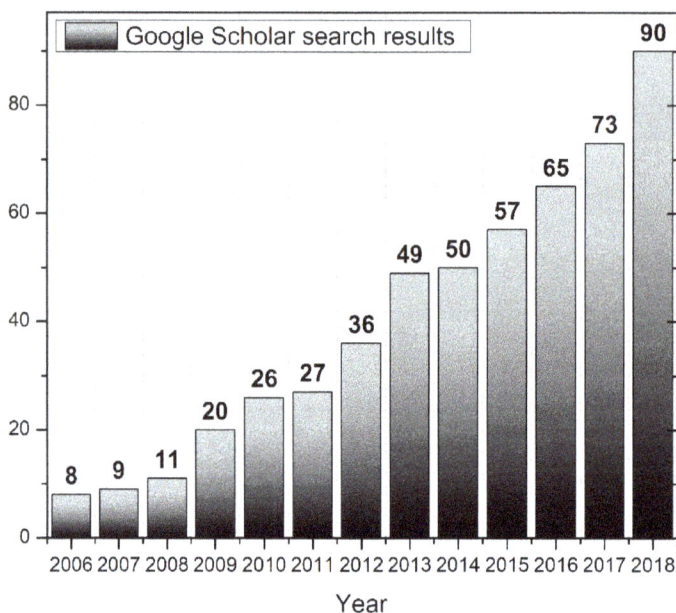

Figure 2. Number of articles published per year mentioning word "aquaphotomics" since 2006. The numbers are obtained using Google Scholar web search engine for articles and patents (excluding citations) containing word "aquaphotomics".

In the following sections, the key ideas of aquaphotomics, which lead paradigm shift of water seen from passive to active component of bio-aqueous systems and how this affected the basis of novel measurement platform—will be explained. Together with the brief illustrations of major contributions to science so far and through the extensive but not exhaustive list of applications, an overview of the huge technological potential of aquaphotomics will be presented.

2. Water Spectrum as a Source of Information

2.1. Water as a Sensor and an Amplifier: The Water-mirror Approach

The fundamental idea of aquaphotomics is that water works as a sensor. This principle is in aquaphotomics—usually expressed in the terms of water being a "collective mirror" (the water-mirror approach) [5,22,25,26]. Every aqueous system is a dynamic arrangement of a water molecular network, hydrogen-bonded between themselves and/or other constituents and influenced by perturbations. As a consequence of the strong potential of water molecules for hydrogen bonding, water changes its absorbance pattern every time it adapts to a physical or chemical change in the system itself or its environment. The spectral pattern extracted through the interaction of light and water hence can be used as an integrative marker or descriptor of the state of aqueous system.

The aquaphotomics approach is complementary to the conventional spectroscopy approaches. In most of the NIR-IR spectroscopy studies, the water absorption bands are considered to be masking the real information. For example, in order to measure proteins or sugars, the samples are usually dried in order to remove water and better observe the absorbance bands related to the structure of these biomolecules. In contrast, in aquaphotomics, the changes of the water spectral pattern are used as a source of information. The change in the concentration of a particular analyte is reflected in the changes of absorbance at several water absorbance bands, which are then used to build the prediction model.

The water-mirror quality of water on a molecular level indirectly permits measurements of small quantities or structural changes of other molecules present in the aqueous system [27]. By tracking the changes in values of absorbance at water absorbance bands in the spectra of aqueous or biological systems, the information is extracted not only regarding the water structure but also of other components present in water, or the state of the system as a whole [4,5,27]. It should be mentioned that this property of water was recognized and utilized very early on in the field of NIR spectroscopy [28–31], but it was only with the development of aquaphotomics that the properties of water as a "collective mirror" were truly explored and the huge potential of it for understanding new phenomena and applications in aqueous systems, in biomeasurements, biodiagnostics and biomonitoring has been truly understood [5].

The fact that changes in concentrations of particular analytes affect many water molecules, and consequently affect many water absorbance bands, has a significant advantage. Traditionally, the quantification limit for NIRS is regarded to be the concentration of 5000 ppm (mgL^{-1}) or 0.5% (*w/v*) [32]. This established limit for the traditional approach to NIRS analysis is based on the utilization of the absorbance bands of respective analytes directly. However, in aquaphotomics, water absorbance bands are used for indirect quantification. The comparison of different approaches— traditional and aquaphotomics and the resulting accuracy of quantification—was performed in one of the proofs of the concept works concerned with the measurement of concentration of polystyrene particles in water [33]. When the first overtone of water (i.e., the aquaphotomics approach) was used to develop a quantification model for polystyrene particles in aqueous suspension (1–0.0001%), the measurements achieved high accuracy—even in the case of very low concentrations. However, when the traditional approach was applied and measurements were based on the polystyrene band near 1680 nm (C-H stretching from aromatic C-H (2v) [34]) a decrease in the concentration of particles led to a substantial decrease in prediction accuracy. Therefore, the two approaches are not equivalent. The possibility of detecting and measuring even low concentrations of analytes—lower than traditionally accepted limit for NIRS—is a result of the different principle of measurement. In all aqueous systems, every molecule of analyte is hydrated with an abundance of water molecules, which adapt to its structure and rearrange, creating a variety of different water molecular species that can be observed based on their respective absorbance bands in the NIR region. Since many water molecules are involved in the hydration of just one molecule of analyte, the water not only acts as a sensor, but also as an amplifier. This means that in aquaphotomics, instead of measuring analytes directly, the information about their concentration is obtained indirectly by measuring changes in always abundant solvent molecules which provides better detection and quantification than it was traditionally assumed the capability of NIRS technique.

Aquaphotomics can thus provide detection and quantification of analytes—even when they are not absorbing near infrared light—and when they are present in low concentrations [5,26,35]. In addition, this approach offers the possibility of using the same water spectra as a source of information about multiple analytes, thus enabling simultaneous measurements of many analytes [5,25,36].

2.2. Water Matrix Coordinates (WAMACS) and Water Spectral Pattern (WASP)

In the NIR region, the water spectrum shows four main bands located approximately around 970, 1190, 1450 and 1940 nm, which are attributed to the second overtone of the OH stretching band ($3v_{1,3}$), a combination of the first overtone of the OH stretching and OH bending band ($2v_{1,3} + v_2$), the first overtone of the OH stretching band ($2v_{1,3}$) and a combination of the OH stretching and OH bending band ($2v_{1,3} + v_2$), respectively [37] (Figure 3). All these main bands are a rich source of information regarding the water structure. However, despite having lower absorbances or being overlapped with absorbance bands of other molecules, it should be noted that from 400 to 2500 nm more than 500 water absorbance bands have been identified [38].

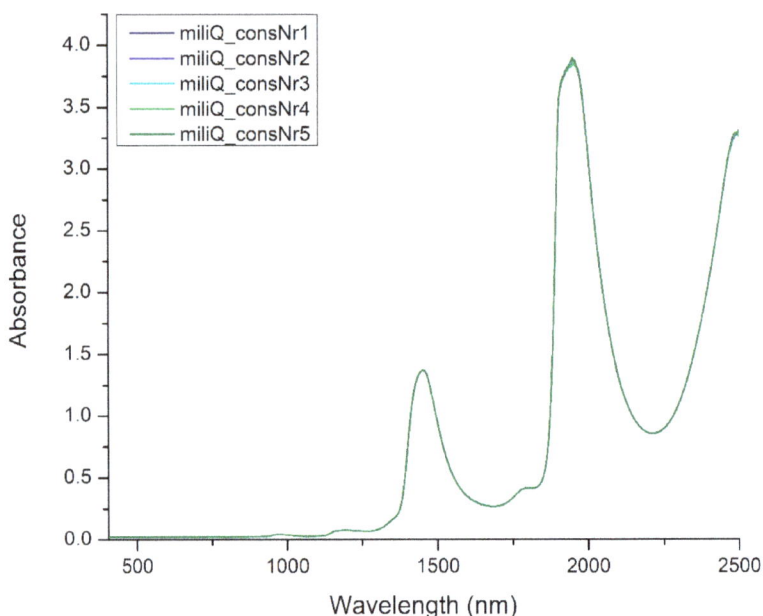

Figure 3. Spectra of pure water (produced by Milli-Q water purification system (Millipore, Molsheim, France) in the visible-near infrared region (400–2500 nm). Five spectra (miliQ_consNr1, miliQ_consNr2, miliQ_consNr5) presented in the figure were acquired by illuminating the same water sample five times consecutively.

Traditionally, water bands in the NIR region around 1450 and 1940 nm have been used for the determination of water content, hydration state [39] and, in particular, the moisture content [40] in various fields (agriculture, food industry, medical and pharmaceutical science).

Conventionally, only few symmetric and asymmetric stretching vibration assignments of water molecules are known in the first overtone of water OH stretching vibrations (Figure 4). This region with its broad band might look completely uninteresting to a classical spectroscopist, and it is often overlooked as informationally poor. In fact, for years, water has been described as the 'greatest enemy' of infrared (IR) and NIR spectroscopy on account of its dominant absorption.

The changes in water spectra in response to any change of water molecular network are very subtle, and require a data mining approach. The recognition of a high information potential of water spectra stimulated the development of novel analytical methods and even new computing tools, in order to meet the needs of the aquaphotomics data analysis [26,41–43]. A step-by-step explanation of aquaphotomics analysis supplemented by analytical tools currently at disposal is provided in Tsenkova et al. (2018) [26]. With so many tools at disposal, the utilization of the richness of NIR water spectra extended its applications far beyond moisture determination.

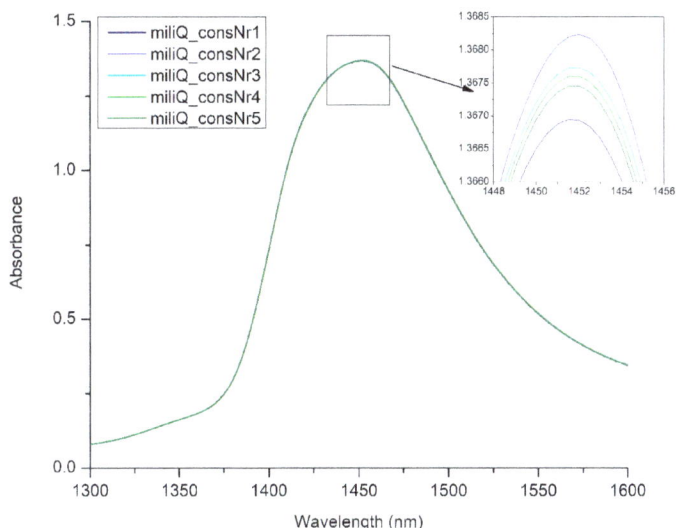

Figure 4. Spectra of pure water subjected to consecutive illuminations (the same spectra presented in Figure 3) in the area of the first overtone of water.

Through extensive experimental research and application of multivariate analysis, the abundance of water absorbance bands in near infrared region was discovered [38]. In the area of the first overtone of water, 12 water absorbance bands corresponding to specific water molecular species were uncovered [5] (Table 1). These 12 absorbance bands, named water matrix coordinates (WAMACS), were found to be consistently important in spectral analysis of different aqueous and biological systems, and under different perturbations. Table 1 provides assignments for the WAMACS of the first overtone of OH stretching vibrations, based on the original publication as a source [5]. This list is only a small part of the full-scale list of water absorbance bands, which is far from being completed, and is continuously expanding as the science progresses. For example, in the region of C5 water matrix coordinate, recent work discovered a subpopulation of quazi-free water molecules—water molecules confined in the local field of ions (1396 to 1403 nm) [44].

Table 1. Water matrix coordinates in the area of the first overtone of water in the near infrared region (1300 to 1600 nm) (based on [5,44,45]).

WAMACS	Range (nm)	Assignment
C1	1336–1348	$2v_3$: H_2O asymmetric stretching vibration
C2	1360–1366	OH-·$(H_2O)_{1,2,4}$: Water solvation shell
C3	1370–1376	$v_1 + v_3$: H_2O symmetrical stretching vibration and H_2O asymmetric stretching vibration
C4	1380–1388	OH-·$(H_2O)_{1,4}$: Water solvation shell O_2-·$(H_2O)_4$: Hydrated superoxide clusters $2v_1$: H_2O symmetrical stretching vibration
C5	1398–1418	Water confined in a local field of ions (trapped water) S_0: Free water Water with free OH-
C6	1421–1430	Water hydration band H-OH bend and O-H ... O
C7	1432–1444	S_1: Water molecules with 1 hydrogen bond
C8	1448–1454	OH-·$(H_2O)_{4,5}$: Water solvation shell
C9	1458–1468	S_2: Water molecules with 2 hydrogen bonds $2v_2 + v_3$: H_2O bending and asymmetrical stretching vibration
C10	1472–1482	S_3: Water molecules with 3 hydrogen bonds
C11	1482–1495	S_4: Water molecules with 4 hydrogen bonds
C12	1506–1516	v_1: H_2O symmetrical stretching vibration v_2: H_2O bending vibration Strongly bound water

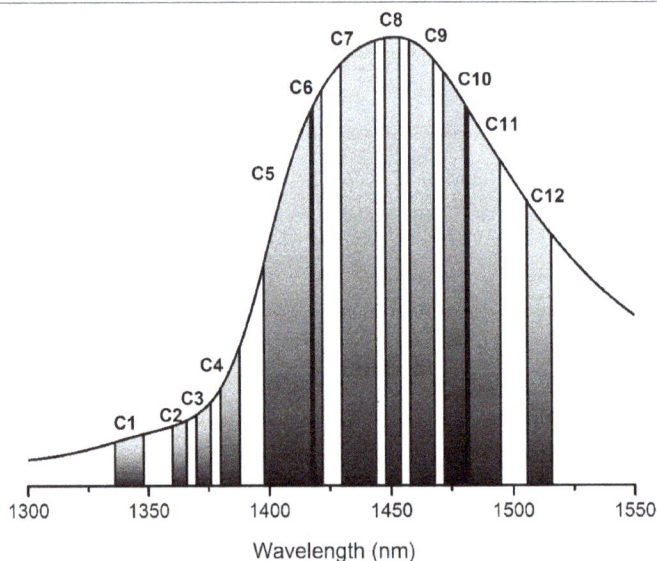

Aquaphotomics starts to build up a "water vocabulary" where the "letters" are the water vibrational frequencies bands (WAMACS) and the water spectral patterns (WASP) are the "words" identifying different water spectral patterns and their relation to functions and phenomena in order to translate findings of water between different disciplines. More information about aquaphotomics terminology are reported in [26]).

The water absorbance spectral pattern WASP is usually presented by aquagrams [46]. There are different types of aquagrams [26]. The simplest form is a classical aquagram—a radar chart that displays normalized absorbance at selected water absorbance bands. For the first overtone of water, the axes of the aquagram are usually based on previously discovered 12 WAMACS. The normalized absorbance is calculated as follows:

$$A'_\lambda = \frac{A_\lambda - \mu_\lambda}{\sigma_\lambda},$$

where A'_λ is normalized absorbance value displayed on radar axis; A_λ is absorbance after scatter correction (multiplicative scatter correction using the mean of the dataset as a reference spectrum or standard normal variate transformation); μ_λ is the mean of all spectra; σ_λ is the standard deviation of all spectra; and λ are the selected wavelengths from WAMACS regions corresponding to the activated water absorbance bands. Water absorbance bands are considered "activated" if they are consistently found to be among highly influential variables in the outputs of aquaphotomics analysis.

The aquagrams are visually very convenient tools that enable quick and comprehensive comparison of different systems or conditions of the same system by comparison of their WASPs.

2.3. Using Perturbation to Elicit Information

One of the most spectacular discoveries from the early years of aquaphotomics development was the observation that the absorbance spectrum of water changed with consecutive measurements [1] (in aquaphotomics, this is called illumination perturbations) (Figure 5). From the example presented in Figure 5, it is evident that every subsequent spectrum after exposure to near infrared light is different. The perturbation in the form of absorbed photons of radiation over sequential illuminations adds the energy to the system and changes the water molecular network.

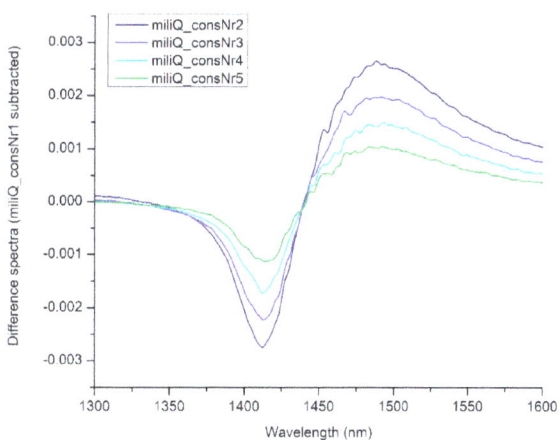

Figure 5. Consecutive illumination of water changes the near infrared spectra. Difference spectra calculated by subtracting the first consecutive spectrum from four subsequently acquired spectra under consecutive illuminations (the same spectra from Figures 3 and 4), show that near infrared light changes the water spectral pattern.

Apart from the most affected bands being located around 1410 and 1488 nm, the subtle troughs and shoulders can be observed throughout the spectra. The bands found to be affected by illumination are located at 1344, 1360, 1376, 1382, 1410, 1418, 1472 and 1482 nm—all within the ranges of 12 WAMACS and corresponding to different water species. These changes in water spectra in response to perturbation due to light serve as a source of additional information. For example, the first publications that presented the influence of light on water spectra, utilized the illumination as perturbation of prion protein solutions in order to discover differences in the functionality of different protein isoforms [1,47]. As the solution evolved with time, the frequencies of the various intramolecular vibration modes fluctuated due to a changing interaction between molecules. Out of three isoforms, only the solution of protein with bound copper ions consistently showed less bulk water despite the light perturbation suggesting it was the most stable form—a finding consistent with the published data.

Similarly to near infrared light, another aquaphotomics study that explored DNA mutation products, showed that exposure to ultraviolet (UV) radiation leads to changes in water spectral pattern of DNA solutions [48]. In addition, this study also showed that it was possible to measure the dose of exposure to irradiation with high accuracy (Figure 6). The regression vector of the developed model for prediction of the irradiation dose (Figure 6B) shows that UV light causes changes of absorbance at C5, C7, C8, C9, C10 and C11 water matrix coordinate. Comparing Figures 5 and 6, the similarities between the influence of NIR and UV light on water spectral pattern can be observed.

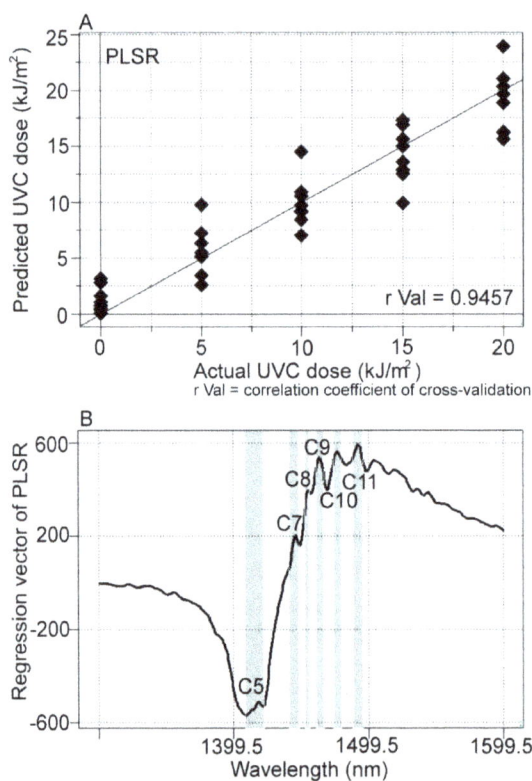

Figure 6. PLSR model for prediction of UV irradiation dose: (**A**) Y-fit curve showing relationship between actual and predicted values; (**B**) regression vector of PLSR model showing water absorbance bands affected by UV light perturbation [48].

The illumination affects the water spectra similar to the temperature; it creates more free water molecules, which are then available to "scan" the rest of the water system and interact with its components. From this interaction, new information can be gained, as explained in the example above.

Similarly to perturbation by light, intentional perturbation by temperature is often used in aquaphotomics as it is possible to use temperature dependent NIR spectra to obtain structural and quantitative information of the aqueous systems [42,43,49,50]. For instance, temperature perturbation was employed to study structural changes of ovalbumin as a model protein in aqueous solutions [51]. Two-dimensional correlation NIR spectroscopy and Gaussian fitting were adopted to investigate the variation of different water species and the sequences of the changes in the structure of protein during gelation. The results showed that in the gelation of protein, the change of S_2 water species (water species with two hydrogen bonds) follows the same phases as the protein; it maintains the stability of the protein in native and molten globule states, while weakening of the hydrogen bond in S_2 caused by high temperature resulted in the destruction of the hydration shell and led to ovalbumin clusters to form a gel structure.

In another work, water was used as a probe to quantify glucose in aqueous glucose solutions and human serum samples [52]. Spectral changes of water were captured from the temperature dependent NIR spectra using multilevel simultaneous component analysis (MSCA). The correlation coefficient for the temperature model was higher than 0.99, and that of the concentration of glucose were 0.99 and 0.84 for aqueous solutions and serum samples, respectively. Even if the changes in the spectra of water caused by temperature or concentration are very subtle, chemometrics provided techniques for the solution of this problem [50].

2.4. Water as a Biomolecule and Water Spectral Pattern as a Collective Biomarker

There are a hundred times as many water molecules in our bodies than the sum of all the other molecules put together. The most abundant molecule in the cell is water. Most biological processes involve water, and the interactions of biomolecules with water affect their structure, function and dynamics [10,24,53,54]. In the last decade, important advances have been made in our understanding of the factors that determine how biomolecules and their aqueous environment influence each other.

In the field of near infrared spectroscopy, however, water is still not considered a molecular network or a biologically relevant matrix, which originates from the general opinion still dominant in life sciences that water is an inert, passive medium. The state of art is so that living processes are described in terms of genes, DNA, proteins, metabolites or other single biomolecules acting as entities isolated from water [53] (Figure 7a). In nature, there are no isolated biomolecules and water is not only the native environment in which all biological processes occur, but also an integral part of all of biological processes [10,24,55,56].

In aquaphotomics, the water spectral pattern is considered as the main source of information. This offers two advantages when analysis of biological systems is performed. First, by focusing on water absorbance, simultaneous measurements of several analytes is possible, and second, which is far more important, the cumulative effect of different biomolecules on water matrix offers opportunity of using water spectral pattern as a novel biomarker. In most conventional spectroscopy studies, quantitative models are made for each separate component to be used to diagnose a system, where combining the models multiplies the errors—thereby producing inaccurate results. In aquaphotomics, despite possibility of measurements of the individual components, using the water spectral pattern as an integrative, global marker provides much more information about the studied systems, because it includes the effect of all components in the system—even the ones that, at the moment, current science does not identify as important and contributing to the system functionality (or disfunctionality, as is the case of diseases).

In aquaphotomics, the 'functionality', the biological state, the biological reaction to a change (dynamics) of the bio aqueous system is the key (focus, objective), instead of the presence of individual molecules. Specific water molecular structures (presented as water spectral patterns) are related to the status, dynamics and 'function' of the bio-aqueous systems studied, thereby building an aquaphotome—a database of water spectral patterns correlating water molecular structures to specific 'perturbations' (disease state, contamination state, reaction to light, change in temperature, and so on).

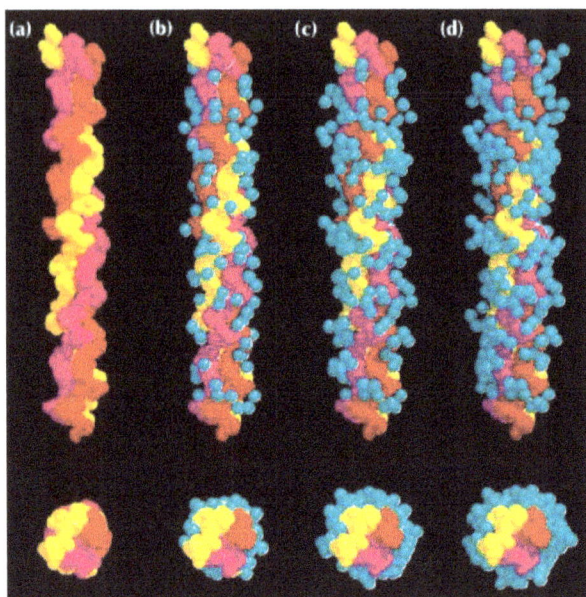

Figure 7. An example of a collagen peptide and its hydration shell: (**a**) in conventional science, biomolecules are usually represented only by this main chain on a black background, as if the biomolecular processes are happening in the vacuum; (**b–d**), a realistic picture, showing water hydration shells as an integral part [57] (Reprinted from Bella J, Brodsky B, Berman HM. Hydration structure of a collagen peptide. Structure 1995; 3:893–906, with permission from Elsevier).

The simple shift in perspective of what water is in biological systems offers novel insights and explanations of certain biological processes or phenomena. For example, DNA damage was detected through changes in the water spectral pattern as nonirradiated and UVC-irradiated DNA solutions were successfully distinguished in the 1488–1543 nm range, corresponding to the first overtone of water [48].

3. Aquaphotomics—Innovative Knowledge Leads to Innovative Applications

Being rapid and non-destructive, NIR spectroscopy is a powerful technique whose horizons have been further expanded by aquaphotomics. Since its establishment, aquaphotomics has grown into a multidisciplinary scientific field, encompassing many research areas and providing a common measurement platform for many applications. Using near infrared spectroscopy in aquaphotomics—in comparison to using light of other frequencies—does offer significant advantage of non-destructive evaluation of aqueous systems, which is of special significance for not only exploration of biological systems, but offers immense potential for biodiagnosis and biomonitoring. This region is furthermore an excellent tool for water observation, which provides an enormous amount of information about water molecular structure [5,58]. Numerous NIR spectra can be obtained in various conditions and states of the systems (under different perturbations)—all in real time.

The work in such a wide variety of applications, with different systems in different conditions led to two significant breakthroughs in aquaphotomics. The first breakthrough is that water spectral pattern can be used as a collective, integrative biomarker—a descriptor of a system's state [4]. The second one is the discovery that the water spectral pattern is related directly to certain functionality of the system. While the first breakthrough is of major significance for applications and provides a novel measurement platform, the second one leads to innovative knowledge of many phenomena. The next sections will illustrate the significance of both.

Contrary to the common understanding of overtone spectroscopy (100 to 1000 times lower absorbance than in the mid-IR range), it has been shown that even very small concentrations of the solutes could be measured with NIR spectroscopy if the aquaphotomics water-mirror approach is applied. Changes in the absorption spectrum of liquid water were used for quantification of the solutes present in water, even when the solutes did not absorb NIR light at all [30,35]. For instance, using very robust experimental design, Gowen et al. performed comprehensive aquaphotomics analysis of aqueous salts solutions (NaCl, KCl, $MgCl_2$, $AlCl_3$) with the aim of establishing limit of detection [35]. This research demonstrated that the best region for the prediction of salt concentration was the first overtone of water, attaining the prediction error of 500-800 ppm. Similar detection limit (1000 ppm) was reported in a research study that explored quantification of different metals (Cu(II), Mn(II), Zn (II) and Fe(III)) in aqueous HNO_3 [59], while another work reported successful prediction of HIV virus concentrations in plasma with the standard error of 23 pg/ml (ppb level) [60]. The water-mirror, indirect approach enables measurements of concentrations previously thought impossible to be measured with NIR spectroscopy at ppm and even at ppb levels under certain experimental conditions [23,25,33,35,59–62]. However, if we look beyond the measurements of individual solutes, what these results illustrate is the sensitivity of water molecular network to the changes in its components. The successful applications list measurements of acidity, pH [63] and effects of mechanical filtration on pure water [64]. Introducing water spectral pattern as an integrative marker represents one step forward from the detection of individual contaminants in water quality monitoring [65] or measurements of single, individual biomarkers in disease diagnostics [4].

This concept is radically novel, because it shifts the perspective of the definition of water quality by a set of physico-chemical and microbiological parameters to the definition of water quality as a water spectrum within some defined spectral limits. The same is true for disease diagnostics, which for many of diseases, especially in the early stage of development, works with very low concentrations of biomarkers in body fluids or does not even have reliable biomarkers. The spectrum of aqueous system integrates the influence of all single markers into one integrative, holistic marker which is a result of cumulative effect of many components and can easily be monitored in real time. The applicability of the proposed concept was evaluated in water quality monitoring [65], food quality monitoring [66] and biodiagnostics [67,68]. Using water as a biomarker, the information on the health status of any organism can thus be acquired in real time and non-destructively, allowing the continuous in vivo monitoring of the same sample.

In plant biology studies, aquaphotomics provided a methodology to follow the impact of a virus infection based on tracking changes in water absorbance spectral patterns of leaves in soybean plants during the progression of the disease [69,70]. Compared to currently used methods such as enzyme-linked immunosorbent assay (ELISA), polymerase chain reaction (PCR), and Western blotting, aquaphotomics was unsurpassable in terms of cost-effectiveness, speed, and accuracy of detection of a viral infection. The diagnosis of soybean plants infected with soybean mosaic virus was done at the latent, symptomless stage of the disease based on the discovery of changes in the water solvation shell and weakly hydrogen-bonded water which resulted from a cumulative effect of virus-induced changes in leaf tissues. A similar study reported the detection of begomovirus in papaya leaves with an aquaphotomics approach [71]. Tracking the cumulative effect of various, most likely unknown, biomarkers of viral infection in leaves provided grounds for successful, early diagnosis based on aquaphotomics principles.

Similarly, different water spectral patterns were found in leaves of genetically modified soybean with different cold stress abilities [69]. This research on the discrimination of soybean cultivars with different cold resistance abilities has proven that resistance to cold stress can be characterized by different water absorbance patterns of the leaves of genetically modified soybean. Different genetic modifications resulted in a multitude of bio-molecular events in response to cold stress, whose cumulative effect was detected as a specific water spectral pattern of leaves; i.e. the higher the cold resistance, the higher was the ability of the cultivar to keep the water structure in less-hydrogen bonded state, providing a supply of "working water" in the conditions of decreased temperature.

In another study, aquaphotomics was applied for exploration of the extreme desiccation tolerance i.e. the ability of some plants—called resurrection plants—to survive extremely long periods in the absence of water and then to quickly and fully recover upon rewatering [72]. Application of aquaphotomics to study one such plant—*Haberlea rhodopensis*—during dehydration and rehydration processes, revealed that in comparison to its biological relative—a non-resurrection plant species, *Deinostigma eberhardtii*—*H. rhodopensis* performs fine restructuring of water in its leaves, preparing itself for the dry period. In the dry state, this plant drastically diminished free water, and accumulated water molecular dimers and water molecules with four bonds (Figure 8). The decrease of free water and increase of bonded water, together with preservation of constant ratios of water species during rapid loss of water, was found to be the underlying mechanism that allows for the preservation of tissues against the dehydration-induced damages and ultimately the survival in the dry state.

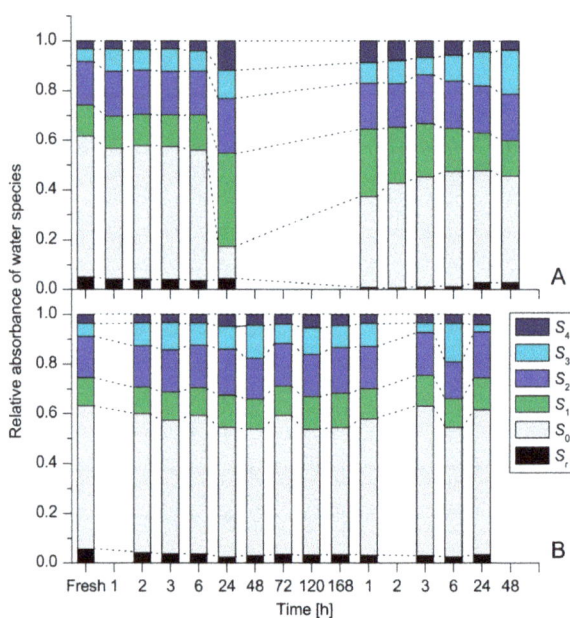

Figure 8. Dynamics of different water species (S_i = water molecules with *i* hydrogen bonds, S_r = protonated water clusters) during dehydration and rehydration of *Haberlea rhodopensis* and *Deinostigma eberhardtii*. Relative absorbance of water species in *Haberlea rhodopensis* (A) and *Deinostigma eberhardtii* (B) during desiccation and subsequent rehydration [72].

In the medical field, aquaphotomics was proposed for in vivo therapy monitoring of topical cream effects [73,74], for monitoring of dialysis efficacy [67] and diagnosis of several diseases: cancer [67], diabetes and coronary heart disease [75]. These applications utilize the concept of a water spectral pattern as an integrative biomarker that offers significant advantage compared to traditional ways of therapy monitoring or diagnostic practices in medicine. For example, monitoring dialysis efficacy is a particularly challenging task that relies on discrete sampling and measurements of only several uremic toxins out of more than 80 currently recognized that contribute to the uremic syndrome (Figure 9). The NIRS method has already been proposed to measure urea in spent dialysate [76]. However, urea is only a single marker and its concentration decreases during dialysis, making the detection harder. By using aquaphotomics approach, individual component measurements were replaced by process monitoring [67]. Instead of measuring waste materials in spent dialysate, their cumulative effect on the water matrix was measured as water spectral pattern changes during the dialysis. In another words, individual component measurement was replaced by monitoring of the process. The water spectral pattern of spent dialysate averaged for all patients after 5, 45, 90 and 135 min of treatment presented as the aquagram in Figure 9 showed, as the therapy progressed there was an increase of free water molecules (1398 and 1410 nm: C5 WAMACS) in the dialysate. In this way, the efficacy of dialysis can be assessed in a simplified way by tracking the changes of the respective dialysate water spectral pattern. The advantage of such an indirect approach of biomonitoring can also be extended to biodiagnostics as the water spectral pattern captures the information regarding all biomolecules that change with the disease—even the biomolecules current science is not aware of.

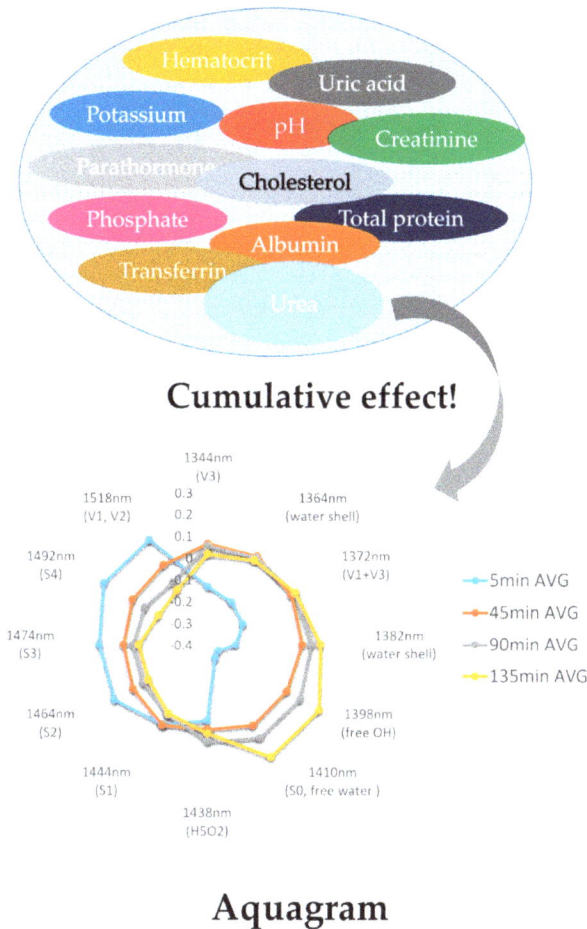

Figure 9. Water spectral pattern of spent dialysate presented on aquagram can be used as a marker of dialysis efficacy. Instead of measurements of different uremic toxins (of which there are more than 80), aquaphotomics provides measurement of their collective cumulative effect on water matrix of spent dialysate. [67].

The works on mastitis [21,77–82] showed that as the various milk components change during the different stages of infection, they influence the water matrix of milk differently. The water spectral patterns of blood, milk, and urine of mastitic cows, revealed that the same water absorbance bands are activated in different body fluids in response to the presence of disease [81]. Similarly, physiological changes such as ovulation can be detected in various body fluids using the same principles such as in the Giant panda [68,83], in the Bornean orangutan [84], in dairy cows [85] and in mares [86].

Aquaphotomics made a significant contribution to the field of microbiology and food engineering by not only providing a fast and nondestructive analysis, but by contributing to better understanding of the mechanism of action of some microorganisms [87–89]. For example, probiotic, non-probiotic and moderate bacteria strains produced a unique water spectral pattern, as shown in aquagrams reported in Figure 10. Probiotic bacteria strains were characterized by a higher number of small protonated water clusters, and free water molecules and water clusters with weak hydrogen bonds [89]. The discovery that strong probiotic bacteria produced more free water and less hydrogen-bonded water species, i.e. they break water structures in a way comparable to an increase in temperature, provides novel insight on their mode of action. Moreover, aquaphotomics was able to distinguish a subdivision into two species within one bacteria strain, where conventional PCR analysis was not enough sensitive [90].

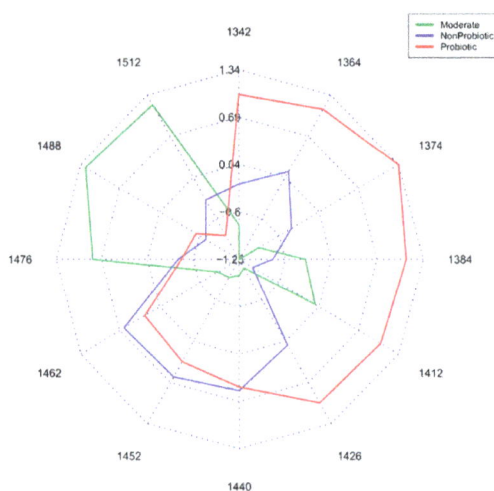

Figure 10. Aquagrams of culture media of groups of probiotic, moderate and non-probiotic strains. Average values of normalized absorbance values of the water matrix coordinates for each group are plotted on each wavelength axis [89].

The aim of studying water interactions on a molecular level was to obtain a better understanding of the relationship between the water structure and a phenomena on a macro scale. For example, one of the novel studies related the sensory texture of apples with particular spectral pattern of fruits: mealy apples had water predominantly in a weakly hydrogen bonded state, while the opposite was true for juicy, firm apples [66]. Another study related the dehydration band (1398 nm) with physical damage in mushrooms [45]. Similarly, wheat kernel hardness was related to specific water absorbance bands (1366 and 1436 nm) [15]. Usually, food texture is not considered a property that stems from certain water structure; however, the above-mentioned works revealed that water structure change with texture. Further studies are needed to better understand the relationships between the water spectral pattern and pectin metabolism in horticultural products.

Interesting findings were obtained in applications of aquaphotomics for basic studies of interaction of biomolecules and water. For example, although many spectroscopic studies have been conducted on glucose, few studies have been carried out on the anomers of glucose despite the fact that spectra—as well as chemical and enzymatic reactions—depend on the specific molecular structure. What aquaphotomics study of glucose isomerism [91] found is that the absorbance band at 1742 nm possess the potential to distinguish glucose anomers qualitatively and quantitatively. What is conventionally regarded as the first overtone of the C-H stretching mode, was confirmed to not be related to glucose—but to water [92,93]. Through work in the field of protein-water interactions, aquaphotomics provided insight into their dynamics and the significant role water plays in their functionality. In a study of prion protein isoforms [47], aquaphotomics analysis of Mn and Cu prion isoforms in water solutions revealed that while binding of copper results in increased protein stability in water, the binding of manganese resulted in less stability—which led to fibril formation, responsible of neurodegenerative disease. The fact that the entire process of protein structural changes in aqueous systems can be monitored indirectly through the water absorbance pattern of the protein solution, was demonstrated in a study of amyloid protein—another protein involved in pathogenesis of neurodegenerative diseases [71]—as well as ovalbumin [51].

Aquaphotomics studies on water-material interaction hold great promise in understanding some of the very complex properties that are of interest for many applications, such as wettability or biocompatibility. A study concerned with investigation of an excellent wettability of titanium dioxide reveled the importance of water species ratios [6]. More recent studies exploring the state of water in hydrogel materials of soft contact lenses [94–96] revealed that the water spectral pattern holds information even about the state of polymer network and protein deposits on the surfaces of worn contact lenses. Other aquaphotomics studies showed how nanomaterials shape the water matrix, as in the case of fullerene-based nanomaterials that act as water structuring elements when present in very low concentrations [73,74,97]. In nanotechnology and nanomedicine, aquaphotomics could lead to novel findings due to the fact that with decreasing size, the available active surface interacting with water playing a significant role increases.

Since its establishment, aquaphotomics has grown into a large, multidisciplinary scientific field, encompassing many research areas and providing common measurement platform for many applications. Table 2 provides an idea of possible fields of applications of aquaphotomics coupled to NIR spectroscopy. These works illustrate the great versatility of this technique and can hopefully inspire novel research and application ideas.

Table 2. Aquaphotomics contribution: from fundamental research to various applications.

Application	Object of Study	Purpose	References
Fundamental research	Sugars	Quantification	[25,43,49,52]
	Glucose	Distinguishing anomers	[91]
	Salts	Quantification and influence on water spectra	[26,35,44,61,98]
	Acids	Quantification, accuracy of prediction depending on acidity	[99]
	Acids and pH	Quantification	[63]
	Ethanol	Quantification, structural analysis	[42,100–103]
	Methanol	Quantification	[98,104]
	Water-ethanol-isopropanol mixture	Quantitative analysis and the effect of temperature	[105]
	Water, methanol, ethanol and ethylenediamine mixture	Quantitative analysis and the effect of temperature	[106]
	Monoethylene-glycol	Quantification	[98]
	Metal ions	Quantification	[107–110]
	Near infrared light	Influence of consecutive irradiation	[1]
	UV light	Measurement of irradiation dose	[48]
	Temperature	Influence of temperature on water spectra	[42,43,111]

Table 2. *Cont.*

Application	Object of Study	Purpose	References
Biomolecules	Oligopeptides	Interaction with water – elucidating the structure, dynamics and function of proteins	[112]
	Prion proteins	Stability of protein structure as a function of metal binding	[47]
	Insulin	Fibrillation phases	[113]
	Albumin and γ-globulin	Quantification	[114]
	Albumin	Structural analysis and hydration properties	[115]
	Ovalbumin	Gelation of globular proteins	[51]
	DNA	Quantification and detection of mutation products	[48]
	Phospholipids	Structural analysis and effect on water	[111]
Water	Water contamination	Quantification of pesticides alachlor and atrazine	[62]
	Water contamination	Detection of contaminants based on salts as model systems	[35]
	Commercial mineral waters	Discrimination	[116]
	Ground water quality	Continuous monitoring based on water spectral pattern as a holistic/integrative marker	[65]
	Pure water	Influence of filtration process	[64]
Food	Honey	Adulteration	[117]
	Mushrooms	Detection of physical damage	[45,118]
	Milk	Components	[36]
	Wafer, coffee, soybean	Water activity and moisture content	[119,120]
	Perches (fish)	Discriminating between wild fishes and raised in the recirculation system	[121]
	Pork loin	Discrimination between fresh and spoiled meat	[121]
	Porcine muscles	Discrimination between fresh and thawed meat	[121]
	Cheese	Ripening process	[122]
	Cheese and winter melon	Influence of packaging material on ripening	[123]
	Salami	Influence of coating on ripening	[124]
	Packaging material	Influence of bioactive compound - propolis	[125]
	Apples	Sensory texture - specific mechanical and structural properties related to water spectral pattern	[66]
	Oilseed Rape	Stem rot detection	[126]
	Rice	Seed vitality	[127]
	Coffee	Roasting degree	[128]
	Wheat kernels	Hardness	[15]

Table 2. *Cont.*

Application	Object of Study	Purpose	References
Materials	Soft contact lenses: hydrogels	Discrimination of hydrogels with different water content	[95,96]
	Soft contact lenses: hydrogels	Discrimination of new and worn contact lenses	[94]
	Titanium dioxide	Wettability	[6]
Environment	Soil	Identification of soil type	[129]
	Water contamination	Monitoring	[65,130]
Nanomaterials	Fullerene based nanomaterials	Hydration properties	[73,74,97]
	Polystyrene	Quantification of particles in water solutions	[33]
Microbiology	Bacteria – metabolites	Contribution to NIR signal from cells and metabolites	[131]
	Bacteria - probiotic	Classification	[87,89]
	HIV virus	Detection and quantification	[60]
	Bacteria	Selection	[88]
Cells and tissues	Somatic cells in milk	Quantification	[21]
	Tissue (mice)	Native state of metals	[132]
	Tissue (mice)	*Ex vivo* discrimination	[133]
Plant biology	Soybean	Detection of mosaic virus infection	[70]
	Soybean	Ability to cope with cold stress in genetically modified cultivars; Detection of mosaic virus infection	[69]
	Resurrection plants	Peculiarities of water structure in leaves of anhydrobiotic organism	[72]
	Papaya leaves	*In vivo* detection of begomovirus infection	[71]
Animal medicine	Mastitis in dairy cows	Disease detection	[21,77–82]
	Estrus detection in urine of giant panda	Finding water spectral pattern as biomarker, quantification of hormone	[68,83]
	Estrus detection in milk of cows	Ovulation period detection and monitoring	[85]
	Estrus detection in urine of Bornean orangutan	Ovulation period detection and monitoring	[84]
	Estrus period detection using serum in mares	Detection of oestrus, metestrus, and diestrus in mares,	[86]
Medicine	DNA mutation products	Detection of DNA damage, quantification of damage products	[48]
	AIDS	HIV virus detection	[60]
	Serum	Serum based diagnosis (diabetes, coronary heart disease)	[75]
	Prion protein disease	Mechanism of disease	[47]
	Skin cream effects	Therapy monitoring	[73,74]
	Dialysis efficacy	Monitoring of spent dialysate	[67]
	Colorectal cancer	Diagnostics based on serum and urine	[67]

4. Future Perspectives

With the theoretical and technological advancements in spectroscopy and data analysis techniques, the development of aquaphotomics as a new science led to development of a steady new knowledge base about water-light interaction and provided a common measurement platform that employs novel measurement principles.

The future of aquaphotomics will be towards building up the aquaphotome database of WAMACs and WASPs for extensive number of systems in our life. It will embrace the rest of the "–omics" data in a collective manner to be used as complementary tool for further understanding new phenomena in science and for the development of feedback systems where the WASP will be the diagnostic tool and the respective individual "–omics" data will provide the information for the regulator in a feedback system to not only monitor and diagnose, but control processes including biological ones.

The innovative knowledge of importance of water as a biologically significant molecule (a biomolecule in its own right [53]) and of water-light interaction as a way of extracting information led the paradigm shift that places different demands on the development of sensing technologies. The advantage of being non-destructive, rapid and capable of comprehensive biomonitoring and biodiagnosis, based on utilization of water spectral pattern as new, more accurate and collective biomarker, aquaphotomics provides great potential to complement conventional technologies used to perform single tasks, while in others it may even lead to the replacement of current ones.

The aquaphotomics based applications vastly extended the possibilities of spectroscopy and especially of the near infrared spectroscopy, while the ever improving sensor technology offers great prospects for high accuracy, real-life applications, being more cost-effective at the same time [134]. Presented here, aquaphotomics works demonstrate outcomes that are presumably just a glimpse of a much larger application potential.

Author Contributions: Conceptualization, J.M. and R.T.; resources, R.T.; writing—original draft preparation, J.M.; writing—review and editing, J.M. and R.T.; visualization, J.M.; supervision, R.T.

Funding: This research received no external funding.

Acknowledgments: J.M. gratefully acknowledges financial support provided by the Japanese Society for Promotion of Science (P17406).

Conflicts of Interest: The authors declare no conflict of interest.

References

1. Tsenkova, R. Visible-Near Infrared perturbation spectroscopy: Water in action seen as a source of information. In Proceedings of the 12th International Conference on Near-infrared Spectroscopy, Auckland, New Zealand, 9–15 April 2005; pp. 514–519.
2. Tsenkova, R. Aquaphotomics: Aquaphotomics and Chambersburg. *NIR News* **2006**, *17*, 12–14. [CrossRef]
3. Tsenkova, R. Aquaphotomics: Exploring water-light interactions for a better understanding of the biological world. Part1: Good yogurt comes from good milk and healthy cows. *NIR News* **2006**, *17*, 10–11. [CrossRef]
4. Tsenkova, R. AquaPhotomics: Water absorbance pattern as a biological marker. *NIR News* **2006**, *17*, 13–20. [CrossRef]
5. Tsenkova, R. Aquaphotomics: Dynamic spectroscopy of aqueous and biological systems describes peculiarities of water. *J. Near Infrared Spectrosc.* **2009**, *17*, 303–313. [CrossRef]
6. Masato, T.; Gianmario, M.; Salvatore, C.; Masakazu, A. Investigations of the structure of H_2O clusters adsorbed on TiO_2 surfaces by near-infrared absorption spectroscopy. *J. Phys. Chem. B* **2005**, *109*, 7387–7391.
7. Tanaka, M.; Hayashi, T.; Morita, S. The roles of water molecules at the biointerface of medical polymers. *Polym. J.* **2013**, *45*, 701–710. [CrossRef]
8. Morita, S.; Tanaka, M.; Kitagawa, K.; Ozaki, Y. Hydration structure of poly(2-methoxyethyl acrylate): Comparison with a 2-methoxyethyl acetate model monomer. *J. Biomater. Sci. Polym. Ed.* **2010**, *21*, 1925–1935. [CrossRef]
9. Morita, S.; Tanaka, M.; Ozaki, Y. Time-resolved in situ ATR-IR observations of the process of sorption of water into a poly(2-methoxyethyl acrylate) film. *Langmuir* **2007**, *23*, 3750–3761. [CrossRef]

10. Chaplin, M. Do we underestimate the importance of water in cell biology? *Nat. Rev. Mol. Cell Biol.* **2006**, *7*, 861–866. [CrossRef]

11. Khesbak, H.; Savchuk, O.; Tsushima, S.; Fahmy, K. The role of water H-bond imbalances in B-DNA substate transitions and peptide recognition revealed by time-resolved FTIR spectroscopy. *J. Am. Chem. Soc.* **2011**, *133*, 5834–5842. [CrossRef]

12. Espinosa-Marzal, R.M.; Fontani, G.; Reusch, F.B.; Roba, M.; Spencer, N.D.; Crockett, R. Sugars communicate through water: Oriented glycans induce water structuring. *Biophys. J.* **2013**, *104*, 2686–2694. [CrossRef]

13. De Simone, A.; Dodson, G.G.; Verma, C.S.; Zagari, A.; Fraternali, F. Prion and water: Tight and dynamical hydration sites have a key role in structural stability. *Proc. Natl. Acad. Sci. USA* **2005**, *102*, 7535–7540. [CrossRef]

14. Erkut, C.; Penkov, S.; Fahmy, K.; Kurzchalia, T.V. How worms survive desiccation. *Worm* **2012**, *1*, 61–65. [CrossRef]

15. Hong, B.H.; Rubenthaler, G.L.; Allan, R.E. Wheat pentosans. II. Estimating kernel hardness and pentosans in water extracts by near-infrared reflectance. *Cereal Chem.* **1989**, *66*, 374–377.

16. Rafsanjani, A.; Brulé, V.; Western, T.L.; Pasini, D. Hydro-responsive curling of the resurrection plant *Selaginella lepidophylla*. *Sci. Rep.* **2015**, *5*, 8064. [CrossRef]

17. Segelstein, D.J. The Complex Refractive Index of Water. Master's Thesis, University of Missouri, Kansas, MO, USA, 1981.

18. Tsenkova, R.; Atanassova, S.; Toyoda, K. Near infrared spectroscopy for diagnosis: Influence of mammary gland inflammation on cow's milk composition measurement. *Near Infrared Anal.* **2001**, *2*, 59–66.

19. Tsenkova, R.; Atanassova, S.; Ozaki, Y.; Toyoda, K.; Itoh, K. Near-infrared spectroscopy for biomonitoring: Influence of somatic cell count on cow's milk composition analysis. *Int. Dairy J.* **2001**, *11*, 779–783. [CrossRef]

20. Pravdova, V.; Walczak, B.; Massart, D.; Kawano, S.; Toyoda, K.; Tsenkova, R. Calibration of somatic cell count in milk based on near-infrared spectroscopy. *Anal. Chim. Acta* **2001**, *450*, 131–141. [CrossRef]

21. Tsenkova, R.; Atanassova, S.; Kawano, S.; Toyoda, K. Somatic cell count determination in cow's milk by near-infrared spectroscopy: A new diagnostic tool. *J. Anim. Sci.* **2001**, *79*, 2550–2557. [CrossRef]

22. Tsenkova, R. Aquaphotomics: Extended water mirror approach reveals peculiarities of prion protein alloforms. *NIR News* **2007**, *18*, 14–17. [CrossRef]

23. Bazar, G.; Kovacs, Z.; Tanaka, M.; Tsenkova, R. Aquaphotomics and its extended water mirror concept explain why NIRS can measure low concentration aqueous solutions. In Proceedings of the Aquaphotomics, "Understanding Water in Biological World", The 5th Kobe University Brussels European Centre Symposium Innovation, Environment, and Globalisation, Brussels, Belgium, 14 October 2014.

24. Biedermannová, L.; Schneider, B. Hydration of proteins and nucleic acids: Advances in experiment and theory. A review. *Biochim. Biophys. Acta Gen. Subj.* **2016**, *1860*, 1821–1835. [CrossRef]

25. Bázár, G.; Kovacs, Z.; Tanaka, M.; Furukawa, A.; Nagai, A.; Osawa, M.; Itakura, Y.; Sugiyama, H.; Tsenkova, R. Water revealed as molecular mirror when measuring low concentrations of sugar with near infrared light. *Anal. Chim. Acta* **2015**, *896*, 52–62. [CrossRef]

26. Tsenkova, R.; Munćan, J.; Pollner, B.; Kovacs, Z. Essentials of aquaphotomics and its chemometrics approaches. *Front. Chem.* **2018**, *6*, 363. [CrossRef]

27. Tsenkova, R. Aquaphotomics: The extended water mirror effect explains why small concentrations of protein in solution can be measured with near infrared light. *NIR News* **2008**, *19*, 13–14. [CrossRef]

28. Hirschfeld, T. Salinity determination using NIRA. *Appl. Spectrosc.* **1985**, *39*, 740–741. [CrossRef]

29. Iwamoto, M.; Uozumi, J.; Nishinari, K. Preliminary investigation of the state of water in foods by near infrared spectroscopy. In Proceedings of the International Near Infrared Spectroscopy/Near Infrared Technology Conference, Budapest, Hungary, 12–16 May 1986; Hello, J., Kaffka, K., Gonczy, J., Eds.; Akademiai Kiado: Budapest, Hungary, 1987; pp. 3–12.

30. Grant, A.; Davies, A.M.C.; Bilverstone, T. Simultaneous determination of sodium hydroxide, sodium carbonate and sodium chloride concentrations in aqueous solutions by near-infrared spectrometry. *Analyst* **1989**, *114*, 819. [CrossRef]

31. Tanaka, M.; Shibata, A.; Hayashi, N.; Kojima, T.; Maeda, H.; Ozaki, Y. Discrimination of commercial natural mineral waters using near infrared spectroscopy and principal component analysis. *J. Near Infrared Spectrosc.* **1995**, *3*, 203–210. [CrossRef]

32. Pasquini, C. Near infrared spectroscopy: A mature analytical technique with new perspectives—A review. *Anal. Chim. Acta* **2018**, *1026*, 8–36. [CrossRef]

33. Tsenkova, R.; Iso, E.; Parker, M.; Fockenberg, C.; Okubo, M. Aqua-Photomics: A NIRS investigation into the perturbation of water spectrum in an aqueous suspension of mesoscopic scale polystyrene spheres. In Proceedings of the 13th International Conference on Near Infrared Spectroscopy, Umea-Vasa, Sweden, 15–21 June 2007.

34. Workman, J. *Handbook of Organic Compounds: NIR, IR, Raman, and UV Spectra Featuring Polymers and Surfaces*; Academic Press: San Diego, CA, USA, 2001.

35. Gowen, A.A.; Marini, F.; Tsuchisaka, Y.; De Luca, S.; Bevilacqua, M.; O'Donnell, C.; Downey, G.; Tsenkova, R. On the feasibility of near infrared spectroscopy to detect contaminants in water using single salt solutions as model systems. *Talanta* **2015**, *131*, 609–618. [CrossRef]

36. Tsenkova, R. Near Infrared Spectroscopy of Raw Milk for Cow's Biomonitoring. Ph.D. Thesis, Hokkaido University, Sapporo, Japan, 2004.

37. Luck, W.A. Structure of water and aqueous solutions. In Proceedings of the International Symposium, Marburg, Germany, 18–28 July 1973; Luck, W.A.P., Ed.; Verlag Chemie: Weinheim, Germany, 1974; pp. 248–284.

38. Tsenkova, R.; Kovacs, Z.; Kubota, Y. Aquaphotomics: Near infrared spectroscopy and water states in biological systems. In *Membrane Hydration*; DiSalvo, E.A., Ed.; Springer: Berlin/Heidelberg, Germany, 2015; pp. 189–211.

39. Ozaki, Y. Applications in Chemistry. In *Near-Infrared Spectroscopy: Principles, Instruments, Applications*; Siesler, H., Ozaki, Y., Kawata, S., Heise, H., Eds.; Wiley: Chichester, UK, 2002; pp. 179–211.

40. Osborne, B.G.; Fearn, T.; Hindle, P.H. *Practical NIR Spectroscopy with Applications in Food and Beverage Analysis*; Browning, D., Ed.; Longman Singapore Publ. Ltd.: Singapore, 1993.

41. Pollner, B.; Kovacs, Z. Multivariate Data Analysis Tools for R Including Aquaphotomics Methods-Aquap2. Available online: https://www.aquaphotomics.com/resources/aquap2/ (accessed on 28 July 2016).

42. Cui, X.; Zhang, J.; Cai, W.; Shao, X. Chemometric algorithms for analyzing high dimensional temperature dependent near infrared spectra. *Chemom. Intell. Lab. Syst.* **2017**, *170*, 109–117. [CrossRef]

43. Shao, X.; Cui, X.; Yu, X.; Cai, W. Mutual factor analysis for quantitative analysis by temperature dependent near infrared spectra. *Talanta* **2018**, *183*, 142–148. [CrossRef]

44. Kojić, D.; Tsenkova, R.; Tomobe, K.; Yasuoka, K.; Yasui, M. Water confined in the local field of ions. *ChemPhysChem* **2014**, *15*, 4077–4086. [CrossRef]

45. Gowen, A.A.; Esquerre, C.; O'Donnell, C.P.; Downey, G.; Tsenkova, R. Use of near infrared hyperspectral imaging to identify water matrix co-ordinates in mushrooms (*Agaricus bisporus*) subjected to mechanical vibration. *J. Near Infrared Spectrosc.* **2009**, *17*, 363–371. [CrossRef]

46. Tsenkova, R. Aquaphotomics: Water in the biological and aqueous world scrutinised with invisible light. *Spectrosc. Eur.* **2010**, *22*, 6–10.

47. Tsenkova, R.N.; Iordanova, I.K.; Toyoda, K.; Brown, D.R. Prion protein fate governed by metal binding. *Biochem. Biophys. Res. Commun.* **2004**, *325*, 1005–1012. [CrossRef]

48. Goto, N.; Bazar, G.; Kovacs, Z.; Kunisada, M.; Morita, H.; Kizaki, S.; Sugiyama, H.; Tsenkova, R.; Nishigori, C. Detection of UV-induced cyclobutane pyrimidine dimers by near-infrared spectroscopy and aquaphotomics. *Sci. Rep.* **2015**, *5*, 11808. [CrossRef]

49. Cui, X.; Cai, W.; Shao, X. Glucose induced variation of water structure from temperature dependent near infrared spectra. *RSC Adv.* **2016**, *6*, 105729–105736. [CrossRef]

50. Cui, X.; Sun, Y.; Cai, W.; Shao, X. Chemometric methods for extracting information from temperature-dependent near-infrared spectra. *Sci. China Chem.* **2019**, *62*, 583–591. [CrossRef]

51. Ma, L.; Cui, X.; Cai, W.; Shao, X. Understanding the function of water during the gelation of globular proteins by temperature-dependent near infrared spectroscopy. *Phys. Chem. Chem. Phys.* **2018**, *20*, 20132–20140. [CrossRef]

52. Cui, X.; Liu, X.; Yu, X.; Cai, W.; Shao, X. Water can be a probe for sensing glucose in aqueous solutions by temperature dependent near infrared spectra. *Anal. Chim. Acta* **2017**, *957*, 47–54. [CrossRef]

53. Ball, P. Water as a Biomolecule. *ChemPhysChem* **2008**, *9*, 2677–2685. [CrossRef]

54. Ball, P. Life's matrix: Water in the cell. *Cell. Mol. Biol.* **2001**, *47*, 717–720.

55. Raschke, T.M. Water structure and interactions with protein surfaces. *Curr. Opin. Struct. Biol.* **2006**, *16*, 152–159. [CrossRef]

56. Ball, P. Water as an active constituent in cell biology. *Chem. Rev.* **2007**, *108*, 74–108. [CrossRef]

57. Bella, J.; Brodsky, B.; Berman, H.M. Hydration structure of a collagen peptide. *Structure* **1995**, *3*, 893–906. [CrossRef]

58. Büning-Pfaue, H. Analysis of water in food by near infrared spectroscopy. *Food Chem.* **2003**, *82*, 107–115. [CrossRef]

59. Sakudo, A.; Tsenkova, R.; Tei, K.; Onozuka, T.; Ikuta, K.; Yoshimura, E.; Onodera, T. Comparison of the vibration mode of metals in HNO_3 by a partial least-squares regression analysis of near-infrared spectra. *Biosci. Biotechnol. Biochem.* **2006**, *70*, 1578–1583. [CrossRef]

60. Sakudo, A.; Suganuma, Y.; Sakima, R.; Ikuta, K. Diagnosis of HIV-1 infection by near-infrared spectroscopy: Analysis using molecular clones of various HIV-1 subtypes. *Clin. Chim. Acta* **2012**, *413*, 467–472. [CrossRef]

61. Gowen, A.A.; Amigo, J.M.; Tsenkova, R. Characterisation of hydrogen bond perturbations in aqueous systems using aquaphotomics and multivariate curve resolution-alternating least squares. *Anal. Chim. Acta* **2013**, *759*, 8–20. [CrossRef]

62. Gowen, A.A.; Tsuchisaka, Y.; O'Donnell, C.; Tsenkova, R. Investigation of the potential of near infrared spectroscopy for the detection and quantification of pesticides in aqueous solution. *Am. J. Anal. Chem.* **2011**, *2*, 53–62. [CrossRef]

63. Omar, A.F.; Atan, H.; MatJafri, M.Z. NIR spectroscopic properties of aqueous acids solutions. *Molecules* **2012**, *17*, 7440–7450. [CrossRef]

64. Cattaneo, T.M.P.; Vero, S.; Napoli, E.; Elia, V. Influence of filtration processes on aqueous nanostructures by NIR spectroscopy. *J. Chem. Chem. Eng.* **2011**, *5*, 1046–1052.

65. Kovacs, Z.; Bázár, G.; Oshima, M.; Shigeoka, S.; Tanaka, M.; Furukawa, A.; Nagai, A.; Osawa, M.; Itakura, Y.; Tsenkova, R. Water spectral pattern as holistic marker for water quality monitoring. *Talanta* **2016**, *147*, 598–608. [CrossRef]

66. Vanoli, M.; Lovati, F.; Grassi, M.; Buccheri, M.; Zanella, A.; Cattaneo, T.M.P.; Rizzolo, A. Water spectral pattern as a marker for studying apple sensory texture. *Adv. Hortic. Sci.* **2018**, *32*, 343–351.

67. Muncan, J.; Mileusnic, I.; Matovic, V.; Sakota Rosic, J.; Matija, L. The prospects of aquaphotomics in biomedical science and engineering. Presented at the 2nd International Aquaphotomics Symposium, Kobe, Japan, 26–29 November 2016.

68. Kinoshita, K.; Miyazaki, M.; Morita, H.; Vassileva, M.; Tang, C.; Li, D.; Ishikawa, O.; Kusunoki, H.; Tsenkova, R. Spectral pattern of urinary water as a biomarker of estrus in the giant panda. *Sci. Rep.* **2012**, *2*, 856. [CrossRef]

69. Jinendra, B. Near infrared Spectroscopy and Aquaphotomics: Novel Tool for Biotic and Abiotic Stress Diagnosis of Soybean. Ph.D. Thesis, Kobe University, Kobe, Japan, 2011.

70. Jinendra, B.; Tamaki, K.; Kuroki, S.; Vassileva, M.; Yoshida, S.; Tsenkova, R. Near infrared spectroscopy and aquaphotomics: Novel approach for rapid in vivo diagnosis of virus infected soybean. *Biochem. Biophys. Res. Commun.* **2010**, *397*, 685–690. [CrossRef]

71. Haq, Q.M.I.; Mabood, F.; Naureen, Z.; Al-Harrasi, A.; Gilani, S.A.; Hussain, J.; Jabeen, F.; Khan, A.; Al-Sabari, R.S.M.; Al-khanbashi, F.H.S.; et al. Application of reflectance spectroscopies (FTIR-ATR & FT-NIR) coupled with multivariate methods for robust in vivo detection of begomovirus infection in papaya leaves. *Spectrochim. Acta Part A Mol. Biomol. Spectrosc.* **2018**, *198*, 27–32.

72. Kuroki, S.; Tsenkova, R.; Moyankova, D.P.; Muncan, J.; Morita, H.; Atanassova, S.; Djilianov, D. Water molecular structure underpins extreme desiccation tolerance of the resurrection plant *Haberlea rhodopensis*. *Sci. Rep.* **2019**, *9*, 3049. [CrossRef]

73. Matija, L.; Tsenkova, R.; Munćan, J.; Miyazaki, M.; Banba, K.; Tomić, M.; Jeftić, B. Fullerene based nanomaterials for biomedical applications: Engineering, functionalization and characterization. *Adv. Mater. Res.* **2013**, *633*, 224–238. [CrossRef]

74. Matija, L.; Muncan, J.; Mileusnic, I.; Koruga, D. Fibonacci nanostructures for novel nanotherapeutical approach. In *Nano-and Microscale Drug Delivery Systems: Design and Fabrication*; Grumezescu, A.M., Ed.; Elsevier: Amsterdam, The Netherlands, 2017; pp. 49–74.

75. Cui, X.; Yu, X.; Cai, W.; Shao, X. Water as a probe for serum–based diagnosis by temperature–dependent near–infrared spectroscopy. *Talanta* **2019**, *204*, 359–366. [CrossRef]

76. Azar, A.T. *Modeling and Control of Dialysis Systems, Volume 2: Biofeedback Systems and Soft Computing Techniques of Dialysis*; Springer-Verlag: Berlin, Germany, 2013.

77. Tsenkova, R.; Atanassova, S.; Toyoda, K.; Ozaki, Y.; Itoh, K. Near infrared spectroscopy for biomonitoring. *J. Jpn. Soc. Agric. Mach.* **1999**, *61*, 529–530.

78. Tsenkova, R.; Murayama, K.; Kawamura, S.; Itoh, K.; Toyoda, K. Near infrared mastitis diagnosis in the process of milking. In *Robotic Milking*; Hogeveen, H., Meijering, A., Eds.; Wageningen Pers: Wageningen, The Netherlands, 2000; p. 127.

79. Tsenkova, R.; Murayama, K.; Kawano, S.; Wu, Y.; Toyoda, K.; Ozaki, Y. Near infrared spectroscopy for mastitis diagnosis: Two-dimensional correlation study in short wavelength region. In *Two-Dimensional Correlation Spectroscopy*; Ozaki, Y., Noda, I., Eds.; AIP: Melville, NY, USA, 1999; pp. 307–311.

80. Meilina, H.; Kuroki, S.; Jinendra, B.M.; Ikuta, K.; Tsenkova, R. Double threshold method for mastitis diagnosis based on NIR spectra of raw milk and chemometrics. *Biosyst. Eng.* **2009**, *104*, 243–249. [CrossRef]

81. Tsenkova, R.; Atanassova, S. Mastitis diagnostics by near infrared spectra of cow's milk, blood and urine using soft independent modelling of class analogy classification. In Proceedings of the Near Infrared Spectroscopy 10th International Conference, Kyongju, Korea, 10–15 June 2001; p. 123.

82. Tsenkova, R.; Yordanova, K.I.; Itoh, K.; Shinde, Y.; Nishibu, J. Near-infrared spectroscopy of individual cow milk as a means for automated monitoring of udder health and milk quality. In Proceedings of the Dairy System for the 21st Century, the Third International Dairy Housing Conference, Orlando, FL, USA, 2–5 February 1994; pp. 82–91.

83. Kinoshita, K.; Morita, H.; Miyazaki, M.; Hama, N.; Kanemitsu, H.; Kawakami, H.; Wang, P.; Ishikawa, O.; Kusunoki, H.; Tsenkova, R. Near infrared spectroscopy of urine proves useful for estimating ovulation in giant panda (*Ailuropoda melanoleuca*). *Anal. Methods* **2010**, *2*, 1671. [CrossRef]

84. Kinoshita, K.; Kuze, N.; Kobayashi, T.; Miyakawa, E.; Narita, H.; Inoue-Murayama, M.; Idani, G.; Tsenkova, R. Detection of urinary estrogen conjugates and creatinine using near infrared spectroscopy in Bornean orangutans (*Pongo Pygmaeus*). *Primates* **2016**, *57*, 51–59. [CrossRef]

85. Takemura, G.; Bázár, G.; Ikuta, K.; Yamaguchi, E.; Ishikawa, S.; Furukawa, A.; Kubota, Y.; Kovács, Z.; Tsenkova, R. Aquagrams of raw milk for oestrus detection in dairy cows. *Reprod. Domest. Anim.* **2015**, *50*, 522–525. [CrossRef]

86. Agcanas, L.A.; Counsell, K.R.; Shappell, N.; Bowers, S.; Ryan, P.L.; Willard, S.T.; Vance, C.K. A novel approach to comparing reproductive stage serum profiles in mares using near-infrared spectroscopy and aquaphotomics. *Reprod. Fertil. Dev.* **2017**, *29*, 117. [CrossRef]

87. Slavchev, A.; Kovacs, Z.; Koshiba, H.; Bazar, G.; Pollner, B.; Krastanov, A.; Tsenkova, R. Monitoring of water spectral patterns of lactobacilli development as a tool for rapid selection of probiotic candidates. *J. Near Infrared Spectrosc.* **2017**, *25*, 423–431. [CrossRef]

88. Kovacs, Z.; Slavchev, A.; Bazar, G.; Pollner, B.; Tsenkova, R. Rapid bacteria selection using Aquaphotomics and near infrared spectroscopy. In Proceedings of the 18th International Conference on Near Infrared Spectroscopy, Copenhagen, Denmark, 11–15 June 2017; IM Publications Open LLP: Chichester, UK, 2019; pp. 65–69.

89. Slavchev, A.; Kovacs, Z.; Koshiba, H.; Nagai, A.; Bázár, G.; Krastanov, A.; Kubota, Y.; Tsenkova, R. Monitoring of water spectral pattern reveals differences in probiotics growth when used for rapid bacteria selection. *PLoS ONE* **2015**, *10*, e0130698. [CrossRef]

90. Remagni, M.C.; Morita, H.; Koshiba, H.; Cattaneo, T.M.P.; Tsenkova, R. Near infrared spectroscopy and aquaphotomics as tools for bacteria classification. In Proceedings of the NIR2013: Picking Up Good Vibrations, La Grande-Motte, France, 2–7 June 2013; p. 602.

91. Tanaka, S.; Kojić, D.; Tsenkova, R.; Yasui, M. Quantification of anomeric structural changes of glucose solutions using near-infrared spectra. *Carbohydr. Res.* **2018**, *463*, 40–46. [CrossRef]

92. Vaidyanathan, S.; Harvey, L.M.; McNeil, B. Deconvolution of near-infrared spectral information for monitoring mycelial biomass and other key analytes in a submerged fungal bioprocess. *Anal. Chim. Acta* **2001**, *428*, 41–59. [CrossRef]

93. Yanmin, Y.; Na, W.; Youqi, C.; Yingbin, H.; Pengqin, T. Soil moisture monitoring using hyper-spectral remote sensing technology. In Proceedings of the 2010 Second IITA International Conference on Geoscience and Remote Sensing, Qingdao, China, 28–31 August 2010; IEEE: Piscataway, NJ, USA, 2010; pp. 373–376.

94. Šakota Rosić, J.; Munćan, J.; Mileusnić, I.; Kosić, B.; Matija, L. Detection of protein deposits using NIR spectroscopy. *Soft Mater.* **2016**, *14*, 264–271. [CrossRef]

95. Munćan, J.; Rosić, J.; Mileusnić, I.; Matović, V.; Matija, L.; Tsenkova, R. The structure of water in soft contact lenses: Near infrared spectroscopy and Aquaphotomics study. In Proceedings of the 18th International Conference on Near Infrared Spectroscopy, Copenhagen, Denmark, 11–15 June 2017; IM Publications Open LLP: Chichester, UK, 2019; pp. 99–104.

96. Muncan, J.; Mileusnic, I.; Sakota Rosic, J.; Vasic-Milovanovic, A.; Matija, L. Water Properties of Soft Contact Lenses: A Comparative Near-Infrared Study of Two Hydrogel Materials. *Int. J. Polym. Sci.* **2016**, *2016*, 1–8. [CrossRef]

97. Matija, L.R.; Tsenkova, R.N.; Miyazaki, M.; Banba, K.; Muncan, J.S. Aquagrams: Water spectral pattern as characterization of hydrogenated nanomaterial. *FME Trans.* **2012**, *40*, 51–56.

98. Haghi, R.K.; Yang, J.; Tohidi, B. Integrated near infrared and ultraviolet spectroscopy techniques for determination of hydrate inhibitors in the presence of NaCl. *Ind. Eng. Chem. Res.* **2018**, *57*, 11728–11737. [CrossRef]

99. Chang, K.; Shinzawa, H.; Chung, H. Concentration determination of inorganic acids that do not absorb near-infrared (NIR) radiation through recognizing perturbed NIR water bands by them and investigation of accuracy dependency on their acidities. *Microchem. J.* **2018**, *139*, 443–449. [CrossRef]

100. Shao, X.; Cui, X.; Wang, M.; Cai, W. High order derivative to investigate the complexity of the near infrared spectra of aqueous solutions. *Spectrochim. Acta Part A Mol. Biomol. Spectrosc.* **2019**, *213*, 83–89. [CrossRef]

101. Dong, Q.; Yu, C.; Li, L.; Nie, L.; Li, D.; Zang, H. Near-infrared spectroscopic study of molecular interaction in ethanol-water mixtures. *Spectrochim. Acta Part A Mol. Biomol. Spectrosc.* **2019**, *222*, 117183. [CrossRef]

102. Kang, J.; Cai, W.; Shao, X. Quantitative determination by temperature dependent near-infrared spectra: A further study. *Talanta* **2011**, *85*, 420–424. [CrossRef]

103. Shao, X.; Cui, X.; Liu, Y.; Xia, Z.; Cai, W. Understanding the molecular interaction in solutions by chemometric resolution of near–infrared spectra. *ChemistrySelect* **2017**, *2*, 10027–10032. [CrossRef]

104. Li, D.; Li, L.; Quan, S.; Dong, Q.; Liu, R.; Sun, Z.; Zang, H. A feasibility study on quantitative analysis of low concentration methanol by FT-NIR spectroscopy and aquaphotomics. *J. Mol. Struct.* **2019**, *1182*, 197–203. [CrossRef]

105. Shan, R.; Zhao, Y.; Fan, M.; Liu, X.; Cai, W.; Shao, X. Multilevel analysis of temperature dependent near-infrared spectra. *Talanta* **2015**, *131*, 170–174. [CrossRef]

106. Shao, X.; Kang, J.; Cai, W. Quantitative determination by temperature dependent near-infrared spectra. *Talanta* **2010**, *82*, 1017–1021. [CrossRef]

107. Putra, A.; Faridah, F.; Inokuma, E.; Santo, R. Robust spectral model for low metal concentration measurement in aqueous solution reveals the importance of water absorbance bands. *J. Sains dan Teknol. Reaksi* **2010**, *8*. [CrossRef]

108. Tsenkova, R.; Fockenberg, C.; Koseva, N.; Sakudo, A.; Parker, M. Aqua-Photomics: Water absorbance patterns in NIR range used for detection of metal ions reveal the importance of sample preparation. In Proceedings of the 13th International Conference on Near Infrared Spectroscopy, Umea-Vasa, Sweden, 15–21 June 2007.

109. Putra, A.; Meilina, H.; Tsenkova, R. Use of near-infrared spectroscopy for determining the characterization metal ion in aqueous solution. In Proceedings of the Annual International Conference, Syiah Kuala University-Life Sciences & Engineering Chapter, Banda Aceh, Indonesia, 22–24 November 2012; Volume 2, pp. 154–158.

110. Vero, S.; Tornielli, C.; Cattaneo, T.M.P. Aquaphotomics: Wavelengths involved in the study of the speciation of metal ions (Zn^{2+}, Pb^{2+} and Ag^+) in aqueous Solutions. *NIR News* **2010**, *21*, 11–13. [CrossRef]

111. Wenz, J.J. Examining water in model membranes by near infrared spectroscopy and multivariate analysis. *Biochim. Biophys. Acta Biomembr.* **2018**, *1860*, 673–682. [CrossRef]

112. Cheng, D.; Cai, W.; Shao, X. Understanding the interaction between oligopeptide and water in aqueous solution using temperature-dependent near-infrared spectroscopy. *Appl. Spectrosc.* **2018**, *72*, 1354–1361. [CrossRef]

113. Chatani, E.; Tsuchisaka, Y.; Masuda, Y.; Tsenkova, R. Water molecular system dynamics associated with amyloidogenic nucleation as revealed by real time near infrared spectroscopy and aquaphotomics. *PLoS ONE* **2014**, *9*, e101997. [CrossRef]

114. Yamada, K.; Murayama, K.; Tsenkova, R.; Wang, Y.; Ozaki, Y. Multivariate determination of human serum albumin and γ-globulin in a phosphate buffer solution by near infrared spectroscopy. *J. Near Infrared Spectrosc.* **1998**, *6*, 375–381.

115. Yuan, B.; Murayama, K.; Wu, Y.; Tsenkova, R.; Dou, X.; Era, S.; Ozaki, Y. Temperature-dependent near-infrared spectra of bovine serum albumin in aqueous solutions: Spectral analysis by principal component analysis and evolving factor analysis. *Appl. Spectrosc.* **2003**, *57*, 1223–1229. [CrossRef]

116. Munćan, J.S.; Matija, L.; Simić-Krstić, J.B.; Nijemčević, S.S.; Koruga, D.L. Discrimination of mineral waters using near-infrared spectroscopy and aquaphotomics. *Hem. Ind.* **2014**, *68*, 257–264. [CrossRef]

117. Bázár, G.; Romvári, R.; Szabó, A.; Somogyi, T.; Éles, V.; Tsenkova, R. NIR detection of honey adulteration reveals differences in water spectral pattern. *Food Chem.* **2016**, *194*, 873–880. [CrossRef]

118. Esquerre, C.; Gowen, A.; O'Donnell, C.P.; Downey, G. Water absorbance pattern of physical damage in mushrooms. In Proceedings of the Biosystems Engineering Research Review 14, Dublin, Ireland, 11 March 2009; Cummins, E., Curran, T., Eds.; University College Dublin: Dublin, Ireland, 2009; pp. 68–71.

119. Gowen, A.A. Water and Food Quality. *Contemp. Mater.* **2012**, *1*, 31–37. [CrossRef]

120. Achata, E.; Esquerre, C.; O'Donnell, C.; Gowen, A.; Achata, E.; Esquerre, C.; O'Donnell, C.; Gowen, A. A study on the application of near infrared hyperspectral chemical imaging for monitoring moisture content and water activity in low moisture systems. *Molecules* **2015**, *20*, 2611–2621. [CrossRef]

121. Veleva-Doneva, P.; Atanassova, S.; Zhelyazkov, G. Innovative engineering methods for quality evaluation and food safety. *Zesz. Nauk. Małopolskiej Wyższej Szkoły Ekon. Tarn.* **2017**, *36*, 13–23.

122. Atanassova, S. Near Infrared Spectroscopy and aquaphotomics for monitoring changes during yellow cheese ripening. *Agric. Sci. Technol.* **2015**, *7*, 269–272.

123. Cattaneo, T.M.P.; Vanoli, M.; Grassi, M.; Rizzolo, A.; Barzaghi, S. The aquaphotomics approach as a tool for studying the influence of food coating materials on cheese and winter melon samples. *J. Near Infrared Spectrosc.* **2016**, *24*, 381–390. [CrossRef]

124. Vanoli, M.; Grassi, M.; Lovati, F.; Barzaghi, S.; Cattaneo, T.M.P.; Rizzolo, A. Influence of innovative coatings on salami ripening assessed by near infrared spectroscopy and aquaphotomics. *J. Near Infrared Spectrosc.* **2019**, *27*, 54–64. [CrossRef]

125. Barzaghi, S.; Cremonesi, K.; Cattaneo, T.M.P. Influence of the presence of bioactive compounds in smart-packaging materials on water absorption using NIR spectroscopy and aquaphotomics. *NIR News* **2017**, *28*, 21–24. [CrossRef]

126. Kong, W.; Zhang, C.; Cao, F.; Liu, F.; Luo, S.; Tang, Y.; He, Y.; Kong, W.; Zhang, C.; Cao, F.; et al. Detection of sclerotinia stem rot on oilseed rape (*Brassica napus* L.) leaves using hyperspectral imaging. *Sensors* **2018**, *18*, 1764. [CrossRef]

127. He, X.; Feng, X.; Sun, D.; Liu, F.; Bao, Y.; He, Y.; He, X.; Feng, X.; Sun, D.; Liu, F.; et al. Rapid and nondestructive measurement of rice seed vitality of different years using near-infrared hyperspectral imaging. *Molecules* **2019**, *24*, 2227. [CrossRef]

128. Chu, B.; Yu, K.; Zhao, Y.; He, Y.; Chu, B.; Yu, K.; Zhao, Y.; He, Y. Development of noninvasive classification methods for different roasting degrees of coffee beans using hyperspectral imaging. *Sensors* **2018**, *18*, 1259. [CrossRef]

129. Mura, S.; Cappai, C.; Greppi, G.F.; Barzaghi, S.; Stellari, A.; Cattaneo, T.M.P. Vibrational spectroscopy and Aquaphotomics holistic approach to determine chemical compounds related to sustainability in soil profiles. *Comput. Electron. Agric.* **2019**, *159*, 92–96. [CrossRef]

130. Bozhynov, V.; Soucek, P.; Barta, A.; Urbanova, P.; Bekkozhayeva, D. Visible aquaphotomics spectrophotometry for aquaculture systems. In Proceedings of the International Conference on Bioinformatics and Biomedical Engineering, Granada, Spain, 25–27 April 2018; Springer: Cham, Switzerland, 2018; pp. 107–117.

131. Nakakimura, Y.; Vassileva, M.; Stoyanchev, T.; Nakai, K.; Osawa, R.; Kawano, J.; Tsenkova, R. Extracellular metabolites play a dominant role in near-infrared spectroscopic quantification of bacteria at food-safety level concentrations. *Anal. Methods* **2012**, *4*, 1389. [CrossRef]

132. Sakudo, A.; Yoshimura, E.; Tsenkova, R.; Ikuta, K.; Onodera, T. Native state of metals in non-digested tissues by partial least squares regression analysis of visible and near-infrared spectra. *J. Toxicol. Sci.* **2007**, *32*, 135–141. [CrossRef]

133. Sakudo, A.; Tsenkova, R.; Tei, K.; Morita, H.; Ikuta, K.; Onodera, T. Ex vivo tissue discrimination by visible and near-infrared spectra with chemometrics. *J. Vet. Med. Sci.* **2006**, *68*, 1375–1378. [CrossRef]

134. Huck, C.W. Theoretical and technical advancements of near-infrared spectroscopy and its operational impact in industry. *NIR News* **2017**, *28*, 17–21. [CrossRef]

Sample Availability: Not available.

![molecules logo] *molecules*

MDPI

Review

Applications of Photonics in Agriculture Sector: A Review

Jin Yeong Tan [1], Pin Jern Ker [1,*], K. Y. Lau [1], M. A. Hannan [1] and Shirley Gee Hoon Tang [2]

[1] Institute of Power Engineering, College of Engineering, Universiti Tenaga Nasional, Kajang 43000, Selangor, Malaysia; adrian_tan_jy@hotmail.com (J.Y.T.); kylau@uniten.edu.my (K.Y.L.); Hannan@uniten.edu.my (M.A.H.)

[2] Microbiology Unit, Department of Pre-clinical, International Medical School, Management and Science University, University Drive, Off Persiaran Olahraga, Seksyen 13, Shah Alam 40100, Selangor, Malaysia; shirley_tang@msu.edu.my

* Correspondence: pinjern@uniten.edu.my

Received: 25 March 2019; Accepted: 12 May 2019; Published: 27 May 2019

Abstract: The agricultural industry has made a tremendous contribution to the foundations of civilization. Basic essentials such as food, beverages, clothes and domestic materials are enriched by the agricultural industry. However, the traditional method in agriculture cultivation is labor-intensive and inadequate to meet the accelerating nature of human demands. This scenario raises the need to explore state-of-the-art crop cultivation and harvesting technologies. In this regard, optics and photonics technologies have proven to be effective solutions. This paper aims to present a comprehensive review of three photonic techniques, namely imaging, spectroscopy and spectral imaging, in a comparative manner for agriculture applications. Essentially, the spectral imaging technique is a robust solution which combines the benefits of both imaging and spectroscopy but faces the risk of underutilization. This review also comprehends the practicality of all three techniques by presenting existing examples in agricultural applications. Furthermore, the potential of these techniques is reviewed and critiqued by looking into agricultural activities involving palm oil, rubber, and agro-food crops. All the possible issues and challenges in implementing the photonic techniques in agriculture are given prominence with a few selective recommendations. The highlighted insights in this review will hopefully lead to an increased effort in the development of photonics applications for the future agricultural industry.

Keywords: agriculture; photonics; imaging; spectral imaging; spectroscopy

---

1. Introduction

Light constitutes a collection of particles known as photons, propagated in the form of waves [1]. In physics, light often relates to radiation in the entire electromagnetic spectrum, encompassing X-rays, ultraviolet, visible light, infrared, and microwaves among others [2]. The unique electromagnetic properties of light have intrigued academics across the globe and the earliest study can be traced back to the early 17th century [3]. As time passes, the accumulation of knowledge and technological advancement have gradually shaped the canvas for light-related research, leading to the establishment of the field of optics and photonics.

Optics can be defined as a branch of physics that studies the behavior and properties of light as well as the interaction of light with other matter [2]. Meanwhile, photonics can be regarded as the application of light through the systematic generation, control and detection of photons [2,4]. Despite the distinction between optics and photonics, both terminologies have often been used interchangeably in the literature to collectively represent the science and application of light [1].

Optics and photonics have influenced various engineering applications, transforming the landscape of various fields and improving the lives of mankind. One of the main applications of optics and photonics can be seen in the field of communications. Knowledge of optics and photonics has been used to develop optical fibers which help to cater for the needs of broadband Internet service in this "data hungry" era. Furthermore, optics and photonics have been used in the manufacturing of modern displays such as liquid crystal display (LCD), organic light-emitting diode (OLED), flexible display and such. Solar cells for energy harnessing too illustrate another application of optics and photonics. Not least, optics and photonics have also been applied in more sophisticated areas such as security surveillance, medical imaging, quantum computing and more [1].

Amidst the modern and complex solutions discussed earlier, it often slipped our minds that optics and photonics can be readily integrated into the field of agriculture. The simplest examples would be the adjustment of plantation direction for optimum sunlight exposure, as well as the usage of incandescent light bulbs in egg incubation and hatching [5]. Over recent decades, academics have been alerted to the potential of optics and photonics in the agricultural industry. This has led to progressive developments that utilize optics and photonic techniques in maximizing the quality and productivity of agricultural products.

This paper aims to review some of the most popular optics and photonic techniques in agriculture, namely imaging, spectroscopy and spectral imaging. In addition, existing applications of each technique in the agricultural industry will also be compiled. A comprehensive discussion will also be made to gauge the potential of exploiting optics and photonic techniques in the agricultural sector with the intention of improving the quality and productivity of the agricultural products at a reduced labor cost.

2. Classification of Photonics Systems in Agriculture

Quantity and quality have always been the primary foci in the field of agriculture. The governing of these attributes is anticipated to be more crucial in the upcoming years. This prediction is based on the constant increase in global population as well as heightened expectations for healthy food sources. However, the agricultural field faces great pressure under globalization. The transformation of the global economic landscape makes agricultural activities seem less profitable in contrast to other industrial activities. The outflow of the workforce makes it increasingly expensive and difficult to meet the demands of agricultural activities.

As a result, modern technology has been integrated into the agricultural field to maximize output efficiency at minimum labor force. Similar to other industries, automation systems have been applied in stages of agricultural activities to reduce a dependency on manual labor [6]. These systems require optics and photonics techniques to complement them, providing the required 'sight' for operations. These vision requirements have been fulfilled by optics and photonic techniques such as imaging, and spectral imaging. These techniques provide machine vision at high dynamic range, high resolution and high accuracy in a non-destructive, non-contact and robust manner [5]. In the subsections below, details of ESS configurations, their classifications and structures have been illustrated.

2.1. Imaging Technique

The imaging technique is analogous to the function of the human eye. It captures the image of the subject for necessary calculations and measurements before performing the final evaluations [7]. The imaging technique is essential for collecting spatial, color [6] and even thermal [8] information of the subject of interest. Therefore, imaging techniques are typically operated in an active manner. The active imaging technique involves image acquisition under two major light sources, namely visible light and infrared sources. Images under visible light can be easily acquired with any standard camera modules. On the other hand, images under exposure to infrared can be acquired with special infrared camera modules [8].

Image acquisition under visible light is similar to our daily photography. The image acquisition process under this light source is straightforward and images captured are usually rich in details and colors. However, complexity often arises while performing analysis on these images due to illumination variations. For instance, images captured outdoors vary under sunny and cloudy conditions. Meanwhile, images captured indoors is categorized by natural light, incandescent and fluorescent conditions [7].

The acquired image will then undergo pre-processing to convert it into an appropriate format before further analysis. Pre-processing tasks may include exposure correction, color balancing, noise reduction, sharpness increase or orientation change. Next, the process of feature detection and matching as well as segmentation is performed on the pre-processed image to extract the object or region of interest. Finally, the subject of interest is analyzed with proper analysis algorithms in the respective area of application [9].

The imaging technique can be easily applied in the simple analysis of static-positioned objects or even in more complex areas which involve moving targets, such as visual navigation and behavioral surveillance. These achievements were made possible by utilizing the spatial information acquired through the imaging technique for position triangulation and motion guidance [7,9]. In image processing, the computer imaging technique has been employed to create, edit, and display graphical images, characters, and objects. The computer image analysis technique is a broad field which consists of computer domains and applications in food quality evaluation [10,11], grading and the sorting of agricultural products [12,13], as well as harvesting the crops [14], and estimating moisture content in the drying stage for the storability of the food product [15]. Computer imaging contributes to the development of digital agriculture. For instance, weed detection and fruit grading systems with digital imaging techniques are cost effective systems in achieving ecological and economically sustainable agriculture [16].

2.2. Spectroscopy Technique

In contrast to the imaging technique, the spectroscopy technique enables the 'sight' of properties that are invisible to the naked eye. The spectroscopy technique functions by extracting spectral information from the sample of interest. The spectral information is obtained when light interacts with the composition of the sample. This interaction leads to changes in the intensity or frequency and wavelength of the initial light source, ultimately defining a spectrum which acts as the fingerprint of the sample [17].

Similar to the imaging technique, variations do exist for spectroscopy. These variations are categorized by the nature of interaction between the light source and the sample when the spectroscopy measurement is conducted. In the agricultural field, the commonly adopted spectroscopy techniques are ultraviolet-visible (UV-VIS) spectroscopy, fluorescence spectroscopy, infrared (IR) spectroscopy, and Raman spectroscopy [17].

2.2.1. Ultraviolet-Visible (UV-VIS) Spectroscopy

The ultraviolet-visible (UV-VIS) spectroscopy is conducted in both the ultraviolet (UV) and visible light (VIS) band, spanning wavelengths from 100 nm to 380 nm (UV) and from 380 nm to 750 nm (VIS). The principle governing the UV-VIS spectroscopy is Beer-Lambert's law, which is expressed by (1) and (2):

$$I = I_0 10^{-\varepsilon c l}, \tag{1}$$

$$\ln \frac{I_0}{I} = \ln \frac{1}{T} = \varepsilon c l = A \tag{2}$$

where I_0 and I are intensity of light entering and leaving a sample respectively, ε is the extinction molar coefficient, c is the molar concentration of substance, l is the thickness of sample (cm), T is transmittance and A is absorbance [18].

A typical model that illustrates Beer-Lambert's law can be seen in Figure 1. It can be observed that as light propagates through a sample, a portion of the incidental light source will be absorbed by the molecules in the sample, while the remaining light rays will transmit and escape across the sample. The ratio between the intensity of the incident and escaped rays defines the absorbance of light by the sample. This value of light absorbance is of main interest in UV-VIS spectroscopy. As in Equation (2), light absorbance is dependent on ε, c, and l [18]. The absorbance value(s) at a single or multiple wavelength(s) will then be used to measure the concentration of compounds in a sample [19–23].

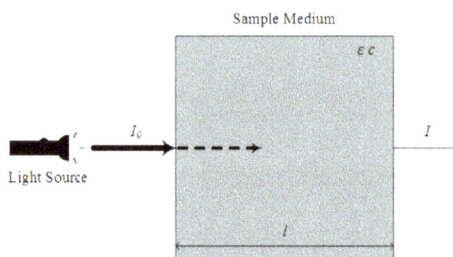

Figure 1. Model of Beer-Lambert's Law [18].

2.2.2. Fluorescence Spectroscopy

Fluorescence spectroscopy is distinct from other spectroscopy techniques in terms of the emission of light when incident rays from an ultraviolet or visible light source is absorbed by fluorescent molecules present in a sample. These fluorescent molecules are known as fluorophores and commonly known examples include quinine, fluorescein, acridine orange, rhodamine B and pyridine 1 [24].

The fluorescence phenomenon can be explained with a Jablonski diagram illustrated in Figure 2. It should first be understood that fluorescence involves the three electronic states of a fluorophore molecule, namely the singlet ground, first and second electronic states. These states are represented by S_0, S_1 and S_2 in Figure 2. The key condition for fluorescence to occur is the excitation of the molecule from the ground state, S_0 to either electronic states S_1 or S_2 upon the absorption of light. If the molecule reaches the S_2 state, internal conversion or vibrational relaxation will occur, returning the molecule to the lower S_1 state without radiation emitted. From here, the molecule will again return to the S_0 while emitting light which has equal energy as the energy difference between S_0 and S_1. This light emission is known as fluorescence and this condition typically occurs 10^{-8} seconds after the initial excitation [17].

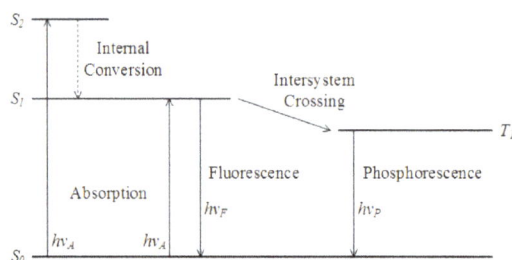

Figure 2. Jablonski diagram [17]. Reproduced with permission from A. Nawrocka, Advances in Agrophysical Research, Published by IntechOpen, 2013.

Fluorescence spectroscopy is highly specific and highly sensitive. The high specificity of the technique arises from the usage of both the excitation and emission spectra; whereas high sensitivity is achieved as radiation measurements are made against absolute darkness. These characteristics however limit the independent usage of the technique [17]. As a result, fluorescence spectroscopy is often combined with high performance liquid chromatography (HPLC) [25]. Variations may also be implemented in the excitation and emission wavelengths, forming the synchronous fluorescence spectroscopy (SFS) [26].

2.2.3. Infrared (IR) Spectroscopy

Infrared (IR) spectroscopy operates within the IR band with wavelengths from 780 nm to 1 mm. The IR band can be further broken down into three sub-bands, namely near-infrared (NIR; 780 nm to 5 µm), mid-infrared (MIR; 5 µm to 30 µm) and far-infrared (FIR; 30 µm to 1 mm). In agriculture-related optics and photonics, the NIR and MIR bands are of greater interest [17].

IR spectroscopy obtains the spectral information of a subject due to molecular vibrations under the excitation of an IR light source. In general, molecular vibrations occur when there exist normal modes of vibrations. A normal mode of vibration (or fundamental) refers to the phenomenon in which every atom in a molecule experiences a simple harmonic oscillation about its equilibrium position. These atoms oscillate in phase at the same frequency while the center of gravity of the molecule remains unchanged. A typical molecule has 3N-6 fundamentals (3N-5 for linear molecules), where N refers to the number of atoms. The diatomic molecular vibrations are illustrated in Figure 3 [27].

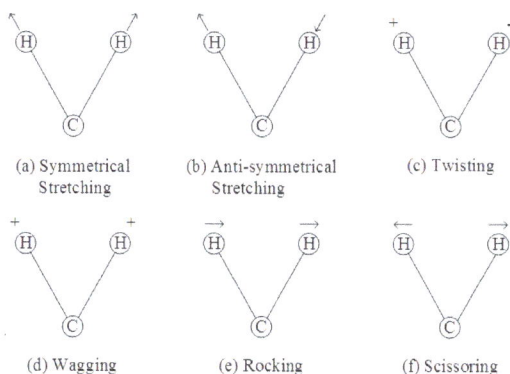

(a) Symmetrical Stretching   (b) Anti-symmetrical Stretching   (c) Twisting

(d) Wagging   (e) Rocking   (f) Scissoring

Figure 3. Vibrations in diatomic molecules [17]. Reproduced with permission from A. Nawrocka, Advances in Agrophysical Research, Published by IntechOpen, 2013.

Molecular vibrations, which occur regardless the presence of IR light source, result in an increase in light absorption. These peaks in absorption form specific bands in the IR spectrum that correspond to the specific frequencies in which molecular vibrations occur. This allows the easy identification of the molecular structure in a sample since different molecules have different vibration frequencies [27]. This unique frequency 'fingerprint' is exceptionally beneficial in the analysis of complex molecules that contains functional groups such as –OH, –NH$_2$, –CH$_3$, C=O, C$_6$H$_5$– and more. For instance, the C$_6$H$_5$– group forms peaks at wavenumbers from 1600 cm^{-1} to 1500 cm^{-1} (wavelengths from 6.25 µm to 6.67 µm) whereas the C=O group exhibits high absorption at wavenumbers from 1800 cm^{-1} to 1650 cm^{-1} (wavelengths from 5.56 µm to 6.06 µm) [28].

2.2.4. Near-Infrared (NIR) Spectroscopy

The near-infrared (NIR) spectroscopy operates within the NIR band with wavelengths from 780 nm to 5 µm. The absorptions within the NIR band exist due to overtones and combinations of the fundamental vibrations. Overtones refer to higher frequencies that are multiples of the fundamental frequency. Meanwhile, combinations involve interactions between two or more vibrations occurring simultaneously, resulting in a frequency which is the sum of multiples of the respective frequencies. A majority of the absorptions in the NIR band are due to vibrations of the C–H, O–H and N–H bands. The S–H and C=O bonds too potentially contribute to these absorptions. Several assignments of the NIR absorption bands can be seen in Table 1 [29].

Table 1. Examples of NIR absorption bands [29].

Wavelength (nm)	Wavenumber (cm^{-1})	Assignment
		Water
1454	6878	1st overtone O–H stretching
1932	5176	O–H combination
		Proteins
1208	8278	2nd overtone C–H stretching
1465	6826	1st overtone N–H and O–H stretching
1734	5767	1st overtone C–H stretching
1932 2058 2180	5176 4859 4587	N–H combination and O–H stretching
2302 2342	4344 4270	C–H stretching combination
		Oil
1210	8264	2nd overtone C–H stretching
1406	7112	1st overtone N–H and O–H stretching
1718 1760	5821 5682	1st overtone C–H stretching
2114	4730	N–H combination and O–H stretching
2308 2346	4333 4263	C–H stretching combination
		Starch
1204	8306	2nd overtone C–H stretching
1464	6831	1st overtone N–H and O–H stretching
1932 2100	5176 4762	N–H combination and O–H stretching
2290 2324	4367 4303	C–H stretching combination

NIR spectroscopy, which is a non-destructive measurement, enables the simultaneous identification of components in a single sample within a short period of time, making it a preferable replacement for various chemical techniques. However, consideration should be taken into account as this technique requires initial calibration with samples of known composition, requiring great expenses of time and resources. Not least, frequent recalibration and issue of instrument interoperability might affect the practicality of the NIR spectroscopy technique [29].

2.2.5. Mid-Infrared (MIR) Spectroscopy

The mid-infrared (MIR) spectroscopy operates within the MIR band with wavelengths from 5 μm to 30 μm (wavenumbers from 4000 cm^{-1} to 400 cm^{-1}; note the presence of slight overlapping with NIR). The absorptions that occur within the MIR band are due to fundamental vibrations and can be segregated into four regions, namely the X–H stretching region (4000 cm^{-1} to 2500 cm^{-1}), triple-bond region (2500 cm^{-1} to 2000 cm^{-1}), double-bond region (2000 cm^{-1} to 1500 cm^{-1}) as well as the fingerprint region (1500 cm^{-1} to 600 cm^{-1}) [27].

The X–H stretching region is due to vibrations from O–H, C–H and N–H stretching. The triple-bond region arises from vibrations of C≡C and C≡N bonds. Besides, the double-bond region relates to C=C,

C=O and C=N vibrations. Lastly, the fingerprint region roots on bending and skeletal vibrations. Table 2 lists some of the common examples of MIR absorption bands [27].

Table 2. Examples of MIR absorption bands [27].

Wavelength (nm)	Wavenumber (cm^{-1})	Assignment
		Water
2.778–3.125	3200–3600	O–H stretching
6.061	1650	H–OH stretching
		Proteins
5.917–6.250	1600–1690	Amide I (C=O stretching)
6.349–6.757	1480–1575	Amide II (C–N stretching and N–H bending)
7.692–8.130	1230–1300	Amide III (C–N stretching and N–H bending)
		Fats
3.333–3.571	2800–3000	C–H stretching
5.731–5.797	1725–1745	C=O stretching
10.309	970	C=C–H bending
		Carbohydrates
3.333–3.571	2800–3000	C–H stretching
7.143–12.500	800–1400	Skeletal stretching and bending

MIR spectroscopy is effective since it provides information on structure-function relationships while performing quantitative analysis. The structure-function relationships are useful in food research and quality control, making MIR spectroscopy a crucial technique in the field of agriculture. The Fourier transform process is often bundled with MIR spectroscopy for data analysis, forming the popular Fourier transform infrared spectroscopy (FTIR) technique [27].

2.2.6. Raman Spectroscopy

Raman spectroscopy (RS), similar to IR spectroscopy, is another form of vibrational spectroscopy technique. RS obtains the spectral information of samples due to the occurrence of Raman effects [30]. Prior to understanding the Raman effects, one should look into the light scattering schemes that occur when incident photons interact with molecules in the sample. The possible light scattering schemes are illustrated in Figure 4. In the case of elastic scattering or Rayleigh scattering, the excited photons experience no change in energy content upon returning to ground state. Alternately, in the case of inelastic scattering or Raman scattering, the excited photons may lose (Stokes' shift) or gain (Anti-Stokes' shift) energy equivalent to the vibrational energy changes in the atoms of the molecules. This affects the motion of the atoms as well as the polarizability of the molecule. The change in molecule polarizability results in increased Raman intensity, ultimately forming the Raman spectrum when plotted across the investigated wavenumbers. However, this effect is weak as the probability of energy exchange is low [30].

The RS technique is gaining popularity as it enables the identification of molecular structure through the characteristic wavenumber in which vibrations occur. Furthermore, samples can be studied in the absence of a solvent as water causes weak Raman scattering. Not least, this technique is instantaneous and may undergo intensity enhancement. However, this technique is not without limitations. Due to the low probability of Raman scattering, this technique requires high concentration of samples. Moreover, sample molecules may experience photo degradation due to excitation of electronic absorption bands. The existence of fluorescence from impurities may disrupt the results obtained as well. These limitations aside, the RS technique can be combined with IR spectroscopy to deliver satisfactory results as summarized in Table 3 [30].

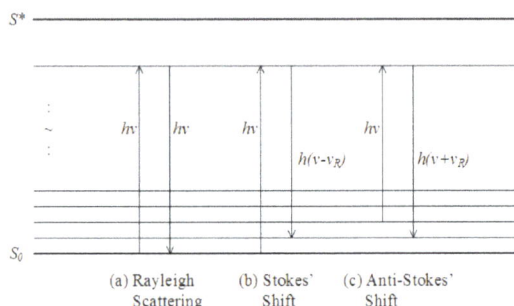

Figure 4. Light scattering schemes [17]. Reproduced with permission from A. Nawrocka, Advances in Agrophysical Research, Published by IntechOpen, 2013.

Table 3. Examples of Raman bands [30].

Wavelength (nm)	Wavenumber (cm^{-1})	Assignment
		Water
2.778–3.125	3200–3600	O–H stretching
		Proteins
19.608	510	
19.048	525	S–S stretching
18.349	545	
14.925–15.873	630–670	C–S stretching
13.423–14.286	700–745	
5.882–6.250	1600–1700	Amide I (C=O stretching and N–H bending)
8.032–8.097	1235–1245	Amide III (C–N stretching and N–H bending)
3.876–3.922	2550–2580	S–H stretching
3.333–3.571	2800–3000	C–H stretching
		Fats
6.940	1441	CH$_2$ bending
6.863	1457	CH$_3$–CH$_2$ bending
6.039	1656	C=C stretching
3.378–3.503	2855–2960	C–H stretching
		Carbohydrates
11.962	836	C–C stretching
9.398	1064	C–O stretching
3.434	2912	C–H stretching
3.397	2944	
2.898	3451	O–H stretching

2.2.7. Additional Spectroscopy Techniques

Apart from the popular spectroscopy techniques discussed earlier, existing studies presented additional variations of spectroscopy techniques which may be more complex in nature. For instance, dielectric spectroscopy has been utilized in agricultural inspections. Dielectric spectroscopy involves the inspection of dielectric properties or permittivity of samples over broad frequency ranges. Dielectric properties or permittivity refers to the ability of samples to store electrical energy in the electric field. In this spectroscopy technique, the permittivity is a complex permittivity relative to the free space, and this complex number is represented by (3):

$$\varepsilon = \varepsilon' + j\varepsilon'' \tag{3}$$

where the real part, ε', is the dielectric constant and the imaginary part, ε'', is the dielectric loss factor which covers losses due to dipolar relaxation and ionic conduction [31]. Another spectroscopy variation is the nuclear magnetic resonance (NMR) spectroscopy technique. The NMR spectroscopy gains spectral information of samples from the interaction between the magnetic moments of nuclei of various atoms and the applied magnetic fields. The two common phenomena that give rise to the NMR spectra are chemical shift and J-coupling [32].

A chemical shift occurs due to different resonant frequencies present in nuclei of the same species. The difference in resonant frequencies is a result of shielding effect from electrons surrounding the nuclei. The shielding effect is sensitive to chemical environments, hence allowing the characteristic identification of specific molecular functional groups [32].

The J-coupling phenomenon is also known as indirect (scalar) spin-spin coupling. This coupling effect results in splitting of spectroscopic lines into multiplets. The J-coupling occurs between two nuclei or groups of nuclei and is governed by the polarization of electrons on the chemical bonds connecting these nuclei. The polarization scheme is in turn dependent on the instant orientation of the nuclear magnetic moments in the presence of a magnetic field [32].

2.2.8. Spectroscopy Processing and Analysis

The raw spectral data undergoes pre-processing or pre-treatment in order to reduce noise and correct baseline variations. The common pre-treatment techniques are multiplicative scattering correction (MSC), standard normal variate (SNV), Savitzky-Golay smoothing as well as first and second derivatives [33,34].

Upon the completion of pre-processing or pre-treatment, the data set undergoes multivariate analysis to select and extract wavelengths that contain useful information. This aids in rectifying issues of collinearity, band overlapping and interaction between spectral variables. The results from multivariate analysis will be used to develop calibration models for calibration and prediction purposes [33,34].

The developed calibration models can be categorized according to the nature of the utilized multivariate analysis such as linear regression or nonlinear regression. Calibration models based on linear regression are built from partial least squares (PLS), interval partial least squares (iPLS), synergy interval partial least squares (SiPLS) or successive projections algorithm (SPA). Meanwhile, calibration models based on nonlinear regression are constructed from principal component analysis (PCA), independent component analysis (ICA), support vector machines (SVM), artificial neural networks (ANN) or a genetic algorithm (GA) [33,34].

The robustness of the final calibration model is evaluated from its ability to perform calibration and prediction. The calibration performance of the model is determined from the root mean square error of calibration (RMSEC) and the correlation coefficient (R_C) in the calibration set. Meanwhile, the prediction performance of the model is identified from the root mean square error of prediction (RMSEP) and the correlation coefficient (R_P) in the prediction set. Ideally, an effective model should register low RMSEC and RMSEP, with minimum difference between RMSEC and RMSEP. Not least, higher R_C and R_P are preferable [33,34].

2.3. Spectral Imaging Technique

The spectral imaging technique is a combination of both imaging and spectroscopy techniques discussed earlier. Being a combinational technique, the spectral imaging technique preserves the best of both worlds, allowing the simultaneous extraction of spatial and spectral information from the inspected sample [35,36].

2.3.1. Classes of Spectral Imaging

The spectral imaging technique acquires multiple images of the same subject at varying wavelengths. The resulting spectral images are three-dimensional (3-D) in nature, consisting of two spatial dimensions (row, x, and column, y) and one spectral dimension (wavelength, λ). Variations of spectral imaging technique are determined by the continuity of data in the wavelength dimension, branching out into hyperspectral imaging and multispectral imaging [35,36].

In general, hyperspectral imaging obtains spectral images in continuous wavelengths, whereas multispectral imaging registers spectral images at discrete wavelengths. Hyperspectral imaging acquires large number of images at high spatial and spectral resolutions. Due to the high volume of data, hyperspectral imaging requires long image acquisition time and involves complex algorithms for image analysis. Despite the complexity, hyperspectral imaging is essential for fundamental research and is the basis for multispectral imaging [35,36].

Multispectral imaging acquires spectral images at a significantly smaller number compared to hyperspectral imaging. Spectral images will only be acquired at optimal wavelengths predetermined from the analysis of dataset obtained through hyperspectral imaging. A smaller number of interested wavelengths allows rapid image acquisition and requires simpler image analysis algorithms. This characteristic of optimum data volume makes multispectral imaging perfectly suited for real-time in-field applications [35,36].

2.3.2. Spectral Image Acquisition Methods

There are several methods in which spectral imaging systems acquire spectral images. The methods are point scan, line scan and area scan as illustrated in Figure 5 [35]. The point scan (whiskbroom) method acquires the spectrum of a single pixel in each scan. A complete hyperspectral cube will be generated as the detector moves from pixel to pixel along the two spatial axes (x and y). The point scan method is similar to a normal spectroscopic approach. Since it cannot cover a large sample area, the point scan method is time consuming and unsuitable for fast image acquisition [35,36].

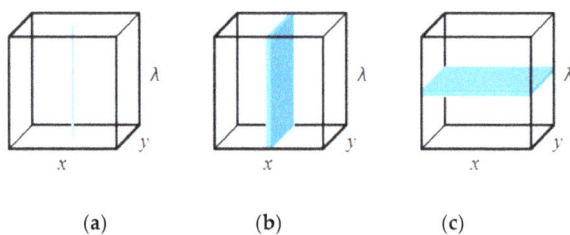

(a) (b) (c)

Figure 5. Methods of spectral image acquisitions with (**a**) point scan, (**b**) line scan, and (**c**) area scan [35]. Reproduced with permission from J. Qin, Journal of Food Engineering, Published by Elsevier, 2013.

The line scan (pushbroom) method, in each scan, acquires a slit (line) of spatial information together with the spectrum of every pixel along the line. A complete hyperspectral cube will be formed when scans are repeated along the direction of motion (*x*). The operation characteristic of the line scan method makes it suitable to acquire spectral images of moving samples. Hence, this method is usually combined with conveyor belt systems, making it a popular method in practical production lines. However, the exposure time should be short and accurately selected to allow uniform exposure at all wavelengths [35,36].

The area scan (band sequential) method, on the other hand, acquires a 2-D grayscale image comprising of complete spatial information in a single wavelength. A complete hyperspectral cube is generated through image stacking when scans are performed along the spectral axis (λ). The nature of the area scan method makes it more suited for the imaging of stationary samples instead of moving samples. In short, among the image acquisition methods discussed, line scan and area scan are greatly preferred over point scan for both hyperspectral and multispectral imaging on the basis of time consumption [35,36].

2.3.3. Spectral Imaging Sensing Modes

Spectral imaging may have varying sensing modes as illustrated in Figure 6. The sensing modes are determined by the positions of the light source and the detector, forming variations such as reflectance, transmittance and interactance modes. In reflectance mode, the detector collects the light reflected off the illuminated surface. This sensing mode is suitable for identifying external features of samples such as size, shape, color, texture and defects. However, when selecting this mode, the detector should be properly positioned to avoid specular reflection [36].

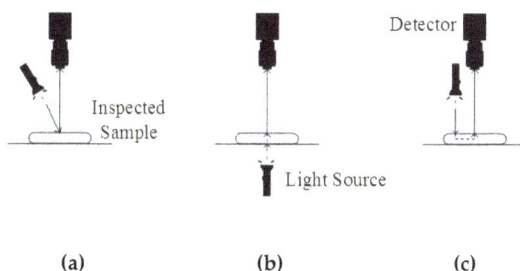

Figure 6. Sensing modes in spectral imaging; (**a**) reflectance, (**b**) transmittance, and (**c**) Interactance [36]. Reproduced with permission from D. Wu, Innovative Food Science & Emerging Technologies, Published by Elsevier, 2013.

The transmittance mode operates by having the detector collect light rays transmitted through inspected samples. In this sensing mode, the light source and the detector will be placed in opposite direction to each other. Due to the absorption of light rays in a sample, the detected signal will be relatively weak and dependent on sample thickness. Hence, the transmittance mode is commonly applied in the internal inspection of relatively transparent samples [36].

Meanwhile, the interactance mode overcomes the limitations of both the reflectance and transmittance modes. This sensing mode exhibits less surface effect compared to reflectance mode. At the same time, it allows detection in deeper layers of a sample without being affected by sample thickness as in transmittance mode. This advantageous sensing mode is set up by installing the light source and the detector at the same side and parallel to each other [36].

2.3.4. Spectral Imaging System Construction

The variations in spectral imaging lead to a diversity of instruments during the construction of a spectral imaging system. In general, a spectral imaging system is made up of a light source, a wavelength dispersive device and an area detector [35,36].

The light source for a spectral imaging system can be classified into illumination and excitation sources. Illumination light source is selected when measurements involve changes in the intensity of the incident rays upon light-sample interaction. The spectral composition of the incident source will not experience any changes. Such interaction is commonly observed in reflectance and transmittance sensing modes. Broadband lights are normally used as illumination sources. An example of illumination light source is the quartz tungsten halogen (QTH) lamp which is capable of generating a smooth spectrum in the visible to infrared range. Besides, the broadband light emitting diode (LED) has gained popularity over time due to its low power consumption, low heat generation, small size and long lifetime [35,36].

Excitation light source is usually selected when measurements involve changes in the frequency and wavelength of the incident rays. Interactions of this nature usually involve fluorescence phenomenon or Raman scattering effect. Narrowband lights are frequently used as excitation sources. A popular excitation light source is the laser which generates powerful monochromatic rays. Not least, UV fluorescent lamp, narrowband LED, high-pressure arc lamp (xeon arc lamp) and low-pressure vapor lamp (mercury vapor lamp) add to the family of excitation light sources [30,31].

The core component of the spectral imaging system is the wavelength dispersive device. A wavelength dispersive device disperses broadband light into different wavelengths to be projected to the area detector. Examples of wavelength dispersive devices include the imaging spectrograph, electronically tunable filter and beam splitting device [35,36].

Compared to traditional spectrograph, an imaging spectrograph extracts both spatial and spectral information. The imaging spectrograph disperses the broadband light illuminated onto different spatial areas of a sample into different wavelengths. This is achieved through diffraction gratings. The two most popular imaging spectrographs are the prism-grating-prism (PGP) imaging spectrograph which uses transmission diffraction gratings and the Offner imaging spectrograph that uses reflection diffraction gratings [35,36]. These variations of imaging spectrographs are commonly applied in line scan acquisitions [37].

An electronically tunable filter utilizes electronic devices to extract the required wavelength. Current electronically tunable filters can be categorized into the acousto-optic tunable filter (AOTF) and liquid crystal tunable filter (LCTF). An AOTF utilizes an acoustic transducer to generate high frequency acoustic waves that change the refractive index of a crystal. The crystal with varied refractive index will only allow the passage of light rays at the specified wavelength. Meanwhile, a LCTF transmits light at the required wavelength through electronically controlled liquid crystal cells [35,36]. These electronically tunable filters allow fast and flexible wavelength switching compared to mechanical filter wheels. They too exhibit advantages of high optical throughput, narrow bandwidth and broad spectral range [38].

Unlike the electronically tunable filter, a beam splitting device allows spectral images to be obtained simultaneously at multiple wavelengths. The beam splitting device divides light into several parts and passes them through bandpass filters which correspond to the required wavelengths. The beam splitting device can be categorized into color splitting and neutral splitting. In color splitting, light rays at a particular waveband are directed to each output, whereas, in neutral splitting, an equal portion of the total light energy is directed to each output [35]. The multiple wavelength acquisition characteristic makes the beam splitting device suitable to be installed in multispectral imaging systems [39].

A spectral imaging system will not be complete without an area detector. The area detector is responsible for collecting light rays which will eventually form the spectral images of the inspected sample. The common categories of area detector are the charge-couple device (CCD) camera and the complementary metal-oxide-semiconductor (CMOS) camera [35,36].

A CCD camera is made up of millions of photodiodes (pixels) that are closely arranged to form an array. These light sensitive photodiodes convert the incident photons into electric charges that correspond to the intensity of the exposed incident rays. The accumulated electric charges at each photodiode will then be moved out of the array to be quantified for spectral image formation [35,36]. One of the common CCD cameras is the silicon CCD camera. The silicon CCD camera exploits the sensitivity of silicon under visible light to perform image acquisition in visible and short-wavelength near-infrared bands [40]. Indium gallium arsenide (InGaAs) CCD camera is another CCD camera variation constructed from InGaAs, an alloy between indium arsenide (InAs) and gallium arsenide (GaAs) which is sensitive in the near-infrared band [41]. Not least, mercury cadmium telluride (MCT or HgCdTe) CCD camera built from HgCdTe, an alloy between mercury telluride (HgTe) and cadmium telluride (CdTe), enables sensing in the long-wavelength near-infrared and mid-infrared band [42].

Comparatively, the CMOS camera is similar to the CCD camera by having a collection of photodiodes (pixels) that convert light rays into electrical charges. The difference, however, lies in the quantification process of the electric charges. Opposed to the remote quantification in the CCD camera, the CMOS camera allows electric charges at each pixel to be independently and instantaneously read by the transistor attached to each photodiode [43]. This unique characteristic allows the CMOS camera to compete with the CCD camera in terms of high imaging acquisition speed, blooming immunity, low cost, low power consumption and small size. However, careful note should be taken as the CMOS camera is susceptible to noise due to on-chip signal transmissions, resulting in lower sensitivity and dynamic range when pitted against CCD camera [36].

2.3.5. Spectral Imaging Processing and Analysis

The raw spectral image data obtained via the spectral imaging technique comes in different formats according to the image acquisition method used. The common formats are Band interleaved by pixel (BIP), band interleaved by line (BIL) and band sequential (BSQ). The BIP format results from the point scan method and stores the complete spectrum of each pixel sequentially. The BIL format comes with the line scan method and stores the complete spectrum of each line in order. Lastly, the BSQ format relates to the area scan method and stacks the spatial image continuously obtained at each wavelength [35,36].

Similar to the imaging and spectroscopy techniques, the raw spectral image data in BIP, BIL and BSQ formats should undergo pre-processing in both the spatial and spectral aspects before being utilized for further analysis. The raw spectral image, which represents detector signal intensity, will first undergo flat-field calibration or reflectance calibration to form useful reflectance or absorbance image. From the spatial aspect, the generated reflectance image can be further improved through image enhancement processes such as edge and contrast enhancement, magnifying, pseudo-coloring and sharpening. Noise reduction can also be achieved through spatial filtering, Fourier transform (FT) and wavelet transform (WT). From the spectral aspect, noise reduction and baseline correction can be performed through algorithms such as MSC, SNV, Savitzky-Golay smoothing, first and second derivatives, FT, WT as well as orthogonal signal correction (OSC) [35,36].

The next step in the analysis flow will be image segmentation. Image segmentation serves to divide the pre-processed spectral image into different regions for the identification of region of interests (ROIs) [44]. In this process, segmentation algorithms are greatly preferred over manual segmentation due to the ease of operation and time saving. The selections of segmentation algorithms include thresholding (global thresholding or adaptive thresholding), morphological processing (erosion, dilation, open, close or watershed algorithm), edge-based segmentation (gradient-based or Laplacian-based methods) and spectral image segmentation [36].

Spatial analysis utilizing spectral image data usually involves quantitative measurement. In this process, gray-level object measurement is performed to quantify the intensity distribution of ROI extracted from image segmentation. Gray-level object measurements can be categorized according to intensity-based or texture-based measurements [45]. Intensity-based measurements are usually first-order measures such as mean [46,47], standard deviation, skew, energy and entropy [36]. Meanwhile, texture-based measurements are second-order measures such as joint distribution functions [36], gray-level co-occurrence matrix (GLCM) [46,48] and 2-D Gabor filter [49].

For spectral analysis, the data set will undergo multivariate analysis to reduce the spectral dimension and select the optimum wavelengths. Similar to the spectroscopy technique, some examples of multivariate analysis algorithms include PLS, linear discrimination analysis (LDA) [35,36], correlation analysis (CA) [50], PCA, ICA [41,51,52], ANN [53], sequential forward selection (SFS) [54] and GA [55]. These results from multivariate analysis will be used to develop calibration models for calibration, validation and prediction purposes [36].

The robustness of the final calibration model is evaluated from its ability to perform calibration and prediction. The calibration performance of the model is determined from the standard error of calibration (SEC), root mean square error of calibration (RMSEC) and the coefficient of determination (r_C^2) in the calibration set. Validation performance is determined via the root mean square error of cross-validation (RMSECV) and the coefficient of determination (r_V^2) in the validation set. Meanwhile, the prediction performance of the model is identified from standard error of prediction (SEP), root mean square error of prediction (RMSEP), residual predictive deviation (RPD) and the coefficient of determination (r_P^2) in the prediction set. Ideally, an effective model should register low SEC, RMSEC, RMSECV, SEP and RMSEP, with a minimum difference between SEC and SEP. Not least, higher r_C^2, r_P^2, r_P^2 and RPD are preferable [36].

2.3.6. Pros and Cons of Spectral Imaging

This technique is advantageous as it omits chemical processes and requires minimum sample preparation. Moreover, the composition of multiple components in a sample can be simultaneously obtained. Upon spectral image acquisition, spectral imaging too allows the flexible selection of region of interest (ROI) for analysis. Furthermore, owing to the rich spatial and spectral information, spectral imaging can easily detect and differentiate subjects even though similar colors, overlapping spectra and morphological characteristics are present [36].

However, the spectral imaging technique does pose several limitations. Hardware speed is a major concern, especially in the case of hyperspectral imaging, due to the massive amounts of data to be acquired and analyzed. Moreover, spectral imaging includes the acquisition of redundant data, resulting in complex data analysis. Spectral imaging systems too require constant calibration in order to maintain their efficiency. The detection limits of spectral imaging are poorer compared to chemical-based analytical methods. Similar to spectroscopy, spectral imaging suffers from multicollinearity and requires multivariate analysis to address the issue. In addition, spectral imaging is inapplicable when the ROI is smaller than the size of a pixel or does not exhibit the characteristic spectral absorption. Lastly, spectral imaging may be irrelevant in the analysis of liquids and homogeneous samples since these samples do not pose distinctive and useful spatial information [36].

2.4. Technique Comparison

Table 4 presents a simple comparison of the optics and photonics techniques in agriculture that have been discussed earlier. From the comparison, the imaging technique is noted to be utilized for the extraction of spatial information only and is sensitive to small-sized objects. In contrast to the imaging technique, the spectroscopy technique allows acquisition of spectral information and is useful in accessing multi-constituent information. The spectral imaging technique covers the benefits of both imaging and spectroscopy techniques, allowing it to obtain spectral and spatial information simultaneously. Apart from this, spectral imaging has the added value of flexible spectral extraction as well as the capability of generating quality-attribute distribution. However, it should be noted that multispectral imaging has poorer access to spectral information compared to hyperspectral imaging due to the acquisition at limited number of wavelengths. Among the compared techniques, the spectral imaging technique can be said to be the most robust. Nonetheless, the area of application should be given the utmost consideration when selecting the best optics and photonic technique in order to avoid the underutilization or overutilization of a particular technique [36].

Table 4. Comparison of optics and photonics techniques in agriculture [36].

Characteristics	Imaging	Spectroscopy	Spectral Imaging
Spectral information	×	✓	✓
Spatial information	✓	×	✓
Multi-constituent information	×	✓	✓
Sensitivity to small-sized objects	✓	×	✓
Flexibility of spectral extraction	×	×	✓
Generation of quality-attribute distribution	×	×	✓

3. Optics and Photonics Applications in Agriculture

The optics and photonics techniques discussed above have been applied in various studies involving agricultural products. The studies will be tabulated in the following sections, enlisting details such as agriculture class, agriculture product, application area, wavelength details and country of applications.

3.1. Applications of Imaging Technique

Table 5 lists some of the agricultural works based on the imaging technique. The imaging technique is performed in the UV-VIS-IR range and involves the acquisition of spatial, color and thermal data from the inspected samples. These works show that the imaging technique is suited for inspection or analysis based on external features of the subject of interest. For instance, bruise detection [56,57] and disease detection [58,59] are performed by inspecting the external damage on the sample. In addition, quantitative analysis [60,61] is performed using the spatial information obtained. The color features extracted are also used for maturity evaluation [57,62,63] and nutrient content detection [64,65]. The thermal data, meanwhile, proves to be useful in similar occasions of bruise detection [66,67], disease detection [68,69] and maturity evaluation [70,71] by analyzing the temperature variations over the inspected sample. Not least, the most significant application of the imaging technique is the development of automated agricultural robots [72–75] and animal behavioral studies [76,77].

Table 5. Applications of imaging technique in agriculture.

Class	Product	Application	Ref.
Fruit	Apple	Bruise detection (thermal)	[66,67,70,78]
	Apple	Maturity evaluation (thermal)	[70]
	Apple	Yield estimation (thermal)	[79]
	Apple	Scab disease detection (thermal)	[68]
	Green apple	Acquisition of segmented fruit region	[80]
	Green apple and orange	Yield estimation	[61]
	Orange	Texture analysis	[81]
	Orange	Bruise detection (thermal)	[67]
	Citrus	Water stress evaluation (thermal)	[82]
	Pear	Maturity evaluation (thermal)	[71]
	Banana	Maturity evaluation	[62]
	Banana	Maturity evaluation	[63]
	Persimmon	Maturity evaluation (thermal)	[71]
	Passion fruit	Mass and volume estimation	[83]
	Blueberry	Bruise detection	[56]
	Grapevine	Pathogen detection (thermal)	[84]
	Tomato	Fruit detection	[85,86]

Table 5. *Cont.*

Class	Product	Application	Ref.
	Tomato	Bruise detection and maturity evaluation	[57]
	Tomato	Bruise detection (thermal)	[87]
	Tomato	Maturity evaluation (thermal)	[71]
	Tomato	Clustered fruit detection	[88]
	Sweet peppers	Peduncle detection	[89]
	Onion	Post-harvest quality assessment (thermal)	[90]
	Lettuce	Segmentation of vegetable	[91]
	Cucumber	Downy mildew disease detection (thermal)	[69,92,93]
Grain	Rice leaf	Nitrogen content detection	[64]
	Wheat	Yield estimation (thermal)	[94–96]
	Corn	Water stress evaluation (thermal)	[97]
	Macadamia nuts	Yield estimation	[60]
	Soybean	Identification of foliar disease	[98]
	Soybean	Identification of leaf disease	[59]
	Maize	Yield estimation (thermal)	[99]
	Maize	Identification of leaf disease	[100]
	Maize	Cultivar identification	[101]
Commercial	Cotton	Water stress evaluation (thermal)	[97,102]
	Silkworm	Gender identification	[103]
Farm and Plantation	Seed	Viability evaluation (thermal)	[104]
	Wheat field	Estimation of nutrient content	[65]
	Cauliflower plantation	Weed detection	[73]
	Asparagus plantation	Crop harvest robot vision	[74]
	Sugar beet and rape plantation	Agriculture robot vision	[75]
	Grapevines	Estimation of intra-parcel grape quantities	[105]
	Cow farm	Behavioural studies	[76,106]
	Goat and sheep farm	Animal species identification	[107]
	Fish aquarium	Behavioural studies	[77,108]
	Baby shrimp farm	Chlorine level detection	[109]
	Orchid farm	Disease and pest detection	[58]
	Surface and ground water	Chemical content detection	[110]

Based on Table 5, bruise detection, yield estimation, and disease identification are the three most common applications with imaging technique in agriculture. In bruise detection, a hyperspectral camera with broad operating wavelength from 400 to 5000 nm [78], a non-destructive and non-contact infrared sensing thermogram [66], and an infrared thermal imaging camera with high temperature resolution of 0.1 K [70] are among the instruments employed as the imaging technique. Moreover, the thermal camera with temperature resolution better than 0.5 °C [79], colour stereo vision camera which creates a 3D environment for further processing [86], and grading machine with a high accuracy of 96.47% [57] are employed for yield estimation. However, the grading machine proposed in [57] has a small capacity in estimation for 300 tomatoes per hour and does not efficiently work for tomato images with high specular reflection. In addition, infrared thermography is a popular device for disease identification due to its non-invasive monitoring and indirect visualization of downy mildew development [69]. This device takes the colour reflectance image for the detection of *V. inaequalis* development on apple leaves [68] and detects the pathogen in grapevines [84]. Additionally, an X-ray computed tomography scanner is utilized to obtain the cross-section of onion inoculated by pathogens [90], whilst an unmanned aerial vehicle (UAV) is presented to track the foliar disease in soybean [98].

Apart from the instrument, numerous types of algorithms are depicted in imaging technique. In bruise detection, PCA and a minimum noise fraction are proposed for 20 apple samples with threshold percentages of success within 86% to 93% and 87% to 97%, respectively [78]. In yield estimation, the fruit detection algorithm is presented for 8–120 apple samples with a correlation coefficient ranging from 0.83 to 0.88 [79]. In addition, the blob detector neural network is demonstrated to detect the yield estimation for both oranges and apples with intersection over union of 81.3% for orange and 83.8% for apple [61]. As for disease identification, a simple linear iterative clustering algorithm is presented for 3624 foliar images with high classification rate of 98.34% for height between 1 to 2 m [98]. The classification rate is reduced for approximately 2% for each meter from the examined height within 1 to 16 m. Moreover, an improved GoogLeNet and Cifar10 models are established for 500 images of maize leaf disease with 4:1 ratio for training and validation which allows the system to have a diversity of sample conditions [100]. The average identification accuracy of GoogLeNet and Cifar10 models is recorded as high as 98.9% and 98.8%, respectively. Apart from bruise detection, yield estimation, and disease identification, algorithms are also shown in maturity evaluation and acquisition of crop segmentation. In maturity evaluation, both a Fuzzy model [62] and medium filter algorithm [85] are employed for 3108 images on banana samples and 100 images on tomato samples with an average identification rate of 93.11%, and within 89% to 98%, respectively. The Fuzzy model is useful in handling ambiguous information for the banana fruit maturity detection using red-green-blue (RGB) components. In the acquisition of crop segmentation, K-mean clustering algorithm is presented for a clustering of apple samples with target acquisition rate of 84% [80]. This algorithm is commonly used in image segmentation whereby crop segmentation can be precisely attained, even with the presence of stems and leaves in the captured images.

3.2. Applications of Spectroscopy Technique

The applications of spectroscopy technique in agriculture are presented in Table 6. The spectroscopy technique is widely applied to inspect internal qualities that are externally invisible. A sizeable amount of research has performed spectroscopy in the UV-VIS-IR region to identify the internal constituents of agricultural products such as pigment compound in apple [111], moisture content in mushroom [112], protein and sugar in potato [113], and caffeine in coffee [114] among others. Within 400 to 1000 nm, 678 nm is sensitive to low chlorophyll content thus the reflectance at 678 ± 30 nm is suggested for the monitoring of the early stage of ripening and the pigment content change with a maximum correlation of closely 0.6 [111]. On the other hand, 590 to 700 nm is recommended for the maturity detection in early stage for yellow colour apple fruits with maximum correlation from 0.7 to 0.9. In the verification of moisture content in mushroom, the spectral region from 600 to 2200 nm gives the lowest standard deviation of cross validation as 0.644% and maximum correlation factor of 0.951 among the investigated wavelengths from 402 to 2490 nm [112]. A high experimental repeatability is presented by a standard deviation of 0.677% and a maximum correlation factor of 0.947 for a separate set of mushrooms of a similar type and treatment. In the protein and sugar content identification in potato, a modified PLS regression model is applied to calculate the relationship between the spectrum and chemical properties of the calibrated samples [113]. Based on the measurement, the standard deviations for crude protein, glucose, fructose, sucrose and red sugar for the 120 potato samples are 0.2%, 0.073%, 0.068%, 0.068%, and 0.122%, respectively. Correspondingly, the squared correlation coefficient for the above five parameters are deduced as 0.96, 0.70, 0.89, 0.62, and 0.82, respectively. In a total of 665 tea leaf samples, NIRS and liquid chromatography is coupled to a diode arrayed detector to determine its content of caffeine [114]. Among 375 calibration sets and 250 validation sets for caffeine in the tea leaf samples, a standard deviation of 8.6 and 8.9, as well as high squared correlation coefficients of 0.97 are acquired for both calibration and validation sets though regression model.

The quality and freshness of fruits [115,116], vegetables [117], and meat [118,119] can be easily inspected using spectroscopy as well. For instance, two wavelengths within 600 to 904 nm of VIS-NIR spectrum are investigated by correlation analysis to discriminate brown core and sound pears [115].

Using eight brown core pears and 32 sound pears, the percentage of soluble solid content achieves a precision of 97.8% and 99% within a standard deviation of 0.5% and 1%, respectively. In addition, NIR spectrum and PLS regression model are used to detect the total anthocyanins content (TAC) and total phenolic compounds (TPC) in jambu fruits [116]. With a total of 50 jambu samples scanning from 1000 to 2400 nm, the correlation coefficients of TAC and TPC are deduced as 0.98 and 0.94, and strong ratio to performance of deviation as 5.19 and 3.27, respectively. Besides that, a 250 to 350 GHz radiation is found to be suitable to distinguish the defective and proper sugar beet seeds [117]. A python package scikit-learn algorithm is used to determine the threshold for these two types of seeds, with 80% detection for proper seed and 94% detection for defective seed. Therefore, the average detection rate of this algorithm is 87%. In addition, meat fraud is injected into bovine meat, aiming to increase the water holding capacity. This issue is characterized with attenuated total reflectance Fourier transform infrared spectrum and the supervision of the 55 meat fraud adulterated samples through PLS square discriminant analysis [118]. The analysis records a precise detection as high as 91% of the adulterated samples. Apart from meat fraud, the freshness of mackerel fish is characterized with auto-fluorescence spectroscopy and analyzed with fluorescence excitation emission matrices (EEM) [119]. The fluorescence EEM data and real freshness values are modelled with PLS regression and an algorithm is developed for this smart system as a predictor with squared correlation coefficient of 0.89.

Furthermore, chemical residues in harvested product [120] or even plantation soil [121–123] can be easily identified, leading to easy detection on contamination of agricultural product. Residual pesticides such as phosmet and thiabendazole in apples are analyzed with surface-enhanced Raman spectroscopy (SERS) coupled with gold nanoparticle [120]. The sensitivities for detectable concentration are 0.5 µg/g for phosmet and 0.1 µg/g for thiabendazole. The PLS regression is also used to correlate the SERS spectrum with the concentration of pesticide in apples with squared correlation coefficient of 0936 for phosmet and 0.959 for thiabendazole. In addition, the effect of drying temperature on the nitrogen detection in soils at four different temperatures of 25 °C for placement, 50, 80, and 95 °C for drying is modelled based on NIR sensor and three successive algorithms, which are multiple linear regression, PLS, and competitive adaptive reweighed square on the spectral information [122]. Based on the three soil samples, loess, calcium soil, and black soil show the correlation coefficients of 0.9721, 0.9588, and 0.9486, respectively at the optimum drying temperature of 80°C. The detection of nitrogen in three types of soils is also alternatively performed in [123], with squared correlation coefficients of 0.95, 0.96 and 0.79 for loess, calcium soil and black soil using PLS regression model. The relatively lower squared correlation coefficient in black soil is due to the interference of high humus content and strong absorption of organic matter in black soil. Lastly, a point worth noting is that the dielectric and NMR spectroscopy are often adopted when the analysis involves more complex chemical compounds [124,125]. These complex chemical compounds include the vulcanization of natural rubber with sulfur-cured and peroxide-cured systems with different dynamics [124] and detection to changes in concentrations of pollutants in agriculture drainage such as heavy metal and heavy oxides [125]. In [124] a sulphur-cured system has features restricted segmented dynamics whereas peroxide-cured system has faster dynamics. In addition, network structure resulting from the vulcanization of both systems also influences the segmental dynamics of natural rubber. The peroxide-cured network is more homogeneous with spatial distribution of cross links than the sulphur-cured network with large inhomogeneity due to the presence of zinc oxide particles and the ionic interaction with the natural rubber chains. In [125], an X-ray fluorescence spectroscopy is employed to investigate the changes of pollutants in dried root and shoot plant parts at a temperature range from 30 to 90 °C. From the measurement, the concentration of pollutant is found to be higher in plant root than plant shoot through the analysis of frequency relaxation process via dielectric modulus measurement. Significantly, the removal of pollutants by plants will be enhanced upon subjecting them to a microwave heating power of 400 W for 30 min. Apart from the aforementioned applications, more applications of the spectroscopy technique in agricultural products with different methods and wavelengths are tabulated in Table 6.

Table 6. Applications of spectroscopy technique in agriculture.

Class	Product	Application	Method	Wavelength (nm)	Ref.
Fruit	Apple	Pigment content change during ripening	UV-VIS-NIR	400–1000	[111]
	Apple	Soluble solid content detection	VIS-NIR	500–1100, 1000–2500	[33]
	Apple	Pesticide residue detection	Raman	5–18 μm	[120]
	Pear	Brown core and soluble solid content detection	UV-VIS-NIR	200–1100	[115]
	Mango	Maturity evaluation	NIR	1200–2200	[126]
	Peach	Peach variety identification	NIR	833–2500	[127]
	Wax jambu	Quality inspection	NIR	1000–2400	[116]
	Grape leaf	Water content estimation	UV-VIS-NIR	350–2500	[128]
Vegetable	Carrot	Carotenoid, fructose, glucose, sucrose and sugar content detection	NIR	1108–2490	[129]
	Potato	Bruise detection	UV-VIS-NIR	250–1750	[130]
	Potato	Protein, fructose, glucose, starch and sucrose content detection	NIR	1100–2500	[113]
	Onion	Soluble solid content detection	VIS-NIR	500–1200	[131]
	Oilseed rape leaf	Aspartic acid content detection	NIR	1100–2500	[132]
	Sugar beet seeds	Quality control	Time-domain spectroscopy	250–350 GHz	[117]
	Mushroom	Moisture content detection	VIS-NIR	600–2200	[112]
Grain	Corn seed	Viability evaluation	NIR Raman	1000–2500 3.125–59 μm	[133]
	Almond	Internal defect detection	VIS-NIR	700–1400	[134, 135]
	Maize	Identification of transgenic ingredients	THz spectral	0–4.5 THz	[136]
	Rice, maize and peanut	Germination and growth of crop	UV-VIS FTIR	380.85–796.62 nm 562.72–3865.11 cm^{-1}	[137]
Meat	Beef	Thermal change inspection	Fluorescence	250–550	[138]
	Beef	Adulteration detection	NIR-MIR	2.5–19 μm	[118]
	Frozen fish	Freshness evaluation	Fluorescence	250–800	[119]
Dairy	Egg	Contamination detection	UV-VIS-NIR	200–860	[139]
	Goat milk	Fatty acid content detection	VIS-NIR	400–2498	[140]
Oil	Edible oil	Stability analysis	NMR	300 MHz (1H)	[141]
	Olive oil	Adulteration detection	Fluorescence	250–720	[142]
	Ocimum essential oil	Antioxidant property identification	NIR-MIR	2.5–18 μm	[143]
Beverage	Tea leaf	Tea polyphenol level detection	UV-VIS-NIR	347–2506	[144]
	Green tea leaf	Caffeine and catechins content detection	VIS-NIR	400–2500	[114]
	Coffee	Geographic and genotypic origin identification	NIR	1100–2498	[145]
	Coffee	Roasting degree and blend composition detection	NIR	800–2857	[146]
	Tomato juice	Quality inspection	NIR-MIR	2.5–14 μm	[147]
	Apple wine	Volatile compound detection	NIR	833–2500	[148]
	Rice wine	Fermentation monitoring	NIR-MIR	2.5–25 μm	[149]

Table 6. Cont.

Class	Product	Application	Method	Wavelength (nm)	Ref.
Commercial	Cotton fibre	Cotton type identification	NIR	800–2500	[150]
	Cotton fibre	Cotton fibre micronaire measurement	VIS-NIR	400–2500	[151]
	Natural rubber	Protein and lipid content detection	NIR-MIR	2.5–25 μm	[152]
	Natural rubber	Chemical interaction during vulcanizing process	NIR-MIR Raman	2.5–25 μm 3.125–100 μm, 6.25–50 μm	[153]
	Natural rubber	Rubber silane reaction	NMR	400 MHz (1H), 100.6 MHz (13C)	[154]
	Natural rubber	Moisture content detection	VIS-NIR	400–1100	[155]
	Natural rubber	Vulcanization system effect	Dielectric NMR	10-1 < Hz < 107 20 MHz (1H)	[124]
	Neem leaf	Pest control	UV-VIS FTIR XRD	200–800 nm 250–4000 cm^{-1} 10–80°	[156]
Farm and Plantation	Soil	Quality inspection	NIR	780–5000	[157]
	Soil	Nitrogen content detection	NIR	800–2564	[158]
	Soil	Chemical and physical property estimation	NIR-MIR	1430–2500, 2.5–27 μm	[159]
	Soil	Nitrogen detection	NIR	900–1700	[122]
	Soil	Nitrogen detection	NIR	900–1700	[123]
	Soil and water	Contaminant detection	VIS-NIR	400–2500	[121]
	Water hyacinth Soybean straw	Pollutant concentration detection Detection of biomass	Dielectric Fluorescence Near infrared spectroscopy	10-1 < Hz < 106 N/A 4000–12,000 cm^{-1}	[125] [160]
	Flower	Plant type identification	VIS	635, 685, 785	[161]

3.3. Applications of Spectral Imaging Technique

As discussed earlier, the spectral imaging technique is a combination of both imaging and spectroscopy techniques. In the agriculture industry, both variations of hyperspectral and multispectral imaging are equally crucial, and some sample applications are compiled in Table 7. As observed from Table 7, hyperspectral imaging involves acquisition over a range of wavelengths while multispectral imaging involved acquisition at fewer selected wavelengths. The robustness of spectral imaging allows its usage in bruise detection [78,162,163], maturity evaluation [164–168], quality evaluation [169–171] and disease detection [172–176]. Internal attributes of samples [177–179] are easily acquired for analysis purposes as well.

In bruise detection of agriculture product, a machine vision system is integrated with optical filter at 740 and 950 nm to detect the bruise in rotating apple with a detection accuracy of 90 to 92% from 54 Pink Lady apples and 60 Ginger Gold apples [162]. In addition, the bruise detection in mushroom is carried out through a line scanning hyperspectral imaging instrument from 400 to 1000 nm with a spectroscopic resolution of 5 nm [163]. The PCA is applied to a set of data comprising of 50 normal and 50 bruise spectra, with standard deviation of 0.025 and 0.055, respectively.

For the maturity evaluation of agriculture products such as peach fruit, a CCD camera is employed at 450, 675 and 800 nm, whereas the fruit ripening is characterized with the increasing in intensity from a histogram with a ratio of red divide with infrared red (R/IR) [164]. The firmness of the peach fruit reduces when the reflectance at 675 nm is increased. An analysis of variance (ANOVA) is presented to access the R/IR clustering which has the highest reflectance at 675 nm, and higher Fisher value as a function of higher R/IR ratio. Apart from the detection of maturity for peach fruit, the maturity stage of strawberry is detected using a hyperspectral imaging system from 380 to 1030 nm and from 874 to

1734 nm [165]. According to the PCA, the optimal wavelengths are from 441.1 to 1013.97 nm and from 941.46 to 1578.13 nm with a classification accuracy of above 85%. Moreover, the maturity of tomato is detected by an electromagnetic spectrum with a continuous 257 bands from 396 to 736 nm [166] and discrete band of 530, 595, 630, and 850 nm using a tomato maturity predictive sensor [167]. Based on the LDA, the classification error is reduced from 51% to 19% [166] and achieves detection accuracy above 85% [167]. The ripening in banana fruit is also characterized with a compact imaging system and an UAV from 500 to 700 nm [168]. The detection is based on the reflectance spectrum whereby in ripe banana, the main element is carotenoid which absorbs less light at 650 nm band. On the other hand, a green banana with a greater amount of chlorophyll than ripe banana absorbs more light at 650 nm band.

For the quality evaluation of agriculture product, the firmness test for two types of apple fruits is conducted with a laser-based multispectral imaging prototype which captures and processes four spectral scattering images at a speed of two fruits per second [169]. The multilinear regression models are developed using to predict the firmness of those two apple types at 680, 880, 905, and 940 nm with a correlation coefficient of 0.86. The quality of grape berries is determined by the reflectance spectrum from a hyperspectral imaging system, whilst the high reflectance at 500 nm, 660 to 700 nm, and 840 nm denotes the chlorophyll content, red-coloured anthocyanin pigment, and sugar content, respectively [170]. A partial least square regression (PLSR) model is applied in order to determine the correlation between the spectral information and the physico-chemical indices. The titratable acidity of the green and black grapes shows a coefficient of determination of 0.95 and 0.82, as well as soluble solid content of 0.94 and 0.93 at pH value of 0.8 and 0.9, respectively. The root mean square error for this method is 0.06 for green grape and 0.25 for black grape. Apart from fruits, the quality of tea leaves is classified by a hyperspectral imaging sensor at 762, 793, and 838 nm, supported by SVM algorithm [171]. Within 700 samples comprising of 500 training samples and 200 prediction sets, the SVM algorithm generates a total classification rate of 98% the training sample and 95% for the prediction set, at result of optimal regularization parameter of 4.37349 and kernel parameter of 13.2131 in SVM model.

For the disease detection in agriculture product, a fruit sorting machine is used to detect the citrus canker at 730 and 830 nm with a bandpass filter installed in the scanning camera [172]. A real-time image processing and classification algorithm is developed based on a two-band ratio (R830/R730) approach, which achieves a detection accuracy of 95.3% for 360 citrus samples. Next, a shortwave infrared hyperspectral imaging system is used to detect the sour skin in onion based on the suitable reflectance spectrum from 1070 to 1400 nm [173]. Two image analysis approaches are utilized based on the log-ratio images at two optimal wavelengths of 1070 and 1400 nm. A global threshold of 0.45 is integrated to segregate sour onion skin infection areas from log-ratio images. With Fisher's discriminant analysis, the detection accuracy of 80% is achieved. The second image analysis approach is the incorporation of three parameters; max, contrast and homogeneity of the log-ratio images as the input features for the SVM. Subsequently, the Gaussian kernel generates higher detection accuracy as 87.14%. Apart from that, the tumorous chicken is detected by the combination of a CCD camera and imaging spectrograph from 420 to 850 nm [174]. Within the wavelength bands, the PCA select the three useful wavelengths of 465, 575, and 705 nm from the tumorous chicken image. Based on the images from 60 tumorous and 20 normal chicken, multispectral image analysis generates the ratio images, which are divided into ROI classified either as tumorous or normal chicken. The image features from ROI such as coefficient of variation, skewness and kurtosis are extracted as the input for the Fuzzy classifier, which generates the detection accuracy of 91% for normal chicken and 86% for tumorous chicken. To detect the nematodes in coffee cultivation, hyperspectral data is used in band simulation of the RapidEye sensor to determine the most sensitive spectral ranges for pathogen discrimination in coffee plants [176]. Multispectral classification identifies the spatial distribution of healthy, moderately infected, and severely infected coffee plants with an overall accuracy of 71%. Apart from the four main applications with the spectral imaging technique in agriculture products, more applications with different scanning methods and various wavelengths are tabulated in Table 7.

Table 7. Applications of spectral imaging technique in agriculture.

Class	Product	Application	Method	Wavelength (nm)	Ref.
Fruit	Apple	Bruise detection	Hyper. line scan	400–2500, 1000–2500	[78,180]
	Apple	Bruise detection timing	Hyper. line scan	400–2500	[181]
	Apple	Bruise detection	Multi. area scan	740, 950	[162]
	Apple	Bruise and faeces detection	Multi. line scan	530, 665, 750, 800	[182]
	Apple	Firmness evaluation	Multi. area scan	680, 880, 905, 940	[169]
	Citrus	Canker detection	Multi. area scan	730, 830	[172]
	Peach	Firmness evaluation	Hyper. line scan	500–1000	[183]
	Peach	Maturity evaluation	Multi. area scan	450, 675, 800	[164]
	Cantaloupe	Faeces detection	Hyper. line scan	425–774	[184]
	Blueberry	Firmness evaluation, soluble solid content detection	Hyper. line scan	400–1000	[177,185]
	Strawberry	Maturity evaluation	Hyper. line scan	380–1030 874–1734	[165]
	Cherry	Pit detection	Hyper. line scan	450–1000	[186]
	Grape	Quality evaluation	Hyper. line scan	400–1000	[170]
	Banana	Maturity evaluation	Hyper. area scan	500–700	[168]
	Tomato	Maturity evaluation	Hyper. line scan	396–736	[166]
	Tomato	Maturity evaluation	Multi. area scan	530, 595, 630, 850	[167]
	Cucumber	Chilling injury detection	Hyper. line scan	447–951	[187]
Vegetable	Freeze-dried broccoli	Glucosinolate detection	Hyper. line scan	400–1700	[188]
	Potato	Cooking time prediction	Hyper. line scan	400–1000	[189]
	Onion	Sour skin disease detection	Hyper. area scan	950–1650	[173]
	Mushroom	Bruise detection	Hyper. line scan	400–1000	[163]
Grain	Rice plant	Nitrogen content detection	Hyper. line scan	400–1000	[190,191]
	Thai jasmine rice	Rice variety identification	Multi. area scan	545, 575	[192]
	Wheat	Fungus detection	Hyper. area scan	1000-1600	[193]
	Wheat	Damage detection	Hyper. line scan	1000–2500	[194]
	Peanut	Tomato spot wilt disease detection	Multi. Area scan	475, 560, 668, 717, 840	[195]
	Corn	Oil and oleic acid content detection	Hyper. area scan	950-1700	[196]
	Corn	Aflatoxin detection	Hyper. line scan	400–600	[197]
Meat	Chicken	Skin tumour detection	Hyper. line scan	420–850	[174]
	Chicken	Heart disease detection	Multi. area scan	495, 535, 585, 605	[198]
	Chicken	Faeces detection	Multi. area scan	520, 560	[199]
	Chicken	Wholesomeness inspection	Multi. line scan	580, 620	[200]
	Beef	Tenderness evaluation	Hyper. line scan	400–1000	[201]
	Beef	Microbial spoilage detection	Hyper. line scan	400–1100	[202]
	Lamb	Lamb variety identification	Hyper. line scan	900–1700	[203]
	Pork meat	*E. coli* detection	Hyper. line scan	470–960	[204]
	Pork meat	Quality inspection	Hyper. line scan	900–1700	[205]
	Fish	Moisture and fat content detection	Hyper. line scan	460–1040	[206]
	Fish	Ridge detection	Hyper. line scan	400–1000	[207]
	Salmon	Microbial spoilage detection	Hyper. line scan	400–1000 880–1720	[208]

Table 7. *Cont.*

Class	Product	Application	Method	Wavelength (nm)	Ref.
	Dehydrated prawn	Moisture content detection	Hyper. line scan	380–1100	[209]
	Prawn	Adulteration detection	Hyper. line scan	380–1030 900–1700	[210]
Dairy	Milk powder	Melamine detection	Hyper. line scan	990–1700	[211]
	Milk	Fat content detection	Hyper. line scan	530–900	[178]
	Milk	Melamine detection	Hyper. point scan	4–98 μm	[212]
Oil	Olive oil	Free acidity, peroxide and moisture content detection	Hyper. line scan	900–1700	[179]
Beverage	Tea	Quality inspection	Hyper. line scan	408–1117	[171]
	Tea	Moisture content detection	Hyper. line scan	874–1734	[213]
	Tea	Tea variety identification	Multi. area scan	580, 680, 800	[214]
Farm and Plantation	Tea bush	Tea variety, growth status and disease identification	Hyper. area scan	325–1075	[175]
	Coffee crop	Detection of disease/infection	Hyper. area scan	440–850	[176]
	Coffee plantation	Monitoring chlorophyll content	Multi. area scan	490–2190	[215]

Note: Hyper. = hyperspectral, multi. = multispectral.

4. Photonics Techniques Implementation in Food Safety Inspection and Quality Control

Food safety inspection and quality control is important for ensuring the high quality of agriculture products. To meet this criterion, photonics techniques have been extensively implemented into numerous applications. For instance, clean drinking water is undeniably one of the most important elements to sustain the organisms' life. Contamination may happen when treated drinking water is travelling in the distribution system to the consumer, whilst the sensitivity to the inhibitor of contamination can be measured by the elevated dissolved organic matter (DOM) at the tap relative to the water leaving the treatment plant [216]. Across a biologically stable drinking water system, humic-like fluorescence (HLF) intensities of less than 2.2% relative standard deviation are measured after accounting for quenching by copper. In addition, a minor infiltration of a contaminant is detectable by sewage with a strong tryptophan-like fluorescence (TLF) signal thus validating the potential of DOM fluorescence in detecting the water quality changes in drinking water system. Moreover, fluorescence spectroscopy was demonstrated in evaluating the microbial quality of untreated drinking water through online monitoring [217]. The DOM peaks are targeted at excitation and emission wavelengths of 280 and 365 nm for TLF, as well as 280 and 450 nm for HLF. Both TLF and HLF are strongly correlated to micro-bacterial cells such as *E. coli* with a correlation coefficient of 0.71 to 0.77. In comparison to turbidity for *E. coli* with correlation coefficient of only 0.4 to 0.48, the DOM sensor appears to be a better indicator for micro-bacterial cells in untreated drinking water. Apart from the DOM sensor, an optical sensor was proposed to differentiate the particles in drinking water as either bacteria or abiotic particles with an accuracy of 90 ± 7% and 78 ± 14% for monotype and fix-type suspensions, respectively, based on a 3D image recognition and classification algorithm [218]. In addition, this optical sensor can detect micro-particles with minimum size of 0.77 μm. Significantly, the aforementioned optical sensors incorporating photonic techniques serve as an early warning for drinking water pollution.

Photonics and optics have also recently gained popularity in the quality inspection of food product. This is because food inspection in the production line needs to be carried out at fast speed and a very fast monitoring system is needed. Food inspection can become even more challenging when it is dealing with large quantities of sample moving very quickly on the conveyor belt. Therefore, high speed and high sensitivity optical system will be very suitable for the online monitoring and inspection of food product. For example, research work on UV-visible-NIR optical spectroscopy have been carried out extensively in the monitoring of extra virgin olive oil [219], honey [220], tea [221], dairy product [222] and alcoholic beverages [223]. However, as these works focus on wavebands below

1100 nm, the results and consistency of the conclusions may be easily affected by ambient lighting conditions and the change in color of the beverages or product. Therefore, more research work shall be carried out to characterize these food products in the NIR (>1100 nm) and MIR wavebands in order to obtain the optical "fingerprint" that correlates to the quality and food safety level of the product.

In addition, food preservative exceeding the allowable limit has been a critical issue in ensuring the health of the public. Butylated Hydroxytoluene (BHT) is commonly used as an antioxidant agent in canned food or bottled beverages. Several optical sensing techniques such as optical spectroscopy and fluorescence may be able to detect the concentration of BHT. BHT is also commonly known as 2,6-ditertiarybutyl-para-cresol (DBPC). Recently, Leong et al. [224] reported the detection of DBPC in transformer oil using optical spectroscopy at waveband near to 1403 nm. This opens up the opportunity of detecting the concentration of BHT in canned food or bottled beverages, leading to an online monitoring system that uses the optical spectroscopy method.

Due to the lack of attention paid during the preparation processes or due to the contamination of water and environment, hazardous residual materials are occasionally found in food. These hazardous materials include heavy metal, pesticides and antibiotics. Conventionally, the screening process or food safety inspections were carried out using laboratory-based equipment or measurement methods such as gas chromatography (GC), GC-mass spectrometry and high-performance liquid chromatography [225]. However, these methods only allow inspection based on sampling due to the high cost and long result waiting time. In this context, optical detection methods such as optical spectroscopy, Raman spectroscopy and fluorescence can be explored for their possible utilization in the online monitoring of food products in order to ensure that they are free of hazardous residual materials.

5. Photonics Techniques Implementation in Tropical Countries Agriculture

Blessed with wide spans of fertile soil, rich marine ecology, abundant rainfall and a tropical climate, tropical countries are exceptionally suited for a myriad of agricultural activities [226–228]. Agriculture activities boost the country's economy by supplying food sources and industrial raw materials. This sector also provides income to farmers, raising their living standards in rural areas. An example of tropical countries with active agricultural activities is Malaysia. Dating back to the early years following Malaysia's independence in 1957, the agriculture sector has been a siginificant driver towards socio-economic development in Malaysia. However, in the early 1980s, the growth of the agricultural field came to an abrupt halt due to the sharp decline in commodity prices, limited technical specialty, volatile rubber prices and lack of incentives [229,230]. Industrialization soon became the leading economic sector, with great focus directed towards manufacturing and services [230]. Fortunately, the agriculture sector is once again emphasized upon the Asian financial crisis in 1997, acting as a measure to minimize external economic shock by first strengthening the domestic economy [227,231]. Since then, agriculture has always been a major agenda item of Malaysian economic plans, with a recent target directed towards modernizing agriculture as drafted in the Eleventh Malaysia Plan [232]. To date, amidst industrial developments, Malaysia has approximately 4.06 million hectares of agricultural land, with 80% allocated for commercial crops such as palm oil, rubber, cocoa, coconut and pepper [229,233], while a portion of the remaining 20% was utilized for the cultivation of agro-food crops [226]. These remarkable statistics have validated the potential of a tropical country to develop its agricultural sector. Apart from Malaysia, other tropical countries such as Indonesia and Thailand are also actively involved in agriculture activities. The following sections will discuss some of the agricultural crops in tropical countries in which optics and photonic techniques can be easily integrated for automated plantation management, yield increment, quality inspection and disease control.

5.1. Implementation in Palm Oil-Related Activities

Palm oil is an extremely valuable commercial crop in tropical countries. Palm oil, which is extracted from oil palm, is often used as raw materials for the production of biofuel, biofertilizers, oleochemicals, biomass products, nutraceuticals and pharmaceuticals. In fact, tropical countries are among the global leaders in the palm oil industry [234]. The implementation of optics and photonic techniques in palm oil-related activities will maintain the competitive power of the tropical countries in the field and help to reap the associated economic benefits.

The implementation of optics and photonics techniques in oil palm related activities can start from the development of agriculture robots. The development of an agriculture robot involves the implementation of the imaging technique in its operation. Spatial and color information attained by the agriculture robot through the imaging technique will greatly improve the efficiency of palm oil plantation management. Automated palm oil fruit harvest is potentially applicable by pinpointing the fruit position as presented in [72,74,75] for other crops. Besides, automated weed detection and removal [73] as well as automated fertilizing can be performed using the developed agriculture robot.

In addition, palm oil quality is governed by fatty acid, moisture and peroxide contents. Microbial or oxidation reactions that take place during the storage of oil palm fruit may modify these contents, resulting in a depreciation of palm oil quality [235]. Under common operations, palm oil plantations are usually distanced further away from refinery factories. Bulk transport of palm fruit upon reaching the necessary processing quota is often practiced for cost savings. As a result, palm fruits that have been harvested earlier will be stored in dedicated storage spaces. The time difference between harvesting and processing greatly increases the risk for microbial or oxidation reaction to take place. In this scenario, spectroscopy or spectral imaging can be implemented in the palm oil extraction stage to perform oil quality segregation. This will greatly prevent contamination of low-quality palm oil in further downstream processes, promoting process efficiency and increasing palm oil yield.

Another area in which optics and photonics techniques may be applied for oil palm activities is disease detection. The most devastating diseases that attack palm oil plantations in South East Asia are basal stem rot (BSR) and upper stem rot (USR). These diseases result in certain death of oil palms if not controlled effectively, resulting in yield loss and disrupting the plantation cycle. These fatal diseases are identified to be caused by the *Ganoderma boninense* (*G. boninense*) fungus. However, the identification of the root cause of these diseases is still insufficient as they cannot be controlled even with the slightest delay in infection detection [236]. In this area, spectral imaging presents itself as one of the possible alternatives to perform early detection of the *G. boninense* fungus [237]. Samples of suspicious fungi in the palm oil plantation can be simultaneously collected and analyzed to identify the presence of disease-causing *G. boninense*. From here, preventive measures can be effectively performed to curb any possible disease spreading.

5.2. Implementation in Natural Rubber Related Activities

Natural rubber is an important commodity that finds it place in the manufacturing of various household, industrial and medical products. Rubber tree plantations have been widely established in the fertile soils of tropical countries. The usage of optics and photonic techniques will again prove to be beneficial in this area.

The simplest idea will again start from the usage of agriculture robots during the plantation stage. In the context of rubber tree plantations, the imaging technique will provide visual guidance for the agriculture robots to perform the scheduled collection of field latex. The usage of these robots will gradually replace manual latex collection done by rubber tappers. This approach will address the decline in manpower to maintain rubber tree plantations.

Meanwhile, the spectroscopy technique can be utilized in the later rubber processing stages. The first application would be rubber quality grading. For instance, cup lump raw rubber, which is an important material in tires, seal strips, conveyor belts and other moulded rubber products, can be graded by using VIS-NIR spectroscopy to inspect the moisture content of the rubber. This spectroscopic approach is fast, accurate and more reliable compared to manual inspection through sight and touch [155]. Similarly, the protein and lipid contents in natural rubber can be detected through NIR-MIR spectroscopy to enable grading [152]. Lastly, spectroscopy variations, such as NIR-MIR, Raman, dielectric or NMR, can be opted to study the structure and properties of rubber during vulcanization. Such studies allow the analysis and selection of accelerators, activators and retarders, leading to improved characteristics in the vulcanized rubber and an optimized vulcanizing process [124,153].

5.3. Implementation in Agro-Food Crops Related Activities

It is important to increase food production and achieve a self-sufficiency level (SSL) for a growing country to become an advanced country. Currently, the agro-food crops in tropical countries comprise of grains, organic fruits and vegetables, herbs and spices, livestock and fisheries [232,234]. By referring to some of the applications stated in Sections 3.1–3.3, optics and photonic techniques can once again improve the overall quality and yield of these crops.

Starting from grains such as rice and corn, crop harvest [72] and weed removal [73] can be easily performed by agriculture robots with imaging capabilities. Thermal imaging can be conducted to evaluate water stress in crops for irrigation control [97]. Moreover, the development of mobile phone application to perform color-based identification of nitrogen content in rice and corn plant is another interesting idea. The usage of such applications promotes the portable and on-site analysis of fertilizer requirements in crop fields [64].

At the same time, all three optics and photonic techniques discussed earlier can be fully utilized to inspect the harvested organic fruits and vegetables for quality evaluation. For instance, imaging in either VIS or IR region is useful in detecting external damage or bruises in mangosteens, wax jambus, cherry tomatoes and more. Spectroscopy may be performed as well to inspect internal features or maturity of fruits and vegetables. Not least, spectral imaging may be considered when spatial and spectral information are required simultaneously for quality evaluation. Meanwhile, the quality inspection of meat products, such as chicken, beef, lamb, and fish among others, is strongly preferred to be performed using spectroscopy or spectral imaging. These two techniques are suitable for identifying the microbial spoilage of meat products due to their ability to obtain spectral information. With the integration and application of optics and photonics in the agriculture industry, it is anticipated that the agricultural products in the tropical countries will meet the public expectation of higher food quality.

5.4. Possible Challenges

The prevailing research challenges of integrating optics and photonics techniques into the agriculture field are the reliability issue of the laser source and sensor, effect of the ambient environmental condition into optics system, and expensive semiconductor materials at operating wavelength from short to mid-IR range. First and foremost, the illumination intensity of the laser and the sensitivity of the sensor may change over time, which leads to the need for recalibration of the system. Therefore, more research is required in terms of the design and fabrication of a more reliable laser source, sensor and optical detector. In addition, the effect of the ambient condition such as humidity, surrounding temperature, and dust particles could be a hindrance in ensuring consistent results obtained from the optical system. Hence, research into the minimization of these effects on the optical system is significant to improve the system performance such as higher sensitivity, lower systematic error and maintenance rate. Moreover, silicon is well-known for its optimum wavelength operation below 1000 nm. From short to mid-IR range, examples of more viable semiconductor materials are gallium antimonide and indium gallium arsenide. The investigation in terms of generating a higher efficiency using these materials for a cost-effective solution creates the research opportunities for further exploration in both simulation and experimental works.

Apart from the research challenges, the main challenge in introducing the discussed optics and photonic techniques into the field of agriculture in tropical countries would be gaining the acceptance of farmers, fisherman and smallholders. The introduction of modern technology and new agriculture practices often raises concerns surrounding their technical and economic feasibilities. Farm and plantation owners will prefer traditional agriculture practices as newly introduced technologies are often regarded to be more suited to a controlled laboratory environment. In this scenario, technology vendors should ensure that complete field testing has been done in the environment where the technology will be introduced. A probationary period may also be set to allow owners to try out and experience the benefits brought forth by the proposed technologies.

The next challenge would be on financial limitations. In general, the cost to fully implement optics and photonics techniques in existing agriculture activities may be a burden to the owners, especially those involving sophisticated optical tools. This deterring factor may be mitigated if financial aids are provided to the owners. In this case, the government of tropical countries should set the right path by providing funds to the owners through attractive policies. For instance, a loan policy of flexible repayment based on harvest cycles is more attractive compared to one of fixed term financing since owners are now presented with flexible loans [232].

Lastly, another challenge lies with the need of technical support. When introducing the optics and photonic techniques, technical training should be provided to farm and plantation workers in order to familiarize them with the operations of new tools. At the same time, advisory and technical services should be easily available in case the agriculture tools experience downtime or require scheduled maintenance.

6. Conclusions

In conclusion, optics and photonics exhibit great benefits if they are integrated into the agricultural industry. A complete knowledge of the behaviors and properties of light upon light-material interaction allows the quantitative and qualitative analysis of agriculture products. In general, optics and photonic techniques for agricultural purposes can be categorized into imaging, spectroscopy and spectral imaging techniques. The imaging technique is effective in collecting spatial, color and thermal information, whereas the spectroscopy technique is essential for collecting spectral information. Meanwhile, spectral imaging is a combination of both imaging and spectroscopy techniques, allowing the collection of a complete data set. These three optics and photonic techniques have been utilized in agriculture categories such as fruits, vegetables, grain, meat, dairy produce, oil, beverages, and commercial crops, as well as farm and plantation management. These works can be referred to and emulated in the agriculture industry of tropical countries, especially in agriculture activities related to oil palm, rubber and agro-food crops. However, challenges in terms of public acceptance, finance and technical support should be overcome before achieving a complete integration of optics and photonics techniques in the agriculture industry.

Thus, the key contribution of this study is the comprehensive analysis of different optics and photonics systems in agricultural applications to provide a detail idea of the advanced techniques and their future deployment in agriculture cultivation and harvesting. The review has proposed important and selective suggestions for the further technological development of optics and photonics in future agricultural applications:

- The incorporation of optical sensors into photonics detection techniques that serve as an early warning for drinking water pollution.
- The characterization of canned food or bottled beverages in the NIR (>1100 nm) and MIR wavebands for their optical "fingerprint" that correlates to the quality and food safety level of the product, such as preservatives concentration.
- The characterization on hazardous residual materials in food using optical spectroscopy, Raman spectroscopy and fluorescence.
- The implementation of an agricultural robot to perform better palm oil plantation management, scheduled collection of field latex and weed removal.
- The spectral imaging provides early detection of disease-causing *G. boninense* in the oil palm.
- Spectroscopy provides moisture content inspection, protein and lipid content detection, as well as improving the rubber vulcanizing process.
- The imaging technique detects external damage or bruises on organic fruits and vegetables.

Molecules **2019**, *24*, 2025

Author Contributions: This review paper is mainly scripted by T.J.Y. and K.P.J., L.K.Y. contributed to the technical detail compilation and references, language proficiency and formatting of the manuscript. H.M.A. contributed in the organization of contents for the review paper and the revision of the manuscript. S.T.G.H. offered expertise in relating the optical sensing technique and their applications in the agriculture industry, and also contributed to the checking of language proficiency.

Funding: This research was funded by Universiti Tenaga Nasional Internal Grant with the project code J510050796.

Conflicts of Interest: The authors declare no conflict of interest.

References

1. National Research Council. *Light: Wave-Particle Duality*; The National Academic Press: Washington, DC, USA, 2013.
2. An Overview of Optics & Photonics, Essential Technologies for Our Nation. Available online: https://www.scribd.com/document/113237529/HLII-Brochure (accessed on 22 November 2018).
3. Brown, L.M.; Pais, A.; Pippard, A.B. *A History of Optical and Optoelectronic Physics in the Twentieth Century*; American Institute of Physics Press: New York, NY, USA, 1995.
4. Sternberg, E. *Photonic Technology and Industrial Policy: U.S. Responses to Technological Change*; State University of New York Press: New York, NY, USA, 1992.
5. Sumriddetchkajorn, S. How Optics and Photonics is Simply Applied in Agriculture? In Proceedings of the International Conference on Photonics Solutions (ICPS), Pattaya City, Thailand, 7 June 2013; p. 888311. [CrossRef]
6. Machine Vision in Agricultural Robotics—A Short Overview. Available online: https://pdfs.semanticscholar.org/ef13/5ac11c38022029da4d607343b33abb033758.pdf (accessed on 22 November 2018).
7. Ji, B.; Zhu, W.; Liu, B.; Ma, C.; Li, X. Review of Recent Machine-Vision Technologies in Agriculture. In Proceedings of the Knowledge Acquisition and Modeling, 2009. KAM'09. Second International Symposium, Wuhan, China, 30 November–1 December 2009; pp. 330–334.
8. Ishimwe, R.; Abutaleb, K.; Ahmed, F. Applications of Thermal Imaging in Agriculture—A Review. *Adv. Remote Sens.* **2014**, *3*, 128. [CrossRef]
9. Szeliski, R. *Computer Vision: Algorithms and Applications*; Springer: London, UK, 2010.
10. Gunasekaran, S.; Ding, K. Using computer vision for food quality evaluation. *Food Technol.* **1994**, *6*, 151–154.
11. Bhargava, A.; Bansal, A. Fruits and vegetables quality evaluation using computer vision: A review. *J. King Saud Univ. Comput. Inf. Sci.* **2018**. [CrossRef]
12. Raj, M.P.; Swaminarayan, P.R.; Istar, A. Applications of image processing for grading agriculture products. *Int. J. Recent Innov. Trends Comput. Commun.* **2015**, *3*, 1194–1201.
13. Mahendran, R.; Jayashree, G.C.; Alagusundaram, K. Application of computer vision technique on sorting and grading of fruits and vegetables. *J. Food Process. Technol.* **2012**, *10*, 2157–7110.
14. Nezhad, M.A.K.B.; Massh, J.; Komleh, H.E. Tomato Picking Machine Vision Using with the Open CV's library. In Proceedings of the 7th Iranian Conference on Machine Vision and Image Processing, Tehran, Iran, 16–17 November 2011; pp. 1–5.
15. Bora, G.C.; Pathak, R.; Ahmadi, M.; Mistry, P. Image processing analysis to track colour changes on apple and correlate to moisture content in drying stage. *Food Qual. Saf.* **2018**, *2*, 105–110. [CrossRef]
16. Digital Agriculture. Available online: https://sites.tufts.edu/eeseniordesignhandbook/2015/digital-agriculture/ (accessed on 29 April 2019).
17. Nawrocka, A.; Lamorska, L. *Advances in Agrophysical Research*; IntechOpen: Rijeka, Croatia, 2013.
18. Swinehart, D.F. The Beer-Lambert Law. *J. Chem. Educ.* **1962**, *39*, 333. [CrossRef]
19. Animal and Vegetable Fats and Oils—Determination of Anisidine Value. Available online: https://www.iso.org/standard/69593.html (accessed on 22 November 2018).
20. Gray, J.I. Measurement of Lipid Oxidation: A Review. *J. Am. Oil Chem. Soc.* **1978**, *55*, 539–546. [CrossRef]
21. Animal and Vegetable Fats and Oils: Determination of Iodine Value. Available online: https://www.researchgate.net/publication/38999704_Animal_and_Vegetable_Fats_and_Oils_Determination_of_Iodine_Value (accessed on 23 November 2018).

22. Mińkowski, K.; Grześkiewicz, S.; Jerzewska, M.; Ropelewska, M. Characteristic of chemical composition of vegetable oil about high contents of linoleic acids (in Polish). *ŻYWNOŚĆ Nauka Technologia Jakość* **2010**, *73*, 146–157.

23. Psomiadou, E.; Tsimidou, M. Pigments in Greek virgin olive oils: Occurrence and levels. *J. Sci. Food Agric.* **2001**, *81*, 640–647. [CrossRef]

24. Lakowicz, J.R. *Principles of Fluorescence Spectroscopy*; Springer: New York, NY, USA, 2006.

25. Albani, J.R. Fluorescence Spectroscopy in Food Analysis. *Encycl. Anal. Chem.* **2006**. [CrossRef]

26. Karoui, R.; Blecker, C. Fluorescence spectroscopy measurement for quality assessment of food systems—A review. *Food Bioprocess Technol.* **2011**, *4*, 364–386. [CrossRef]

27. Stuart, B.H. *Infrared Spectroscopy: Fundamentals and Applications*; John Wiley and Sons Ltd.: Chichester, UK, 2004.

28. Shurvell, H. *Spectra-Structure Correlations in the Mid- and Far-Infrared*; John Wiley and Sons Ltd.: Chichester, UK, 2006.

29. Li-Chan, E.C.Y.; Ismail, A.A.; Sedman, J.; Voort, F.R. *Vibrational Spectroscopy of Food and Food Products*; John Wiley and Sons Ltd.: Chichester, UK, 2006.

30. Niaura, G. *Raman Spectroscopy in Analysis of Biomolecules*; John Wiley and Sons: New York, NY, USA, 2006.

31. Nelson, S.O. Dielectric spectroscopy in agriculture. *J. Non-Cryst. Solids* **2005**, *351*, 2940–2944. [CrossRef]

32. Mlynárik, V. Introduction to nuclear magnetic resonance. *Anal. Biochem.* **2016**, *529*, 4–9. [CrossRef]

33. Guo, Z.; Huang, W.; Peng, Y.; Chen, Q.; Ouyang, Q.; Zhao, J. Color compensation and comparison of shortwave near infrared and long wave near infrared spectroscopy for determination of soluble solids content of 'Fuji' apple. *Postharvest Biol. Technol.* **2016**, *115*, 81–90. [CrossRef]

34. Guo, Z.; Chen, Q.; Chen, L.; Huang, W.; Zhang, C.; Zhao, C. Optimization of Informative Spectral Variables for the Quantification of EGCG in Green Tea Using Fourier Transform Near-Infrared (FT-NIR) Spectroscopy and Multivariate Calibration. *Appl. Spectrosc.* **2011**, *65*, 1062–1067. [CrossRef]

35. Qin, J.; Chao, K.; Kim, M.S.; Lu, R.; Burks, T.F. Hyperspectral and multispectral imaging for evaluating food safety and quality. *J. Food Eng.* **2013**, *118*, 157–171. [CrossRef]

36. Wu, D.; Sun, D. Advanced applications of hyperspectral imaging technology for food quality and safety analysis and assessment: A review—Part I: Fundamentals. *Innov. Food Sci. Emerg. Technol.* **2013**, *19*, 1–14. [CrossRef]

37. Kim, M.S.; Chao, K.; Chan, D.E.; Jun, W.; Lefcourt, A.M.; Delwiche, S.R.; Kang, S.; Lee, K. Line-scan hyperspectral imaging platform for agro-food safety and quality evaluation: System enhancement and characterization. *Trans. ASABE* **2011**, *54*, 703–711. [CrossRef]

38. Morris, H.R.; Hoyt, C.C.; Treado, P.J. Imaging spectrometers for fluorescence and Raman microscopy–acousto-optic and liquid-crystal tunable filters. *Appl. Spectrosc.* **1994**, *48*, 857–866. [CrossRef]

39. Kise, M.; Park, B.; Heitschmidt, G.W.; Lawrence, K.C.; Windham, W.R. Multispectral imaging system with interchangeable filter design. *Comput. Electron. Agric.* **2010**, *72*, 61–68. [CrossRef]

40. Kim, M.S.; Chen, Y.R.; Mehl, P.M. Hyperspectral reflectance and fluorescence imaging system for quality and safety. *Trans. ASAE* **2001**, *44*, 721–729.

41. Lu, R. Detection of bruises on apples using near-infrared hyperspectral imaging. *Trans. ASAE* **2003**, *46*, 523–530.

42. Manley, M.; Williams, P.; Nilsson, D.; Geladi, P. Near infrared hyperspectral imaging for the evaluation of endosperm texture in whole yellow maize (*Zea maize* L.) kernels. *J. Agric. Food Chem.* **2009**, *57*, 8761–8769. [CrossRef]

43. Litwiller, D. CMOS vs. CCD: Maturing technologies, maturing markets. *Photonics Spectra* **2005**, *39*, 54–61.

44. ElMasry, G.; Wang, N.; Vigneault, C. Detecting chilling injury in Red Delicious apple using hyperspectral imaging and neural networks. *Postharvest Biol. Technol.* **2009**, *52*, 1–8. [CrossRef]

45. Ngadi, M.O.; Liu, L. *Hyperspectral Image Processing Techniques*; Academic Press/Elsevier: Cambridge, MA, USA, 2010; pp. 99–127.

46. ElMasry, G.; Wang, N.; ElSayed, A.; Ngadi, M. Hyperspectral imaging for non-destructive determination of some quality attributes for strawberry. *J. Food Eng.* **2007**, *81*, 98–107. [CrossRef]

47. Qiao, J.; Wang, N.; Ngadi, M.O.; Gunenc, A.; Monroy, M.; Gariepy, C. Prediction of drip-loss, pH, and color for pork using a hyperspectral imaging technique. *Meat Sci.* **2007**, *76*, 1–8. [CrossRef] [PubMed]

48. Qin, J.W.; Burks, T.F.; Ritenour, M.A.; Bonn, W.G. Detection of citrus canker using hyperspectral reflectance imaging with spectral information divergence. *J. Food Eng.* **2009**, *93*, 183–191. [CrossRef]

49. Daugman, J.G. Uncertainty relation for resolution in space, spatial-frequency, and orientation optimized by two-dimensional visual cortical filters. *J. Opt. Soc. Am. A Opt. Image Sci. Vis.* **1985**, *2*, 1160–1169. [CrossRef]

50. Lee, K.; Kang, S.; Delwiche, S.R.; Kim, M.S.; Noh, S. Correlation analysis of hyperspectral imagery for multispectral wavelength selection for detection of defects on apples. *Sens. Instrum. Food Qual. Saf.* **2008**, *2*, 90–96. [CrossRef]

51. Kim, M.S.; Lefcourt, A.M.; Chao, K.; Chen, Y.R.; Kim, I.; Chan, D.E. Multispectral detection of fecal contamination on apples based on hyperspectral imagery: Part I–Application of visible and near-infrared reflectance imaging. *Trans. ASAE* **2002**, *45*, 2027.

52. Park, B.; Lawrence, K.C.; Windham, W.R.; Buhr, R.J. Hyperspectral imaging for detecting fecal and ingesta contaminants on poultry carcasses. In *2001 ASAE Annual Meeting*; American Society of Agricultural and Biological Engineers: St. Joseph, MI, USA, 1998. [CrossRef]

53. Bajwa, S.G.; Bajcsy, P.; Groves, P.; Tian, L. Hyperspectral image data mining for band selection in agricultural applications. *Trans. ASAE* **2004**, *47*, 895. [CrossRef]

54. Nakariyakul, S.; Casasent, D.P. Hyperspectral waveband selection for contaminant detection on poultry carcasses. *Opt. Eng.* **2008**, *47*, 087202.

55. Xing, J.; Guver, D.; Ariana, D.; Lu, R. Determining optimal wavebands using genetic algorithm for detection of internal insect infestation in tart cherry. *Sens. Instrum. Food Qual. Saf.* **2008**, *2*, 161–167. [CrossRef]

56. Leiva-Valenzuela, G.A.; Aguilera, J.M. Automatic detection of orientation and diseases in blueberries using image analysis to improve their postharvest storage quality. *Food Control* **2013**, *33*, 166–173. [CrossRef]

57. Arakeri, M.P.; Lakshmana. Computer Vision Based Fruit Grading System for Quality Evaluation of Tomato in Agriculture industry. *Procedia Comput. Sci.* **2016**, *79*, 426–433. [CrossRef]

58. Sumriddetchkajorn, S.; Somboonkaew, A.; Chanhorm, S. Mobile Device-Based Digital Microscopy for Education, Healthcare, and Agriculture. In Proceedings of the 9th International Conference on Electrical Engineering/Electronics, Computer, Telecommunications and Information Technology (ECTI-CON), Phetchaburi, Thailand, 16–18 May 2012; IEEE: Piscataway, NJ, USA, 2012; pp. 1–4. [CrossRef]

59. Kaur, S.; Pandey, S.; Goel, S. Semi-automatic leaf disease detection and classification system for soybean culture. *IET Image Process.* **2018**, *12*, 1038–1048. [CrossRef]

60. Billingsley, J. *The Counting of Macadamia Nuts*; Research Studies Press Ltd.: Baldock, UK, 2002.

61. Chen, S.W.; Shivakumar, S.S.; Dcunha, S.; Das, J.; Okon, E.; Qu, C.; Taylor, C.J.; Kumar, V. Counting Apples and Oranges with Deep Learning: A Data-Driven Approach. *IEEE Robot. Autom. Lett.* **2017**, *2*, 781–788. [CrossRef]

62. Intaravanne, Y.; Sumriddetchkajorn, S.; Nukeaw, J. Cell phone-based two-dimensional spectral analysis for banana ripeness estimation. *Sens. Actuators B-Chem.* **2012**, *168*, 390–394. [CrossRef]

63. Marimuthu, S.; Roomi, S.M.M. Particle Swarm Optimized Fuzzy Model for the Classification of Banana Ripeness. *IEEE Sens. J.* **2017**, *17*, 4903–4915. [CrossRef]

64. Intaravanne, Y.; Sumriddetchkajorn, S. Android-based rice leaf color analyzer for estimating the needed amount of nitrogen fertilizer. *Comput. Electron. Agric.* **2015**, *116*, 228–233. [CrossRef]

65. Sulistyo, S.B.; Woo, W.L.; Dlav, S.S. Regularized Neural Networks Fusion and Genetic Algorithm Based On-Field Nitrogen Status Estimation of Wheat Plants. *IEEE Trans. Ind. Inform.* **2017**, *13*, 103–114. [CrossRef]

66. Varith, J.; Hyde, G.; Baritelle, A.; Fellman, J.; Sattabongkot, T. Non-Contact Bruise Detection in Apples by Thermal Imaging. *Innov. Food Sci. Emerg. Technol.* **2003**, *4*, 211–218. [CrossRef]

67. Danno, A.; Miyazato, M.; Ishiguro, E. Quality Evaluation of Agricultural Products by Infrared Imaging Method: Grading of Fruits for Bruise and Other Surface Defects. *Mem. Fac. Agric. Kagoshima Univ.* **1978**, *14*, 123–138.

68. Oerke, E.; Fröhling, P.; Steiner, U. Thermographic Assessment of Scab Disease on Apple Leaves. *Precis. Agric.* **2011**, *12*, 699–715. [CrossRef]

69. Oerke, E.; Steiner, U.; Dehne, H.; Lindenthal, M. Thermal Imaging of Cucumber Leaves Affected by Downy Mildew and Environmental Conditions. *J. Exp. Bot.* **2006**, *57*, 2121–2132. [CrossRef] [PubMed]

70. Hellebrand, H.J.; Linke, M.; Beuche, H.; Herold, B.; Geyer, M. *Horticultural Products Evaluated by Thermography*; The Leibniz Institute for Agricultural Engineering Potsdam-Bornim: Potsdam, Germany, 2000.

71. Danno, A.; Miyazato, M.; Ishiguro, E. Quality Evaluation of Agricultural Products by Infrared Imaging Method: Maturity Evaluation of Fruits and Vegetables. *Mem. Fac. Agric. Kagoshima Univ.* **1980**, *16*, 157–164.

72. Wu, G.; Tan, Y.; Zheng, Y.; Wang, S. Walking Goal Line Detection Based on Machine Vision on Harvesting Robot. In Proceedings of the 2011 Third Pacific-Asia Conference on Circuits, Communications and System (PACCS), Wuhan, China, 17–18 July 2011; IEEE: Piscataway, NJ, USA, 2011; pp. 1–4.

73. Tillett, N.D.; Hague, T.; Miles, S.J. A field assessment of a potential method for weed and crop mapping on the basis of crop planting geometry. *Comput. Electron. Agric.* **2001**, *32*, 229–246. [CrossRef]

74. Irie, N.; Taguchi, N.; Horie, T.; Ishimatsu, T. Asparagus harvesting robot coordinated with 3-D vision sensor. In Proceedings of the IEEE International Conference on Industrial Technology (ICIT 2009), Gippsland, VIC, Australia, 10–13 February 2009; IEEE: Piscataway, NJ, USA, 2009; pp. 1–6.

75. Astrand, B.; Baerveldt, A.J. A vision based row-following system for agricultural field machinery. *Mechatronics* **2005**, *15*, 251–269. [CrossRef]

76. Songa, X.Y.; Lerova, T.; Vrankena, E.; Maertens, W.; Sonck, B.; Berckmans, D. Automatic detection of lameness in dairy cattle Vision-based trackway analysis in cow's locomotion. *Comput. Electron. Agric.* **2008**, *64*, 39–44. [CrossRef]

77. Kane, A.S.; Salierno, J.D.; Gipson, G.T.; Molteno, T.C.A.; Hunter, C. A video-based movement analysis system to quantify behavioural stress responses of fish. *Water Res.* **2004**, *38*, 3993–4001. [CrossRef]

78. Baranowski, P.; Mazurek, W.; Wozniak, J.; Majewska, U. Detection of early bruises in apples using hyperspectral data and thermal imaging. *J. Food Eng.* **2012**, *110*, 345–355. [CrossRef]

79. Stajnko, D.; Lakota, M.; Hocevar, M. Estimation of Number and Diameter of Apple Fruits in an Orchard during the Growing Season by Thermal Imaging. *Comput. Electron. Agric.* **2004**, *42*, 31–42. [CrossRef]

80. Lv, J.; Shen, G.; Ma, Z. Acquisition of Fruit Region in Green Apple Image Based on the Combination of Segmented Regions. In Proceedings of the 2nd International Conference on Image, Vision and Computing, Chengdu, China, 2–4 June 2017.

81. Dunn, M.; Billingsley, J. A Machine Vision System for Surface Texture Measurements of Citrus. In Proceedings of the 11th IEEE conference on Mechatronics and Machine Vision in Practice, Macau, China, 30 November–2 December 2004; pp. 73–76.

82. Ballester, C.; Castel, J.; Jiménez-Bello, M.; Castel, J.; Intrigliolo, D. Thermographic Measurement of Canopy Temperature Is a Useful Tool for Predicting Water Deficit Effects on Fruit Weight in Citrus Trees. *Agric. Water Manag.* **2013**, *122*, 1–6. [CrossRef]

83. Bonilla, J.; Prieto, F.; Pérez, C. Mass and Volume Estimation of Passion Fruit using Digital Images. *IEEE. Lat. Am. Trans.* **2017**, *15*, 275–282. [CrossRef]

84. Stoll, M.; Schultz, H.R.; Loehnertz, B.B. Exploring the Sensitivity of Thermal Imaging for Plasmopara viticola Pathogen Detection in Grapevines under Different Water Status. *Funct. Plant Biol.* **2008**, *35*, 281–288. [CrossRef]

85. Xiao-Lian, L.; Xiao-Rong, L.; Bing-Fu, L. Identification and Location of Picking Tomatoes Based on Machine Vision. In Proceedings of the 2011 International Conference on Intelligent Computation Technology and Automation (ICICTA), Shengzhen, Guangdong, China, 28–29 March 2011; pp. 101–107.

86. Yang, L.; Dickinson, J.; Wu, Q.M.J.; Lang, S. A fruit recognition method for automatic harvesting. In Proceedings of the 14th International Conference on Mechatronics and Machine Vision in Practice (M2VIP2007), Xiamen, China, 3–5 December 2007; pp. 152–157.

87. Vanlinden, V.; Vereycken, R.; Ramon, H.; Baerdemaeker, J.D. Detection technique for tomato bruise damage by thermal imaging. *Acta Hortic.* **2003**, *599*, 389–394. [CrossRef]

88. Quan, Q.; Lanlan, T.; Xiaojun, Q.; Kai, J.; Qingchun, F. Selecting Candidate Regions of Clustered Tomato Fruits under Complex Greenhouse Scenes Using RGB-D Data. In Proceedings of the 3rd International Conference on Control, Automation and Robotics, Nagoya, Japan, 22–24 April 2017; pp. 389–393.

89. Sa, I.; Lehnert, C.; McCool, C.; Dayoub, F.; Upcroft, B.; Perez, T. Peduncle Detection of Sweet Pepper for Autonomous Crop Harvesting-Combined Color and 3-D Information. *IEEE Robot. Autom. Lett.* **2017**, *2*, 765–772. [CrossRef]

90. Speir, R.A.; Heidekker, M.A. Onion postharvest quality assessment with X-ray computed tomography— A pilot study. *IEEE Instrum. Meas. Mag.* **2017**, *20*, 15–19. [CrossRef]

91. Shi-Gang, C.; Heng, L.; Xing-Li, W.; Yong-Li, Z.; Lin, H. Study on segmentation of lettuce image based on morphological reorganization and watershed algorithm. In Proceedings of the IEEE Chinese Control and Decision Conference (CCDC), IEEE, Shenyang, China, 9–11 June 2018; pp. 6595–6597.

92. Lindenthal, M.; Steiner, U.; Dehne, H.; Oerke, E. Effect of Downy Mildew Development on Transpiration of Cucumber Leaves Visualized by Digital Infrared Thermography. *Phytopathology* **2005**, *95*, 233–240. [CrossRef]

93. Oerke, E.; Lindenthal, M.; Fröhling, P.; Steiner, U. Digital Infrared Thermography for the Assessment of Leaf Pathogens. In Proceedings of the 5th European Conference on Precision Agriculture, Uppsala, Sweden, 9–11 June 2005; pp. 91–98.

94. Smith, R.; Barrs, H.; Steiner, J.; Stapper, M. Relationship between Wheat Yield and Foliage Temperature: Theory and Its Application to Infrared Measurements. *Agric. For. Meteorol.* **1985**, *36*, 129–143. [CrossRef]

95. Du, W.Y.; Zhang, L.D.; Hu, Z.F.; Shamaila, Z.; Zeng, A.J.; Song, J.L.; Liu, Y.J.; Wolfram, S.; Joachim, M.; He, X.K. Utilization of Thermal Infrared Image for Inversion of Winter Wheat Yield and Biomass. *Spectrosc. Spectr. Anal.* **2011**, *31*, 1476–1480.

96. Hu, Z.; Zhang, L.; Wang, Y.; Shamaila, Z.; Zeng, A.; Song, J.; Liu, Y.; Wolfram, S.; Joachim, M.; He, X. Application of BP Neural Network in Predicting Winter Wheat Yield Based on Thermography Technology. *Spectrosc. Spectr. Anal.* **2013**, *33*, 1587–1592.

97. Wanjura, D.; Upchurch, D.R. Water Status Response of Corn and Cotton to Altered Irrigation. *Irrig. Sci.* **2002**, *21*, 45–55. [CrossRef]

98. Tetila, E.C.; Machado, B.B.; de Souza Belete, N.A.; Guimarães, D.A.; Pistori, H. Identification of Soybean Foliar Diseases Using Unmanned Aerial Vehicle Images. *IEEE Geosci. Remote Sens. Lett.* **2017**, *14*, 2190–2194. [CrossRef]

99. Zhou, C.; Yang, G.; Liang, D.; Yang, X.; Xu, B. An Integrated Skeleton Extraction and Pruning Method for Spatial Recognition of Maize Seedlings in MGV and UAV Remote Images. *IEEE Geosci. Remote Sens. Lett.* **2018**, *56*, 4618–4632. [CrossRef]

100. Zhang, X.; Qiao, Y.; Meng, F.; Fan, C.; Zhang, M. Identification of Maize Leaf Diseases Using Improved Deep Convolutional Neural Networks. *IEEE Access* **2018**, *6*, 30370–30377. [CrossRef]

101. Lu, H.; Cao, Z.; Xiao, Y.; Fang, Z.; Zhu, Y. Toward good practices for fine-grained maize cultivar identification with filter-specific convolutional activations. *IEEE Trans. Autom. Sci. Eng.* **2018**, *15*, 430–442. [CrossRef]

102. Padhi, J.; Misra, R.; Payero, J. Estimation of Soil Water Deficit in an Irrigated Cotton Field with Infrared Thermography. *Field Crops Res.* **2012**, *126*, 45–55. [CrossRef]

103. Kamtongdee, C.; Sumriddetchkajorn, S.; Chanhorm, S.; Kaewhom, W. Noise reduction and accuracy improvement in optical-penetration-based silkworm gender identification. *Appl. Opt.* **2015**, *54*, 1844–1851. [CrossRef]

104. Kranner, I.; Kastbergerb, G.; Hartbauerb, M.; Pritcharda, H.W. Noninvasive Diagnosis of Seed Viability Using Infrared Thermography. *Proc. Natl. Acad. Sci. USA* **2010**, *107*, 3912–3917. [CrossRef]

105. Henry, D.; Aubert, H.; Véronèse, T.; Serrano, É. Remote estimation of intra-parcel grape quantity from three-dimensional imagery technique using ground-based microwave FMCW radar. *IEEE Instrum. Meas. Mag.* **2017**, *20*, 20–24. [CrossRef]

106. Tosi, M.V.; Ferrante, V.; Mattiello, S.; Canali, E.; Verga, M. Comparison of video and direct observation methods for measuring oral behaviourin veal calves. *Ital. J. Anim. Sci.* **2006**, *5*, 19–27. [CrossRef]

107. Dunn, M.; Billingsley, J.; Finch, N. *Machine Vision Classification of Animals*; Research Studies Press Ltd.: Baldock, UK, 2003.

108. Stien, L.H.; Brafland, S.; Austevollb, I.; Oppedal, F.; Kristiansen, T.S. A video analysis procedure for assessing vertical fish distribution in aquaculture tanks. *Aquac. Eng.* **2007**, *37*, 115–124. [CrossRef]

109. Sumriddetchkajorna, S.; Chaitavonb, K.; Intaravanne, Y. Mobile-platform based colorimeter for monitoring chlorine concentration in water. *Sens. Actuators B-Chem.* **2014**, *191*, 561–566. [CrossRef]

110. Iqbal, Z.; Bjorklund, R.B. Colorimetric analysis of water and sand samples performed on a mobile phone. *Talanta* **2011**, *84*, 1118–1123. [CrossRef] [PubMed]

111. Nagy, A.; Riczu, P.; Tamás, J. Spectral evaluation of apple fruit ripening and pigment contentalteration. *Sci. Hortic.* **2016**, *201*, 256–264. [CrossRef]

112. Roy, S.; Anantheswaran, R.C.; Shenk, J.S.; Beelman, R. Determination of moisture content of mushrooms by Vis-NIR spectroscopy. *J. Sci. Food Agric.* **1993**, *63*, 355–360. [CrossRef]

113. Hartmann, R.; Büning-Pfaue, H. NIR determination of potato constituents. *Potato Res.* **1998**, *41*, 327–334. [CrossRef]

114. Lee, M.; Hwang, Y.; Lee, J.; Choung, M. The characterization of caffeine and nine individual catechins in the leaves of green tea (*Camellia sinensis* L.) by near-infrared reflectance spectroscopy. *Food Chem.* **2014**, *158*, 351–357. [CrossRef]

115. Sun, X.; Liu, Y.; Li, Y.; Wu, M.; Zhu, D. Simultaneous measurement of brown core and soluble solids content in pear by on-line visible and near infrared spectroscopy. *Postharvest Biol. Technol.* **2016**, *116*, 80–87. [CrossRef]

116. Viegas, T.R.; Mata, A.L.M.L.; Duarte, M.M.L.; Lima, K.M.G. Determination of quality attributes in wax jambu fruit using NIRS and PLS. *Food Chem.* **2016**, *190*, 1–4. [CrossRef]

117. Gente, R.; Busch, S.F.; Stübling, E.; Schneider, L.M.; Hirschmann, C.B.; Balzer, J.C.; Koch, M. Quality control of sugar beet seeds with THz time-domain spectroscopy. *IEEE Trans. Terahertz Sci. Technol.* **2016**, *6*, 754–756. [CrossRef]

118. Nunes, K.M.; Andrade, M.V.O.; Filho, A.M.P.S.; Lasmar, M.C.; Sena, M.M. Detection and characterisation of frauds in bovine meat in natura by non-meat ingredient additions using data fusion of chemical parameters and ATR-FTIR spectroscopy. *Food Chem.* **2016**, *205*, 14–22. [CrossRef] [PubMed]

119. ElMasry, G.; Nagai, H.; Moria, K.; Nakazawa, N.; Tsuta, M.; Sugiyama, J.; Okazaki, E.; Nakauchi, S. Freshness estimation of intact frozen fish using fluorescence spectroscopy and chemometrics of excitation-emission matrix. *Talanta* **2015**, *143*, 145–156. [CrossRef] [PubMed]

120. Luo, H.; Huang, Y.; Lai, K.; Rasco, B.A.; Fan, Y. Surface-enhanced Raman spectroscopy coupled with gold nanoparticles for rapid detection of phosmet and thiabendazole residues in apples. *Food Control* **2016**, *68*, 229–235. [CrossRef]

121. Cozzolino, D. Near infrared spectroscopy as a tool to monitor contaminants in soil, sediments and water–State of the art, advantages and pitfalls. *Trends Environ. Anal. Chem.* **2016**, *9*, 1–7. [CrossRef]

122. Nie, P.; Dong, T.; He, Y.; Xiao, S. Research on the effects of drying temperature on nitrogen detection of different soil types by near infrared sensors. *Sensors* **2018**, *18*, 391. [CrossRef]

123. Xiao, S.; He, Y.; Dong, T.; Nie, P. Spectral Analysis and Sensitive Waveband Determination Based on Nitrogen Detection of Different Soil Types Using Near Infrared Sensors. *Sensors* **2018**, *18*, 523. [CrossRef]

124. Hernández, M.; Valentín, J.L.; López-Manchado, M.A.; Ezquerra, T.A. Influence of the vulcanization system on the dynamics and structure of natural rubber: Comparative study by means of broadband dielectric spectroscopy and solid-state NMR spectroscopy. *Eur. Polym. J.* **2015**, *68*, 90–103. [CrossRef]

125. Mahani, R.; Atia, F.; Neklawy, M.M.A.; Fahem, A. Dielectric spectroscopic studies on the water hyacinth plant collected from agriculture drainage. *Spectrochim. Acta A Mol. Biomol. Spectrosc.* **2016**, *162*, 81–85. [CrossRef]

126. Jha, S.N.; Narsaiah, K.; Jaiswal, P.; Bhardwaj, R.; Gupta, M.; Kumar, R.; Sharma, R. Nondestructive prediction of maturity of mango using near infrared spectroscopy. *J. Food Eng.* **2014**, *124*, 152–157. [CrossRef]

127. Guo, W.; Gu, J.; Liu, D.; Shang, L. Peach variety identification using near-infrared diffuse reflectance spectroscopy. *Comput. Electron. Agric.* **2016**, *123*, 297–303. [CrossRef]

128. González-Fernández, A.B.; Rodríguez-Pérez, J.R.; Marabel, M.; Álvarez-Taboada, F. Spectroscopic estimation of leaf water content in commercial vineyards using continuum removal and partial least squares regression. *Sci. Hortic.* **2015**, *188*, 15–22. [CrossRef]

129. Schulz, H.; Drews, H.; Quilitzsch, R.; Krüger, H. Application of near infrared spectroscopy for the quantification of quality parameters in selected vegetables and essential oil plants. *J. Near Infrared Spectrosc.* **1998**, *6*, A125–A130. [CrossRef]

130. Evans, S.D.; Muir, A.Y. Reflectance Spectrophotometry of Bruising in Potatoes. I. Ultraviolet to Near Infrared. *Int. Agrophys.* **1999**, *13*, 203–210.

131. Birth, G.S.; Dull, G.G.; Renfroe, W.T.; Kays, S.J. Nondestructive Spectrophotometric Determination of Dry Matter in Onions. *J. Am. Soc. Hortic. Sci.* **1985**, *110*, 297–303.

132. Zhang, C.; Kong, W.; Liu, F.; He, Y. Measurement of aspartic acid in oilseed rape leaves under herbicide stress using near infrared spectroscopy and chemometrics. *Heliyon* **2016**, *2*, e00064. [CrossRef]

133. Ambrose, A.; Lohumi, S.; Lee, W.; Cho, B.K. Comparative nondestructive measurement of corn seed viability using Fourier transform near-infrared (FT-NIR) and Raman spectroscopy. *Sens. Actuators B-Chem.* **2016**, *224*, 500–506. [CrossRef]

134. Pearson, T.C. Spectral Properties and Effect of Drying Temperature on Almonds with Concealed Damage. *LWT-Food Sci. Technol.* **1999**, *32*, 67–72. [CrossRef]

135. Pearson, T.C. Use of Near Infrared Transmittance to Automatically Detect Almonds with Concealed Damage. *LWT-Food Sci. Technol.* **1999**, *32*, 73–78. [CrossRef]

136. Lian, F.; Xu, D.; Fu, M.; Ge, H.; Jiang, Y.; Zhang, Y. Identification of Transgenic Ingredients in Maize Using Terahertz Spectra. *IET Nanobiotechnol.* **2017**, *7*, 378–384. [CrossRef]

137. Prasad, T.N.; Adam, S.; Rao, P.V.; Reddy, B.R.; Krishna, T.G. Size dependent effects of antifungal phytogenic silver nanoparticles on germination, growth and biochemical parameters of rice (*Oryza sativa L*), maize (*Zea mays L*) and peanut (*Arachis hypogaea L*). *IET Nanobiotechnol.* **2016**, *11*, 277–285. [CrossRef]

138. Sahar, A.; Rahman, U.; Kondjoyan, A.; Portanguen, S.; Dufour, E. Monitoring of thermal changes in meat by synchronous fluorescence spectroscopy. *J. Food Eng.* **2016**, *168*, 160–165. [CrossRef]

139. Liu, M.; Yao, L.; Wang, T.; Li, J.; Yu, C. Rapid determination of egg yolk contamination in egg white by VIS spectroscopy. *J. Food Eng.* **2014**, *124*, 117–121. [CrossRef]

140. Núñez-Sánchez, N.; Martínez-Marín, A.L.; Polvillo, O.; Fernández-Cabanás, V.M.; Carrizosa, J.; Urrutia, B.; Serradilla, J.M. Near Infrared Spectroscopy (NIRS) for the determination of the milk fat fatty acid profile of goats. *Food Chem.* **2016**, *190*, 244–252. [CrossRef] [PubMed]

141. Almoselhy, R.I.M.; Allam, M.H.; El-Kalyoubi, M.H.; El-Sharkawy, A.A. 1H NMR spectral analysis as a new aspect to evaluate the stability of some edible oils. *Ann. Agric. Sci.* **2014**, *59*, 201–206. [CrossRef]

142. Mabood, F.; Boqué, R.; Folcarelli, R.; Busto, O.; Jabeen, F.; Al-Harrasi, A.; Hussain, J. The effect of thermal treatment on the enhancement of detection of adulteration in extra virgin olive oils by synchronous fluorescence spectroscopy and chemometric analysis. *Spectrochim. Acta Part A: Mol. Biomol. Spectrosc.* **2016**, *161*, 83–87. [CrossRef]

143. Hzounda, J.B.F.; Jazet, P.M.D.; Lazar, G.; Răducanu, D.; Caraman, I.; Bassene, E.; Boyom, F.F.; Lazarca, I.M. Spectral and chemometric analyses reveal antioxidant properties of essential oils from four Cameroonian *Ocimum*. *Ind. Crops Prod.* **2016**, *80*, 101–108. [CrossRef]

144. Dutta, D.; Das, P.K.; Bhunia, U.K.; Singh, U.; Singh, S.; Sharma, J.R.; Dadhwal, V.K. Retrieval of tea polyphenol at leaf level using spectral transformation and multi-variate statistical approach. *Int. J. Appl. Earth Obs. Geoinf.* **2015**, *36*, 22–29. [CrossRef]

145. Marquetti, I.; Link, J.V.; Lemes, A.L.G.; dos Santos Scholz, M.B.; Valderrama, P.; Bona, E. Partial least square with discriminant analysis and near infrared spectroscopy for evaluation of geographic and genotypic origin of arabica coffee. *Comput. Electron. Agric.* **2016**, *121*, 313–319. [CrossRef]

146. Bertone, E.; Venturello, A.; Giraudo, A.; Pellegrino, G.; Geobaldo, F. Simultaneous determination by NIR spectroscopy of the roasting degree and Arabica/Robusta ratio in roasted and ground coffee. *Food Control* **2016**, *59*, 683–689. [CrossRef]

147. Ayvaz, H.; Sierra-Cadavid, A.; Aykas, D.P.; Mulqueeney, B.; Sullivan, S.; Rodriguez-Saona, L.E. Monitoring multicomponent quality traits in tomato juice using portable mid-infrared (MIR) spectroscopy and multivariate analysis. *Food Control* **2016**, *66*, 79–86. [CrossRef]

148. Ye, M.; Gao, Z.; Li, Z.; Yuan, Y.; Yue, T. Rapid detection of volatile compounds in apple wines using FT-NIR spectroscopy. *Food Chem.* **2016**, *190*, 701–708. [CrossRef] [PubMed]

149. Kim, D.; Cho, B.; Lee, S.H.; Kwon, K.; Park, E.S.; Lee, W. Application of Fourier transform-mid infrared reflectance spectroscopy for monitoring Korean traditional rice wine 'Makgeolli' fermentation. *Sens. Actuators B-Chem.* **2016**, *230*, 753–760. [CrossRef]

150. Fortier, C.; Rodgers, J. Preliminary Examinations for the Identification of U.S. Domestic and International Cotton Fibers by Near-Infrared Spectroscopy. *Fibers* **2014**, *2*, 264–274. [CrossRef]

151. Liu, Y.; Delhom, C.; Campbell, B.T.; Martin, V. Application of near infrared spectroscopy in cotton fiber micronaire measurement. *Inf. Process. Agric.* **2016**, *3*, 30–35. [CrossRef]

152. Rolere, S.; Liengprayoon, S.; Vaysse, L.; Sainte-Beuve, J.; Bonfils, F. Investigating natural rubber composition with Fourier Transform Infrared (FT-IR) spectroscopy: A rapid and non-destructive method to determine both protein and lipid contents simultaneously. *Polym. Test.* **2015**, *43*, 83–93. [CrossRef]

153. Musto, P.; Larobina, D.; Cotugno, S.; Straffi, P.; Florio, G.D.; Mensitieri, G. Confocal Raman imaging, FTIR spectroscopy and kinetic modelling of the zinc oxide/stearic acid reaction in a vulcanizing rubber. *Polymer* **2013**, *54*, 685–693. [CrossRef]

154. Yrieix, M.; Cruz-Boisson, F.D.; Majesté, J. Rubber/silane reaction sand grafting rates investigated by liquid-state NMR spectroscopy. *Polymer* **2016**, *87*, 90–97. [CrossRef]

155. Suchat, S.; Theanjumol, P.; Karrila, S. Rapid moisture determination for cup lump natural rubber by near infrared spectroscopy. *Ind. Crops Prod.* **2015**, *76*, 772–780. [CrossRef]

156. Avinash, B.; Venu, R.; Prasad, T.N.; Rao, K.S.; Srilatha, C. Synthesis and characterisation of neem leaf extract, 2, 3-dehydrosalanol and quercetin dihydrate mediated silver nano particles for therapeutic applications. *IET Nanobiotechnol.* **2016**, *11*, 383–389. [CrossRef]

157. Cécillon, L.; Barthès, B.; Gomez, C.; Ertlen, D.; Génot, V.; Hedde, M.; Stevens, A.; Brun, J. Assessment and monitoring of soil quality using near infrared reflectance spectroscopy (NIRS). *Eur. J. Oral Sci.* **2009**, *60*, 770–784.

158. Zhang, Y.; Li, M.; Zheng, L.; Zhao, Y.; Pei, X. Soil nitrogen content forecasting based on real-time NIR spectroscopy. *Comput. Electron. Agric.* **2016**, *124*, 29–36. [CrossRef]

159. Ludwig, B.; Linsler, D.; Höper, H.; Schimdt, H.; Piepho, H.; Vohland, M. Pitfalls in the use of middle-infrared spectroscopy: Representativeness and ranking criteria for the estimation of soil properties. *Geoderma* **2016**, *268*, 165–175. [CrossRef]

160. Wang, Y.; Jiang, F.; Gupta, B.B.; Rho, S.; Liu, Q.; Hou, H.; Jing, D.; Shen, W. Variable Selection and Optimization in Rapid Detection of Soybean Straw Biomass Based on CARS. *Cellulose* **2018**, *144*, 28–51. [CrossRef]

161. Symonds, P.; Paap, A.; Alameh, K.; Rowe, J.; Miller, C. A real-time plant discrimination system utilising discrete reflectance spectroscopy. *Comput. Electron. Agric.* **2015**, *117*, 57–69. [CrossRef]

162. Bennedsen, B.S.; Peterson, D.L.; Tabb, A. Identifying defects in images of rotating apples. *Comput. Electron. Agric.* **2005**, *48*, 92–102. [CrossRef]

163. Gowen, A.A.; O'Donnell, C.P.; Taghizadeh, M.; Cullen, P.J.; Frias, J.M.; Downey, G. Hyperspectral imaging combined with principal component analysis for bruise damage detection on white mushrooms (*Agaricus bisporus*). *J. Chemom.* **2008**, *22*, 259–267. [CrossRef]

164. Lleo, L.; Barreiro, P.; Ruiz-Altisent, M.; Herrero, A. Multispectral images of peach related to firmness and maturity at harvest. *J. Food Eng.* **2009**, *93*, 229–235. [CrossRef]

165. Zhang, C.; Guo, C.; Liu, F.; Kong, W.; He, Y.; Lou, B. Hyperspectral imaging analysis for ripeness evaluation of strawberry with support vector machine. *J. Food Eng.* **2016**, *179*, 11–18. [CrossRef]

166. Polder, G.; van der Heijden, G.W.; Young, I.T. Spectral image analysis for measuring ripeness of tomatoes. *Trans. ASAE* **2002**, *45*, 1155–1161. [CrossRef]

167. Hahn, F. Multi-spectral prediction of unripe tomatoes. *Biosyst. Eng.* **2002**, *81*, 147–155. [CrossRef]

168. Chen, J.; Cai, F.; He, R.; He, S. Experimental Demonstration of Remote and Compact Imaging Spectrometer Based on Mobile Devices. *Sensors* **2018**, *18*, 1989. [CrossRef] [PubMed]

169. Lu, R.; Peng, Y. Development of a multispectral imaging prototype for real-time detection of apple fruit firmness. *Opt. Eng.* **2007**, *46*, 123201.

170. Baiano, A.; Terracone, C.; Peri, G.; Romaniello, R. Application of hyperspectral imaging for prediction of physico-chemical and sensory characteristics of table grapes. *Comput. Electron. Agric.* **2012**, *87*, 142–151. [CrossRef]

171. Zhao, J.; Chen, Q.; Cai, J.; Ouyang, Q. Automated tea quality classification by hyperspectral imaging. *Appl. Opt.* **2009**, *48*, 3557–3564. [CrossRef] [PubMed]

172. Qin, J.; Burks, T.F.; Zhao, X.; Niphadkar, N.; Ritenour, M.A. Development of a two-band spectral imaging system for real-time citrus canker detection. *J. Food Eng.* **2012**, *108*, 87–93. [CrossRef]

173. Wang, W.; Li, C.; Tollner, E.W.; Gitaitis, R.D.; Rains, G.C. Shortwave infrared hyperspectral imaging for detecting sour skin (*Burkholderia cepacia*)-infected onions. *J. Food Eng.* **2012**, *109*, 38–48. [CrossRef]

174. Chao, K.; Mehl, P.M.; Chen, Y.R. Use of hyper- and multi-spectral imaging for detection of chicken skin tumors. *Appl. Eng. Agric.* **2002**, *18*, 113. [CrossRef]

175. Kumar, A.; Manjunath, K.R.; Meenakshi; Bala, R.; Sud, R.K.; Singh, R.D.; Panigrahy, S. Field hyperspectral data analysis for discriminating spectral behavior of tea plantations under various management practices. *Int. J. Appl. Earth Obs. Geoinf.* **2013**, *23*, 352–359. [CrossRef]

176. Martins, G.D.; Galo, M.D.L.B.T.; Vieira, B.S. Detecting and Mapping Root-Knot Nematode Infection in Coffee Crop Using Remote Sensing Measurements. *IEEE J. Sel. Top. Appl. Earth Obs. Remote Sens.* **2017**, *10*, 5395–5403. [CrossRef]

177. Leiva-Valenzuela, G.A.; Lu, R.; Aguilera, J.M. Prediction of firmness and soluble solids content of blueberries using hyperspectral reflectance imaging. *J. Food Eng.* **2013**, *115*, 91–98. [CrossRef]

178. Qin, J.; Lu, R. Measurement of the absorption and scattering properties of turbid liquid foods using hyperspectral imaging. *Appl. Spectrosc.* **2007**, *61*, 388–396. [CrossRef] [PubMed]

179. Gila, D.M.M.; Marchal, P.C.; García, J.G.; Ortega, J.G. On-line system based on hyperspectral information to estimate acidity, moisture and peroxides in olive oil samples. *Comput. Electron. Agric.* **2015**, *116*, 1–7. [CrossRef]

180. Keresztes, J.C.; Goodarzi, M.; Saeys, W. Real-time pixel based early apple bruise detection using short wave infrared hyperspectral imaging in combination with calibration and glare correction techniques. *Food Control* **2016**, *66*, 215–226. [CrossRef]

181. Baranowski, P.; Mazurek, W.; Pastuszka-Woźniak, J. Supervised classification of bruised apples with respect to the time after bruising on the basis of hyperspectral imaging data. *Postharvest Biol. Technol.* **2013**, *86*, 249–258. [CrossRef]

182. Kim, M.S.; Lee, K.; Chao, K.; Lefcourt, A.M.; Jun, W.; Chan, D.E. Multispectral line-scan imaging system for simultaneous fluorescence and reflectance measurements of apples: Multitask apple inspection system. *Sens. Instrum. Food Qual. Saf.* **2008**, *2*, 123–129. [CrossRef]

183. Lu, R.; Peng, Y. Hyperspectral scattering for assessing peach fruit firmness. *Biosyst. Eng.* **2006**, *93*, 161–171. [CrossRef]

184. Rajkumar, P.; Wang, N.; EImasry, G.; Raghavan, G.S.V.; Gariepy, Y. Studies on banana fruit quality and maturity stages using hyperspectral imaging. *J. Food Eng.* **2012**, *108*, 194–200. [CrossRef]

185. Leiva-Valenzuela, G.A.; Lu, R.; Aguilera, J.M. Assessment of internal quality of blueberries using hyperspectral transmittance and reflectance images with whole spectra or selected wavelengths. *Innov. Food Sci. Emerg. Technol.* **2014**, *24*, 2–13. [CrossRef]

186. Qin, J.; Lu, R. Detection of pits in tart cherries by hyperspectral transmission imaging. *Trans. ASAE* **2005**, *48*, 1963–1970. [CrossRef]

187. Liu, Y.; Chen, Y.R.; Wang, C.; Chan, D.E.; Kim, M.S. Development of a simple algorithm for the detection of chilling injury in cucumbers from visible/near-infrared hyperspectral imaging. *Appl. Spectrosc.* **2005**, *59*, 78–85. [CrossRef]

188. Hernández-Hierro, J.M.; Esquerre, C.; Valverde, J.; Villacreces, S.; Reilly, K.; Gaffne, M.; González-Miret, M.L.; Heredia, F.J.; O'Donnell, C.P.; Downey, G. Preliminary study on the use of near infrared hyperspectral imaging for quantitation and localisation of total glucosinolates in freeze-dried broccoli. *J. Food Eng.* **2014**, *126*, 107–112. [CrossRef]

189. Trong, D.; Nyugen, N.; Tsuta, M.; Nicolaï, B.M.; Baerdemaeker, J.D.; Saeys, W. Prediction of optimal cooking time for boiled potatoes by hyperspectral imaging. *J. Food Eng.* **2011**, *105*, 617–624. [CrossRef]

190. Onoyama, H.; Ryu, C.; Suguri, M.; Iida, M. Estimation of Nitrogen Contents in Rice Plant at the Panicle Initiation Stage Using Ground-Based Hyperspectral Remote Sensing. *IFAC Proc. Vol.* **2010**, *43*, 166–171. [CrossRef]

191. Onoyama, H.; Ryu, C.; Suguri, M.; Iida, M. Potential of Hyperspectral Imaging for Constructing a Year-invariant Model to Estimate the Nitrogen Content of Rice Plants at the Panicle Initiation Stage. *IFAC Proc. Vol.* **2013**, *46*, 219–224. [CrossRef]

192. Suwansukho, K.; Sumriddetchkajorn, S.; Buranasiri, P. Demonstration of a single-wavelength spectral-imaging- based Thai jasmine rice identification. *Appl. Opt.* **2011**, *50*, 4024–4030. [CrossRef]

193. Zhang, H.; Paliwal, J.; Jayas, D.S.; White, N.D.G. Classification of fungal infected wheat kernels using near-infrared reflectance hyperspectral imaging and support vector machine. *Trans. ASABE* **2007**, *50*, 1779–1785. [CrossRef]

194. Xing, J.; Huang, P.; Symons, S.; Shahin, M.; Hatcher, D. Using a short wavelength infrared (SWIR) hyperspectral imaging system to predict alpha amylase activity in individual Canadian western wheat kernels. *Sens. Instrum. Food Qual. Saf.* **2009**, *3*, 211. [CrossRef]

195. Patrick, A.; Pelham, S.; Culbreath, A.; Holbrook, C.C.; De Godoy, I.J.; Li, C. High throughput phenotyping of tomato spot wilt disease in peanuts using unmanned aerial systems and multispectral imaging. *IEEE Instrum. Meas. Mag.* **2017**, *20*, 4–12. [CrossRef]

196. Weinstock, B.A.; Janni, J.; Hagen, L.; Wright, S. Prediction of oil and oleic acid concentrations in individual corn (*Zea mays* L.) kernels using near-infrared reflectance hyperspectral imaging and multivariate analysis. *Appl. Spectrosc.* **2006**, *60*, 9–16. [CrossRef] [PubMed]

197. Yao, H.; Hruska, Z.; Kincaid, R.; Brown, R.; Cleveland, T.; Bhatnagar, D. Correlation and classification of single kernel fluorescence hyperspectral data with aflatoxin concentration in corn kernels inoculated with *Aspergillus flavus* spores. *Food Addict. Contam. Part A Chem.* **2010**, *27*, 701–709. [CrossRef] [PubMed]

198. Chao, K.; Chen, Y.R.; Hruschka, W.R.; Park, B. Chicken heart disease characterization by multi-spectral imaging. *Appl. Eng. Agric.* **2001**, *17*, 99. [CrossRef]

199. Kise, M.; Park, B.; Lawrence, K.C.; Windham, W.R. Design and calibration of a dual-band imaging system. *Sens. Instrum. Food Qual. Saf.* **2007**, *1*, 113–121. [CrossRef]

200. Chao, K.; Yang, C.C.; Kim, M.S.; Chan, D.E. High throughput spectral imaging system for wholesomeness inspection of chicken. *Appl. Eng. Agric.* **2008**, *24*, 475–485. [CrossRef]

201. Naganathan, G.K.; Grimes, L.M.; Subbiah, J.; Calkins, C.R.; Samal, A.; Meyer, G.E. Visible/near-infrared hyperspectral imaging for beef tenderness prediction. *Comput. Electron. Agric.* **2008**, *64*, 225–233. [CrossRef]

202. Peng, Y.; Zhang, J.; Wang, W.; Li, Y.; Wu, J.; Huang, H.; Gao, X.; Jiang, W. Potential prediction of the microbial spoilage of beef using spatially resolved hyperspectral scattering profiles. *J. Food Eng.* **2011**, *102*, 163–169. [CrossRef]

203. Kamruzzaman, M.; ElMasry, G.; Sun, D.; Allen, P. Application of NIR hyperspectral imaging for discrimination of lamb muscles. *J. Food Eng.* **2011**, *104*, 332–340. [CrossRef]

204. Tao, F.; Peng, Y. A method for nondestructive prediction of pork meat quality and safety attributes by hyperspectral imaging technique. *J. Food Eng.* **2014**, *126*, 98–106. [CrossRef]

205. Barbin, D.; Elmasry, G.; Sun, D.; Allen, P. Near-infrared hyperspectral imaging for grading and classification of pork. *Meat Sci.* **2012**, *90*, 259–268. [CrossRef] [PubMed]

206. ElMasry, G.; Wold, J.P. High-speed assessment of fat and water content distribution in fish fillets using online imaging spectroscopy. *J. Agric. Food Chem.* **2008**, *56*, 7672–7677. [CrossRef] [PubMed]

207. Sivertsen, A.H.; Chu, C.K.; Wang, L.C.; Godtliebsen, F.; Heia, K.; Nilsen, H. Ridge detection with application to automatic fish fillet inspection. *J. Food Eng.* **2009**, *90*, 317–324. [CrossRef]

208. Wu, D.; Sun, D. Potential of time series-hyperspectral imaging (TS-HSI) for non-invasive determination of microbial spoilage of salmon flesh. *Talanta* **2013**, *111*, 39–46. [CrossRef]

209. Wu, D.; Shi, H.; Wang, S.; Hea, Y.; Bao, Y.; Liu, K. Rapid prediction of moisture content of dehydrated prawns using online hyperspectral imaging system. *Anal. Chim. Acta* **2012**, *726*, 57–66. [CrossRef] [PubMed]

210. Wu, D.; Shi, H.; He, Y.; Yu, X.; Bao, Y. Potential of hyperspectral imaging and multivariate analysis for rapid and non-invasive detection of gelatin adulteration in prawn. *J. Food Eng.* **2013**, *119*, 680–686. [CrossRef]

211. Lim, J.; Kim, G.; Mo, C.; Kim, M.S.; Chao, K.; Qin, J.; Fu, X.; Baek, I.; Cho, B. Detection of melamine in milk powders using near-infrared hyperspectral imaging combined with regression coefficient of partial least square regression model. *Talanta* **2016**, *151*, 183–191. [CrossRef]

212. Qin, J.; Chao, K.; Kim, M.S. Raman chemical imaging system for food safety and quality inspection. *Trans. ASABE* **2010**, *53*, 1873–1882. [CrossRef]

213. Deng, S.; Xu, Y.; Li, X.; He, Y. Moisture content prediction in tealeaf with near infrared hyperspectral imaging. *Comput. Electron. Agric.* **2015**, *118*, 38–46. [CrossRef]

214. Wu, D.; Yang, H.; Chen, X.; He, Y.; Li, X. Application of image texture for the sorting of tea categories using multi-spectral imaging technique and support vector machine. *J. Food Eng.* **2008**, *88*, 474–483. [CrossRef]

215. Chemura, A.; Mutanga, O.; Odindi, J. Empirical Modeling of Leaf Chlorophyll Content in Coffee (Coffea Arabica) Plantations with Sentinel-2 MSI Data: Effects of Spectral Settings, Spatial Resolution, and Crop Canopy Cover. *IEEE J. Sel. Top. Appl. Earth Obs. Remote Sens.* **2017**, *10*, 5541–5550. [CrossRef]

216. Heibati, M.; Stedmon, C.A.; Stenroth, K.; Rauch, S.; Toljander, J.; Säve-Söderbergh, M.; Murphy, K.R. Assessment of drinking water quality at the tap using fluorescence spectroscopy. *Water Res.* **2017**, *125*, 1–10. [CrossRef]

217. Sorensen, J.P.R.; Vivanco, A.; Ascott, M.J.; Gooddy, D.C.; Lapworth, D.J.; Read, D.S.; Rushworth, C.M.; Bucknall, J.; Herbert, K.; Karapanos, I.; et al. Online fluorescence spectroscopy for the real-time evaluation of the microbial quality of drinking water. *Water Res.* **2018**, *137*, 301–309. [CrossRef] [PubMed]

218. Højris, B.; Christensen, S.C.B.; Albrechtsen, H.J.; Smith, C.; Dahlqvist, M. A novel, optical, on-line bacteria sensor for monitoring drinking water quality. *Sci. Rep.* **2016**, *6*, 23935. [CrossRef] [PubMed]

219. Mignani, A.G.; Ciaccheri, L.; Ottevaere, H.; Thienpont, H.; Conte, L.; Marega, M.; Cichelli, A.; Attilio, C.; Cimato, A. Visible and near-infrared absorption spectroscopy by an integrating sphere and optical fibers for quantifying and discriminating the adulteration of extra virgin olive oil from Tuscany. *Anal. Bioanal. Chem.* **2011**, *399*, 1315–1324. [CrossRef]

220. Woodcock, T.; Downey, G.; O'Donnel, C. Near infrared spectral fingerprinting for confirmation of claimed PDO provenance of honey. *Food Chem.* **2009**, *114*, 742–746. [CrossRef]

221. Li, X.; He, Y.; Wu, C.; Sun, D.W. Non desctructive measurement and fingerprint analysis of soluble content of tea soft drink based on Vis/NIR spectroscopy. *J. Food Eng.* **2007**, *82*, 316–323. [CrossRef]

222. Fagan, C.C.; Castillo, M.; O'Donnel, C.P.; Callaghan, D.J.; Payne, F.A. Online prediction of cheese making indices using backscatter of near infrared light. *Int. Dairy J.* **2008**, *18*, 120–128. [CrossRef]

223. Egidio, V.D.; Oliveri, P.; Woodcock, T.; Downey, G. Confirmation of brand identity in foods by near infrared transflectance spectroscopy using classification and class-modelling chemometric techniques—The example of a Belgian beer. *Food Res. Int.* **2011**, *44*, 544–549. [CrossRef]

224. Leong, Y.S.; Ker, P.J.; Jamaludin, M.Z.; Nomanbhay, S.M.; Ismail, A.; Abdullah, F.; Looe, H.M.; Shukri, C.N.S. New near-infrared absorbance peak for inhibitor content detection in transformer insulating oil. *Sens. Actuators B Chem.* **2018**, *266*, 577–582. [CrossRef]

225. Li, T.L.; Chung-Wang, Y.J.; Shih, Y.C. Determination and confirmation of chloramphenicol residues in swine muscle and liver. *J. Food Sci.* **2002**, *67*, 21–28. [CrossRef]

226. Overview: Malaysian Agricultural Biotechnology. BiotechCorp, 2009. Available online: http://www.bioeconomycorporation.my/wp-content/uploads/2011/11/publications/White_Paper_Agricultural.pdf (accessed on 25 November 2018).

227. Ahmad, T.T.M.A.; Suntharalingam, C. Transformation and Economic Growth of the Malaysian Agricultural Sector. *Econ. Technol. Manag. Rev.* **2009**, *4*, 1–10.

228. Matahir, H. The Empirical Investigation of the Nexus between Agricultural and Industrial Sectors in Malaysia. *Int. J. Bus. Manag. Soc. Res.* **2012**, *3*, 225–231.

229. Onn, F.C. Small and Medium Industries in Malaysia: Economic Efficiency and Entrepreneurship. *Dev. Econ.* **1990**, *28*, 152–179.

230. Rahman, A.A.Z. Economic Reforms and Agricultural Development in Malaysia. *ASEAN Econ. Bull.* **1998**, *15*, 59–76. [CrossRef]

231. Shaffril, M.H.A.; Asmuni, A.; Ismail, A. The Ninth Malaysian Plan and Agriculture Extension Officer Competency: A Combination for Intensification of Paddy Industry in Malaysian. *J. Int. Soc. Res.* **2010**, *3*, 450–457.

232. *Eleventh Malaysia Plan 2016–2020 Anchoring Growth on People*; Percetakan Nasional Malaysia: Kuala Lumpur, Malaysia, 2015; Available online: https://www.mkma.org/Notice%20Board/2015/MP11Book.pdf (accessed on 25 November 2018).

233. Murad, M.W.; Mustapha, N.H.; Siwar, C. Review of Agricultural Policies with Regards to Sustainability. *Am. J. Environ. Sci.* **2008**, *4*, 608–614. [CrossRef]

234. *Tenth Malaysia Plan 2011–2015*; Percetakan Nasional Malaysia: Kuala Lumpur, Malaysia, 2010. Available online: http://www.pmo.gov.my/dokumenattached/RMK/RMK10_E.pdf (accessed on 25 November 2018).

235. Tagoe, S.M.A.; Dickinson, M.J.; Apetorgbor, M.M. Factors influencing quality of palm oil produced at the cottage industry level in Ghana. *Int. Food Res. J.* **2012**, *19*, 271–278.

236. Hushiarian, R.; Yusof, N.A.; Dutse, S.W. Detection and control of *Ganoderma boninense*: Strategies and perspectives. *SpringerPlus* **2013**, *2*, 555. [CrossRef] [PubMed]

237. Lelong, C.C.; Roger, J.; Brégand, S.; Dubertret, F.; Lanore, M.; Sitorus, N.; Raharjo, D.; Caliman, J. Evaluation of Oil-palm fungal disease infestation with canopy hyperspectral reflectance data. *Sensors* **2010**, *10*, 734–747. [CrossRef] [PubMed]

MDPI

St. Alban-Anlage 66

4052 Basel

Switzerland

Tel. +41 61 683 77 34

Fax +41 61 302 89 18

www.mdpi.com

Molecules Editorial Office

E-mail: molecules@mdpi.com

www.mdpi.com/journal/molecules

www.ingramcontent.com/pod-product-compliance
Lightning Source LLC
Chambersburg PA
CBHW051701210326
41597CB00032B/5335